The Laboratory Mouse

The Handbook of Experimental Animals

Editor-in-Chief

Peter Petrusz
Department of Cell Biology and Anatomy
University of North Carolina
Chapell Hill, NC
USA

List of Editorial Advisory Board

The Laboratory Mouse

Edited by

Professor Hans J Hedrich

Institute for Laboratory Animal Science
Hannover Medical School
Hannover, Germany

and

Professor Gillian Bullock

University Hospital
Department of Pathology
Ghent, Belgium

ELSEVIER
ACADEMIC
PRESS

Amsterdam • Boston • Heidelberg • London • New York • Oxford
Paris • San Diego • San Francisco • Singapore • Sydney • Tokyo

Elsevier Academic Press
84 Theobald's Road, London WC1X 8RR, UK
http://books.elsevier.com

Elsevier Academic Press
525 B Street, Suite 1900 San Diego, California 92101-4495, USA
http://books.elsevier.com

ISBN 0–12–336425–6

A catalogue record for this book is available from the British Library

Library of Congress Catalog Number: 2003115957

Typeset by Newgen Imaging Systems (P) Ltd, Chennai, India

Printed and bound in the United Kingdom
Transferred to Digital Print 2010

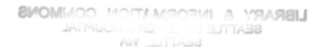

Contents

Part 2 Anatomy and Developmental Biology

Part 3 Pathophysiology (Including Non-Infectious Diseases)

Colour plates appear between pages 300 and 301

List of Contributors

Adler, A
Institute of Experimental Genetics,
GSF – National Research Center for
    Environment and Health, GmbH
Neuherberg,
Germany

Anver, M R
SAIC Frederick, Inc.,
National Cancer Institute,
Frederick,
Maryland,
USA

Beckers, J
Institute of Experimental Genetics,
GSF – National Research Center for
    Environment and Health, GmbH
Neuherberg,
Germany

Berdanier, C D
Nutrition and Cell Biology,
University of Georgia,
Athens,
Georgia,
USA

Boggess, D
The Jackson Laboratory,
Bar Harbor,
Maine,
USA

Bonhomme, F
Université Montpellier II,
Laboratoire Génome, Populations,
    Interactions,
Montpellier,
France

Braun, A
Fraunhofer Institut fur Toxikologie und
    Aerosolforschung,
Hannover,
Germany

Brayton, C
Comparative Pathology Laboratory,
Baylor College of Medicine,
Houston,
Texas,
USA

Buerge, T
ROCH/Laboratory Animal Services,
Novartis Institutes for BioMedical
    Research,
Basel,
Switzerland

Davisson, M T
The Jackson Laboratory,
Bar Harbor,
Maine,
USA

de Angelis, M H
Institute of Experimental Genetics,
GSF – National Research Center for
    Environment and Health, GmbH
Neuherberg,
Germany

de Leeuw, W A
Department for Veterinary Public Health,
DEN HAAG,
The Netherlands

Dixon, A K
Ethology Research Unit,
Research Institute Wander,
Neuenegg,
Switzerland

Dorsch, M M
Zentrales Tielaboratorium,
Medizinische Hochschule,
Hannover,
Germany

Entman, M L
Baylor College of Medicine,
Houston,
Texas,
USA

Ernst, H
Fraunhofer Institut fur Toxikologie Und
    Aerosolforschung,
Hannover,
Germany

Everds, N
American College of Veterinary
    Pathology,
Newark,
Delaware,
USA

Frangogiannis, N G
Baylor College of Medicine,
Houston,
Texas,
USA

Fukuta, K
Graduate School of Bioagricultural
    Sciences,
Nagoya,
Japan

Gailus-Durner, V
Institute of Developmental Genetics,
GSF – National Research Center for
    Environment and Health, GmbH
Neuherberg,
Germany

Gärtner, K
Zentrales Tierlaboratorium,
Medizinische Hochschule Hannover,
Hannover,
Germany

Gossler, A
Institut für Molekularbiologie OE5250,
Medizinische Hochschule Hannover,
Hannover,
Germany

Guénet, J-L
Institut Pasteur,
Unité de Génétique des Mammifères,
Paris,
France

Hafner, M
GBF German Research Centre for
    Biotechnology,
Braunschweig,
Germany

Haines, D C
SAIC Frederick, Inc.,
National Cancer Institute,
Frederick,
Maryland,
USA

Hardy, P
Charles River Laboratories,
Lyon,
France

Harlemann, J H
Novartis Pharma AG,
Basel,
Switzerland

Hartley, C J
Baylor College of Medicine,
Houston,
Texas,
USA

Hedrich, H J
Hannover Medical School,
Hannover,
Germany

Holmdahl, R
Medical Inflammation Research,
Lund University,
Lund,
Sweden

Houdebine, L-M
Laboratoire de Biologie Cellulaire et
Moleculaire,
Institut National de la Recherche
Agronomique,
Jouy-en-Josas,
France

Hoymann, H G
Fraunhofer Institute of Toxicology and
Experimental Medicine,
Hannover,
Germany

Ichiki, T
The Jackson Laboratory,
Bar Harbor,
Maine,
USA

Imai, K
Institute of Developmental Genetics,
GSF – National Research Center for
Environment and Health, GmbH
Neuherberg,
Germany

Jilge, B
Laboratory Animal Research Unit,
University of Ulm,
Germany

Kannan, Y
Laboratory of Integrative Physiology,
Graduate School of Agriculture and Biological
Sciences,
Osaka Prefecture University,
Osaka,
Japan

Kispert, A
Institut für Molekularbiologie OE5250,
Medizinische Hochschule Hannover,
Hannover,
Germany

Komárek, V
Sidlistni 212,
Lysolaje,
Czech Republic

Krinke, G J
Syngenta AG,
Stein,
Switzerland

Kunz, E
Laboratory Animal Research Unit,
University of Ulm,
Germany

Linder, C C
The Jackson Laboratory,
Bar Harbor,
Maine,
USA

Mähler, M
Biomedical Diagnostics,
Hannover,
Germany

Michael, L H
Baylor College of Medicine,
Houston,
Texas,
USA

Mikaelian, I
The Jackson Laboratory,
Bar Harbor,
Maine,
USA

Mossmann, H
Max-Planck-Institut für Immunbiologie,
Freiburg i. Br,
Germany

Müller, W
Department of Experimental
Immunology,
GBF German Research Centre for
Biotechnology,
Braunschweig,
Germany

Nicklas, W
Central Animal Laboratories,
German Cancer Research Centre,
Heidelberg,
Germany

Otto, K
Zentrales Tierlaboratorium,
Medizinische Hochschule Hannover,
Hannover,
Germany

Price, R E
American College of Veterinary Pathologists,
Houston,
Texas,
USA

Ritskes-Hoitinga, M
Laboratory Animal Science and Comparative
Medicine,
University of Southern Denmark,
Odense,
Denmark

Rittinghausen, S
Fraunhofer Institute of Toxicology and
Experimental Medicine,
Hannover,
Germany

Seymour, R
The Jackson Laboratory,
Bar Harbor,
Maine,
USA

Shimizu, S
National Institute of Animal Health,
Tsukuba,
Japan

Silva, K A
The Jackson Laboratory,
Bar Harbor,

Maine,
USA

Soewarto, D
Institute of Experimental Genetics,
GSF – National Research Center for
Environment and Health, GmbH
Neuherberg,
Germany

Sundberg, J P
The Jackson Laboratory,
Bar Harbor,
Maine,
USA

Taffet, G E
Baylor College of Medicine,
Houston,
Texas,
USA

Wagner, S
Institute of Experimental Genetics,
GSF – National Research Center for
Environment and Health, GmbH
Neuherberg,
Germany

Ward, J M
Veterinary and Tumor Pathology Section
CCR, NCI,
Frederick,
Maryland,
USA

Weiss, T
ROCK/LAS/Veterinary Services
Novartis Institutes for BioMedical Research,
Novartis Pharma AG,
Basel,
Switzerland

Foreword

*R*ESEARCH in the biological sciences is dramatically changing. The driving force is, like in many cases, technology development. Genomic and proteomic data can now be collected with a rate and a quality which for many of us was completely outside our imagination less than 20 years ago. This development is not so different from what we have witnessed in the computer industry. Automatization and miniaturization have moved into the center stage of biosciences.

Navigation with the click of a mouse through the 30 billion base pairs of the recently finished mouse genome seems to become as important for our students and postdocs than navigation through the complexities of mouse anatomy, physiology or endocrinology. The real excitement comes when you are able to do both. Interdisciplinarity has been preached for decades. Implementing interdisciplinarity, however, is much more difficult. This is where the present book will make its mark.

There is a saying among geneticists: "Your genetics is only as good as your phenotype". This requires in-depth, standardized and high-quality phenotyping of mutant mice. In order to be able to judge the relevance of an apparently pathophysiological mouse trait you have to know the normal physiology. You should also be aware of the genetic heterogeneity between different inbred and wild mouse strains. Not so much different from the genetic variation between human individuals and populations.

There are still many papers published where we read: "This mouse mutant has no phenotype". Apart from the trouble of thinking about the literal meaning of such a statement, we all know that very often abnormalities in physiological or biochemical parameters are missed in a first line phenotyping of mouse mutants.

The Laboratory Mouse edited by Hans Hedrich and Gillian Bullock fills an important gap for anybody working with mice. Rather than describing other people's experiments, this book really helps you to do your experiments. It guides us through the origins of mouse genetics and its historical foundation. It gives us an introduction into the toolbox of generating mouse mutants and into the development of mouse embryos. It provides an impressive journey through the majority of organs and how you should analyze physiological and pathophysiological processes, and illustrates how the mouse is an excellent model for the study of human genetics, disease processes and general biology.

The Laboratory Mouse will not run into the danger of becoming outdated before the ink of the 1st edition is dry. Good laboratory practice on how to handle mice, how to administer compounds or collect body fluids from mice, is here to stay and professional knowledge in this area needs to enter the curriculum of any student and researcher working with mice, young or old.

The book comes at the right time. It will help to switch our attention to the importance of phenotyping, where "genetics meets physiology". In 1999, Hartwell wrote: "The next generation of students should learn to look for amplifiers and logic circuits, as well as to describe and look for molecules and genes" (Hartwell, L.H., Hopfield, J.J., Leibler, S. and Murray, A.W. (1999) *Nature* 402, C47-C52). I would like to extend this by saying: "The next generation of students has to read this book".

Rudi Balling
June 2004

Preface

WITH the immense amount of research that has taken place using the mouse as an experimental tool, one may well ask 'why another book about mice?'. This, however, is not just another book. Immense care has been taken to select topics and authors to define and illustrate the most important features of this species.

The volume has been divided up in such a way that the new and/or established researcher can easily track down the most up-to-date information in any one area. While headline-grabbing topics such as mouse genomics and the generation of mouse mutants sit comfortably with the analysis of the total mouse genome, equal importance has been given to the basics of mouse development, pathological anatomy and pathophysiology. Further consideration is paid to husbandry, methodological aspects, alleviation of pain as well as legal aspects.

On practical grounds the coverage of certain systems and topics as well as the extent of data had to be limited to some extent; for instance, rather than covering viral, bacterial and parasitic diseases within the part on *Infectious Diseases*, we decided to cover only viral infections. Since murine viruses pose the greatest risk for any mouse facility it was decided that it would be beneficial to have this aspect dealt with in a comprehensive manner.

I am extremely grateful to all my friends and colleagues who have helped me to put together this volume. As with the previous volumes in this series – The Laboratory Rat and The Laboratory Fish – the authors, being associated with both, universities and applied research organisations, came from a wide range of countries, thus providing a global, well-balanced approach.

Finally, we would like to thank Jenny Taylor and Sara Purdy from the publishing house for their patience and support irrespective of the transition of Academic Press from Harcourt to Elsevier during the development of this volume.

Professor Hans Hedrich
Volume Editor

Professors Gillian Bullock and Peter Petrusz
Editors-in-Chief
The Handbook of Experimental Animals

PART 1

History, Development and Genetics of the Mouse as a Laboratory Model

Contents

Origin of the Laboratory Mouse and Related Subspecies

Jean-Louis Guénet
Institut Pasteur, Unité de Génétique des Mammifères,
Paris, France

François Bonhomme
Université Montpellier II, CNRS UMR 5000, Laboratoire
Génome, Populations, Interactions, Montpellier,
France

Introduction

Based on paleontological data it seems that men and mice have been in contact since the early Pleistocene (Berry, 1970), which means for over a million years (Myrs), and numerous historical records (Keeler, 1931; Staats, 1966; Morse, 1978; Berry, 1987; Moriwaki *et al.*, 1994) indicate that mice were already bred as pets 3 millennia ago: it was then logical that these small mammals, as well as the rat and some small sized pet-birds, be used by scientists of the early days for performing their experiments. However, if this choice was more opportunistic rather than based on scientific considerations, it nevertheless appears to be an excellent one in the context of modern biomedical research where the house mouse has become a model of predilection.

Mice are easy to keep. Because they are rodents, they eat a rather large quantity of food but do not have very specific or expensive nutritional requirements. They breed all year round, with a short generation time; they deliver relatively large progenies and tolerate inbreeding rather well compared to other mammalian species. With the passing years, hundreds of mutations, most of them with deleterious alleles, have been collected that all have contributed and still contribute to the identification of genes by their function(s), and several programs of intensive mutagenesis have been developed worldwide to increase further this invaluable resource. Another very important advantage to be credited to the mouse is that it seems to be one of the rare, maybe the only species, where it is possible to grow totipotent embryonic stem (ES) cells *in vitro*, which can be genetically engineered in a number of ways and still retain the capacity to participate in

The Laboratory Mouse
Copyright 2004 Elsevier
ISBN 0-1233-6425-6

the formation of the germ line once re-injected into a developing embryo. Finally, and this is not the slightest of the advantages, the complete sequence of the mouse genome is now available (Waterston *et al.*, 2002), which will allow comparisons with other mammalian genomes and annotations concerning the function of the genes to be made. In short, the mouse is the only mammalian species whose genomic sequence is known and for which technical procedures exist for the generation of a virtually unlimited number of genetic alterations.

In this chapter we will describe the origins of laboratory mice, starting with their phylogenetic relationships with the other mammalian species. We will also discuss the advantage of strains established from recently trapped wild specimens as a source of polymorphisms for scientific research.

The phylogenetic relationships of the house mouse

The position of rodents among mammalian species

Mice are rodents. They belong to the most abundant (around 40%) and diversified order of living placental mammals, with slightly over 2000 species grouped in 28 families (Huchon *et al.*, 2002). Because of their great diversity, the phylogenetic relationships between the different species of this order has been a matter of controversy for many years, especially when morphological markers were the only criteria available for the establishment of phylogeny. Nowadays, with the use of various molecular (mostly DNA) markers and possible references to the complete genomic sequence of numerous orthologous genes, the situation is much clarified and Figure 1.1 represents the most likely phylogenetic tree for a sample of 40 different eutherian mammals. Based on comparisons at the level of nuclear DNA sequences, the divergence between man and murid rodents (*Mus* or *Rattus* genus) has been set somewhere between 65 and 75 Myrs ago (Waterston *et al.*, 2002).

Mice among rodents

The rodent family of *Muridae* encompasses at least 1326 species grouped in 281 genera (Musser and Carleton,

1993). The establishment of the evolutionary systematics in this group has also been disputed but, this time, it was because many mammals in this family are very similar in size and shape. Here again studies making use of DNA sequences of various types (Michaux *et al.*, 2001; Lundrigan *et al.*, 2002) have greatly contributed to clarify the situation and Figure 1.2 represents the evolutionary relationships among a sample of 21 rodent species anchored into the broader phylogeny of eutherian mammals. The divergence between the *Mus* and *Rattus* genus has been estimated at around 10–15 Myrs ago (Jaeger *et al.*, 1986; Murphy *et al.*, 2001), while the divergence of these two genera with *Peromyscus maniculatus*, the deer mouse (subfamily *Sigmodontinae*), occurred at around 25 Myrs ago. This is to be remembered because deer mice, which are abundantly used as laboratory models, are often considered close relatives of the laboratory mice while, in fact, they are no more related to them than hamsters.

Systematics in the genus Mus

Figure 1.3 (Guénet and Bonhomme, 2003 and references therein) summarizes the phylogenetic relationships within the genus *Mus* (subfamily *Murinae*). The individualization of the subgenus *Mus sensu stricto* occurred around 5 Myrs ago with the split of three other different subgenera, the African *Nannomys* and the Asian *Coelomys* and *Pyromys*.

The subgenus *Mus* comprises several species that are extremely similar in size and shape but never hybridize in the wild. Among the Asian species are *Mus cervicolor*, *Mus cookii*, and *Mus caroli* as well as the group of Indian pigmy mice related to *Mus dunni*. *Mus famulus* from India should also be cited as well as the recently discovered species *Mus fragilicauda* (Auffray *et al.*, 2003) from Thailand.

Mus spicilegus and *Mus macedonicus* are short tailed mice that are found in central Europe and the eastern Mediterranean, respectively, while mice belonging to the species *Mus spretus* are common in the western Mediterranean regions (south east France, Spain, Portugal and North Africa).

Mice of the *Mus musculus* complex are closely related. They have their evolutionary origins in the Indian subcontinent (Bonhomme *et al.*, 1994) but are now spread over the five continents. The best known representatives of the complex are the three *Mus musculus* subspecies: *Mus m. domesticus*, common in western Europe, Africa, the near-East, and transported by man to the Americas and Australia; *Mus m. musculus*, whose habitat spans from eastern Europe to Japan, across

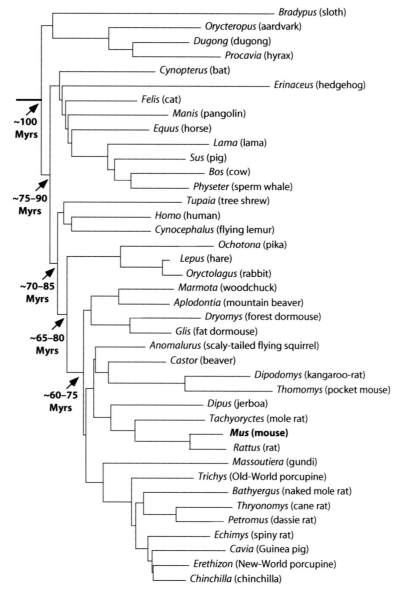

Figure 1.1 Evolutionary tree concerning 40 mammalian species including 21 rodent species, with an estimated time of divergence in Myrs (from Huchon *et al.* (2002). *Mol. Biol. Evol.* **19**, 1053–1065).

Russia, and northern China, and *Mus m. castaneus*, which is found from Sri Lanka to south east Asia including the Indo-Malayan archipelago. Various molecular criteria discriminate easily between these different species (Figure 1.4; Boursot *et al.*, 1993; Moriwaki *et al.*, 1994)

Mouse interspecific hybridization

Hybrids between mice of the genus *Mus* and mice of the subgenera *Nannomys, Coelomys* or *Pyromys*

have never been reported and probably never occur. Hybrids between wild mice of the species *Mus cervicolor, Mus caroli, Mus dunni*[1] and mice of the *Mus musculus* complex have never been found in the wild but hybrids between the former three wild species and laboratory mice have been produced by artificial insemination (West *et al.*, 1977). In these experiments, hybrids generated by insemination of female laboratory mice with *Mus cervicolor* sperms failed to complete more than a few cleavage divisions. Hybrids generated from *Mus dunni* sperms and laboratory female oocytes implanted but died *in utero* at a very

[1]See legend to Figure 1.3.

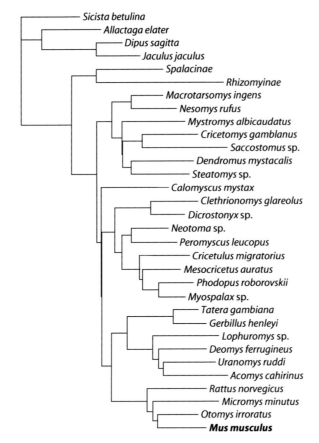

Figure 1.2 Phylogenetic relationships between 32 species of rodents representing 14 subfamilies of Muridae (redrawn from Michaux *et al.* (2001). *Mol. Biol. Evol.* **18**, 2017–2031).

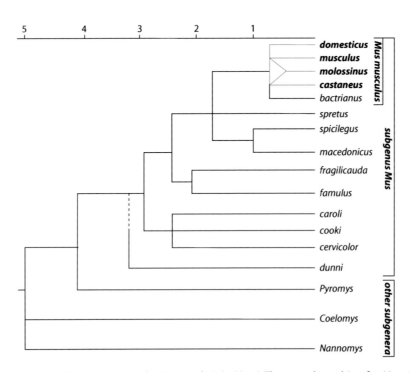

Figure 1.3 Evolutionary tree of the genus *Mus* (the time scale is in Myrs). The exact branching for *Mus dunni* is not precisely known. The four species at the origin of the classical laboratory strains are highlighted in bold (from Guénet, J.L. and Bonhomme, F. (2003). *Trends Genet.* **19**, 24–31).

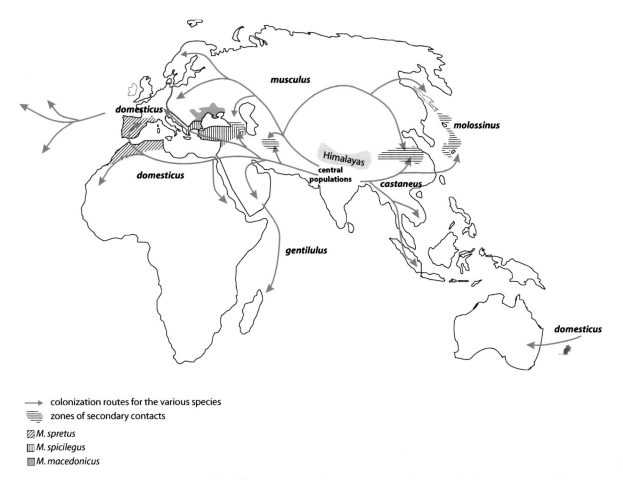

Figure 1.4 Geographical distribution of the different species of the genus *Mus* and routes of colonization. Mice of the American and Australian continents were imported by man during colonization (from Guénet, J.L. and Bonhomme, F. (2003). *Trends Genet.* **19**, 24–31).

early developmental stage. Hybrids generated from the same laboratory females and sperm from *Mus caroli* completed fetal development and a very low percentage of them even survived to maturity but none reproduced.

Although they share the same range (they are sympatric) with some *Mus musculus* subspecies, the short tailed species *Mus spretus, Mus spicilegus* and *Mus macedonicus* rarely produce hybrids in nature. However, evidence from studies on mtDNA (Orth *et al.*, 2002) and LINE transposable elements (Greene-Till *et al.*, 2000) indicate that exchanges can occur sporadically that would allow alleles with a selective advantage to circulate outside the species in which they originated. The three species mentioned above produce viable offspring with laboratory mice but male offspring of these crosses are sterile in compliance with Haldane's rule. Male hybrids born from a *Mus musculus* × *Mus spretus* cross, for example, are invariably sterile regardless of the direction of the cross. This sterility is controlled by

a relatively small number of genes since fertile males are frequently observed in the backcross progeny of F1 females with a male of either of the parental species (Guénet *et al.*, 1990; Forejt, 1996; Pilder *et al.*, 1997; Elliott *et al.*, 2001).

Mice of the *Mus musculus* complex are not genetically isolated and, in those locations where they meet, there is evidence of gene exchanges ranging from limited introgression to complete blending (Boursot *et al.*, 1993). The best-documented cases of such gene exchanges are those occurring between *M. m. musculus* and *M. m. domesticus* in Europe, along a narrow hybrid zone, and between *M. m. musculus* and *Mus m. castaneus* in Japan. In this archipelago, the two subspecies have hybridized extensively, giving rise to a unique population often referred to as *Mus m. molossinus* (Yonekawa *et al.*, 1988). These gene exchanges, which indicate that the speciation process is in progress but not yet completed, explain the use of Latin trinomens for the designation of the different subspecies in the *M. musculus* complex.

The house mouse as a laboratory model: a historical perspective

Mice, rats and other small vertebrates have been used in biomedical research since the middle of the sixteenth century when biology progressively shifted from a descriptive to an experimental science. Morse (1981), reported that William Harvey used mice for his fundamental studies on reproduction and blood circulation and, according to Berry (1987), the earliest record of the use of mice in scientific research seem to have been in England, in 1664, when Robert Hooke used mice to study the biological consequences of an increase in air pressure. Much later, Joseph Priestly (1733–1804) and his intellectual successor, Antoine Lavoisier (1743–94), both used mice repeatedly in their experiments on respiration.

During the nineteenth century several fanciers in Europe and the United States were breeding and exchanging pet mice segregating for a variety of coat color or behavioral mutations. According to Grüneberg (1957), one of these fanciers, M. Coladon, who was established as a pharmacist in Geneva, reported results from his breeding experiments that were in perfect agreement with the Mendelian expectations ... but this was 36 years before the publication of Mendel's own results on peas. As mentioned by Paigen (2003a,b) in his notes about a century of mouse genetics, it seems that Mendel's first experiments on the transmission of characters were made using mice segregating for coat color markers but Mendel was rapidly asked by his ecclesiastical superior to stop breeding in his cellule awfully smelly creatures that, in addition, had sex and copulated. Mendel changed his experimental material for peas and published in 1866 his observations in a botanical journal where they had a much lower impact and remained virtually ignored until the very beginning of the twentieth century. Once rediscovered by De Vries, Correns and von Tschermark, the three of them working independently with plants, it was really tempting to check whether the so-called Mendel's laws were also valid for animals and experiments were published in 1902 by L. Cuénot (1902), indicating that this was indeed the case. Cuénot's observations were shortly confirmed and extended to other species as well as for other genetic traits by G. Bateson, E.R. Saunders, A. Garrod, W.E. Castle and C.C. Little (Paigen, 2003a).

Even if mice have been extensively used during the twentieth century in most areas of biomedical research, animals of this species have played an instrumental role in research in immunology, oncology, and genetics because the breeding systems[2] which are used to produce them allow the establishment of highly standardized strains whose characteristics are precisely known and monitored generation after generation. Most laboratory strains have their origins from a few pet dealers who progressively became suppliers of 'laboratory' mice. For many years, most of the albino strains used in laboratories were collectively designated 'Swiss' mice to recapitulate their Helvetian origin.

Strain DBA/2 (formerly dba, then DBA) is the most ancient of all inbred strains since it was established by C.C. Little in 1909 (Russell, 1978), by intercrossing mice homozygous for the coat color markers non agouti (*a*), brown (formerly *b* now *Tyrp1*), and dilute (formerly *d* and now *Myo5a*). About 10 years later, strain C57BL/6 was established by Miss Lathrop (Granby, Massachusetts, USA) intercrossing the 'black' offspring of her female 57, while strains C3H, CBA, and A were created by L.C. Strong, a cancer geneticist established at Cold Spring Harbor Laboratory (Strong, 1978). At this point, it is interesting to note that, among the strains established by L.C. Strong, strains CBA and C3H stemmed from the offspring of an outcross with a few wild specimens trapped in a pigeon coop in Cold Spring Harbor. This explains how the wild allele at the agouti locus (*A*) was re-introduced in laboratory strains.

With a few exceptions, historical records concerning the genealogy of most laboratory inbred strains are well documented and several interesting reviews on this subject are available (Morse, 1978; Festing, 1979). A chart concerning the genealogy of these strains, including the recently established ones, has been published (Beck *et al.*, 2000) and regularly up-dated information is available at the website <http://www.informatics.jax.org>.

In addition to its contribution to the re-discovery of Mendel's laws, the mouse has been closely associated with many important discoveries in biology during the twentieth century. To cite just a few among the most

[2]Among the genetically standardized strains, inbred strains are the most widely used. They result from the systematic and uninterrupted mating of brothers to their sisters, which leads to complete homozygosity for the same allele at all loci.

Figure 1.5 Mice of the *Mus spretus* species (left, with an agouti coat color) and C57BL/6 (right, with a non agouti or solid black coat color). In spite of their great similarity in size and body shape these mice are distantly related species but can still produce viable and fertile hybrids (female only). Mice of the *Mus spretus* species have been extensively used for the development of the mouse genetic map. **(See also Color Plate 1.)**

important, we could say that our understanding of the genetic determinism underlying the success or failure of tissue transplantations is a consequence of the many experiments performed with inbred mouse strains by P.A. Gorer (Gorer *et al.*, 1948), then by G.D. Snell and co-workers (Snell, 1978) who developed a series of congenic resistant strains, which were all genetically identical to the C57BL/10Sn background strain, with the exception of single short-sized chromosomal regions determining graft rejection. The discovery and genetic interpretation of the phenomenon of X-inactivation in female mammals, by M.F. Lyon (Lyon, 1961), has been facilitated by the existence and use of several X-linked mutations in the mouse and the observation of variegation in the coat color. The first chimeric organisms produced by A.K. Tarkowski, in Warsaw (Tarkowski, 1961), and B. Mintz, in Philadelphia (Mintz, 1962), were mice. The observation of a particularly high frequency of testicular terato-carcinomas in strain 129 (Stevens and Little, 1954; Stevens, 1970) and the *in vitro* culture of cell lines derived from these tumors, which represented for almost a decade a material of choice for investigating the processes at work in tissue differentiation (Jacob, 1983), undoubtedly opened the way to the establishment of the so-called embryonic stem (ES) cells, by G.R. Martin (1981) and, simultaneously, by M.J. Evans and M.H. Kaufman (1981). The discovery of parental imprinting of some chromosomal regions has been a consequence of

experiments performed by J. McGrath and D. Solter (1984) and M.A. Surani and co-workers (1984), who demonstrated that a normal mouse embryo can only result from the fusion of a male and a female pronucleus, while B.M. Cattanach and M. Kirk (1985) demonstrated that the parental origin for the two elements of a given chromosome pair was sometimes not genetically equivalent. The first transgenic mammal created by pronuclear injection of a cloned DNA, was a mouse (Gordon *et al.*, 1980; Brinster *et al.*, 1981; Costantini and Lacy, 1981; Gordon and Ruddle, 1981; Harbers *et al.*, 1981; Wagner *et al.*, 1981a,b) and the first *in vitro* genetically engineered mammalian organism was also a mouse (Kuehn *et al.*, 1987). Only the first cloned mammal was not a mouse although this type of uniparental procreation has been achieved also in the mouse (Wakayama and Yanagimachi, 1999).

Information concerning many aspects of the biology of the mouse considered as a laboratory model, in particular about its genetics, has been published in the 95 issues of the *Mouse News Letters*. First issued in 1949 and published regularly every semester till 1997,[3] this informal publication, edited by the scientists from the MRC at Harwell and distributed at low cost, has been for several decades the major medium for the dissemination of information among the community of mouse geneticists. In this sense, the *Mouse News Letters* will for ever remain the best place to get information about the history of mouse genetics, and in particular, about the history and location of most inbred strains, the progressive development and refinement of the linkage map and the discovery of hundreds of spontaneous mutations.

The Jackson Laboratory (the JAX-Lab), which was founded in Bar Harbor (Maine, USA), in 1929, by C.C. Little, has played a pivotal role in the promotion of the mouse as a laboratory model and still is a unique center for mouse genetics. The Jax-Lab, a non-profit organization entirely dedicated to basic research on the genetics of mammals, is nowadays almost exclusively dedicated to the mouse. It is, at the same time, a top ranked research institution, a meeting place where courses and conferences are organized on the various aspects of mouse genetics, and the world largest genetic repository where a great variety of genotypes and biological samples of all kind are stored, under the form of frozen embryos or sperm cells, for distribution to the community. Several other Institutions, like the

[3]The name *Mouse News Letter* was changed for *Mouse Genome* in 1990 when this publication became a peer reviewed journal. From 1998, *Mouse Genome* and *Mammalian Genome* have merged in one and a single journal.

Oak Ridge National Laboratory in Tennessee (USA) and the MRC center at Harwell in England have also played a very important role in the development of the mouse as a laboratory model for researches in genetics, oncology and immunology. Recently, the European Union has decided to support the establishment of a network of genetic repositories (the so-called European Mouse Mutant Archive or EMMA), with major nodes in Italy (EMMA head quarters is in Monteretondo, near Roma), in England (Harwell), in France (Orléans-la-Source) and in Germany (Munich). Finally, more recently, Japan has established a bio-resource center at the RIKEN, in Tsukuba. More information about all these centers is available at the websites provided at the end of this chapter.

The house mouse and its wild relatives

As discussed above, the classical laboratory inbred strains of mouse have many advantages that are related to their great genetic homogeneity. After all, a population of F1 hybrid mice, born from an intercross set between two highly inbred strains, can be considered identical from the genetic point of view to a population of cloned mice. Unfortunately, the coin has another side and these inbred strains, because they are derived from a relatively small pool of ancestors do not exhibit a great variety of genetic polymorphisms of natural origin. This relative genetic homogeneity is well reflected in the fact that most of the classical strains possess the same maternally inherited molecule of mitochondrial DNA (Yonekawa, 1980; Ferris et al., 1982) and relatively reduced polymorphisms for the Y chromosome (Bishop et al., 1985; Tucker et al., 1992). Aside from this relative genetic homogeneity, a careful analysis of the genetic polymorphism also indicates that laboratory strains have a mosaic genome derived from more than one species (Bonhomme et al., 1987; Wade et al., 2002) and today's classical laboratory strains should be regarded as interspecific recombinant strains derived (in unequal percentages) from three parental components: Mus m. domesticus, Mus m. musculus and Mus m. castaneus. For this reason, and to point to a relatively unnatural genetic constitution, it would probably be more appropriate to designate them as Mus m. 'laboratorius'!

Over the last 20 years a variety of strains, derived from small breeding nuclei of wild specimens, trapped in well-defined geographical areas and belonging to well characterized species, have been established in various laboratories. A list of the strains that are completely inbred, i.e. that have been propagated by strictly unrelaxed brother × sister matings for more than 20 generations, is given in Bonhomme and Guénet (1996). Other useful stocks of wild mice are also maintained in various laboratories and a complete description of these stocks has been published by Potter (1986). These 'new' inbred strains have been extremely important over these last years because they represent a virtually unlimited reservoir of genetic polymorphisms. Wild mice have been useful in providing geneticists with polymorphisms such as electrophoretic variants, restriction fragment length polymorphisms (RFLPs), or more generally single nucleotide polymorphisms (SNPs) that were much less numerous in standard inbred strains. With the introduction of strains derived from wild progenitors, in particular, from Mus spretus, the genetic map of mouse has dramatically increased its resolution (Guénet, 1986). Comparisons of non-coding orthologous regions at the sequence level indicate that any inbred strain derived from Mus spretus exhibits, on average, one SNP at every 80–100 bp when compared with any of the classical laboratory strains. This means that virtually any DNA sequence of 100–200 bp, can be used as a molecular marker in assays such as denaturing gradient gel electrophoresis (DGGE) or single strand confirmation polymorphism (SSCP). This high density of polymorphisms represents a considerable advantage in experiments when the aim is positional cloning of a gene identified only by phenotype because it allows an accurate delineation of the targeted locus. In fact, one can consider that there is no upper limit to the density in molecular markers when a genetic map is established from an interspecific or intersubspecific cross (Breen et al., 1994). The high density of polymorphisms turns out to be an even greater advantage when quantitative traits are mapped, because every animal with a relevant phenotype can be genotyped for a very large number of markers. In this respect, the mouse is unique since the frequency of SNPs between humans is roughly one order of magnitude lower than that of Mus spretus compared to laboratory strains (Flint and Mott, 2001; Matin and Nadeau, 2001). The high frequency of SNPs in coding regions means that the genome of Mus spretus or Mus m. musculus is full of ready-made 'quantitative trait loci (QTL)-point mutations' waiting for functional genomic studies!

Wild mice have also been invaluable in providing cytogeneticists with a large collection of Robertsonian translocations (or centric fusions) recovered from the many populations of *Mus m. domesticus* where they occur in homozygous conditions (Gropp and Winking, 1981). These translocations are characterized by the fusion of two acrocentric chromosomes by their centromeres and they often interfere with the normal process of meiosis resulting in the production of gametes with an aneuploid (unbalanced) complement. Using carefully designed crosses involving these centric fusions, it has been possible to produce and study trisomies and monosomies for all mouse chromosomes (Epstein, 1986).

In addition to their homogeneity in terms of chromosome morphology, laboratory strains have only long telomeres while, for instance, strains of *Mus spretus* origins, have both long and short telomeres (Coviello-McLaughlin and Prowse, 1997; Zhu *et al.*, 1998). This peculiarity might be helpful for investigating the significance of the still mysterious variations in telomere size found in mammalian cells.

When infectious agents of various kinds are injected into mice it is common to observe that some strains are more susceptible than others and that wild derived strains are in general more resistant than classical laboratory strains. A commonly accepted explanation, although not demonstrated, is that some alleles of laboratory strains, which are essential for determining innate or acquired mechanisms of defense, have been by chance replaced by a defective mutant allele without any consequences for the mice because these animals are kept in protected environments. Even if in most cases the level of susceptibility or resistance is controlled by several genes interacting together (QTLs) or having an additive effect, the situation is sometimes under the control of a single gene making its analysis relatively simple. This is the case, for example, when mice of most laboratory strains die after an injection with an appropriate dose of orthomyxoviruses while most wild strains are resistant (Haller *et al.*, 1998). This phenotype is controlled by a single gene (*Mx1* – chromosome 16) with two alleles: $Mx1^+$ (resistant, dominant) and $Mx1^-$ (susceptible, recessive) and the discrepancy between wild mice and laboratory mice in terms of susceptibility indicates that the mutated allele of *Mx1* is over-represented in laboratory strains, probably due to a sampling effect. A similar example exists with experimental flavivirus infections where all laboratory inbred strains, except strain PL/J, are susceptible while most wild derived inbred strains are resistant (Sangster *et al.*, 1998). Here again, it seems that Flv^s, the allele responsible for susceptibility at the *Flv* locus (chromosome 5), has been fortuitously selected in laboratory strains while it is rare or absent in wild mice.

Similar phenotypes of resistance/susceptibility have also been reported for a variety of pathogens (Sebastiani *et al.*, 2000; Lengeling *et al.*, 2001) and, even if in most instances genetic differences have been observed among classical laboratory strains, these differences also exist between laboratory and wild derived strains making the genetic analysis much easier. Wild animals also proved particularly useful for investigating the biology of murine leukemia viruses and both new *Fv* loci and new alleles at the *Fv1* and *Fv2* loci have been discovered in wild mice (Gardner *et al.*, 1991; Qi *et al.*, 1998).

The comments addressed concerning the susceptibility of mice to infectious diseases also apply to carcinogenesis and comparisons between classical laboratory strains allowed the complex influences of genetic background on tumor susceptibility to be unraveled and several genes modifying tumor susceptibility have been identified. However, while the phenotype of F1 hybrids between any two classical laboratory strains is generally intermediate between the two parental strains, it is often identical to the phenotype of the wild parent when crosses are performed with wild inbred strains, indicating dominance of the wild derived allele. (Nagase *et al.*, 2001).

Besides their use in mapping, interspecific crosses also offer an opportunity for analyzing the effects of bringing together the products of genes separated by divergent evolution in the cells of an offspring. This can help identify the genetic functions that are subject to rapid divergence and to pinpoint the functions that eventually promote speciation. Those functions that are mostly unaffected during the evolution of the taxa are most likely to be basic functions that are under more constraint. This last point will be important in the comparison of orthologous regions between human and mouse genomes now that sequencing is completed for both species.

Questions concerning epistatic interactions can also be addressed by investigating offspring of interspecific crosses at the genomic level. So far, we have no clear answers to this question but data exist indicating that some combinations of alleles are strongly counter-selected in the offspring of some interspecific crosses (Montagutelli *et al.*, 1996).

A less dramatic but still very interesting situation is frequently observed when wild mice are used for the mapping of mutations with deleterious phenotypic effects. In this case, the interspecific offspring, homozygous for the mutant allele, often exhibit a wide range of variations in the degree of severity of their phenotypes with severe forms and weaker ones. In these cases, genes or loci with a modifying effect can be identified, mapped and eventually cloned (Upadhya *et al.*, 1999; Sawamura *et al.*, 2000). Genes of this kind, which are of potentially great value cannot be recognized in an animal with a normal genotype.

Because many different inbred strains belonging to several more or less related taxa of the genus *Mus* are now easily available, it is possible to address questions aimed at a better understanding of genome structure and functions. For example: are all the genes present in one strain also present in the others, or are there differences and/or variations in the copy number? If the answer is that a particular gene exists in one strain and not in a closely related one, then what use is the gene in question? Examples of that kind have already been reported (Ye *et al.*, 2001) and have allowed fundamental questions to be answered in a very elegant way.

It would also be interesting to study certain categories of orthologous genes in closely related species to see how their pattern of spatio-temporal expression evolves and in what sort of sequence variation this evolution is involved. This can be particularly interesting when adaptive traits are concerned.

Investigations at the genomic level using inbred strains derived from wild mice are bound to become very popular in the near future because they can be achieved with a high level of refinement and can be correlated in a very reliable way to the phenotype of the living animal. At this point it is no exaggeration to say that this new type of mouse strain is bound to be of expanding interest and it is predictable that, in the future, the house mouse and its related species will be even more useful for scientific research than it has been over the last centuries, especially when more than one complete genome will be available for comparative purposes in the genus *Mus*.

List of relevant URL for Mouse Resource Centers:

Mouse Genome Informatics: http://www.informatics.jax.org

Mammalian Genetics Unit, Harwell, UK: http://www.mgu.har.mrc.ac.uk

RIKEN Bioresource Center: http://www.brc.riken.jp/en/

European Mouse Mutant Archive: http://www.emma.rm.cnr.it/

References

Auffray, J.C., Orth, A., Catalan, J., *et al.* (2003). *Zoologica Scripta* **32**, 119.

Beck, J.A., Lloyd, S., Hafezparast, M., *et al.* (2000). *Nat. Genet.* **24**, 23–25.

Berry, R.J. (1970). *Field Studies* **3**, 219–262.

Berry, R.J. (1987). *Biologist* **34**, 177–186.

Bishop, C.E., Boursot, P., Baron, B., *et al.* (1985). *Nature* **325**, 70–72.

Bonhomme, F. and Guénet, J.-L. (1996). In *Genetic Variants and Strains of the Laboratory Mouse* (eds M.F. Lyon, S. Rastan, and S.D.M. Brown), pp. 1577–1596. Oxford University Press, Oxford, New York, Tokyo.

Bonhomme, F., Guénet, J.-L., Dod, B., *et al.* (1987). *Biol. J. Linn. Soc. Lond.* **30**, 51–58.

Bonhomme, F., Anand, R., Darviche, D., *et al.* (1994). In *Genetics in Wild Mice* (eds K. Moriwaki, T. Shiroishi, and H. Yonekawa), pp. 13–23. Japan Sci. Soc. Press, Tokyo/S.Karger, Basel.

Boursot, P., Auffray, J.C., Britton-Davidian, J., *et al.* (1993). *Ann. Rev. Ecol. Syst.* **24**, 119–152.

Breen, M., Deakin, L., Macdonald, B., *et al.* (1994). *Hum. Mol. Genet.* **3**, 621–627.

Brinster, R.L., Chen, H.Y., Trumbauer, M., *et al.* (1981). *Cell* **27**, 223–231.

Cattanach, B.M. and Kirk, M. (1985). *Nature* **315**, 496–498.

Costantini, F. and Lacy, E. (1981). *Nature* **294**, 92–94.

Coviello-McLaughlin, G.M. and Prowse, K.R. (1997). *Nucleic Acids Res.* **25**, 3051–3058.

Cuénot, L. (1902). *Arch. Zool. Exp. Gén., 3e sér.* **10**, xxvii–xxx.

Elliott, R.W., Miller, D.R., Pearsall, R.S., *et al.* (2001). *Mamm. Genome* **12**, 45–51.

Epstein, C.J. (1986). *The Consequences of Chromosome Imbalance: Principle, Mechanisms and Models.* Cambridge University Press, New York.

Evans, M.J. and Kaufman, M.H. (1981). *Nature* **292**, 154–156.

Ferris, S.D., Sage, R.D. and Wilson, A.C. (1982). *Nature* **295**, 163–165.

Festing, M.F. (1979). *Inbred Strains in Biomedical Research.* The MacMillan Press Ltd., London and Basingstoke.

Flint, J. and Mott, R. (2001). *Nat. Rev. Genet.* **2**, 437–445.

Forejt, J. (1996). *Trends Genet.* **12**, 412–417.

Gardner, M.B., Kozak, C.A. and O'Brien, S.J. (1991). *Trends Genet.* **7**, 22–27.

Gordon, J.W. and Ruddle, F.H. (1981). *Science* **214**, 1244–1246.

Gordon, J.W., Scangos, G.A., Plotkin, D.J., *et al.* (1980). *Proc. Natl. Acad. Sci. USA* **77**, 7380–7384.

Gorer, P.A., Lyman, S. and Snell, G.D. (1948). *Proc. R. Soc. Lond. B* **135**, 499–505.

Greene-Till, R., Zhao, Y. and Hardies, S.C. (2000). *Mamm. Genome* **11**, 225–230.

Gropp, A. and Winking, H. (1981). *Zool. Soc. Lond. Symp.* **47**, 141–181.

Grüneberg, H. (1957). *Genes in Mammalian Development*. Lewis, H.K., London.

Guénet, J.L. (1986). In *The Wild Mouse in Immunology* (eds M. Potter, J.H. Nadeau, and M.P Cancro), pp. 109–113. Springer Verlag, Berlin, Heidelberg, New York, Tokyo.

Guénet, J.L. and Bonhomme, F. (2003). *Trends Genet.* **19**, 24–31.

Guénet, J.-L., Nagamine, C., Simon-Chazottes, D., *et al.* (1990). *Genet. Res.* **56**, 163–165.

Haller, O., Frese, M. and Kochs, G. (1998). *Rev. Sci. Tech.* **17**, 220–230.

Harbers, K., Jahner, D. and Jaenisch, R. (1981). *Nature* **293**, 540–542.

Huchon, D., Madsen, O., Sibbald, M.J., *et al.* (2002). *Mol. Biol. Evol.* **19**, 1053–1065.

Jacob, F. (1983). *Cold Spring Harbor Conf. Cell Proliferation* **10**, 683–687.

Jaeger, J.J., Tong, H. and E., B. (1986). *C. R. Acad. Sci. Paris* **302**, 917–922.

Keeler, C.E. (1931). *The Laboratory Mouse: Its Origin, Heredity, and Culture*. Harvard University Press, Cambridge, MA.

Kuehn, M.R., Bradley, A., Robertson, E.J., *et al.* (1987). *Nature* **326**, 295–298.

Lengeling, A., Pfeffer, K. and Balling, R. (2001). *Mamm. Genome* **12**, 261–271.

Lundrigan, B.L., Jansa, S.A. and Tucker, P.K. (2002). *Syst. Biol.* **51**, 410–431.

Lyon, M.F. (1961). *Nature* **190**, 372–373.

Martin, G.R. (1981). *Proc. Natl. Acad. Sci. USA* **78**, 7634–7638.

Matin, A. and Nadeau, J.H. (2001). *Trends Genet.* **17**, 727–731.

McGrath, J. and Solter, D. (1984). *Cell* **37**, 179–183.

Michaux, J., Reyes, A. and Catzeflis, F. (2001). *Mol. Biol. Evol.* **18**, 2017–2031.

Mintz, B. (1962). *Am. Zool.* **2**, 432.

Montagutelli, X., Turner, R. and Nadeau, J.H. (1996). *Genetics* **143**, 1739–1752.

Moriwaki, K., Shiroishi, T. and Yonekawa, H. (1994). *Genetics in Wild Mice. Its Application to Biomedical Research*. Japan Scientific Societies Press, Tokyo.

Morse, H.C., 3rd (1978). *Origins of Inbred Mice*. Academic Press, New York.

Morse, H.C., 3rd (1981). In *The Mouse in Biomedical Research* (eds H.L. Foster, J.D. Small and J.G. Fox), pp. 1–16. Academic Press, New York.

Murphy, W.J., Eizirik, E., Johnson, W.E., *et al.* (2001). *Nature* **409**, 614–618.

Musser, G.G. and Carleton, M.D. (1993). In *Mammal Species of the World. A Taxonomic and Geographic Reference* (eds D.E. Wilson and D.M. Reeder), pp. 501–755. Smithsonian Institution Press, Washington, D.C. and London.

Nagase, H., Mao, J.H., de Koning, J.P., *et al.* (2001). *Cancer Res.* **61**, 1305–1308.

Orth, A., Belkhir, K., Britton-Davidian, J., *et al.* (2002). *Comptes Rendus Biologies* **325**, 89–97.

Paigen, K. (2003a) *Genetics* **163**, 1–7.

Paigen, K. (2003b) *Genetics* **163**, 1227–1235.

Pilder, S.H., Olds-Clarke, P., Orth, J.M., *et al.* (1997). *J. Androl.* **18**, 663–671.

Potter, M. (1986). In *Current Topics in Microbiology and Immunology. The Wild Mouse in Immunology* (eds M. Potter, J.H. Nadeau, and M.P. Cancro). Springer-Verlag, New York.

Qi, C.-F., Bonhomme, F., Buckler-White, A., *et al.* (1998). *Mamm. Genome* **9**, 1049–1055.

Russell, E.S. (1978). In *Origins of Inbred Mice* (ed. H.C. Morse, 3rd), pp. 33–44. Academic Press, New York.

Sangster, M.Y., Mackenzie, J.S. and Shellam, G.R. (1998). *Arch. Virol.* **143**, 697–715.

Sawamura, K., Davis, A.W. and Wu, C.I. (2000). *Proc. Natl. Acad. Sci. USA* **97**, 2652–2655.

Sebastiani, G., Leveque, G., Lariviere, L., *et al.* (2000). *Genomics* **64**, 230–240.

Snell, G.D. (1978). In *Origins of Inbred Mice* (ed. H.C. Morse, 3rd), pp. 119–156. Academic Press, New York.

Staats, J. (1966). In *Biology of the Laboratory Mouse* (ed. E.L. Green). McGraw-Hill, New York and London.

Stevens, L.C. (1970). *Dev. Biol.* **21**, 364–382.

Stevens, L.C. and Little, C.C. (1954). *Proc. Natl. Acad. Sci. USA* **40**, 1080–1087.

Strong, L.C. (1978). In *Origins of Inbred Mice* (ed. H.C. Morse, 3rd), pp. 45–68. Academic Press, New York.

Surani, M.A., Barton, S.C. and Norris, M.L. (1984). *Nature* **308**, 548–550.

Tarkowski, A.K. (1961). *Nature* **184**, 1286–1287.

Tucker, P.K., Lee, B.K., Lundrigan, B.L., *et al.* (1992). *Mamm. Genome* **3**, 254–261.

Upadhya, P., Churchill, G., Birkenmeier, E.H., *et al.* (1999). *Genomics* **58**, 129–137.

Wade, C.M., Kulbokas, E.J., 3rd, Kirby, A.W., *et al.* (2002). *Nature* **420**, 574–578.

Wagner, E.F., Stewart, T.A. and Mintz, B. (1981a) *Proc. Natl. Acad. Sci. USA* **78**, 5016–5020.

Wagner, T.E., Hoppe, P.C., Jollick, J.D., *et al.* (1981b) *Proc. Natl. Acad. Sci. USA* **78**, 6376–6380.

Wakayama, T. and Yanagimachi, R. (1999). *Nat. Genet.* **22**, 127–128.

Waterston, R.H., Lindblad-Toh, K., Birney, E., *et al.* (2002). *Nature* **420**, 520–562.

West, J.D., Frels, W.I., Papaioannou, V.E., *et al.* (1977). *J. Embryol. Exp. Morphol.* **41**, 233–243.

Ye, X., Zhu, C. and Harper, J.W. (2001). *Proc. Natl. Acad. Sci. USA* **98**, 1682–1686.

Yonekawa, H. (1980). *Jpn. J. Genet.* **55**, 289–296.

Yonekawa, H., Moriwaki, K., Gotoh, O., *et al.* (1988). *Mol. Biol. Evol.* **5**, 63–78.

Zhu, L., Hathcock, K.S., Hande, P., *et al.* (1998). *Proc. Natl. Acad. Sci. USA* **95**, 8648–8653.

CHAPTER 2

Historical Foundations

Muriel T Davisson and Carol C Linder
The Jackson Laboratory, Bar Harbor, Maine, USA

Introduction

The laboratory mouse (derived from the common house mouse) has played a key role in mammalian genetic and biomedical research. Research using the mouse spans the twentieth century, from 1902 to 2002, from the birth of mammalian genetics to having the whole genome sequence. The ability to add to and selectively alter the mouse genome increased the power of the mouse as a research tool for understanding the genetic basis of human health and disease (Gordon et al., 1980; Mansour et al., 1988; Capecchi, 1989). Mutant and inbred mice frequently have syndromes similar to human inherited diseases because of their close metabolic and internal anatomical similarities to human beings. Genes in the mouse and human genomes are >99% conserved (Consortium, 2002). Hence, the mouse provides models for research not only on mammalian biology but also on a wide variety of human diseases including cancer, diabetes, aging, atherosclerosis, endocrine diseases, immunological diseases, autoimmunity, neurological dysfunction, and numerous others. Inbred and mutant mice are universally accepted as the primary model for analyzing and understanding inherited human disorders (Paigen, 1995, 2002; Davisson, 1999; O'Brien and Woychik, 2003).

The ultimate recognition of this value of the mouse was its selection as the first model organism to have its genome sequenced in the Human Genome Initiative (Battey et al., 1999; Consortium, 2002).

Historical foundations

The laboratory mouse originates from ancestors in the Middle East in the area that is now Pakistan. A commensal organism, mice have emigrated to most corners of the world as human beings' traveling companions. For a detailed history of the origins of the house mouse see Silver (1995). On the tiny island of Tenedos at the mouth of the Dardanelles stands a temple to Apollo God of Mice that predates the Trojan War. Albino mice were used in auguries for Egyptian rulers (Keeler, 1978). The earliest drawings of mice may be seen in Chinese prints as early as AD 300 and mutant mice, such as albino and waltzer, appear in eighteenth- and nineteenth-century Asian prints. Mouse fanciers of the late nineteenth and early twentieth centuries were the origin of most laboratory mice of today. The mouse fancy hobby originated in Asia and later spread to Europe and from there to America.

The Laboratory Mouse
Copyright 2004 Elsevier
ISBN 0-1233-6425-6

Because of their origins in the mouse fancy trade, laboratory mouse strains are a genetic mix of four different subspecies: *Mus musculus musculus* (eastern Europe), *Mus musculus domesticus* (western Europe), *Mus musculus castaneus* (Southeast Asia), and *Mus musculus molossinus* (Japan). The latter is thought to be a hybrid between *castaneus* and *musculus* (Moriwaki *et al.*, 1990). Genome analysis confirms that the laboratory mouse is a blend of these four different species or subspecies of the genus *Mus* (Wade *et al.*, 2002). Phylogenetically the house mouse (*Mus*) belongs to the family Muridae along with several other species of mice and the common rat.

Many inbred strains derive from Miss Abbie Lathrop, a mouse fancier who bred and sold mice in Granby, Massachusetts, from ~1900 to her death in 1918. She obtained her mice from dealers, European fanciers, and those captured in the wild. Although Miss Lathrop is often mentally pictured as a little old lady who collected fancy mice, she was an experimentalist and keen observer. She carried out cancer research experiments, collaborating with Dr Leo Loeb at the University of Pennsylvania (Lathrop and Loeb, 1915a,b, 1918). This collaboration grew out of her observation of tumor growths in her mice and her curiosity to learn more (Morse, 1978). She also carried out breeding experiments in collaboration with William Castle, and later Clarence Cook ('C.C.') Little, at the Bussey Institute at Harvard. Miss Lathrop's breeding records and notebooks, including such observations, are preserved in the library at The Jackson Laboratory.

History of mouse genetics and research with the laboratory mouse

Genetics

Ten major milestones mark the history of mouse genetics. The history of mouse genetics might have begun in the 1860s if an Augustinian bishop had not forbidden the breeding of mice within the monastery where Gregor Mendel did his classic genetic studies with sweet peas (Paigen, 2003). Thus, the first milestone in mouse genetics was the proof that mice, like sweet peas, had genes when French geneticist Lucien Cuénot

demonstrated that mammals show Mendelian inheritance using the inheritance of coat colors in mice (1902). He went on to demonstrate that a gene can have multiple alleles (1904) and that some alleles can be lethal, using the yellow allele of the agouti gene, A^y (1905). In 1903 William Castle at Harvard also published a paper on coat color genetics in mice (Castle and Allen, 1903). He and his student, C.C. Little, are often credited with the first cogent report and explanation of a lethal allele, also A^y (Castle and Little, 1910).

The second milestone was when mouse genetic research was initiated at the Bussey Institute for Research in Applied Biology at Harvard. William Ernest Castle directed the mammalian research program (Morse, 1985; Snell and Reed, 1993). Two of Castle's first students were Sewall Wright and C.C. Little. Most of the well-known names in the history of mouse genetics can be traced to the Bussey Institute. Examples include L.C. Dunn (developmental biology), L.C. Strong, C.C. Little, and Lloyd Law (cancer genetics), Clyde Keeler (behavior genetics), Paul Sawin (quantitative biology), and George Snell (histocompatibility genetics). A 'genealogical' tree of mouse geneticists drawn by Elizabeth ('Tibby') Russell and modified by Sandy Morse shows the extent to which students of the Bussey determined the future of mouse genetics (Morse, 1978). Genetic research using mice in the early twentieth century centered around coat color genetics, cancer, and tumor transplantability.

The third major milestone was the development of inbred strains of laboratory mice. C.C. Little is credited with conceiving of and creating the first inbred strains, although others including Miss Lathrop and Leonell Strong, were inbreeding mice at the same time. Breeders from C.C. Little's Line C (derived from Ms Lathrop's mice) founded the C57/C58 family of strains; females 57 and 58 were mated to male 52 to give rise to the C57BL, C57BR, and C58 inbred strains.

Most principles of genetics in the mid 1900s were established in nonmammalian species. Studies in *Drosophila* and micro-organisms led to the understanding of chromosomal theory, the nature of mutation, the discovery of the genetic code, and gene structure and function. During this period, however, the study of mouse genetics and gene mapping grew steadily, creating a solid foundation for the future of mouse genetics. In the 1960s and 1970s mouse genetics became prominent again with the recognition of the need for a mammalian model for biomedical research, the development of efficient genetic mapping tools in mice, and the realization of the high degree of genomic conservation between the mouse and human genomes. In the 1980s and 1990s, of course, mouse genomics burst again into the limelight at center stage with the development of powerful methods to manipulate the mouse genome.

The genetic map of the mouse

Genetic maps, the road maps of genetics, are of two types: linkage and physical. The 'sign posts' on the maps are loci, any location or marker in the genome that can be detected by genetic or DNA analysis. The term 'gene' is more restrictive than loci and refers to DNA segments that encode proteins or can be linked to phenotypes. Linkage maps are recombinational maps and are constructed by carrying out linkage crosses that measure the recombination frequency between genes or loci on the same chromosome. The first genetic linkage in the mouse (and first autosomal linkage in mammals) was described in 1915 in the classic paper on the linkage of pink-eyed dilution and albino (Haldane *et al.*, 1915). This fourth milestone of mouse genetics was the beginning of the genetic linkage mapping effort that continued through the rest of the twentieth century and into the twenty-first. Until the early 1970s, the mouse genetic map was composed of such linkage groups in which two or more genes were linked together. Linkage groups were not assigned to specific mouse chromosomes until the early 1970s (see next section on Physical Mapping). Composite linkage maps were compiled at The Jackson Laboratory (Margaret Dickie, Margaret Green, James Womack, Thomas Roderick, and Muriel Davisson) and at the Medical Research Council (MRC) Genetics Unit at Harwell in England (Colin Beechey, J. Butler, Susan Hawkes, and R. Meredith) by statistically combining data from all scientists' linkage crosses. As more and more genes were identified and mapped, the number of linkage groups and genes mapped within them grew increasingly rapidly until it was impossible to graphically depict the whole mouse linkage map in print publications. The last such published map was by Davisson *et al.* (1990).

Most mouse genetic and biological research in the middle of the twentieth century was carried out at a triumvirate of mouse research centers: the Harwell MRC Genetics Unit, the Biology Unit at the Atomic Energy Commission's facility in Oak Ridge, Tennessee, and The Jackson Laboratory in Bar Harbor, Maine. During the 1940s, the two major focuses for genetic mapping were identification of histocompatibility genes and discovery and characterization of visible, morphological markers resulting from spontaneous mutations. Study of the latter also provided the first mouse models of human inherited diseases. Genetic mapping with spontaneous mutations that created visible phenotypes, such as

changes in coat color/texture (e.g. albino, Tyr^c; piebald, $Ednrb^s$; satin, sa; fuzzy, fz) or behavior (e.g. waltzer, v; reeler, $Reln^{rl}$; shiverer, Mbp^{shi}) was laborious and sometimes took years. This was because crosses between mice carrying recessive mutations yielded so few informative progeny, and genes on only one or two chromosomes could be scored in each cross. Determining linkage demanded sophisticated statistical analysis and large numbers of progeny were required to obtain statistical significance (Green, 1963). It was not uncommon to generate thousands of intercross (F2) progeny – but the results did not lead to the high resolution maps that such crosses with today's genetic markers provide. During the 1950s and 1960s, linkage testing stocks (e.g. V/Le, SB/Le) combining multiple visible markers were created to speed the mapping process.

The first real breakthrough in linkage mapping, enabling the scoring of many test markers and chromosomes in the same cross, was the discovery and use of co-dominant biochemical (isoenzyme) genes (e.g. glucose phosphate isomerase 1, $Gpi1$; Hutton, 1969; Hutton and Coleman, 1969; Hutton and Roderick, 1970). Thus, the fifth milestone was the transition from visible markers to polymorphisms. In the 1980s and 1990s, DNA markers revolutionized genetic mapping. Their use was greatly facilitated by the development of the polymerase chain reaction (PCR) in 1983. DNA polymorphic markers, restriction fragment length polymorphisms (RFLPs) (Elliott, 1996), and later simple sequence length polymorphisms (SSLPs), such as the MIT markers (Dietrich *et al.*, 1994), are widespread throughout the genome. One of the biggest advantages of DNA markers for mapping is that newly discovered markers can be typed in indefinitely stored DNAs from linkage crosses or mapping panels. In 2003 came the identification and use of single nucleotide polymorphisms (SNPs) of which millions are present in the mouse genome (Lindblad-Toh *et al.*, 2000).

During this period as well, several scientists developed inbred strains from wild populations to increase genetic variability in mapping crosses (Kozak *et al.*, 1984; Guenet, 1986). Verne Chapman (Roswell Park), Michael Potter (National Institutes of Health), Jean-Louis Guenet (Pasteur Institute, France), and Eva Eicher and Tom Roderick (The Jackson Laboratory) developed wild-derived inbred strains. *Mus musculus castaneus* (e.g. CAST/Ei), *Mus spretus* (e.g. SPRET/Ei), and *Mus musculus molossinus* (e.g. MOLD/Rk) are the most widely used. Johnson *et al.* (1994) improved mapping mutations with RFLP loci by combining intersubspecific intercrosses (F2) using inbred CAST/Ei with RFLPs for gene families, allowing multiple genes to be detected on a single Southern blot. Taylor *et al.* (1994) further improved the efficiency of mapping mutations with polymorphic markers by pooling F2 progeny DNAs for the initial genome screen with PCR. Panels of backcross DNAs have been developed for efficient mapping of polymorphic markers that can be typed in DNA. The C57BL/6Ros × *Mus spretus* panel of Neal Copeland and Nancy Jenkins (Ceci *et al.*, 1994) and the C57BL/6J × SPRET/Ei backcross panel at The Jackson Laboratory are mapping resources available to investigators (Rowe *et al.*, 1994).

Identification (or cloning) of mutated genes in the last decades of the twentieth century was made possible by the development of libraries of artificial chromosomes containing inserts of mammalian DNA: plasmid clones (P1-derived artificial chromosomes, PACs), bacterial artificial chromosomes (BACs), and yeast artificial chromosomes (YACs), listed in order of size. High resolution genetic crosses with sometimes thousands of progeny are used to narrow the chromosomal interval harboring the gene of interest. Once a mutant gene is mapped to a segment less than a fraction of a centiMorgan, contigs of overlapping clones can be constructed across the region by hybridizing PACs, BACs, and/or YACs to each other. Although laborious, this approach to identifying mutated genes became increasingly successful as DNA sequencing technologies and quality of libraries improved. With the electronic publication of the entire mouse genome sequence (CDS, 2001; Consortium, 2002), candidate gene cloning became more common and has yielded gene identification more rapidly than positional cloning did. It is now a simple matter to examine electronically the chromosomal interval identified by the high resolution genetic cross for candidate genes whose mutation might lead to the phenotype observed. With both methods, candidate genes are tested by assessing RNA expression levels or examining the gene itself by Southerns for large DNA alterations and, ultimately, by sequencing exons. In either case, that the mutated gene identified causes the phenotype must be proved by showing mutations in the same gene in multiple alleles (if available) or by creating transgenic mice with the mutated gene.

The physical map of the mouse

Of course, a major milestone for all genetics, the sixth for the mouse, was the determination of DNA's physical structure in 1953 by Watson and Crick (1953) using X-ray crystallography images produced by Rosalind Franklin. The seventh major milestone in mouse genetics was the development of chromosomal banding techniques in the late 1960s. This was particularly important because all laboratory mouse chromosomes are telocentric (the centromere is located at one end) and identification of most individual chromosomes prior to banding technology was virtually impossible. Development of fluorescent Q-banding (Q- for quinacrine; Francke and Nesbitt, 1971) and G-banding (Giemsa; Buckland *et al.*, 1971) enabled the assignment of linkage groups to physical chromosomes by combining knowledge of linkage groups associated with reciprocal translocations with cytological identification of the chromosomes involved based on alterations in banding patterns (Miller *et al.*, 1971). This approach is credited to John Hutton but drew on the accumulated data of many laboratories. However, prior to chromosomal banding, Eva Eicher assigned the first linkage group to a chromosome, based on the size and unusual unbanded cytological appearance of Chromosome 19 and linkage analysis of Linkage Group XII (Eicher, 1971). The first composite linkage map showing both chromosome numbers and linkage groups was published by Margaret Green (1973). Davisson and Roderick (1978) published the first linkage map in which linkage groups were adjusted to chromosome size based on physical measurements of the chromosomes (Committee, 1972). The final advance in the cytological physical map was the development of fluorescent *in situ* hybridization (FISH; Kranz *et al.*, 1985). FISH allows mapping of single genes to cytological bands on the physical chromosome and identification of chromosomal rearrangements using paints (Rabbitts *et al.*, 1995; Liyanage *et al.*, 1996).

Because genetic crosses are possible in mice, somatic cell hybrid panels, enabling the assignment of genes to chromosomes, were never used in mouse gene mapping to the extent they were in human mapping. Somatic cells of two species are fused and one species' chromosomes are segregated out during cell line propagation, creating a panel of cell lines, each with one or a few chromosomes of the species of interest. Radiation hybrid (RH) panels, in which the chromosomes of the species of interest have been fragmented by irradiation prior to fusion, have been used effectively in the mouse as well as human to physically locate and order genes in chromosomal segments. The T31 mouse RH mapping panel also is available through The Jackson Laboratory mapping resource (Rowe *et al.*, 2003).

Mary Lyon at Harwell maintained, and published in *Mouse News Letter* for many years, the composite 'Chromosome Atlas' map, which combined linkage data and physical mapping by FISH or cytological location of chromosomal rearrangement breakpoints. Updates of the Atlas were published for several of the final issues of *Mouse News Letter*.

Comparative mapping

The mouse is a powerful model organism for research on human disease because it is a mammal and because of the high degree of conservation between the mouse and human genomes. Comparative mapping began in the early 1970s and gained momentum until it culminated with the sequencing of the two genomes in 2001 and 2002. The first conserved mouse and human autosomal linkage was reported in 1978 (Lalley *et al.*, 1978). At the 4th Human Gene Mapping (HGM) Workshop, six autosomal linkages and the X Chromosome, were known (Pearson and Roderick, 1978). In 1984, Nadeau and Taylor (1984) identified 13 conserved autosomal segments and estimated 178 ± 39 chromosomal rearrangements between mouse and human chromosomes. By the last formal meeting of the HGM Comparative Mapping Committee, 105 conserved linkages were known (Andersson *et al.*, 1996). Sequencing of the two genomes revealed that 95% of the coding sequence is conserved at the DNA level (Consortium, 2002).

Bioinformatics

The development of large, comprehensive databases must be recognized as a milestone, the eighth, because without their development the rapidly increasing accumulation of genetic and biological data in the last three decades of the twentieth century would be overwhelming and impossible to manage. In the late 1970s, the linkage data Margaret Green had accumulated on 4×6 in. cards was entered into a computer program known as GBASE (the Genetic Database of the Mouse) developed by Thomas Roderick, Muriel Davisson, and Carolyn Blake at The Jackson Laboratory. The data were proof read by Mary Lyon while she was on an extended visit to the Laboratory. GBASE became the first on-line database of mouse genetic information. Subsequently, Margaret Green's catalog describing mouse genes (Green, 1979) was added as *Mouse Locus Catalog*, which was maintained for many years by Donald Doolittle. At about this time, Thomas Roderick coined the term 'genomics' for the new journal of that name, providing a name for the expanded science that encompasses genetic mapping, sequencing, and genome analysis. In 1995, GBASE was combined with a linkage analysis database developed by Janan Eppig and homology database developed by Joseph

Nadeau to become the predecessor of the Mouse Genome Database (MGD). Today's Mouse Genome Informatics program at The Jackson Laboratory encompasses the MGD of genomic and phenotype information, a gene expression database (GXD), a mouse genome sequence (MGS) analysis database, a tumor biology database, and related analysis tools (MGI, 2003). The advent of DNA sequencing, culminating in the sequencing of entire genomes, has generated sophisticated bioinformatics systems beyond the scope of this chapter to describe. The many sequence databases and analysis software packages that are available are valuable tools for the mouse geneticist. A critical aspect of all genetic mapping and the bioinformatics programs that support it is the use of controlled genetic nomenclature. Conventions of genetic nomenclature are described in Chapter 3 on Strains, Stocks, and Mutant Mice.

Genetic manipulation of the mouse genome

The ninth milestone in mouse genetics – genetic engineering – has catapulted the mouse into the lead mammalian model organism for biomedical research. With transgenesis and gene targeting, it is possible to selectively modulate the amount or composition of a gene product (see Chapter 5 on Generation of Mouse Mutants by Sequence Information Driven and Random Mutagenesis for a detailed explanation). The first technology introduced was the insertion of foreign genes into mouse chromosomes to produce gain-of-function mutants. The first transgenic mouse was created and described by Jon Gordon in Frank Ruddle's laboratory in 1980 (Gordon *et al.*, 1980). Creation of loss-of-function mutations followed the pioneering demonstration by Leroy Stevens that embryonic teratocarcinoma cell lines could give rise to differentiated tissues that led to the discovery that pluripotent embryonic stem (ES) cells could be grown in culture (Evans and Kaufman, 1981). In 1988, two research groups reported the first successful alteration of a mouse gene by homologous recombination or targeting (Doetschman *et al.*, 1988; Mansour *et al.*, 1988; Capecchi, 1989). The early efforts with this technology basically created null mutations or 'knockouts'. The discovery that many such mutations led to embryonic lethality instigated the development of targeting technology that makes it possible to determine tissue and temporal specificity using conditional mutation systems. The first developed was the *creLox* system. Mice carrying a transgene containing the gene for the prokaryotic *cre* recombinase enzyme linked to a tissue-specific promotor are mated with mice carrying insertions of the LoxP target sequence flanking the gene to be removed (Utomo *et al.*, 1999; Nagy, 2000). Tissue-specific expression of the *cre* enzyme enhances recombination between the LoxP sites and deletes the targeted gene's function in that tissue. A similar system can be created with the Flp recombinase (Dymecki, 1996; Rodriguez *et al.*, 2000). Temporal control is achieved using Tetracycline-inducible mutations (Saam and Gordon, 1999; Schonig and Bujard, 2003). Finally, it is possible to replace an endogenous gene with another functional gene (Hanks *et al.*, 1995). We can now go from gene to phenotype (reverse genetics) as well as from phenotype to gene (forward genetics). Nevertheless, it should be noted that being able to manipulate specific genes is still a long way from being able to generate specific phenotypes; frequently targeted mutations cause an unexpected phenotype or, sometimes, no detectable phenotype at all.

High-throughput mutagenesis is making it possible to increase the mutation rate and screen for subtle phenotypes that will be the key to identifying novel genes. The widespread use of the powerful mutagen ethylnitrosourea (ENU) has developed from the research of William Russell's group at Oak Ridge National Laboratory (Russell *et al.*, 1979). Several large-scale mutagenesis centers or multi-center programs were established in Europe and North America during the latter 1990s (e.g. Hrabé de Angelis *et al.*, 2000; Justice, 2000; Nolan *et al.*, 2000). Insertional mutagenesis can be used to randomly mutate genes by insertion of a DNA sequence which can be subsequently used to identify the mutated gene. The first of these was developed by Rick Woychik, then at Oak Ridge; using insertion of transgenes carrying a selectable construct (Moyer *et al.*, 1994). This approach has largely been replaced by gene-trapping technology using sequences that integrate only into gene-specific genomic regions such as promotors (Friedrich and Soriano, 1991) or other gene-specific sequences in ES cells (Skarnes *et al.*, 1995; Stanford *et al.*, 2001).

Sequencing the mouse genome

In 1999, three major sequencing centers, the Wellcome Trust Sanger Institute, the Whitehead Center for Genome Research, and Washington University Genome Sequencing Center, combined efforts to form the Mouse Genome Sequencing Consortium (MGSC) to sequence the mouse genome. C57BL/6J was chosen as the first inbred strain to be sequenced. The ultimate physical map

is the sequence of the genome. This tenth major milestone was achieved in the year 2001, only 1 year after this achievement for the human genome and 2 years ahead of schedule, when Celera (2001) announced completing first pass sequence in May of 2001 of a mixed mouse genome derived from the following inbred strains: 129S1/SvImJ, 129X1/SvJ, A/J, C57BL/6, and DBA/2J (CDS, 2001). In December 2002, the MGSC public consortium announced the complete draft sequence of C57BL/6J (Consortium, 2002). One hundred years after Mendel's principles were shown to operate in the mouse, we can relate the genetic map to the cytological map to the ultimate physical map of the genome. Yet having the sequence reinforces the need for the genetic map because we need to genetically map traits in order to locate their positions in the sequence and identify candidate genes.

Biomedical research

The earliest biomedical research using the mouse involved the genetics of tumor transplantability and cancer susceptibility. Initial studies suggested that the genetic component in these traits was weak or non-existent because of the use of outbred mice and the complexity of the trait (Loeb, 1902; Tyzzer, 1909). The development of defined genetic backgrounds by inbreeding encouraged the continuation and growth of research on cancer.

C.C. Little and others continued to develop inbred strains to analyze the genetics of susceptibility and resistance to cancer and tumor transplantation (Strong, 1978). Haldane suggested in 1933 that cancer had a genetic component (Haldane, 1933). Jacob Furth at the University of Pennsylvania developed the high leukemia strain AKR (Furth, 1978). Leonell C. Strong studied the genetics of susceptibility and resistance to tumor transplantation. Howard Andervont studied the genetics and viral etiology of cancer and went on to head the National Cancer Institute until 1961 (Deringer, 1978). Walter Heston began his early work on lung cancer (Heston, 1978). In 1933 the Staff of The Jackson Laboratory published a paper describing maternal inheritance of mammary tumors in mice (Staff, 1933). This phenomenon was later shown to be nonmendelian transmission by a factor in the dam's milk (Bittner, 1936), which was subsequently identified as a virus (Bryan *et al.*, 1942; Visscher *et al.*, 1942). Ultimately, it was shown that the virus could be integrated into the mouse genome and transmitted as a gene by both females and males.

George Snell, inspired by Little and his early work on tumor transplantation, began his lifelong study of the genetics of transplantation (Snell, 1978) that became the basis for all histocompatibility and tissue transplantation research. He joined the staff at The Jackson Laboratory in 1935. During the 1940s, Snell created congenic strains (strains differing at a locus of interest and a linked chromosomal segment carried over during backcrossing) to isolate, identify, and map genes involved in tissue rejection or acceptance (Snell, 1948). He rediscovered the mouse major histocompatibility complex (MHC), described by Peter Gorer in 1938 as a red blood cell antigen affecting transplantation. Snell and Gorer's joint research on histocompatibility genes (Gorer *et al.*, 1948) became the foundation for the medical field of tissue and organ transplantation in humans today. Because Gorer died prematurely, it was Snell who was awarded the Nobel prize in 1980 for this research.

During the 1940s and 1950s, spontaneous and induced mutations created in radiation risk assessment studies following the Second World War were used to map genes in the mouse. As the effects of these mutations were studied, the potential of mutant mice to provide research tools for studying human inherited disease became evident and the types of biomedical research broadened. Areas of research expanded from cancer genetics and histocompatibility to hematopoietic stem cell research, hematological disorders, skeletal abnormalities, neurological and neuromuscular diseases, kidney disease, and many more.

In the 1980s and 1990s molecular technology advances made it possible (1) to more rapidly identify mutated genes and (2) to genetically manipulate the mouse genome to alter genes shown to be mutated in human diseases (see section on Genetic Manipulation of the Mouse Genome). The transgenic strain overexpressing the promoter region and exon 1 of the human Huntington's disease gene, B6CBA-TgN(HDexon1) 62Gpb/J, causes disease symptoms that mimic the human condition beginning at 9–11 weeks of age (Mangiarini *et al.*, 1996). A transgenic mouse carrying the human superoxide dismutase 1 gene with the mutation associated with amyotrophic lateral sclerosis (ALS, Lou Gehrig's disease) provides a good model for that disease (Gurney *et al.*, 1994). One of the early targeted mutations created was in the gene encoding the transmembrane protein mutated in human cystic fibrosis (*Cftr*; Koller *et al.*, 1991), but the mutant mice died early in life because of intestinal abnormalities. Combining the *Cftr* knockout with a functional human *CFTR* transgene expressed in the intestine allowed cystic fibrosis null mice to survive long enough to provide a model to study and test therapies for the debilitating lung phenotype that affects human

patients (Manson *et al.*, 1997). Recently created conditional targeted mutations allow control of the tissue specificity of the mutation or onset of gene expression (temporal control; Utomo *et al.*, 1999; Nagy, 2000; Schonig and Bujard, 2003).

Completion of the draft sequence of the mouse genome (Consortium, 2002) provides the most comprehensive map of the mouse genome to date. Direct sequence analysis enables more rapid identification of mutated genes and sequence mining to identify as yet unrecognized genes. The mouse is expected to be a major player in the post genome era of identifying the functions of these genes.

Large-scale mutagenesis programs using mice are taking advantage of powerful mutagens and sophisticated bioimaging and behavioral screening methods to provide tools for understanding how mouse, and ultimately mammalian, genes function and what disorders occur when their function goes awry. A selected list of electronic resources available to gain entry into the broad areas of mouse genetics and biomedical research is provided in Chapter 3 on Strains, Stocks, and Mutant Mice.

The future

The future of the mouse in genome analysis and as a model organism seems virtually unlimited. Whole genome sequence and gene prediction programs make it quite feasible to knock out or genetically modify every mouse gene. Almost certainly, ES cells will play a key role in future mouse genomics because they can be manipulated in culture. More phenotype screens that can detect mutations prior to making live mice will be needed to increase the number and types of mutations that can be detected in the ES cells themselves. ES cell deletion banks will allow the recovery of recessive mutations in ES cells (You *et al.*, 1997). More than 5000 knockouts and more than 9000 gene-trapped mouse genes already were estimated in 2003 (Nagy *et al.*, 2003). In 1998, Cumulina the mouse joined Dolly in being a mammal that can be cloned from somatic cells (Wakayama *et al.*, 1998), making possible the creation of mice from somatic mutations. This technology is only likely to be exploited to its full potential if the efficiency and success rate of whole mouse cloning equals or exceeds that of mutating ES cells and turning them into mice. The whole mouse will continue to be the final test bed for determining how genes function, the science of functional genomics. Ironically, one of the most rapidly growing areas of mouse genomic research, the genetic analysis of complex traits and diseases, combines the latest molecular technology, such as SNPs for gene mapping, with breeding technologies developed in the mid-twentieth century. Initiatives have begun to develop larger panels of recombinant inbred strains and their variant recombinant congenic strains, and panels of consomic (chromosome substitution) strains (Nadeau *et al.*, 2000; Threadgill *et al.*, 2002). In the latter, one can search for complex trait genes on individual chromosomes and then narrow the analysis with segmental congenic strains. Thus, while other model organisms, such as *Drosophila*, yeast, worms (*C. elegans*), and zebrafish may be easier to manipulate and allow analyses that require hundreds or thousands of animals, the mouse is likely to continue to be the premier mammalian model for understanding human inherited diseases.

Acknowledgments

We are grateful to Herbert 'Sandy' Morse, III, Elizabeth 'Tibby' Russell, Ken Paigen, Lee Silver, Eva Eicher, and Margaret Green from whose previous writings on the history of the laboratory mouse and mouse genetics we have borrowed heavily.

References

Andersson, L., Archibald, A., Ashburner, M., Audun, S., Barendse, W., Bitgood, J., Bottema, C., Broad, T., Brown, S., Burt, D., Charlier, C., Copeland, N., Davis, S., Davisson, M., Edwards, J., Eggen, A., Elgar, G., Eppig, J.T., Franklin, I., Grewe, P., Gill, T., 3rd, Graves, J.A.M., Hawken, R., Hetzel, J., Hillyard, A., Jacob, H., Jaswinska, L., Jenkins, N., Kunz, H., Levan, G., Lie, O., Lyons, L., Maccarone, P., Mellersh, C., Montgomery, G., Moore, S., Moran, C., Morizot, D., Neff, M., Nicholas, F., O'Brien, S., Parsons, Y., Peters, J., Postlethwait, J., Raymond, M., Rothschild, M., Schook, L., Sugimoto, Y., Szpirer, C., Tate, M., Taylor, J., VandeBerg, J., Wakefield, M., Wienberg, J. and Womack, J. (1996). *Mamm. Genome* 7, 717–734.

Battey, J., Jordan, E., Cox, D. and Dove, W. (1999). *Nat. Genet.* 21, 73–75.

Bittner, J.J. (1936). *Science* 84, 162.

Bryan, W.R., Kahler, H., Shimkin, M.B. and Andervont, H.B. (1942). *J. Natl. Cancer Inst.* 2, 451.

Buckland, R.A., Evans, H.J., Sumner, A.T. (1971). *Exp. Cell Res.* 69, 231–236.

Capecchi, M.R. (1989). *Science* 244, 1288–1292.

Castle, W.E. and Allen, G.M. (1903). *Proc. Am. Acad. Arts Sci.* 38, 602–622.

Castle, W.E. and Little, C.C. (1910). *Science* 32, 868–870.

Ceci, J.D., Matsuda, Y., Grubber, J.M., Jenkins, N.A., Copeland, N.G. and Chapman, V.M. (1994). *Genomics* 19, 515–524.

Celera Discovery System (CDS) (2001). http://cds.celera.com

Committee on Standardized Genetic Nomenclature for Mice. (1972). *J. Hered.* 63, 69–72.

Consortium (Mouse Genome Sequencing Consortium). (2002). *Nature* 420, 520–562.

Cuénot, L. (1902). *Arch. Zool. Exp. Gen.* 3, 27–30.

Cuénot, L. (1904). *Arch. Zool. Exp. Gen.* 4, 33–38.

Cuénot, L. (1905). *Arch. Zool. Exp. Gen.* 4, 123–132.

Davisson, M.T. (1999). *Lab. Anim.* (Millenium Issue) 28, 2–5.

Davisson, M.T. and Roderick, T.H. (1978). *Cytogenet. Cell Genet.* 22, 552–557.

Davisson, M.T., Roderick, T.H., Doolittle, D.P., Hillyard, A.L. and Guidi, J.N. (1990). In *Genetic Maps*, 5th edn (ed. S.J. O'Brien), pp. 4.3–4.14. Cold Spring Harbor Laboratory Press, Cold Spring Harbor.

Deringer, M.K. (1978). In *Origins of Inbred Mice*, Proceedings of a Workshop, Bethesda, MD, February 1978 (ed. H.C. Morse, III), pp. 99–102. Academic Press, New York.

Dietrich, W.F., Miller, J.C., Steen, R.G., Merchant, M., Damron, D., Nahf, R., Gross, A., Joyce, D.C., Wessel, M., Dredge, R.D., Marquis, A., Stein, L.D., Goodman, N., Page, D.C. and Lander, E.S. (1994). *Nat. Genet.* 7, 220–245.

Doetschman, T., Maeda, N. and Smithies, O. (1988). *Proc. Natl. Acad. Sci. USA* 85, 8583–8587.

Dymecki, S.M. (1996). *Proc. Natl. Acad. Sci. USA* 93, 6191–6196.

Eicher, E.M. (1971). *Genetics* 69, 267–271.

Elliott, R.W. (1996). In *Genetic Variants and Strains of the Laboratory Mouse*, 3rd edn (eds M.F. Lyon, S. Rastan and S.D.M. Brown), pp. 1312–1423. Oxford University Press, New York.

Evans, M.J. and Kaufman, M.H. (1981). *Nature* 292, 154–156.

Francke, U. and Nesbitt, M.N. (1971). *Cytogenetics* 10, 356–366.

Friedrich, G. and Soriano, P. (1991). *Genes Dev.* 5, 1513–1523.

Furth, J. (1978). In *Origins of Inbred Mice*, Proceedings of a Workshop, Bethesda, MD, February 1978 (ed. H.C. Morse, III), pp. 69–97. Academic Press, New York.

Gordon, J.W., Scangos, G.A., Plotkin, D.J., Barbosa, J.A. and Ruddle, F.H. (1980). *Proc. Natl. Acad. Sci. USA* 77, 7380–7384.

Gorer, P.A., Lyman, S. and Snell, G.D. (1948). *Proc. R. Soc. Lond. (Biol.)* **135**, 499.

Green, M.C. (1963). In *Methodology in Mammalian Genetics* (ed. W.J. Burdette), pp. 56–82. Holden-Day, San Francisco.

Green, M.C. (1973). *Mouse News Lett.* **49**, 17.

Green, M.C. (1979). In *Genetic Variants and Strains of the Laboratory Mouse* (ed. M.C. Green), pp. 8–278. Oxford University Press, New York.

Guenet, J.L. (1986). *Curr. Top. Microbiol. Immunol.* **127**, 109–113.

Gurney, M.E., Pu, H., Chiu, A.Y., Dal Canto, M.C., Polchow, C.Y., Alexander, D.D., Caliendo, J., Hentati, A., Kwon, Y.W., Deng, H.X., *et al.* (1994). *Science* **264**, 1772–1775.

Haldane, J.B.S. (1933). *Nature* **132**, 265–267.

Haldane, J.B.S., Sprunt, A.D. and Haldane N.M. (1915). *J. Genet.* **5**, 133–135.

Hanks, M., Wurst, W., Anson-Cartwright, L., Auerbach, A.B. and Joyner, A.L. (1995). *Science* **269**, 679–682.

Heston, W.E. (1978). In *Origins of Inbred Mice*, Proceedings of a Workshop, Bethesda, MD, February 1978 (ed. H.C. Morse, III), pp.109–117. Academic Press, New York.

Hrabé de Angelis, M.H., Flaswinkel, H., Fuchs, H., Rathkolb, B., Soewarto, D., Marschall, S., Heffner, S., Pargent, W., Wuensch, K., Jung, M., Reis, A., Richter, T., Alessandrini, F., Jakob, T., Fuchs, E., Kolb, H., Kremmer, E., Schaeble, K., Rollinsk, B., Roscher, A., Peters, C., Meitinger, T., Strom, T., Steckler, T., Holsboer, F., Klopstock, T., Gekeler, F., Schindewolf, C., Jung, T., Avraham, K., Behrendt, H., Ring, J., Zimmer, A., Schughart, K., Pfeffer, K., Wolf, E. and Balling, R. (2000). *Nat. Genet.* **25**, 444–447.

Hutton, J.J. (1969). *Biochem. Genet.* **3**, 507–515.

Hutton, J.J. and Coleman, D.L. (1969). *Biochem. Genet.* **3**, 517–523.

Hutton, J.J. and Roderick, T.H. (1970). *Biochem. Genet.* **4**, 339–350.

Johnson, K.R., Cook, S.A. and Davisson, M.T. (1994). *Mamm. Genome* **5**, 670–687.

Justice, M.J. (2000). *Nat. Rev. Genet.* **1**, 109–115.

Keeler, C. (1978). In *Origins of Inbred Mice*, Proceedings of a Workshop, Bethesda, MD, February 1978 (ed. H.C. Morse, III), pp. 179–193. Academic Press, New York, 719 pp.

Koller, B.H., Kim, H.S., Latour, A.M., Brigman, K., Boucher, R.C. Jr., Scambler, P., Wainwright, B. and Smithies, O. (1991). *Proc. Natl. Acad. Sci. USA* **88**, 10730–10734.

Kozak, C.A., Hartley, J.W. and Morse, H.C. III. (1984). *J. Virol.* **51**, 77–80.

Kranz, D.M., Saito, H., Disteche, C.M., Swisshelm, K., Pravtcheva, D., Ruddle, F.H., Eisen, H.N. and Tonegawa, S. (1985). *Science* **227**, 941–945.

Lalley, P.A., Francke, U. and Minna, J.D. (1978). *Proc. Natl. Acad. Sci. USA* **75**, 2382–2386.

Lathrop, A.E.C. and Loeb, L. (1915a). *J. Exp. Med.* **22**, 646–673.

Lathrop, A.E.C. and Loeb, L. (1915b). *J. Exp. Med.* **22**, 713–731.

Lathrop, A.E.C. and Loeb, L. (1918). *J. Exp. Med.* **28**, 475–500.

Lindblad-Toh, K., Winchester, E., Daly, M.J., Wang, D.G., Hirschhorn, J.N., Laviolette, J.P., Ardlie, K., Reich, D.E., Robinson, E., Sklar, P., Shah, N., Thomas, D., Fan, J.B., Gingeras, T., Warrington, J., Patil, N., Hudson, T.J. and Lande, E.S. (2000). *Nat. Genet.* **24**, 381–386.

Liyanage, M., Coleman, A., du Manoir, S., Veldman, T., McCormack, S., Dickson, R.B., Barlow, C., Wynshaw-Boris, A., Janz, S., Wienberg, J., Ferguson-Smith, M.A., Schrock, E. and Ried, T. (1996). *Nat. Genet.* **14**, 312–315.

Loeb, L. (1902). *J. Med. Res.* **3**, 44–82.

Mangiarini, L., Sathasivam, K., Seller, M., Cozens, B., Harper, A., Hetherington, C., Lawton, M., Trottier, Y., Lehrach, H., Davies, S.W. and Bates, G.P. (1996). *Cell* **87**, 493–506.

Manson, A.L., Trezise, A.E., MacVinish, L.J., Kasschau, K.D., Birchall, N., Episkopou, V., Vassaux, G., Evans, M.J., Colledge, W.H., Cuthbert, A.W. and Huxley, C. (1997). *EMBO J.* **16**, 4238–4249.

Mansour, S.L., Thomas, K.R. and Capecchi, M.R. (1988). *Nature* **336**, 348–352.

Miller, O.J., Miller, D.A., Kouri, R.E., Allderdice, P.W., Dev, V.G., Grewal, M.S. and Hutton, J.J. (1971). *Proc. Natl. Acad. Sci. USA* **68**, 1530–1533.

Moriwaki, K. Sagai, T., Shiroishi, T., Bonhomme, F., Wang, C.-H., He, X.-Q., Jin, M-L. and Wu, Z-A. (1990). *Biol. J. Linnean Soc.* **41**, 125–139.

Morse, H.C., III. (1978). In *Origins of Inbred Mice*, Proceedings of a Workshop, Bethesda, MD, February 1978 (ed. H.C. Morse, III), pp. 3–21. Academic Press, New York.

Morse, H.C., III. (1985). *Immunogenetics* **21**, 109–116.

Mouse Genome Informatics (MGI), The Jackson Laboratory, Bar Harbor, Maine. (2003). http://www.informatics.jax.org/

Moyer, J.H., Lee-Tischler, M.J., Kwon, H.Y., Schrick, J.J., Avner, E.D., Sweeney, W.E., Godfrey, V.L., Cacheiro, N.L., Wilkinson, J.E., and Woychik, R.P. (1994). *Science* **264**, 1329–1333.

Nadeau, J.H., Singer, J.B., Matin, A. and Lander, E.S. (2000). *Nat. Genet.* **24**, 221–225.

Nadeau, J.H. and Taylor, B.A. (1984). *Proc. Natl. Acad. Sci. U.S.A.* **81**, 814–818.

Nagy, A. (2000). *Genesis* **26**, 99–109.

Nagy, A., Perrimon, N., Sandmeyer, S. and Plasterk, R. (2003). *Nat. Genet.* **33** (Suppl.), 276–284.

Nolan, P.M., Peters, J., Strivens, M., Rogers, D., Hagan, J., Spurr, N., Gray, I.C., Vizor, L., Brooker, D., Whitehill, E., Washbourne, R., Hough, T., Greenaway, S., Hewitt, M., Liu, X., McCormack, S., Pickford, K., Selley, R., Wells, C., Tymowska-Lalanne, Z., Roby, P., Glenister, P., Thornton, C., Thaung, C., Stevenson, J.A., Arkell, R., Mburu, P.,

Hardisty, R., Kiernan, A., Erven, A., Steel, K.P., Voegeling, S., Guenet, J.L., Nickols, C., Sadri, R., Nasse, M., Isaacs, A., Davies, K., Browne, M., Fisher, E.M., Martin, J., Rastan, S., Brown, S.D. and Hunter, J. (2000). *Nat. Genet.* **25**, 440–443.

O'Brien, T. and Woychik, R. (2003). *Nat. Genet.* **33**, 1–2.

Paigen, K. (1995). *Nat. Med.* **1**, 215–220.

Paigen, K. (2002). *ILAR J.* **43**, 123–135.

Paigen, K. (2003). *Genetics* **163**, 1–7.

Pearson, P.L. and Roderick, T.H. (1978). *Cytogenet. Cell Genet.* **22**, 150–162.

Rabbitts, P., Impey, H., Heppell-Parton, A., Langford, C., Tease, C., Lowe, N., Bailey, D., Ferguson-Smith, M. and Carter, N. (1995). *Nat. Genet.* **9**, 369–375.

Rodriguez, C.I., Buchholz, F., Galloway, J., Sequerra, R., Kasper, J., Ayala, R., Stewart, A.F. and Dymecki, S.M. (2000). *Nat. Genet.* **25**, 139–140.

Rowe, L.B., Nadeau, J.H., Turner, R., Frankel, W.N., Letts, V.A., Eppig, J.T., Ko, M.S., Thurston, S.J. and Birkenmeier, E.H. (1994). *Mamm. Genome* **5**, 253–274.

Rowe, L.B., Barter, M.E., Kelmenson, J.A. and Eppig, J.T. (2003). *Genome Res.* **13**, 122–133.

Russell, W.L., Kelly, E.M., Hunsicker, P.R., Bangham, J.W., Maddux, S.C. and Phipps, E.L. (1979). *Proc. Natl. Acad. Sci. USA* **76**, 5818–5819.

Saam, J.R. and Gordon, J.I. (1999). *J. Biol. Chem.* **274**, 38071–38082.

Schonig, K. and Bujard, H. (2003). In *Transgenic Mouse Methods and Protocols* (eds M. Hofker and J. van Deursen), pp. 69–104. Humana Press, Totowa, New Jersey.

Silver, L.M. (1995). *Mouse Genetics, Concepts and Applications*, pp. 3–31. Oxford University Press, New York.

Skarnes, W.C., Moss, J.E., Hurtley, S.M. and Beddington, R.S. (1995). *Proc. Natl. Acad. Sci. USA* **92**, 6592–6596.

Snell, G.D. (1948). *J. Genet.* **49**, 87.

Snell, G.D. (1978). In *Origins of Inbred Mice*, Proceedings of a Workshop, Bethesda, MD, February 1978 (ed. H.C. Morse, III), pp. 119–156. Academic Press, New York.

Snell, G.D. and Reed, S. (1993). *Genetics* **133**, 751–753.

Staff of the Roscoe B. Jackson Memorial Laboratory. (1933). *Science* **78**, 465–466.

Stanford, W.L., Cohn, J.B. and Cordes, S.P. (2001). *Nat. Rev. Genet.* **2**, 756–758.

Strong, L.C. (1978). In *Origins of Inbred Mice*, Proceedings of a Workshop, Bethesda, MD, February 1978 (ed. H.C. Morse, III), pp. 45–67. Academic Press, New York.

Taylor, B.A., Navin, A. and Phillips, S.J. (1994). *Genomics* **21**, 626–632.

Threadgill, D.W., Hunter, K.W. and Williams, R.W. (2002). *Mamm. Genome* **13**, 175–178.

Tyzzer, E.E. (1909). *J. Med. Res.* **21**, 519.

Utomo, A.R., Nikitin, A.Y. and Lee, W.H. (1999). *Nat. Biotechnol.* **17**, 1091–1096.

Visscher, M.B., Green, R.G. and Bittner, J.J. (1942). *Proc. Soc. Exp. Biol. Med.* **49**, 94.

Wade, C.M., Kulbokas, E.J., 3rd, Kirby, A.W., Zody, M.C., Mullikin, J.C., Lander, E.S., Lindblad-Toh, K. and Daly, M.J. (2002). *Nature* **420**, 574–578.

Wakayama, T., Perry, A.C., Zuccotti, M., Johnson, K.R. and Yanagimachi, R. (1998). *Nature* **394**, 369–374.

Watson, J. and Crick, F. (1953). *Nature* **171**, 737–738.

You, Y., Bergstrom, R., Klemm, M., Lederman, B., Nelson, H., Ticknor, C., Jaenisch, R. and Schimenti, J. (1997). *Nat. Genet.* **15**, 285–288.

CHAPTER 3

Strains, Stocks, and Mutant Mice

Carol C Linder and Muriel T Davisson
The Jackson Laboratory, Bar Harbor, Maine, USA

Introduction

Genetically defined inbred mice are universally accepted as the principal model for analyzing and understanding inherited human disorders (Paigen, 1995, 2002; Davisson, 1999; Zambrowicz and Sands, 2003). The ultimate recognition of their value was when the mouse was selected as the first model organism to have its genome sequenced in the Human Genome Initiative. The advantages for using mice for research purposes are numerous. Despite their obvious physical differences, genes from mice and humans are approximately 99% identical (Mouse Genome Sequencing Consortium, 2002). In addition, genes in the mouse and humans function in virtually the same way in a biological context. Unlike many mammalian model organisms, the genome of the mouse is easily manipulated through numerous genetic engineering technologies. Their small size and ease of maintenance reduce the costs of research. Their accelerated lifespan (1 mouse year = ~30 human years) allows all life stages to be studied. Their short gestation time (~3 weeks) and large litter size quickly provide a large sample population and enable rapid genetic and pathophysiologic characterizations.

Genetically defined backgrounds and controlled environmental factors (e.g. diet, exercise, reproduction, and specific pathogens) minimize variability. This latter point is especially important for defining effects due to mutant genes because variability in human diseases is frequently due to the heterogeneity of the genetic background. The ability to place mutations on different genetic backgrounds makes it possible to study the effect of modifying genes. Developmental and invasive studies not possible or ethical with human subjects can be done under controlled conditions in mice. Diet, drug, and gene therapies can be tested and all tissues are accessible. Finally, knowledge of the mouse genome is far more advanced than that of any other experimental mammal. Databases containing the entire DNA sequences for both the human and mouse genomes are available from both private (Celera, celera.com) and public (international human and mouse sequencing consortia, ensembl.org) sources. Direct DNA sequence comparisons can now be used to delineate exact regions of gene conservation between the two genomes. This chapter describes the types of strains available for research, standardized genetic strain and gene nomenclature, and mouse and bioinformatics resources available to the research community.

Laboratory mouse strains

Inbred strains

Inbred strains are defined as those derived from 20 or more consecutive sister × brother matings; a strain's degree of inbreeding is designated by the letter F followed by the number of generations of filial breeding (e.g. F20). Residual heterozygosity will essentially be eliminated by F60 (Bailey, 1978). Most commonly used inbred strains have been inbred for more than 200 generations. Continual inbreeding produces mice that are genetically uniform, being homozygous at virtually all of their loci.

Inbred strains offer several advantages over outbred or random bred mice. In addition to their genetic and phenotypic uniformity, commonly used inbred strains are well characterized. There is a wealth of information about how inbred strains respond to experimental perturbations, specific strain characteristics, and what to expect in terms of pathobiology as mice age (Festing, 2003; Mouse Phenome Database (MPD), 2003). The uniformity of inbred strains reduces the number of mice needed for experiments. This is because experimental variability in phenotype is limited to variations in epigenetic, extragenetic, and/or varying uncontrolled environmental factors. The advantages of using inbred strains outweigh their lack of robustness, often-overlooked strain-specific characteristics, low fecundity, and the relatively high cost of some inbred strains.

A number of classical inbred strains were developed from fancy mice in the first third of the twentieth century. These include BALB/c, the C57 series, C3H, DBA, and 129 parent strains, all of which have become the standards for research in most areas of mouse biology. Individual inbred strains exhibit specific characteristics, passed on from generation to generation, that make them ideally suited for specific types of research (see Table 3.1). Unfortunately, often the characteristics of some of the classical inbred strains are overlooked during experimental design or data analysis. For example, mice of several strains (e.g. C3H, FVB/N, and SJL/J) are blind due to homozygosity for the retinal degeneration 1 mutation, $Pde6b^{rd1}$ (Chang et al., 2002); some strains (e.g. 129, A/J, C57BLKS/J, DBA/2J) exhibit early hearing loss (Zheng et al., 1999); DBA/2J mice are prone to **audiogenic** seizures prior to their early hearing loss (Fuller and Sjursen, 1967). Genetic differences among 129 strains can substantially impact their value as background strains for targeted mutation experiments (Simpson et al., 1997; Threadgill et al., 1997). See next section on Inbred Strain Nomenclature for more details on 129 strain differences. In addition, many inbred strains carry recessive genes affecting coat color (e.g. BALB/c mice are homozygous for both the brown, $Tyrp1^b$, and albino, Tyr^c, loci). When designing experiments, it is critical to select the appropriate strain and then to determine whether mice of the strain have any characteristics that might confound the experimental results.

Hybrid mice

The deliberate crossing together of mice of two inbred strains generates hybrid mice. As long as the parental strains exist, they can be repeatedly produced. F1 hybrids (e.g. B6D2F1) are similar to inbred strains in that they are genetically and phenotypically uniform. In contrast to most inbred strains, F1 hybrids display an overall hybrid vigor (i.e. increased disease resistance, better survival under stress, greater natural longevity, larger litters). Thus, they provide the advantage of genetic uniformity with more robustness than the average inbred strain. They are useful as hosts for tissue transplants from mice of either parental strain. Because of the combination of hybrid vigor with genetic and phenotypic uniformity, F1 hybrid mice are often preferred over random bred or outbred mice in a wide variety of research endeavors, including radiation research, behavioral research, and bioassays for nutrients, drugs, pathogens, and hormones.

Some deleterious mutations (e.g. the osteopetrosis mutation, Csf^{op}) cause inviability on an inbred background but can be maintained and provided for research by breeding mice carrying the mutation to an F1 hybrid. The progeny from this cross are not F1 hybrids, but rather are segregating for alleles of the two parental strains; i.e. for any locus that differs between the parental strains, a mouse may be heterozygous or homozygous for either parental allele.

Many strains carrying targeted mutations (knockouts) are maintained by homozygous matings for reasons of economy and scale. For targeted mutations maintained homozygously on a mixed C57BL/6 and 129 genetic background (designated B6;129), there are no matching wild-type controls. F2 hybrids (e.g. B6129SF2) are frequently used as physiological controls for these strains. Like the targeted mutant mice, the genetic background composition of F2 hybrid mice varies among littermates because of allelic segregation in gametes of the F1 hybrid parents. F2 hybrid mice provide only an approximate genetic match to the B6;129 background.

Random bred and outbred stocks

A large number of researchers rely on outbred mice in experiments when the precise genetic makeup is not considered crucial. Commonly used stocks include CD-1, Swiss Webster, Black Swiss, ICR, and NIH Swiss. Most outbred stocks of mice exhibit hybrid vigor similar to or exceeding that of F1 hybrids. Compared to inbred strains, they have longer life spans, higher disease resistance, earlier fertility and higher overall reproductive performance, lower neonatal mortality, and are considerably less expensive. They are ideally suited as stud males or foster mothers. Using outbred mice for any experiment, including test treatments that may lead to future genetic studies (e.g. susceptibility vs. resistance), will decrease the value of the results and may not be cost effective in the long term.

Random bred and outbred stocks are often confused. Outbred mouse stocks are maintained by non-sibling matings. For random bred stocks, a random numbers table or computer program must be used to select breeder pairs. Also, unless a closed colony of either type of stock has sufficient numbers (typically >25 pairs), the mice will gradually become increasingly genetically homogeneous.

Most commercial mouse suppliers use breeding schemes that avoid crosses between closely related individuals in order to maintain a maximal level of heterozygosity in progeny of outbred stocks. They are not really 'random'. However, it is a common misconception that outbred stocks of mice are more representative of the genome of human populations than inbred strains. Outbred stocks are essentially closed colonies and many were originally derived from a very limited gene pool. Their lack of genetic homogeneity frequently confounds results rather than mimicking the human condition.

Recombinant inbred strains, recombinant congenic strains, and advanced intercross lines

Recombinant inbred (RI) strains are derived by systematic inbreeding from a cross of two distinct inbred strains. Donald Bailey (1971) first developed them at The Jackson Laboratory. Most RI strain sets result from randomly mated pairs of F2 mice followed by at least 20 generations of inbreeding. Each strain within an RI set is equally likely to have inherited either the maternal or paternal progenitor strain allele at each autosomal locus. Alleles of unlinked loci are randomized in the F2 generation. Thus, parental and recombinant allelic combinations should be fixed with equal probability in RI strains. Linked genes will tend to remain linked and will become fixed in parental combinations in the strains of an RI set at frequencies directly proportional to the genetic distance between them. Thus, sets of RI strains are useful for segregation and linkage analysis of both polymorphic markers and phenotypic traits that differ between the progenitor strains. They can be used to distinguish between single and multiple gene traits; the former will appear in the progeny strains only in the two parental types, while the latter will appear as parental and intermediate phenotypes. In addition, individual RI strains may exhibit a more robust (e.g. granulosa cell tumorigenesis in SWXJ9/BmJ mice; Tennent et al., 1993) or novel (mu opioid receptor gene defects in CXB7/ByJ (CXBK) mice; Ikeda et al., 2001) phenotype when compared to either one of the progenitor strains. Recombinant inbred sets are limited in their ability to map polygenic traits because they do not exhibit enough genetic recombination or heterogeneity. Intercrossing mice of individual RI strains of the same set, producing RIX strains, has been proposed to increase the number of recombinants and make them more powerful for mapping complex traits (Threadgill et al., 2002).

Recombinant congenic (RC) strains are a variation on recombinant inbred strains. Following the initial outcross of mice of two inbred strains, F1 hybrid progeny mice are backcrossed to mice of one of the parental strains for one or two generations prior to sibling mating for at least 14 generations. In contrast to RI strains that are a ~50:50 mixture of the progenitor strains, the genome of RC strains will be derived predominantly from one parent (the proportion depending upon the number of backcross generations before incrossing). Sets of RC strains (e.g. NONcNZO1–NONcNZO10) have unique characteristics and are valuable for dissecting polygenic diseases like Type 2 diabetes (Reifsnyder and Leiter, 2002).

Advanced intercross lines (AILs) are initiated in the same way as RI strain sets; however, mice of the F2 generation and each subsequent generation are intercrossed by avoiding sibling matings (Darvasi and Soller, 1995). The purpose of AILs is to increase the frequency of recombination between tightly linked genes.

TABLE 3.1: Commonly used inbred strains, origins, and research applications

Parent strain	Origin[a]	Commonly used substrains[b]	Research applications[c]
129	101 mice L.C. Dunn, Bussey Institute	129P1/J, 129P2/OlaHsd, 129S1/SvImJ, 129S2/SvPas, 129S4/SvJae, 129S6/SvEvTac, 129X1/SvJ	General purpose, targeted mutagenesis (embryonic stem (ES) cell lines), low background incidence of testicular teratomas, sensorineural research (early hearing loss), neurodevelopmental defects (callosal agenesis, incomplete penetrance)
A	Cold Spring Harbor albino crossed with a Bagg albino L.C. Strong, Bussey Institute	A/J, A/He, A/WySn	General purpose, cancer research, immunology research, sensorineural research (early hearing loss), developmental biology research (low background incidence cleft palate). High susceptibility to cortisone-induced cleft palate and carcinogen-induced lung adenomas, the latter making it useful for carcinogen testing
AKR	Detwiler stock, Norristown PA H. Furth, Rockefeller Institute	AKR/J	Cancer Research (high leukemia strain)
BALB/c	Albino stock H. Bagg, Memorial Hospital, NY	BALB/c, BALB/cAn, BALB/cBy	General purpose, immunology research (production of monoclonal antibodies and hybridomas), neurodevelopmental defects (callosal agenesis, incomplete penetrance)
C3H	Bagg albino female crossed with a DBA male L.C. Strong, Bussey Institute	C3H/He, C3H/HeOu, C3H/HeSn	General purpose, cancer research (mammary tumor development enhanced by presence of exogenous mouse mammary tumor virus, MMTV), sensorineural research (homozygous for *Pde6b*[rd1])
C57BL	Lathrop stock C.C. Little, Bussey Institute	C57BL/6, C57BL/6By, C57BL/Ei, C57BL/10, C57BL/10Sn	General purpose, cardiovascular biology research, background strain for most transgenes, spontaneous or targeted mutations, C57BL/10Sn background strain for histocompatibility congenics
DBA	Coat color stock C.C. Little, Bussey Institute	DBA/1, DBA/1Lac, DBA/2	General purpose, DBA/2: cardiovascular biology, neurobiology (audiogenic seizures), and sensorineural research (early hearing loss), often contrasted with C57BL/6 because of many polymorphic differences between the two. DBA/1: autoimmunity and arthritis research

FVB	Outbred N:GP (National Institutes of Health) General Purpose) Swiss mice	FVB/N	Transgenic production (large male pronuclei, good breeder, background strain for many transgenes), sensorineural research (homozygous for $Pde6b^{rd1}$)
NOD	Outbred Jc1:ICR S. Makino, Shionogi Research Laboratories	NOD/LtJ, NOD/MrkTac, NOD/Shi	Type 1 Diabetes research, autoimmunity research
NZB	Outbred mice from Imp. Cancer Research Fund M. Bielschowsky, University of Otago Medical School	NZB/B1NJ	Autoimmunity research. F1 hybrids of NZB/B1NJ and NZW/LacJ (NZBWF1/J) are widely used as a model for autoimmune disease resembling human systemic lupus erythematosus
SJL	Outbred Swiss Webster J. Lambert, TJL	SJL/J	Cancer research (reticulum cell sarcomas), autoimmunity (experimental allergic encephalomyelitis, EAE), sensorineural research (homozygous for $Pde6b^{rd1}$)
SWR	Swiss mice from A. de Coulon of Lausanne, Switzerland C.J. Lynch, Rockefeller Institute	SWR/J	General purpose, cancer research (high incidence of lung and mammary gland tumors in aging mice), metabolic disease research (nephrogenic diabetes insipidus with increasing age), autoimmunity research (EAE)

[a] Festing (2003).

[b] Distinct substrains are available from various mouse vendors (e.g. Bom, M & B A/S; Crl, Charles River; Ico, IFFA-CREDO; J, The Jackson Laboratory; Hsd, Harlan Sprague Dawley, Inc.; Tac, Taconic, etc.). Important note: substrain differences exist due to residual heterozygosity, genetic drift, or spontaneous mutations that occur after separation from the parent strain; these may affect overall strain characteristics.

[c] For more information on research applications (including references), refer to Table 3.5, General information resources on the Web.

Congenic strains

Congenic strains are inbred strains carrying a mutant gene or 'foreign' polymorphic allele from a different strain or stock. George Snell (1948, 1978) first developed congenic strains. They are derived by successively mating mice carrying a mutant gene or foreign locus of interest from a donor strain to mice of a recipient inbred strain. A fully congenic strain and its inbred partner are expected to be identical at virtually all loci except for the transferred locus and a linked segment of chromosome. A strain is considered congenic after 10 generations of backcrossing to a recipient inbred strain (N10). Strains carrying mutations that have been backcrossed onto the background strain less than 10 generations are considered incipient congenics.

Congenic strains may be created in <10 generations using marker-assisted technologies, creating so-called speed congenics. This technology takes advantage of the extensive mapping of DNA microsatellite markers in numerous inbred strains (Mouse Genome Database (MGD), 2003). Microsatellite markers are **dinucleotide repeats** present in noncoding regions of the genome. Inbred strains frequently differ from each other in the number of dinucleotide repeats at a specific site that are amplified by microsatellite marker primers. These marker differences are called simple sequence length polymorphisms (SSLPs). SSLPs between inbred strains are the basis of speed congenic development (Markel et al., 1997, Wakeland et al., 1997).

Speed congenic strains are created using a panel of mapped SSLP markers that span the entire genome (except the X and Y chromosomes) and that are polymorphic between the donor and recipient strain genomes. The speed congenic approach takes advantage of the fact that progeny following the second backcross generation (N2) have a range of genomic identities. Progeny that contain the highest percentage of the recipient genome, or preferably the most nonrecombinant recipient chromosomes, are selected for the next round of backcrossing. By selecting such optimal breeders that are heterozygous for the gene(s) of interest at each subsequent backcross generation, it is possible to reach 99% recipient strain genomic identities after five generations (N5). Selecting mice that are carrying recipient-like markers close to the transferred gene can reduce the interval spanning the gene of interest, still of significant size at N5. A simple breeding strategy accomplishes fixation of the recipient sex chromosomes. The use of at least one heterozygous female to a recipient inbred male followed by a subsequent backcross using a male carrier to a recipient inbred

female insures that both the X and Y chromosomes are 100% recipient genome.

Consomic (or chromosome substitution) strains are a variation of congenic strains. An entire chromosome is transferred to a new recipient background by repeated backcrossing. Traditionally most **consomic strains** involved transferring the Y chromosome from one strain to another (e.g. BALB/cByJ-ChrY^{B6By}). More recently, however, complete sets of chromosome substitution strains (CSS panels) have been generated in the mouse (Nadeau et al., 2000). A CSS panel includes individual recipient strains that have had each of the 19 autosomes and the X and Y chromosomes replaced by a donor strain. CSS panels facilitate quantitative trait loci (QTL) mapping of polygenic traits that differ between two progenitor strains. First, the CSS panel is screened for a phenotype of interest. If a CSS strain differs from the recipient strain, this indicates there must be at least one QTL located on the donor chromosome. In contrast, traditional QTL mapping requires an initial large scale two-generation cross and the production of perhaps thousands of recombinant progeny, followed by development of multiple congenic strains carrying QTL of interest. These first steps are avoided with the use of CSS panels. However, like traditional QTL analysis, finer structure mapping to delineate genes/loci of interest requires additional crosses. Generation of CSS panels between strains that demonstrate numerous polygenic trait differences (e.g. A/J and C57BL/J) have widespread utility.

Wild-derived inbred strains

Wild-derived inbred mice are descendants of captured wild mice and represent several different *Mus* species from around the world. Many such strains were created as genetic mapping tools by Thomas Roderick and Eva Eicher at The Jackson Laboratory, Jean-Louis Guenet at the Institut Pasteur, and Verne Chapman at the Roswell Park Cancer Institute, as well as several Japanese investigators. Examples of frequently used wild-derived strains include *Mus musculus castaneus* from Thailand (CASA/RkJ and CAST/EiJ), *M. m. molossinus* from Japan (MOLC/RkJ, MOLD/RkJ, MOLF/EiJ), *Mus caroli* and *Mus pahari* (both from Thailand), *Mus hortulanus* from Serbia (PANCEVO/EiJ), and *Mus spretus* from Spain (SPRET/EiJ). Inbreeding of wild mice to create the laboratory strains listed above was commenced from one pair or trio mating. The large number of genetic differences in progeny from the interspecific crosses with common inbred laboratory mice makes wild-derived inbred mice valuable tools for gene mapping, evolution, and systematics research.

Several of the wild-derived inbred strains (e.g. RBA/DnJ, TIRANO/EiJ, and ZALENDE/EiJ) naturally carry multiple Robertsonian chromosomes, a fusion of two nonhomologous telocentric chromosomes to form a single metacentric chromosome. **Robertsonian chromosomes** are useful as tissue or cell markers for chimera and transplantation studies, chromosome specific trisomy production, and mapping genes by fluorescent *in situ* hybridization (FISH) of gene probes.

Mice with chromosomal aberrations

The diploid chromosomal complement of standard inbred laboratory strains is $2N = 40{:}19$ autosomes, X and Y sex chromosomes. The autosomes and the X chromosomes are telocentric (i.e. the centromere is at one end of a single-armed chromosome) while the Y chromosome is acrocentric (i.e. it has a short p arm as well as the longer q arm, use of 'p' and 'q' is patterned on human chromosomal nomenclature). The sex-determining genes reside in the short arm of the Y chromosome.

Strains of mice whose chromosomal complement deviates from the normal chromosomal makeup are designated chromosomal aberration strains. Chromosomal aberrations can include intra- and interchromosomal rearrangements or aneuploidy. These include (1) inversions and transpositions, rearrangements of DNA segments within chromosomes; (2) reciprocal translocations, Robertsonian chromosomes, and insertions, exchanges of DNA segments between chromosomes; and (3) aneuploidy, deviations from the normal diploid number of chromosomal arms in somatic cells (e.g. trisomies). Some chromosomal deletions and duplications also may be cytologically detectable.

The ability to detect chromosomal aberrations cytologically makes them useful as dominant markers for linkage studies and for marking tissues in chimera and transplantation experiments. Many chromosomal aberrations are useful in FISH gene mapping and meiotic nondisjunction studies.

Mutant mice

A large number of mouse models are the result of spontaneous single gene mutations (MGD, 2003). Many of these occurred within The Jackson Laboratory's large production breeding colonies and have been developed into models by the Laboratory's Mouse Mutant Resource (MMR) program or by members of the research staff.

Many mouse mutations (both spontaneous and induced) have come from the radiation/chemical risk assessment programs at the Oak Ridge National Laboratory (ORNL) in the U.S. and the Medical Research Council Genetics Program at Harwell in the U.K.

Traditionally, the detection of spontaneous mutations in an animal colony has been limited to alterations of observable phenotypes. These include mutations that cause changes in coat color (e.g. yellow, leaden), growth defects (e.g. dwarf, pigmy), abnormal morphology (e.g. limb deformity, legless), or alterations in behavior or motor coordination (e.g. ataxia, circling). Large-scale phenotypic screening for desired traits that are not easily observed or measured is time consuming and not cost effective, given the rarity of spontaneous mutations.

Mice carrying spontaneous mutations provide a rich source of animal models for human genetic diseases. Advances in gene cloning technology and the sequencing of the mouse genome have led to the molecular identification of the genes underlying many mouse mutations and their homologous human inherited conditions. Spontaneous mutations complement the ever-increasing numbers of transgenic and targeted mutations made possible by recent advances in genetic engineering. Spontaneous mouse mutations offer several advantages for the study of human genetic diseases. Spontaneous mutations are identified on the basis of a biomedically relevant phenotype first and the gene identified by reverse genetics later, whereas the phenotype of targeted mutations cannot be accurately predicted. Spontaneous mutations usually lead to the discovery of new genes, whereas targeted mutations can be engineered only for those genes that already have been identified and cloned. Spontaneous mutations also are more likely to resemble the naturally occurring mutations that cause human inherited diseases (Bruneau *et al.*, 2001). Recent advances in positional cloning technology, available tools, such as YAC, BAC, P1, **cosmid libraries**, and availability of the complete mouse genome draft sequence have dramatically improved the ability to identify the mutated genes.

Hundreds of spontaneous mutations that cause genetic disease have been identified in mice (MGD, 2003). Many of these have been proposed to be homologs of known human diseases; some are known to be mutations in homologous genes and others are postulated on the basis of phenotypic similarities or comparative mapping or both (Andersson *et al.*, 1996; Bedell *et al.*, 1997; MGD, 2003). Table 3.2 lists some examples of mouse models of human disease, developed and/or currently maintained at The Jackson Laboratory whose underlying genes have been identified. Because of the

TABLE 3.2: Selected cloned mouse genes with orthologous human disorders

Gene/ Allele Symbol	Chr	Allele name	Gene name	Reference	Human ortholog	Human map location	Human disorder (OMIM (Online Mendelian Inheritance in Man) number[a])
Ar^{Tfm}	X	testicular feminization	androgen receptor	Charest et al. (1991)	AR	X (q11.2–q12)	Androgen insensitivity syndrome (AIS) (#300068)
$Galc^{twi}$	12	twitcher	Galactosylceramidase	Sakai et al. (1996)	GALC	14 q31	Krabbe disease (#245200)
$Ghrh^{lit}$	6	little	growth hormone releasing hormone receptor	Godfrey et al. (1993)	GHRHR	7(p15-p14)	growth hormone deficiency, isolated (*139191)
Gus^{mps}	5	mucopolysaccharidosis VII	beta-glucoronidase	Sands & Birkenmeier (1993)	GUSB	7 q22	mucopolysaccharidosis type VI (*253220)
$Hps4^{le}$	5	light ear	Hermansky–Pudlak syndrome (HPS) 4 homolog	Suzuki et al. (2002)	HPS4	22 q11.2–q12.2	HPS (*606682; #203300)
Lep^{ob}	6	obese	leptin	Zhang et al. (1994)	LEP	7 q32.1	obesity, leptin deficiency, hypogonadism (*164160)
$Lepr^{db}$	4	diabetes	leptin receptor	Chen et al. (1996)	LEPR	1 p31	obesity, morbid, with hypogonadism (*601007)

Pit1[dw]	16	dwarf	pituitary-specific transcription factor 1	Li et al. (1990)	POU1F1	3p11	Pituitary Hormone Deficiency (CPHD) (#173110)
Pou4f3[ddl]	18	dreidel	POU domain, class 4, transcription factor 3	Frankel et al. (1999)	POU4F3	5 q31	deafness, autosomal dominant nonsyndromic sensorineural 15 (DFNA15) *602460, (#602459)
Rab27a[ash]	9	ashen	RAB27A, member RAS oncogene family	Wilson et al. (2000)	RAB27A	15 (q15–q21.1)	Griscelli syndrome (#214450)
Tgn[cog]	15	congenital goiter	thyroglobulin	Kim et al. (1998)	TG	8 (q24.2–q24.3)	goiter, familial, with hypothyroidism AR (*188450)
Tnfsf6[gld]	1	generalized lymphoproliferative disease	tumor necrosis factor (ligand) superfamily, member 6	Takahashi et al. (1994)	TNFSF6	1 q23	autoimmune lymphoproliferative syndrome (ALPS), type 1B (*134638, #601859)

[a]An asterisk (*) before an OMIM number means that the phenotype determined by the gene at the given locus is separate from those represented by other asterisked entries and that the mode of inheritance of the phenotype has been proved (in the judgment of the authors and editors). In general, an attempt has been made to create only one asterisked entry per gene locus. A number symbol (#) before an OMIM number means that the phenotype can be caused by mutation in any of two or more genes. The #-labeled entries are considered useful for avoiding repetition of the same phenotypic information in several entries and necessary because it is often unknown which genetic type is referred to in a particular report.

high degree of gene conservation between the mouse and human genomes, such models are valuable for identifying human disease genes.

Mutant mice do not need to be exact copies of human diseases to be valuable for research. Even when phenotypically different, mouse models increase our understanding of basic underlying mechanisms and functional roles of genes during development. Indeed, differences between mouse and human phenotypes can provide information about the underlying biology of a disease.

The generation and use of mice carrying induced or genetically engineered mutations has increased over the past 15 years. Random mutagenesis protocols, such as treating mouse gametes or embryonic stem (ES) cells with chemical mutagens (Schimenti and Bucan, 1998) and gene trapping with retroviral vectors (Friedrich and Soriano, 1993), are used to produce valuable new models. Random mutagenesis produces both dominant and recessive mutations, although most efforts to date have concentrated on identifying the more easily detected dominant mutations. To obtain maximum value from random mutagenesis approaches, rapid and systematized protocols for phenotypic screening must be available, as well as sufficient resources for mapping and cloning genes and subsequently distributing these new models. Increasing the mutation frequency by chemical mutagenesis, such as ethyl nitrosourea (ENU), when coupled with screening protocols, enables the detection of mutations that cause subtle phenotypes to model specific categories of diseases (Russell et al., 1979). Several large-scale ENU mutagenesis projects are currently underway (Table 3.3; Hrabé de Angelis et al., 2000; Nolan et al., 2000).

Two broad areas of technology – transgenesis and targeted mutagenesis using homologous recombination – are currently used to create genetically engineered strains of mice (see Chapter 5 on Generation of Mouse Mutants by Sequence Information Driven and Random Mutagenesis for more information on these technologies). Transgenic mice have genetic material randomly added to their genomes (Gordon et al., 1980). Thousands of transgenic strains have been used to study gene function and expression and have resulted in many important disease models (MGD, 2003; Transgenic/ Targeted Mutation Database (TBASE), 2003). Since transgene insertion is a random event, the phenotype of the mouse may vary depending on the site of integration and/or the copy number of transgenes integrated. Transgene integration may cause disruption in an endogenous gene (insertional mutation), creating an inherited phenotype (usually recessive) unrelated to

transgene expression. In these cases, the transgenic animal provides a vehicle for gene discovery through the mapping and subsequent cloning of the disrupted gene (e.g. the pigmy locus was identified as an allele of the high mobility group AT-hook 2 (*Hmga2*) gene as a result of an insertional mutation caused by a human globin transgene; Xiang et al., 1990).

Targeted mutations are created using homologous recombination to alter or replace a specific locus or gene (Smithies et al., 1985; Mansour et al., 1988; Capecchi, 1989). Currently, the majority of strains created by gene targeting carry a null mutation for the gene in question. Increasingly, however, conditional targeted mutations are being created that allow control of both the tissue specificity of the mutation (Gu et al., 1994; Utomo et al., 1999; Nagy 2000) or the temporal onset of gene expression (Kistner et al., 1996, Schonig and Bujard, 2003). Gene targeting produces strains used to study gene function and to create models for human genetic diseases for which the offending gene is known.

Transgenesis and targeted mutagenesis technologies often produce unexpected results, creating mice with either no observable phenotype or an unexpected phenotype, one outside the researcher's area of expertise or interest. Thus, this gene-based approach may lead to the discovery of novel pathways for an already known gene.

Importance of standardized nomenclature

A standardized nomenclature system for both strains and genes is essential to provide unique identifiers for each and to establish relationships among laboratory mice from different sources. Cross-species standardization of gene names and symbols facilitates access, comparison, and interpretation of the vast amount of data being generated by various genome and gene expression projects. Understanding and use of standardized nomenclature greatly improves the retrieval and analysis of information. The essential tenets of mouse nomenclature will be addressed in this section. The complete rules set forth by the International Committee on Standardized Genetic Nomenclature for Mice are available on the Web through the MGD (2003).

A set of rules for naming and referring to strains was deemed necessary to effectively communicate results from mouse-based experimental research as early as 1921

TABLE 3.3: Selected ENU mutagenesis centers	
The Jackson Laboratory Neuroscience Mutagenesis Facility (JAX-NMF) www.jax.org/nmf	JAX-NMF is planned as a progressive scientific resource that each year will provide 50 novel mouse models useful for the study of important human neurological disorders. Valuable models will be identified using an extensive phenotype-driven approach that targets a broad range of recessive neurological phenotypes and characterizes them sufficiently to attract further detailed study. Mice are undergoing vision screening, primarily for retinal degeneration and mutations that lead to progressive vision impairment. Member of the NeuroMice.org Consortium.
The Jackson Laboratory Heart, Lung, Blood and Sleep Disorders Mutagenesis Facility pga.jax.org/index.html	The goal of the Mouse Heart, Lung, Blood, and Sleep Disorders Center is to link genetic variation to biological function and dysfunction. Using a phenotype-driven approach, the program will identify single genes and interacting gene networks that underlie the physiology and pathophysiology of atherosclerosis, hypertension, lung function, blood formation, thrombosis, obesity, and sleep function.
MRC Mammalian Genetics Unit (MGU, Harwell): ENU mutagenesis program www.mgu.har.mrc.ac.uk/mutabase	The ENU mutagenesis program at the MRC MGU is screening mice for a wide variety of phenotypes including defects in hearing, eye development, and vision.
ENU-Mouse Mutagenesis Screen Project www.gsf.de/isg/groups/enu-mouse.html	The German Human Genome Project (DHGP) has established a research center to perform a large ENU mutagenesis screen in mouse. The research center consists of a core facility (GSF, Gene Center) and several associated laboratories. The core facility generates mutagenized F1 and G3 mice that are analyzed by the associated laboratories. Furthermore, other research groups that are not participants of the German Human Genome Project can perform additional screens.
NeuroMice.org Consortium www.neuromice.org	The NeuroMice.org consortium effort is a NIH supported effort to maintain, characterize, and distribute recently-identified mouse lines with alterations in nervous system function and/or behavior. Consortium members include the JAX-NMF, the Neurogenomics Project at Northwestern University, and the Tennessee Mouse Genome Consortium (TMGC).
Neurogenomics Project at Northwestern University www.genome.northwestern.edu/neuro/	Neurogenomics Project at Northwestern University is funded by NIH to use forward genetic strategies to identify genes involved in nervous system function and behavior. An additional Center is dedicated to the genetics of development. Member of the NeuroMice.org Consortium.
TMGC tnmouse.org	The TMGC is a collaborative effort that includes Tennessee researchers from East Tennessee State University, Meharry Medical College, Oak Ridge National Laboratory, St Jude Children's Research Hospital, University of Memphis, The University of Tennessee, and Vanderbilt University. Researchers contribute their expertise for the study of complex biological systems using mouse model systems. Member of the NeuroMice.org Consortium.

(Little, 1921). The first formal nomenclature committee convened in 1939 (Dunn *et al.*, 1940). The International Committee on Standardized Genetic Nomenclature for Mice continually develops and revises nomenclature conventions so mouse strain names will be informative unique identifiers. For strains carrying engineered mutations, a name may also convey the technology used to generate the strain (i.e. transgene vs. targeted mutation). Proper nomenclature provides information about the specific substrain used and its producer and is critical to assist investigators to ensure that they obtain the correct mouse for their experiments.

Inbred strain nomenclature

Inbred strain nomenclature is a combination of parent strain and substrain designations. A parent strain is designated by a brief symbol made up of upper case letters or numbers or combination of letters and numbers. Inbred strain names may be rooted in their coat color, origin, or a defining characteristic. For example, C.C. Little's first inbred strain DBA, originally called dba, is named for its coat color genes; DBA mice are homozygous for dilute (*Myo5a^d^*/*Myo5a^d^*), brown (*Tyrp1^b^*/*Tyrp1^b^*), and nonagouti (*a/a*) recessive mutations. C57BL and C57BR parent strains derived from inbreeding black (BL) and brown (BR) progeny from a mating of female #57 to male #52 in Little's Line C (Russell, 1978). Inbred strains also may be named for more application-based phenotypes like the non-obese diabetic (NOD) strain and the strains with high (BPH/J), normal (BPN/J), and low (BPL/J) blood pressure.

Substrains are strains of mice that have diverged from their parent strain for 20 or more generations (10 generations each from a common ancestor), have demonstrated residual heterozygosity left over from the time of separation, or carry new mutations not found in the parent strain. Substrain designations are appended to the parent strain symbol following a forward slash (e.g. C57BL/6J). Substrain designations are typically a Laboratory Registration Code ('lab code'), but may include a number if there are multiple related substrains. A lab code usually consists of 3–4 letters (first letter is upper case) and identifies the particular institute, laboratory, or investigator that produces and/or maintains a mouse strain (e.g. J for The *Jackson* Laboratory, Mcw for *Medical College of Wisconsin*, Crl for *Charles River Laboratories*). The Institute for Laboratory Animal Research (ILAR; http://dels.nas.edu/ilar/index.asp) assigns lab codes and maintains a master registry.

Inbred strain names may change as more information is obtained about their genetic makeup and relationship to other strains. The nomenclature for 129 strains provides a prime example. L.C. Dunn originally produced the 129 inbred strain at Columbia University. The Jackson Laboratory obtained the mice in 1945, and again in 1948 following the 1947 Bar Harbor Fire. L.C. Stevens used 129 mice as a background strain for research on testicular teratomas and several different 129 substrains were subsequently distributed to outside investigators over the next 50 years. The 129 strains can be classified into three lineages: parental, steel, and teratoma. In some 129 strains, investigators have bred out coat color mutations (e.g. pink-eyed dilution and albino) and bred in selectable markers (e.g. *Hprt*, *Gpi1*; Simpson *et al.*, 1997). Work with teratocarcinomas and early gene transfer experiments in the late 1970s and early 1980s led to the eventual development of ES cells lines derived from several different 129 strains. Extensive analysis of the numerous 129 strains and their ES cell lines reveals more genetic variability than can be explained solely by genetic drift (Simpson *et al.*, 1997, Threadgill *et al.*, 1997). This variation has considerable impact on experimental strategies centered on targeted mutagenesis technologies. Matching the DNA library used for obtaining the targeting construct to the ES cell line can increase homologous recombination efficiency. Additionally, by matching the ES cell line to the appropriate 129 strain, it is possible to create a pure strain that would differ from its inbred strain control only by the targeted gene. Given the importance of understanding the origin of 129 strains, their nomenclature was modified to create separate parent strains (Festing *et al.*, 1999; The Jackson Laboratory, 2001a). These include strains derived from the original 129 parental strains (129P), strains derived from lines carrying the steel-J mutation (129S), strains derived from lines carrying the teratoma mutation (129T), and 129X (the X denotes the documented genetic contamination of the 129X1/SvJ strain). Despite the genetic contamination in the strain's history, 129X1/SvJ mice currently are completely inbred (Simpson *et al.*, 1997; Threadgill *et al.*, 1997; The Jackson Laboratory, 2001b).

The specialized types of strains and stocks described in the first section each have a specific format or symbol that identifies the strain type. Hybrid mice are designated by the abbreviated symbols of the female parent followed by the male parent; e.g. B6D2F1 means the offspring of a C57BL/6J female mated to a DBA/2J male. Random or outbred stocks are designated by writing the lab code first, followed by a stock symbol, e.g. Crl:CD-1.

Recombinant inbred strains are written similarly but are distinguished from hybrids by the inclusion of an 'X' in the symbol and single letter strain abbreviations, e.g. BXD1/Ty is the first of a series of RI strains created from C57BL/6 and DBA/2 by Benjamin Taylor (Ty). Recombinant congenic RC strains are designated like RI strains except that a lower case 'c' is inserted between the strain abbreviations. The host strain symbol is followed by the donor strain symbol, e.g. NONcNZO1 is the first in a series of RC strains in which NON × NZO F1 mice were backcrossed to NON mice prior to inbreeding (Reifsnyder and Leiter, 2002). Symbols for AIL sets contain the lab code of the laboratory that produced the lines, followed by a colon and the two parental inbred strain abbreviations, separated by a comma; with the generation number included in the symbol following a hyphen. Generations are designated G3, G4, etc., beginning with the first non-sibling cross after the F2 generation. For example, Pri:B6,D2-G# is an AIL stock created at *Pri*nceton from the inbred strains C57BL/6 × DBA/2. The G number increases with each generation.

Congenic strains are designated by the abbreviated symbol for the host strain, followed by a period, the abbreviated symbol for the donor strain, a hyphen, and the symbol for the transferred locus. If the donor mouse is unknown, noninbred, or of mixed origin, 'Cg' may be used to signify congenic. For example, the yellow mutation (A^y) occurred in non-inbred 'fancy' mice and was subsequently backcrossed to the C57BL/6J strain, designated B6.Cg-A^y.

Wild-derived inbred strains often are given symbols that identify the species, e.g. SPRET/Ei is an inbred strain of *Mus spretus* created by Eva Eicher (Ei). Chromosomal aberrations are designated by a letter abbreviation identifying the type of aberration (e.g. In for inversion, Ts for trisomy, Rb for Robertsonian) followed by the number(s) of the chromosome(s) involved in parentheses and a series symbol (e.g. Rb(6.16)24Lub).

Gene nomenclature

Mouse gene symbols serve the sole purpose of providing short unique identifiers, although they may provide information secondarily. They are typically composed of 3–5 characters that are usually acronyms for the gene names. The symbols (but not gene names) are italicized and the first letter is capitalized, except when the symbol is for a gene that is recognized only by a recessive mutation. Human gene symbols are designated by all capital letters in italics. Mouse and human protein symbols are designated by all capital letters and are not italicized, distinguishing them from gene symbols but making the protein symbols of the two species indistinguishable.

Spontaneous mutations are alleles of initially unknown genes and are given allele names and symbols based on their phenotype (e.g. diabetes, *db*). Recessive mutations (i.e. requiring two copies of the mutated allele to manifest the phenotype) are represented by all lower case letters while dominant (i.e. one or two copies of the mutated allele produces the phenotype) and semidominant (i.e. one mutant allele produces an intermediate phenotype) spontaneous mutations are represented by an uppercase first letter, followed by lower case letters. Once the gene responsible for the mutant phenotype has been identified, the allele symbol is superscripted to an approved gene symbol (e.g. the diabetes mutation is a point mutation in the leptin receptor gene, *Lepr*^{db}). The Mouse Genomic Database Nomenclature Committee approves and assigns gene names and symbols, which may be registered online (http://www.informatics.jax.org) or requested by email (nomen@informatics.jax.org). Gene names and symbols may change as the function of a gene is better understood or to better correspond with gene symbols of other species (primarily human).

The International Committee for Standardized Genetic Nomenclature for Mice distinguishes between transgenes (insertion of exogenous DNA, usually leading to an overexpression of the transgene and a semidominant phenotype) and targeted mutations that 'knockout' or alter an endogenous gene product. Transgenes are designated by Tg, followed by a designation for the DNA insert in parentheses (preferably the gene symbol), then a number indicating the founder line, and finally a lab code. Transgene symbols are not italicized. For example, Tg(CD8)1Jwg is a transgene containing the human *CD8* gene, the first transgenic line using this construct, described by the laboratory of Jon W. Gordon (Jwg). The promoter also may be designated within the parentheses to clarify the transgene expression pattern. Tg(Zp3-cre)3Mrt designates the cre transgene with a *Zp3* promoter, the third transgenic line from the laboratory of Gail Martin (Mrt).

Targeted alleles of genes are designated by the approved gene symbol followed by a superscript containing tm (for targeted mutation), an allele number, and the lab code (all in italics). For example, *Apoe*^{tm1Unc} represents the first targeted mutation in the apolipoprotein E (*Apoe*) gene made in the laboratory of Nobuyo Maeda at the University of North Carolina at Chapel Hill (Unc).

In a mutant strain designation, the genetic background of mice with spontaneous or induced mutations is given prior to the gene, transgene designation, or allele symbol. Many strains are maintained on a mixture of C57BL/6 and 129 genetic backgrounds (e.g. B6;129-*Alox5^{tm1Fun}*) because 129-derived ES cell lines are commonly used in gene targeting and chimeric mice are mated to C57BL/6 to determine germline transmission. Mutations transferred from a mixed to an inbred background by repeated backcrossing are designated using congenic nomenclature. For example, B6.129S2-*Alox5^{tm1Fun}* indicates that the *Alox5^{tm1Fun}* mutation originated in the 129S2 parent strain (via the D3H ES cell line) and was subsequently backcrossed to the C57BL/6J inbred strain for at least five generations. A strain is considered fully congenic at N10; however, the guidelines permit the use of congenic nomenclature for incipient congenics (backcrossing $\geq$N5 to the host strain; The Jackson Laboratory, 2000; MGD, 2003). Although generation numbers (e.g. N10 for B6.129S2-*Alox5^{tm1Fun}*) are not part of a strain name, the backcross generation should be obtained and considered prior to use of a congenic strain. Care must be taken in deciphering symbols used in strain nomenclature; the semicolon used to denote a mixed background versus a period used to denote a congenic background is a subtle but critical distinction. In the former, the strain background is a mix of alleles from the two designated parental strains; in the latter, the genetic background contains primarily host strain alleles.

The use of genetically defined inbred mice has increased dramatically with the ability to genetically manipulate the mouse genome. Be aware that the strain nomenclature used by many investigators and mouse suppliers in publications and in product literature is improper or incomplete. Scientists need complete information to make informed decisions about selecting and using appropriate mouse strains. Just as critical is the use of proper nomenclature in the mouse room on cage cards and in breeding records. Lack of attention to this type of detail may compromise entire research endeavors.

Mouse resources and repositories

The number of different mutant mouse strains currently numbers in the thousands. With the sequencing of the mouse genome and the creation of new induced, targeted, and transgenic mutations, we are slated to reach tens of thousands. Witness the words of Dr Allen Bradley of the Wellcome Trust Sanger Institute.

The Mouse genome encodes an experimentally tractable organism. This means that it is now truly possible to determine the function of each and every component gene by experimental manipulation and evaluation, in the context of the whole organism. (Bradley, 2002)

How will the research community manage this vast number of mice? There is an absolute requirement for additional resources and repositories to produce, house, and distribute these models. Substrains of many of the basic, standard inbred strains are available from a variety of commercial and non-profit distributors. Specialty and mutant strains are more typically available from repositories, which are the best solution for maintaining and distributing strains that require supplemental funding. The functions of the international repositories are to (1) identify, (2) select and (3) import important genetically engineered and mutant strains of mice, (4) cryopreserve embryos or gametes from these strains, (5) transfer mutations onto defined genetic backgrounds when appropriate, (6) maintain and distribute these strains, and (7) provide information on them to the scientific community.

Since its inception in 1929, The Jackson Laboratory has served as a central repository, developing and distributing a wide variety of inbred strains and mice carrying mutations. Originally held in individual research laboratories it became clear by the late 1940s that it was important to consolidate the growing number of mutant mouse strains into central colonies. Early repositories like the Mouse Mutant Stocks Center (MMSC) and the Mouse Mutant Gene Resource (MMGR) were combined to form the MMR in 1983. This resource program is funded by the National Center for Research Resources (NCRR) at the NIH. Several special emphasis colonies within the MMR are supported by other institutes and agencies that fund the identification, characterization, and distribution of mice for specific disease areas. For example, the Eye Mutant Resource at The Jackson Laboratory is funded by NIH's National Eye Institute and the Foundation Fighting Blindness; the Neural Tube Defect and Cytogenetic Models Resources are funded by the NIH's National Institute for Child Health and Human Development. Other specialty resources include models for Type 1 diabetes, complex disease gene mapping, and the Alzheimer's Disease Mouse Model Resource.

Beginning in the late 1990s several large ENU mutagenesis programs were funded to generate induced mutations for specific research areas. Two of these also are located at The Jackson Laboratory: the JAX-NMF and the Heart, Lung, Blood, Sleep Disorder Center. A large cooperative mutagenesis program was established in Europe at about the same time, the principal participants being the U.K. (MRC), Germany, and France. The MRC, German Human Genome Project, and JAX-NMF ENU mutagenesis programs are all currently screening ENU-induced mutant mice for neurological disorders, vision impairments or defects in eye development, and hearing deficits.

Mice carrying genetically engineered mutations are available from a variety of mouse vendors and special repositories (Table 3.4). The Jackson Laboratory established the IMR in 1993. As the first repository for genetically engineered mouse models, the IMR has served as the prototype for the establishment of newer repositories like EMMA, MMRRC, and MMHCC. These federally and privately funded repositories have many functions, including the identification and selection of important mouse models, their importation, cryopreservation, and distribution.

Information resources

The World Wide Web offers a virtual library of information resources to the scientist interested in finding out what mouse strains are available to the research community. Searching the Web can be overwhelming. It is useful to take advantage of the as yet limited number of information warehouses and databases available (Table 3.5) when trying to select a new mouse model or learn more specific information about the nature of the model (e.g. genetic and phenotypic characteristics, homology data, animal husbandry and maintenance, etc.). The Jackson Laboratory and the MRC Genome Center at Harwell have established the International Mouse Strain Resources (IMSR) that lists all strains held at these two institutions (www.jax.org/pub-cgi/imsrlist). Now that the prototype has been tested, other institutions are being encouraged to join to provide a 'one-stop-shopping' website for mouse strains.

As part of its mission to the research community, The Jackson Laboratory is an essential knowledge center for information about mice. The MGI Project was initiated in 1992 to integrate and centralize the numerous mouse-related databases to facilitate new data integration and analysis. Released on the Web in June 1994,

the first version incorporated all the data sets from the Genomic Database of the Mouse (GBASE) the Mouse Locus Catalog (MLC), Mouse Linkage Database and Programs (MLDP), Homology Database and Programs (HMDP), and the probes and PCR databases describing molecular reagents and polymorphisms (MusProb and MusPCR). Since its inception, the Web site has undergone major construction to incorporate new databases and data sets, to improve functionality and query mechanisms, and to develop and enhance analysis tools.

At present, the MGI Web site (www.informatics. jax.org) provides a unique integrated access to various sources for information on the genetics and biology of the laboratory mouse. It is comprised of the MGD (Blake *et al.*, 2003), GXD (Ringwald *et al.*, 2001), MGS Project, and related resources including the MTB database (Naf *et al.*, 2002), the Rat Data resource, Michael Festing's Listing of Inbred Strains of Mice and Rats, and MouseBLAST. MGI also provides links to a wide variety of other scientific web sites.

The MGI group, as a member of Gene Ontology™ Consortium (www.geneontology.org), recently implemented a new classification system for genes and gene products (The Gene Ontology Consortium, 2001). Defined, structured vocabularies, called **gene ontologies**, provide an annotation pipeline across eukaryotic species to describe the 'molecular function', 'biological process', and 'cellular component' aspects of a given gene. Current annotation data sets include the following model organisms: Human, *Drosophila melanogaster*, *Caenorhabditis elegans*, *Mus musculus*, *Arabidopsis thaliana*, *Rattus norvegicus*, *Danio rerio*, *Saccharomyces cerevisiae*, and several others. In addition to the gene ontology (GO) terms, MGI is constructing a Mammalian Phenotype browser to provide standard terms for annotating phenotypic data. More information about the Gene Ontology Project and the use of the GO terms in MGI can be found on their Web site (www.informatics.jax.org/mgihome/GO/ontology.shtml).

The Jackson Laboratory is also the home of the recently established international MPD Project. The goal of MPD is 'to establish a collection of baseline phenotypic data on commonly used and genetically diverse inbred mouse strains through a coordinated international effort' (Bogue, 2003). There is a tremendous need for comprehensive phenotypic information on inbred mouse strains. The laboratory mouse is the primary genetic model for exploring normal human biology and disease. In addition, spontaneous and induced mutations are frequently maintained on an inbred background with the mutant phenotype usually resulting from the combination of both the mutation and the

TABLE 3.4: Mouse vendors, special repositories, and genetic resources

Charles River Laboratories, Inc. www.criver.com	Charles River is a commercial distributor of outbred and inbred mice, rats, guinea pigs, hamsters, gerbils, swine, cats, and dogs for the biomedical research community.
Comparative Mouse Genomics Centers Consortium (CMGCC) www.niehs.nih.gov/cmgcc	CMGCC was initiated by the Environmental Genome Project (EGP) to develop transgenic and knockout mouse models based on human DNA sequence variants in environmentally responsive genes. These mouse models are tools to improve understanding of the biological significance of human DNA polymorphism. CMGCC is focusing on variation in genes involved in DNA repair or cell cycle control, because many of these are well-characterized environmentally responsive genes. Environmentally responsive genes also play roles in cell division, cell signaling, cell structure, gene expression, apoptosis, and metabolism.
European Mouse Mutant Archive (EMMA) www.emma.rm.cnr.it	EMMA is a non-profit repository for transgenic mouse strains essential for basic biomedical research and models for research into complex diseases. Qualified research scientists can obtain these strains. EMMA consists of several nodes in different locations in European countries, the main archival site being located in Monterotondo (Rome, Italy). Other centers in the EMMA system are at the Harwell MGU in England, the GSF Institute of Experimental Genetics in Germany, and in France, Sweden, and Portugal.
Harlan Sprague Dawley www.harlan.com	Harlan Sprague Dawley is a commercial distributor of outbred and inbred mice, hybrids, mutant and transgenic mice, as well as a large number of other laboratory animal stocks and strains. Member of Mouse Mutant Regional Resource Centers (MMRRC).
MRC MGU, Harwell www.mgu.har.mrc.ac.uk/	The largest of the EMMA-associated centers with mouse strains and a mouse embryo bank dating back to the 1940s.
International Mouse Strain Resources (IMSR) www.jax.org/pub-cgi/imsrlist	The IMSR is an international multicenter collaboration to promote the use of the mouse as a model organism for research into human diseases, as well as into mammalian physiology and genetics. The ultimate aim of this project is to provide one Web site at which scientists can find the location of any publicly available mouse strain.
JAX® Mice (The Jackson Laboratory) www.jax.org/jaxmice	The Jackson Laboratory is a not-for-profit private research institution that distributes the broadest range of mice including inbred strains, hybrids, and JAX® GEMM™ strains (Genetically Engineered and Mutant Mice), recombinant inbred and recombinant congenic strains, wild-derived mice, and mice with chromosomal aberrations. Resources include federally and privately funded repositories like the MMR, the Induced Mutant Resource (IMR), Eye Mutant Resource Program, Cytogenetics Models Resource, and Mouse Mutant Informatics Coordinating Center (MMRRC-ICC) for the MMRRC.
JAX® Mice DNA Resource www.jax.org/dnares/index.html	The Jackson Laboratory Mouse DNA Resource provides genomic DNAs from most of The Laboratory's diverse genetic stocks. High molecular weight DNA is prepared by phenol–chloroform extracting

	nuclei from the tissues of individual mice. The DNA is suitable for Southern blotting and amplification by polymerase chain reaction (PCR). Most extractions are from spleen or brain and spleen of pedigreed male mice; however, occasionally DNA is extracted from other tissues and some DNA is preserved from female mice.
Mouse Models for Human Cancer Consortium (MMHCC) Repository web.ncifcrf.gov/researchresources/mmhcc	The MMHCC Repository is a National Cancer Institute-funded repository for mouse cancer models and associated strains. The repository makes strains available to all members of the scientific community.
MMRRC www.mmrrc.org	The MMRRC Program is a national network of regional breeding and distribution centers. The goals of the MMRRC include: (1) distribution of selected mouse models to the scientific community; (2) storage of mouse embryos or gametes by cryopreservation and subsequent recovery of cryopreserved strains; (3) phenotypic characterization (structural and behavioral) of laboratory mice; (4) genetic quality control of laboratory mice (including genetically modified mice) (5) maintenance of an MMRRC mouse resource database MMRRC members include Harlan Sprague Dawley, University of North Carolina, Chapel Hill, Taconic Farms, Inc., and UC Davis. The MMRRC Distribution Centers are coordinated through the MMRRC Informatics Coordinating Center (ICC). The ICC is responsible for designing and developing the necessary software and Web site, installing an electronic network linking the MMRRC Distribution Centers, receiving and coordinating requests for mice, and maintaining the MMRRC database.
Mymouse.org www.mymouse.org	The goal of mymouse.org is to facilitate international collaborative research between phenotype screening centers and the mouse mutagenesis community, and to promote interactions among scientists interested in analyzing the biological mechanisms linking genes, function, and phenotypes. Mymouse.org features: (1) a list of screening laboratories with specific details on the assay performed; (2) direct email links to the principal investigator to discuss and arrange for the analysis of a new mutant mice; and (3) an interactive database where individuals can enter new mutants or search for a specific mutant mouse model for collaborative analysis.
ORNL lsd.ornl.gov/htmouse/mmdmain.htm	ORNL has a collection of several hundred mouse stocks, most of which propagate mutations induced (over a period of several decades) by radiations or chemicals in various stages of male or female gametogenesis. These mutations may range from single base-pair changes to rearrangements of various sizes, depending on mutagen and germ cell stage.
Taconic www.taconic.com	Taconic is a commercial distributor of a wide range of mice and rats for biomedical research with new initiatives in Emerging Models Program. Member of the MMRRC.

TABLE 3.5: General information resources on the Web

Dysmorphic Human–Mouse Homology Database (DHMHD) www.hgmp.mrc.ac.uk/DHMHD/ dysmorph.html	DHMHD is a composite of three separate databases of human and mouse malformation syndromes together with a database of mouse/human conserved regions. The database can be searched by (1) specifying specific malformations, clinical features, or chromosome locations; (2) homology; (3) asking for human syndromes located at a chromosome region conserved with a specific mouse chromosome region (and vice versa from human to mouse).
EMICE emice.nci.nih.gov	EMICE contains a wealth of information on different kinds of cancers and mouse models available through the National Cancer Institute. The site includes the Cancer Models Database, searchable links to PubMed, and a collection of information resources (i.e. gene chip protocols, funding agencies, and mouse suppliers).
Edinburgh Mouse Atlas Project (emap) and gene expression database (emage) genex.hgu.mrc.ac.uk	The UK MRC Human Genetics Unit in Edinburgh is developing a digital atlas of mouse development (emap) to be a resource for spatially mapped data such as *in situ* gene expression and cell lineage. The project is in collaboration with the Department of Anatomy, University of Edinburgh. The gene expression database (emage) is part of the Mouse Gene Expression Information Resource (MGEIR) in collaboration with The Jackson Laboratory.
Festing's Inbred Strain Characteristics www.informatics.jax.org	M.F.W. Festing's Inbred Strains Characteristics is a searchable, hypertext listing of inbred strains of mice and rats with associated strain characteristics. It is a valuable resource for basic characteristics of inbred strains including incidence of spontaneous diseases and which commonly used inbred strain carry mutations affecting eye development and vision.
Genetic Atlas of the Mammary Gland mammary.nih.gov	The goal of the Mammary Genome Anatomy Project (MGAP) is to discover isolate, and understand the genes and signaling pathways that regulate normal development and neoplastic transformation of the breast. The National Institute of Diabetes, Digestive and Kidney Diseases, NIH supports this site. It contains a list of experimental models important for breast cancer and mammary gland development research as well as extensive data on the biology of the mammary gland.
The Jackson Laboratory Web site www.jax.org	The Jackson Laboratory Web site provides an overview of basic mammalian genetics research being conducted at this private research facility, courses and conferences, pre-doctoral and post-doctoral training and educational opportunities, the mouse genome informatics consortium, numerous research resources, the JAX® Mice website (www.jax.org/jaxmice), and much more. The JAX® Mice website contains a wealth of information about JAX® Mice and Services, available research models, strain datasheets, and other helpful information. New areas include research models pages (e.g. Cardiovascular Biology, Immunobiology and inflammation, Neurobiology, etc.) with special emphasis on specific diseases (e.g. Alzheimer's, Cancer, Diabetes, etc.).
Mouse Genetics: Concepts and Applications www.informatics.jax.org	L.M. Silver's Mouse Genetics book (Oxford University Press, 1995) is now available online through the Mouse Genome Informatics (MGI) Web site. Contents include an (1) introduction to mice and the origins of laboratory strains; (2) overview of animal husbandry, reproduction, and breeding techniques; (3) overview of the mouse genome, mutagenesis, and mapping strategies.

MGI Project www.informatics.jax.org	The MGI Database provides integrated access to data on the genetics, genomics, and biology of the laboratory mouse. Projects included in this resource: MGD Project includes data on gene characterization and nomenclature, mapping, gene homologies among mammals, sequence links, phenotypes, allelic variants and mutants, and strain data. Gene Expression Database (GXD) Project integrates different types of gene expression information from for the laboratory mouse, and provides an electronic index of published experiments on endogenous gene expression during mouse development. Mouse Genome Sequence (MGS) Project is integrating mouse genomic sequence data with the genetic and biological data available in MGD and GXD. Mouse Tumor Biology (MTB) Database Project catalogs information relevant to using mouse models in cancer research including tumor names and classifications, tumor incidence and latency data, tumor pathology reports and images, genetic factors, and references.
Mouse Knockout and Mutation Database (MKMD) www.biomednet.com/db/mkmd	Established in 1995, MKMD is BioMedNet's fully searchable database of phenotypic information related to knockout and classical mutations in mice.
MPD www.jax.org/phenome	An information repository for protocols and raw data related to the phenotypic characteristics of 40 commonly used and genetically diverse inbred mouse strains. Provides current project information including opportunities for community participation in this coordinated international effort.
Online Mendelian Inheritance in Man (OMIM). www3.ncbi.nlm.nih.gov/Omim	OMIM catalogs human phenotypes and genotypes and relevant mouse models. Authored and edited by Dr. Victor A. McKusick and his colleagues and developed for the Web by the National Center for Biotechnology Information (NCBI). The database contains textual information, pictures, and reference information. It also contains copious links to NCBI's Entrez database of MEDLINE articles and sequence information.
TBASE tbase.jax.org	TBASE is an information database of mice carrying transgenes and targeted mutations ('knockouts') generated by the worldwide research community.
Trans NIH Mouse Initiative www.nih.gov/science/models/mouse	The Trans-NIH Mouse Initiative was initiated in 1998. A coordinating group of scientists convened to establish priorities for mouse genomics and genetics resources. This Web site provides information on: funding opportunities for mouse genomics and genetics resources; major resources-producing grants funded in response to the initiative; progress toward meeting the goals of the initiative; major mouse genomics and genetics resources; courses and scientific meetings related to the mouse initiative; and selected reports and publications.
Whole Mouse Catalog www.rodentia.com/wmc	This Web site serves as an information warehouse scientific researchers using mice or rats in their work.

TABLE 3.6: Electronic listservers	
COMPMED www.aalas.org/association/ links/compmed.htm	COMPMED is a comparative Medicine Discussion List – Mailing list for discussing the topics of comparative and laboratory animal medicine.
MGI email lists www.informatics.jax.org/ lmgihome/lists/lists.shtml	MGI maintains several E-mail lists for the genetics research community. The list service uses the Lyris list service software package, version 3.0. MGI-LIST – Forum for topics in mouse genetics, MGI news updates RAT-LIST – Forum for rat genetics PHENOME-LIST – Forum for strain characterization and phenotypic data collection.
Transgenic-List lists.man.ac.uk/mailman/ listinfo/transgenic-list	Transgenic-list is an email forum focusing on issues surrounding the generation, care and use, of genetically engineered mice.

genetic background. The MPD Project collaborates with scientists to gather phenotypic information on a diverse panel of inbred strains. Currently the base panel consists of 40 strains, but the MPD is being expanded to include information on strain sets (such as RI sets) derived from these basic strains.

The MPD Web site (www.jax.org/phenome) helps researchers select appropriate strains for a wide variety of research applications including physiological testing, gene identification, drug discovery, toxicology studies, mutagenesis, disease onset and susceptibility, and QTL analyses.

Electronic listservers, or electronic bulletin boards, provide a valuable and often overlooked service to the research community (Table 3.6). Most listservers are geared toward a specific research area. With thousands of subscribers, a researcher can often find a source for a rare mouse model or even gather information on whether an observed phenotype is the result of the mutation or rather the environment.

Conclusion

Tremendous accomplishments in mouse genetics were made during the twentieth century (Paigen, 2003). Accompanying the major milestones (see Chapter 2 on Historical Foundations), was the rapidly increasing number of mouse strains, stocks, and mutants available to the biomedical research community. Beginning with the creation of the first inbred strain in 1909, the number and different types of mouse models has increased from numbering in the dozens to the current exponential growth and generation of thousands in the last 20 years. The sequencing of the mouse genome is certain to ensure a continued need for additional resources to produce, house, characterize, and distribute these models. Continual improvements and centralization of Web-based information resources and databases will make information about these models more accessible.

Ideally, researchers prefer to have central repositories maintaining large colonies of mice in order to immediately meet any anticipated demand of new and classic mouse models. However, this option is not feasible, due to both monetary and capacity resource constraints. Compromises on availability level have already become a reality to many individual investigators looking for mouse models. Many repositories have limited distribution, only providing a couple of breeder pairs; customers then are able to contract with mouse providers to breed larger numbers of mice as required.

Cryopreservation provides the most efficient and economical option for strains that are in low demand. The recovery of mice from the cryopreserved state takes longer than if they were maintained on the shelf. However, implementation of artificial reproductive strategies, such as in vitro fertilization and intracytoplasmic sperm injection (ICSI), can dramatically improve strain recovery and colony expansion times.

It is evident that without careful planning, efficient utilization of space and personnel, and innovation and use of new technologies, existing and future repositories will struggle to keep pace with the creation of new mouse models for biomedical research.

Acknowledgments

The authors would like to thank The Jackson Laboratory colleagues within Genetic Resources, Technical Information Services, and the Mouse Genome Informatics Consortium for their tireless and continuous efforts to generate and maintain quality information and resources pertaining to the mouse. Their efforts contributed greatly to the writing of this chapter.

References

Andersson, L., Archibald, A., Ashburner, M., Audun, S., Barendse, W., Bitgood, J., Bottema, C., Broad, T., Brown, S., Burt, D., Charlier, C., Copeland, N., Davis, S., Davisson, M., Edwards, J., Eggen, A., Elgar, G., Eppig, J.T., Franklin, I., Grewe, P., Gill, T., 3rd, Graves, J.A.M., Hawken, R., Hetzel, J., Hillyard, A., Jacob, H., Jaswinska, L., Jenkins, N., Kunz, H., Levan, G., Lie, O., Lyons, L., Maccarone, P., Mellersh, C., Montgomery, G., Moore, S., Moran, C., Morizot, D., Neff, M., Nicholas, F., O'Brien, S., Parsons, Y., Peters, J., Postlethwait, J., Raymond, M., Rothschild, M., Schook, L., Sugimoto, Y., Szpirer, C., Tate, M., Taylor, J., VandeBerg, J., Wakefield, M., Wienberg, J. and Womack, J. (1996). *Mamm. Genome* **7**, 717–734.

Bailey, D.W. (1971). *Transplantation* **11**, 325–327.

Bailey, D.W. (1978). In *Origins of Inbred Strains*, Proceedings of a Workshop, Bethesda, MD, February 1978 (ed. H.C. Morse, III), p. 199. Academic Press, New York.

Bedell, M.A., Largaespada, D.A., Jenkins, N.A. and Copeland, N.G. (1997). *Genes Dev.* **11**, 11–43.

Blake, J.A., Richardson, J.E., Bult, C.J., Kadin, J.A. and Eppig, J.T. and Mouse Genome Database Group. (2003). *Nucleic Acids Res.* **31**, 193–195.

Bogue, M. (2003, March). Mouse Phenome Database website, The Jackson Laboratory, Bar Harbor, Maine. http://www.jax.org/phenome.

Bradley, A. (2002). *Nature* **420**, 512–514.

Bruneau, S., Johnson, K.R., Yamamoto, M., Kuroiwa, A. and Duboule, D. (2001). *Dev. Biol.* **237**, 345–353.

Capecchi, M.R. (1989). *Science* **244**, 1288–1292.

Chang, B., Hawes, N.L., Hurd, R.E., Davisson, M.T., Nusinowitz, S. and Heckenlively, J.R. (2002). *Vision Res.* **42**, 517–525.

Charest, N.J., Zhou, Z.X., Lubahn, D.B., Olsen, K.L., Wilson, E.M. and French, F.S. (1991). *Mol. Endocrinol.* **5**, 573–581.

Chen, H., Charlat, O., Tartaglia, L.A., Woolf, E.A., Weng, X., Ellis, S.J., Lakey, N.D., Culpepper, J., Moore, K.J.,

Breitbart, R.E., Duyk, G.M., Tepper, R.I. and Morgenstern, J.P. (1996). *Cell* **84**, 491–495.

Darvasi, A. and Soller, M. (1995). *Genetics* **141**, 1199–1207.

Davisson, M.T. (1999). *Lab. Anim.* **28**, 53–56.

Dunn, L.C., Gruneberg, H. and Snell, G.D. (1940). *J. Hered.* **31**, 505–550.

Festing, M.F.W. (2003, March). http://www.informatics. jax.org/

Festing, M.F.W., Simpson, E.M., Davisson, M.T. and Mobraaten, L.E. (1999). *Mamm. Genome* **10**, 836.

Frankel, W.N., Mahaffey, C.L. and Bartlett, F.S. 2nd. (1999). MGI Direct Data Submission to Mouse Genome Database (MGD), MGI:1349786. http://www.informatics. jax.org

Friedrich, G. and Soriano, P. (1993). *Methods Enzymol.* **225**, 681–701.

Fuller, J.L. and Sjursen, F.H. (1967). *J. Hered.* **58**, 135–140.

Godfrey, P., Rahal, J.O., Beamer, W.G., Copeland, N.G., Jenkins, N.A. and Mayo, K.E. (1993). *Nat. Genet.* **4**, 227–232.

Gordon, J.W., Scangos, G.A., Plotkin, D.J., Barbosa, J.A. and Ruddle, F.H. (1980). *Proc. Natl. Acad. Sci. USA.* **77**, 7380–7384.

Gu, H., Marth, J.D., Orban, P.C., Mossmann, H. and Rajewsky, K. (1994). *Science* **265**, 103–106.

Hrabé de Angelis, M.H., Flaswinkel H., Fuchs, H., Rathkolb, B., Soewarto, D., Marschall, S., Heffne, S., Pargent, W., Wuensch, K., Jung, M., Reis, A., Richter, T., Alessandrini, F., Jakob, T., Fuchs, E., Kolb, H., Kremmer, E., Schaeble, K., Rollinski, B., Roscher, A., Peters, C., Meitinger, T., Strom, T., Steckler, T., Holsboer, F., Klopstock, T., Gekeler, F., Schindewolf, C., Jung, T., Avraham, K., Behrendt, H., Ring, J., Zimmer, A., Schughart, K., Pfeffer, K., Wolf, E. and Balling, R. (2000). *Nat. Genet.* **25**, 444–447.

Ikeda, K., Kobayashi, T., Ichikawa, T., Kumanishi, T., Niki, H. and Yano, R. (2001). *J. Neurosci.* **21**, 1334–1339.

Kim, P.S., Hossain, S.A., Park, Y.N., Lee, I., Yoo, S.E. and Arvan, P. (1998). *Proc. Natl. Acad. Sci. USA* **95**, 9909–9913.

Kistner, A., Gossen, M., Zimmermann, F., Jerecic, J., Ullmer, C., Lubbert, H. and Bujard, H. (1996). *Proc. Natl. Acad. Sci. USA* **93**, 10933–10998.

Li, S., Crenshaw, E.B. 3rd, Rawson, E.J., Simmons, D.M., Swanson, L.W. and Rosenfeld, M.G. (1990). *Nature* **347**, 528–533.

Little, C.C. (1921). *Amer. Nat.* **55**, 175–178.

Mansour, S.L., Thomas, K.R. and Capecchi, M.R. (1988). *Nature* **336**, 348–352.

Markel, P., Shu, P., Ebeling, C., Carlson, G.A., Nagle, D.L., Smutko, J.S. and Moore, K.J. (1997). *Nat. Genet.* **17**, 280–284.

Mouse Genome Database (MGD), Mouse Genome Informatics website, The Jackson Laboratory, Bar Harbor, Maine. (2003, March). http://www.informatics. jax.org/

Mouse Genome Sequencing Consortium. (2002). *Nature* **420**, 520–562.

Mouse Phenome Database (MPD) website, The Jackson Laboratory, Bar Harbor, Maine. (2003, March). http://www.jax.org/phenome

Nadeau, J.H., Singer, J.B., Matin, A. and Lander, E.S. (2000). *Nat. Genet.* **24**, 221–225.

Naf, D., Krupke, D.M., Sundberg, J.P., Eppig, J.T. and Bult, C.J. (2002). *Cancer Res.* **62**, 1235–1240.

Nagy, A. (2000). *Genesis* **26**, 99–109.

Nolan, P.M., Peters, J., Strivens, M., Rogers, D., Hagan, J., Spurr, N., Gray, I.C., Vizor, L., Brooker, D., Whitehill, E., Washbourne, R., Hough, T., Greenaway, S., Hewitt, M., Liu, X., McCormack, S., Pickford, K., Selley, R., Wells, C., Tymowska-Lalanne, Z., Roby, P., Glenister, P., Thornton, C., Thaung, C., Stevenson, J.A., Arkell, R., Mburu, P., Hardisty, R., Kiernan, A., Erven, A., Steel, K.P., Voegeling, S., Guenet, J.L., Nickols, C., Sadri, R., Nasse, M., Isaacs, A., Davies, K., Browne, M., Fisher, EM, Martin, J., Rastan, S., Brown, S.D. and Hunter, J. (2000). *Nat. Genet.* **25**, 440–443.

Paigen, K. (1995). *Nat. Med.* **1**, 215–220.

Paigen, K. (2002). *ILAR J.* **43**, 123–135.

Paigen, K. (2003). *Genetics* **163**, 1–7.

Reifsnyder, P.C. and Leiter, E.H. (2002). *Diabetes* **51**, 825–832.

Ringwald, M., Eppig, J.T., Begley, D.A., Corradi, J.P., McCright, I.J., Hayamizu, T.F., Hill D.P., Kadin, J.A. and Richardson, J.E. (2001). *Nucleic Acids Res.* **29**, 98–101.

Russell, E.S. (1978). In *Origins of Inbred Mice*, Proceedings of a Workshop, Bethesda, MD, February 1978 (ed. H.C. Morse, III), pp. 33–43. Academic Press, New York.

Russell, W.L., Kelly, E.M., Hunsicker, P.R., Bangham, J.W., Maddux, S.C. and Phipps, E.L. (1979). *Proc. Natl. Acad. Sci. USA* **76**, 5818–5819.

Sakai, N., Inui K, Tatsumi, N., Fukushima, H., Nishigaki, T., Taniike, M., Nishimoto, J., Tsukamoto, H., Yanagihara, I., Ozono, K., and Okada, S. (1996). *J. Neurochem.* **66**, 1118–1124.

Sands, M.S. and Birkenmeier, E.H. (1993). *Proc. Natl. Acad. Sci. USA* **90**, 6567–6571.

Schimenti, J. and Bucan, M. (1998). *Genome Res.* **8**, 698–710.

Schonig, K. and Bujard, H. (2003). *Transgenic Mouse Methods and Protocols* (eds M. Hofker and J. van Deursen) pp. 69–104. Humana Press, Totowa, New Jersey.

Simpson, E.M., Linder, C.C., Sargent, E.E., Davisson, M.T., Mobraaten, L.E. and Sharp, J.J. (1997). *Nat. Genet.* **16**, 19–27.

Smithies ,O., Gregg, R.G., Boggs, S.S., Koralewski, M.A. and Kucherlapati, R.S. (1985). *Nature* **317**, 230–234.

Snell, G.D. (1948). *J. Genet.* **49**, 87–99.

Snell, G.D. (1978). In *Origins of Inbred Strains*, Proceedings of a Workshop, Bethesda, MD, February 1978 (ed. H.C. Morse, III), pp. 1–31. Academic Press, New York.

Suzuki, T., Li, W., Zhang, Q., Karim, A., Novak, E.K., Sviderskaya, E.V., Hill, S.P., Bennett, D.C., Levin, A.V., Nieuwenhuis, H.K., Fong, C.T., Castellan, C., Miterski, B., Swank, R.T. and Spritz, R.A. (2002). *Nat. Genet.* **30**, 321–324.

Takahashi, T., Tanaka, M., Brannan, C.I., Jenkins, N.A., Copeland, N.G., Suda, T. and Nagata, S. (1994). *Cell* **76**, 969–976.

Tennent, B.J., Shultz, K.L. and Beamer, W.G. (1993). *Cancer Res.* **53**, 1059–1063.

The Gene Ontology Consortium. (2001). *Genome Res.* **11**, 1425–1433.

The Jackson Laboratory. (2000). JAX Bulletin, No. 4, The Jackson Laboratory, Bar Harbor, Maine. http://jaxmice.jax.org/info/bulletin/bulletin04.html

The Jackson Laboratory. (2001a). JAX Bulletin, No. 1, The Jackson Laboratory, Bar Harbor, Maine. http://jaxmice.jax.org/html/nomenclature/nomen_129.shtml

The Jackson Laboratory. (2001b). JAX Notes 481,4, The Jackson Laboratory, Bar Harbor, Maine. http://jaxmice.jax.org/library/notes/481.pdf

Threadgill, D.W., Yee, D., Matin, A., Nadeau, J.H. and Magnuson, T. (1997). *Mamm. Genome* **8**, 390–393.

Threadgill, D.W., Hunter, K.W. and Williams, R.W. (2002). *Mamm. Genome* **13**, 175–178.

Transgenic and Targeted Mutation Database (TBASE), The Jackson Laboratory, Bar Harbor, Maine. (2003, March). http://tbase.jax.org

Utomo, A.R., Nikitin, A.Y. and Lee, W.H. (1999). *Nat. Biotechnol.* **17**, 1091–1096.

Wakeland, E., Morel, L., Achey, K., Yui, M. and Longmate, J. (1997). *Immunol. Today* **18**, 472–477.

Wilson, S.M., Yip, R., Swing, D.A., O'Sullivan, T.N., Zhang, Y., Novak, E.K., Swank, R.T., Russell, L.B., Copeland, N.G. and Jenkins, N.A. (2000). *Proc. Natl. Acad. Sci. USA* **97**, 7933–7938.

Xiang, X., Benson, K.F. and Chada, K. (1990). *Science* **247**, 967–969.

Zambrowicz, B.P. and Sands, A.T. (2003). *Nat. Rev. Drug Discov.* **2**, 38–51.

Zhang, Y., Proenca, R., Maffei, M., Barone, M., Leopold, L. and Friedman, J.M. (1994). *Nature* **372**, 425–432.

Zheng, Q.Y., Johnson, K.R. and Erway, L.C. (1999). *Hear Res.* **130**, 94–107.

CHAPTER 4

Mouse Genomics

M Hrabé de Angelis

Institute for Mammalian Genetics, GSF – National
Research Center for Environment and Health,
Neuherberg, Germany

A Adler, J Beckers, D Soewarto and S Wagner

Institute of Experimental Genetics, GSF – National
Research Center for Environment and Health,
Neuherberg, Germany

V Gailus-Durner and K Imai[1]

Institute of Developmental Genetics, GSF – National
Research Center for Environment and Health,
Neuherberg, Germany

ENU mutagenesis

History

Genetic variation is fundamental to all genetic experiments, and phenotypic variation, in particular, is used in surveys to uncover the normal function of a wild-type allele. Characterization of spontaneous and induced mutants, the analysis of transgenic and gene-targeted phenotypes in model organisms such as *Drosophila*, zebra fish, or rodents, are common tools to obtain insight into the biological function of genes. In general, there are two different strategies, which can be used for the systematic production of mutant phenotypes in the mouse, called gene-driven and phenotype-driven approach.

The gene-driven approach is based on mouse embryonic stem (ES) cell technology in which mouse

[1]K. Imai wrote the section on Positional Cloning.

The Laboratory Mouse
Copyright 2004 Elsevier
ISBN 0-1233-6425-6

mutants generated for any targeted mutation engineered through homologous recombination (Thomas and Capecchi, 1987; Ramirez-Solis *et al.*, 1993). Mostly, the engineered construct alters the targeted DNA region of the wild-type gene and interrupts its function resulting in many cases in a null allele or a knocked out gene. Prerequisite for the production of the targeting construct is the gene of interest, which involves an engineered DNA fragment containing the desired mutation.

The transgenic mouse system is another important genetic tool for studying gene function and gene regulation. In this case, a DNA construct, typically containing a promoter and regulatory unit and a gene of interest or a reporter gene, is injected into one of the pro-nuclei of the fertilized egg (Hogan *et al.*, 1994). Classically, such transgenic animals are used to study the effects of regulated over-expression of a functional gene product. In addition, the transgenic mouse system has successfully and frequently been used to study the functional potentials of regulatory DNA sequences.

Both technologies, gene-targeting in ES cells and the transgenic mouse system, require that the gene of interest has been isolated and that the gene structure is known, at least partially (see also Chapter 5 on Generation of Mouse Mutants by Sequence Information Driven and Random Mutagenesis).

Another gene-driven approach is the **gene trap** strategy (Evans *et al.*, 1997), where a selectable insert is introduced into ES cell DNA. By building a large number of cell lines carrying disrupted units, a valuable ES cell mutation bank is established for generating mouse mutants (Wiles *et al.*, 2000).

Complementary to such gene-driven approaches, in which mutants are produced for already known genes, the phenotype-driven approach identifies new genes, gene products and their relevant biological pathways by systematically recovering mouse mutants with a novel phenotype (Hrabé de Angelis and Balling, 1998).

In early studies of genetics, the house mouse became an ideal organism because of the availability of many variant and visible phenotypes caused by spontaneous mutations. Now, it is known that the spontaneous mutation frequency in mouse spermatogonia varies between 0.008 and 1.8×10^{-5} mutations per locus and gamete depending on the genetic endpoints (Favor, 1994). More than 1200 mutant phenotypes could be collected. Very soon it became clear that the number of different variants was still limited due to the fact that the detectable mutation frequency in the gametes was so low. Therefore, there was a deep interest in developing methods to induce mutations by physical or chemical agents. In 1927, Muller showed that the exposure to ionizing radiation (X-rays) induced heritable mutations. Since that time, several geneticists who studied all of the major experimental organisms have used ionizing radiation and various chemicals to assess the induced mutation frequencies and, in parallel, to reveal novel alleles as tools to understand gene function.

Different protocols of exposure could show, that X-irradiation can induce mutations in mouse spermatogonia at a rate of 0.5 to 13.3×10^{-5} per locus, which is 1–2 orders of magnitude higher than the spontaneous frequency (Favor, 1994). Ionizing radiation induces mainly large deletions or other large lesions such as translocations or complex rearrangements within the genome.

In addition to ionizing radiation, more and more chemicals were assessed for their ability to induce mutations in germ cells of different model organisms (from fruit fly to mouse). Chemical mutagens were characterized by differential spermatogenic response (Ehling *et al.*, 1972). Oakberg (1956) determined the duration of the different stages of spermatogenesis in the mouse which formed the basis for breeding schemes in mutagenicity tests. The frequencies and types of mutations induced by chemical mutagens have been analysed by employing the specific locus test (SLT; Russell, 1951), which will be described in more detail in this chapter (see section on Testing for Mutation Frequencies). Depending on the sensitivity of the germ cell stages, chemical mutagens can be divided into three groups: the first group consists of agents such as N-ethyl-N-nitrosourea (ENU), methylnitrosourea (MNU), procarbazine and triethylenemelamine (TEM), which induce mutations predominantly in stem cell spermatogonia. Most of these stem cell mutagens are also characterized by a sterile period, which results from the cytotoxic effect of the chemical to differentiating spermatogonia. The stem cell mutagens induce mostly intragenic lesions (base pair substitutions and small deletions). The second category includes chemicals affecting early spermatids. Chlorambucil (CHL) and melphalan belong to this group. The third group includes chemicals such as ethylmethanesulfonate (EMS), methyl methanesulfonate (MMS), cyclophosphamide (CP) and diethylsulfate (DES). They affect spermatozoa and late spermatids. Chemicals that affect post-spermatogonial stages of spermatogenesis mostly induce large lesions (intragenic and intergenic deletions and structural chromosomal aberrations; L.B. Russell *et al.*, 1990). In contrast to chemical mutagens, ionizing radiation induces mostly large lesions also in stem cell spermatogonia.

In 1979, Russell *et al.* (1979) demonstrated that ENU was the most powerful mutagen in mouse spermatogonial stem cells using the SLT. In 1981, Johnson and Lewis applied electrophoretic procedures to detect ENU-induced mutations in mouse proteins. They found a haemoglobin mutant in which a single amino acid substitution was discovered, and it was proposed that ENU had induced an A to T transversion in a histidine codon (Popp *et al.*, 1983). Subsequently, more mutations in offspring from ENU-treated stem cell spermatogonia of mice were molecularly characterized. Of the 27 mutations, 24 showed base-pair substitutions at A/T sites (Favor, 1990). This suggests that the relevant ENU-induced DNA adducts in mouse spermatogonia are ethylated adenine or thymine. ENU was the first chemical to be used for saturation mutagenesis of specific chromosomal regions, which were defined by induced deletions such as the *albino* deletion and the *t*-complex (Bode, 1984; Justice and Bode, 1986, 1988; Shedlovsky *et al.*, 1986, 1988). Genome-wide large-scale mutagenesis screens have been successfully carried out in other organisms such as *Drosophila melanogaster* (Nüsslein-Vollhardt *et al.*, 1984; Wieschaus *et al.*, 1984), *Caenorhabditis elegans* (Jansen *et al.*, 1997), and *Danio rerio* (Driever *et al.*, 1996; Haffter *et al.*, 1996). They demonstrated the validity and success of these approaches by giving insight into biological mechanisms of pattern formation, organogenesis and other processes in developmental pathways.

Using mice as animal models for human diseases, a few genome-wide screens for specific dominant mutations (cataracts, blood enzymes) were launched recognizing the advantage of ENU-induced mutations for studying gene function in mammals as well (see also classical phenotyping in "Testing for mutation frequencies" of this chapter). In several cases, a number of different alleles of a single locus was isolated and used to study protein function (Jungblut *et al.*, 1998; Klopp *et al.*, 1998; Graw *et al.*, 2002).

Two large-scale genome-wide projects were launched utilizing systematic ENU mutagenesis to recover mutant of neuropathological and allelic series that would be particularly relevant for inherited diseases in humans. The main challenge was to establish appropriate procedures to assess mutant mouse phenotypes of interest. Some assays that are routinely used for medical examination in humans were adopted for the mouse in the Munich protocol (GSF Research Center and University of Munich, Germany; Hrabé de Angelis *et al.*, 2000; Soewarto *et al.*, 2000). Additionally, new assays and protocols were developed to recover novel phenotypes with neuropathological and behavioural abnormalities.

This project was termed SHIRPA (*S*mithKline Beecham Pharmaceuticals; *H*arwell, MRC Mouse Genome Centre and Mammalian Genetics Unit; *I*mperial College School of Medicine at St Mary's; *R*oyal London Hospital, St Bartholomew's and the Royal London School of Medicine; *P*henotype *A*ssessment) protocol (Rogers *et al.*, 1997). Both these protocols were developed to assess mutant phenotypes for specific, postnatal abnormalities comprising congenital malformations, clinical chemical, biochemical, haematological, immunological defects, and complex traits such as allergy and behaviour. To date, these projects have uncovered a rich diversity of mutations. Since the start of the projects 6 years ago, a higher number of induced mutants have been recovered in comparison to the number of spontaneous mutants, which occurred in the last 100 years.

A number of genome-wide recessive screens to isolate new recessive mutations have been set up in the last few years beside other screens for recessive mutations in specific chromosomal regions using either deletions or inversions (Table 4.1).

Nevertheless, it became obvious that huge efforts are needed to obtain a mutant for each single gene or pathophysiological event, and even more so, for each informative allele. In order to tackle this goal, an international effort has been launched, which includes mouse centres from Asia, Australia, Europe and North America (Nadeau *et al.*, 2001).

Function and mechanism of ENU mutagenesis

N-ethyl-N-nitrosourea is an alkylating agent that can transfer its ethyl group to oxygen or nitrogen atoms at a number of reactive sites in the DNA bases (Table 4.2). N-ethyl-N-nitrosourea does not require metabolic activation.

If the alkylated sites are not repaired, they result in mispairing of the two DNA strands and base pair substitutions during DNA replication (Shibuya and Morimoto, 1993).

ENU mainly induces point mutations, namely base pair substitutions such as A/T → T/A transversions and A/T to G/C transitions, covering about 82%

TABLE 4.1: Overview of current large-scale ENU mutagenesis projects

ENU screen center/ location	Mutagenesis approach	Mouse regions	Summary of screens	Contact (URL/email)
Mammalian Genetics Unit, Medical Research Council, Harwell, UK	Dominant Recessive	Genome-wide Chr 13 36H del	SHIRPA, LMA, Startle, PPI and clinical chemistry	http://www.mgu.har.mrc.ac.uk/mutabase
GSF, Neuherberg, Germany	Dominant Recessive	Genome-wide Genome-wide	dysmorphology, clinical chemistry, immunological	http://www.gsf.de/ieg/groups/enu-mouse.html
Jackson Laboratory, Bar Harbor, Maine	Dominant Recessive	Genome-wide Genome-wide	eye defects, hearing, seizures, anxiety, rotarod startle, PPI and heart, lung and blood	htttp://www.jax.org/
University of Pennsylvania, Philadelphia, USA	Recessive	Chr 5	circadian behavior and sleep	http://www.uphs.upenn.edu/cnb/
Oak Ridge National Laboratory, Oak Ridge, Tennessee, USA	Recessive Recessive Recessive Recessive	Chr 7 Chr 15 Chr X Chr 10	social behavior, drug abuse, alcohol, eye, ageing, auditory, nociception, learning and memory	http://www.bio.ornl.gov/htmouse
Neurobiology and Genetics Department, Case Western Reserve University, Cleveland, USA	Dominant Sensitized	Genome-wide	circadian, general behaviour, ataxia	http://www.northwestern.edu/neurobiology/faculty/takahashi.html
RIKEN Genomic Sciences Centre, Yokohama, Japan	Dominant Recessive	Genome-wide Genome-wide	late-onset phenotype in a variety of phenotypic areas	http://www.gsc.riken.go.jp/Mouse/
University of Toronto, Mouse Models Initiative, Toronto, Canada	Dominant Sensitized	Genome-wide	bone, cardiovascular, learning and memory, developmental	http://www.cmhd.ca/sub/mutagenesis.htm
Medical Genome Centre, Australia National University., Canberra, Australia	Recessive	Genome-wide	Immunological phenotpyes and other visibles	http://jcsmr.anu.edu.au/group_pages/mgc/research.htm
Mouse Genome Centre, Baylor College of Medicine, Houston, Texas, USA	Recessive	Chr 11 Chr 4	developmental phenotypes	http://www.mouse-genome.bcm.tmc.edu/ENU/MutagenesisProj.asp

Source: Modified from Brown and Balling, 2001.

TABLE 4.2: Reactive sites that are susceptible for alkylation	
Base	**Reactive sites**
Adenine	N1-, N3-, N7-
Guanine	O6-, N3-, N7-
Thymine	O2-, O4-, N3-
Cytosine	O2-, N3-

of all induced mutations (Justice *et al.*, 1999), or small deletions. Base pair substitutions at A/T sites result in missense, nonsense or mRNA splicing defects.

ENU is a highly effective mutagen in mammalian germ cells and has been extensively used to study the influence of a number of variables on germ cell mutagenicity such as gender (Ehling and Neuhäuser-Klaus, 1988), sensitivity of germ cell stages (Favor *et al.*, 1990) and dose fractionation (Russell *et al.*, 1982; Hitotsumachi *et al.*, 1985; Favor *et al.*, 1988). In the mouse, the mutagenic action of ENU is most potent in pre-meiotic spermatogonial stem cells, which allows large numbers of mutant offspring from permanent mating of ENU-treated male mice to be recovered.

Optimal doses of ENU can induce mutations with a frequency of about $6–1.5 \times 10^{-3}$, which is equivalent to one mutation per gene in every 175–655 gametes (Hitotsumachi *et al.*, 1985; Shedlovsky *et al.*, 1993). A threshold dose has been calculated to be 38.75 mg ENU per kilogram of body weight which is probably due to saturable repair processes (Favor, 1990). One of those DNA repair enzymes is O^6-alkylguanine-DNA alkyltransferase (AGT), which was initially studied in hamster spermatogonial cells (Seiler *et al.*, 1997). O^6-alkylguanine-DNA alkyltransferase can efficiently repair alkylated guanine adducts by transferring the alkyl group from the affected nucleotide to its own cysteine moiety (Lindahl *et al.*, 1988). Hence, only 13% G/C alterations were detected among all reported ENU induced DNA lesions (Justice *et al.*, 1999).

Due to the nature of ENU-induced mutations, the mutant phenotypes represent a unique mutant resource because they (i) reflect the consequences of single gene changes independent of position effects; (ii) provide a fine-structure dissection of protein function; (iii) display a range of mutation types from complete or partial loss of function alleles to hypermorphic and dominant negative alleles; and (iv) discover gene functions in an unbiased manner.

Testing for mutation frequencies

Initially, the tests described in the following section were developed to evaluate the mutagenicity of chemicals and ionizing radiation and to study the mechanisms of mutation induction. In the early 1950s, W.L. Russell developed the SLT (Russell, 1951), which was initially employed to quantify the mutagenic effects of ionizing radiation in mouse germ cells. The homozygous *Tester*-stock was constructed for seven recessive, viable mutations affecting easily recognizable external traits: *a* (non-agouti, Chr. 2), *b* (brown, Chr. 4, tyrosine related protein 1 *Tyrp1^b*), *c^{ch}* (chinchilla at albino, Chr. 7, tyrosinase *Tyr^{c-ch}*), *d* (dilute, Chr. 9, myosin Va *Myo5a^d*), *p* (pink-eyed dilution, Chr. 7), *s* (piebald spotting, Chr. 14, endothelin-B receptor *Ednrb^s*) and *se* (short ear, Chr. 9, bone morphogenetic protein 5 *Bmp5^{se}*). Another tester-stock was constructed in Harwell, England using different loci: *bp* (brachypodism, Chr. 2, *Gdf5^{bp-H}*), *fz* (fuzzy, Chr. 1), *ln* (leaden, Chr. 1, melanophilin *Mlph^{ln}*), *pa* (pallid, Chr. 2, pallidin *Pldn^{pa}*), and *pe* (pearl, Chr. 13, *Ap3b1^{pe}*; Lyon and Morris, 1966). The Harwell tester stock was especially created to test if some genes of Russell's SLT are more mutable than others in the genome.

In Russell's SLT, untreated *T*-stock females are mated to mutagen-treated wild-type males and to another set of untreated wild-type males as a control group. In the absence of any mutation, offspring derived from this cross will display the wild-type phenotype. An induced mutation to a recessive allele in a germ cell of the treated wild-type male will generate an offspring homozygous for the respective recessive allele so that the phenotype of the specific locus is immediately visible among the otherwise wild-type progeny. This approach allows the calculation of mutation rates by simply scoring the number of mutant offspring. All mutants were subjected to allelism tests to confirm the mutation.

In the early 1980s, Ehling developed another phenotypic assay for dominant cataract mutations in mice (Ehling *et al.*, 1982). At the age of 4–6 weeks, F1 offspring of mutagen-treated males were examined biomicroscopically, with the aid of a slit lamp, for lens opacities. The alterations in the lens were categorized into seven phenotypic classes: *total opacity, nuclear and zonular cataract, nuclear cataract, anterior pyramidal cataract, anterior polar cataract, anterior capsular cataract,* and *vacuolated lens*. Variants exhibiting lens opacity were bred further to confirm the genetic nature of its variant phenotype.

A number of different biochemical methods have been established to detect and quantify dominant mutations such as the erythrocyte enzyme-activity assay (Pretsch et al., 1982; Charles and Pretsch 1987). In this assay, 10 enzymes from the glycolytic, pentose phosphate pathways and the citric acid cycle, which represent a set of structural, regulatory and operating loci, controlling the activity of each particular enzyme, were analysed. The enzymes included Lactate dehydrogenase (LDH), Triosephosphate isomerase (TPI), Malate dehydrogenase (MDH), Glucosephosphate isomerase (GPI), 3-phosphoglycerate kinase (PGK), Phosphoglycero mutase (PGAM), Glyceraldehyde-3-phophate dehydrogenase (GAPDH), Glucose-6-phosphate dehydrogenase (G6PD), Pyruvate kinase (PK) and Glutathione reductase (GR). At the age of 4–6 weeks, about 100 µl of blood was taken from F1 offspring of mutagenized males and from wild-type controls. The enzyme assays were performed in the afternoon, when the biological variability of the enzyme activities is low. Variants with an altered enzyme activity lying beyond 3 SD of the mean were defined as outliers, and these were genetically confirmed to be accepted as mutants.

The three endpoints – specific recessive genes, dominant cataracts and enzyme activity mutations – were combined in the multiple endpoint approach to increase the number of loci tested from 7 to 70 and to compare the mutability of different loci (Ehling et al., 1985). Other mutation tests were performed by screening for dominant visible traits in combination with specific recessive loci (Lyon, 1983) or by screening for dominant skeletal mutations (Ehling, 1966) and for histocompatibility mutations (Melvold and Kohn, 1975).

The tests described above meet all criteria for an ideal mutagenicity test which should be simple in its use, fast in carrying out and unambiguous, because large numbers of animals have to be examined to detect an increase over the spontaneous mutation rate. To perform mutagenicity tests in order to generate mouse mutants as models for human genetic diseases the numbers of phenotypic endpoints screened have to be increased considerably which was achieved in the large scale ENU mutagenesis screens (Table 4.1).

Breeding schemes

The breeding scheme for dominant phenotypes is simple and fast (Figure 4.1a), since it includes one single breeding generation. Fertile, mutagen-treated males of a certain inbred strain are mated to wild-type females of the same or another genetic background to produce F1 animals, among which may be some that are heterozygous for dominant or semi-dominant mutations. These F1 animals can be assessed in different phenotype assays (see section on Phenogenetics). If one of the F1 offspring displays an altered phenotype in comparison to the wild-type phenotype it is designated as a variant. Inheritance has to be tested by producing offspring from crosses of the variant animal with wild-type animals of opposite sex. Theoretically, 50% of these offspring will show the same phenotype as the variant parent and will be genetically heterozygous for the dominant mutation (+/−). As soon as the phenotype is confirmed to be a mutation, a new mutant line can be established by crossing affected heterozygous (+/−) animals to the original inbred strain in order to produce an **isogenic** line. If F1 hybrids between two inbred lines were initially produced, the aim is to develop a **congenic** mutant line on a certain inbred background.

Semi-dominant mutations can be determined if two affected, heterozygous (+/−) animals of opposite sex are mated to produce homozygous (−/−) offspring. If the homozygous (−/−) offspring display a more severe phenotype in comparison to their heterozygous siblings, a semi-dominant trait can be concluded.

For recessive phenotypes, a two-step breeding scheme is required (Figure 4.1b). To date, all reported recessive screens have used male F1 mice as founders, since it is much faster than using female F1 mice. A male F1 animal is expected to have heterozygous mutations and is crossed to wild-type females of the same inbred strain to produce G2 offspring. Initially, dominant visible phenotypes in the F1 animal should be excluded in order to avoid any interference of dominant and recessive phenotypes in the G3 generation, which could complicate the phenotypic screening. Only female G2 animals (+/− or +/+) are collected and subsequently crossed back to the male F1 founder to produce G3 animals, out of which one eighth might be homozygous for a specific recessive phenotype of interest. A minimum number of, for example, 20 G3 animals can be assessed for phenotypic alterations. A phenotypic variant of interest has to be tested for inheritance by producing homozygous offspring showing the same phenotype. This is performed by back-crossing to the G1 founder (in case of a G3 female), or to the respective G2 female (in case of a G3 male). If the parental animals are not available any more, the respective G3 animal is out-crossed to a wild-type animal to produce heterozygous G4 offspring. By setting up G4 inter-crosses, one quarter of the G5 offspring should be homozygous for the recessive mutation of interest.

Beside large-scale mutagenesis screens targeting a broad range of mutations, smaller labs are also able to perform recessive screens for specific chromosomal regions, as initially performed by E. Rinchik (1990) for the *albino* locus. In this case (Figure 4.1c), a mutagen-treated male is mated to a female that carries a dominant coat colour marker in a homozygous state. The G1 mice are then mated to animals of the opposite sex carrying a deletion of interest and marked with a different dominant coat colour marker.

Those G2 animals, which obtain the null allele from the mother and the other allele carrying a point mutation from the father, are then hemizygous for the affected recessive gene of interest. The absence of the hemizygous mutant class indicates the recovery of a presumably linked lethal mutation. If the hemizygous mutant class survives it can be screened for a novel phenotype.

New strategies have been developed to introduce defined chromosomal rearrangements such as deletions (Ramirez-Solis *et al.*, 1993) or inversions by using ES cells in combination with the *Cre/loxP* site-specific recombination system, thus creating balancer chromosomes (Zheng *et al.*, 1999). These balancer chromosomes are used in screens that are designed to isolate region-specific, recessive lethal or detrimental mutations by screening for the presence or absence of the coat colour marker that they confer. In general, a balancer chromosome suppresses recombination over the length of the inverted chromosomal region.

The first balancer chromosome in the mouse was constructed on Chr. 11, in which one endpoint of the inversion breaks in the *Wnt3* gene, which is an essential gene in early embryonic development and lethal in a homozygous state. It can be tagged, for example, with a vector that contains the *K14*-agouti transgene conferring dominantly a yellow coat colour. In this scheme (Figure 4.1d), mutagen-treated C57BL/6 males are mated to females carrying the inversion. G1 animals, which are yellow, are mated to animals carrying both a balancer chromosome and another dominant coat colour marker on the non-mutated chromosome, *Rex* (*Re*) on Chr. 11 causing curly fur (mottled), which are then yellow and mottled. The coat colour markers allow differentiation of all classes of G2 offspring (explained in more detail in Figure 4.1e). The G2 animals of interest carry the balancer chromosome and the mutated chromosome (yellow and straight fur) and are inter-crossed to look for new mutations in the G3 generation. The G3 animals that are homozygous for the inversion are not viable. The absence of the homozygous mutant class indicates the recovery of a presumably linked lethal mutation. If the homozygous mutant class survives it can be screened for a novel phenotype.

Modifier or sensitized screens are other strategies to study genetic interaction in a particular biological pathway. In 1993, Dietrich *et al.* identified the first modifying locus of the *Min* mutation, which carries a mutant mouse Apc gene and develops many intestinal adenomas. Mice carrying *Min* provided a model system for studying human familial adenomatous polyposis, and for identifying genes that can modify the phenotype by mating the carrier either to different genetic backgrounds, as in Figure 4.1e or to ENU treated animals as in Figure 4.1f. The F1 animals derived from a different parental strain (AKR/J) displayed a lower number of tumours than the original congenic strain C57BL/6-*Min/* + suggesting that the AKR strain carries alleles that can act in a dominant way to modify the tumourigenic effect of *Min* (Figure 4.1e). Alternatively, a sensitized screen against a drug or an environmental modification can be carried out. In the G1 generation, mutant phenotypes may be recovered (1) that would show a completely new dominant phenotype owing to the presence of a dominant mutation that is independent from the pathway of interest (Figure 4.1e and f) (2) that show a less (suppressor) or more severe (enhancer) appearance compared to the original mutant phenotype due to a dominant mutation, which directly affects and modifies the pathway of interest (Figure 4.1e and f); and (3) that show a mutant phenotype caused by new alleles at the locus of the original recessive (Figure 4.1f) mutation. The aim of such a screen is to detect new alleles that modify (in a non-allelic, non-complementary manner) the phenotype of the original mutation or transgenic model by enhancement or suppression.

Unlike gene targeting or insertional mutagenesis, ENU induced mutations isolated by their phenotype are not molecularly tagged. In general, these mutations are mapped in meiotic back-crosses (back-cross or inter-cross) and identified by either positional cloning or the candidate gene approach (see also section on Genetic Mapping in this chapter, KI).

Gene-driven approaches using ENU mutagenesis

The rapid development and refinement of DNA analysis and sequencing methods promoted the establishment of efficient, high through-put, sequence-based genetic screens.

In a mutagenesis screen, routine freezing of sperm from F1 and mutant male mice provides the possibility of maintaining valuable mutant lines for future scientific projects. Thus, a combination of a sperm and

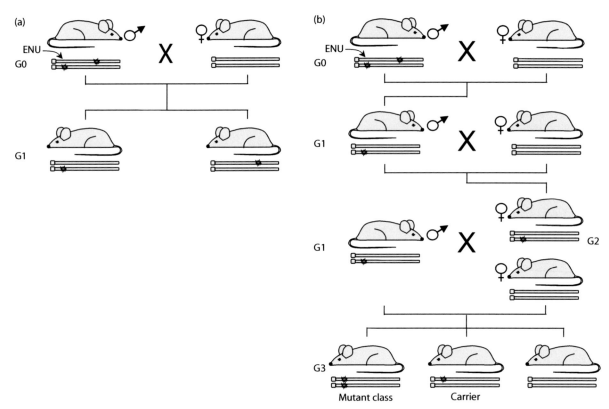

Figure 4.1 (a) Dominant genome-wide screen: G0 represents the parental generation. An ENU-treated male is mated to a wild-type female. All F1 offspring of the G1 generation are heterozygous for dominant mutations; (b) recessive genome-wide screen: ENU-treated males are mated to wild-type females (G0 generation). In the resultant G1 generation male animals (F1) probably heterozygous for a mutation, are mated to wild-type females to produce heterozygous or wild-type G2 offspring. Only the female G2 mice are selected and mated to their respective F1 fathers to produce the G3 generation; (c) region-specific recessive screen using chromosomal deletions: ENU-treated males are mated to wild-type females carrying a dominant yellow coat colour marker (K-14-*agouti*). Animals of the G1 generation are mated to animals of the opposite sex that are dominant for a different coat colour and carry a deletion of interest. The G2 offspring may be classified in a first step by its coat colour as a mutant animal or a carrier of the mutant gene. Offspring with the yellow coat colour of the G0 female will be uninformative; (d) region-specific recessive screen using balancer chromosomes: ENU-treated C57BL/6 males are mated to females carrying a balancer chromosome containing an inversion, which is marked with the agouti transgene (K-14-*agouti*) being dominant for a yellow coat colour. The offspring in the G1 generation have yellow fur. G1 males which are now carriers of the inversion, are mated to females being heterozygous for the balancer chromosome and being dominant for a different coat colour and structure (yellow mottled) on the non-mutated chromosome (*Rex* on Chr. 11). The coat colours allow one to distinguish the different classes that occur in the G2 generation. Since in homozygous animals with yellow fur and with the inversion localized on both chromosomes the mutation will be lethal, yellow mice with straight hair will carry both the mutation and the inversion and are useful for further inter-crossing. Dark mice with curly fur do not carry the inversion. In the G3 generation the offspring will easily be classified as the (dark) mutant class which is carrying the balanced point mutation and the yellow carrier class. The wild-type mutant class indicates whether the likelihood of a linked lethal mutation is missing; (e) dominant modifier screen: In the G0 generation, a mutant line with a dominant mutation on a specific genetic background (e.g. C57BL/6) is mated to other different inbred strains. The F1 hybrids are examined for modifier alleles that could act in a dominant fashion to confer a partial suppression or enhancement of the original mutant phenotype; (f) recessive modifier screen: ENU-treated males are mated to females already carrying a recessive mutation in the pathway of interest. In the G1 generation, offspring with a new dominant phenotype may be created. The second group of offspring might show the recessive mutant phenotype to a greater extent compared with the original phenotype, which is classified as a dominant modifier as the pathway of interest is directly affected and modified. In the third group new alleles at the locus of the original recessive mutation cause a different mutant phenotype.

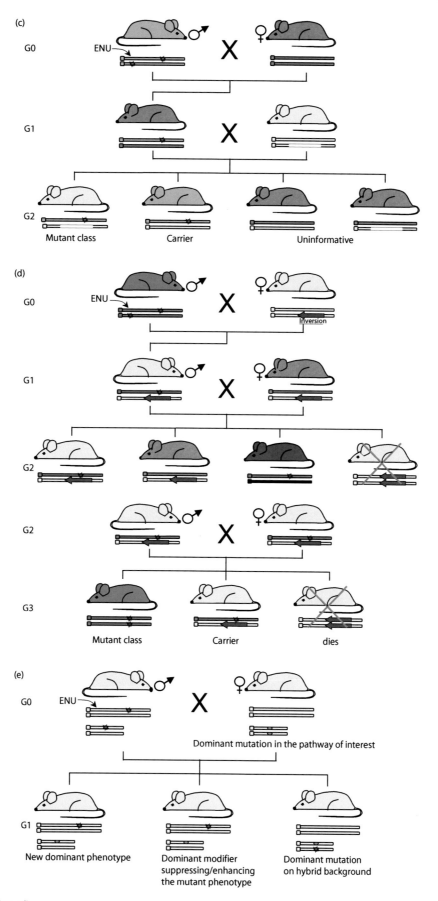

Figure 4.1 (continued)

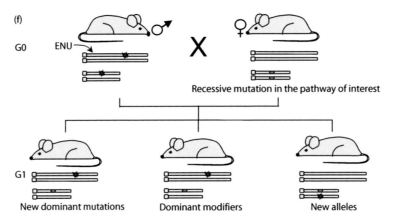

Figure 4.1 (continued)

DNA archive from mutagen-treated males with appropriate methods to screen for mutations of a gene of interest (Coghill *et al.*, 2002) provides a powerful tool for the detection of new mutations on the DNA level. Rapid screening of DNA sequences from the archive is then followed by mutant recovery of identified mutations by *in vitro* fertilization. Phenotypic and functional analysis has to be performed as a second step.

Another recently developed strategy for the generation of mutants is a screen for chemically induced mutations in mouse ES cells (Chen *et al.*, 2000; Munroe *et al.*, 2000), which can be applied by either a phenotype- or genotype-driven approach. In this strategy, ES cells are exposed to EMS, ENU or other mutagens to establish an ES cell bank. Subsequently, these mutagen-treated ES cells are injected into mouse blastocysts and re-implanted into pseudo-pregnant females to generate chimeric mice in which the germ line is affected. Subsequently, in the phenotype-driven approach, a large number of chimeras are generated in order to assess novel phenotypes revealing new genes or alleles of known genes as already described for the whole-animal mutagenesis screens. Alternatively, the genotype-driven approach utilizes the ES cell bank in the first step as a DNA bank in order to screen for ES cell clones with a mutation in a gene of interest by using polymerase chain reaction (PCR)- and sequence-based analyses. In the second step, chimeric mice are generated only from those ES cell clones, which carry a mutation of interest. Some PCR- and sequence-based methods will be briefly described in the following section.

Mutation detection

Restriction fragment length polymorphism

A *r*estriction *f*ragment *l*ength *p*olymorphism (RFLP) is defined by the existence of alternative alleles associated with restriction fragments that differ in size from each other (Silver, 1995). RFLPs can be visualized by digesting the DNA from a mutant and a wild-type animal with a restriction enzyme, followed by separating the fragments in a gel electrophoresis system according to size, then blotting and hybridization to a labelled probe that identifies the locus of interest. Since each restriction enzyme cuts at a very specific recognition site, an alteration in the investigated DNA sequence can be immediately detected. A mutant animal that carries, for example, a point mutation in the investigated locus would display another pattern of DNA fragments in a Southern blot analysis than a wild-type animal if the mutation sets a new restriction site or alters the existing one.

As a tool to detect DNA variations, RFLPs were predominantly used for linkage analysis until the invention of the PCR, which is (i) the ultimate in sensitivity; (ii) the ultimate in resolution, since all kind of polymorphisms can be distinguished (from single base changes to large rearrangements) and (iii) extremely rapid in its use. The advantage of RFLPs over PCR-based protocols is that no prior sequence information or oligonucleotide synthesis is required to detect DNA variation and large regions can be analysed.

Single-strand confirmation polymorphism

DNA that normally exists in double strands can be separated by heating above the melting temperature and, if cooled down slowly, will re-anneal to the double-stranded molecule. If cooled very rapidly, however, the single strands obtain a unique three-dimensional shape, which is highly dependent on their sequence. Each single strand will assume a most-favoured confirmation based on the lowest free energy state in which a large number of bases form hydrogen bonds with each other. This phenomenon is known as *s*ingle *s*trand *c*onfirmation *p*olymorphism (SSCP; Orita *et al.*, 1989; Beier *et al.*, 1992) and is highly sensitive, such that a single base change will alter the shape that can be visualized on a gel. In its basic form, DNA fragments of interest (100–300 bp fragments) are amplified by PCR, then denatured at 94 °C and cooled on ice followed by gel electrophoresis on standard non-denaturing polyacrylamide gels. The two single DNA strands from an animal carrying a point mutation in the gene of interest will move with different speeds and can be easily distinguished from the wild type DNA strands. The identified mutated DNA fragments then have to be confirmed by sequencing.

Thermal gradient gel electrophoresis

The *t*hermal *g*radient *g*el *e*lectrophoresis (TGGE) is a method of mutation analysis without a chemical gradient (compared with denaturing gradient gel electrophoresis (DGGE)). DNA fragments with point mutations exhibit different melting behaviour and therefore a different conformation compared with the wild-type DNA. By varying the denaturing conditions with a temperature gradient, TGGE achieves unparalleled separation and sensitivity. When a wild-type DNA sequence is compared with a mutant DNA sequence in a denaturing gel simultaneous temperature gradient, the migration of the two fragments will differ based on the difference in the melting temperatures of the DNA sequences. Typically, DNA fragments of up to 1000 bases in length can be investigated by TGGE (Anderson *et al.*, 1978; Arimondo *et al.*, 2001).

Heteroduplex analysis

Denaturing High Performance Liquid Chromatography (DHPLC) identifies mutations by detecting sequence variation in re-annealed DNA-strands (heteroduplexes; Liu *et al.*, 1998). With this technique, heteroduplex fragments are eluted differently from the chromatography column to the homoduplex fragments and result in a readily distinguishable peak of ultraviolet (UV) absorbance. This method efficiently detects single nucleotide alterations and insertion/deletion events in crude PCR products directly without DNA sequencing. Therefore, DHPLC is useful for finding novel mutations and polymorphisms in genes and other genomic regions, for screening clones to identify candidates for sequencing, for scoring samples for known mutations, and for performing quality control on oligonucleotides.

Sequencing

Sequencing by synthesis is based on the detection of incorporation of di-deoxy NTPs chain termination method by using a primer-directed polymerase extension. The sequence can be deduced by repeated cycles of nucleotide incorporation. The di-deoxy method of DNA sequencing, which used radioactive-labelled dNTPs was originally developed by Sanger *et al.* (1977).

Meanwhile, one of the most rapid and reliable identification of sequence polymorphisms is the automated fluorescent di-deoxy sequencing method of DNA (Voss *et al.*, 1989). The combination of optimized protocols for the sequencing of PCR amplified fragments of about 1 kb and a direct evaluation of the sequence traces on a gel makes it feasible to test directly for the presence of mutations in the offspring of mutagen-treated mice.

Capillary electrophoresis (CE) is a technique for DNA sequencing where electrophoretic separation of fluorescence-labelled Sanger fragments takes place in a very small capillary tube. Capillary electrophoresis separates molecules based on differences in their electrophoretic mobility in the buffer, which is equal to the charge to mass ratio of the molecule (Luckey *et al.*, 1990).

A completely new method for sequencing DNA has been devised recently that does not use electrophoresis, radioactivity, or fluorescence. The method works by measuring inorganic pyrophosphate (PPi) generated by the DNA polymerase reaction and is termed as Pyrosequencing (Ronaghi *et al.*, 1996, 1998). The biotinylated target DNA and DNA polymerase are immobilized onto streptavidin-coated paramagnetic beads in a column and solutions containing the four different dNTPs are pumped through. The detection is based on the PPi, which is released during the DNA polymerase reaction and the quantitative conversion of PPi to ATP by the enzymatic activity of sulfurylase. The concentration of ATP is sensed by luciferase, which produces visible light. A unique pattern of signals is obtained in a so-called pyrogram, which determine the exact sequence of different nucleotide positions. Pyrosequencing allows accurate determination of 20 bases for a large number of PCR products.

Database

Establishing a large-scale mutagenesis screen requires a database that combines the functions for animal management, for managing sample tracking and workflow, for the documentation of results and, not at least, for the evaluation and analysis of data (Hrabé de Angelis and Strivens, 2001). Such a system should normally be based in a centralized data storage system, which can be a proprietary relational database management system (RDGMS) such as Oracle 8i, Sybase Adaptive Server Enterprise, Microsoft SQL server or MySQL (Hrabé de Angelis and Strivens, 2001). The system should provide a user interface that is practical and easily operated, as a broad range of personnel, from animal caretakers, students to scientists will use the database for animal and data administration (Figure 4.2). For the installation of a database system, one has to consider that the computer equipment has to be installed in the animal facility itself and that various net connections in the mouse rooms will facilitate entering all data via, for example, a laptop as soon as a procedure with the animals or samples is completed. In general, each animal needs a unique identification number in combination with, for example, an ear-mark or a transponder and is assigned to a certain room, rack and cage number supplied by the animal management system (AMS).

The AMS is also necessary for the collection of all physical data of mouse breeding such as mating, weaning and disposal procedures within a certain colony. The mating type, line code or generation of each animal should be printable on a cage card (for an example of a cage card see Figure 4.3) to facilitate the organization in the mouse facility. The AMS is connected to a sample tracking system (STS), which is able to optimize the selection of appropriate animals with respect to line, age or type of sample. Moreover, it supplies the workflow by creating working lists for internal and external use by tracking each single sample and provides an automated ordering system for the screening lab or investigator and a scheduling system for the respective animal facility. A result documentation system (RDS) supports the accurate recording and storage of observational material and quantitative analysis by named individuals. The combination with a data analysis system that offers appropriate statistical tools and the knowledge of specialists for the interpretation of evaluated data sets results in a comprehensive picture of each mutant line and may recover new aspects for the pathogenesis of diseases.

Finally it will be necessary for international collaboration to present the results obtained in the world

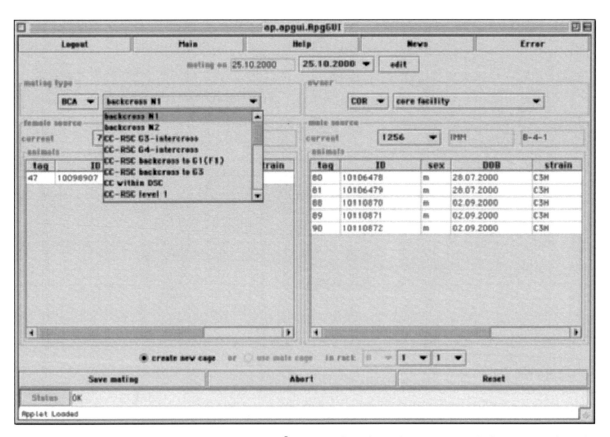

Figure 4.2 Screen shot of a DB user interface (MouseNet©). This interface shows the mating panel of MouseNet where the day of mating, the mating type (depending on the breeding scheme), the Ids of the female/male mice and the destination of the created cage number are recorded. Additionally, all mating steps of an individual mouse are recorded in log files and can be queried at any time.

Figure 4.3 Example of a cage card. Each animal has a unique identification number in combination with an earmark. All important information about an animal such as strain or line, sex, date of birth, generation, parent information, cage number, location of the cage are printed on the cage card.

wide web where data of all contributing research centres may be found.

Considerations before getting started

In general, the outcome of a comprehensive phenotype-driven screen is influenced by several factors, (a) the choice of a strain, (b) a defined genetic predisposition, (c) the response to ENU treatment and (d) the logistical assistance by cryopreservation.

The first important prerequisite for effective mutagenesis screens is the choice of appropriate mouse strains. It is known that certain outbred and inbred strains have a different sensitivity to chemical mutagens presumably due to a variable capability for DNA repair in their germ cells (van Zeeland, 1988; Vogel *et al.*, 1996). If an out-bred strain is selected for a mutagenesis experiment, one has to consider the mixed genetic background, which may complicate further analysis of the mutant phenotype (Justice, 2000). When inbred strains are used for mutagenesis, one can avoid the difficulties of a polymorphic genetic background for phenotypic analysis and characterization, since the mutant phenotype can be clearly defined in comparison with the wild-type phenotype. Since a different inbred strain is used for backcrossing, one has to consider that different strain dependent alleles at the same locus or modifying genes may still interfere with the mutated allele. Modifiers or modifying genes can be located on the same or on a completely different chromosome, where they can act as regulatory elements by alteration or compensation of the mutated allele.

A second issue is the choice of an appropriate strain for mutagenesis according to the category of phenotypes and parameters, which have to be assessed (Rogers *et al.*, 1997). In a large-scale phenotype-driven screen, inbred strains with a genetic predisposition might not be appropriate where this genetic predisposition interferes with the mutated loci. For example, the C3H strain should be used with care in behavioural studies since it carries the *rd* (retinal degeneration) gene, which leads to blindness after about 6 weeks.

The third issue is the determination of the optimal dose of ENU for that (inbred) strain in order to perform an effective mutagenesis screen. Effectiveness of ENU is measured by the highest possible dose considering (i) the period of sterility, (ii) the percentage of males that regain fertility, (iii) the mortality rate before or after regaining fertility and (iv) the mutation frequency in the F1 or G3 generation. Weber *et al.* (2000) determined optimal doses of ENU for several inbred strains. They came to the conclusion that, for example, a fractionated dose of 3×100 mg/kg for BALB/cJ, 3×80 mg/kg for C3HeB/FeJ or 3×100 mg/kg for C57BL/6J mice, given as three weekly injections, results in a high number of fertile males ($>60\%$) and a satisfactory mutation frequency.

For a project to create a large number of new mouse mutant lines it is beneficial to establish an archive of cryopreserved embryos or germ cells in liquid nitrogen. It became obvious that a sperm or embryo archive optimizes the logistics of an animal facility with respect to space and breeding time. Moreover, it provides a valuable platform for genotype- and phenotype-based screens and facilitates the systematic and detailed study of gene function.

Embryo freezing is a well-established method for the maintenance of mutant lines by cryopreservation (Nakagata, 1994; Marschall and Hrabé de Angelis, 1999). However, 100–500 embryos are required to ensure a reliable re-establishment of a breeding colony. The cryopreservation of sperm is much faster than collecting embryos, and in addition, sperm from a single male can provide up to 20 times more offspring than embryos from a single female. If an important mutant line is to be maintained exclusively via females, cryopreservation of ovaries can be performed as an alternative. Moreover, spermatozoa of F1 males or of males from mutant lines can be archived in liquid nitrogen for almost unlimited time and if needed, a sufficient number of offspring can be recovered by *in vitro* fertilization and embryo transfer at any time to establish a new breeding colony. Nevertheless limitations for sperm freezing are set as only the male haplotype is preserved.

Phenogenetics

The need for standardized phenotyping

The phenotype of an individual organism comprises its observable traits and physical characteristics, as distinguished from its genetic constitution, the genotype. The presence of a phenotypic trait, such as a disease, may or may not be genetic. Non-genetic factors of phenotypic traits may include ecological, environmental and social determinants. Phenogenetics is defined as the dependence of phenotypic characteristics from genetic determinants. However, the phenotyping methods that we describe below, may also be applied to assess the influence of environmental and other non-genetic parameters on the phenotype.

Standardized phenotyping protocols are required for the efficient and reliable analysis of mouse mutant resources across laboratories. An overview will be given of recent initiatives to standardize the diagnostic methods for mouse phenotyping and different screening methods will be described in more detail. Some emphasis will be given to the potential of recent technologies for the analysis of transcript and protein expression profiles in mouse tissues and brief reference will be given to some major public database repositories for phenotypes of wildtype and mutant mouse strains.

In order to reach the goal of systematically and comprehensively annotating every gene of the mouse genome with at least one described function (*one mutation in every gene*), the International Mouse Mutagenesis Consortium (IMMC) has called for a coordinated worldwide endeavour to integrate the research efforts of academic and commercial laboratories (Nadeau *et al.*, 2001). Specifically, it has been recognized that the establishment of networks for phenotyping centres will be one of the major milestones of this enterprise. It will be necessary to establish standard operating procedures and detailed phenotyping protocols that are made public so that phenotyping conditions can be reproduced and are directly comparable in laboratories world-wide.

International repositories for mouse mutant lines such as the archive of *The Jackson Laboratory* in the USA, the *European Mouse Mutant Archive* (*EMMA*), European projects to support and co-ordinate mouse mutagenesis projects and standardization of phenotyping methods (such as *EUmorphia* and *EuroComp*) and similar organizations in Asia (such as the Riken

FANTOM initiative), Australia and Canada, represent some of the initiatives that have adopted the objectives of the *IMMC*. As one part of this challenge, a phenotyping centre for mouse mutants has been established at the GSF in Munich (Germany). The goal of the German Mouse Clinic is to develop and offer a large-scale standardized phenotype analysis of mouse mutants from various sources (including transgenic mice, knock-out mice, mutants from ENU or other mutagenesis screens and gene-trap approaches), to ensure an efficient, reliable and extensive analysis of the growing numbers of mouse mutants.

This phenotyping centre has a modular structure of 11 laboratories for specific pathophysiological tests of mouse mutants. Each unit is run by specialists in their field and contains a laboratory directly adjacent to a mouse room equipped with heating ventilation and air conditioning (HVAC)-connected IVC racks and class II changing stations.

The units are designated to different areas. These include units specialized in the analysis of clinical-chemical parameters, metabolism, behaviour, neurological disorders, toxicological and pharmacological parameters, as well as immmunological and allergological parameters. Additional units focus on bone and cartilage disease and development, dysmorphologies, lung and respiration function, eye development and vision as well as pain perception. The basic phenotypic assessment of a mutant mouse line is organized as a workflow of successive primary examinations, followed by subsequent secondary and tertiary phenotypic screens (see Figure 4.4) – representing the workflow.

A 'hotelling-system' for visiting scientists has been established in which guest groups are hosted for a limited time to contribute their specific phenotyping expertise for the identification of new mouse models. Maintenance of animal health and avoidance of cross infections are a challenging task in such multi-user units, due to the constant influx of mice from different institutions with varying health status.

Phenotyping methods for the identification of mouse models

Establishing phenotyping methods

The general concept of phenogenetics and phenotype-driven approaches to identify new mouse models is to determine phenotypic parameters of mouse mutant lines that are distinct from established baselines of the corresponding wild-type strain. Using classical genetic

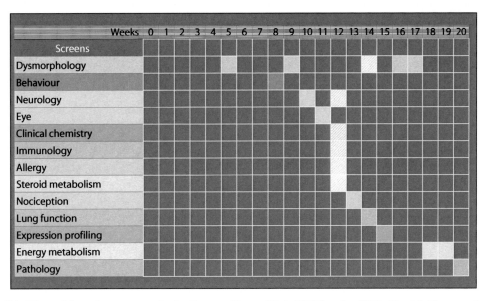

Figure 4.4 Workflow of the primary screen in the German Mouse Clinic (GMC) at the GSF in Munich (Germany).

approaches, such as back-crosses, the phenotypic alterations are genetically confirmed and the mutated gene may be identified using candidate gene approaches.

A comprehensive analysis of mutant phenotypes is crucial for realizing the full utility of the information generated in mouse genomics and mutagenesis. Although many different methods are already established, new tools for the assessment of mouse mutant phenotypes have to be developed and adapted mainly from clinical setups. This effort requires close interaction between experts of various fields, including mouse biologists, physicians, engineers, technicians and computer scientists. Not only the analytical methods and assays have to be adapted to the mouse, but also the technical equipment and instruments must be designed for animals as small as mice.

In general, an extensive mutant phenotypic analysis starts with a measurement of basic parameters to reveal traits of interest. These primary measurements are usually easy to perform and are not invasive in order to permit repeated measurements on individual mice and on the other hand to allow efficient analysis of a large number of animals. In a secondary screen interesting outliers uncovered by the primary assays are subjected to more detailed analysis. These measurements are more time and cost intensive and may require surgery or other invasive methods. Therefore, an in-depth analysis may only be performed with selected animals.

As a first approach towards phenotyping mice quantitatively rather than qualitatively, a systematic routine screening protocol was developed (SHIRPA; Rogers et al., 1997). The primary screen utilizes standard methods to provide a behavioural and functional profile by observational assessment. In the secondary screen, a comprehensive behavioural assessment battery and pathological analysis is performed. The tertiary screening stage is tailored to the assessment of models of neurological disease, as well as the assessment of phenotypic variability. SHIRPA has been developed to evolve with experience. Therefore, new tests are constantly being developed and added to this protocol (Hatcher et al., 2001) which is used as the basis for the development and adaptation of standard procedures in behaviour, neurology and dysmorphology screens (Fuchs et al., 2000).

Dysmorphology and anatomy

The main goal of the dysmorphology screen is to isolate mouse mutants that exhibit malformations in organ systems most relevant to human congenital dysmorphologies. Phenotypic abnormalities of sense-organs, limbs, axial skeleton, pigmentation and central nervous system (CNS) are analysed in the screen. For example, tail malformations (kinked tails, short tails) indicate abnormalities of the development of axial structures during embryogenesis. Abnormal behaviour, tremors, or seizures are indicative of distinct defects of the CNS.

For screening dysmorphological abnormalities a standard procedure was developed (Fuchs et al., 2000). The primary screen for morphological abnormalities is carried out in subsequent steps depending on the age of the mice. First, newborns are examined for obvious abnormalities and general appearance. The weight of the animals is determined from the time of weaning

TABLE 4.3: Parameters for the primary dysmorphological screening of adult mice

Body
Body Size, Body Shape, Right–Left-Abnormality, Skull Shape, Vibrissae, Tooth Length, Tooth Shape, Tooth Colour, Ear Size, Ear Form, Eye Size, Cataracts, Limbs, Double Digits, Crippled Digits, Syndactylism, Nails, Kinked Tail, Coiled Tail

Physical Appearance
General Physical Condition, Weight, Coat Colour, Coat Structure, Skin Colour, Skin Structure, Swellings, Carcinoma/Tumours, Strength, Gait, Trembling, Cramping, Paralysis, Seizures, Respiration

Behaviour
General Behaviour, Activity, Aggressiveness, Exploring, Head Tossing, Head Shaking, Circling, Click-Box Test, Landing Behaviour, Hanging Behaviour, Articulation

Environment
Social Structure in Cage, Cage Cleanness, Urine, Faeces

TABLE 4.4: Clinical chemical parameters

Basic haematology (blood counter)
  Parameters (measured)
    White blood cell (WBC) count, red blood cell (RBC) count, haematocrit (HCT), haemoglobin (HGB), platelets (PLT)
  Parameters (calculated)
    mean corpuscular volume (MCV)
    mean corpuscular haemoglobin (MCH), mean corpuscular haemoglobin concentration (MCHC),
Differential blood count (microscopy)
  Segmented and not segmented neutrophils (SEG), lymphocytes (LYM), monocytes (MOC), eosinophils (EOS), basophils (BAS)
Plasma enzymes activities (blood autoanalyser)
  Alkaline phosphatase (AP), α-Amylase (AMY), creatine kinase (CK), AST, ALT
Plasma concentrations of specific metabolites
  Glucose (GLS), cholesterol (CHO), triglycerides (TGL), total protein (TP), uric acid (HS), urea (HST), creatinine (CREA), ferritin (FER), transferrin (TRF), Calcium (Ca), inorganic phosphate (P)
Plasma concentrations of electrolytes
  Potassium (K), sodium (Na), chloride (CI)

until the 10th week of age. Adult mice are examined in a standardized protocol for dysmorphological abnormalities (see Table 4.3). To detect age-dependent abnormalities, the screen focuses on dysmorphologies that appear after 5–6 months of age.

New developments in electronics, physics and computer science have made it possible to design and adapt instruments for non-invasive imaging techniques like computer tomography and X-ray for the application to mouse phenotyping (Paulus *et al.*, 2000). In a primary screen, these techniques can be used to scan the whole body of individual mice for potential bone and cartilage abnormalities with resolution down to less than 50 μm for (computer tomography <50 μm). In a secondary screen detailed scanning of body parts or tissues using higher resolution or different views may be performed. A bone densitometer that is routinely used for osteoporosis check-ups in a clinical setup can be used to measure the bone density, a non-invasive method that might be included in the primary screen.

Clinical-chemical and biochemical metabolite assays

The clinical-chemical screen focuses on the identification and characterization of mouse mutants with defects of various organ systems, changes in metabolic pathways, haematological parameters and electrolyte homeostasis. For example, abnormalities in glucose metabolism (elevated fasting glucose levels) are indicative for diabetes mellitus, serum uric levels for gout, and hypoproteinaemia is associated with gastrointestinal tumours, severe liver damage and other human metabolic diseases (Rathkolb *et al.*, 2000). The set-up used is based on routine laboratory diagnostic procedures (e.g. blood autoanalyser). Therefore large numbers of mice can be screened for a broad spectrum of clinical-chemical and haematological parameters.

A general search spectrum for clinical-chemical parameters and blood count is presented in Table 4.4. In a secondary screen, specific profiles of organs are obtained if parameters of the primary screen show significant deviations from the baseline. For example, a liver profile consists of the measurement of aspartate-aminotransferase (AST/old: GOT), alanine-aminotransferase (ALT/old: GPT), alkaline phosphatase, bilirubin (total), bilirubin (direct), and cholesterol. To include the field of renal diseases SDS-PAGE might be used in the secondary screen in order to qualitatively analyse urinary proteins (detection of kidney lesions).

Figure 4.5 BALB/c mouse engaging in grooming behaviour and defecation (see Table 4.5, behavioural parameters) close to the partition in a version of the modified hole board test which also measures social behaviour. Courtesy of Dr. Sabine Hölter, German Mouse Clinic, GSF Neuherberg, Germany. **(See also Colour Plate 2.)**

method allows comprehensive metabolite scanning in a wide range of pathways from small amounts of blood (Rolinski *et al.*, 2000).

Behavioural screens

Classical behavioural measurements

The behavioural screen focuses on emotional and cognitive aspects such as anxiety, learning, and memory. As a primary screen all animals may be tested in the modified hole board (see Figure 4.5). This introductory test offers a first overview over the mouse's behaviour, since behavioural parameters such as anxiety, risk assessment, exploration, locomotion, food-intake, novelty seeking, and arousal are scored simultaneously in just one short test (see Table 4.5, Ohl *et al.*, 2001). The mice used in the primary screen should be as experimentally naive as possible, because apparently 'small' manipulations may already influence the test results. For example, food shortage which is a common experience in the natural environment can modify the behavioural differences of inbred strains (Cabib *et al.*, 2000). In the secondary screen specific confirmation tests are carried out. These may include specific tests for anxiety-related, exploratory and locomotor behaviour as well as tests for social behaviour and aggression. The tertiary

Another approach for identifying mouse mutants with defects in metabolic pathways is to analyse metabolites of key pathways (organic acids, amino acids), e.g. changes in the threonine and alanine levels which are associated with vitamin B6 deficiency and lactic acidosis, respectively. Electrospray-tandem-masspectrometry (ESI-MSMS) as a high throughput and cost-effective

TABLE 4.5: Behavioural parameters measured in the modified hole board		
Behavioural parameter	**Abbreviation**	**Behavioural dimension**
Percentage of time on board	Time	Anxiety
Latency until first board entry	Latency board	
Board entries	Entries	
Stretched attends (stretched body posture)	Stretched attends	Risk assessment
Latency until the first hole visit	Latency hole	Exploration
Holes visited	Holes visited	
Rearings on the board	Rear board	
Rearings in the box (i.e. not on the board)	Rear board	
Latency until the first exploration of familiar food	familiar food (latency exploration)	Novelty seeking
Latency until the first exploration of unfamiliar food	Unfamiliar food (latency exploration)	
Latency until the first intake of familiar food	Familiar food (latency intake)	Food intake inhibition
Line crossings	Line crossings	Locomotion
Self-grooming	Groom	
Number of boli	Defecation	Arousal

Source: Modified from Ohl *et al.*, 2001.

screen may serve as differentiated analysis of learning and memory abilities. For example, in the Morris–Water–Maze test spatial learning abilities can be assessed by having the mice swim in a water tank where they have to find a hidden platform as the only safe place in the tank.

In general, the behavioural screen should be carried out in combination with other screens such as the dysmorphology or metabolic screens to exclude major non-specific behavioural alterations.

Neurological parameters

In the neurological screen, mutant mice are examined for neurological phenotypes. These comprise neuromuscular (e.g. myopathy) and mitochondrial disorders, peripheral neuropathies, basal ganglia and movement disorders (e.g. Parkinson), cerebellar and brainstem, vascular, seizure and demyelinating diseases. As a primary screen, a clinical examination by SHIRPA-protocols for sensory/motor function may be performed as stated above. In a secondary screen, muscle strength, coordination and performance are analysed in more detail, e.g. in the rotarod test sensorimotor control is tested by the ability to remain on a rotating rod as the speed of rotation increases. Positive phenotypes such as the presence of seizures (indicating epilepsy) or neuromuscular impairments can then be explored further using methods such as telemetric electro encephalography (EEG) or electromyography (EMG), respectively. Moreover, examinations by magnetic resonance imaging, myopathology, and neuropathology may be performed. Additionally, in a tertiary screen somatosensory phenotypes may be assessed for spinal reflex function using an *in vitro* spinal cord preparation. With this method, mono- and polysynaptic neurotransmission in the spinal cord can be characterized quantitatively (assessment of synaptic connections). Anatomical analysis of the sensory nerves and standard pharmacology supplement the physiological assessment for mutants with defects in the somatosensory system.

Nociceptive parameters

The nociceptive screen measures pain sensitivity at the spinal and supra-spinal level. The animals are tested in several behavioural paradigms to evaluate different nociceptive circuits, as well as different modalities. Initially, spinal and supra-spinal responses to acute thermal stimuli may be tested with the hot plate assay. Additional tests in the primary screen used are the tail-flick-test (lightbeam heating of the tail), Hargreaves-test (heating of the hind paws), paw pressure test (mechanical pressure of the hind paws), von Frey Filament-test (poking with a filament with increasing pressure) and phenylquinone-induced writing test (injection of phenylquinone intraperitonally). These experiments are followed by the evaluation of tonic pain responses in the formalin-induced writhing-test (injection of formalin in hind paw subcutaneously). In the tertiary screen, inflammation-induced hyperalgesia and allodynia can be determined after the acetic acid, carrageenan or capsaicin injection using the van-Frey-filament or the Hargreaves tests.

Immunology and allergy

The immunology/allergy screen focuses on the detection and characterization of abnormal immune responses (e.g. immune dysfunction associated with autoimmune disease) in mouse mutants. As is true for the clinical-chemical and metabolite screen, mutations that affect the immune system or immune cell populations are not easy to identify upon gross examination. Therefore, high throughput and semi-automated methods were developed and used for the assessment of immune dysfunction in the primary screen (Flaswinkel et al., 2000). Defects of the immune system or the immune cell population can be analysed by determination of basal immunology parameters in peripheral blood using flow cytometry analysis (FACS) and enzyme-linked immunosorbance assay (ELISA). With FACS analysis the composition of cell lineages of the peripheral immune system (e.g. B-cells and their subpopulations) can be determined. Moreover, the quantification of expression levels of selected surface proteins that are involved in immune function is possible. Analysis by ELISA allows detection of levels of the different immunoglobulin subclasses and anti-DNA autoreactive autoantibodies in order to detect dysfunction in the humoral immunocompetence and autoimmune diseases, respectively. Defects in the immune system that lead to inflammation and allergic reactions can be assessed in the primary screen by the determination of the T-helper cell population (the T-cell response plays a pivotal role in the pathogenesis of atopic disorders) and IgE levels (indicative for the susceptibility to protein allergens – type I allergy).

In the secondary screen, specific aspects of immunity can be analysed, e.g. antigen-specific antibody response and differentiation of antigen-specific cell populations after immunization (e.g. ovalbumin). Challenging experiments with different pathogens and allergens are performed with outliers identified in the primary screen, because they require complex set-ups

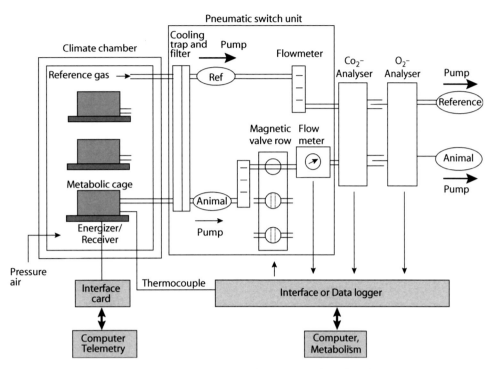

Figure 4.6 System for automatic recording of oxygen consumption and carbon dioxide production. (Courtesy of Dr. R. Elvert and G. Heldmaier, German Mouse Clinic, GSF Neuherberg and University of Marburg, Germany)

(e.g. analysis of broncho-pulmonary inflammatory reaction in broncho-alveolar lavage), specialized laboratories and mouse husbandry (depending on pathogen) or long-term experiments.

Metabolic screen

In the metabolic screen, a comprehensive analysis of bioenergetics in mice is performed, aiming towards identification of phenotypic variation in the mechanisms regulating body mass and composition, energy balance, metabolism, locomotor activity and thermal balance. Metabolic phenotyping includes measurements of growth and body composition, energy expenditure, respiratory exchange ratio, daily food intake and metabolizable energy.

The energy value of food intake and excreta, primarily faeces, can be determined with a calorimetric system based on a dry measurement principle to measure gross calorific values. Since some parameters during digestion, e.g. metabolism of microorganisms in the intestines, are difficult to control, the dry calorimetric system is only one method to approach measurement of energy metabolism. To complete the metabolic phenotyping the amount of oxygen used in oxidation processes may be measured. The automatic monitoring of oxygen consumption and carbon dioxide production as an estimate for metabolic rate requires the development

of a complex system (Figure 4.6). It includes metabolic chambers, a pneumatic switch unit, gas analysers and an interface connected to a computer to record and control the entire system. The prolonged analysis of mice in this open respiratory system is possible and parameters measured include ambient temperature, basal and resting metabolic rate, daily energy expenditure, heat production and respiratory exchange ratio. Upgrading the metabolic system with a telemetrical system enables parallel recording of body temperature and activity pattern. By implanting a temperature sensitive transmitter into the abdominal cavity, prolonged measurements in unrestrained mice are possible.

Lung function tests

The goal of the lung screen is to identify mouse mutants with defects in respiratory and non-respiratory lung functions, The susceptibility to environmental air pollutants may be studied by exposure to ultra-fine particles or aerosols. In the primary screen the spontaneous breathing patterns in unrestrained animals can be analysed by the determination of parameters such as **tidal volume**, breathing rate, inspiratory and expiratory time. Airway responsiveness to an unspecific challenge (e.g. methacholine) can be assessed by measuring changes in 'enhanced pause' (parameter for airway obstruction) as a function of the amount of nebulized substance. In the secondary screen respiratory mechanics (e.g. lung volume, airway resistance) and intrapulmonary and alveolo-capillary gas transfer can be measured in a miniaturized lung function unit. This requires **narcosis** of the animal, and a complex and elaborate experimental set-up. In combination with the immunology screen, the method of broncho-alveolar lavage can be used to determine inflammatory parameters in the broncho-alveolar lavage fluid (e.g. determination of total cell count, cell vitality etc.) of untreated and exposed animals (e.g. nebulized methacholine, oxygen, ultra-fine model aerosols, etc.)

Eye defects

The eye screen focuses on the identification of mouse mutants with morphological and physiological defects of the visual system to generate models for inherited ocular diseases such as cataracts. Obvious morphological defects of the eye or lens are detected during the general dysmorphological examination. Methods used are slit lamp biomicroscopy (cataracts), funduscopy (retinal degenerations glaucoma) and measurement of intraocular pressure and nystagmus analysis (opticus atrophy). Part of the morphological study is the histological examination. For the assessment of physiological features of the visual system methods such as electroretinography (test of retinal function) and the visual evoked potential (VEP, a measure of the mouse's responsiveness to a spatial feature of a stimulus) may be used (Pinto and Enroth-Cugell, 2000). To distinguish environmental from genetic factors that affect the visual system the mice have to be reared under tightly controlled lighting conditions during the course of the experiments.

Cardiovascular

Atheriosclerosis, hypertension, and other cardiovascular diseases are major causes for mortality in Western societies. Cardiovascular diseases are in many cases of multifactorial origin due to complex interactions between genetically determined homeostatic control mechanisms and environmental factors. Therefore, the screening for cardiovascular disorders to establish new animal models for these diseases involves the study of cardiac function, blood pressure regulation, cardiovascular hormones and lipoprotein metabolism. Methods used in a primary screen may be echocardiography (non-invasive assessment of left ventricle performance), electrocardiography (abnormal conduction and arhythmogenesis), noninvasive plethysmographic blood pressure measurement, blood chemistry (e.g. cholesterol, free fatty acids, triglycerides etc.) and exercise capacity (for more detail see Hoit and Nadeau, 2001). In a secondary screen more invasive methods may be used, such as the implementation of **telemetric transmitters** to monitor cardiovascular variables (e.g. heart rate, blood pressure) or left ventricular catheterization with high fidelity devices. Moreover, electrophysiological studies *in vivo* (intravascular placement of multipolar catheters) and *in vitro* (using heart cell preparations) or studies in combination with other screens such as the metabolic screen (characterization of kidney function with metabolic chambers) and lung function screen (respiratory rate) may be performed.

Pathology

For a complete assessment of phenotypic traits a comprehensive pathological examination may be particularly informative and may reveal traits of disease models that would otherwise be overlooked. Although necropsy is a terminal examination, it has successfully been used as a final first line screen to identify novel animal models. In this case, germ cells (sperms or ovaries) need to be archived before or during necropsy. This allows re-derivation of mutant lines using *in vitro* fertilization technologies in case of positive pathological findings. For established mouse mutant lines, animals meant for necropsy may be bred for this purpose.

For a comprehensive phenotypical analysis of mouse mutant lines it is of fundamental importance to execute a technically correct necropsy with a careful external and internal examination of the body. Any particular character of organs such as their position, the relationship to other organs, their volume, weight, shape, colour and consistency may be important evidence found in the course of the autopsy. This information together with the history of the animal and symptoms that preceded the autopsy will be required for the pathologist to reach a conclusive judgment of the particular phenotype. It will not be possible here to give an exhaustive description of good practice in autopsy and secondary examinations that may include, for example, histopathological diagnosis, immunohistological techniques, *in situ* hybridization, laser guided microdissection, quantitative PCR, or ultrastructural investigations.

In the mouse in particular, post-mortem autolysis of organs is very fast and begins as soon as cadaveric rigidity is over. It is therefore necessary to perform autopsies immediately after death or, if that is not possible, to store cadavers in a refrigerator ($+2$ °C to

+4 °C) as soon as possible after death. Appropriate equipment and a series of systematic operations to examine the animal in the best possible way, without altering any particular characteristics, are required for an exact appraisal of normal and pathological traits. This is best done with the help of pathology cards, which are a useful tool for the objective and systematic evaluation of macroscopic data. The examination begins with an external assessment of the body. The skin is inspected for the presence of wounds, infectious processes, tumours of the skin or of mammary glands. A rough, dry and **hirsute** fur may indicate a chronic disease or the presence of parasites. Crusty and dry areas, depilated areas, variation of fur colour or depigmentation is information that is important for pathological evaluation. Anomalies of the natural orifices, as the mouth, nose and anus should be reported. Modifications may be observed in the colouring of the labial or oral mucosa. Lesions, erosion or fracture of teeth, in particular of the incisors, may be a cause for malnutrition or even death. The examination of the anal opening may provide evidence, for example, of diarrhoea or prolapsed intestine.

Following external examination a complete internal necropsy is advisable, starting with the report of, for example, oedema or haematoma in subcutaneous tissues. The morphology of superficial and deep lymph nodes must be examined carefully. A complete examination of the skeleton is possible by X-raying. Formation of bone tumours, thinning or atrophic lesions, or developmental defects such as homeotic transformations or fusions of vertebral elements may be detected with this technique.

After the examination of the superficial organs, the necropsy proceeds with the examination of inner cavities, starting with the abdomen, the thorax and finally the skull. After opening the abdominal cavity, the investigator will first observe the position of organs, the presence of adhesions or liquids in the cavity. Following the general observation, the investigator may proceed to the extraction of abdominal viscera, starting with the spleen, pancreas, intestine and stomach (and its contents), liver and the associated gall bladder, and finally the urinary and genital apparatus. Colour, size, shape and anomalies of these organs need to be recorded carefully.

The investigator will then continue with the examination of the thorax. Before extracting thymus, lungs, heart and thyroid, it is necessary again to pay attention to some characteristics that may be changed as a consequence of the extraction techniques. The position of organs (situs inversus, congenital malformations), the presence of adhesions (as a result of pleuritis or pericarditis) or hydro- and haemothorax must be considered. As with all previous organs the shape, colour, volume and consistency of the thoracic viscera are crucial characteristics for a proper assessment of pathological phenotypes.

Finally, the investigator will proceed to the examination of the cranial cavity. The necropsy may start with an examination of eyes and the Harderian (retro-orbital) glands. Having opened the skull, typical diseases of the meningus, such as diffuse haemorrhage caused by rupture of meningeal vessels, or brain enlargements as in the case of oedema or tumours may be observed. After removal of the encephalon the general necropsy of the mouse is completed.

Histological examinations may be required either for all organs or only for organs with pathological findings. Organs should be fixed as soon as possible, for example, in 10% buffered formalin, 5% Zenker formolic acid, or Bouin's fixative. Fixation should not exceed 1 or 2 days. Generally, large organs should be dissected into smaller pieces prior to fixation (see Chapter 30 on Necropsy Methods).

For practical advice and for detailed and systematic pathological examinations see:

- http://www.ita.fhg.de/reni/index.htm (advice on histological preparation for each mouse organ)
- http://www.eulep.org/Necropsy_of_the_Mouse/index.php (guidelines for a comprehensive and systematic necropsy of the mouse).
- http://www.geocities.com/virtualbiology/necropsy (a beginner's guide to necropsy, more cartoon oriented than the previous web-site)
- http://eulep.anat.cam.ac.uk/Pathology_Ontology/index.php (an ontology of pathology terms)
- PathBase, a web-site of pathological tissue sections is currently under construction. It will be of major value for the evaluation of histological data. Access will be provided through the EULEP (European Late Effects Project) homepage: http://www.eulep.org/.

Molecular phenotyping: gene expression profiling

Novel tools for the molecular analysis of mutant phenotypes have emerged from recent technologies that allow the genome wide monitoring of expression at the transcript and protein level. Applied to expression analysis, DNA-chip technology facilitates the measurement of RNA levels for the complete set of transcripts of an organism. Technically the DNA-chip is similar to the procedures of the classical dot-blot: Gene specific oligonucleotides or double-stranded cDNAs are immobilized in defined positions on a solid support and hybridized to complex mixtures of expressed nucleic acids. Using the current standards of microarray spotters, approximately 20,000 spots can easily be fitted on a standard chip of the size of an ordinary histological slide. An important advantage of using glass as a transparent, solid support is that it allows the simultaneous, competitive hybridization of test and reference samples labelled with different fluorescent dyes. Relative expression levels are analysed directly by comparing each fluorescent signal on every spot. Production, hybridization, and scanning of such DNA-chips can be automated to a great extent allowing for high-throughput approaches.

The microarray technology has a substantial effect on the questions that can be addressed and answered, particularly for model organisms whose genomes are already well characterized (*Nature Genetics* Vol. 21 (Supplement), 1999). For example, a comprehensive transcriptome analysis in a compendium of yeast mutants has led to the identification of new gene functions and co-regulated synexpression groups of genes (Hughes *et al.*, 2000). For the fruit fly, DNA-chip technology has been used to study molecular pathways during metamorphosis (White *et al.*, 1999), and in human cancer research expression profiling has provided new insights into underlying molecular changes and into the classification of tumours (Elek *et al.*, 2000; Dhanasekaran *et al.*, 2001; Pomeroy *et al.*, 2002) and of inflammatory diseases (Heller *et al.*, 1997).

It has been suggested that comprehensive genome wide expression profiling ought to be one of the tools in the worldwide effort to annotate biological functions to the mouse genome (Nadeau *et al.*, 2001; Beckers and Hrabé de Angelis, 2002). Whereas the current knowledge of gene function is usually limited to single pathways or a small set of target genes, transcript profiling of mouse mutant lines (their organs or derived cell lines) or of mice challenged by infectious diseases allows a comprehensive analysis of regulatory interactions in global molecular networks. Several publications have successfully used DNA microarray technologies for transcriptome analysis in mice. For example, the transcriptional response to ageing in the mouse brain has significant similarities to that in human neurodegenerative disorders, such as Alzheimer's disease (Lee *et al.*, 1999, 2000). The differential gene expression in several brain regions and the response to seizure has also been analysed and the results provided

evidence that particular differences in gene expression may account for distinct phenotypes in mouse inbred strains (Sandberg *et al.*, 2000). These and other papers (Livesey *et al.*, 2000; Campbell *et al.*, 2001; Porter *et al.*, 2001) have provided the proof-of-principle that despite the complexity of mammalian organs, expression profiling is a useful tool to identify pathways associated with particular biological processes in the mouse model system.

Other novel technologies applied for the assessment of genome wide gene expression include serial analysis of gene expression (SAGE). This methodology is based on the finding that short sequence tags of 10–14 bp are sufficient to identify the gene from which the cDNA tag has been isolated and that the number of times a particular tag is observed provides the expression level of the corresponding gene. To obtain this information, SAGE tags are isolated from RNA pools, linked to concatamers, cloned and finally sequenced. This technology has successfully been used for the evaluation of differential gene expression and the identification of new genes in complex RNA isolations from organ samples.

Technologies for the assessment of gene expression are under constant refinement. For example, a recent technology aims at the identification specifically of differentially expressed genes that may be expressed even at very low levels. The analysis of such weakly expressed genes is rather difficult using the DNA-chip technology or SAGE. Using specific probes immobilized on microbeads and fluorescently labelled cDNAs from two samples, over- and under-expressed cDNAs can be sorted using FACS. The sorting is based exclusively on the ratio of fluorescent signals and is not dependent on absolute gene expression levels. The isolated pools of differentially expressed genes are then analysed by high-parallel sequencing approaches.

A comprehensive understanding of molecular networks requires in addition the analysis of the cellular **proteome**. Most biochemical processes within and between cells are realized by the interaction of proteins, or of proteins and their substrates. The proteome of a cell is the result of controlled biosynthesis, and hence largely (but not exclusively) regulated by gene expression. In turn, gene expression can be regarded as a sensitive read-out of the biochemical state of the cell, or in other words the proteome. **Transcriptome** and proteome feedback to each other in a highly complex and somehow controlled way. Thus, the regulatory context is a crucial part of gene function (Fessele, 2002). Beyond the analysis of the transcriptome, it will therefore be required to improve technologies further in order to analyse the proteome with similar efficiencies. At least two prerequisites make this effort a major challenge for functional genomics. First, the proteome is probably an order of magnitude more complex than the entirety of transcripts. Second, proteins are biochemically highly diverse and cannot be amplified as easily as nucleic acids. Despite these difficulties efforts are being undertaken to improve the sensitivity and reproducibility of 2D-gel electrophoresis and to automate the subsequent identification of proteins. Prefractionation of protein samples and the availability of comprehensive sets of protein specific antibodies and antibodies against characteristic functional domains of protein families will be an integral part of proteome analysis.

Phenome databases

The IMMC has recognized that efficient and reliable methods for the dissemination of complex phenotype data are a major goal for the comprehensive, functional annotation of the mammalian genome. This is a particularly challenging task since, in contrast to genomic and sequence databases, there are no existing large-scale phenotype databases on which to model databases for phenotype data. To support the integration of relevant phenome databases and to assist queries of diverse aspects of the biology of laboratory mice, the development of standardized phenotype vocabularies, similar to the Gene Ontology program (http://www.geneontology.org/), will be a prerequisite. The Mouse Phenome Database (MPD) housed at The Jackson Laboratory serves as the central community resource for the integration of phenotypic information and genetic information (http://aretha.jax.org/pub-cgi/phenome/mpdcgi?rtn = docs/home).

As a first step towards a comprehensive description of phenotypes, the Mouse Phenome Project is establishing a collection of baseline phenotypic data on commonly used and genetically diverse inbred mouse strains through a coordinated international effort with both academic and industrial participation. The data collected includes phenotypic information on gross and micro anatomy, heart, lung, and blood parameters, immunology, physiology, neurobiology, behaviour and pain, sensory function, cell biology and biochemistry, disease susceptibility, aging studies and gene expression. Other phenome databases focus on the data acquisition either of distinct organ systems and disease models, or on particular sets of mutated genes. One such example is the Mouse Brain Library (MBL; http://www.mbl.org) which consists of high-resolution images and databases of brains from many genetically characterized strains of mice. The goal of the MBL initiative is to map and characterize genes that modulate the architecture of the mammalian CNS systematically. A more comprehensive list of similar phenome databases can be found under http://www.utu.fi/erill/ kek/ phenolinks.html. The Online Mendelian Inheritance in Man (OMIM) database is the most comprehensive online catalogue of human genes and genetic disorders developed by the National Center for Biotechnology Information (NCBI) (http://www.ncbi.nlm.nih.gov/ Omim/). The database contains textual information, pictures, and reference information on human disease phenotypes. A similar database, the Mouse Locus Catalog (MLC), has been developed at The Jackson Laboratory. In addition, an integrated OMIM/MLC query form is available which allows the user to enter one search string to execute a query in both MLC and OMIM (http://www.informatics.jax.org/searches/ noforms_mlc_omim.cgi).

Positional cloning: from mutation to gene through mapping

Introduction

Analysis of abnormal phenotypes in mutants can provide insight into the mechanisms that underlie the biological process affected by the mutation. Nevertheless, the primary defect caused by the mutation is best understood by the identification of the gene whose mutation is causally related to the mutant phenotype. As a next step to the large-scale production of new mouse mutant lines from ENU mutagenesis projects described in the previous sections, the issue to be discussed here is how the causal gene can be identified. Some phenotypes may directly reflect a mutation in a certain gene. For example, changes in amount or in biochemical nature of a protein as phenotype may be due to mutations in the gene encoding the protein itself. However, such cases are rather exceptional, and usually there is an unknown degree of gap between observed mutant phenotypes and the primary defects caused by a mutation in the causal gene. Thus, in a majority of the cases, the identification of the causal gene usually relies on information about the chromosomal position of the mutant locus. This section describes the procedure called 'positional cloning' through which the gene responsible for the mutation is identified. This procedure normally starts from the localization of a mutant locus to a chromosomal segment by genetic mapping. However, conceptual and technical details of genetic mapping cannot be provided here, because it would require considerably more space to describe them in full depth. Therefore, one should consult more appropriate references, such as *Mouse Genetics* by L. Silver (Oxford University Press, 1995; the electronic version is freely accessible online at www.informatics.jax.org/silver/), which includes over 150 pages for comprehensive descriptions of concepts and methodologies. Here, basic steps of the positional cloning procedure are outlined, and only general and practical issues to be considered for positional cloning will be discussed, in connection with the recent advances in the mouse genome project.

Nowadays, one can obtain a mutation and its causal gene as a molecular entity, either by gene targeting or by positional cloning. The former and the latter approaches are also referred to as 'reverse genetics' and 'forward genetics', respectively. It is a coincidence that the general feasibility of the reverse and forward genetic approaches was shown at almost the same time by the first targeted inactivation of a non-selectable (i.e. ordinary) gene, *c-abl* (Schwartzberg *et al.*, 1989), and by the cloning of the *T* gene, encoding a T-box transcription factor, as the causal gene of the *Brachyury* mutation (Herrmann *et al.*, 1990), respectively. In general, positional cloning is a more time-consuming task than gene targeting. This fact is reflected by large imbalance in numbers between the successful cases of targeted mutations in cloned genes and those of positional cloning of existing mutations. According to the Mouse Genome Database (MGD; Phenotypic Alleles, including 6523 alleles from 2848 genes as of mid-2003), a total of 2281 targeted alleles have been created in 1372 genes, while natural (spontaneous) or induced (physically or chemically) mutations are identified in about 300 genes. Among the latter cases, there are about 100 cases of gene identification by executing the entire procedures of positional cloning as described below. In a number of these cases, mutations in causal genes were identified by so-called '(positional) candidate cloning' where some way of making shortcuts in the procedure was possible (see below). In some cases, positional cloning in other species (mostly in humans) led to the identification of the mouse counterpart affected in existing mouse mutations: for example, a mutation in

the mouse counterpart of human Duchenne muscular dystrophy (*DMD*) in *mdx* mice (Heilig *et al.*, 1987) and a mutation in the ectodermal dysplasia (*EDA*) gene in *Tabby* mice (Srivastava *et al.*, 1997). Incidental disruption of a gene by transgenic insertion could lead to the creation of a new allele of existing mutations. In this case, the locus may be molecularly tagged, thereby leading to the identification of the causal gene. Examples of this kind include a *dystonin* mutation in *dystonia musculorum* (*dt*) mice (Brown *et al.*, 1995) and a *formin* mutation in *limb deformity* (*ld*) mice (Mass *et al.*, 1990).

Table 4.6 lists selected examples of positional cloning works in the mouse, including some pioneering cases and also ones where positional cloning was particularly difficult.

The numbers of meioses analysed in genetic mapping partly reflect the difficulty with the positional cloning works. Nevertheless, it is also true that the identification of a gene responsible for a mutation is not always that difficult. In late 1980s and early 1990s, when a number of so-called 'developmental control genes' were cloned based on the sequence similarities to those originally found in *Drosophila*, mouse geneticists and molecular biologists became excited by witnessing several pioneering examples where classical mouse mutants were shown to carry mutations in such developmental control genes. Such examples include mutations of *Pax1* in *undulated* (Balling *et al.*, 1988), *Pit1* in *dwarf* (Li *et al.*, 1990), *Pax3* in *Sploch* (Epstein *et al.*, 1991), *Pax6* in *Small eye* (Hill *et al.*, 1991) and *Wnt1* in *swaying* (Thomas *et al.*, 1991). In these cases, the genes were considered to be good candidates, because they were expressed in the tissues affected in the corresponding mutants, and because the gene and mutant loci were co-localized by genetic linkage studies with low resolutions. Certainly, these examples can be regarded as the simplest cases of positional cloning, and often such cases are referred to as 'positional candidate cloning' to reflect the fact that the most time consuming steps normally needed for 'real' positional cloning could be skipped. Nonetheless, there is no objective way to discriminate between 'positional candidate cloning' and 'positional cloning', provided that in either case it is information on gene/locus location that leads towards the identification of the causal gene, and that the principle difference is only in the size of the critical interval determined by mapping studies with either low or high resolution. In any case, all genes in the critical interval are considered as 'positional candidates' and their candidacy is assessed for the same criteria (see section on Candidate Gene Search).

TABLE 4.6: Selected examples of positional cloning works in the mouse

Mutation	Symbol	Gene	Meiosis	Reference
Brachyury	T	T	[a]	Herrmann et al., 1990
short ear	se	Bmp5	[a]	Kingsley et al., 1992
obese	ob	Lep	1606	Zhang et al., 1994
natural resistance to infection	Bcg/Lsh/Ity	Slc11a1	1000[b]	Vidal et al., 1995
tottering	tg	Cacna1a	2800	Fletcher et al., 1996
tubby	tub	tub	1232	Kleyn et al., 1996
clock	Clock	Clock	2400	King et al., 1997
vibrator	vb	Pitpn	2600	Hamilton et al., 1997
Lurcher	Lc	Grid2	504	Zuo et al., 1997
syndactylism	sm	Jag2	5766	Sidow et al., 1997
pudgy	pu	Dll3	2264	Kusumi et al., 1998
shaker-2	sh2	Myo15	500[b]	Probst et al., 1998
shaker-2	sh2	Myo15	1305	Wakabayashi et al., 1998
Dactylaplasia	Dac	Fbxw4	7182	Sidow et al., 1999
mahogany	mg	Atrn	2437	Nagle et al., 1999
mahogany	mg	Atrn	1727	Gunn et al., 1999
progressive ankylosis	ank	ank	1846	Ho et al., 2000
fidget	fi	Fign	2400	Cox et al., 2000
dreher	dr	Lmx1a	738	Millonig et al., 2000
fatty liver dystrophy	fld	Lpin1	706	Peterfy et al., 2000
flexed-tail	f	Sfxn1	1000	Fleming et al., 2001
cytomegalovirus resistance 1	Cmv1	Klra8	1967	S.H. Lee et al., 2001
Loop tail	Lp	Ltap	753	Kibar et al., 2001
waltzer	v	Cdh23	3830[b]	Di Palma et al., 2001
Waltzer	v	Cdh23	1648	Wada et al., 2001
hypolipidemia	hypl	Angptl3	3344	Koishi et al., 2001

[a]The critical interval in these cases was determined mainly by the analysis of a series of deletion alleles.

[b]For mapping data of Bcg/Lsh/Ity, shaker-2 and waltzer, see the references by Malo et al. (1993), Liang et al. (1998) and Bryda et al. (2001), respectively.

The mouse is one of the best model vertebrate organisms for genetic studies. Nevertheless, for the purpose of positional cloning, our knowledge about genes in the genome had been very limited until recently. However, now in the post-genome era, the efficiency of positional cloning procedures can be improved significantly by the use of recently developed resources and techniques. Furthermore, ever improving information about genes in the genome (i.e. sequence-based genome map/annotation) greatly helps finding good candidates for the causal gene, even with a moderate resolution in genetic mapping. Therefore, we can expect to see an increasing number of successful cases of 'positional candidate cloning' in the next years. With the help of the sequence-based genome annotation, even in the form that is available now, a full version of positional cloning, which is described in the following sections, can be dramatically simplified as compared to how it had been done until very recently.

General flow of positional cloning procedure

The standard procedures of positional cloning include the following steps: (1) fine genetic mapping, (2) physical mapping of a critical interval, (3) search for (candidate) genes in the critical interval, (4) identification of a molecular lesion in a candidate gene, and (5) confirmation of the causal relationship between the gene mutation and the mutant phenotype. For each step, it normally took 1 or 2 years in the past. In the post-genome era, at least the steps (2) and (3) may be significantly shortened (see later sections).

Genetic mapping

Linkage analysis

The first step to localize a mutant locus is carried out by genetic linkage analysis with a panel of progeny from a cross segregating for the mutation and marker loci. Linkage analysis is the way in which to measure genetic distances between loci of interest and to determine the order of more than two loci on a chromosomal segment. The frequency of meiotic recombination

events between two loci is used to express a genetic distance, such that 1 recombination in 100 meiotic events is defined as 1 centi-Morgan (cM). The order of three or more loci is determined by statistical evaluations on haplotype data. Haplotype is a set of allele typing data for multiple loci in a chromosomal segment in individual samples. Among all theoretically possible locus orders by permutation, the best or most likely locus order is determined as one with the minimal number of recombination events. The use of computer programs such as MapManager (K. Manley, available at mapmgr.roswellpark.org), or MAPMAKER (Lander *et al.*, 1987) greatly facilitates this statistical analysis.

Intercross or backcross

There are two major ways for setting up a cross for a linkage study: intercross (Figure 4.7) and backcross (Figure 4.8). In either case, the whole procedure goes in two steps (generations), starting with a generation of heterozygous animals at the F1 (first filial) generation by crossing mutant animals to wild-type mating partners

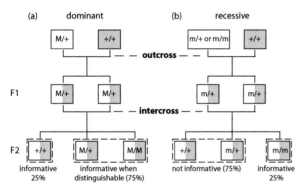

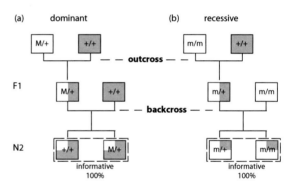

Figure 4.7 Scheme of intercross for a dominant mutation (a) or a recessive mutation (b). In either case, mutant animals are first crossed to wild-type mating partners of a different strain to generate F1 animals (outcross). Then F1 animals are mated together to generate F2 animals (intercross). The colours (white or grey) in the boxes represent average genetic compositions (genomic contributions from parental strains) of individual animals in different generation. (a) If the penetrance is 100% and heterozygotes (M/+) can be distinguished from homozygotes (M/M) by phenotyping (i.e., semidominance), then all F2 animals are informative (100%). If the penetrance is 100%, but heterozygotes are indistinguishable from homozygotes by phenotyping, only wild-type (+/+) F2 animals are informative (25%). If the penetrance (either M/+ or M/M) is not complete, none of the F2 animals are informative. In such a case, backcross mapping with affected N2 animals is the only possibility to solve the issue (see Figure 4.2). (b) By definition, heterozygotes (m/+) are indistinguishable from wild-type (+/+) animals by phenotyping. Regardless of the degree of penetrance, only affected F2 mice (all or a fraction of the homozygotes) are informative. If the penetrance is 100%, the efficiency is as high as 25%. If homozygotes (m/m) are lethal (but recoverable for genotyping) or sterile, this strategy is the only way to set up a linkage cross.

Figure 4.8 Scheme of backcrosses for a dominant mutation (a), or a recessive mutation (b). In either case, mutant animals are first crossed to wild-type mating partners of a different strain to generate F1 animals (outcross). Then F1 animals are mated with either +/+ (dominant) or m/m (recessive) animals to generate N2 animals (backcross). The colours (white or grey) in the boxes represent genetic average compositions (genomic contributions from parental strains) of individual animals in different generations. (a) In this strategy, no homozygotes are produced in the N2 generation. If the penetrance is 100%, all N2 animals are informative (100%). If the penetrance is not complete, only affected N2 animals (a fraction of the heterozygotes) are informative (i.e. less than 50%). (b) For a recessive mutation, all N2 animals are either heterozygous (m/+) or homozygous (m/m). For this strategy, viable and fertile homozygotes have to be available. If the penetrance is 100%, all N2 animals are informative (100%). If the penetrance is not complete, only affected N2 animals (a fraction of the homozygotes) are informative (i.e. less than 50%).

of a different mouse strain (outcross). In intercrosses, heterozygous F1 animals are crossed to each other, and meiotic recombination events (meioses) in the germ cells of both maternal and paternal F1 animals can be scored in animals of the next generation (F2), expecting wild-type (+/+), heterozygous (+/−) and homozygous (−/−) animals at 1 : 2 : 1 ratio. For backcrosses, heterozygous F1 (or N1) animals are mated with either wild-type (for a dominant mutation) or homozygous (for a recessive mutation) animals. In this case, only meiotic recombination events in germ cells of F1 parents can be scored in the N2 generation and only wild-type and heterozygous (dominant cross), or heterozygous and homozygous offspring (recessive cross) are expected at a 1 : 1 ratio.

The apparent advantage with the intercross is in that the numbers of meioses analysed would be double for the same number of progeny scored. Nevertheless, the following fact about the intercross should be noted. In the case of recessive mutation (m), where in principle heterozygous (m/+) animals (50% of the F2 progeny) cannot be distinguished from wild-type (+/+) animals (25%) on the phenotypic basis, only homozygous (m/m) F2 animals (25%) are informative. Consequently, in order to analyse, for example, 100 meioses by intercross, one needs 50 homozygous F2 animals, which is one fourth of the total intercross progeny (i.e. $n = 200$). On the other hand, the same number of meioses ($n = 100$) can be analysed from 100 backcross N2 progeny, which is half the size of the required intercross progeny, provided that heterozygous and homozygous animals can be undoubtedly distinguished by phenotype. This means that backcrossing could be more efficient than intercrossing, in terms of mouse colony size and animal production. In the case of dominant mutations (M), if heterozygous (M/+) and homozygous (M/M) animals cannot be distinguished phenotypically, only wild-type animals are informative, and the same considerations as above on the efficiencies between intercross and backcross apply. However, if heterozygous (M/+) and homozygous (M/M) animals are distinguishable (ideal cases of semi-dominant mutations), then all animals are informative and intercross mapping would be twice as efficient as backcross mapping.

The discussions above are more or less hypothetical. In reality, there are several factors that can affect the strategy and efficiency of linkage studies. One needs to consider several points in setting up a genetic cross. They include, (1) the viability and fertility of the mutants, (2) the penetrance rate of the mutant phenotype, and (3) the effects of the genetic background on the phenotype. It is often the case that a mutation leads to lethality (but recoverable for genotyping) or to sterility in homozygous mice. In such a case, intercrossing is the only way to analyse recessive mutations. Phenotypic consequences of a mutation can be variable even on an inbred background. If it is conceivable that a significant fraction of the mutants does not show the phenotype, then only phenotypically affected animals are informative in both intercross and backcross mapping. In such a case, intercrossing may be advantageous for recessive mutations, and the safest method would be backcrossing for dominant mutations. The genetic background can sometimes modulate the mutant phenotype to a considerable extent. In extreme cases, the original phenotype can completely disappear on a mixed genetic background. It is also possible that on a certain genetic background a viable mutation can be lethal. These genetic background effects can be advantageous in special occasions (e.g. modifier locus screening), but they are usually problematic. This issue is important in considering which strain of mice to use as the mating partner, which is discussed in the next section.

When one has to decide between intercrossing and backcrossing, in general neither one of the two strategies can be regarded as superior a priori. Instead, one has to choose one of the two alternative strategies, depending on the nature of the mutation of interest.

Mating partner strain

The best mating partner strain would be the one with a genetic background that does not modify the mutant phenotype and with a high level of genetic polymorphisms as compared with the strain on which the mutation is maintained. If the nature of a mutation on the original genetic background is already problematic in terms of viability, fertility and penetrance, certain strains as the mating partner could improve these aspects. How the genetic composition of a mouse strain affects a mutation on a different genetic background is unpredictable. Therefore, in each case it has to be tested which strain of mice is suitable as the mating partner strain. Another important consideration for choosing mating partner strains is the species or subspecies of mice. To change the genetic polymorphisms between original and partner strains, the use of different species or subspecies of mice as the mating partner may be advantageous. Genetic polymorphism is the key nature of a marker in a linkage study. For the best use of available polymorphic markers, it has been recommended choosing either *Mus spretus*, or *Mus musculus castaneous*, or *Mus musculus molossinus* as mating partners (most laboratory strains are of either *Mus musculus domesticus* or *Mus musculus musculus* origin, or a mixture of these species). This idea may be outdated soon, as the use of single-nucleotide polymorphism (SNP) markers replaces that of microsatellite markers.

Microsatellite vs. SNP markers

The introduction of microsatellite markers in the early 1990s has dramatically improved the efficiency of genetic studies. All so-called 'high-resolution' mapping studies incorporated a number of microsatellite markers. Microsatellite mapping, based on the difference (polymorphism) in the number of CA repeats among mouse strains, is easily carried out by PCR in any laboratory. A total of 6000 microsatellite markers are commonly used in mouse genetic mapping. The genetic and physical sizes of the mouse genome are about 1500 cM and 3000 Mb, respectively. Therefore, the average spacing of microsatellite markers is estimated to be about 0.25 cM or 500 kb. This is a rough approximation, because a genetic distance does not necessarily reflect a physical distance, since in certain genome intervals meiotic recombination is either over- or under-represented, i.e. recombination hot or cold spots, respectively. Microsatellite mapping can certainly provide a considerable resolution, but it may not be high enough for the purpose of positional cloning in many occasions. One should note that not all markers in a cross are polymorphic (informative) and that microsatellite markers may be sparsely distributed in certain genome intervals. Therefore, the availability of even higher numbers of markers that can be easily typed is desirable. This is true for SNP markers. Currently, the number of available SNP markers for the mouse genome is still modest (2848 in the MIT SNP database), but the potential of the presence of SNP between mouse strains is enormously high. In humans, over 3 million SNP markers are already known. As mouse genome sequencing proceeds with more than one strain, the situation with mouse SNP markers will improve dramatically. One million SNP markers evenly distributed in the mouse genome would mean that they are positioned only 3 kb apart. Such a density is certainly high enough for any type of genetic mapping. With the improvement in SNP detection techniques, it is likely that SNP mapping will replace microsatellite mapping in the near future.

How many meioses to look at?

Regardless of the number and types of markers to be used in a linkage study, one may discuss how many

meiotic events should be analysed for the purpose of positional cloning. The author would like to suggest that 1000 meioses be looked at in a positional cloning experiment, unless a short cut by candidate cloning or by any other means is possible. What resolution can one expect through the analysis of 1000 meioses? One recombination in 1000 meioses means a genetic distance of 0.1 ± 0.1 cM (± one standard error), which can be translated into up to 400 kb of physical distance. If a marker does not recombine with the mutant locus in N meioses, a maximum genetic distance (d) between the mutant locus and the non-recombinant marker can be calculated with 95% confidence as follows:

$$(1 - p)^N = 0.05$$

where p is the maximum probability of recombination between the two loci under this confidence limit ($d = 100 \times p$). When solving the formula for p with $N = 1000$:

$$p = 0.003$$

Therefore, the mutation is expected to be located within ±0.3 cM or 600 kb from the non-recombining marker. This is a reasonable size of the genome, for which one can start to think of more direct approaches to identify molecular lesion. When $N = 500$, then p is 0.006 and d is 0.6 cM (or 1.2 Mb). This region might be too big, depending on the gene density in the interval. When $N = 1500$, then p is 0.002 and d is 0.2 cM (or 400 kb). Narrowing down the critical interval to this range could be significant, in some occasions (high gene-density regions), but it may not be so in other cases (low gene-density regions). The amount of work for production and typing of additional 500 meioses could differ significantly from one mutation to another. Therefore, one has to decide whether it pays off to analyse additional meioses for the expected degree of reducing the interval range. Considering the average size of bacterial artificial chromosome (BAC) clones (200 kb), which is often used for transgenic rescue experiments (see below), 1000 meioses would be a good number for a linkage study, and at this point one can think of an alternative strategy to narrow down further the critical interval. With great luck, of course, there is always a chance to hit on an excellent candidate gene that turns out to be the right one (candidate cloning). This could happen at any time once the mutant locus is chromosomally mapped, even with very limited numbers of meioses (less than 100).

Physical mapping

BAC- and sequence-based maps

As sequence-based genome maps become publicly available online at Ensembl (www.ensembl.org) and NCBI (www.ncbi.nlm.nih.gov), the need for the isolation and analysis of genomic clones with large inserts, such as BAC, P1-derived artificial chromosome (PAC), or P1 clones, is decreasing. In most cases, there is no need to screen BAC libraries to isolate clones of interest and to analyse them for assembly into a contig: the step called 'physical mapping'. It is now possible to find *in silico* BAC clones that contain key (recombinant or non-recombinant) markers defined by the preceding genetic mapping study. Furthermore, on many occasions, it is possible to obtain the entire genome sequence of the interval between proximal and distal recombinant markers (the critical interval) in the public databases. At present, mouse genome information on a particular interval may still be limited. In such a case, looking at a corresponding human genome interval(s) (syntenic regions) may help to reveal which genes probably located in the corresponding mouse interval.

At this step, one can get the information about the physical size of the critical interval that has been genetically defined and about genes in this interval. If recombinant markers that define the critical interval hit more than one recombination event in the cross(es), it is possible to narrow down further the critical interval by refining the genetic mapping data with additional sequence-based markers (sequence-tagged sites) within the initial critical interval. For example, one will find a number of segments containing CA-repeats, which may serve as new microsatellite markers polymorphic between the parental strains used. When good candidate genes can be found for certain criteria (see section on Candidate Gene Search), one can examine them for a molecular lesion (mutation; see section on Mutation Detection). When this attempt fails, and/or when there are still too many positional candidates in the interval, one can consider the BAC transgenic rescue approach to narrow down further the critical interval.

BAC transgenic rescue

A BAC transgenic rescue experiment could significantly narrow down the critical interval beyond the practical limit of genetic mapping. For example, when a single BAC clone as a transgene can rescue a mutant

phenotype, the critical interval is this BAC interval. If this is 200 kb in size, this critical interval would correspond to what one can expect from a linkage study with about 6000 meioses at 95% confidence (see the theoretical calculations in previous section). This scale of genetic study in the mouse is not totally unrealistic, as exemplified in the cases of *syndactylism* (Sidow et al., 1997) and *dactyloplasia* (see Table 4.6; Sidow et al., 1999), but it is practically beyond the capacity of many laboratories. If each of two overlapping BAC clones can rescue a mutant phenotype, the overlapping region defines the critical interval. Nevertheless, the success with BAC rescue procedures does not necessarily mean the identification of the causal gene, because one BAC interval (~200 kb) could contain more than one gene. Provided that the total number of protein-coding genes in the mouse genome is 30,000, the average interval for a gene is 100 kb. Apparently, genes are not distributed evenly in the genome. Therefore, a 200-kb interval could contain several genes.

There are certain limitations with the applicability of the BAC transgenic rescue approach. This approach is based on an overexpression of a normal copy of a gene, therefore, loss-of-function mutations (hypomorphic or null alleles) may well be rescued by supplying additional copies of a wild-type gene as a transgene. However, a rescue experiment may not be useful for certain types of mutations such as strong dominant negative, or hyper-active alleles. Simply based on its phenotype, it is difficult to know the nature of a mutation in advance. Another principle limitation exists for genes that are larger than the average size of BAC inserts. There are a number of examples of such large genes. In such cases, BAC transgenic rescue will not be successful.

Candidate gene search: use of genome annotation databases

It is very important to define early defects caused by the mutation at the molecular and cellular levels as much as possible, in order to be able to make good assumptions for potential functions and possible expression profiles of the causal gene. These assumptions provide the basis for a candidate gene search in the critical interval. Genome and gene annotation is the on-going effort of international collaborative projects in the post-genome era (see for example the Gene Ontology Consortium web site at www.geneontology. org). There are two points to be considered: (1) whether genes in an interval are comprehensively listed, and (2) whether annotations for (potential) functions of genes are reliable, or available at all. At present, the status of genome annotation databases is far from completion, especially with respect to annotations for potential functions of uncharacterized genes in the current context. Regarding to the first point, one has to keep in mind the possibility of the presence of so-far uncovered genes in the genome. Aside from protein-coding genes, most of which are believed to have been discovered, there is a great possibility for the presence of a considerable number of non-coding RNA genes, which might have some regulatory functions in controlling transcription and/or translation of specific protein-coding genes (Eddy, 2001). Furthermore, *cis*-regulatory elements of the genes are to a large extent uncovered, and other regulatory sequences that have important functions to control gene expression at the chromatin or chromosomal level are poorly understood. These are all potential targets for mutations, leading to abnormal gene functions.

With respect to the second point, the number of genes with DNA sequence information is 28,378 for the mouse, and less than half of them (13,436) are associated with protein sequence information. The number of known genes with well-characterized functions is less than 10,000. The annotations for the rest of the genes (usually represented by EST clones) remain hypothetical. If such uncharacterized genes show the structural similarities to known genes with well-characterized functions, predictions can be made for their potential functions with certain confidence. On the other hand, when a totally novel gene that has no structural basis for predictable function is to be identified, positional cloning may have to be performed in all its detailed steps. Nevertheless, in most cases, gene annotation databases provide useful information to derive a priority list of candidate genes for further analysis in the next step: mutation detection.

Other large-scale efforts like systematic analysis of gene expression and of protein interactions (see, for example RIKEN databases at www.genome.gsc.riken.go.jp) are underway. In addition to the sequence similarity as a basis for putative functions, information from such projects may provide further possibilities for the deduction of potential functions of novel genes. Integration of information from different databases will be one of the keys to successful positional cloning.

Mutation detection

Once having candidate genes, several points should be tested for each candidate gene: both quantitative and qualitative aspects of genome DNA and mRNA (or cDNA). They include: (1) whether the gene is present in the mutant chromosome at all, (2) any changes in transcript level and in transcript size(s), (3) whether any mutation is present in the cDNA sequences from mutants as compared with those from appropriate control samples, and (4) whether any changes exist in sequences of genome segments including exons together with splice signals.

A candidate gene might be ablated either completely or partially by deletion. PCR or Southern analysis on genomic DNA from homozygous carriers of the mutation may provide the answer to the first point. However, very small deletions may not be detected by either way. If a gene spans over a wide genomic interval with many exons, it may not be easy to scan through the whole gene interval. As a first step for point (4), PCR amplification of genome segments including exons may be of help for this type of scanning.

The level of gene expression and transcript size(s) can be assessed most conveniently by Northern analysis, although subtle changes in the amount or in size may not be detected. Disturbance of gene expression might occur in a tissue-specific manner. In such a case, *in situ* hybridization or Northern analysis with tissue-specific blots might detect some changes. One should keep in mind, however, that a gene mutation does not necessarily change the expression level of the gene. This means that negative data (i.e. no change) do not exclude the candidate gene. Thus, sequencing of cDNA segments (normally RT-PCR products) obtained from mutant RNA samples is required, unless the expression of the candidate gene is totally downregulated. In parallel, sequencing of genome segments including exons and flanking splice signals is performed, as long as the gene is present in the mutant genome.

If any of the alterations is detected, the gene in question is a very good candidate. One should always compare data between mutant animals and appropriate wild-type control animals, ideally mice of the strain in which the mutation arose originally. It is always helpful to test further wild-type controls of other strains to assess whether the observed change(s) is strain-specific (functionally silent polymorphism) or mutant-specific (mutation). A molecular lesion may reside in an intron, or even in regulatory regions. Detection of such a molecular lesion may be more difficult. Nevertheless, in such a case, one can usually detect some changes at the transcript level, which is encouraging enough to search through all possible regions for a mutation.

Formal proof: more than one allele?

When a molecular lesion in a gene from the critical interval is found in the genome of a mutant strain, does this provide a formal proof for the gene to be responsible for the mutation of interest? The answer is 'no', even if there is support from gene expression data. Some functional tests on the product of the gene may provide strong support. In principle, such a line of evidence is required, but not sufficient to prove the causal relationship between the mutation in the gene and the mutant phenotype. If only one allele is available, this issue becomes very critical. If multiple alleles are present and the same gene is mutated in a different way in each allele, then the causal relationship of the gene mutation to the phenotype is usually granted. Otherwise, evidence from genetic complementation by either transgenic rescue experiments with the gene, or targeted inactivation of the gene will be required to obtain a formal proof. There are several examples for the former case (Wilkinson *et al.*, 1990; King *et al.*, 1997; Probst *et al.*, 1998). As examples for the latter case, see the references by Vidal *et al.* (1995) and Cox *et al.* (2000).

Final remarks

Even with the possibility of skipping steps that are needed for physical mapping and gene search in the candidate interval, positional cloning can still be a hard task, if one has to go through all the other steps. However, there is always a chance to make a shortcut by 'candidate cloning', and this chance is ever increasing. Furthermore, as a result of large-scale ENU mutagenesis projects being performed worldwide in parallel, multiple alleles may well become available for many genes. This will significantly help the positional cloning experiments during the final phase. The end of a positional cloning project automatically marks the beginning of a functional study, in a new dimension, of the identified gene and on the pathway in which the gene product plays a role with clear knowledge of the molecular basis of primary defects caused by the mutation.

Cytogenetics

Introduction

In mice as in other experimental rodents, meiotic and mitotic chromosomes can be subjected to cytogenetic studies. The tissues of choice are testicular cells (spermatocytes) for meiotic chromosomal studies or bone-marrow cells and spermatogonia for mitotic chromosomal studies. If necessary, any other proliferating tissue may be used, such as hepatocytes after partial hepatectomy, gut epithelium, or splenocytes and lymphocytes after *in vitro* stimulation of cell division. The cytogenetic studies may have two purposes, detection of mutagenicity (clastogenicity) induced by chemical or physical agents or diagnosis of karyotype alterations in mutant mice or cell lines. Furthermore, localization of novel genes in the mouse genome may be performed by fluorescence *in situ* hybridization (FISH) of c-DNA in 4'-6-diamino-2-phenylindole (DAPI)-banded somatic mouse chromosomes. Recently, mouse karyotyping has been facilitated by the development of multicolour FISH. Competitive genomic hybridization (CGH) for detection of loss and gain of genetic material in tumour tissues is presently being developed for the mouse as well.

Mutagenicity testing

Meiotic chromosome studies during diakinesis/metaphase I of the first meiotic division will reveal the presence of reciprocal translocations by the formation of ring or chain multivalents (Figure 4.9). The preparation

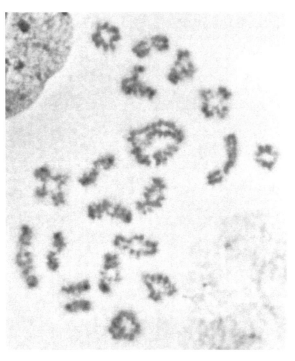

Figure 4.9 Mouse spermatocyte at diakinesis/metaphase I of the first meiotic division with a reciprocal translocation forming a ring multivalent.

procedure for meiotic chromosomes in mouse spermatocytes was described by Evans *et al.* (1964). Meiotic chromosomal analysis was performed in mutagen-treated animals after exposure of stem cell spermatogonia to ionizing radiation or chemical mutagens (Cattanach and Williams, 1971; Lyon *et al.*, 1972; Brewen and Preston, 1973; Brewen *et al.*, 1973; Adler, 1974). Individual mice carrying a reciprocal translocation induced in the heritable translocation assay can also be diagnosed by analysis of multivalents during the first meiotic division (Ford *et al.*, 1969; Adler, 1978).

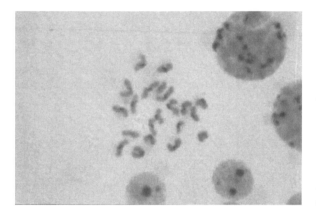

Figure 4.10 Mouse spermatocyte at meiotic metaphase II with 21 c-banded chromosomes (hyperhaploid).

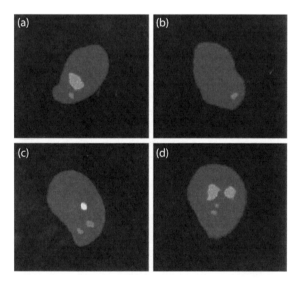

Figure 4.11 Mouse sperm with fluorescence signals for chromosomes to detect aneuploidy. (a) Normal sperm with chromosomes 8 (labelled with CY3, red) and Y (labelled with FITC, green); (b) hypohaploid sperm with only chromosome 8 (red); (c) hyperhaploid (disomic) sperm with two chromosomes 8 (red) and one X chromosome (labelled with FITC plus CY3, white); (d) diploid mouse sperm with two chromosomes 8 and two Y chromosomes. Images digitized with the software ISIS (MetaSystems, Altlussheim, Germany). **(See also Colour Plate 3.)**

To detect chemical induction of aneuploidy in mouse male germ cells, the chromosomes of second meiotic divisions can be counted in testicular preparations after specific giemsa-staining for centromeric heterochromatin (c-banding) (Russo *et al.*, 1984).

To detect aneuploid sperm, fluorescent signals can be counted in sperm (Figure 4.10) after FISH with a small number (up to three) of chromosomal-specific DNA probes labelled with different fluorochromes such as FITC (green), Texas red (Vector) or CY3 (red), Spectrum orange® (Vysis; Figure 4.11; Lowe *et al.*, 1995; Schmid *et al.*, 1999). Small chromosomal-specific DNA probes are commercially available for all mouse chromosomes except the Y chromosome from Biomar Diagnostic Systems GmbH, Marburg, Germany.

Somatic chromosome studies of structural chromosome aberrations in bone marrow cells of mice or other rodents belong to the prescribed battery of *in vivo* mutagenicity tests to determine chemical safety (Figure 4.12; EU Federal Register, 1985).

Indirectly, such events can also be assessed in the *in vivo* micronucleus assay (Schmid, 1975). In the common version of the *in vivo* micronucleus assay, polychromatic erythrocytes are analysed for the presence of micronuclei shortly after expulsion of the main nucleus. Variations in the *in vivo* rodent micronucleus assay have been published recently by an international committee (Hayashi *et al.*, 2000). Micronuclei can be formed by acentric chromosome fragments due to the clastogenic effect of a chemical. Micronuclei can also be formed by lagging chromosomes due to impairment of the mitotic chromosome segregation (Figure 4.13). Complemented by centromere labelling with centromere-specific antibodies (CREST) or centromeric DNA probes, the micronucleus assay is also capable of detecting chemicals that induce aneuploidy (chromosomal loss; Miller *et al.*,

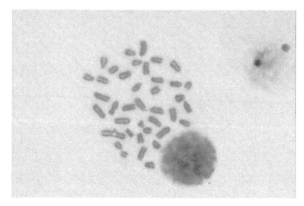

Figure 4.12 Mouse cell at mitotic metaphase with a structural chromosomal aberration (chromatid interchange).

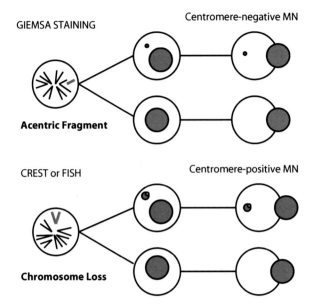

GIEMSA STAINING

Centromere-negative MN

Acentric Fragment

CREST or FISH

Centromere-positive MN

Chromosome Loss

Figure 4.13 Schematic diagram of the formation of micronuclei (MN) scored in polychromatic mouse erythrocytes. The induction of acentric fragments indicates a clastogenic effect and chromosome loss indicates induction of aneuploidy.

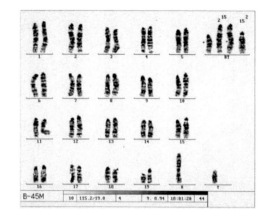

Figure 4.14 Trypsin–giemsa-banded mouse bone marrow chromosomes karyotyped with the software IKAROS (MetaSystems, Altlussheim, Germany). The picture shows a reciprocal translocation (RT) between chromosomes 2 and 15 (upper right).

1991). Recently, two-colour primed *in situ* DNA synthesis (PRINS) was applied to label simultaneously telomeric and centromeric (minor satellite DNA) sequences in micronuclei induced by chemicals in the murine cell line C6 (Basso and Russo, 2000).

Diagnostic cytogenetics

The diploid cell of the mouse contains 40 acrocentric chromosomes that do not differ much in length, i.e. between 7.2% and 2.7% of total haploid length of the female chromosome complement. With the method of the trypsin–giemsa-banding (G-banding), pairs of homologous mouse chromosomes can be discriminated (Figure 4.14). Various banding techniques developed for human chromosomes were also applied to mouse chromosomes (Buckland *et al.*, 1971; Schnedl, 1971; Sperling and Wiesner, 1972; Wurster, 1972). The mechanisms involved in producing bands of euchromatic and heterochromatic areas along the axes of the chromosomes were discussed by several authors (Kato and Moriwaki, 1972; Sumner, 1973; Sumner and Evans, 1973a,b). The basic principle is that the fixed mitotic chromosomes are pre-treated in diluted warm sodium chloride–tri-sodium citrate (SSC) solution followed by short treatment with a trypsin solution and then stained by giemsa stain or other suitable chromatin stains. A nomenclature for mouse chromosome bands was first suggested by Nesbitt and Franke

(1973) and was standardized by the Committee on Standardized Genetic Nomenclature for Mice (1985). The standard idiogram of the mouse was described by E.P. Evans (1989). A compilation of chromosomal variants of the mouse was given by A.G. Searle (1989). With the action of actinomycin D, a high-resolution G-banded karyotype can be produced (Sawyer and Hozier, 1986).

The new molecular technique of FISH has opened new dimensions of cytogenetic analyses. Gene localization within the genome of a specific c-DNA is possible with FISH (Hagenbuch *et al.*, 2000). The c-DNA probe is labelled with a fluorochrome by nick translation. The probe is then hybridized to mitotic chromosome preparations and the mouse chromosomes are banded by the DAPI–actinomycin procedure (Tucker *et al.*, 1988). Whole chromosome painting probes have been generated using the sequence-independent PCR to amplify sequences contained in DNA extracted from flow sorted chromosomes (Miyashita *et al.*, 1994). Painting probes for mouse chromosomes 1, 2, 3, 4, 5, 7, 9, 16 and Y are commercially available from Biomar Diagnostic Systems GmbH, Marburg, Germany. Painting probes were employed for multicolour FISH to produce karyotypes of mice in which each chromosome was entirely labelled by a specific colour. Spectral karyotyping (SKY) has been developed for the mouse by Liyanage *et al.* (1996) and a filter-based multiplex FISH (M-FISH) was described for the mouse by Jentsch *et al.* (2001). These techniques permit scanning of the entire mouse genome for chromosomal rearrangements (Adler *et al.*, 2002). However, assignment of breakpoints of translocations or insertion points of genes to specific chromosome bands is only possible by a combination of chromosomal banding

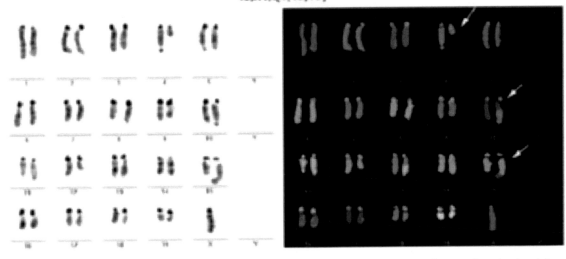

AAD101
40,XY,t(4;10;15)

Figure 4.15 DAPI-banded and multicolour karyotype (M-FISH) of a mouse carrying a complex translocation involving 3 chromosomes T(4;10;15) (obtained from Isabelle Jentsch, GSF-Institute of Human Genetics). **(See also Colour Plate 4.)**

(e.g. DAPI-banding) and chromosomal painting. An example is shown in Figure 4.15.

Specific gains or losses of genetic material in tumour cells can be assessed by CGH. In CGH, normal genomic DNA (labelled by a green fluorochrome) and tumour-derived DNA (labelled by a red fluorochrome) is competitively hybridized to normal metaphase chromosomes (Kallionemi *et al.*, 1992). Green chromosomal domains indicate loss of genetic material (deletions) in the tumour cell DNA and red chromosomal domains indicate gains (i.e. gene amplifications). Even though CGH is presently only reported for human tumour samples, the development of CGH for the mouse is in progress (Zitzelsberger, GSF-Institute of Radiation Biology, pers. comm.)

References

Adler, I.-D. (1974). *Mutat. Res.* **23**, 369–379.

Adler, I.-D. (1978). *Biol. Zentralblatt* **97**, 441–451.

Adler, I.-D., Kliesch, U., Jentsch, I. and Speicher, M.R. (2002). *Mutagenesis* **17**, 383–389.

Anderson, N.L., Eisler, W.J. and Anderson, N.G. (1978). *Anal. Biochem.* **91**(2), 441–445.

Arimondo, P.B., Garestier, T., Helene, C. and Sun, J.S. (2001). *Nucleic Acids Res* **29**(4), E15.

Balling, R., Deutsch, U. and Gruss, P. (1988). *Cell* **55**, 531–535.

Basso, K. and Russo, A. (2000). *Mutagenesis* **15**, 349–356.

Beckers, J. and Hrabé de Angelis, M. (2002). *Curr. Opin. Chem. Biol.* **6**(1), 17–23.

Beier, D.R., Dushkin, H. and Sussman, D.J. (1992). *Proc. Natl. Acad. Sci. USA* **89**(19), 9102–9106.

Bode, V.C. (1984). *Genetics* **108**(2), 457–470.

Brewen, G.J. and Preston, R.J. (1973). *Environ. Health Persp.* **6**, 157–166.

Brewen, G.J., Preston, R.J., Jones, K.P. and Gosselee, D.G. (1973). *Mutat. Res.* **17**, 245–254.

Brown, A., Bernier, G., Mathieu, M., Rossant, J. and Kothary, R. (1995). *Nat. Genet.* **10**, 301–306.

Brown, S.D. and Balling, R. (2001). *Curr. Opin. Genet. Dev.* **11**(3), 268–273.

Brown, S.D. and Peters, J. (1996). *Trends Genet.* **12**(11), 433–435.

Bryda, E.C., Kim, H.J., Legare, M.E., Frankel, W.N. and Noben-Trauth, K. (2001). *Genomics* **73**, 338–342.

Buckland, R.A., Evans, H.J. and Sumner, A.T. (1971). *Exp. Cell Res.* **69**, 231–236.

Cabib, S., Orsini, C., Le Moal, M., and Piazza, P.V. (2000). *Science* **289**(5478), 463–465.

Campbell, W.G., Gordon, S.E., Carlson, C.J., et al. (2001). *Am. J. Physiol. Cell Physiol.* **280**, C763–768.

Cattanach, B.M. and Williams, C.E. (1971). *Mutat. Res.* **13**, 371–375.

Charles, D.J. and Pretsch, W. (1987). *Mutat. Res.* **176**(1), 81–91.

Chen, Y., Yee, D., Dains, K., Chatterjee, A., Cavalcoli, J., Schneider, E., Om, J., Woychik, R.P. and Magnuson, T. (2000). *Nat. Genet.* **24**(3), 314–317.

Coghill, E.L., Hugill, A., Parkinson, N., Davison, C., Glenister, P., Clements, S., Hunter, J., Cox, R.D. and Brown, S.D. (2002). *Nat. Genet.* **30**(3), 255–256.

Committee on Standardized Genetic Nomenclature for Mice. (1985). *Mouse Newslett.* **72**, 13–15.

Cox, G.A., Mahaffey, C.L., Nystuen, A., Letts, V.A. and Frankel, W.N. (2000). *Nat. Genet.* **26**, 198–202.

Dhanasekaran, S.M., Barrette, T.R., Ghosh, D., et al. (2001). *Nature* **412**, 822–826.

Dietrich, W.F., Lander, E.S., Smith, J.S., Moser, A.R., Gould, K.A., Luongo, C., Borenstein, N. and Dove, W. (1993). Cell **75**(4), 631–639.

Di Palma, F., Holme, R.H., Bryda, E.C., Belyantseva, I.A., Pellegrino, R., Kachar, B., Steel, K.P. and Noben-Trauth, K. (2001). *Nat. Genet.* **27**, 103–107.

Driever, W., Solnica-Krezel, L., Schier, A.F., Neuhauss, S.C., Malicki, J., Stemple, D.L., Stainier, D.Y., Zwartkruis, F., Abdelilah, S., Rangini, Z., Belak, J. and Boggs, C. (1996). *Development* **123**, 37–46.

Eddy, S.R. (2001). *Nat. Rev. Genet.* **2**, 919–929.

Ehling, U.H. (1966). *Genetics* **54**, 1381–1389.

Ehling, U.H. (1974). *Arch. Toxicol.* **32**(1), 19–25.

Ehling, U.H., Charles, D.J., Favor, J., Graw, J., Kratochvilova, J., Neuhäuser-Klaus, A. and Pretsch, W. (1985). *Mutat. Res.* **150**, 393–401.

Ehling U.H., and Neuhäuser-Klaus, A. (1984). In Problems of Threshold in Chemical Mutagenesis (eds Y. Tazima *et al.*), pp. 15–35. Nissan Science Foundation, Tokyo.

Ehling, U.H. and Neuhäuser-Klaus, A. (1988). *Mutat. Res.* **202**, 139–146.

Ehling, U.H., Doherty, D.G. and Malling, H.V. (1972). *Mutat. Res.* **15**(2), 175–184.

Ehling, U.H., Favor, J., Kratochvilova, J. and Neuhauser-Klaus, A. (1982). *Mutat. Res.* **92**(1–2), 181–192.

Elek, J., Park, K.H., and Narayanan, R. (2000). *In Vivo* **14**, 173–182.

Epstein D.J., Vekemans, M. and Gros, P. (1991). *Cell* **67**, 767–774.

Evans, E.P. (1989). In *The Laboratory Mouse* (eds M.F. Lyon and A.G. Searle), pp. 576–581.

Evans, E.P., Breckon, G. and Ford, C.E. (1964). *Cytogenetics* **3**, 289–294.

Evans, M.J., Carlton, M.B.L. and Russ, A.P. (1997). *Trends Genet.* **13**, 370–374.

EU Federal Register. (1985). *Rules and Regulations* **50** (188).

Favor, J. (1998). *Mutat. Res.* **405**, 221–226.

Favor J., Neuhäuser-Klaus, A., and Ehling, U.H. (1988). *Mutat. Res.* **198**, 269–275.

Favor, J., Neuhäuser-Klaus, A., Ehling, U.H., Wulff, A. and van Zeeland, A.A. (1997). *Mutat. Res.* **374**(2), 193–199.

Favor, J. (1990). *Biology of Mammalian Germ Cell Mutagenicity, Banbury Rep.* **34**, 221–235.

Favor, J. (1994). In *Male-Mediated Developmental Toxicity* (eds D.R. Mattison and A.F. Olsham), pp. 23–36. Plenum Press, New York.

Favor J., Neuhäuser-Klaus, A., and Ehling, U.H. (1990). *Mutat. Res.* **229**, 105–114.

Fessele, S., Maier, H., Zischek, C., et al. (2002). *Trends Genet.* **18**, 60–63.

Flaswinkel, H., Alessandrini, F., Rathkolb, B., Decker, T., Kremmer, E., Servatius, A., Jakob, T., Soewarto, D., Marschall, S., Fella, C., Behrendt, H., Ring, J., Wolf, E., Balling, R., Hrabé de Angelis, M., and Pfeffer, K. (2000). *Mamm. Genome* **11**(7), 526–527.

Fleming, M.D., Campagna, D.R., Haslett, J.N., Trenor, C.C. 3rd and Andrews, N.C. (2001). *Genes Dev.* **15**, 652–657.

Fletcher, C.F., Lutz, C.M., O'Sullivan, T.N., Shaughnessy, J.D. Jr, Hawkes, R., Frankel, W.N., Copeland, N.G. and Jenkins, N.A. (1996). *Cell* **87**, 607–617.

Ford, C.E., Hilary, M. and Clegg, H.M. (1969). *Br. Med. Bull.* **25**, 110–114.

Fuchs, H., Schughart, K., Wolf, E., Balling, R. and Hrabé de Angelis, M. (2000). *Mamm. Genome* **11**(7), 528–530.

Graw, J., Neuhäuser-Klaus, A., Loster, J. and Favor, J. (2002). *Invest. Ophthalmol. Vis. Sci.* **43**(1), 236–240.

Gunn, T.M., Miller, K.A., He, L., Hyman, R.W., Davis, R.W., Azarani, A., Schlossman, S.F., Duke-Cohan, J.S. and Barsh, G.S. (1999). *Nature* **398**, 152–156.

Haffter, P., Granato, M., Brand, M., Mullins, M.C., Hammerschmidt, M., Kane, D.A., Odenthal, J., van Eeden, F.J., Jiang, Y.J., Heisenberg, C.P., Kelsh, R.N., Furutani-Seiki, M., Vogelsang, E., Beuchle, D., Schach, U., Fabian, C. and Nüsslein-Volhard, C. (1996). *Development* **123**, 1–36.

Hagenbuch, B., Adler, I.-D. and Schmid, T.E. (2000). *Biochem. J.* **345**, 115–120.

Hamilton, B.A., Smith, D.J., Mueller, K.L., Kerrebrock, A.W., Bronson, R.T., van Berkel, V., Daly, M.J., Kruglyak, L., Reeve, M.P., Nemhauser, J.L., Hawkins, T.L., Rubin, E.M. and Lander, E.S. (1997). *Neuron* **18**, 711–722.

Hatcher, J.P., Jones, D.N., Rogers, D.C., Hatcher, P.D., Reavill, C., Hagan, J.J. and Hunter, A.J. (2001). *Behav. Brain Res.* **125**(1–2), 43–47.

Hayashi, M., MacGregor, J.T., Gatehouse, D.G. Adler, I.-D., Blakey, D.H. Dertinger, S.D., Krishna, G., Morita, T., Russo, A. and Sutou, S. (2000). *Environ. Mol. Mutagen.* **35**, 234–252.

Heilig, R., Lemaire, C., Mandel, J.L., Dandolo, L., Amar. L. and Avner, P. (1987). *Nature* **328**, 168–170.

Heller, R.A., Schena, M., Chai, A., et al. (1997). *Proc. Natl. Acad. Sci. USA* **94**, 2150–2155.

Herrmann, B.G., Labeit, S., Poustka, A., King, T.R. and Lehrach, H. (1990). *Nature* **343**, 617–622.

Hill, R.E., Favor, J., Hogan, B.L., Ton, C.C., Saunders, G.F., Hanson, I.M., Prosser, J., Jordan, T., Hastie, N.D. and van Heyningen, V. (1991). *Nature* **354**, 522–525.

Hitotsumachi, S., Carpenter, D.A. and Russell, W.L. (1985). *Proc. Natl. Acad. Sci. USA* **82**(19), 6619–6621.

Ho, A.M., Johnson, M.D. and Kingsley, D.M. (2000). *Science* **289**, 265–270.

Hogan, B., Bedington, R., Constantini, F. and Lacy, E., (1994). *Manipulating the Mouse Embryo.* Cold Spring Harbor Laboratory Press, NY.

Hoit, B.D. and Nadeau, J.H. (2001). *Trends Cardiovasc. Med.* **11**(2), 82–89.

Hrabé de Angelis, M., Flaswinkel, H., *et al.* (2000). Nat. Genet. 25, 444–447.

Hrabé de Angelis, M. and Balling, R. (1998). *Mutat Res.* **400**, 25–32.

Hrabé de Angelis, M. and Strivens, M. (2001). *Brief Bioinform.* **2**(2), 170–180.

Hughes, T.R., Marton, M.J., Jones, A.R., et al. (2000). *Cell* **102**, 109–126.

Ichimori, Y., Sasano, K., Itoh, H., Hitotsumachi, S., Kimura, Y., Kaneko, K., Kida, M. and Tsukamoto, K. (1985). *Biochem. Biophys. Res. Commun.* **129**(1), 26–33.

Jansen, G., Hazendonk, E., Thijssen, K.L. and Plasterk, R.H. (1997). *Nat. Genet.* **17**(1), 119–121.

Jentsch, I., Adler, I.-D. Carter N. and Speicher M.R. (2001). *Chromosome Res.* **9**, 211–214.

Johnson, F.M. and Lewis, S.E. (1981). *Proc. Natl. Acad. Sci. USA* **78**, 3138–3141.

Jungblut, P.R., Otto, A., Favor, J., Lowe, M., Muller, E.C., Kastner, M., Sperling, K. and Klose, J. (1998). *FEBS Lett.* **435**(2–3), 131–137.

Justice, M.J. (2000). *Nat. Rev. Genet.* **1**(2), 109–115.

Justice, M.J. and Bode, V.C. (1986). *Genet. Res.* **47**(3), 187–192.

Justice, M.J. and Bode, V.C. (1988). *Genet. Res.* **51**(2), 95–102.

Justice, M.J., Noveroske, J.K., Weber, J.S., Zheng, B. and Bradley, A. (1999). *Hum. Mol. Genet.* **8**(10), 1955–1963.

Kallionemi, A., Kallionemi, O.-P., Sudar, D., Rutovitz, D., Gray, J.W., Waldman, F. and Pinkel, D. (1992). *Science* **258**, 818–821.

Kato, H. and Moriwaki, K. (1972). *Chromosoma* **38**, 105–120.

Kibar, Z., Vogan, K.J., Groulx, N., Justice, M.J., Underhill, D.A. and Gros, P. (2001). *Nat. Genet.* **28**, 251–255.

King, D.P., Zhao, Y., Sangoram, A.M., Wilsbacher, L.D., Tanaka, M., Antoch, M.P., Steeves, T.D., Vitaterna, M.H., Kornhauser, J.M., Lowrey, P.L., Turek, F.W. and Takahashi, J.S. (1997). *Cell* **89**, 641–653.

Kingsley, D.M., Bland, A.E., Grubber, J.M., Marker, P.C., Russell, L.B., Copeland, N.G. and Jenkins, N.A. (1992). *Cell* **71**, 399–410.

Kleyn, P.W., Fan, W., Kovats, S.G., Lee, J.J., Pulido, J.C., Wu, Y., Berkemeier, L.R., Misumi, D.J., Holmgren, L., Charlat, O., Woolf, E.A., Tayber, O., Brody, T., Shu, P., Hawkins, F., Kennedy, B., Baldini, L., Ebeling, C., Alperin, G.D., Deeds, J., Lakey, N.D., Culpepper, J., Chen, H., Glucksmann-Kuis, M.A., Carlson, G.A., Duyk, G.M. and Moore, K.J. (1996). *Cell* **85**, 281–290.

Klopp, N., Favor, J., Loster, J., Lutz, R.B., Neuhäuser-Klaus, A., Prescott, A., Pretsch, W., Quinlan, R.A., Sandilands, A., Vrensen, G.F. and Graw, J. (1998). *Genomics* **52**(2), 152–158.

Koishi, R., Ando, Y., Ono, M., Shimamura, M., Yasumo, H., Fujiwara, T., Horikoshi, H. and Furukawa, H. (2001). *Nat. Genet.* **30**, 151–157.

Kusumi, K., Sun, E.S., Kerrebrock, A.W., Bronson, R.T., Chi, D.C., Bulotsky, M.S., Spencer, J.B., Birren, B.W.,

Frankel, W.N. and Lander, E.S. (1998). *Nat. Genet.* **19**, 274–278.

Lander, E.S., Green, P., Abrahamson, J., Barlow, A., Daly, M.J., Lincoln, S.E. and Newburg, L. (1987). *Genomics* **1**, 174–181.

Lee, C.K., Klopp, R.G., Weindruch, R., and Prolla, T.A. (1999). *Science* **285**, 1390–1393.

Lee, S.H., Girard, S., Macina, D., Busa, M., Zafer, A., Belouchi, A., Gros, P. and Vidal, S.M. (2001). *Nat. Genet.* **28**, 42–45.

Lee, W.R., Beranek, D.T., Byrne, B.J. and Tucker, A.B. (1990). *Mutat. Res.* **231**(1), 31–45.

Li, S., Crenshaw, E.B. 3rd, Rawson, E.J., Simmons, D.M., Swanson, L.W. and Rosenfeld, M.G. (1990). *Nature* **347**, 528–533.

Liang, Y., Wang, A., Probst, F.J., Arhya, I.N., Barber, T.D., Chen, K.S., Deshmukh, D., Dolan, D.F., Hinnant, J.T., Carter, L.E., Jain, P.K., Lalwani, A.K., Li, X.C., Lupski, J.R., Moeljopawiro, S., Morell, R., Negrini, C., Wilcox, E.R., Winata, S., Camper, S.A. and Friedman, T.B. (1998). *Am. J. Hum. Genet.* **62**, 904–915.

Lindahl, T., Sedgwick, B., Sekiguchi, M. and Nakabeppu, Y. (1988). *Annu. Rev. Biochem.* **57**, 133–157.

Liu, W., Smith, D.I., Rechtzigel, K.J., Thibodeau, S.N. and James, C.D. (1998). *Nucleic Acids Res.* **26**(6), 1396–1400.

Livesey, F.J., Furukawa, T., Steffen, M.A. et al. (2000). *Curr. Biol.* **10**, 301–310.

Liyanage M., Coleman, E., du Manoir, S., Veldman, T., McCormack, S., Dickson, R.B., Barlow, C., Wynshaw-Boris, A., Janz, S., Wienberg, J., Ferguson-Smith, M.A., Schröck, E. and Ried, T. (1996). *Nat. Genet.* **14**, 312–315.

Luckey, J.A., Drossman, H., Kostichka, A.J., Mead, D.A., D'Cunha, J., Norris, T.B. and Smith, L.M. (1990). *Nucleic Acids Res.* **18**(15), 4417–4421.

Lyon, M.F. and Morris, T. (1966). *Genet. Res.* **7**, 12–17.

Lyon, M.F. (1983). *INSERM* **119**, 323–345.

Lyon, M.F., Phillips, R.J.S. and Glenister, P.H. (1972). *Mutat. Res.* **15**, 191–195.

Malo, D., Vidal, S.M., Hu, J., Skamene, E. and Gros, P. (1993). *Genomics* **16**, 655–663.

Marschall, S. and Hrabé de Angelis, M. (1999). *Trends Genet.* **15**(4), 128–131.

Mass, R.L., Zeller, R., Woychik, R.P., Vogt, T.F. and Leder, P. (1990). *Nature* **346**, 853–855.

Melvold, R.W. and Kohn, H.I. (1975). *Mutat. Res.* **27**, 415–418.

Miller, B.M., Zitzelberger, H.F., Weier, H.U. and Adler, I.-D. (1991). *Mutagenesis* **6**, 297–302.

Millonig, J.H., Millen, K.J. and Hatten, M.E. (2000). *Nature* **403**, 764–769.

Mills, E., Kuhn, C.M., Feinglos, M.N. and Surwit, R. (1993). *Am. J. Physiol.* **264**(1 Pt 2), R73–R78.

Miyashita, K., Vooijs, M.A., Tucker, D.J., Lee, D.A., Gray, J.W. and Pallavicini, M.G. (1994). *Cytogenet. Cell Genet.* **66**, 54–57.

Munroe, R.J., Bergstrom, R.A., Zheng, Q.Y., Libby, B., Smith, R., John, S.W., Schimenti, K.J., Browning, V.L. and Schimenti, J.C. (2000). *Nat. Genet.* **24**(3), 318–321.

Nadeau, J.H., Balling, R., Barsh, G., Beier, D., Brown, S.D., Bucan, M., Camper, S., Carlson G., Copeland, N., Eppig, J., Fletcher, C., Frankel, W.N., Ganten, D., Goldowitz, D., Goodnow, C., Guenet, J.L., *et al.* (2001). *Science* **291**(5507), 1251–1255.

Nagle, D.L., McGrail, S.H., Vitale, J., Woolf, E.A., Dussault, B.J. Jr, DiRocco, L., Holmgren, L., Montagno, J., Bork, P., Huszar, D., Fairchild-Huntress, V., Ge, P., Keilty, J., Ebeling, C., Baldini, L., Gilchrist, J., Burn, P., Carlson, G.A. and Moore, K.J. (1999). *Nature* **398**, 148–152.

Nakagata, N. (1994). *Jikken Dobutsu* **43**(1), 11–8.

Nesbitt, M.N. and U. Franke (1973). *Chromosoma* **41**, 145–158.

Nishina, P.M., Wang, J., Toyofuku, W., Kuypers, F.A., Ishida, B.Y. and Paigen, B. (1993). Lipids **28**(7), 599–605.

Nüsslein-Vollhardt, C., Wieschaus, E. and Kluding, H. (1984). *Roux's Arch. Dev. Biol.* **193**, 267–282.

Oakberg, E.F. (1956). *Am. J. Anat.* **99**, 507–516.

Ohl, F., Sillaber, I., Binder, E., Keck, M.E. and Holsboer, F. (2001). *J. Psychiatr. Res.* **35**(3), 147–154.

Orita, M., Suzuki, Y., Sekiya, T. and Hayashi, K. (1989). *Genomics* **5**(4), 874–879.

Paulus, M.J., Gleason, S.S., Kennel, S.J., Hunsicker, P.R. and Johnson, D.K. (2000). *Neoplasia* **2**(1–2), 62–70.

Peterfy, M., Phan, J., Xu, P. and Reue, K. (2000). *Nat. Genet.* **27**, 121–124.

Pinto, L.H., Enroth-Cugell, C. (2000). *Mamm. Genome* **11**(7), 531–536.

Pomeroy, S.L., Tamayo, P., Gaasenbeek, M., et al. (2002). *Nature* **415**, 436–442.

Popp, R.A., Bailiff, E.G., Skow, L.C., Johnson, F.M. and Lewis, S.E. (1983). *Genetics* **105**(1), 157–167.

Porter, J.D., Khanna, S., Kaminski, H.J., et al. (2001). *Proc. Natl. Acad. Sci. USA* **98**, 12062–12067.

Pretsch, W., Charles, D.J. and Natarayan, K.R. (1982). *Electrophoresis* **3**, 142–145.

Probst, F.J., Fridell, R.A., Raphael, Y., Saunders, T.L., Wang, A., Liang, Y., Morell, R.J., Touchman, J.W., Lyons, R.H., Noben-Trauth, K., Friedman, T.B. and Camper, S.A. (1998). *Science* **280**, 1444–1447.

Ramirez-Solis, R., Davis, A.C. and Bradley, A. (1993). *Methods Enzymol.* **225**, 855–878.

Ramirez-Solis, R., Liu, P. and Bradley, A. (1995). *Nature* **378** (6558), 720–724.

Rathkolb, B., Decker, T., Fuchs, E., Soewarto, D., Fella, C., Heffner, S., Pargent, W., Wanke, R., Balling, R., Hrabé de Angelis, M., Kolb, H.J. and Wolf, E. (2000). *Mamm. Genome* **11**(7), 543–546.

Rinchik, E.M., Carpenter, D.A. and Selby, P.B. (1990). *Proc. Natl. Acad. Sci. USA* **87**(3), 896–900.

Rogers, D.C., Fisher, E.M., Brown, S.D., Peters, J., Hunter, A.J. and Martin, J.E. (1997). *Mamm. Genome* **8**(10), 711–713.

Rolinski, B., Arnecke, R., Dame, T., Kreischer, J., Olgemoller, B., Wolf, E., Balling, R., Hrabé de Angelis, M., and Roscher, A.A. (2000). *Mamm. Genome* **11**(7), 547–551.

Ronaghi, M., Karamohamed, S., Pettersson, B., Uhlen, M. and Nyren, P. (1996). *Anal. Biochem.* **242**(1), 84–89.

Ronaghi, M., Uhlen, M. and Nyren, P. (1998). *Science* **281**(5375), 363, 365.

Russell, L.B., Russell, W.L., Rinchik, E.M. and Hunsicker, P.R. (1990). *Banbury Rep.* **34**, 271–289.

Russell, W.L. (1951). *X-Ray-Induced Mutations in Mice*, pp. 327–336: Cold Spring Harbor Symp. Quant. Biol.

Russell, W.L., Hunsicker, P.R., Raymer, G.D., Steele, M.H., Stelzner, K.F. and Thompson, H.M. (1982). *Proc. Natl. Acad. Sci. USA* **79**(11), 3589–3591.

Russell, W.L., Kelly, P.R. and Hunsicker, P.R. (1979). *Proc. Natl. Acad. Sci. USA* **76**, 5918–5922.

Russo, A., Pacchierotti, F. and Metalli, P. (1984). *Environ. Mutagen.* **6**, 695–703.

Sanger, F., Nicklen, S. and Coulson, A.R. (1977). *Proc. Natl. Acad. Sci. USA* **74**(12), 5463–5467.

Sanger, F., Nicklen, S. and Coulson, A.R. (1977). *Proc. Natl. Acad. Sci. USA* **74**(12), 5463–5467.

Sawyer, J.R. and J.C. Hozier (1986). *Science* **232**, 1632–1635.

Schmid, W. (1975). *Mutat. Res.* **31**, 9–15.

Schmid, T.E., Wang Xu and Adler, I.-D. (1999). *Mutagenesis* **14**, 173–179.

Schnedl, W. (1971). *Chromosoma* **35**, 111–116.

Schwartzberg, P.L., Goff, S.P. and Robertson, E.J. (1989). *Science* **246**, 799–803.

Seiler, F., Kamino, K., Emura, M., Mohr, U. and Thomale, J. (1997). *Mutat. Res.* **385**(3), 205–211.

Shedlovsky, A., Guenet, J.L., Johnson, L.L. and Dove, W.F. (1986). *Genet. Res.* **47**(2), 135–142.

Shedlovsky, A., McDonald, J.D., Symula, D. and Dove, W.F. (1993). *Genetics* **134**(4), 1205–1210.

Shedlovsly, A., King, T.R. and Dove, W.F. (1988). *Proc. Natl. Acad. Sci. USA* **85**(1), 180–184.

Shibuya, T. and Morimoto, K. (1993). *Mutat. Res.* **297**(1), 3–38.

Shibuya, T., Murota, T., Horiya, N., Matsuda, H. and Hara, T. (1993). *Mutat. Res.* **290**(2), 273–280.

Sidow, A., Bulotsky, M.S., Kerrebrock, A.W., Bronson, R.T., Daly, M.J., Reeve, M.P., Hawkins, T.L., Birren, B.W., Jaenisch, R. and Lander, E.S. (1997). *Nature* **389**, 722–725.

Sidow, A., Bulotsky, M.S., Kerrebrock, A.W., Birren, B.W., Altshuler, D., Jaenisch, R., Johnson, K.R. and Lander, E.S. (1999). *Nat. Genet.* **23**, 104–107.

Silver, L.M. (1995). Mouse Genetics: Concepts and Applications. Oxford University Press, New York, Oxford.

Soewarto, D., Fella, C., Teubner, A., et al. (2000). *Mamm. Genome* **11**, 507–510.

Sperling, K. and Wiesner, R. (1972). *Humangenetik* **15**, 349–353.

Srivastava, A.K., Pispa, J., Hartung, A.J., Du, Y., Ezer, S., Jenks, T., Shimada, T., Pekkanen, M., Mikkola, M.L., Ko, M.S., Thesleff, I., Kere, J. and Schlessinger, D. (1997). *Proc. Natl. Acad. Sci. USA* **94**, 13069–13074.

Sumner, A.T. (1973). *Exp. Cell Res.* **83**, 438–442.

Sumner, A.T. and Evans, J.H. (1973a). *Exp. Cell Res.* **81**, 214–222.

Sumner, A.T. and Evans, J.H. (1973b). *Exp. Cell Res.* **81**, 223–236.

Thomas, K.R. and Capecchi, M.R. (1987). *Cell* **51**(3), 503–512.

Thomas, K.R., Musci, T.S., Neumann, P.E. and Capecchi, M.R. (1991). *Cell* **67**, 969–976.

Tucker, J.D., Christensen, M.L. and Carrano, A.V. (1988). *Cytogenet. Cell Genet.* **48**, 103–106.

van Zeeland, A.A. (1988). Mutagenesis **3**(3), 179–191.

Vidal, S., Tremblay, M.L., Govoni, G., Gauthier, S., Sebastiani, G., Malo, D., Skamene, E., Olivier, M., Jothy, S. and Gros P. (1995). *J. Exp. Med.* **182**, 655–666.

Vogel, E.W., Nivard, M.J., Ballering, L.A., Bartsch, H., Barbin, A., Nair, J., Comendador, M.A., Sierra, L.M., Aguirrezabalaga, I., Tosal, L., Ehrenberg, L., Fuchs, R.P., Janel-Bintz, R., Maenhaut-Michel, G., Montesano, R., Hall, J., Kang, H., Miele, M., Thomale, J., Bender, K., Engelbergs, J. and Rajewsky, M.F. (1996). *Mutat. Res.* **353**(1–2), 177–218.

Voss, H., Schwager, C., Wirkner, U., Sproat, B., Zimmermann, J., Rosenthal, A., Erfle, H., Stegemann, J. and Ansorge, W. (1989). *Nucleic Acids Res.* **17**(7), 2517–2527.

Wada, T., Wakabayashi, Y., Takahashi, S., Ushiki, T., Kikkawa, Y., Yonekawa, H. and Kominami, R. (2001). *Biochem. Biophys. Res. Commun.* **283**, 113–117.

Wakabayashi, Y., Takahashi, Y., Kikkawa, Y., Okano, H., Mishima, Y., Ushiki, T., Yonekawa, H. and Kominami, R. (1998). *Biochem. Biophys. Res. Commun.* **248**, 655–659.

Weber, J.S., Salinger, A. and Justice, M.J. (2000). *Genesis* **26**(4), 230–233.

Wieschaus, E., Nüsslein-Volhard, C. and Kluding, H. (1984). *Dev. Biol.* **104**(1), 172–186.

Wiles, M.V., Vauti, F., Otte, J., Fuchtbauer, E.M., Ruiz, P., Fuchtbauer, A., Arnold, H.H., Lehrach, H., Metz, T., von Melchner, H. and Wurst, W. (2000). *Nat. Genet.* **24**(1), 13–14.

Wilkinson, D.G., Bhatt, S., and Herrmann, B.G. (1990). *Nature* **343**, 657–659.

Wurster, D.H. (1972). *Cytogenetics* **11**, 379–387.

Zhang, Y., Proenca, R., Maffei, M., Barone, M., Leopold, L. and Friedman, J.M. (1994). *Nature* **372**, 425–432.

Zheng, B., Sage, M., Cai, W.W., Thompson, D.M., Tavsanli, B.C., Cheah, Y.C. and Bradley, A. (1999). *Nat. Genet.* **22**(4), 375–378.

Zuo, J., De Jager, P.L., Takahashi, K.A., Jiang, W., Linden, D.J. and Heintz, N. (1997). *Nature* **388**, 769–773.

CHAPTER 5

Generation of Mouse Mutants by Sequence Information Driven and Random Mutagenesis

Martin Hafner and Werner Müller
Department of Experimental Immunology, GBF
German Research Centre for Biotechnology,
Braunschweig, Germany

This article discusses methods for generating mouse mutants either by sequence driven or by random mutagenesis and gives a general idea of the advantages and disadvantages of the various approaches. Time, lab and housing space, mouse numbers, manpower and the amount of money required to generate the respective mutants as well as additional resources to analyse phenotypes will be discussed. References to books and articles describing the methods in more detail are provided. The comparison of the different technologies will include a discussion of their deep intellectual and scientific differences (forward vs. reverse genetics). After reading the chapter, it will be up to the reader to decide which method is the best possible way to reach a certain goal.

Choosing the sequence driven mutagenesis approach

The mouse genome sequence was finished in December 2002 (Waterson *et al.*, 2002). With the availability of the genome sequence information, the design and generation of gene targeting vectors is now much easier and faster compared with the time when the gene of interest had to be cloned, mapped and partially sequenced prior to the generation of the targeting vector. The fastest way to obtain the technical information (exon/intron structure, intron phase, restriction map) needed for the design of the targeting vector is to visit the EnsEMBL Internet webpage (http://www.ensembl.org). At this site one can view the gene of interest and download the sequences necessary for the construction of the targeting vector. In addition, although not completed and mostly based on automatic sequence analysis tools, the sequence annotation provided in EnsEMBL is quite informative already and artefacts like accidental inactivating genes encoded on the opposite strand of the gene of interest without notice are easily avoidable. Once inside the EnsEMBL database the user is connected to other databases via cross-references. It should be noted that until the mid-1990s the genomic sequence as such did not drive the decision to create a targeted mutant of a particular gene. It was more likely that such decisions were based on other information or knowledge that had been gained in unrelated experiments (RNA or protein expression studies, homology searches, functional data of **orthologs** and **paralogs** in other species, structural data, results obtained with mice over expressing the respective cDNA etc.). Thus, researchers applying sequence driven mutagenesis approaches already had some hypotheses on the function of the respective gene. However, during the last decade as the targeting methodology improved and access to biological information became easier and faster, targeted mutagenesis was used to unravel the function of newly cloned genes. Still, even today, in most cases the choice for a target gene originates from very specific scientific questions, which try to verify, prove or disprove hypotheses or which ask the very basic question about the function of a gene which was not yet studied by targeted mutation. In the past a number of genetic screens have been performed in less complex eucaryotic organisms such as yeast, C. elegans, Drosophila and fish species. In many cases, where a phenotype was observed in these species, it may now be worthwhile to repeat such mutations in mice using the sequence driven mutagenesis approach. However, results obtained with other species may be misleading. For example, while antibodies against the SPARC protein disrupt the normal embryogenesis of Xenopus laevis and inactivation of its ortholog in C. elegans also causes a severe morphological phenotype, ablation of the orthologous gene in mice did not affect embryogenesis and the animals only developed

an eye disease later during adulthood (Bradshaw and Sage, 2001). Still, in many other cases data obtained with non-mammalian species fed developments in mammalian genetics.

The basic procedure for inactivating a target gene in the mouse germline requires the generation of a gene-targeting vector, which contains genomic sequences of the target gene, a resistance marker gene, and, if wanted, small sequence motives of target sequences (loxP or frt sites) for recombinases (cre or flp). These allow the elimination of the resistance marker gene in the genome, the exchange of DNA segments, the removal or inversion of sequences located between two recombinase target sequences.

Targeting vectors are used to transfect murine embryonic stem (ES) cells. After transfection, stably transfected clones of the ES cells are selected *in vitro* and the genomic DNA of such stable clones are analysed either by Southern blotting or polymerase chain reaction (PCR) procedures in order to identify the clones in which the wanted homologous recombination event occurred. Murine ES cells with the correct mutation are used to generate chimeric mice. The latter are then further used for breeding and in some of them ES cells will contribute to the mouse germline. Thus, the mutation introduced into the ES cell is transmitted to the offspring. Intercrossing siblings will finally lead to the generation of a mouse line in which both alleles of the target gene are mutated.

Embryonic stem cells were first isolated from the 129 mouse strain. Today ES cells are available for other inbred mouse strains such as C57BL/6 or BALB/c. When the chimeric mice are crossed with mice of the same strain from which the employed ES cell line was originally derived, inbred mouse lines can be established from the chimeras directly, thereby saving time, labour and financial resources.

In order to identify chimeric mice and to identify offspring, which are potential carriers of the mutations, coat colour genes are used. For example, when ES cells of C57BL/6 mice are injected into blastocysts of BALB/c mice, the resulting chimeras are black and white (with stripes or patches). In order to analyse whether the ES cells contribute to the germline, the chimeras are crossed with C57BL/6 mice. In the next generation mice derived from the ES cells are black; mice derived from the blastocyst cells are brown. According to Mendel's laws, about 50% of the black mice will carry the mutation that was introduced into the ES cell (with the exception of sex chromosome linked genes).

How to choose genes for a sequence driven approach?

The starting point of the sequence driven approach is the knowledge about a gene or genomic region of interest. It is also important to verify that a mutant mouse was not produced so far. Information about the published mouse mutants can be gathered at http://tbase.org (free) or at http://research.bmn.com/mkmd (subscription). It is also useful to use a general Internet search tool like http://www.google.com/ to search for pages on the Internet, which may point to new mouse mutants. To choose the best possible target for the mutagenesis, it is necessary to collect as much information about the gene as possible, prior to designing the best possible gene-targeting vector. Simply go through the list of the following questions and answer simply yes or no.

Does the target gene exist with multiple copies in the genome?
Does the target gene belong to a gene family?
Do pseudo genes of the target gene exist in the genome?

If your answer is three times no, you can proceed to the following questions, otherwise the experimental approaches for studying the gene functions become difficult and the discussion of these approaches is outside the scope of this article.

Are natural mutations known in the target gene of the mouse or in other species? Information on natural mutation in disease related human genes is found in the OMIM (Online Mendelian Inheritance in Men) database maintained at the National Centre for Biotechnology Information (http://www.ncbi.nlm.nih. gov). Relevant mouse mutants might be identified in the Mouse Genome Informatics (MGI) database maintained at The Jackson Lab (http://www.informatics.jax. org/).

If a mutation is well documented, what is the phenotype of such mutants and what will you learn new if you generate a new mutation in such a target gene?

Where and when is the target gene expressed in the mouse? Public gene expression data may be obtained from the Genexpression Omnibus (GEO) database maintained at the National Centre for Biotechnology Information (http://www.ncbi.nlm.nih.gov/geo/).

If it is expressed in many tissues and/or early in development there is a chance that inactivation of your target gene may result in an early phenotype or in a very complex phenotype. In such cases it may be better to follow a conditional gene targeting approach (see below). Is it likely that the complete inactivation of the target gene will result in embryonic lethality? If yes, again you may consider the conditional gene targeting approach (see below).

Assuming the gene of interest is a single copy gene specifically expressed in the cell type of interest and is not a member of a gene family, the complete knockout of the target gene is the best possible way to get insight into the biological relevance of the gene. In the traditional way, the targeting vector can be constructed in a relatively short time and checking of the construct can be done in a very crude way as interrupting the gene structure is the main goal. Such a construct would cover about 4–12 kb of your target gene with a simple resistance marker cassette gene, located somewhere in the middle of the targeting vector, replacing an essential part of your target gene. In addition, you need a DNA probe in order to analyse the gene-targeting event by Southern blotting. Ideally you would like to have two DNA probes, one at the 5′ end of your targeting vector and one at the 3′ end of your targeting vector. Using both probes you can verify that in case of the correct targeting event, you really replaced part of your target gene by the resistance marker gene (homologous recombination) and did not simply insert the resistance marker gene into the target locus, thereby partially duplicating the target gene. Such recombination events occasionally occur in ES cells and the result could be gene targeting without or only little effect on the gene function.

The main drawback of the traditional gene targeting approach is that the resistance marker gene remains in the genome. It has been shown, that such resistance marker genes, which consist of a promoter, an open reading frame encoding the resistance marker and a polyA signal, may not only disrupt the target gene but may also influence the expression of neighbouring genes, either by enhancing or silencing mechanisms. Therefore the use of this approach is not recommended. Today a better approach is to use resistance marker genes, which carry loxP or frt sites at their ends. The use of targeting vectors with such cassettes allows, after the homologous recombination event, the removal of the resistance marker gene in the ES cell prior to the generation of chimera by transfecting the targeted ES cell clone with the respective recombinase (cre or flp). Alternatively, if chimeras are generated first, the resistance marker can be removed by crossing the chimeras with **deleter** mice, which express the respective recombinases in the germline. By the latter approach one has to take into account that one introduces genetic material from the deleter strain and thereby may risk an unnecessary genetic impurity of the mutant strain. However, in the future this situation will improve, since recombinase expressing mice are being backcrossed to all important inbred strains used biomedical research. While these simple targeting vectors are easy to construct, one may consider immediately designing targeting vectors for a conditional gene targeting approach, a method that allows the generation of the complete knockout as well as the conditional mutant (see below).

Conditional gene targeting

The conditional gene targeting approach (Rajewsky et al., 1996) allows the introduction of much more sophisticated changes of the mouse germline compared with the simple gene modification approach described in the previous paragraph. The method depends on the use of site-specific recombinases and their target sequences as tools to improve the site-specific mutagenesis in the genome. This approach allows the introduction of point mutations, gene replacement, and gene inactivation in the absence of resistance marker genes in the mutants. In addition, it offers the opportunity of introducing mutations in somatic cells by the use of recombinase transgenic mice. This is probably the most useful application of conditional gene targeting, because it allows the predefined gene modification to be introduced in particular cell types and/or at particular time points. The procedure can be divided into two separate steps. The first step implies the introduction of the recombination target sites into the mouse genome by homologous recombination, a procedure similar to the complete gene inactivation approach described above. The construction of the mutation requires more careful planning of the targeting vector. In addition, the construction of the targeting vector and the mutation in the ES cells have to be carefully monitored in order to make sure that the introduction of the recombination sites do not disturb the target gene. In addition, one has to make sure that the removal of the gene segment flanked by recombination sites leads to complete inactivation of the target gene.

One can easily inactivate the target gene in the targeted ES cell clone by transiently transfecting a suitable recombinase expression vector to generate a mouse line in which the target gene is completely ablated. The spatial inactivation of the target gene in certain cell types or the ablation of the respective gene at defined time points requires breeding of the newly generated mouse line with transgenic mice specifically expressing the respective recombinase. About 30 cre-transgenic mouse lines have been published. An Internet page collecting cre-transgenic mouse lines is maintained by Andras Nagy (http://www.mshri.on.ca/nagy/cre.htm). If a suitable recombinase transgenic mouse line is not available a second mouse engineering step is required for the generation of such a mouse line. The generation of useful recombinase transgenic lines is much more demanding compared with the transgenic mice one generates to overproduce other gene products. The use of recombinase transgenic mice requires complete activity in all cells of a particular cell type. Variegation, leading to mosaic expression of the recombinase, will lead to the generation of chimeric mice made up of cells either with or without a deletion of the target sequence. In addition, the recombinase may be transiently expressed in other cell types during mouse development thus leading to gene modification in other cell types than initially anticipated. In particular, in situations where mutant cells would have a disadvantage, the mutant cells will be counter selected and overgrown by wild type cells. This process could lead to the complete loss of mutant cells, an experimental situation which would make it impossible to study the function of the target gene in such mice.

In practice, the combination of the mouse line with the targeted gene and the recombinase expressing mouse line will determine the success of a particular experimental setup. It is impossible to predict the outcome prior to the experiment. Due to the requirement of a suitable recombinase expressing mouse strain and due to the necessity of more careful planning and construction of the mouse mutant, more time and larger mouse numbers are required for the analysis of the gene function. Since transgenic mice are usually generated by microinjection of expression constructs into fertilised oocytes of outbred crossings, additional time for backcrossing the transgenic lines to a clean background has to be taken into account. However, alternative methods such as transgene transfection in suitable ES cell lines or site specific introduction of the recombinases via homologous recombination (referred to as knock in technology) within the respective ES cells are available today.

On the other hand, by the use of conditional mutants one is able to pose very sophisticated questions, which cannot be answered with conventional mutants. For example, it is possible to study the function of genes which are required for the development of mice and which may have additional or different functions in mature cell type. It is also possible to identify cell autonomous processes.

Many articles on the conditional gene targeting approach have been published during recent years. The most commonly used system is that of the P1 bacteriophage derived recombinase Cre and its recombinase target sequence loxP. The Cre recombinase is not the only suitable site-specific recombinase used. The second most often used is the FLP recombinase and its target sequence FRT. This recombinase, although less efficient compared with the Cre recombinase, is of particular interest, as it allows the cassette exchange of genomic regions in ES cells thereby allowing the fast generation of allelic mutants of the gene of interest. Other recombinases have been identified and proven to work *in vitro* and *in vivo*.

Choosing the random mutagenesis approach

In contrast to the sequence driven approaches described above, random mutagenesis approaches do not rely on the knowledge of target sequences. In Drosophila, random mutagenesis has been applied since the first half of the twentieth century. Indeed, many models of modern molecular biology are based on observations first made in this species. In mice, two different basic procedures are used to randomly mutate the genome, random insertion of DNA pieces (gene trap approach; Wiles *et al.*, 2000) and chemical induction of point mutations (ethylnitrosourea (ENU) mutagenesis; Nolan *et al.*, 2000a,b; Soewarto *et al.*, 2000). With both techniques, the initial event is random. Therefore, in a later step, particular events have to be selected. The key issue is to define a very selective screening procedure, which allows the identification of individual alterations significantly different to the natural population. Large screens have been set up in big centres where several phenotyping screens are conducted around a central mutagenesis platform. Thus,

one given mutant is tested for as many parameters as possible. Besides phenotyping of mutants generated by random mutagenesis the infrastructures of these 'mouse clinics' are very useful platforms for the analysis of targeted mouse mutants without apparent phenotypes.

Generation of gene trap mouse mutants

Gene trap mutagenesis starts with the transfection of murine ES cells, as described in the first part of this chapter. DNA pieces (gene trap vectors) carrying an impaired selection marker gene are randomly inserted into the genomic DNA of murine ES cells (Wiles *et al.*, 2000). Gene trap vectors are designed in a way that the resistance marker gene is transcribed if the construct was inserted in the promoter regions of any gene. Alternatively gene trap vectors, activated after the insertion upstream of poly adenylation sites, are in use. Thus, only ES cells, in which the resistance marker genes are inserted in a functional genomic region that is transcribed in ES cells, are selected in the culture. Individual clones are picked and the insertion points are determined by sequencing the region neighbouring the gene trap vector. Thus, with knowledge of the mouse genome sequence the insertion point can determined precisely and one can predict how the insertion will alter the function of the trapped gene. This is important if one has to deal with genes that are transcribed from alternate promoters or are alternatively spliced, respectively. Due to the limited sequence information such analyses have not been possible before the mouse genome sequence was completed and researchers were lucky if they could identify a trapped known gene.

Three major problems have to be envisioned with the gene trap approach. Firstly, only ES cells with a single insertion point are really useful and sometimes clones are selected with multiple gene trap vector insertions. Secondly, the targeted gene may not be inactivated but only partially affected by the gene trap vector insertion (e.g. alternate promoters might be used to transcribe the trapped gene). Thirdly, the insertion of gene trap vectors in the mouse genome is not really random but favours particular genes. Therefore one often finds the same gene targeted over and over again while other genes are never found in the clone collection. On the Internet, several collections of gene trap mutants can be searched in order to find mutants of interest. For example see the German Genetrap Consortium at http://www.genetrap.de (alternatively use: http://tikus.gsf.de/) or visit the following sites: http://baygenomics.ucsf, http://socrates.berkeley.edu/%7Eskarnes/resource.html.

Generation of ENU mutants

The mutagen ENU allows the introduction of point mutations in the mouse genome (e.g. http://www.gsf.de/ieg/groups/enu-mouse.html). Only male mice are treated with the mutagen. Thus spermatocytes carry randomly introduced mutations. The mutation procedure introduces a specific mutation at a rate of about 10^{-3}. This means that in every genome more than one gene is mutated and that about 1/1000 of the offspring carries a mutation in a given gene. The mutagenised males are then used for breeding with untreated females and mutant mice are selected in the next generation. Since only male gametes carry induced mutations only dominant mutations and Y-linked mutations can be identified in the offspring directly. To a degree it is also possible to identify animals in which the mutation causes gene dosage effects.

The most widely used approach is to use a first line phenotyping, which allows the identification of mutants by markers, which can be accessed, in the living animal. Due to the great number of mutants the applied phenotyping techniques have to be either fast or have the potential of automation.

Due to the applied phenotyping techniques dominant mutants are detected relatively easily. In these mutants, a single point mutation in a gene would lead to a change in the phenotype of the mice. However, it is well known that the majority of human inherited diseases are caused by recessive mutations. In addition, targeted mutations induced in mice normally do not cause a phenotype in the heterozygous state. Thus efforts are made to identify recessive mutations induced by random mutagenesis. This is of course more complicated, but not impossible. For this purpose offspring from individual mice have to be intercrossed. According to Mendel's laws, one quarter of the offspring will harbour the introduced mutation in a homozygous fashion and display a phenotype of interest.

With ENU mutagenesis the generation of mutants is fast. The first line screening requires a good logistic and a good baseline in the screening procedure. The real work starts when the mutation causing the phenotype has to be identified. At the time ENU projects were set up, the goal to identify a single point mutation linked to the phenotype was quite ambitious. Even today with the availability of polymorphic markers this goal remains time and cost intensive, because every mutant has to be backcrossed to other inbred strains to identify markers which cosegregate with the mutation. Based on the genome sequence one may then identify a candidate gene that might cause the disease phenotype. It has to be stressed that the identification of the point mutation *per se* is not sufficient and other experiments are required to make sure that the mutation isolated is really the cause of the phenotype. The most extreme view would be to generate a targeted mutation of the ENU mutant to verify the phenotype observed. The main drawback of this method is that mutations are introduced randomly and it is hard to exclude the fact that more than one mutation have been introduced into the mutant. While mutations located on different chromosomes will segregate relatively fast during the process of backcrossing, the segregation rate of mutations on the same chromosome is a reciprocal function of the distance of two mutations. Thus, the number of backcross generations increases when the mutations are closely neighboured. In addition, it is hard to generate the same mutant twice. To a degree this problem is solved by new methods such as sequence driven ENU mutagenesis (Beier, 2000; Chen *et al.*, 2000), which allow the generation of allelic series of mutants, but again with the danger of introducing more than one mutation into the genomic region of interest.

The most ideal situation, which however is rarely seen, is achieved when the same type of mutation is seen in different species. For example the Beethoven Mutation introduced by ENU mutagenesis into mice (Vreugde *et al.*, 2002) mimics a mutation of the corresponding gene in humans (Kurima *et al.*, 2002). Only these findings together allowed the unequivocal association of the phenotype with the mutation identified.

The forgotten mutants

In most cases scientists are working with inbred mouse strains. These strains are generated by intercrossing outbred (or wild) mice for many generations by brother/sister interbreeding. After a very long time of inbreeding, stable and genetically uniform inbred mouse strains are generated. During the inbreeding process, a constant selection for mouse mutants is continuing and some genes, not required for the survival in the animal house, are lost from the inbred population. In addition, due to the relatively small population size some alleles will be lost by genetic drift. As a byproduct selection and genetic drift during the inbreeding process, all genes become homozygous for a particular allele. All inbred mice therefore represent a genetically homogenous pool of a particular mouse mutant. If one compares the genes present in inbred mice with those of wild type mice, one detects for example that many inbred mice lack a functional MX gene required for the immediate early immune response against viral infections. Comparing different inbred mice for a particular biological feature or response may be a good starting point to search for alleles of relevant genes, even without any mutagenesis approach. It is sometimes even worth comparing the same inbred line after a long time of separate breeding by two independent breeders.

Spontaneous Mutations causing phenotypic changes occur during the normal housing and breeding of mice. Since inbred mouse lines are maintained by intercrossing siblings there is a good chance of identifying recessive mutations in such mouse colonies. Thus, interesting mouse mutants appear in a way automatically all the time, although with a much lower frequency compared with the ENU mutants. In contrast to the point mutations introduced by the chemical agents, the types of spontaneous mutations are not predictable. For example point mutations, chromosomal aberrations or retroviral insertions have been observed in such cases. Like with ENU mutants the key issue is to identify the mutants while keeping the mice. Coat colour genes or gross skeletal defects are easily seen and the mutants can be identified. Other, more subtle phenotypic changes may only be noticed by chance. In addition, the animals in a normal mouse-breeding house are only kept for a limited time and mutations which only manifest at older ages, will remain unnoticed. Nevertheless, with a breeding colony of about 10,000 mice one could on the average pick up one mutation/year.

Transgenic mice are generated by injecting foreign DNA into the nucleus of fertilised eggs. The DNA integrates randomly and may influence other genes in the vicinity of the integration site. If one tries to make transgenic mice homozygous for the transgene, one will notice that about 10% of the transgenic lines cannot be bred to homozygosity, indicating that the inserted DNA has hit a vital region of the genome that contains genes involved in embryogenesis. By identifying the integration point and by a close follow up of the embryos one may learn about the function of the target gene. In addition, unexpected phenotypes only observed in adult animals homozygous for the transgene, may give a hint to a recessive mutation caused by the integration of the foreign DNA.

The generation of informative mutations is not limited to the mouse germline. Interesting target genes can also be identified by mutagenesis of adult cells. For example, proto-oncogenes have been identified by retroviral infection of haematopoietic stem cells and by the subsequent identification of the retrovirus insertion point in tumours that developed from the infected cells in living mice.

Thus retroviral infection procedures may be considered an interesting alternative to the other methods currently used for random mutagenesis. Since hereby foreign genetic material is introduced into the mouse genome, the identification of the mutated locus should be much easier compared with ENU mutants (Table 5.1). Also, in comparison with gene trap mutants, this method seems to be advantageous since it is not limited to the identification of genes expressed in ES cells.

The Gedanken-experiment

The following discussion tries to roughly calculate the resources needed to generate a medium size number of mouse mutants in a given time period. At the end of the experiment, the mutants should fulfil the following criteria: the mutation in the germline is identified, the initial phenotype is characterised and the mutation is verified in a second independent mutant.

The system, which requires the smallest number of resources is the gene trap mutagenesis approach, since the vector needed to introduce the mutations has to be constructed only once. The target gene is however random and the nature of the mutation not 100% predictable. The cost of 100 Mutants in 5 Years (including the generation of a second independent mutant by the same method) is set to 1.

The production of a simple gene targeted mutant requires more effort and resources than a gene trap mutant. An individual targeting vector has to be constructed and the homologous targeting event has to

TABLE 5.1: Comparison of resources required for targeted mutation and for random methods

Method	Time to generate (months)	Time to identify (months)	Time to analyse (months)	Time to verify	Number of mice	Overall time
Gene targeting	6	6	6	Part of the generation	100	1.5 year
Conditional gene targeting	12	6	12	Part of the generation	400	2.5 years
Gene Trap	3	1	6	Unknown	100/mutant	10 months
ENU	3	6–12	24	Unknown	At least 500/mutant	2–4 years

be identified in a series of transfected ES cell clones. However, usually more than one homologously recombined clone is identified and a second independent mouse line can be generated within the same experiment. Compared with gene trap mutants, the amount of additional time and effort increases the cost per mutant by about a factor of 2. The advantage compared with the gene trap approach is based on the fact, that the mutation is already clearly defined at the stage of the construction of the targeting vector.

The generation of a conditional mouse mutant is again more demanding compared with the simple targeted mutation. However, this approach allows the introduction of much more sophisticated mutations and in addition offers the option of limiting the gene inactivation to somatic cells. The additional resources for this approach, assuming the use of a one cell type specific recombinase transgenic mouse line, are roughly again a factor of 2. The nice thing about the conditional approach is that the control is built in the mutation itself, because the initial mutation just carries the recombination signals for the respective recombinase and leaves the targeted allele functionally unaffected. Thus, only after recombination the phenotype linked to the mutation should appear. Once the conditional mutation is generated, the phenotypic analysis can be continued almost limitlessly by the use of different recombinase transgenic mouse lines with other specificities. In cases where a severe phenotype is observed in gene trap mutants or complete knock outs, conditional mutants offer the opportunity for identifying subtle phenotypes in certain tissues that would remain hidden in animals in which the respective target gene is completely inactivated. It should be noted that many of such experiments could be conducted with recombinase

transgenic mice already available. Therefore costs for the generation of such animals can be omitted.

The generation of ENU Mutation appears to be quite cheap at the first glance. As the screening is based on the phenotype of the mutant, it may require expensive screening of negative mice until the identification of a desired mutant, especially if one aims to identify recessive mutations. In addition, during the breeding needed to pinpoint the mutation in the genome, all the offspring have to be checked for the respective phenotype. Thus the molecular characterisation of the mutation requires huge resources in terms of breeding, phenotyping and sequencing to identify the causative mutation. As the mutations in the ENU screen are generated randomly, each mutant is unique and the phenotype cannot be easily verified by another independent ENU mutant. Therefore other methods like the analysis of gene trap mutants of the ENU mutated gene or the targeting mutation approach must complement the ENU mutagenesis screen. As the mutant is unique, the strain must be cryopreserved at an early stage. Of course this is also advisory for mutants generated by gene trapping or targeted mutation. However, ES cells altered by either method can be stored frozen, thus allowing regeneration of the respective mouse line in cases where the allele was lost during breeding. With targeting approaches the same mutation might even be introduced a second time if necessary.

On the other hand, ENU mutagenesis centres have been established together with large phenotyping platforms, which allow the screening of mouse mutants for a particular phenotype. To achieve a particular phenotyping by sequence driven approaches is nearly impossible. Also, the combination of sequencing of gene segments in sperm of mutant mice allows the selection of allelic series of mutations in a particular gene. This approach would reduce the screening costs dramatically and the genotype of the mutation becomes known very early with this procedure. The phenotype however, remains uncertain. We assume that the resources required for one ENU mutant are a factor of 8 higher compared with gene trap mutants. These costs do not include the verification by an independent method.

In order to describe the resources required to find a point mutation introduced into the mouse germline, one publication, describing the identification of a point mutation in an intermediate mutant of the *Hyplip1* gene, will be discussed. The gene has a size of 160 kb, the mutation that leads to hyperlipidaemia in mice. The mutation was a spontaneous mutation, therefore the chance of several mutations in a certain region is minimal. The first screen to map the gene required the analysis of 259 wildtype, 489 heterozygous and 214 homozygous mutant mice. The fine mapping, done in a recombinant congenic strain to the evolutionarily distant strain CAST/Ei required a further 183 F2 mice. Finally, after screening 2700 F2 mice, 184 mutants were identified which showed a crossing-over between two linkage markers flanking the candidate gene. By this, the number of candidate genes could be reduced to 13 genes. By gene expression analysis, the difference was linked to the Txnip gene. After sequencing 175 kb twice, the point mutation was identified in Exon 2 of the Tcnip gene, leading to the introduction of a stop codon.

Taken together, the ranking in terms of resources is shown in Table 5.2.

TABLE 5.2: Ranking of resources

Type of mutation	Resources required for one mutant
Gene trap mutant	1
Gene targeting	1.5
Conditional gene targeting (complete knockout)	2
Conditional gene targeting	4
ENU mutant	8
Forgotten mutant	8

Notes: Or in other words, for one ENU Mutant one could generate two conditional mouse mutants, four gene targeted mutations or eight gene trap mutants.

Conclusion

To get an impression of the relation of the resources needed for sequence driven and random mutagensis approaches and the scientific yield of the respective methods, it is worthwhile to roughly survey the literature of the last 15 years. The first mouse mutant generated by gene targeting was published in 1988. It took until 1991 for the number of published mutants to increase significantly. Since 1991 the numbers of new mouse mutants published per year is growing exponentially. It should be noted that not only does the generation of new mutants continue, but also the data generated with such mutants have a very strong impact on the formulation of scientific hypotheses. Almost every model to date has been either found or verified by the analysis of mutants, preferably in targeted mouse mutants (Mak *et al.*, 2001). In the field it is very important to follow the description of mutants for a longer time as more and more information gets accumulated and more than one mutant is generated for a particular gene. To exemplify, initially two groups inactivated the prion gene. One group found a neurological phenotype, while another group could not verify this finding. It became clear that the group, which did not describe a phenotype of the prion gene, is correct, as it turned out that the other group unwillingly inactivated a neighbouring gene, the doppel gene, by their targeting vector. Based on such experience obtained by targeted mutations, it becomes questionable if researchers will be able to correlate a single gene function with a point mutation. The ENU mutagenesis approach will therefore miss many genes. On the other hand, one could argue that the ENU mutagenesis approach will only reveal the 'relevant' genes.

The first conditional mouse mutant based on a biological questions was published in 1998 (Betz *et al.*, 1998). Soon the power of this technology was proven in the many other publications which followed. Time will tell whether the increased resources for the conditional targeting method will pay off. It is however quite clear that certain biological question can only be answered by the conditional gene targeting approach.

The ENU mutagenesis approach was evangelised in 2000 in a special issue of the journal *Mammalian Genome* (Vol 11(7), 2000). The different ENU screens were presented in overview articles. One of the mutants from this mutagenesis approach is the Beethoven mouse, a mutant already discussed above (Vreugde *et al.*, 2002). The time from the start of the ENU mutagenesis programme until today may be too short to demonstrate the efficiency of the method. However, one other series of ENU mutants shows the very selective power of the method. In this screen mouse mutants were analysed for the presence of cataracts in the eye of the mouse. The turbidity in the lenses was visible very early on in the mutants. The screening for this particular phenotype was therefore straightforward. In the Munich screen, after analysing 456,890 individual mice, 203 dominant mutants were recovered and many mutations could be mapped to the crystalline genes (Favor and Newhauser-Klaus, 2000). This approach was therefore very useful for the generation of an allelic series of mutants in the crystalline gene.

References

Beier, D.R. (2000). *Mamm. Genome* **11**, 594–597.

Betz, U.A., Bloch, W., van den Broek, M., Yoshida, K., Taga, T., Kishimoto, T., Addicks, K., Rajewsky, K. and Müller, W. (1998). *J. Exp. Med.* **188**, 1955–1965.

Bradshaw, A.D. and Sage, E.H. (2001). *J. Clin. Invest.* **107**, 1049–1054.

Chen, Y., Yee, D., Dains, K., Chatterjee, A., Cavalcoli, J., Schneider, E., Om, J., Woychik, R.P. and Magnuson, T. (2000). *Nat. Genet.* **24**, 314–317.

Favor, J. and Neuhauser-Klaus, A. (2000). *Mamm. Genome* **11**, 520–525.

Kurima, K., Peters, L.M., Yang, Y., Riazuddin, S., Ahmed, Z.M., Naz, S., Arnaud, D., Drury, S., Mo, J., Makishima, T., Ghosh, M., Menon, P.S., Deshmukh, D., Oddoux, C., Ostrer, H., Khan, S., Deininger, P.L., Hampton, L.L., Sullivan, S.L., Battey, J.F., Jr., Keats, B.J., Wilcox, E.R., Friedman, T.B. and Griffith, A.J. (2002). *Nat. Genet.* **30**, 277–284.

Mak, T.W., Penninger, J.M. and Ohashi, P.S. (2001). *Nat. Rev. Immunol.* **1**, 11–19.

Mammalian Genome (2000). Special issue: Chemical mutagenesis in mice. *Mamm. Genome* **11(7)**, 471–607.

Nolan, P.M., Peters, J., Strivens, M., Rogers, D., Hagan, J., Spurr, N., Gray, I.C., Vizor, L., Brooker, D., Whitehill, E., Washbourne, R., Hough, T., Greenaway, S., Hewitt, M., Liu, X., McCormack, S., Pickford, K., Selley, R., Wells, C., Tymowska-Lalanne, Z., Roby, P., Glenister, P., Thornton, C., Thaung, C., Stevenson, J.A., Arkell, R., Mburu, P., Hardisty, R., Kiernan, A., Erven, A., Steel, K.P., Voegeling, S., Guenet, J.L., Nickols, C., Sadri, R., Nasse, M., Isaacs, A., Davies, K., Browne, M., Fisher, E.M., Martin, J., Rastan, S., Brown, S.D. and Hunter, J. (2000a). *Nat. Genet.* **25**, 440–443.

Nolan, P.M., Peters, J., Vizor, L., Strivens, M., Washbourne, R., Hough, T., Wells, C., Glenister, P., Thornton, C., Martin, J., Fisher, E., Rogers, D., Hagan, J., Reavill, C., Gray, I., Wood, J., Spurr, N., Browne, M., Rastan, S., Hunter, J. and Brown, S.D. (2000b). *Mamm. Genome.* **11**, 500–506.

Rajewsky, K., Gu, H., Kühn, R., Betz, U.A., Müller, W., Roes, J. and Schwenk F. (1996). *J. Clin. Invest.* **98**, 600–603.

Soewarto, D., Fella, C., Teubner, A., Rathkolb, B., Pargent, W., Heffner, S., Marschall, S., Wolf, E., Balling, R. and Hrabé de Angelis, M. (2000). *Mamm. Genome* **11**, 507–510.

Vreugde, S., Erven, A., Kros, C.J., Marcotti, W., Fuchs, H., Kurima, K., Wilcox, E.R., Friedman, T.B., Griffith, A.J., Balling, R., Hrabé de Angelis, M., Avraham, K.B. and Steel, K.P. (2002). *Nat. Genet.* **30**, 257–258.

Waterston, R.H., Lindblad-Toh, K., Birney, E., Rogers, J., Abril. J.F., *et al.* (2002). *Nature* **420**, 520–562.

Wiles, M.V., Vauti, F., Otte, J., Fuchtbauer, E.M., Ruiz, P., Fuchtbauer, A., Arnold, H.H., Lehrach, H., Metz, T., von Melchner, H. and Wurst, W. (2000). *Nat. Genet.* **24**, 13–14.

Suggested reading

Clarke, Alan R. (ed.) (2002). *Transgenesis Techniques. Principles and Protocols (Methods in Molecular Biology*, 2nd edn, Vol. 180). Humana Press, ISBN 0896036960.

Crawley, Jaqueline N. (2000). *What's Wrong with my Mouse?: Behavioral Phenotyping of Transgenic and Knockout Mice*, 1st edn. Wiley-Liss, ISBN 0471316393.

Hofker, Marten and van Deursen, Jan. (eds) (2002). *Transgenic Mouse Methods and Protocols (Methods in Molecular Biology*, Vol. 209). Humana Press, ISBN 0896039153.

Hogan, Brigid, Beddington, Rosa, Costantini, Frank and Lacy, Elizabeth. (eds) (1994). *Manipulating the Mouse Embryo: A Laboratory Manual*, 2nd edn. Cold Spring Harbour Laboratory, ISBN 0879693843.

Houdebine, Louis-Marie. (2003). *Animal Transgenesis and Cloning*, 1st edn. John Wiley & Sons, Inc., ISBN 0-470-84827-8.

Jackson, Ian J. and Abbott, Catherine M. (eds) (2000). *Mouse Genetics and Transgenics: A Practical Approach*, 1st edn. Oxford University Press, ISBN 0199637083.

Joyner, Alexandra L. (ed.) (2000). *Gene Targeting: A Practical Approach*, 2nd edn. Oxford University Press, ISBN 019963792X.

Kaufmann, Matthew H. (1992). *Atlas of Mouse Development*, 1st edn. Academic Press, ISBN 0124020356.

Kaufmann, Matthew H. and Bard, Jonathan B.L. (1999). *The Anatomical Basis of Mouse Development*, 1st edn. Academic Press, ISBN 0124020607.

Kmiec, Eric B. and Gruenert, Dieter C. (eds) (2000). *Gene Targeting Protocols* (*Methods in Molecular Biology*, 1st edn, Vol. 133). Humana Press, ISBN 0896033600.

Mak, Tak W., Penninger, Josef, Roder, John, Rossant, Janet and Saunders, Mary. (eds) (1998). *The Gene Knockout Factsbook (2-volume set)*. Academic Press, ISBN 0124660444.

Sukov, Mark A., Danneman, Peggy and Brayton, Cory. (2000). *Laboratory Mouse*, 1st edn. Lewis Publishers, Inc., ISBN 0849303222.

Torres, Raul M. and Kühn, Ralf. (eds) (1997). *Laboratory Protocols for Conditional Gene Targeting*, 1st edn. Oxford University Press, ISBN 019963677X.

Tymms, Martin J. and Kola, Ismail. (eds) (2000). *Gene Knockout Protocols* (*Methods in Molecular Biology*, 1st edn, Vol. 158). Humana Press, ISBN 0896035727.

Ward, Jerrold Michael. (ed.), Mahler, Joel, Maronpot, Robert R. and Sundberg, John P. (2000). *Pathology of Genetically-Engineered Mice*, 1st edn. Iowa State University Press, ISBN 0813825210.

Mouse genome related Internet resources

EnsEMBL. http://www.ensebl.org/

Genexpression Omnibus. http://www.ncbi.nlm.nih.gov/geo/

TBASE. http://tbase.org/

Mouse knockout and mutant database. http://research.bmn.com/mkmd/

Cre transgenic mouse lines. http://www.mshri.on.ca/nagy/cre.htm

EMMA. http://www.emma.rm.cnr.it/

Jackon Lab. http://www.informatics.jax.org/

The Mouse as an Animal Model for Human Diseases

Louis-Marie Houdebine

Biologie du Développement et Reproduction,
Cellulaire et Moleculaire, Institut National de la
Recherche Agronomique, Jouy-en-Josas, France

Introduction

For decades, animals have been used as models to study human diseases. Essentially small laboratory animals and more particularly mice have been taken as a reference. This is clearly due to the fact that the mouse is a small mammal which is highly prolific and relatively cheap to breed. This small size is sometimes a limitation for experimenters. On the other hand, the mouse is one species among others and some of its biological functions are not so close to their counterparts in human. Other species such as the rat or rabbit are therefore preferred and used whenever possible.

Mice are extensively used to study reproduction and embryo development. These reasons converge and help to explain why the mouse is the species of choice to generate **transgenic animals**. The first animals harboring a functional transgene having a phenotypic effect were mice showing enhanced growth (Palmiter *et al.*, 1982). These pioneer experiments revealed that, although laborious, foreign DNA transfer by microinjection into pronuclei was possible. About ten years later (Capecchi, 1989) it was shown that precise gene replacement by homologous recombination was possible in the mouse genome using transformed embryonic stem (ES) cells to generate mutated embryos.

These two techniques are extensively used. Innumerable gene constructs have been transferred to mice and it is acknowledged that almost 5000 genes have been knocked out in this species. This experimental approach is expected to be still more extensively followed in the future. Indeed, the conventional methods for cloning genes coding for known proteins have been followed by the use of genetic markers, namely of **microsatellites**, to tentatively identify unknown genes responsible for human diseases. The complete sequencing of several genomes, including human and mouse genomes, is almost complete and will provide experimenters with a huge number of genes to be studied. Transgenesis is certainly going to have a central position in deciphering the mechanisms which control gene expression, and in defining the role of a number of newly identified genes in normal cell differentiation. Transgenesis has already proved to be a unique tool for studying a number of human diseases. The generation

The Laboratory Mouse
Copyright 2004 Elsevier
ISBN 0-1233-6425-6

of mutated mice by gene transfer is complementary to the conventional selection and breeding of spontaneously mutated mice or to the systematic induction of gene mutation by chemical compounds such as ethylnitrosourea (ENU; Brown and Balling, 2001; this volume).

Transgenesis in the mouse is also complementary to similar experiments carried in several other model species and even of invertebrates such as *Caenorhabditis elegans* or *Drosophila* (Bernards and Hariharan, 2001). Indeed, in these two latter species, the whole genome has been sequenced. Many genes show high structural and biological homology to human ones. On the other hand, gene addition and specific gene extinction by knockout or RNA interference (RNAi) can be performed in these two species.

The cloning technique by nuclear transfer has recently opened new avenues to generate transgenic mammals other than mice, including gene addition and gene knockout (Readout *et al.*, 2000). This will complete experiments carried out in the mouse.

The recent demonstration that human pluripotent embryonic cells can be cultured and differentiated into many cell types *in vitro* has provided attractive new clues for cell therapy. Within just a few years, animal cloning, transgenesis, cell therapy and gene therapy have become closely interconnected.

Although important progress has been made to add and knock out genes and to improve control of transgene expression, several technical problems remain to be solved and further improvements of the currently used methods would be welcome. The present chapter is an overview of the recent progress which have been made to generate transgenic animals and to control gene expression. Examples of transgenic mouse models obtained to study various types of human diseases are depicted in the last part of the chapter.

Tools used to generate transgenic mice and to express transgenes

Problems still remaining to be solved for creating animal models for the study of human diseases or more generally of biological functions are numerous. In some cases,

foreign genetic information is added to the animals. In other cases an endogenous gene is replaced by a mutant. An important issue is to block the expression of the endogenous gene. This can be achieved by interfering at different levels of gene expression. On the other hand, ideally the expression of the transgene should occur in tissues and at the stages decided by experimenters.

Recent technical advances have made the creation of relevant mouse models easier.

Methods for adding and replacing genes

The protocol to microinject DNA into embryo pronuclei has not essentially been modified since the initial success (Gordon *et al.*, 1980; Palmiter *et al.*, 1982). The technique proved to be efficient not only for common short DNA fragments (up to 20 kb) but also for chromosome fragments cloned in P1 phage, bacterial artificial chromosome (BAC) or yeast artificial chromosome (YAC) (Giraldo and Montoliu, 2001).

In order to enhance the efficiency of gene transfer or to make it easier, several alternatives to DNA microinjection have been proposed.

Lentiviral vectors have been designed to transfer up to 8 kb into a mouse single cell embryo efficiently (Lois *et al.*, 2002). The use of such vectors raises some biosafety problems but the protocol has been standardized and this may appear more attractive than classical DNA microinjection.

The Sleeping Beauty retrotransposon is also an efficient tool for transfering foreign genes into mouse embryos and a possible alternative to naked DNA microinjection (Dupuy *et al.*, 2002).

Foreign DNA has been transferred to mouse embryos via sperm using different approaches. Retroviral vectors were able to transfer genes into sperm precursors in the testis, leading to the generation of transgenic mice (Nagano *et al.*, 2001). Sperm membrane alteration by a detergent followed by incubation with naked DNA and *in vitro* fertilization using intracytoplasmic sperm injection (ICSI) is another way to transfer foreign DNA via sperm. The most attractive method in this field seems to be the use of a complex formed by a monoclonal antibody bound to foreign DNA and recognizing a specific sperm antigen. The sperm kept its capacity to fertilize mouse oocytes *in vitro* and generated a large number of transgenics (Qian *et al.*, 2002).

Episomal vectors capable of replicating autonomously would theoretically be highly efficient in the generation of transgenic mice and would also avoid the influence of chromatin currently observed on the expression of integrated foreign genes (Co *et al.*, 2000; Bok Hee *et al.*, 2002).

Gene replacement by homologous recombination can be achieved using different vectors designed to specifically knock out or knock in endogenous genes (Viville *et al.*, 1997; this volume). This laborious but potent technique can still be implemented, for unclear reasons, with just the ES cell from only two mouse lines. In no other species have the ES-like cells the capacity to form chimeric embryos with a subsequent transmission of the mutation to progeny.

The frequency of homologous recombination is naturally low. It can be markedly enhanced by cleaving genomic DNA using restriction enzymes such as Isce 1 which recognize the corresponding site previously added to the genome (Cohen-Tannoudji *et al.*, 1998).

In a number of cases, gene knockout is lethal during embryo development preventing the creation of a model. A way to avoid this problem consists of triggering the gene knockout in a restricted number of cell types and only at a chosen stage of development or differentiation. To reach this goal, the DNA fragment to be withdrawn to knock the gene out is previously bordered by two LoxP sequences. Their recombination eliminating the DNA fragment is induced by the Cre recombinase which is driven by a promoter active only in the chosen cell types. This conditional knockout is used more and more (Nebert and Duffy, 1997).

A recent study has shown that ES cells are unstable and this seems due to a quite variable epigenetic state of the cells. A number of genes essential for development are more or less inactivated by a methylation process. This might explain why functional ES cell lines are so rare and why the established lines sometimes have an unpredictable capacity to generate chimera (Humphreys *et al.*, 2001).

The use of tetraploid embryos as recipients for genetically modified ES cells gives rise to the generation of non-mosaic animals. Indeed, the tetraploid cells can help the ES cells to divide but not participate in the development of the embryo. This saves time and contributes to reduce the number of experimental animals (Moore, 2001).

Animal cloning by nuclear transfer has proved to be an attractive alternative to microinjection for gene addition (Schnieke *et al.*, 1997) and the only way in which to replace genes by homologous recombination in ruminants (McCreath *et al.*, 2000) and other large animals. However, cloning remains too hard a task, even in mice, to become an alternative to microinjection for gene addition and to the use of ES cells for gene replacement in the near future.

Methods for targetting the integration of foreign genes

Most transgenes are sensitive to the position effect. Transgenes are thus either extinguished by the formation of heterochromatin (see section on General Rules for the Design of Vectors for Transgene Expression) or unduly stimulated by neighbor enhancers. This leads to the generation of numerous useless animals and in some cases to the establishment of lines having unexpected, unique and quite interesting biological properties.

To reduce this wastage, appropriate sites of integration may be defined. In these conditions, the position effect does not exist or, at least, is reproducible in all the transgenic mice.

Homologous recombination depicted above is a good but laborious way to reach this goal. Alternative methods have been defined. They rely essentially on the implementation of the Cre–LoxP system (Nagy, 2000). A 34 bp sequence from P1 phage is specifically and quite efficiently recognized by the Cre recombinase from the same phage. This property is used to eliminate DNA fragments previously bordered by two LoxP sequences. This may lead to an inactivation of the gene or to an activation of the gene if the deleted fragment was a transcription inhibitor.

A LoxP site introduced into a genome by microinjection or homologous recombination may recombine with another LoxP sequence added in the gene construct. In practice, the transgene embryo harboring the integrated LoxP is microinjected with the LoxP construct and with a circular plasmid expressing the Cre recombinase.

The efficiency of this method is limited by the fact that the two LoxP sites present in the genome after the integration of the construct can easily recombine and eliminate the insert. A selection of the ES cells or embryos in which the insert was not ejected then has to be made.

Several methods greatly reduce this unwanted elimination of the insert. Two different mutants of LoxP can be added, one in the mouse genome and the other in the construct. These mutants allow an efficient integration of the construct by the Cre recombinase but not its elimination (Araki *et al.*, 1997). A wild and the LoxP 511 mutant can be added on both sides of the gene construct and in the mouse genome. This double Lox recombination system made it possibly an irreversible targeting of the foreign gene (Sonkharev *et al.*, 1999). Another more efficient approach also implies a double recombination but with wild LoxP sequences added in inverted orientation and a selection gene (Feng *et al.*, 1999). These methods known as recombinase-mediated cassette exchange (RMCE) are being used to integrate various mutant constructs in a chosen site (Schübeler *et al.*, 2000; Baer and Bode, 2001).

The expression of Cre recombinase specifically in male germ cells by a cell specific promoter (Sage *et al.*, 1998) proved to be highly efficient in recombining two LoxP located in two different chromosomes. The recombination then takes place during meiosis and this might explain its high efficiency. This method called transallelic targeted meiotic recombination (TAMERE) seems the best currently available for the engineering of chromosomes (Herault *et al.*, 1998). Interestingly, the expression of the Cre recombinase gene in these conditions allows an exceptionally high frequency of targeted integration at a given LoxP site in the mouse genome.

Interchromosomal recombination can also be mediated by the Cre–LoxP system leading to relevant biological models (Hoike *et al.*, 2002).

The choice of the chromosomal site where the LoxP was initially integrated is of paramount importance for the expression of the targeted transgenes. The LoxP sequence may be added randomly by microinjection. The appropriate sites must then be identified and retained, or not, after the targeted integration of a foreign reporter gene. Alternatively, the LoxP site may be inserted at a chosen site by homologous recombination. Another simpler and potentially efficient way may consist of randomly inserting a LoxP sequence surrounded by insulators allowing a reproducible expression of the targeted transgenes irrespective of the chromosomal site where the LoxP sequence was integrated.

Although the LoxP sequence is not present in mammalian genomes, the Cre recombinase at high concentrations is cytotoxic. This has been attributed to the fact that the enzyme recognizes degenerated LoxP sequences in the genome. This generates DNA cleavage and possibly recombination. The use of Cre recombinase, whose activity is controlled by estrogen analogs, limits the enzyme activity at a level which is sufficient to recombine LoxP but insufficient to induce genome alteration (Loonstia *et al.*, 2001).

Methods for conditional transgene expression

Expressing a transgene only in a given tissue and at precise periods cannot be achieved easily. The promoters generally added in gene constructs drive a more or less leaky expression of the transgenes. On the other hand,

natural inducers stimulate not only the transgene but also several host genes. This sometimes reduces considerably the value of a model for studying human diseases.

A specific control of the transgenes can be achieved by using exogenous inducers which are recognized only by the promoter engineered for that purpose and added before the gene of interest. Several systems are currently available. The system based on the use of tetracycline and analogs is presently the most popular. In its original version, this system (and most of the others) suffered from a fundamental limitation. The background expression of the transgene is substantial even in the absence of the inducers. The simultaneous use of a transcription inhibitor present on the promoter only, in the absence of the inducers, and of a transcription stimulator present only after addition of the inducers reduces quite significantly the background level of expression (Rossi *et al.*, 1998; Blau and Rossi, 1999; Forster *et al.*, 1999; Freundlieb *et al.*, 1999). The other systems are based on the use of various inducers such as rapamycin (Rivera *et al.*, 1996), ecdysone (No *et al.*, 1996), RU 486 (Wang *et al.*, 1997) and streptogramin (Fussenegger *et al.*, 2000). All these systems rely on minimum promoters enhanced by transcription factors activated by specific inducers. Another approach consists of engineering a promoter containing repressor elements. Lac Z repressor binding sequences have been added in the tyrosinase promoter. The hybrid promoter becomes tightly controlled by the inducers isopropyl beta-D-thiogalactopyranoside (IPTG) (Cronin *et al.*, 2001). The advantage of this system is that the gene of interest is highly inducible with no background expression in the absence of the inducers. Such a system can theoretically be extended to other promoters. However, it may be difficult to engineer a hybrid promoter being both functional and inducible.

Methods for the inhibition of endogenous gene expression

Inactivating an endogenous gene is as important as adding exogenous genetic information in order to create relevant models for the study of human diseases. The most frequently used method to inactivate a gene is to knock it out. Although efficient, this approach is laborious and it has the major drawback of being irreversible. In the best conditions, gene knockout can be induced in a given tissue at a precise period of the animal's life by using the Cre–LoxP system and a tissue specific promoter. Engineered Cre recombinase activated by steroids can contribute to the induction of gene knockout in a conditional way (Wunderlich et al., 2001). The Flp–FRT system may be used as well, especially in its improved version (Umana et al., 2001). This protocol has met with success in some cases. Yet, it remains complex and it does not always offer all the flexibility required to create a relevant model.

A more attractive theoretical possibility consists of using transgenes whose products (RNA or proteins) inhibit a given endogenous gene. Several systems have given satisfactory results in some cases but they must all still be explored before being routinely used in transgenesis.

Formation of triple helix with DNA or RNA

Pyrimidine rich in DNA or RNA can form a triple helix with purine rich sequences in a precise orientation. A number of studies have shown that oligonucleotides can enter cells and form a triple helix with DNA. This can prevent gene transcription if the triple helix is in a CpG rich sequence of the promoter or within the transcribed region (Giovannangeli and Helene, 2000). Triple helices can be formed by an mRNA and a complementary double strand RNA (Upegui-Gonzales et al., 2000). It is therefore conceivable that a given endogenous gene at the DNA or RNA level could be inhibited by expressing a single-strand or double-strand RNA respectively from a transgene. This method has met with success in cultured cells (Upegui-Gonzales et al., 2000) but has not yet been successfully used in transgenic animals.

Use of single-strand antisense RNA

RNAs complementary to mRNAs are used in nature to block their translation or to destabilize them. The overexpression of an antisense RNA coded by a transgene specifically recognizing an mRNA can significantly reduce the synthesis of the corresponding protein (L'Huillier et al., 1996). In practice, it appears that the method is not very efficient. This is probably due to the fact that the targeted mRNA and the antisense RNA are both structured and associated with proteins which considerably reduces the chance for the two partners to meet. A systematic search of the single-strand regions in a given mRNA may define which antisense RNAs have the best chance of efficiently targeting the mRNA.

Use of double-strand RNAi

In plants but also in lower vertebrates and mammals, a double-stranded RNA can trigger a specific and very potent destruction of the complementary RNA. This mechanism which is not fully understood has been called RNAi. In *Caenorhabditis elegans* (Zipperlen et al., 2001) as well as in *Drosophila*, the expression of double-stranded RNAs from transgenes is being used systematically to inhibit endogenous gene expression.

It is acknowledged that the RNAi mechanism is triggered when double-stranded RNAs of at least 300 bp are present in a cell. The double-stranded RNAs are degraded in 21–23 bp fragments which specifically degrade RNAs having the same sequence. Long double-stranded RNAs are known to induce interferon and cell death in higher vertebrates. Expression of foreign genes coding for long double-stranded RNA in mammalian cells led to confusing results. The RNAi was clearly observed when short double-strand RNAs with an overhanging 3′ end were chemically synthesized (Elbashir et al., 2001) or obtained by transcription with in vitro T7 polymerase (Yu et al., 2001; Paddison et al., 2002).

Quite interestingly, short double-stranded RNAs synthesized by templates under the control of RNA polymerase III promoters showed quite efficient RNAi effects in various mammalian cells (Tuschl, 2002). This observation should be extrapolated to transgenic mice. This method could replace or complement the classical knockout to inhibit the expression of a gene in mice.

Use of ribozymes

Ribozymes are antisense RNAs which have a catalytic site capable of specifically cleaving complementary RNAs. The efficiency of ribozymes is strongly limited by the fact (mentioned in the section on use of Single-Stranded Antisense RNA) that the catalytic RNA only has a small chance of meeting its target. Quite interestingly, a recent study showed that a hybrid molecule containing a **helicase** and a ribozyme can reach its target RNA much more easily. The helicase acts by rendering the targeted region accessible to the ribozyme (Warashina et al., 2001; Kawasaki and Taira, 2002).

Use of transdominant negative proteins

Transdominant negative proteins are mutated proteins which have lost one of their essential biological properties. The overexpressed transdominant negative proteins act as competitive inhibitors of the normal proteins. The efficiency of this approach is limited only by the availability of the proteins having a transdominant effect.

Use of intrabodies

Natural antibodies are normally secreted but they can be engineered to remain in the intracellular space. Such molecules are known as intrabodies. An intrabody recognizing a given cellular protein can block its biological activity (Jones and Marasco, 1997). To be fully efficient, intrabodies must be quite specific, to be present at a relatively high concentration in the cells and sometimes targeted to a specific cell compartment.

All these systems designed to inhibit endogenous gene expression can potentially be used in transgenic animals. Their flexibility may be high, especially if the promoters used to direct transgene expression are tissue specific or activated by exogenous inducers.

Identification of genes by gene trap

The systematic identification of genes by genome sequencing and the establishment of gene expression pattern by DNA chips may be insufficient to determine the role of a gene in a given biological function. Gene trapping is currently being used to identify genes of interest. For this purpose, vectors containing a marker gene, an intron splicing site and a transcription terminator, but devoid of promoter, are introduced by **electroporation** in ES cells. The cells are then used to generate chimeric mice. The expression of the marker gene (often a hybrid containing β-galactosidase and neomycin resistant genes) indicates that the vector was randomly integrated within a gene. This gene can easily be identified. Interesting models can be obtained in this way if a correlation can be established between the expression of the gene trap vector in a given tissue and particular health problems in the animals.

A series of gene trap vectors have been designed to identify genes expressed in ES cells or in differentiated tissues, coding for secreted proteins or activated by specific inducers such as hormones or growth factors (Cecconi and Meyer, 2000; Jackson, 2001; Medico et al., 2001; Mitchell et al., 2001). Banks of ES cells in which gene trap vectors have been introduced are available (Goodwin et al., 2001).

General rules for the design of vectors for transgene expression

A gene is composed of multiple signals which interact in a subtle manner to control its expression. Engineered genes have often lost some of their signals and gained others. The gene constructs may thus be far from optimal to direct the expression of the transgene.

A few rules have progressively emerged and they may help experimenters to avoid leaking but mainly silencing of the transgenes. This rules have been depicted in more detail elsewhere (Houdebine et al., 2002).

A transgene may be badly expressed if it contains cDNA rather than genomic DNA, if it is integrated as multicopies in tandem, if it is too rich in CpG motives and if it is integrated in inactive chromatin regions (centromere and telomere). A gene construct should therefore ideally contain genomic DNA or at least one intron added before the cDNA. Its copy number may be reduced by the Cre–LoxP system (Whitelaw and Martin, 2001). The addition of long stretch of CpG rich DNA must be avoided. Other sequences or mutated AT rich sequences should be used whenever possible. Gene construct must be used surrounded by long genomic DNA fragments containing insulators or by identified insulators (Taboit-Dameron et al., 1999) which prevent the formation of inactive heterochromatin induced by the transgene itself (West et al., 2002).

Examples of transgenic models for human diseases

The transgenic models to study human diseases are very numerous and it is impossible to depict all of them. It is therefore of interest to summarize those studies which have been carried out in order to understand diseases which are particularly complex or generally which could not have been studied with the single cell systems or natural mouse mutants.

Models for genetic diseases

At least 6000 human diseases have been shown to be of genetic origin. In some cases, a clear gene mutation is responsible for the disease. This is true for cystic fibrosis which is triggered when the CFTR gene is mutated. However, several hundreds of mutations have been found in this gene, but only some of them cause a severe pathogenicity. On the other hand, it is known that the most potent mutations, namely the ΔF 508 mutation, have not the same effect in all patients. Clearly, the cooperation with other genes may enhance or attenuate the effect of the mutation. Transgene mice harboring the ΔF 508 mutation in their CFTR gene have been independently generated by several groups. In no single case, did the models appear fully relevant. CFTR is a chloride channel which is less active and not properly transported to plasma membrane in mutants. Mice have other chloride channels and their lungs are only weakly affected by the CFTR gene knockout. Yet, these animals show intestinal disorders typical of cystic fibrosis. Rabbit and sheep are expected to be better models for cystic fibrosis since their CFTR genes are more similar to the human CFTR gene. Experiments are underway to generate rabbits and sheep with the mutated CFTR gene, but the major hurdle is that the cloning technique by nuclear transfer must be implemented.

In many cases, a mutated gene increases the probability of a disease developing but is insufficient in itself. This situation is seen with the BRCA 1 and BRCA 2 genes in human breast cancer. Several transgenes must then be expressed simultaneously to generate the pathology. The cooperative effect of two genes is sometimes quite unexpected. The transgenic mice used to study amyotrophic lateral sclerosis exemplify this point. The crossing of two lines of mice, one harboring the superoxide dismutase-1 gene and the other gene for a neurofilament subunit gave much more relevant models than each line separately (Kong and Xu, 2000).

In some cases, the effect of a knockout gene may be compensated by the action of a quite different gene. This may reveal a non-anticipated role of the second gene. Mice, in which the LAM 2 gene has been knocked out, suffer from congenital muscular dystrophy. The addition of the agrin gene known to be involved in the formation of neuromuscular junctions restored muscular function (Moll et al., 2001).

Models for infectious diseases

Many pathogens have species specificity and conventional laboratory animals may not be relevant models. The limitation is often due to the lack of appropriate receptors for the pathogens in the animal cells. Receptor genes can then be transferred to animals to be preferentially expressed in tissues which are the normal targets of the pathogens.

A model for listeriosis

Listeria monocytogenes is responsible for severe foodborne infections. A surface protein of this bacteria,

internalin, recognizes the intestinal host receptor E-cadherin Human and guinea-pig E-cadherin are recognized by internalin. Transgenic mice expressing the human E-cadherin gene exclusively in the enterocytes of their small intestine have been generated. This was achieved by using the promoter from the rat *FABP-1* gene. These mice were sensitive to infection by *Listeria monocytogenes* and constitute an excellent model for the study of this infectious pathogen. Interestingly, E-cadherin is not expressed in the tight junction of the intestinal epithelium which is accessible only when enterocytes divide. The mouse model correctly reproduced this phenomenon. They are also good candidates for internalizing foreign molecules as soon as they are associated with internalin (Lecuit *et al.*, 2001). *L. monocytogenes* is known to impact cells other than enterocytes in the organism. These mechanisms could be studied by this type of tool.

Models for three viral infections

Mice sensitive to measles have been obtained by transferring to them the viral receptor CD 46 gene (Oldstone *et al.*, 1999). Similarly, mice harboring the gene for the polymyelitis receptor are used as models to study the viral infection (Ren *et al.*, 1990).

The case of hepatitis C infection is more complex. This virus does not infect cells *in vitro* and *in vivo* models are currently strictly needed. Severe combined immunodeficiency (SCID; *Prkd^{scid}*) mice harboring human hepatocytes can be infected by serum from patients suffering from hepatitis C. This model has been greatly improved by using homozygous transgenic mice expressing the U-plasminogen activator (uPA) gene in their liver under the control of an albumin promoter. The grafted human hepatocytes survived a longer time in the mice (Fausto, 2001).

Models for AIDS study

The HIV-1 virus is known to have two major receptors in human cells CD4 and CCR5. Transgenic rabbits expressing the human CD4 gene were shown to become seropositive after HIV-1 infection (Dunn *et al.*, 1995). However although the virus replicated in rabbit cells, it was unable to generate any disease. The CCR5 receptor could have been added to the CD4 rabbits. Due to the difficulty in maintaining rabbits in appropriate confinement and for other reasons, many potential mouse models were generated (Cohen, 2001). It finally appeared that transgenic rats expressing all the HIV-1 genes except gag and pol showed pathogeny having many similarities to human AIDS (Reid *et al.*, 2001).

Models for prion diseases

It is no exaggeration to say the **prion** diseases would be much less understood without the contribution of transgenic mice (Moore and Melton, 1997; Prusiner *et al.*, 1998). Creutzfeldt–Jakob and related diseases are typically pathologies which incubate for very long periods of time and require the cooperation of several cell types, especially when infection occurs by the oral route.

It was essential to show that the PrP $-/-$ ($Prnp^{tm}$) mice were insensitive to any prion infection which then demonstrated that the PrP gene has a major role in prion diseases.

Transgenic mice overexpressing the PrP genes from different species become more sensitive to the prion disease of the same species. This phenomenon is amplified when the endogenous mouse PrP gene is knocked out. PrP $-/-$ mice expressing the bovine PrP gene are sensitive to bovine spongiform encephalopathy (Scoot *et al.*, 1999). PrP $-/-$ mice expressing the hamster PrP gene specifically in astrocytes are sensitive to hamster scrapie (Raeber *et al.*, 1997). Transgenic mice also revealed that some epitopes of the PrP protein are essential for the transmission of the disease (Scott *et al.*, 1997), that some PrP alleles are more favorable for infection than others and that genetic environmental factors modify the incubation period of bovine spongiform encephalopathy (Manolakou *et al.*, 2001). Transgenic mice also helped to define the putative existence of a protein X, the role of which is essential for infection. This protein X is supposed to bind directly to the PrP protein.

The prion disease appears more rapidly when the PrP gene of the mouse and the cell extract used for infection come from more closely related species. It was also surprising to observe that the cell extracts become increasingly infectious after repeated passages to animals. This adaptation has not yet been explained at the molecular level.

All these studies *in vivo* have led to the definition of more flexible *in vitro* systems in which cultured cells can transform the PrP^c into PrP^{sc} (Vilette *et al.*, 2001).

Various transgenic mice are invaluable models for the evaluation of the therapeutic effects of chemical compounds. Encouraging results showing that different drugs can delay or even partly cure scrapie have recently been obtained.

Models for Alzheimer's disease

Alzheimer's disease (AD) is complex and it occurs after a long period of incubation. AD is characterized by the extracellular accumulation of polymerized tau protein. β-amyloid is cleaved from amyloid precursor protein (ADP), the cleavage being performed by three secretases α, β and γ (Esler and Wolfe, 2001).

Transgenic mice, in which the ADP gene had been knocked out, kept their sensitivity to the disease (Lewis *et al.*, 2001). This is due to the existence of another gene called ADP-like gene. The knockout of the genes is lethal.

Transgenic mice having knocked-out ApoE and IL 1 genes are less sensitive to the disease indicating that these two genes contribute to the AD development.

Different models have been obtained. Some of them overexpress normal ADP and prenisilin genes while others express mutated forms of the genes. None of these models reflect all the aspect of AD. Fewer neurons in general are destroyed in the models than in patients. Correlation between behavioral disorders and accumulation of β-amyloid is not always strict. Yet, the models show a similar disease pattern and they are used to define treatments. An anti-inflammatory molecule, ibuprofen, was shown to reduce the number of β-amyloid plaques. Inhibitors of secretase-β also show some therapeutic effect. Vaccination against β-amyloid can delay the development of AD. These models may therefore help to identify active molecules and to design analogs capable of reaching the brain (Chapman *et al.*, 2001).

Other brain degenerative diseases are being studied using transgenic models. Transgenic mice which overexpress the α-synuclein gene in their brain are used to study Parkinson's disease (Betarbet *et al.*, 2002).

Huntington's disease is also studied in transgenic mice overexpressing the huntingtin gene variants having oligoglutamines of variable length in their N terminus (Rubinsztein, 2002).

Models for cell death

Apoptosis has a major role in many biological events and is highly conserved in evolution. It is indispensable for development and organogenesis which imply specific cell elimination. Defaults in apoptosis may lead to autoimmune diseases, tumor formation and neurodegeneracy. At least 25 genes involved in apoptosis have been knocked out (Ranger *et al.*, 2001). This includes genes for caspases, adaptors, regulators, bcl2 family members and mitochondrial proteins. Genes involved in the control of apoptosis are redundant and their role is not easily identified in some cases. Ill-controlled apoptosis may be lethal for animals. A finely tuned expression of the transgenes is therefore required to point out their effect *in vivo*.

Models for aging

Aging is a complex phenomenon which has only been partially described. Defects in genomes appear to be a major cause of aging (Hasty and Vijg, 2002). A growing number of transgenic models are being used to study aging. Mice in which the XPD gene has been knocked out are more sensitive to oxidative DNA damage. This sensitivity was increased further when the XPA gene was also knocked out (De Boer *et al.*, 2002). These models reflect some of the aging syndromes in human.

Models for atherosclerosis

Atherosclerosis results from complex disorders in lipid metabolism. Many genes are involved in this process. The disease has a genetic component and its development is favored by cholesterol-rich diets.

Numerous genes coding for different apolipoproteins, lipases and other factors have been added or knocked out in mice (Miller and Rubin, 1997). These studies have given important information as to the role of the different genes. However, lipid metabolism is quite different in mouse and human. Mice must be submitted to diets very rich in cholesterol to mimic even partially atherosclerosis. The rabbit is much closer to humans for this biological function and this animal is used extensively in addition to mice for the study of atherosclerosis (Brousseau and Hoeg, 1999). Some transgenic rabbits are also used to define new drugs capable of protecting humans against arteriosclerosis.

Models for cancer

Cancer occurs under multiple forms and it is the result of several steps occurring sequentially. Cells are first **immortalized**. They later get the capacity to multiply at a high rate and more or less independently of normal inducers. They finally colonize different tissues forming metastases. All these steps result from alterations of oncogene and antioncogene expression and activity. Mutations, chromosomal translocations and environmental factors are thus responsible for tumor formation (Elenbaas *et al.*, 2001).

Transgenic mice were soon used to try to generate models for cancer study. The first oncomouse expressed c-myc gene in the mammary gland. This was sufficient to trigger the formation of mammary tumors. Further studies made it possible to identify additional genes involved in mammary cancer (Siegel *et al.*, 2000). Genes whose expression is amplified in mammary tumors (c-myc, erb B2 and cyclin D1) have an oncogenic effect when used as transgenes. Crossing mice harboring different oncogenes and having knocked out genes has made it possible to determine the cooperative actions of some of these genes.

A recent study pointed out the role of cyclin-D1 in mammary cancer (Bartek and Lukas, 2001; Yu *et al.*, 2001). Mice having knocked out cyclin D1 gene are insensitive to *neu* and ras oncogenes whereas wint and c-myc genes have kept their capacity to trigger mammary tumors. These data help to search for new drugs capable of inhibiting neu, ras and cyclin D1 genes.

The stability of mammary epithelial cells is highly dependent on the extracellular matrix. Development of mammary tumors and metastasis is correlated with local degradation of some of the extracellular matrix components. Degradation of the extracellular matrix is achieved by metalloproteinases which are controlled by specific inhibitors. The role of different genes involved in extracellular matrix synthesis and degradation is being studied using transgenic mouse models. A recent study has pointed out the role of Akt gene which delays mammary gland involution. This effect is mediated by the prolonged presence of one of the metalloproteinase inhibitors (Schwertfeger *et al.*, 2001). A better understanding of these phenomena may lead to the identification of new drugs preventing metastasis.

It is acknowledged that tumor cells derive from a single cell in which all the mutations required for cancer development occurred. This explains in part why cancer development is a slow process at the early steps and becomes rapid in time. The classical transgenic models do not take into account the fundamental clonal origin of cancers. Specific gene constructs capable of sporadically activating oncogenes have been designed (Berns, 2001). The technique makes it possible to activate k-ras oncogene randomly and at a low frequency (Johnson *et al.*, 2001). This new approach is expected to reflect more precisely the process of tumor formation under natural conditions.

Models for xenografting

The growing shortage of cells and organs of human origin for patients led scientists to envisage the use of animal material. The first attempts to graft animal organs into humans were made one century ago. This pioneer work was a surgical success but organs were strongly rejected and destroyed after only a few hours or days. The use of immosuppressors did not improve the situation. This clearly indicated that phenomena other than rejection which occurs after allografting have a potent negative effect on **xenoorgans**. Studies performed *in vitro* and *in vivo* revealed that complement attack is one of the major mechanisms involved in xenograft rejection. Several genes, including DAF and CD59 genes were shown to have anti-complement effects. These genes and others were transferred to cells which became resistant to human complement. Transgenic mice, rats, rabbits and pigs harboring these two human genes have been generated. Their cells and some of their organs (kidney and heart) are protected for several days when grafted to primates.

The first success encouraged study of the other rejection mechanisms and to identify the genes involved in this phenomenon or being capable of blocking it.

It is acknowledged that xenoorgan rejection is due to at least three types of mechanisms acting sequentially (Platt, 2002). The first known as acute rejection is rapid, (occurring within minutes or hours), potent and essentially due to complement action. It is considered to be controlled by the DAF and CD59 transgenes. The second mechanism is delayed acute rejection which occurs during the week following grafting. The cellular events responsible for this phenomenon are not all known. The third rejection mechanism occurs later. It seems to be similar but more intense than that observed after allografting.

The pig has been retained as the best candidate for the supply of cells and organs for humans. This implies that genes will be added to pigs to block the rejection mechanisms. In addition, pig genes coding for strong antigens or for retroviruses will have to be inactivated by knockout or any other methods. Although the mouse is too small an animal for experimental organ transplantation to primates, it remains essential to define which genes should be transferred to pigs (Houdebine and Weill, 2002; Platt,2002).

Models for reproduction and endocrinological studies

Fetal gonad development

The development of gonads in the fetus is known to be different in males and females. In mice, gonad differentiation starts at day 10.5 post mating. The sexual difference is known to be controlled by the SRY gene which is located in the Y chromosome. SRY activates a cascade of gene expression. Sox 9 gene is activated soon after SRY and it precedes the expression of the Mullerian inhibiting substance (MIS) which directly determines the morphology of the nascent gonad.

Transgenic mice harboring the Sox 9 gene have been generated. The Sox 9 was inserted into a YAC containing the Wit gene (Wilm's tumor suppressor) by a knock-in. The regulatory region of the Wt 1 gene was chosen since it is known to be active in the urogenital ridge at day 10.5 post coitum, in both male and female fetuses.

All the fetuses developed a male gonad, even those which were XX. In the XX transgenic fetus, male reproductive ducts developed normally and Sertoli cells were present. At day 13.5, the male and female gonads appeared similar. Adult testes, seminiferous tubules and Sertoli cells were observed while Leydig cells were present in the interstitial tissue. However, germ cells were absent, as was already found in naturally masculinized XX animals.

This result points out the importance of the Sox 9 gene and it suggests that the number of steps between the activation of SRY and Sox 9 genes is small (Vidal *et al.*, 2001).

Role of progesterone in mammary gland development

Several hormone receptor genes have been overexpressed or knocked out to determine their role in tissue growth and differentiation.

The mammary gland is composed of two major categories of cells. The parenchyme contains the secretory epithelial and the myoepithelial cells. The stroma which surrounds the parenchyme is formed by fibroblasts, adipocytes and connective tissue. Progesterone is one of the hormones favoring mammary gland growth *in vivo* and preventing milk secretion during pregnancy until parturition. The progesterone receptor gene is particularly present in the parenchyme. To determine its role, the progesterone receptor gene was knocked out. The development of the mammary gland was impaired in these animals. The grafting of the fetal parenchyme from knocked out mice into the stroma of normal mice did not lead to a development of the mammary gland. Conversely, normal parenchyme grafted into the stroma of knocked out mice developed to become a normal mammary gland. These experiments confirm the role of progesterone in mammary gland development and they point out the different roles of parenchyme and stroma.

A number of other gene modifications in mice have been carried out to decipher the different mechanisms in the development and the differentiation of the mammary gland (Shillingford and Hennighausen, 2001).

Models for the study of insulin resistance

Non-insulin-dependent diabetes mellitus is a frequent and complex polygenic disease. Numerous genetic modifications have been carried out in mice. They revealed the involvement of the role of several insulin transduction mechanisms in the disease. These data also suggest that disorders in insulin action may be important in tissues not considered as direct insulin targets (Baudry *et al.*, 2002; Hribal *et al.*, 2002).

Conclusions and perspectives

Transgenic mice for the study of human diseases are an essential tool and this methodology is expected to be even more extensively used in the coming decades.

Despite its numerous successes, genetic modification in mouse is sometimes disappointing. The available tools which are steadily being improved will further reduce the number of animals giving no interpretable results. The increased possibility to add and replace genes in the animals will be welcome. The introduction of a series of transgenes in a chosen site may greatly simplify the interpretation of the data. The addition of insulators may reduce the number of lines expressing the transgenes in an inappropriate manner. The possibility of stimulating a transgene or to inactivate reversibly an endogenous gene with specific inducers having no effect on the other genes of the host will increase the generation of relevant models. Not only isolated genes, but group of genes on a given chromosome must be studied simultaneously in some cases. The YAC and the BAC vectors are invaluable tools for this purpose. Episomal circular vectors (Kelleher *et al.*, 1998; Bok Hee *et al.*, 2002) and chromosomal vectors (Voet *et al.*, 2001) might provide experimenters with still more potent tools in the future.

Several points will limit the use of mouse models. Other species namely rats (Charreau *et al.*, 1996), rabbits (Fan *et al.*, 1999) and pigs (Malassagne *et al.*, 2003) are and will be used for specific studies in which mice are inappropriate for various reasons.

Transgenes as endogenous genes are subjected to epigenetic mechanisms which may greatly modify their expression and in an unpredictable manner. One of the mechanisms involved in the variation of gene and transgene expression is mediated by DNA methylation (Rakyan *et al.*, 2001; Whitelaw and Martin, 2001). This means that individuals of the same line of animals may show variable levels of expression for a given transgene in some cases.

Up to 30% of the knocked-out gene is not accompanied by any observable phenotypic effect. This may be due to redundant genes which compensate for the inactivated genes. In these cases, the redundant genes must be found and knocked-out. It should be kept in mind that some of the knockout effects are not seen because experimenters do not proceed to appropriate investigations. A number of genes are highly conserved and play an essential role in animals living free but not in those kept in cages. The animals with additional or knocked-out genes must be subjected to various standard tests, namely behavioral tests, capable of revealing phenotypic alterations not visible in animals kept in cages.

The interpretation of the data obtained with transgenics is and will remain difficult in some cases. This may be due to the intrinsic highly complex interactions between genes (Szathmary *et al.*, 2001). Transgenesis and alternatives to transgenesis such as cell culture may not be able to unravel such a mechanism.

Establishing relevant models to study human disease is more complicated than realized by many experimenters. Animals showing only some of the syndromes of a disease may be of no help to really understand the disease and to design new therapies. The experimenters must also take into account that the effect of a transgene may be markedly influenced by the genetic background of the animals (Carvallo *et al.*, 1997). The choice of one mouse line rather than another must therefore be done as soon as possible.

Generating transgenic animal models is a problem. Keeping the transgenic lines available for other experimenters with the relevant information is another problem. The Jackson Laboratory in USA, the European Mouse Mutant Archive (EMMA) in EU and companies are becoming more and more organized structures with which research laboratories collaborate.

Another problem that must be taken into account is the fact that animal experimentation is not well-tolerated by some activists. The experimenters must be able to justify their use of the animals and keep them in perfect condition (Moore, 2001).

References

Araki, K., Araki, M. and Yamamura, K. (1997). *Nucleic Acids Res.* **25**, 868–872.

Baer, A. and Bode, J. (2001). *Curr. Opin. Biotechnol.* **12**, 473–480.

Bartek, J. and Lukas, J. (2001). *Nature* **411**, 1001–1002.

Baudry, A., Leroux, L., Jackerott, M. and Joshi, R.L. (2002). *EMBO Rep.* **3**, 323–328.

Bernards, A. and Hariharan, I.K. (2001). *Curr. Opin. Genet. Dev.* **11**, 274–278.

Berns, A. (2001). *Nature* **410**, 1043–1044.

Betarbet, R., Sherer, T.B. and Greenamyre, J.T (2002). *BioEssays* **24**, 308–318.

Blau, H.M. and Rossi, F.M. (1999). *Proc. Natl. Acad. Sci. USA* **96**, 797–799.

Bok Hee, C., Fetzer, C.P., Stehle, I.M., Jönsson, F., Fackelmayer, F.O., Conradt, H., Bode, J. and Lipps, H.J. (2002). *EMBO Rep.* **3**, 349–354.

Brousseau, M.E. and Hoeg, J.M. (1999). *J. Lipid Res.* **40**, 365–375

Brown, S.D. and Balling, R. (2001). *Curr. Opin. Genet. Dev.* **11**, 268–273.

Capecchi, M.R. (1989). *Science* **244**, 1288–1292.

Carvallo, G., Canard, G. and Tucker, D. (1997). In *Transgenic Animal Generation and Use* (ed. L.M. Houdebine), pp. 403–410. Harwood Academic Publishers, Amsterdam.

Cecconi, F. and Meyer, B.I. (2000). *FEBS Lett.* **480**, 63–71.

Chapman, P.F., Falinska, A.M., Knevett, S.G. and Ramsay, M.F. (2001). *Trends Genet.* **1**, 254–261.

Charreau, B., Tesson, L., Soulillou, J.P., Pourcel, C. and Anegon, I. (1996). *Transgenic Res.* **5**, 223–234.

Co, D.O., Borowski, A.H., Leung, J.D., van der Kaa, J., Hengst, S., Platenburg, G.J., Pieper, F.R., Perez, C.F., Jirik, F.R. and Drayer, J.I (2000). *Chromosome Res.* **8**, 183–191.

Cohen, J. (2001). *Science* **29**, 1034–1046.

Cohen-Tannoudji, M., Robine, S., Choulika, A., Pinto, D., El Marjou, F., Babinet, C., Louvard, D. and Jaisser, F. (1998). *Mol. Cell. Biol.* **18**, 1444–1448.

Cronin, C.A., Gluba, W. and Scrable, H. (2001). *Genes Dev.* **15**, 1506–1517.

De Boer, J., Andressoo, J.O., De Wit, J., Huijmans, J., Beems, R.B., Van Steeg, H., Weeda, G., Van der Horst, G.T.J., Van Leueuwen, W., Themmen, A.P.N., Meradji, M. and Hoeijmakers, J.H.J. (2002). *Science* **296**, 1276–1279.

Dunn, C.S., Mehtali, M., Houdebine, L.M., Gut, J.P., Kirn, A. and Aubertin, A.M. (1995). *J. Gen. Virol.* **76**, 1327–1336.

Dupuy, A.J., Clark, K., Carlson, C.M., Fritz, S., Davidson, A.E., Markley, K.M., Finley, K., Fletcher, C.F., Ekker, S.C., Hackett, P.B., Horn, S. and Largaespada, D.A. (2002). *Proc. Natl. Acad. Sci. USA* **99**, 4495–4499.

Elbashir, S.M., Harborth, J., Lendeckel, W., Yalcin, A., Weber, K. and Tuschl, T. (2001). *Nature* **411**, 494–498.

Elenbaas, B., Spirio, L., Koerner, F., Fleming, M.D., Zimonjic, D.B., Donaher, J.L., Popescu, N.C., Hahn, W.C. and Weinberg, R.A. (2001). *Genes Dev.* **15**, 50–65.

Esler, W.P. and Wolfe, M.S. (2001). *Science* **293**, 1449–1454.

Fan, J., Challah, M. and Watanabe, T. (1999). *Pathol. Int.* **49**, 583–594.

Fausto, N. (2001). *Nat. Med.* **7**, 890–891.

Feng, Y.Q., Seibler, J., Alami, R., Eisen, A., Westerman, K.A., Leboulch, P., Fiering, S. and Bouhassira, E.E. (1999). *J. Mol. Biol.* **292**, 779–785.

Forster, K., Helbl, V., Lederer, T., Urlinger, S., Wittenburg, N. and Hillen, W. (1999). *Nucleic Acids Res.* **27**, 708–710.

Freundlieb, S., Schirra-Muller, C., Bujard, H. (1999). *J. Gene Med* **1**, 4–12.

Fussenegger, M., Morris, R.P., Fux, C., Rimann, M., von Stockar, B., Thompson, C.J. and Bailey, J.E. (2000). *Nat. Biotechnol.* **18**, 1203–1208.

Giovannangeli, C. and Helene, C. (2000). *Nat. Biotechnol.* **18**, 1245–1246.

Giraldo, P. and Montoliu, L. (2001). *Transgenic Res.* **10**, 83–103.

Goodwin, N.C., Ishida, Y., Hartford, S., Wnek, C., Bergstrom, R.A., Leder, P. and Schimenti, J.C. (2001). *Nat. Genet.* **28**, 310–311.

Gordon, J.W., Scangos, G.A., Plotkin, D.J., Barbosa, J.A. and Ruddle, F.H. (1980). *Proc. Natl. Acad. Sci. USA* **77**, 7380–7384.

Harris, D. (1997). *Human Mol. Genet.* **6**, 2191–2193.

Hasty, P. and Vijg, J. (2002). *Science* **296**, 1250–1251.

Herault, Y., Rassoulzadegan, M., Cuzin, F. and Duboule, D. (1998). *Nat. Genet.* **20**, 381–384.

Houdebine, L.M., Attal, J. and Vilotte, J.L. (2002). In *Transgenic Animal Technology*, 2nd edn (ed. Carl A. Pinkert), pp. 419–458. Academic Press, San Diego.

Hribal, M.L., F., O. and Domenico, A. (2002). *J. Physiol. Endocrinol. Metab.* **282**, 977–981.

Humpherys, D., Eggan, K., Akutsu, H., Hochedlinger, K., Rideout, W.M., 3rd, Biniszkiewicz, D., Yanagimachi, R. and Jaenisch, R. (2001). *Science* **293**, 95–97.

Jackson, I.J. (2001). *Nat. Genet.* **28**, 198–200.

Jenke, B.H., Fetzer, C.P., Stehle, I.M., Jonsson, F., Fackelmayer, F.O., Conradt, H., Bode, J. and Lipps, H.J. (2002). *EMBO Rep.* **3**, 349–354.

Johnson, L., Mercer, K., Greenbaum, D., Bronson, R.T., Crowley, D., Tuveson, D.A. and Jacks, T. (2001). *Nature* **410**, 1111–1116.

Jones, S.D. and Marasco, W.A. (1997). In *Transgenic Animal Generation and Use* (ed. L.M. Houdebine), pp. 501–506. Harwood Academic Publishers, Amsterdam.

Kawasaki, H. and Taira, K. (2002). *EMBO Rep.* **3**, 443–450.

Kelleher, Z.T., Fu, H., Livanos, E., Wendelburg, B., Gulino, S. and Vos, J.M. (1998). *Nat. Biotechnol.* **16**, 762–768.

Koike, H., Horie, K., Fukuyama, H., Kondoh, G., Nagata, S. and Takeda, J. (2002). *EMBO Rep.* **3**, 433–437.

Kong, J. and Xu, Z. (2000). *Neurosci. Lett.* **281**, 72–74.

Lecuit, M., Vandormaeel-Pournin, S., Lefort, J., Huerre, M., Gounon, P., Dupuy, C., Babinet, C., Cossart, P. (2001). *Science* **292**, 1722–1725.

Lewis, J., Dickson, D.W., Lin, W.L., Chisholm, L., Corral, A., Jones, G., Yen, S.H., Sahara, N., Skipper, L., Yager, D., Eckman, C., Hardy, J., Hutton, M. and McGowan, E. (2001). *Science* **293**, 1487–1491.

L'Huillier, P.J., Soulier, S., Stinnakre, M.G., Lepourry, L., Davis, S.R., Mercier, J.C. and Vilotte, J.L. (1996). *Proc. Natl. Acad. Sci. USA* **93**, 6698–6703.

Lois, C., Hong, E.J., Pease, S., Brown, E.J. and Baltimore, D. (2002). *Science* **295**, 868–872.

Loonstra, A., Vooijs, M., Beverloo, H.B., Allak, B.A., van Drunen, E., Kanaar, R., Berns, A. and Jonkers, J. (2001). *Proc. Natl. Acad. Sci. USA* **98**, 9209–9214.

Malassagne, B., Regimbeau, J.M., Taboit, F., Troalen, F., Chereau, C., Moire, N., Attal, J., Batteux, F., Conti, F., Calmus, Y., Houssin, D., Boulard, C., Houdebine, L.M.., Weill, B. (2003). *Xenotransplantation* **10**, 267–277

Manolakou, K., Beaton, J., McConnell, I., Farquar, C., Manson, J., Hastie, N.D., Bruce, M. and Jackson, I.J. (2001). *Proc. Natl. Acad. Sci. USA* **98**, 7402–7407.

McCreath, K.J., Howcroft, J., Campbell, K.H., Colman, A., Schnieke, A.E. and Kind, A.J. (2000). *Nature* **405**, 1066–1069.

Medico, E., Gambarotta, G., Gentile, A., Comoglio, P.M. and Soriano, P. (2001). *Nat. Biotechnol.* **19**, 579–582.

Miller, M.W. and Rubin, E.M. (1997). In *Transgenic Animal Generation and Use*, (ed. L.M. Houdebine), pp. 445–448. Harwood Academic Publishers, Amsterdam.

Mitchell, K.J., Pinson, K.I., Kelly, O.G., Brennan, J., Zupicich, J., Scherz, P., Leighton, P.A., Goodrich, L.V., Lu, X., Avery, B.J., Tate, P., Dill, K., Pangilinan, E., Wakenight, P., Tessier-Lavigne, M. and Skarnes, W.C. (2001). *Nat. Genet.* **28**, 241–249.

Moll, J., Barzaghi, P., Lin, S., Bezakova, G., Lochmuller, H., Engvall, E., Muller, U. and Ruegg, M.A. (2001). *Nature* **413**, 302–307.

Moore, A. (2001). *EMBO Rep.* **2**, 554–558.

Moore, R.C. and Melton, D.W. (1997). *Mol. Hum. Reprod.* **3**, 529–544.

Nagano, M., Brinster, C.J., Orwig, K.E., Ryu, B.Y., Avarbock, M.R. and Brinster, R.L. (2001). *Proc. Natl. Acad. Sci. USA* **98**, 13090–13095.

Nagy, A. (2000). *Genesis* **26**, 99–109.

Nebert, D.W. and Duffy, J.J. (1997). *Biochem. Pharmacol.* **53**, 249–254.

No, D., Yao, T.P. and Evans, R.M. (1996). *Proc. Natl. Acad. Sci. USA* **93**, 3346–3351.

Oldstone, M.B., Lewicki, H., Thomas, D., Tishon, A., Dales, S., Patterson, J., Manchester, M., Homann, D., Naniche, D. and Holz, A. (1999). *Cell* **98**, 629–640.

Paddison, P.J., Caudy, A.A. and Hannon, G.J. (2002). *Proc. Natl. Acad. Sci. USA* **99**, 1443–1448.

Palmiter, R.D., Brinster, R.L., Hammer, R.E., Trumbauer, M.E., Rosenfeld, M.G., Birnberg, N.C. and Evans, R.M. (1982). *Nature* **300**, 611–615.

Perry, A.C., Wakayama, T., Kishikawa, H., Kasai, T., Okabe, M., Toyoda, Y. and Yanagimachi, R. (1999). *Science* **284**, 1180–1183.

Platt, J.L. (2002). *Nat. Biotechnol.* **20**, 231–232.

Prusiner, S.B., Scott, M.R., DeArmond, S.J. and Cohen, F.E. (1998). *Cell* **93**, 337–348.

Qian, J., Lin, Y., Jiang, M., Hung, H. and Wang, K. (2002). *Transgenic Res.* **11**, 92.

Rakyan, V.K., Preis, J., Morgan, H.D. and Whitelaw, E. (2001). *Biochem. J.* **356**, 1–10.

Ranger, A.M., Malynn, B.A. and Korsmeyer, S.J. (2001). *Nat. Genet.* **28**: 113–118.

Readeout, W.M.III, W.T., Wutz A., Eggan K., Jackson-Grubsy L., Dausman J., Yanagimachi R. and Jaenisch R. (2000). *Nat. Genet.* **24**, 109–110.

Reid, W., Sadowska, M., Denaro, F., Rao, S., Foulke, J., Jr., Hayes, N., Jones, O., Doodnauth, D., Davis, H., Sill, A., O'Driscoll, P., Huso, D., Fouts, T., Lewis, G., Hill, M., Kamin-Lewis, R., Wei, C., Ray, P., Gallo, R.C., Reitz, M. and Bryant, J. (2001). *Proc. Natl. Acad. Sci. USA* **98**, 9271–9276.

Ren, R.B., Costantini, F., Gorgacz, E.J., Lee, J.J. and Racaniello, V.R. (1990). *Cell* **63**, 353–362.

Rivera, V.M., Clackson, T., Natesan, S., Pollock, R., Amara, J.F., Keenan, T., Magari, S.R., Phillips, T., Courage, N.L., Cerasoli, F., Jr., Holt, D.A. and Gilman, M. (1996). *Nat. Med.* **2**, 1028–1032.

Rossi, F.M. and Blau, H.M. (1998). *Curr. Opin. Biotechnol.* **9**, 451–456.

Rossi, F.M., Guicherit, O.M., Spicher, A., Kringstein, A.M., Fatyol, K., Blakely, B.T. and Blau, H.M. (1998). *Nat. Genet.* **20**, 389–393.

Rubinsztein, D.C. (2002). *Trends Genet.* **18**, 202–209.

Sage, J., Martin, L., Rassoulzadegan, M. and Cuzin, F. (1998). *Gene* **221**, 85–92.

Schnieke, A.E., Kind, A.J., Ritchie, W.A., Mycock, K., Scott, A.R., Ritchie, M., Wilmut, I., Colman, A. and Campbell, K.H. (1997). *Science* **278**, 2130–2133.

Schübeler, D., Lorincz, M.C., Cimbora, D.M., Telling, A., Feng, Y.Q., Bouhassira, E.E. and Groudine, M. (2000). *Mol. Cell. Biol.* **20**, 9103–9112.

Schwertfeger, K.L., Richert, M.M. and Anderson, S.M. (2001). *Mol. Endocrinol.* **15**, 867–881.

Scott, M.R., Safar, J., Telling, G., Nguyen, O., Groth, D., Torchia, M., Koehler, R., Tremblay, P., Walther, D., Cohen, F.E., DeArmond, S.J. and Prusiner, S.B. (1997). *Proc. Natl. Acad. Sci. USA* **94**, 14279–14284.

Scott, M.R., Will, R., Ironside, J., Nguyen, H.O., Tremblay, P., DeArmond, S.J. and Prusiner, S.B. (1999). *Proc. Natl. Acad. Sci. USA* **96**, 15137–15142.

Shillingford, J.M. and Hennighausen, L. (2001). *Trends Endocrinol. Metab.* **12**, 402–408.

Siegel, P.M., Hardy, W.R. and Muller, W.J. (2000). *BioEssays* **22**, 554–563.

Silver, D.P. and Livingston, D.M. (2001). *Mol. Cell.* **8**, 233–243.

Soukharev, S., Miller, J.L. and Sauer, B. (1999). *Nucleic Acids Res.* **27**, e21.

Szathmary, E., Jordan, F. and Pal, C. (2001). *Science* **292**, 1315–1316.

Taboit-Dameron, F., Malassagne, B., Viglietta, C., Puissant, C., Leroux-Coyau, M., Chereau, C., Attal, J., Weill, B. and Houdebine, L.M. (1999). *Transgenic Res.* **8**, 223–235.

Tuschl, T. (2002). *Nat. Biotechnol.* **20**, 446–448.

Umana, P., Gerdes, C.A., Stone, D., Davis, J.R., Ward, D., Castro, M.G. and Lowenstein, P.R. (2001). *Nat. Biotechnol.* **19**, 582–585.

Upegui-Gonzalez, L.C., Francois, J.C., Ly, A. and Trojan, J. (2000). *Adv. Exp. Med. Biol.* **465**, 319–332.

Vidal, V.P., Chaboissier, M.C., de Rooij, D.G. and Schedl, A. (2001). *Nat. Genet.* **28**, 216–217.

Vilette, D., Andreoletti, O., Archer, F., Madelaine, M.F., Vilotte, J.L., Lehmann, S. and Laude, H. (2001). *Proc. Natl. Acad. Sci. USA* **98**, 4055–4059.

Viville, B., Charnock-Jones, D.S., Sharkey, A.M., Wetzka, B. and Smith, S.K. Hum. (1997). *Reproduction* **12**, 815–822.

Voet, T., Vermeesch, J., Carens, A., Durr, J., Labaere, C., Duhamel, H., David, G. and Marynen, P. (2001). *Genome Res.* **11**, 124–136.

Wang, Y., DeMayo, F.J., Tsai, S.Y. and O'Malley, B.W. (1997). *Nat. Biotechnol.* **15**, 239–243.

Warashina, M., Kuwabara, T., Kato, Y., Sano, M. and Taira, K. (2001). *Proc. Natl. Acad. Sci. USA* **98**, 5572–5577.

West, A.G., Gaszner, M. and Felsenfeld, G. (2002). *Genes Dev.* **16**, 271–288.

Whitelaw, E. and Martin, D.I. (2001). *Nat. Genet.* **27**, 361–365.

Wunderlich, F.T., Wildner, H., Rajewsky, K. and Edenhofer, F. (2001). *Nucleic Acids Res.* **29**, E47.

Yu, Q., Geng, Y. and Sicinski, P. (2001). *Nature* **411**, 1017–1021.

Zipperlen, P., Fraser, A.G., Kamath, R.S., Martinez-Campos, M. and Ahringer, J. (2001). *EMBO J.* **20**, 3984–3992.

The Mouse in Preclinical Safety Studies

Johannes H Harlemann

Novartis Pharma, Preclinical Safety, Basel, Switzerland

The mouse has been one of the main mammalian species used in preclinical studies ranging from pharmacology and safety assessment. Its' use came into effect for various reasons, some of which were quite mundane such as being considered a pest, ease of breeding and small size. The latter made it a favorite animal for mammalian geneticists. Between the 1920s and the 1950s numerous inbred strains were developed and characterized. The Jackson Institute was a key player and an example of the work carried out, such as providing breeding and research facilities for this work. Another boost for the mouse as a model was given by the immunologists, especially after the pioneering work to develop monoclonal antibodies in this species. Recently genetics have again made major inroads with the development of **knockout** and **knock-in** mouse models and strains (Rosenberg, 1997; Rudmann and Durham, 1999; Hopley and Zimmer, 2001; Lesch, 2001).

The latter technology, by deleting specific genes out of animals or respectively inserting e.g. human genes into the mouse genome has expanded exponentially. Transgenic animals are nowadays a major source of research material (Rudolph and Möhler, 1999). The variations and extent of genetic manipulation is only limited by one's imagination, resources and ethics. In the first 6 months of 2002 alone, a literature search restricted to this period using as search criteria 'transgenic mice' resulted in a listing of over 1400 citations (Sigmund, 2000). Much of this work involves basic biology and pharmacology. A complete review would go beyond the scope of this section. Transgenic animal technology has recently been combined with another powerful tool gene chips, in which mice are used as a model. Expression profiling of mRNA is expanding rapidly our basic understanding of regulatory pathways in mammalian systems.

This section will focus however mainly on the mouse in preclinical safety testing. Here the use of this species is historic and for the same reasons as above. The use of mice as models in safety evaluations is currently required in international guidelines for both chemicals and pharmaceuticals.

Mice used in these safety studies are often either random bred albino mice, frequently with a Swiss strain origin, or hybrids like the B6C3F1 (an F1 hybrid from the C57Bl and Ch3 strains).

In oncogenicity testing in particular the use of transgenic animals is being examined and is already favored by some. These strains have either certain repair deficiencies and/or carry one or more particular oncogenes making them potentially more sensitive to the effects of tumor induction and promotion (Ashby, 2001; Cohen, 2001; Venkatachalam *et al.*, 2001). Transgenic mice are also been used in special cases for the testing of biotechnology products in which no other models exist except man itself or higher apes (Dayan, 1995; Rosenberg, 1997; Griffiths and Lumley, 1998; Bugelski *et al.*, 2000; Goodman, 2001). These special applications will be discussed further in the respective sections. Mice are used in safety testing in the following types of studies: acute toxicity, subacute/chronic toxicity, carcinogenicity, mutagenicity and immune toxicity.

Acute toxicity

Acute toxicity testing is a frequently underestimated model. The acute toxicity test gives an estimate for the therapeutic margin and absolute safety of a drug or chemical. In the past, LD50 tests were used to quantify exactly and calculate accurately this endpoint. This, however, used large numbers of animals, which nowadays is avoided by approximative dose regimes. The latter is also used in the classification of chemicals (Chan and Hayes, 1994).

Acute toxicity testing has one drawback in that a detailed examination is often not carried out e.g. histopathology, clinical pathology and kinetics. This can lead to an underestimation of the toxicity of a test article and identification of target organs with those compounds in which repeat dose toxicity dosing is limited to markedly lower doses e.g. gastrointestinal toxicity with non-steroidal anti-inflammatory drugs (NSAIDs).

The routes of exposure in acute toxicity tests is generally similar to the expected exposure in man e.g. oral as well as parenteral e.g. intravenous or intraperitoneal.

Repeat dose subacute/chronic toxicity

Whereas in pharmacology the mouse is a preferred species because of the plethora of models available and its inherent small size, this limits the amount of compound used. In toxicology the mouse is in contrast to the rat, which is not the standard species for toxicity testing, mostly because of its size, which limits sampling to obtaining clinical pathology samples as well as kinetic samples only. Hence most repeat dose toxicity tests use the rat as the standard rodent species. The mouse is used in certain cases when, for example, the pharmacological target is only present in man and not in any other species. In such cases knock-in transgenic mice have been used as a model (Dayan, 1995; Thomas, 1995; Griffiths and Lumley, 1998). In repeat dose studies non-rodent species showed a higher prediction rate of adverse effects in humans (Olson *et al.*, 1998, 2000). However experience in mice is limited because most studies involve rats. Based on results in carcinogenicity studies, a lower rate of prediction may be expected.

Repeat dose mouse studies are most often carried out relatively late in the development or safety testing of a drug or chemical and then usually as dose range finding studies for the oncogenicity studies. In these studies the mixamal tolerated dose (MTD) after repeat dosing is determined for use in lifetime exposure in the subsequent carcinogenicity study.

Carcinogenicity studies

The mouse and the rat are the standard species for the conduct of carcinogenicity studies (Haseman *et al.*, 2001). Although these species have been able to help identify potential human carcinogens, they may also identify many rodent specific carcinogens (Huff, 1994; Battershill and Fielder, 1998). The prediction rate of the mouse model has been widely criticized (Infante, 1993; Alden *et al.*, 1996). It has been proposed that these models should be replaced by the transgenic models (Cannon *et al.*, 2000; Carmichael *et al.*, 2001; Cohen *et al.*, 2001; Usui *et al.*, 2001; Van Kreijl *et al.*, 2001; Van Steeg *et al.*, 2001). Whatever model is used for the testing, it is of utmost importance to use a standardized nomenclature (Keenan *et al.*, 2002). Progress has been made in international standardization of this nomenclature, most notably by the RITA-group. Their nomenclature has been published and is available on the Internet.

The most widely used models are the p53 strain, which is primarily used for compounds in a positive mutagenicity test and the TgH2RAS model which is promoted for both mutagenic and non-mutagenic compounds (Carmichael *et al.*, 2001; Storer *et al.*, 2001; Tamaoki, 2001; Usui *et al.*, 2001; Van Kreijl *et al.*, 2001; Van Steeg *et al.*, 2001; Venkatachalam *et al.*, 2001). The TgAC model is only favored for dermally applied compounds (Spalding *et al.*, 1999; Eastin *et al.*, 2001; Tennant *et al.*, 1998, 2001; Morton *et al.*, 2002). Experience with these compounds is still limited, but these models have been accepted by the regulatory authorities as alternative models for the mouse lifetime 2-year carcinogenicity testing (Haseman *et al.*, 2001). Initially it was thought that these models would give more specific answers, because of their genetic back ground, predisposing them to tumor development. This does not always appear to be the case. Some investigators suggest increasing the number of animals in each dose group and extend the duration of these studies from 6 to 9 months. Another issue is the use of positive control groups in these studies. As positive controls p-cresidine and benzidine most frequently are used. However several groups have reported difficulties in the use of these compounds either not leading to positive results or having mixed toxicities.

The handling of these known carcinogens within a testing facility is also not favored. The results of testing by the ILSI group and the NTP indicated that the use of these models can identify correctly many known human carcinogens and also had a lower false positive rate compared with the 2-year models (Van der Laan and Spindler, 2002). Only a few indirect carcinogens such as cyclosporin, estradiol, phenobarbital and chloroform appear to be non-responding in the models tested (Van der Laan and Spindler, 2002; Van der Laan *et al.*, 2002).

Mutagenicity studies

The mouse is used in one standard mutagenicity study, the micronucleus test (Brusick, 1994). In this test high doses of test material are administered and the effect on the formation of micronuclei in the bone marrow is evaluated. This method is either evaluated manually or by automated morphometric analysis.

Another model, which is used occasionally to detect mutation in certain organ tissues is the Mutamouse or Big-blue system (Gossen *et al.*, 1989; Short *et al.*, 1990; Myhr, 1991). Here a bacterial genome is inserted into the mouse genes. After treatment with a test material,

the mutation frequency in these inserts is evaluated and compared with controls indicating a direct effect where mutations are found.

Immunotoxicity

The mouse is used in two models for the testing for the immunotoxic potential of a test material. These models are proposed as a replacement for the Magnusson Kligman or Buehler test in guinea pigs.

One model, the local lymph node test, has been widely validated and has been accepted by the regulatory authorities as an alternative for the above guinea pig models (Ulrich and Vohr, 1996; Basketter *et al.*, 2000; Dean *et al.*, 2001; Kimber, 2001; Ulrich *et al.*, 2002). The other test, the popliteal lymph node test has been proposed in the newest Food and Drug Administration (FDA) guidelines as a possible test for autoimmunity and allergy. Both tests use the primary response in the draining lymph node after exposure to the test material. In the local lymph node this is applied to the skin of the ear and in the popliteal lymph node test injected in the foot pad (Gleichmann *et al.*, 1989; Bloksma *et al.*, 1995; Pieters, 2001; Pieters *et al.*, 2002). With a modification of the local **lymph mode test (LLN)** one can also check compounds for their photo-sensitization potential (Ulrich *et al.*, 1998; Vohr *et al.*, 2000).

References

Abelson, P.H. (1990). *Science* **249**(4975), 1357.

Alden, C.L., Smith, P.F., Piper, C.E. and Brej, L. (1996). Acritical appraisal of the value of the mouse cancer bioassay safety assessment. *Toxcol. Pathol.* **24**, 722–725.

Ashby, J. (2001). *Toxicol. Pathol.* **29**(Suppl.) 177–182.

Basketter, D.A., Balikie, L., Dearman, R.J., Kimber, I., Ryan, C.A., Gerberick, G.F., Harvey, P., Evans, P., White, I.R. and Rycroft, R.J. (2000). *Contact Dermatitis* **42**, 344–348.

Battershill, J.M. and Fielder, R.J. (1998). *Hum. Exp. Toxicol.* **17**, 193–205.

Bloksma, N., Kubicka-Muranyi, M., Schuppe, H.C., Gleichmann, E. and Gleichmann, H. (1995). *Crit. Rev. Toxicol.* **25**, 369–396.

Brusick, D. (1994). In *Principles and Methods of Toxicology* (ed. W.A. Hayes), pp. 545–577. Raven Press, New York.

Bugelski, P.J., Herzyk, D.J., Rehm, S., Harmsen, A.G., Gore, E.V., Williams, D.M., Maleeff, B.E., Badger, A.M., Truneh, A., O'Brien, S.R., Macia, R.A., Wier, P.J., Morgan, D.G. and Hart, T.K. (2000). *Hum. Exp. Toxicol.* **19**, 230–243.

Cannon, R.E., Graves, S., Spalding, J.W., Trempus, C.S. and Tennant, R.W. (2000). *Mol. Carcinog.* **29**, 229–235.

Carmichael, P.L., Mills, J.J., Campbell, M., Basu, M. and Caldwell, J. (2001). *Toxicol. Pathol.* **29**(Suppl.), 155–160.

Chan, P.K. and Hayes, A.W. (1994). In *Principles and Methods of Toxicology* (ed. W.A. Hayes), pp. 579–647. Raven Press, New York.

Cohen, S.M. (2001). *Toxicol. Pathol.* **29**(Suppl.), 183–190.

Cohen, S.M., Robinson, D. and MacDonald, J. (2001). *Toxicol. Sci.* **64**, 14–19.

Dayan, A.D. (1995). *Toxicology* **105**, 59–68.

Dean, J.H., Twerdok, L.E., Tice, R.R., Sailstad, D.M., Hattan, D.G. and Stokes, W.S. (2001). *Regul. Toxicol. Pharmacol.* **34**, 258–273.

Eastin, W.C., Mennear, J.H., Tennant, R.W., Stoll, R.E., Branstetter, D.G., Bucher, J.R., McCullough, B., Binder, R.L., Spalding, J.W. and Mahler, J.F. (2001). *Toxicol. Pathol.* **29**(Suppl.), 60–80.

Gleichmann, E., Klinkhammer, C. and Gleichmann, H. (1989). *Arch. Toxicol. Suppl.* **13**, 188–190.

Goodman, J.I. (2001). *Toxicol. Pathol.* **29**(Suppl.), 173–176.

Gossen, J., de Leeuw, W., Tan, C., *et al.* (1989). *Proc. Natl. Acad. Sci. USA* **86**, 7971–7975.

Griffiths, S.A. and Lumley, C.E. (1998). *Hum. Exp. Toxicol.* **17**, 63–83.

Haseman, J., Melnick, R., Tomatis, L. and Huff, J. (2001). *Food Chem. Toxicol.* **39**, 739–744.

Hopley, R. and Zimmer, A. (2001). *BioTechniques* **30**, 130–132.

Huff, J. (1994). In *Carcinogenesis* (eds M.P. Waalkes and J.M. Ward), pp. 25–37. Raven Press, New York.

Infante, P.F. (1993). *Environ. Health Perspect.* **101**(Suppl. 5), 143–148.

Keenan, C., Hughes-Earle, A., Case, M., Stuart, B., Lake, S., Mahrt, C., Halliwell, W., Westhouse, R., Elwell, M., Morton, D., Morawietz, G., Rittinghausen, S., Deschl, U. and Mohr, U. (2002). *Toxicol. Pathol.* **30**(1), 75–79.

Kimber, I. (2001). *Toxicology* **158**, 59–64.

Lesch, K.P. (2001). *Pharmacogenomics J.* **1**, 187–192.

Morton, D., Alden, C.L., Roth, A.J. and Usui, T. (2002). *Toxicol. Pathol.* **30**(1), 139–146.

Myhr, B. (1991). *Environ. Mol. Mutagen.* **18**, 308–315.

Olson, H., Betton, G., Stritar, J. and Robinson, D. (1998). *Toxicol. Lett.* **102–103**, 535–538.

Olson, H., Betton, G., Robinson, D., Thomas, K., Monro, A., Kolaja, G., Lilly, P., Sanders, J., Sipes, G., Bracken, W., Dorato, M., van Deun, K., Smith, P., Berger, B. and Heller, A. (2000). *Regul. Toxicol. Pharmacol.* **32**, 56–67.

Pieters, R. (2001). *Toxicology* **158**, 65–69.

Pieters, R., Ezendam, J., Bleumink, R., Bol, M. and Nierkens, S. (2002). *Toxicol. Lett.* **127**, 83–91.

Popp, J.A. (2001). *Toxicol. Pathol.* **29**(Suppl.), 20–23.

Rosenberg, M.P. (1997). *Mol. Carcinog.* **20**, 262–274.

Rudmann, D.G. and Durham, S.K. (1999). *Toxicol. Pathol.* **27**(1), 111–114.

Rudolph, U. and Möhler, H. (1999). *Eur. J. Pharmacol.* **375**, 327–337.

Sigmund, C.D. (2000). *Arterioscler. Thromb. Vasc. Biol.* **20**, 1425–1429.

Spalding, J.W., French, J.E., Tice, R.R., Furedi-Machacek, M., Haseman, J.K. and Tennant, R.W. (1999). *Toxicol. Sci.* **49**, 241–254.

Short, J.M., Kohler, S.W., Provost, G.S., Ferik, A. and Kretz, P.L. (1990). In *Mutation and the Environment* (eds M. Mendelsohn and R. Albertini), pp. 355–367. Wiley and Liss, New York.

Storer, R.D., French, J.E., Haseman, J., Hajian, G., LeGrand, E.K., Long, G.G., Mixson, L.A., Ochoa, R., Sagartz, J.E. and Soper, K.A. (2001). *Toxicol. Pathol.* **29** (Suppl.), 30–50.

Tamaoki, N. (2001). *Toxicol. Pathol.* **29**(Suppl.), 81–89.

Tennant, R.W., Tice, R.R. and Spalding J.W. (1998). *Toxicol. Lett.* **102–103**, 465–471.

Tennant, R.W., Stasiewicz, S., Eastin, W.C., Mennear, J.H. and Spalding J.W. (2001). *Toxicol. Pathol.* **29**(Suppl.), 51–59.

Thomas, J.A. (1995). *Toxicology* **105**, 7–22.

Ulrich, P. and Vohr, H.W. (1996). *Eur. Cytokine Netw.* 7, 401–407.

Ulrich, P., Homey, B. and Vohr, H.W. (1998). *Toxicology* **125**, 149–168.

Ulrich, P., Grenet, O., Bluemel, J., Vohr, H.W., Wiemann, C., Grundler, O. and Suter, W. (2002). *Arch. Toxicol.* **76**, 62.

Usui, T., Mutai, M., Hisada, S., Takoaka, M., Soper, K.A., McCullough, B. and Alden, C. (2001). *Toxicol. Pathol.* **29**(Suppl.), 90–108.

Van der Laan, J.W. and Spindler, P. (2002). *Regul. Toxicol. Pharmacol.* **35**, 122–125.

Van der Laan, J.W., Lima, B.S. and Snodin, D. (2002). *Toxicol. Pathol.* **30**(1), 157–159.

Van Kreijl, C.F., McAnulty, P.A., Beems, R.B., Vynckier, A., van Steeg, H., Fransson-Steen, R., Alden, C.L., Forster, R., van der Laan, J.W. and Vandenberghe, J. (2001). *Toxicol. Pathol.* **29**(Suppl.), 117–127.

Van Steeg, H., de Vries, A., van Oostrom, C.T.M., van Benthem, J., Beems, R.B. and van Kreijl, C.F. (2001). *Toxicol. Pathol.* **29**(Suppl.), 109–116.

Venkatachalam, S., Tyner, S.D., Pickering, C.R., Boley, S., Recio, L., French, J.E. and Donehower, L.A. (2001). *Toxicol. Pathol.* **29**(Suppl.), 147–154.

Vohr, H.W., Blumel, J., Blotz, A., Homey, B., and Ahr, H.J. (2000). *Arch. Toxicol.* **73**, 501–509.

PART 2

Anatomy and Developmental Biology

Contents

CHAPTER 8

Gross Anatomy

Vladimír Komárek
Sidlistni, Lysolaje, Czech Republic

Introduction

This chapter presents illustrations likely to be of practical importance to those working with laboratory mice. They include the body regions, a simple demonstration of the skeleton, the muscles and a dissection of the body cavities with description of major organs. More detailed information is provided in publications by Cook (1965), Hummel *et al.* (1975), Feldman and Seely (1988), Popesko *et al.* (1990), and Iwaki *et al.* (2001).

The terminology used here is based on the international veterinary anatomical nomenclature published by Schaller *et al.* (1992). In the figure captions, XY denotes male and XX female.

Acknowledgement

With her kind consent, the figures presenting the myology and most of those presenting the splanchnology were drawn following the concept of Professor Dr Viera Rajtova (Popesko *et al.*, 1990).

References

Cook, M.J. (1965). *The Anatomy of the Laboratory Mouse.* Academic Press, London.

Feldman, D.B. and Seely, J.C. (1988). *Necropsy Guide: Rodents and the Rabbit.* CRC Press, Boca Raton, Florida.

Hummel, K.P., Richardson, F.L. and Fekete, E. (1975). *In Biology of the Laboratory Mouse*, 2nd edn (eds E.L. Green and E.U. Fahey), pp. 247–307. Dover Publications, New York.

Iwaki, T., Yamashita, H. and Hayakawa, T. (2001). *A Color Atlas of Sectional Anatomy of the Mouse.* Adthree, Tokyo.

Popesko, P., Rajtova, V. and Horak, J. (1990). *Atlas anatomie malych laboratornych zvierat*, Vol. 2. Priroda, Bratislava. (English version published by Wolfe Publishing Ltd, London, 1992.)

Schaller, O., Constantinescu, G.M., Habel, R.E., Sack, W.O., Simoens, P. and de Vos, N.R. (1992). *Illustrated Veterinary Anatomical Nomenclature.* Enke Verlag, Stuttgart.

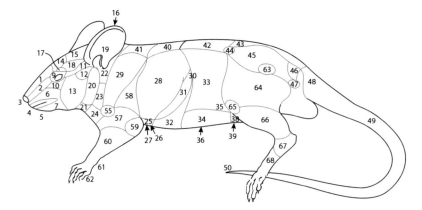

Figure 8.1 The regions of the body (regiones corporis), lateral view.

Regions of the face (regiones faciei)
1. regio dorsalis nasi
2. regio lateralis nasi
3. regio naris et apex nasi
4. regio oralis
5. regio mentalis
6. regio buccalis
7. regio mandibularis
8. regio intermandibularis
9. regio orbitalis
10. regio infraorbitalis
11. regio zygomatica
12. regio articulationis temporomandibularis
13. regio masseterica

Regions of the skull (regiones cranii)
14. regio frontalis
15. regio parietalis
16. regio occipitalis
17. regio supraorbitalis
18. regio temporalis
19. regio auricularis et auricula

Regions of the neck (regiones colli)
20. regio parotidea
21. regio subhyoidea
22. regio colli dorsalis
23. regio colli ventralis
24. regio trachealis

Regions of the chest (regiones pectoris)
25. regio presternalis
26. regio sternalis
27. regio mammaria thoracica (see Figure 8.4)

28. regio costalis
29. regio scapularis
30. arcus costalis

Regions of the cranial abdomen (regiones abdominis craniales)
31. regio hypochondriaca
32. regio xiphoidea

Regions of the middle abdomen (regiones abdominis mediae)
33. regio abdominis lateralis
34. regio umbilicalis
35. regio plicae genus
36. regio mammaria abdominalis (see Figure 8.4)

Regions of the caudal abdomen (regiones abdominis caudales)
37. regio inguinalis
38. regio pubica (scrotalis et preputialis in XY)
39. regio mammaria inguinalis (see Figure 8.4)

Regions of the back (regiones dorsi)
40. regio vertebralis thoracis
41. regio interscapularis
42. regio lumbalis

Regions of the pelvis (regiones pelvis)
43. regio sacralis
44. regio tuberis coxae
45. regio glutea
46. regio clunis
47. regio tuberis ischiadici
48. regio radicis caudae
49. regio corporis caudae

50. regio apicis caudae
51. regio perinealis
52. regio analis
53. regio vulvae
54. regio clitoridis

Regions of the forelimb (regiones membri thoracici)
55. regio articulationis humeri
56. regio axillaris
57. regio brachii
58. regio tricipitalis
59. regio cubiti
60. regio antebrachii (cranialis, lateralis, caudalis, medialis)
61. regio carpi (cranialis, lateralis, caudalis, medialis)
62. regio manus (metacarpi et digiti, cranialis, lateralis, volaris/palmaris, medialis)

Regions of the hindlimb (regiones membri pelvini)
63. regio articulationis coxae
64. regio femoris (cranialis, lateralis, caudalis, medialis)
65. regio genus
66. regio cruris (cranialis, lateralis, caudalis, medialis)
67. regio tarsi (cranialis, lateralis, caudalis, medialis)
68. regio pedis (metatarsi et digiti, dorsalis, lateralis, plantaris, medialis)

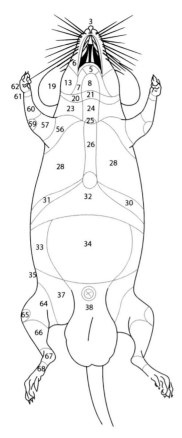

Figure 8.2 The regions of the body (regiones corporis), ventral view XY. (For labelling see Figure 8.1.)

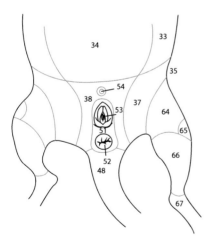

Figure 8.3 The regions of the body (regiones corporis), ventral view XX. (For labelling see Figure 8.1.)

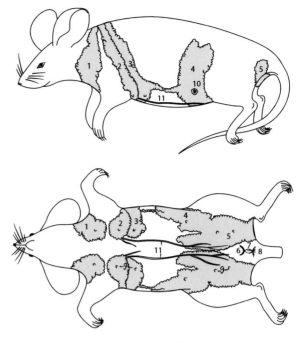

Figure 8.4 The mammary gland (XX).

1. pars cervicalis
2. pars thoracica cranialis
3. pars thoracica caudalis
4. pars abdominalis
5. pars inguinalis
6. clitoris et orificium urethrae externum
7. introitus vaginae
8. anus
9. papillae mammae
10. lymphonodus subiliacus
11. vena epigastrica cranialis superficialis

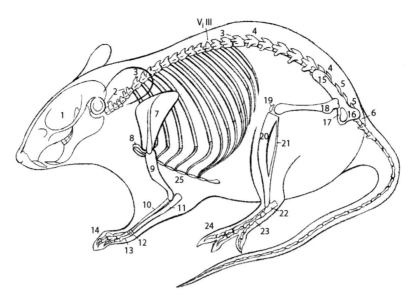

Figure 8.5 The skeleton.

1. skeleton capitis
2. vertebrae cervicales (7)
3. vertebrae thoracicae (about 13, 12–14)
4. vertebrae lumbales (5–7)
5. vertebrae sacrales (3–4)
6. vertebrae caudales (about 28, 27–30)
7. scapula
8. clavicula
9. humerus

10. radius
11. ulna
12. ossa carpi
13. ossa metacarpi
14. phalanges digitorum
15. os ilium
16. os ischii
17. os pubis (15, 16 and 17 considered to form 'innominate bone' of the pelvic girdle)
18. femur

19. patella
20. tibia
21. fibula
22. calcaneus et ossa tarsi
23. ossa metatarsi
24. phalanges digitorum
25. sternum

V_IIII vertebra inflexa III sive vertebra anticlinalis

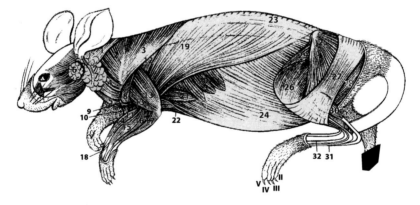

Figure 8.6 The muscles of the body (m. = musculus).

1. m. sphincter colli superficialis
2. m. trapezius, pars cervicalis
3. m. trapezius, pars thoracica
4. m. cleidocephalicus
5. pars scapularis musculi deltoidei
6. glandula lacrimalis extraorbitalis
7. m. paritidoauricularis et glandula parotidea
8. m. sternooccipitalis
9. pars clavicularis musculi deltoidei
10. m. biceps brachii
11. pars acromialis musculi deltoidei
12. m. teres major

13. m. triceps brachii, caput longum
13'. caput laterale
14. m. extensor carpi radialis longus
15. m. extensor digitorum communis
16. m. extensor digitorum lateralis
17. m. extensor carpi ulnaris
18. m. abductor digiti I. (pollicis) longus
19. m. latissimus dorsi
20. m. serratus ventralis
21. and 22. pars abdominalis m. pectoralis majoris
23. fascia thoracolumbalis

24. m. obliquus externus abdominis
25. m. gluteus superficialis
26. m. rectus femoris – m. quadriceps
27. m. biceps femoris
28. m. semitendinosus
29. m. tensor fasciae latae
30. caput laterale musculi gastrocnemii
31. m. extensor digitorum lateralis
32. m. extensor digitorum longus

II–V digitus secundus, tertius, quartus, quintus

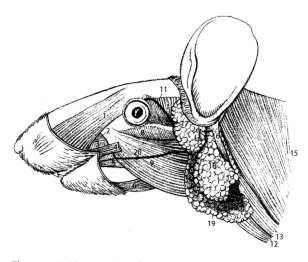

Figure 8.7 The muscles of the head.

1. m. levator nasolabialis
2. m. levator labii superioris proprius
3. m. buccinatorius, pars buccalis
4. m. zygomaticus
5. m. depressor labii inferioris
6. m. digastricus, venter rostralis
7. m. digastricus, venter caudalis
8. m. masseter, pars profunda
9. m. masseter, pars superficialis
10. m. buccinatorius, pars molaris
11. m. temporalis
12. m. sternooccipitalis
13. m. cleidooccipitalis
14. m. sternohyoideus
15. m. trapezius, pars cervicalis
16. m. parotidoauricularis
17. glandula lacrimalis extraorbitalis et eius ductus
18. glandula parotidea
19. glandula submandibularis
20. ductus parotideus

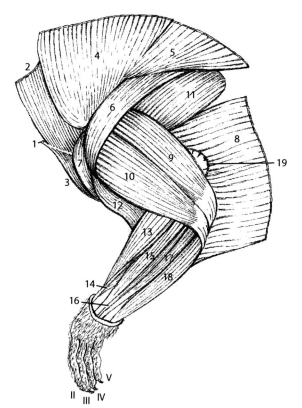

Figure 8.9 The muscles of the forelimb, lateral view.

1. clavicula
2. m. cleidocephalicus
3. m. cleidobrachialis
4. m. trapezius, pars cervicalis
5. m. trapezius, pars thoracica
6. m. deltoideus, pars scapularis
7. m. deltoideus, pars acromialis
8. m. cutaneus trunci
9. m. triceps brachii, caput longum
10. m. triceps brachii, caput laterale
11. m. infraspinatus
12. m. biceps brachii
13. m. extensor carpi radialis longus
14. m. abductor digiti I. (pollicis) longus
15. and 16. m. extensor digitorum communis
17. m. extensor digitorum lateralis
18. m. extensor carpi ulnaris
19. lymphonodus axillaris accessorius
II–V digitus secundus, tertius, quartus, quintus

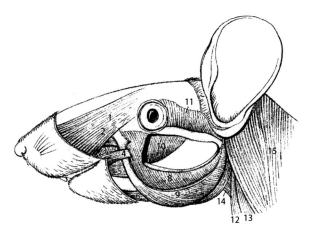

Figure 8.8 The muscles of the head. (For labelling see Figure 8.7.)

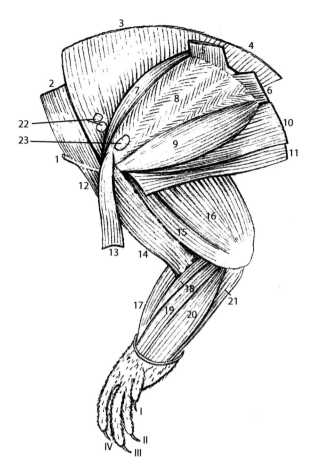

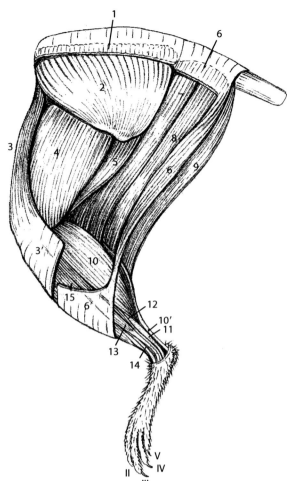

Figure 8.10 The muscles of the forelimb, medial view.

1. clavicula
2. m. cleidocephalicus
3. m. trapezius, pars cervicalis
4. m. trapezius, pars thoracica
5. m. rhomboideus, pars cervicalis
6. m. rhomboideus, pars thoracica
7. m. supraspinatus
8. m. subscapularis
9. m. teres major
10. m. latisimus dorsi
11. m. cutaneus trunci
12. m. cleidobrachialis
13. m. pectoralis ascendens
14. m. biceps brachii
15. m. triceps brachii, caput mediale
16. m. triceps brachii, caput longum
17. m. extensor carpi radialis
18. m. pronator teres
19. m. flexor carpi radialis
20. m. flexor digitorum profundus
21. m. flexor carpi ulnaris
22. lymphonodi cervicales superficiales
23. lymphonodus axillary proprius
I–IV digitus primus, secundus, tertius, quartus

Figure 8.11 The muscles of the hindlimb, lateral view.
1. m. gluteus superficialis
2. m. gluteus medius
3. m. tensor fasciae latae, 3! fascia lata
4. m. rectus femoris
5. m. vastus lateralis
6. m. biceps femoris, 6! fascia cruris
7. m. adductor
8. m. semimebranosus
9. m. semitendinosus
10. caput laterale musculi gastrocnemii, 10! tendo musculi
    tricipitis surae
11. m. flexor digiti I, (hallucis) longus
12. m. extensor digitorum lateralis
13. m. extensor digitorum longus
14. tendo musculi peronei longi
15. m. tibialis cranialis
II–V digitus secundus, tertius, quartus, quintus

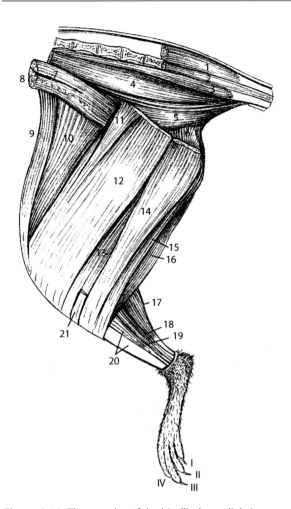

Figure 8.12 The muscles of the hindlimb, medial view.

1. m. lumbosacrocaudalis dorsalis lateralis
2. musculi intertransversarii
3. m. lumbosacrocaudalis ventralis lateralis
4. m. coccygeus dorsalis
5. m. coccygeus ventralis
6. m. obturator externus, pars intrapelvina
7. m. psoas minor
8. m. psoas major
9. m. tensor fasciae latae
10. m. rectus femoris
11. m. pectineus
12. m. vastus medialis
13. m. adductor
14. m. gracilis
15. m. semimembranosus
16. m. semitendinosus
17. caput mediale musculi gastrocnemii
18. m. tibialis caudalis
19. m. flexor digiti I. (hallucis) longus
20. m. flexor digitorum longus et tibia
21. insertio musculi sartorii
I–IV digitus primus, secundus, tertius, quartus

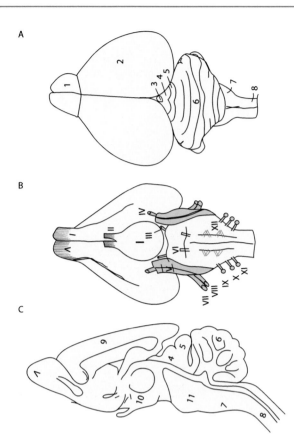

Figure 8.13 The brain.

A dorsal view
B ventral view
C midline section
1. bulbus olfactorius
2. hemispherium cerebri
3. corpus pineale
4. colliculi rostrales (tectum mesencephali)
5. colliculi caudales (tectum mesencephali)
6. cerebellum
7. medulla oblongata
8. medulla spinalis
9. cortex telencephali
10. hypothalamus
11. pons
 I n. (= nervus) olfactorius (termination in the bulbus)
 II n. opticus
III n. oculomotorius
IV n. trochlearis
 V n. trigeminus
VI n. abducens
VII n. facialis
VIII n. vestibulocochlearis
IX n. glossopharyngeus
 X n. vagus
XI n. accesorius
XII n. hypoglossus

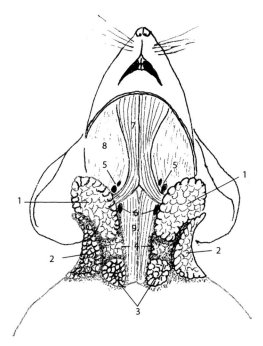

Figure 8.14 The salivary glands.

1. glandula submandibularis
2. glandula parotidea
3. pars cervicalis thymi

4. glandula sublingualis
5. lymphonodi mandibulares
6. lymphonodus retropharyngeus lateralis

7. m. digastricus
8. m. masseter
9. m. sternohyoideus et m. sternothyroideus

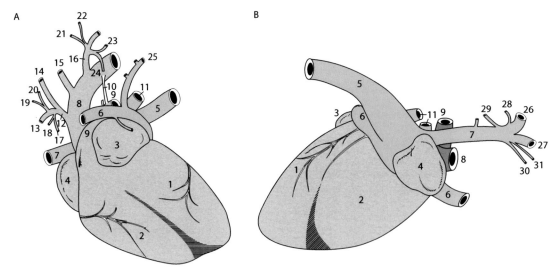

Figure 8.15 The heart.

A left lateral view, branching of the aorta
B right lateral view, branching of the vena cava cranialis dextra

1. ventriculus sinister
2. ventriculus dexter
3. auricula sinistra
4. auricula dextra
5. vena cava caudalis
6. vena cava cranialis sinistra
7. vena cava cranialis dextra
8. arcus aortae

9. truncus pulmonalis
10. ligamentum arteriosum (Botalli)
11. venae pulmonales
12. truncus brachiocephalicus
13. arteria subclavia sinistra
14. arteria carotis communis sinistra
15. arteria carotis communis dextra
16. arteria subclavia dextra
17. arteria thoracica interna sinistra
18. ramus thymicus sinister
19. arteria cervicalis superficialis sinistra
20. arteria vertebralis sinistra

21. arteria vertebralis dextra
22. arteria cervicalis superficialis dextra
23. ramus thymicus dexter
24. arteria thoracica interna dextra
25. vena azygos sinistra
26. vena subclavia dextra
27. vena jugularis externa dextra
28. vena thymica dextra
29. vena thoracica interna dextra
30. vena jugularis interna dextra
31. vena cervicalis superficialis dextra

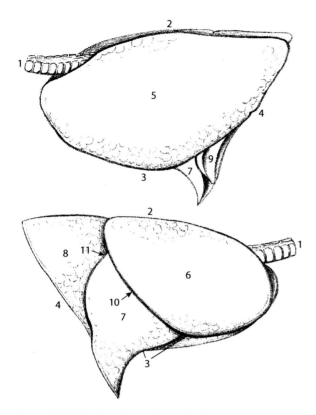

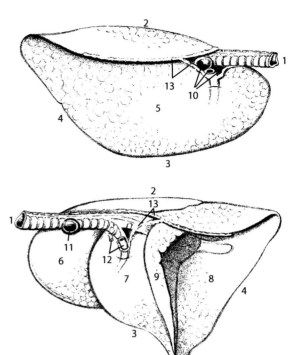

Figure 8.16 The lung, lateral (costal) view.

1. trachea
2. margo obtusus
3. margo acutus
4. margo basalis
5. pulmo sinister
6. lobus cranialis pulmonis dextri
7. lobus medius pulmonis dextri
8. lobus caudalis pulmonis dextri
9. lobus accessorius pulmonis dextri
10. incisura cardiaca
11. fissura interlobaris

Figure 8.17 The lung, medial (mediastinal) view.

1. trachea
2. margo obtusus
3. margo acutus
4. margo basalis
5. pulmo sinister
6. lobus cranialis pulmonis dextri
7. lobus medius pulmonis dextri
8. lobus caudalis pulmonis dextri
9. lobus accessorius pulmonis dextri
10. bronchus principalis dexter et rami arteriae et venae pulmonalis
11. bronchus principalis sinister
12. rami arteriae et venae pulmonalis
13. ligamentum pulmonale

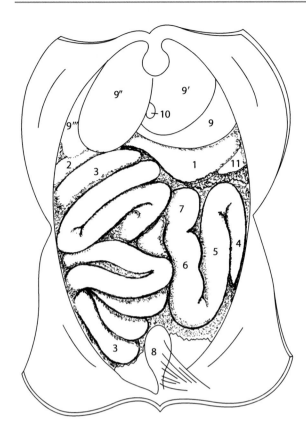

Figure 8.18 The abdominal situs viscerum.

1. gaster
2. duodenum ascendens
3. jejunum
4. apex ceci
5. corpus ceci
6. ampulla coli
7. colon ascendens
8. vesica urinaria et ligamenta
9. lobus sinister hepatis lateralis, 9ʹ. lobus sinister hepatis medialis, 9ʺ.lobus dexter hepatis medialis, 9ʺʹ. lobus dexter hepatis lateralis
10. vesica fellea
11. lien

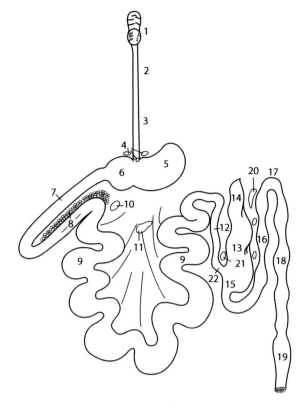

Figure 8.19 The digestive tract.

1. lingua
2. pars cervicalis esophagei
3. pars thoracica esophagei
4. pars abdominalis esophagei et lymphonodi gastrici
5. pars cardiaca ventriculi (saccus cecus, forestomach)
6. pars fundica et pylorica ventriculi (glandular stomach)
7. pars descendens duodeni
8. pars ascendens duodeni et pancreas
9. jejunum
10. lymphonodus pancreaticuduodenalis
11. lymphonodus jejunalis
12. ileum
13. corpus ceci
14. apex ceci
15. ampulla coli
16. colon ascendens
17. colon transversum
18. colon descendens
19. rectum
20. lymphonodi colici
21. lymphonodus ileocolicus
22. sacculus rotundus

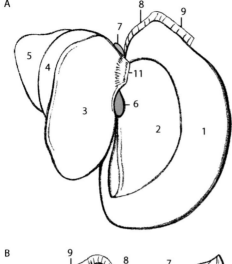

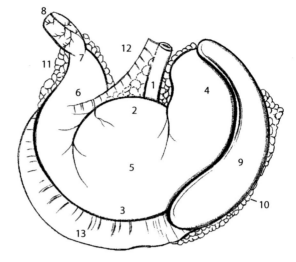

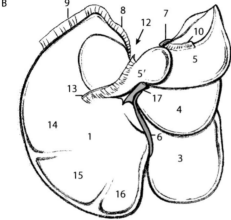

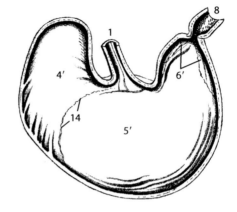

Figure 8.20 The liver.

A facies diaphragmatica
B facies visceralis
1. lobus sinister lateralis hepatis
2. lobus sinister medialis hepatis
3. lobus dexter medialis hepatis
4. lobus dexter lateralis hepatis
5. lobus caudatus hepatis, 5.′ processus papillaris hepatis
6. vesica fellea
7. vena cava caudalis
8. ligamentum coronarium sinistrum
9. ligamentum triangulare sinistrum
10. ligamentum hepatorenale
11. ligamentum falciforme et ligamentum teres hepatis
12. impressio esophagica
13. omentum minus
14. impressio ventricularis
15. impressio duodenalis
16. impressio jejunalis
17. vena portae, arteria hepatica, ductus choledochus

Figure 8.21 The stomach.

1. oesophagus
2. curvatura minor
3. curvatura major
4. saccus cecus ventriculi, 4.′ pars cardiaca tunicae mucosae
5. fundus ventriculi, 5.′ pars fundica tunicae mucosae
6. pars pylorica ventriculi, 6.′ pars pylorica tunicae mucosae
7. pylorus
8. duodenum
9. lien
10. and 11. pancreas
12. omentum minus
13. omentum majus
14. margo plicatus

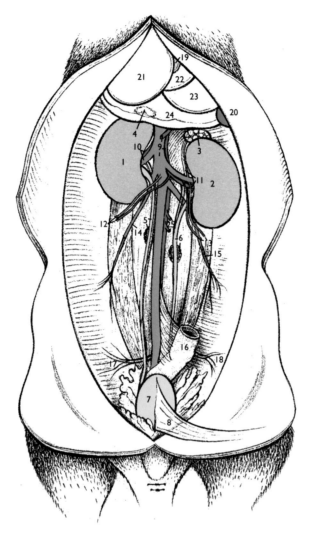

Figure 8.22 The kidneys *in situ*.

1. ren dexter
2. ren sinister
3. glandula adrenalis sinistra
4. glandula adrenalis dextra (under the stomach)
5. ureter dexter et lymphonodus lumbalis aorticus dexter
6. ureter sinister et lymphonodus lumbalis aorticus sinister
7. vesica urinaria
8. ligamentum vesicae medianum
9. vena cava caudalis, aorta abdominalis, arteria mesenterica cranialis
10. arteria et vena renalis dextra
11. arteria et vena renalis sinistra, arteria et vena adrenalis sinistra

12. arteria et vena ovarica dextra (in XX)
13. arteria et vena ovarica sinistra (in XX)
14. ramus muscularis dorsalis dexter
15. ramus muscularis dorsalis sinister
16. colon descendens
17. arteria et vena circumflexa ilium dextra
18. arteria et vena circumflexa ilium sinistra
19. vesica fellea
20. lien
21. lobus dexter medialis hepatis
22. lobus sinister medialis hepatis
23. lobus sinister lateralis hepatis
24. curvatura major ventriculi
25. arteriae mesentericae caudales

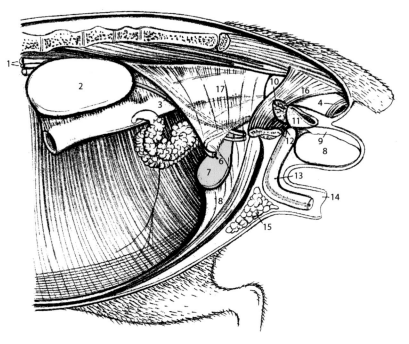

Figure 8.23 The male genital organs *in situ*.

1. aorta descendens et vena cava
   caudalis
2. ren sinister
3. ureter sinister
4. rectum et m. sphincter ani externus
5. glandula vesiculosa
6. vas deferens sinister
7. vesica urinaria

8. testis sinister
9. epididymis
10. glandula bulbourethralis
    (partim resecta)
11. diverticulum glandulae
    bulbourethralis
12. symphysis pelvina et m. urethralis
    et radix penis

13. penis
14. preputium
15. glandula preputialis
16. m. coccygeus
17. ligamentum vesicae laterale
18. ligamentum vesicae medianum

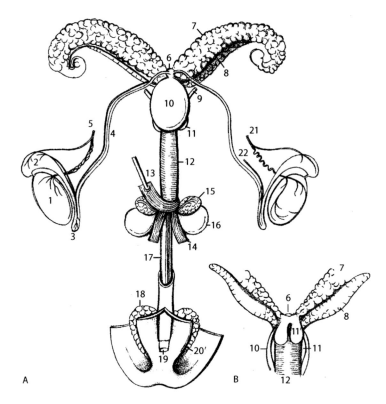

Figure 8.24 The male genital organs.

A ventral view
B dorsal view

1. testis dexter
2. caput epididymidis
3. cauda epididymidis
4. vas deferens dexter
5. vena testicularis dextra
6. glandula ampullaris
7. glandula vesiculosa
8. pars anterior prostatae (glandula coagulationis)

9. ureter sinister
10. vesica urinaria
11. prostata (pars ventralis),
    11'. prostata (pars dorsalis)
12. pars membranacea urethrae et m. urethralis
13. m. bulboglandularis
14. m. ischiocavernosus
15. glandula bulbourethralis

16. diverticulum glandulae bulbourethralis
17. penis
18. glandula preputialis
19. glans penis
20. preputium
21. arteria testicularis sinistra
22. arteria vas deferentis sinistra

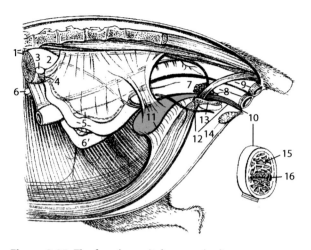

Figure 8.25 The female genital organs *in situ*.

1. aorta abdominalis et vena cava caudalis
2. ren sinister
3. ovarium sinistrum
4. oviductus sinister et mesosalpinx sinister
5. cornu uteri sinistrum, ligamentum latum uteri
6. ovarium dextrum, 6.' cornu uteri dextrum
7. cervix uteri
8. vagina et m. constrictor vulvae
9. anus et m. sphincter ani externus
10. clitoris et seccio transversalis clitoridis
11. vesica urinaria et ligamentum vesicae laterale et ligamentum vesicae medianum
12. urethra
13. symphysis pelvina
14. glandula clitoridis (preputialis feminina)
15. corpus cavernosum clitoridis
16. urethra

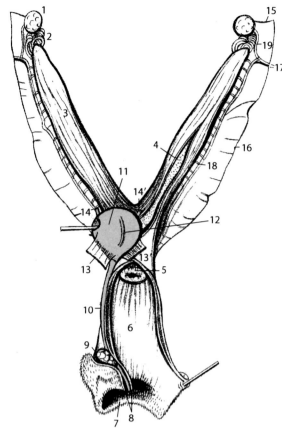

Figure 8.26 The female genital organs, ventral view.

1. ovarium dextrum
2. oviductus dexter
3. cornu uteri dextrum
4. cornu uteri sinistrum (partim resectum)
5. portio vaginalis uteri, cervix
6. fornix vaginae
7. vestibulum vaginae
8. clitoris et preputium clitoridis
9. glandula clitoridis (preputialis feminina)
10. urethra
11. vesica urinaria
12. ligamentum vesicae medianum
13. ligamentum vesicae laterale dextrum, 13.' ligamentum vesicae laterale sinistrum
14. and 14.' ureter dexter et sinister
15. mesovarium
16. mesometrium
17. arteria et vena ovarica sinistra
18. ramus uterinus arteriae et venae ovaricae sinistrae
19. ramus ovaricus arteriae et venae ovaricae sinistrae

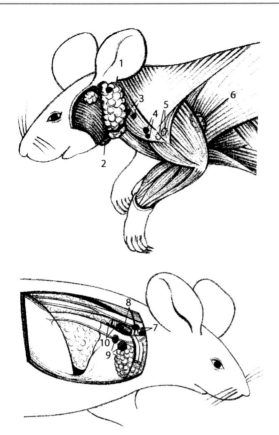

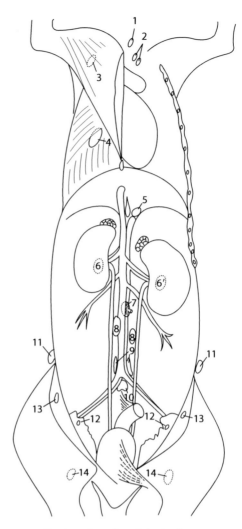

Figure 8.27 The lymph nodes of the head, neck and thorax (In. = lymphonodus, Inn. = lymphonodi).

1. In. parotideus
2. In. madibularis
3. Inn. cervicales profundi
4. Inn. cervicales superficiales
5. In. axillaris proprius
6. In. axillaris accessorius
7. In. mediastinalis cranialis
8. Inn. tracheobronchiales
9. Inn. mediastinales medii et pars thoracica thymi
10. Inn. mediastinales caudales

Figure 8.28 The lymph nodes of the body.

1. In. cervicalis profundus caudalis
2. Inn. mediastinales craniales
3. In. axillaris proprius
4. In. axillaris accessorius
5. In. aorticus
6. and 6´. Inn. renales
7. In. mesentericus caudalis
8. Inn. lumbales aortici
9. Inn. iliaci externi
10. In. iliacus internus
11. In. subiliacus
12. In. iliofemoralis
13. In. inguinalis superficialis
14. In. popliteus

Normative Histology of Organs

Georg J Krinke

Eggstrasse 26, CH-4402 Frenkendork BL, Switzerland

Introduction

This chapter provides concise descriptions and microscopic illustrations of the organs, arranged in alphabetical order. The 'normative' descriptions focus on features characteristic for the mouse in general. The abundance of mouse strains, spontaneous mutants and genetically engineered models does not allow excursions to their detailed features.

The following rules were applied. For bilateral organs the word 'paired' is consistently used. Bilaterally symmetrical organs such as the teeth, or the organs located in the body midline, such as the brain or the nasal cavity, are not considered as 'paired'. The description of hollow organs generally adheres to a concept of three-layered wall: inner lining, such as intima or mucosa, middle layer such as muscularis, and outer covering such as adventitia. Adventitia exposed to the abdominal cavity is called serosa. For description of the glands consisting of smaller units whose excretory ducts progressively join to form main ducts the term 'compound' is used.

The morphological features described here are generally those of young adult, healthy individuals. Further details can be found in publications by Hummel et al. (1975), Gude et al. (1982), Mohr et al. (1996), Maronpot et al. (1999) and Iwaki et al. (2001). Various research projects, mostly devoted to 'phenotyping' and correlation of genotypic with phenotypic features are displayed on the internet: a selection is presented as 'Internet Addresses' under the References.

Changes occurring with aging, as well as pathology of spontaneous diseases including the hyperplastic and neoplastic changes, are dealt with in a number of publications, especially those by Bannoasch and Gössner (1994, 1997), Cotchin and Roe (1967), Faccini et al. (1990), Maronpot et al. (1999), Mohr et al. (1996), Mohr (2001) and Turusov and Mohr (1994).

All microphotographs in this chapter show hematoxylin and eosin-stained paraffin sections from young adult mice of CD-1 strain, except where noted otherwise. Corresponding organ weights are given in Table 9.1.

Organ weights

Organ	Males		Females	
	Absolute (g)	Relative (%)	Absolute (g)	Relative (%)
Carcass	38.480	100.00	28.200	100.00
Adrenals	0.010	0.03	0.018	0.06
Brain	0.550	1.43	0.570	2.02
Epididymides	0.180	0.47	—	—
Heart	0.210	0.55	0.190	0.67
Kidneys	0.640	1.66	0.430	1.53
Liver	2.180	5.67	1.640	5.82
Ovaries	—	—	0.054	0.19
Spleen	0.100	0.26	0.100	0.36
Testes	0.290	0.75	—	—
Thymus	0.050	0.13	0.048	0.17
Thyroid	0.010	0.03	0.011	0.04

TABLE 9.1: Organ weights[a]

Comprehensive data for mice of various ages are given by Iwaki *et al.* (2001).

[a]Average organ weights of young adult CD-1 mice (about 60 males and 60 females, 4 months old, carcass = exsanguinated body)

Adrenal gland (glandula suprarenalis, glandula adrenalis; Figures 9.1–9.3)

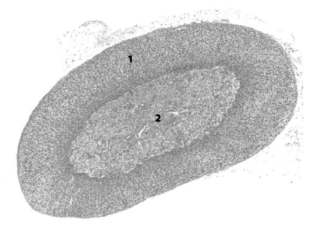

Figure 9.1 Adrenal gland (male).
1. cortex   2. medulla

The paired adrenal glands are located cranially to the kidneys. The gland is composed of the cortex, derived from the coelomic epithelium, and the medulla, derived from the neural crest. In the mouse, accessory adrenal cortical nodules are commonly found attached to the cortex or dispersed in the retroperitoneal fat. Mouse adrenal cortex is capable of regenerating by downgrowth from the subcapsular area. Spontaneously proliferating subcapsular cells may be fusiform, so called 'A cells', or more rounded 'B cells', resembling normal cortical cells. The cortical cells contain lipid droplets and produce steroid hormones such as mineralocorticoids, glucocorticoids and sex hormones. The cortex is composed of superficial zona glomerulosa and deep zona fasciculata. The zona reticularis, which occurs in other species, is not recognizable in mice. At the junction of the cortex and medulla there is the so called 'X-zone' which represents a specific feature of mouse adrenal gland. This zone is fully developed during the first postnatal weeks, until weaning, and regresses

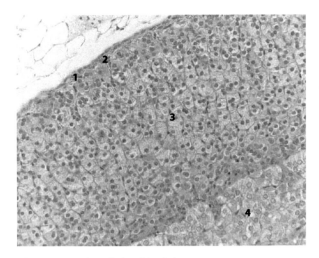

Figure 9.2 Adrenal gland (male).

1. capsule
2. zona glomerulosa
3. zona fasciculata
4. medulla

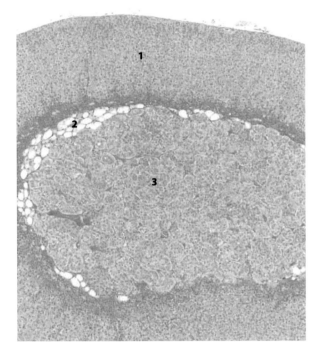

Figure 9.3 Adrenal gland (female).

1. cortex
2. 'X-zone' (showing lipid vacuolation)
3. medulla

in adult life. The function of the 'X-zone' is not well understood. In males the zone disappears by the age of puberty (about 5 weeks) without undergoing lipid vacuolization, whereas in females it continues to increase in size until the age of about 9 weeks and then regresses rapidly during the first pregnancy. In virgin females, however, it regresses slowly and undergoes lipid vacuolization, which in some cases may persist until advanced age. Sexual dimorphism of the mouse adrenal gland is expressed by a difference in size: female glands

are generally larger than male. The medulla produces biogenic amines noradrenalin and adrenalin (also called norepinephrine, epinephrine), and a number of regulatory peptides. The medullary cells contain 'dense core vesicles' for storage of biogenic amines. Adrenal medullary cells are often referred to as 'chromaffin' since the oxidation of their biogenic amines by chromate solutions results in red-brown coloration. The cells react positively with antibody for tyrosine hydroxylase, chromogranin and synaptophysin.

Ampullary gland (glandula ampullaris)

The paired ampullary glands are male accessory genital glands. They form groups of branched tubular glands lined by low cuboidal epithelium with large, oval nuclei. The glands open into the ampullae, near the seminal collicle.

Aorta (Figure 9.4)

The wall of the aorta is composed of the intima with an endothelial lining, the thick media formed predominantly by elastic fibers with smooth muscle fibers, and the adventitia.

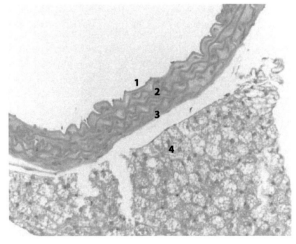

Figure 9.4 Aorta.

1. intima
2. media, with prominent wavy elastic fibers
3. adventitia
4. mediastinal brown fat (the cytoplasm contains multiple small lipid droplets)

Bone (Os; Figure 9.5)

The mouse skeleton is composed of bones which, according to their shape and structure are tubular (extremities) or flat (cranium, scapula, ribs). Some bones, such as the vertebrae, have both tubular and flat portions. Long tubular bones have a shaft (diaphysis), ending on both ends as metaphysis. The metaphysis is connected by epiphyseal cartilage (the so called 'growth plate') to the most peripheral part, the epiphysis. The surface of the bones is covered by a connective tissue membrane, the periost. The bones grow by 'endochondral' ossification in which precursor cartilage such as the epiphyseal growth plate is converted to bone, or from the periost by 'periosteal' ossification. Most of the diaphysis is formed by compact cortical bone, whereas the inside of the metaphysis and the epiphysis, as well as of short tubular and flat bones contains trabecular (cancellous) bone tissue. The bones form cavities in which the hematopoietic tissue, the bone marrow is located. The bone tissue contains the bone lining cells, osteoblasts, osteocytes and osteoclasts. The bone-lining cells are resting pre-osteoblasts, lining the bone surface as flattened cells. The osteoblasts are also located at the bone surface, but are larger and polyhedral. They actively deposit osteoid and progressively become incorporated in osteoid matrix and differentiate to osteocytes. The osteocytes are completely surrounded by mineralized bone. The osteoclasts are multinucleated macrophages which resorb bone and enable bone remodeling. The bone contains collagen fibers which may be arranged in parallel layers (lamellar bone) or in a random pattern (non-lamellar, woven bone). The woven bone is considered to represent immature bone, whereas the lamellar bone is more differentiated and forms both the cortical (compact) and the trabecular (cancellous) bone tissue. In the mouse, the cortical bone does not have distinct Haverian systems, which occur in other species. Growth and modeling of various long bones in mice is complete by the age of about 26 weeks and further remodeling serves to the maintenance or occurs in response to changing external forces acting on the bones, or to a disease.

Bone marrow (medulla ossium; Figure 9.6)

The bone marrow consists of a highly vascular, loose connective tissue stroma and the hematopoietic cells. In the mouse, nearly all bony cavities are filled with active marrow, leaving little reserve space for extending hematopoietic activity. This lack of marrow reserve is compensated for by extramedullary hematopoietic activity, especially in the spleen. In decalcified, hematoxylin and eosin-stained paraffin sections an estimate of general hematopoietic activity (cellularity) and

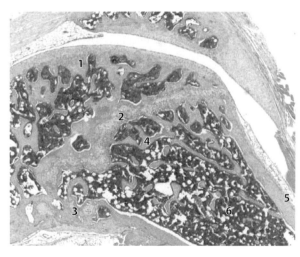

Figure 9.5 Bone, long tubular (femur).

1. epiphysis
2. epiphyseal growth plate
3. metaphysis
4. trabecular (cancellous) bone
5. cortical (compact) bone
6. bone marrow

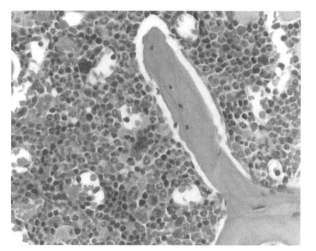

Figure 9.6 Bone marrow (sternum). Bone marrow exhibits large megakaryocytes with multilobulated nuclei, erythroid elements with deeply basophilic nuclei, myeloid elements with larger pale nuclei which differentiate to doughnut form and then to segmental form. The mature erythrocytes are red.

myeloid/erythroid ratio can be made. Finer differentiation requires special stains and preparation techniques.

Brain (cerebrum, encephalon; Figures 9.7 and 9.8)

The mouse brain is lissencephalic, since the surface of cerebral hemispheres is devoid of gyri and sulci. The caudate nucleus and the putamen form a continuous structure, the caudatoputamen. The major parts of the brain are the forebrain (including the cerebral cortex, hippocampus and olfactory bulbs), the upper brain stem (including the basal ganglia, septum, epithalamus, thalamus and hypothalamus), the midbrain (including the tectum, tegmentum and pedunculi cerebri), the cerebellum with the pons and the medulla oblongata. Atlases showing selected brain parts or the complete mouse brain have been published by Kovac and Denk (1968), Krueger (1971), Montemurro and Dukelow (1972) and Sidman *et al.* (1971). More recently, a comprehensive stereotaxic atlas was prepared by Paxinos and Franklin (2000). The brain tissue consist of functional cells (nerve cells, neurons) and the supporting cells, macroglia and microglia. The macroglia are oligodendrocytes, which are the central myelin-forming cells, and astrocytes, which occur both in the gray and the white matter. The ependymal cells line the walls of brain ventricles, are ciliated and may react positively with astrocytic markers such as GFAP. The choroid plexus epthelium forms microvilli and reacts positively with epithelial markers. The central nervous system is covered by meninges (leptomenix, pachymeninx) and surrounded by cerebrospinal fluid.

Brown fat (Figure 9.4)

This type of adipose tissue is found especially between the scapulae (so called 'hibernating gland'), in the axillae, along the jugular veins, adjacent to the thymus, along

Figure 9.7 Brain (cerebellar cortex, cresyl violet and luxol blue).

1. molecular layer
2. Purkinje cell layer
3. granular cell layer
4. white matter

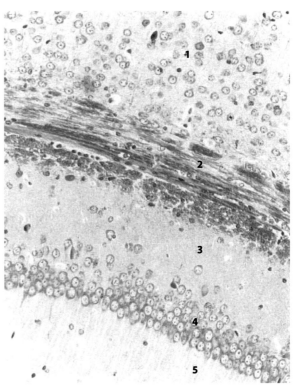

Figure 9.8 Brain (forebrain, cresyl violet and luxol blue).

1. neocortex (gray matter)
2. neocortex, white matter (corpus callosum)
3. molecular layer of hippocampus
4. pyramidal layer of hippocampus
5. polymorphic layer of hippocampus

the aorta, at the renal hilum and along the urethra. The tissue is composed of polygonal cells with multiple lipid droplets in the cytoplasm and centrally located nuclei.

Bulbourethral gland (glandula bulbourethralis; Figures 9.9, 9.10)

The paired bulbourethral glands are male accessory genital glands (also called Cowper's glands). They are located

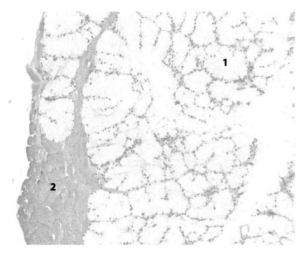

Figure 9.9 Bulbourethral gland (body).
1. epithelium in secretory state (foamy)
2. skeletal muscle

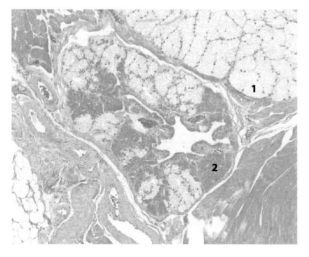

Figure 9.10 Bulbourethral gland (tail).
1. epithelium in secretory state (foamy)
2. epithelium in resting state (eosinophilic, fine granular)

at the base of the penis, and consist of the body and tail. The body is buried in the bulbocavernosus muscle, the tail is covered by ischiocavernosus muscle. The excretory ducts open into the urethra immediately cranially to the urethral diverticulum. The mucosal epithelium has abundant foamy cytoplasm considered a secretory state, and eosinophilic, fine granular cytoplasm considered a resting state. The 'resting' cells occur in the tail area.

Cecum

See intestine

Colon

See intestine

Clitoral gland (glandula clitoridis)

'Female preputial gland' (paired), – *see preputial gland*

Coagulating gland

Anterior prostate, *see prostate gland*

Duodenum

See intestine

Ear (auris; Figure 9.11)

The paired ears are composed of three parts, the external ear, the middle ear and the inner ear. The external ear is

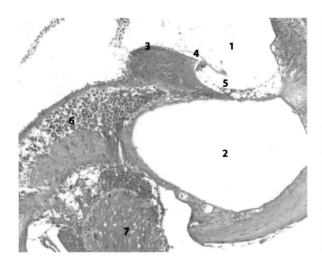

Figure 9.11 Ear (cochlea with organ of Corti).

1. scala media
2. scala tympani
3. limbus
4. tectorial membrane
5. hair cells
6. spiral ganglion
7. cochlear nerve

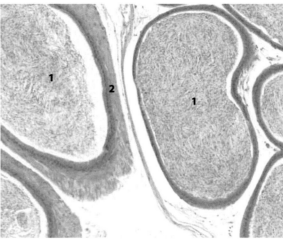

Figure 9.12 Epididymis.

1. spermatozoa in the lumen
2. epididymal ducts formed by epithelial lining and smooth muscle

formed by concha auriculae (auricula, the pinna) and the external auditory canal (meatus). The pinna is an elastic cartilaginous structure covered by skin on both sides. In the mouse, the lateral portion exhibits hair and some sebaceous glands, the medial part has much less hair and more sebaceous glands. The middle ear consists of the tympanic cavity with the tympanic membrane and the Eustachian tube, connecting the middle ear to the pharynx. Within the tympanic cavity lie the auditory ossicles malleus, incus and stapes. The inner ear consists of the labyrinth (organ of equilibrium) and the cochlea (organ of audition). In the cochlea there is the organ of Corti with sensory hair cells which are connected to spiral ganglion cells. The mouse cochlea has one and a half turns, but its length varies among some strains (Stejskal, 1996).

Epididymis (Figure 9.12)

The paired epididymides consist of the head (caput), body (isthmus) and the tail (cauda). The organ contains ducts lined by columnar to cuboidal epithelium and occasional specialized 'clear cells' which contain lysosomal bodies and exert enzymatic activity. The head receives testicular ductuli efferentes, and in the tail begins the vas deferens. The ductal wall contains smooth muscle which becomes more prominent at the transition to vas deferens.

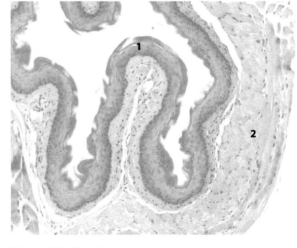

Figure 9.13 Esophagus.

1. stratified squamous epithelium
2. striated muscle arranged in longitudinal and circular direction

Esophagus (Figure 9.13)

The esophagus is located dorsally to the trachea, slightly to the left of the medial level. It is lined by stratified squamous epithelium, and the muscular wall is formed by longitudinal and circular striated muscle fibers.

Estrous cycle

See uterus and vagina

Eye (oculus; Figures 9.14 and 9.15)

The paired eyes are nearly spherical in shape. The lens is also spherical and relatively large. Owing to the nocturnal way of life, the mouse retina does not have areas of increased visual acuity such as a 'central round area' or 'horizontal streak'. The mouse eye is atapetal, i.e. the tapetum lucidum is not developed. The outer fibrous tunic is formed by the sclera and the cornea. The cornea consists of the external layer of stratified squamous epithelium and of the stroma formed by collagen fibers, fibroblasts and a few elastic fibers. The 'Bowman's membrane' is not recognizable in mice. 'Descement's membrane' lines the inner surface of the cornea. The uvea consists of the iris, ciliary body and choroid. Except in albino mice, these layers are pigmented, as well as the retinal pigment epithelium. The lens consists of laminated fibers formed by modified epithelial cells, enclosed by a capsule. The retina is formed by photoreceptor cells, predominantly the rods,

lined by retinal pigment epithelium. There are three layers of cell nuclei arranged in the outer and inner nuclear layer and the innermost ganglion cell layer. The outer nuclear layer is formed by photoreceptors, the inner nuclear layer contains specialized 'bipolar', 'horizontal' and 'amacrine' cells. The glial cells of the retina are the astrocyte-like Müller cells which traverse all retinal layers and have the nuclei located in the inner nuclear layer. The ganglion cells form the axons of the optic nerve, conducting the visual impulses towards the brain.

Gallbladder (vesica biliaris, vesica fellea; Figure 9.16)

The mouse gallbladder is located at the base of the deep bifurcation of the median lobe of the liver. It consists of a fundus, a body and a neck which continues into the

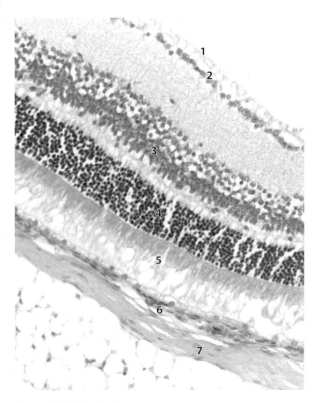

Figure 9.15 Eye (retina).

1. nerve fibers formed by the axons of ganglion cells
2. ganglion cells
3. inner nucler layer
4. outer nuclear layer
5. photoreceptor layer
6. pigment epithelium and choroid
7. sclera

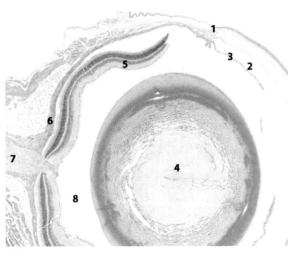

Figure 9.14 Eye.

1. cornea
2. anterior chamber
3. iris
4. lens
5. retina
6. choroid and sclera
7. optic nerve
8. vitreous body

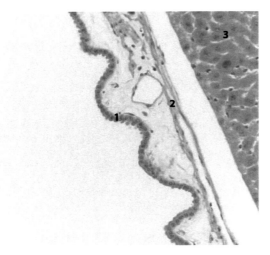

Figure 9.16 Gall bladder.

1. mucosa
2. muscularis and serosa
3. liver

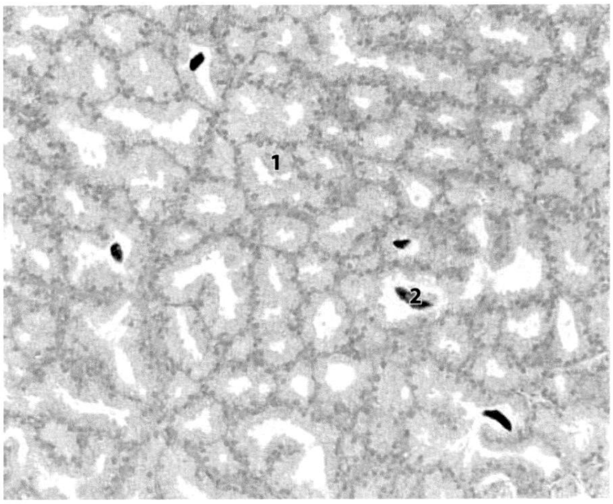

Figure 9.17 Harderian gland.

1. secretory cells with finely vacuolated cytoplasm
2. occasional porphyrin accretions

cystic duct. The cystic duct unites with the hepatic duct to form the common bile duct which opens at the duodenal papilla. The wall of gallbladder is formed by mucous membrane, thin smooth muscle and serosa. The mucosa is lined by cuboidal epithelium and is folded when the bladder is empty.

Harderian gland (Figure 9.17)

The paired Harderian gland is located deep within the orbit, surrounding the optic nerve and several external ocular muscles from the dorsal, medial and ventral direction. The gland has a tubuloalveolar structure. On each side, a single excretory duct resulting from connection of alveolar, lobular and lobar lumina opens at the base of the surface of the nictitating membrane. There are no intralobular or interlobular ducts. The excretory duct is lined by columnar epithelium except at the opening on the nictitating membrane where stratified squamous epithelium occurs. The cytoplasm of secretory cells appears finely vacuolated owing to the presence of lipid droplets containing mainly glyceryl ester diesters and phospholipids. Major secretory products are porphyrins controlling the amount and quality of light reaching the retina and providing photoprotection to the eye. Occasional porphyrin accretions occur in the glandular lumina. In many mouse strains the amount of porphyrins is significantly higher in female

than in male gland. Between the secretory cells and the basement membrane are located the myoepithelial cells, which enable release of secret in response to nervous, especially cholinergic, stimuli. (Krinke *et al.*, 1996).

Heart (cor; Figures 9.18 and 9.19)

The heart is located in the thoracic cavity, surrounded by the pericardium. It is a hollow muscular organ containing left and right atrium and left and right ventricle. The heart wall consists of the endocardium, myocardium and epicardium. The myocardium has striated fibers with centrally located nuclei. At the basis of the heart there is a supportive 'skeleton' formed by fibrous connective tissue. The valves between the atria and ventricles (right tricuspid and left bicuspid) are formed by connective tissue and covered by endocardium. They have a pale, myxomatous appearance. Certain mouse strains are genetically predisposed to develop spontaneous myocardial calcification (Vargas *et al.*, 1996).

Ileum

See intestine

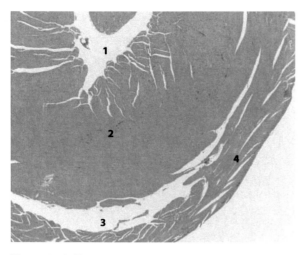

Figure 9.18 Heart.

1. left ventricle
2. interventricular septum
3. right ventricle
4. wall of the right ventricle

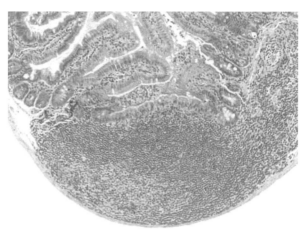

Figure 9.20 Intestine (Peyer's patch). The Peyer's patch consists of lymphoid tissue located between the muscularis and the mucosal epithelium.

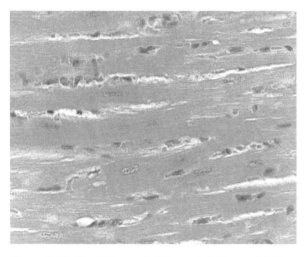

Figure 9.19 Heart. Myocardial fibers exhibit cross striation and centrally located nuclei.

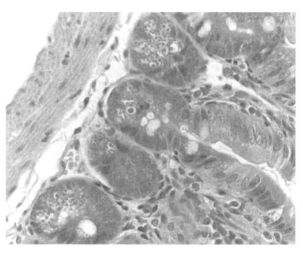

Figure 9.21 Intestine (Paneth cells). Paneth cells contain large eosinophilic granules and are located at the base of intestinal crypts.

Intestine (intestinum; Figures 9.20–9.27)

The intestine is divided into the small and large intestine. Throughout its length the intestine has three principle layers: the mucosa with submucosa, the muscularis and the serosa. The mucosa has an epithelial lining and fibrovascular stroma called the 'lamina propria mucosae'. The lamina propria is separated from the submucosa by the 'lamina muscularis mucosae', a

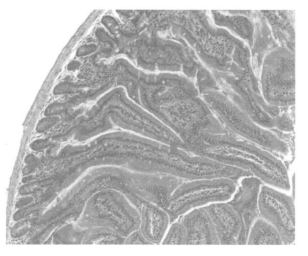

Figure 9.22 Intestine (duodenum).

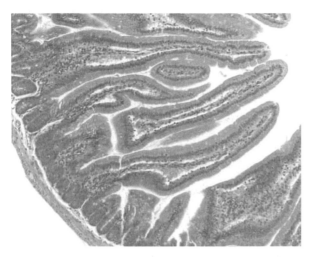

Figure 9.23 Intestine (jejunum).

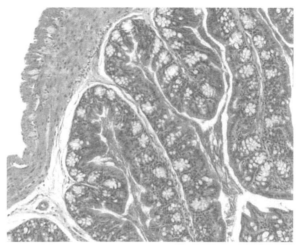

Figure 9.26 Intestine (colon).

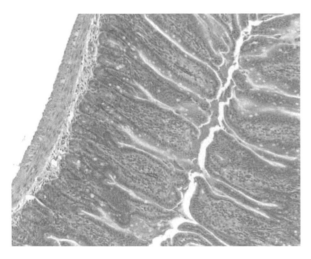

Figure 9.24 Intestine (ileum).

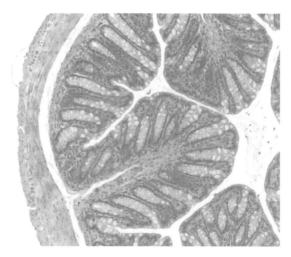

Figure 9.27 Intestine (rectum).

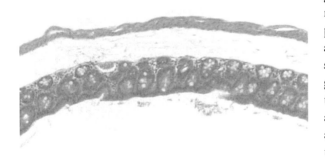

Figure 9.25 Intestine (cecum).

thin layer of smooth muscle. The submucosa consists of connective tissue surrounding blood and lymphatic vessels and nerves. The muscularis consists of an outer longitudinal layer and an inner circular layer of smooth muscle. The serosa is formed by a thin layer of visceral peritoneum. The lymphoid tissue (GALT, gut-associated lymphatic tissue) forms nodules scattered in the submucosa and the lamina propria. The larger aggregates form 'Peyer's patches', which in the small intestine are located opposite the mesenteric attachment, in an 'antimesenteric' position. In the large intestine they are not strictly antimesenteric. The most common cell type of the mucosal epithelium are the absorptive cells with a luminal cell membrane forming microvilli. The mucous goblet cells are scattered between other cell types. The Paneth cells contain brightly eosinophilic cytoplasmic granules, especially large in the mouse, which contain lysozyme and antimicrobial peptides. They occur in the small intestine, especially the jejunum and become conspicuous after several hours of fasting. The enteroendocrine cells are polypeptide-producing endocrine cells, diffusely distributed along the gastrointestinal tract. On the surface of the Peyer's patches the epithelium forms so called 'M cells' which serve to sample antigens. So called 'caveolated cells' are considered to represent intestinal chemoreceptors.

Small intestine (intestinum tenue)

The small intestine is formed by the duodenum, jejunum and ileum. The mucosal surface of the small intestine of the mouse lacks the folds (plicae) that are found in larger species. The mucosa forms villi consisting of epithelium and lamina propria, which project into the intestinal lumen. Each villus contains a central lymph vessel, the lacteal. The length of the villi decreases from the duodenum to the ileum. Among the villi are mucosal protrusions in opposite direction, beneath the mucosal surface, forming so called 'crypts' or 'intestinal glands'. The initial portion of duodenum is equipped by special tubuloalveolar duodenal glands, the so called 'Brunner's glands'. One or more main pancreatic ducts and the common bile duct open at the duodenal papilla.

Large intestine (intestinum crassum)

The large intestine consists of the cecum, colon and rectum. The mucous membrane of the large intestine contains a larger proportion of goblet cells than that of the small intestine. It forms crypts, but no villi. The mouse cecum has a corpus and apex. The entrance of the ileum forms the 'sacculus rotundus', and the exit of the colon the 'ampulla coli'. The cecal mucosa forms transverse folds. The colon has an ascending, a transverse and a descending part. The mucosa of the ascending and transverse colon forms transverse folds, whereas the descending colon and the rectum have prominent longitudinal folds, protruding into the lumen, formed by mucosa and submucosa. The muscularis mucosae is more prominent in the rectum than in the colon. At the transition of rectum to anus the surface epithelium becomes stratified squamous. Around the anus are the modified sebaceous 'circumanal glands'.

Jejunum

See intestine

Joint (articulatio)

The bones are fixed to each other by different types of joints. Some joints are quite rigid and fibrous (such as sutures in the skull), or cartilaginous (such as between the vertebrae or the sternebrae). The synovial joints connect the bones more loosely, allowing for movement. The bones in synovial joints are equipped with articular cartilage, sometimes with additional cartilaginous menisci, ligaments and capsule. The inside of the synovial joints is covered by the synovial membrane and filled with synovial fluid.

Kidney (ren; Figures 9.28–9.30)

The paired kidneys are located in the dorsal part of abdominal cavity, the right kidney slightly more cranially then the left. The right kidney is usually larger than the left, and male kidneys are relatively larger than female. The mouse kidney is unilobar with a single papilla. It consists of the cortex and the medulla. The cortex contains 'cortical tubular labyrinths' (mainly proximal convoluted tubules), and 'medullary rays' extending from the outer medulla. The medulla is subdivided in an outer and inner zone. The outer zone has an 'outer and inner stripe', the inner zone forms the papilla. The functional unit is the nephron, consisting of the glomerulus, convoluted and straight portions of the proximal tubule, the descending and ascending portions of the loop of Henle, and the straight and convoluted portions of the distal tubule. The nephrons are connected to the collecting ducts, which run into papillary ducts. The papillary ducts open at the tip of the renal papilla into the renal pelvis. The renal pelvis is lined by a transitional cell epithelium and its continuation forms the ureter. The mouse renal papilla may be very long and protrude into the initial portion of the ureter. The glomerulus is surrounded by the Bowman's capsule, which in most mouse strains is considered to exhibit sexual dimorphism: the parietal

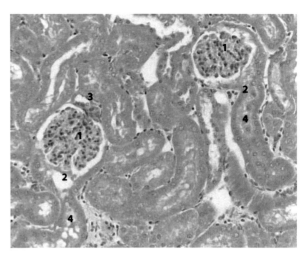

Figure 9.30 Kidney (cortex).

1. glomerulus   3. vascular pole with macula densa
2. urinary pole   4. proximal convoluted tubule

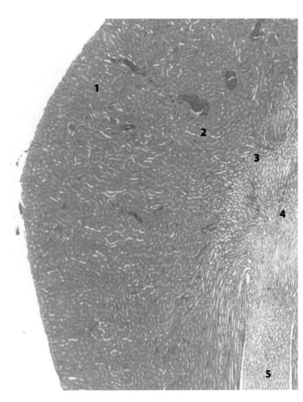

Figure 9.28 Kidney (cortex, medulla).

1. cortex   4. inner medulla
2. outer stripe of outer medulla   5. papilla
3. inner stripe of outer medulla

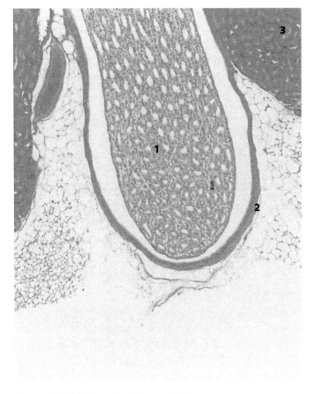

Figure 9.29 Kidney (pelvis, papilla).

1. papilla   2. pelvis   3. cortex

epithelial cells are cuboidal in males and flattened in females. This difference could not be demonstrated in the sections of CD-1 mouse kidneys used to illustrate this text. Regardless of the gender, the parietal cells of Bowman's capsule were flattened at the vascular pole and cuboidal at the urinary pole of glomeruli. Reportedly the sexual dimorphism occurs under influence of testosterone only in mature males. The exact statement regarding this dimorphism says that 'many (but obviously not all) of the Bowman's capsules of the adult male are lined by cuboidal cells' (Hummel *et al.*, 1975).

The proximal tubules are found mainly in the cortex. They have cuboidal cells with prominent brush border (microvilli). The descending and ascending portions of the loop of Henle are found in the medulla. They are lined by flattened epithelium resembling the endothelium of blood vessels. The distal tubules reenter the cortex and have a cuboidal epithelium similar to that of the proximal tubules, but devoid of a brush border. The straight portion of the distal tubules leads to the macula densa at the vascular pole of glomerulus, where renin is produced by specialized cells. The mouse renal vasculature is similar to other species. The branches of the renal artery form the arcuate arteries at the corticomedullary border. Interlobular branches of the arcuate arteries supply the afferent arterioles of glomeruli. The efferent arterioles supply the cortex and form the descending vasa recta which supplies the medulla. The venous blood collects in the ascending vasa recta and interlobular and arcuate veins. Spontaneously occurring vacuolation, probably of lysosomal origin, in the renal tubular epithelium of the outer medulla in CD-1 mice was described by Johnson *et al.* (1998).

Lacrimal gland (glandula lacrimalis; Figure 9.31)

The mouse possesses two pairs of lacrimal glands. The exorbital glands are located subcutaneously ventral and anterior to the external ear. The intraorbital glands are located at the outer canthus, where the joint excretory ducts of both ipsilateral glands open. The lacrimal glands are tubuloacinar, consisting of lobes and lobules. The serous secretory cells have basophilic cytoplasm near the basally located nuclei and more pale cytoplasm at the lumen. The cells are larger than in the parotid salivary gland, which otherwise is similar. Myoepithelial cells are found between the epithelium and the basement membrane. The intralobular ducts are lined by cuboidal cells and the excretory ducts by stratified columnar epithelium. (Krinke et al., 1996).

Larynx

The larynx is located between the pharynx and the trachea. The entrance of the larynx is bordered by the epiglottis. The laryngeal wall is formed by three layers: the epithelial lining, the cartilage with striated muscles and vocal cords, and the outer loose connective tissue. The epithelial lining varies from stratified squamous epithelium cranially on the epiglottis, to pseudostratified

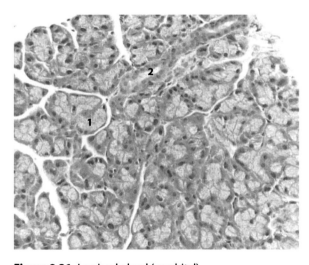

Figure 9.31 Lacrimal gland (exorbital).
1. secretory cells   2. intralobular duct

ciliated columnar respiratory epithelium caudally, at the transition to the trachea. At the base of epiglottis there are subepithelial seromucous glands. Specific areas of the laryngeal wall have intermediate types of epithelium. The vocal folds and vocal processes are covered by low cuboidal to squamous epithelium, the ventral laryngeal poach (diverticulum) is lined by a mixture of ciliated columnar and cuboidal cells. Recommendations about the appropriate processing for detailed histological examination of the mouse larynx have been published by Renne et al. (1992) and Sagartz et al. (1992).

Leydig cells (Figures 9.60 and 9.61)

The endocrine Leydig cells are located between the seminiferous tubules of the paired testes. They are also called 'interstitial cells'. Their cytoplasm is abundant, eosinophilic, and can be finely vacuolated. The cells produce testosterone under the regulation of pituitary luteinizing hormone (LH).

Liver (hepar; Figures 9.32 and 9.33)

The mouse liver consists of the left lateral lobe, the median lobe subdivided into left and right portions, the right lateral lobe subdivided horizontally into anterior and posterior portions, and the caudal lobe subdivided to two portions, located dorsally and ventrally to the esophagus. The posterior surface of the caudal lobe forms the papillary process. This particular pattern of hepatic lobulation is most frequent, however at least 13 different patterns have been described (Hummel et al., 1975). The surface of the liver is covered by a fibrous capsule forming connective tissue septa within the liver tissue. The liver tissue is arranged in lobules with portal triads at the periphery and the central vein in the middle. The portal triads consist of branches of the hepatic artery and portal vein, as well as intrahepatic bile ducts. The blood flows from the perilobular area towards the central vein from where it is conducted over large hepatic veins to the vena cava. The liver cells, hepatocytes, are arranged in plates radiating from the central vein towards lobular periphery. A characteristic feature of the mouse liver is normally occurring anisocytosis and anisokaryosis, e.g. great variation in size of the liver cells and their nuclei. The hepatocytes have a bile canalicular surface, which together with the surfaces of other hepatocytes forms the bile canaliculus, and a perisinusoidal surface, which is separated by the space of Disse from the sinusoidal wall formed by fenestrated endothelial cells. Specialized hepatic cells are the Kupffer cells, fixed macrophages attached to the sinusoidal wall, Ito cells containing cytoplasmic lipid droplets and storing vitamin A, and pit cells which are large granular lymphocytes with activity of natural killer cells. During the first few weeks of postnatal life, megakaryocytes can be seen in the mouse liver. In contrast to the anatomical hepatic lobules, the functional units, defined as acini, have their center at the portal triads and the periphery at the central vein.

Lung (pulmo; Figures 9.34 and 9.35)

The lung is located in the thoracic cavity, covered by visceral and surrounded by the parietal pleura. The

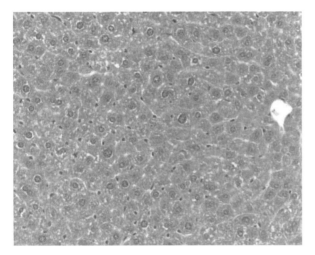

Figure 9.32 Liver. Anisocytosis, an uneven size of hepatocytes, and anisokaryosis, an uneven size of hepatocellular nuclei, are characteristic features of the mouse liver.

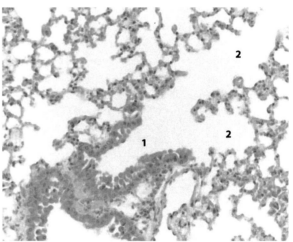

Figure 9.34 Lung.
1. terminal bronchiole   2. alveolar duct

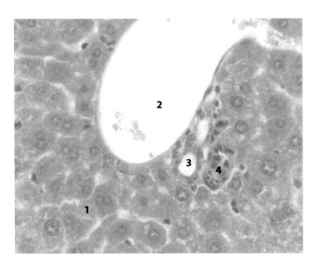

Figure 9.33 Liver.
1. hepatocytes, some of them binucleated
2. portal triad: vein
3. portal triad: bile duct
4. portal triad: artery

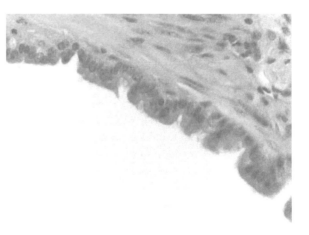

Figure 9.35 Lung (main bronchus). Epithelial cell lining of the main bronchus consists mainly of Clara cells and ciliated columnar cells.

left lung forms a single lobe, whereas the right lung is subdivided into cranial, middle, caudal and accessory lobes. At least nine patterns of pulmonary lobulation have been described (Hummel *et al.*, 1975). The lung is entered by the main bronchi produced by tracheal bifurcation. The main bronchi branch to form intra-pulmonary bronchi to terminate as bronchioles. The larger airways are lined by columnar epithelial cells, mainly nonciliated Clara cells, ciliated cells, and some neuroendocrine cells, mucous cells, and brush cells. The mucous cells are of a small granule type, goblet cells with a prominent mucous cytoplasm do not occur in the mouse lung. The smallest airway of the mouse lung is the terminal bronchiole which opens into the alveolar ducts. The alveolar ducts lead to alveolar sacs and alveoli. The alveolar epithelium consists of thin pulmonary type I cells and cuboidal type II cells. The lymphoid tissue (BALT, bronchial-associated lym-phatic tissue) is rarely seen in healthy mouse lung. The lung receives nutritive blood from the systemic circula-tion via the bronchial arteries and venous blood via the pulmonary arteries from the heart. The pulmonary arteries and veins follow the bronchial tree. The walls of pulmonary veins contain cardiac (striated) muscle.

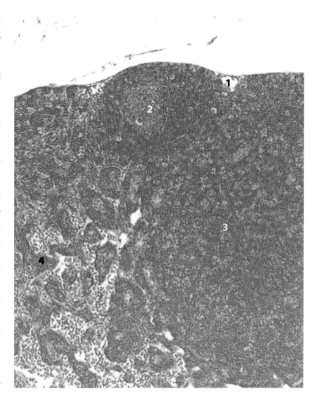

Figure 9.36 Lymph node (mesenteric).

1. subcapsular sinus   3. paracortex
2. lymphoid follicle   4. medullary cords

Lymph nodes (lymphonodi, nodi lymphatici; Figure 9.36)

The lymph nodes are connected with the lymphatic system and distributed through the whole body. Some are paired (e.g. axillary) but most of the visceral nodes are not (e.g. mesenteric). Lymphatic tissue incorpo-rated in selected organs is commonly known as NALT (nose-associated lymphatic tissue), BALT in the lung, and GALT in the intestine. The lymph nodes are lym-phatic structures separate from the organs. Each lymph node is covered by a connective tissue capsule which can form septae within the node. The lymph nodes consist of numerous endothelial sinuses and reticular tissue, arranged in a meshwork filled with lymphatic cells. The lymph enters the lymph node through vasa afferentia which penetrate the capsule and reaches the subcapsular sinus, which is connected to paratrabecular and medullary sinuses. It exits through the vas efferens

in the hilum. The superficial lymphatic tissue forms the cortex which contains lymphatic follicles, mainly com-posed of B cells. Primary follicles are nonstimulated and contain dense aggregates of small lymphocytes. Secondary follicles are produced in response to immune stimulation. They are larger and have germi-nal centers with numerous large pale lymphoblasts and some macrophages with cell debris. The follicles are surrounded by paracortex, composed mainly of T-cells. The periphery of the paracortex has the highest con-centration of specialized 'high endothelial venules'. The deep lymphatic tissue forms the medulla which is arranged in medullary cords spreading towards the hilum.

Mammary gland (glandula lactifera, mamma; Figure 9.37)

Female mice have five pairs of mammary glands, three pairs in the cervicothoracic region and two pairs in the inguinoabdominal region. Male mice have only four

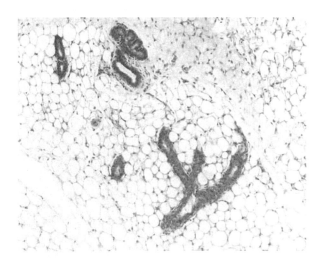

Figure 9.37 Mammary gland (virgin female). Resting mammary gland in a virgin female consists essentially of lactiferous ducts embedded in adipose tissue.

pairs of glands and no nipples. The mammary gland is a compound tubuloalveolar gland. In prepubertal mice, there is a branched system of lactiferous ducts, embedded in adipose tissue. Postpubertal development reaches the maximum in virgin females of 4–7 months and consists of further duct proliferation and formation of some secretory alveoli. Complete lobuloalveolar development occurs only during pregnancy and lactation. The lactiferous ducts are lined by pseudostratified low columnar or cuboidal epithelium, the alveoli by low cuboidal epithelium. The secretory cells are surrounded by myoepithelial cells. Details of mouse mammary gland biology have been reviewed by Cardiff (2002).

Nasal cavity (cavum nasi; Figures 9.38–9.41)

The nasal cavity is separated by cartilaginous septum. From the walls so called turbinates (conchae) project into the lumen these being formed by bone and covered by mucous membrane. In the anterior nasal cavity there are nasoturbinates and maxilloturbinates, whereas the posterior nasal cavity contains ethmoturbinates. Among the turbinates are the air passages, the dorsal, middle and ventral meatus. The vomeronasal organ occurs medioventrally in the nasal cavity, and is an organ of chemoreception for pheromones and food flavor.

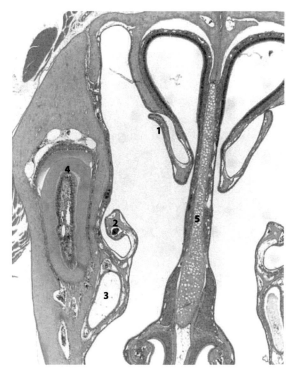

Figure 9.38 Nasal cavity.

1. nasoturbinate
2. maxilloturbinate
3. nasolacrimal duct
4. incisor tooth
5. nasal septum

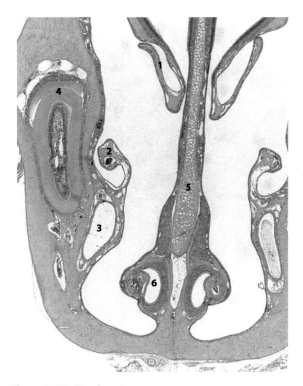

Figure 9.39 Nasal cavity.

1. nasoturbinate
2. maxilloturbinate
3. nasolacrimal duct
4. incisor tooth
5. nasal septum
6. vomeronasal organ

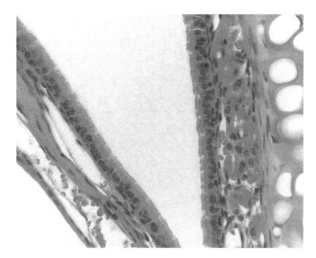

Figure 9.40 Nasal cavity. Pseudostratified columnar respiratory epithelium.

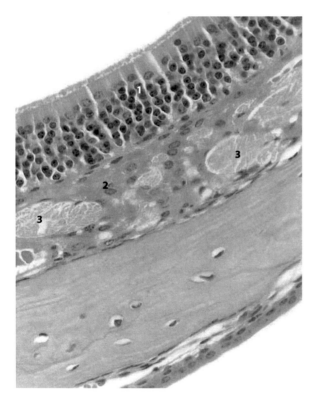

Figure 9.41 Nasal cavity.

1. olfactory epithelium
2. lamina propria with Bowman's glands
3. bundles of nerve fibers

The paired nasolacrimal ducts connect the medial canthus of the eye with the nasal cavity and pass through the bony nasolacrimal canal. The anterior portion of the nasal cavity (vestibule) is lined by stratified squamous epithelium which extends through the ventral meatus into the pharynx. The nasoturbinates, maxilloturbinates, cranioventral portion of ethmoturbinates and most of the nasal septum are covered by respiratory epithelium which contains ciliated and nonciliated columnar cells, cuboidal cells, goblet cells, brush cells and basal cells. The lamina propria of the respiratory epithelium contains serous glands at the anterior nasal septum and mucous glands at the posterior septum. The lateral walls of air passages and the naso- and maxilloturbinates are lined by 'transitional respiratory epithelium' consisting of cuboidal cells, unciliated columnar cells, brush cells and basal cells. The dorsal wall of the nasal cavity and the ethmoturbinates are covered by olfactory epithelium. This epithelium is pseudostratified columnar and consists of specialized bipolar olfactory neurons, sustentacular (supporting) cells and basal cells. The axons of olfactory neurons form bundles of unmyelinated nerve fibers which synapse with the neurons in the olfactory bulb. The lamina propria of olfactory epithelium contains tubuloalveolar Bowman's glands. Mery *et al.* (1994) have developed 'nasal diagrams' which demonstrate the nasal topography at various section levels and serve as a basis for assessment of histopathological findings.

Optic nerve (nervus opticus; Figure 9.14)

The paired optic nerves are formed by the processes of retinal ganglion cells connecting the eyes to the brain. The optic nerve tissue belongs to the central nervous tissue, the myelinating cells are oligodendrocytes and the outer sheaths investing the optic nerve are a continuation of cerebral meninges. The mouse optic nerve contains about 65,000 nerve fibers.

Oral cavity (cavum oris)

The upper lip is split, exposing the two upper incisor teeth. The lip folds close the space between the incisor and the molar teeth (the diastema). The hard palate has eight rows of ridges formed by dense connective tissue. The mucosa is formed by keratinizing stratified squamous epithelium. There is no distinct submucosa. The soft palate forms the roof of the posterior oral cavity and the floor of the nasopharynx.

Ovary (ovarium; Figures 9.42 and 9.43)

The paired ovaries lie caudally and laterally to the kidneys. They are surrounded by the ovarian bursa and connected to the uterine horns by the convoluted oviducts. The ovary is covered by simple cuboidal to columnar epithelium and consists of numerous follicles, yellow bodies and clusters of polygonal interstitial cells. Rudiments of the rete ovarii may persist near the hilum as blind tubules or cords of epithelial cells. The follicles contain eggs and develop from small primordial follicles to larger growing follicles and large preovulatory 'Graafian follicles', which, owing to the presence of a cavity belong to 'antral' or 'vesicular' follicles. Mature follicles contain an egg (oocyte) surrounded by granulosa cells and outer layer of fusiform theca cells. Polyovular follicles may occur in young mice or in certain strains. During the life time, only about 20% of the available follicles ovulate, the majority undergo 'follicular atresia' characterized by cell death (apoptosis) of granulosa cells. The mature follicles produce hormones such as estradiol, inhibin, progesterone and androgens. Follicular growth is regulated by pituitary follicle stimulating hormone (FSH) and LH and ovulation is stimulated by release of LH. After ovulation, the granulosa and theca cells form a progesterone-producing yellow body (corpus luteum). The presence of a regular number of oocytes and follicles in different stages of development is considered an indicator of intact fertility and is tested in special experimental procedures. A classification scheme for mouse ovarian follicles was proposed by Pedersen and Peters (1968). They further classify small, medium and large follicles to 10 types or subtypes in all. For practical applications, Bolon *et al.* (1997) and Buci *et al.* (1997) developed a simplified classification categorizing the follicles as small, growing and antral.

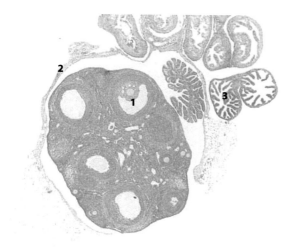

Figure 9.42 Ovary.

1. large antral follicle   2. ovarian bursa   3. oviduct

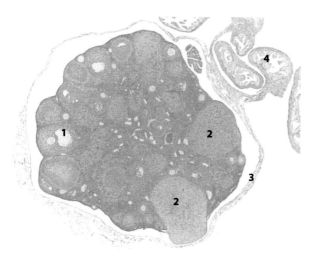

Figure 9.43 Ovary.

1. antral follicle   3. ovarian bursa
2. corpus luteum   4. oviduct

Oviduct (oviductus, salpinx uterina, tuba uterina, tuba fallopii; Figures 9.42 and 9.43)

The paired oviducts are convoluted tubes connecting the ovaries to the uterine horns. The oviduct begins in the ovarian bursa by infundibulum and widened ampulla, it continues as a narrow, coiled isthmus and terminates as an intramural portion in the uterine wall. The epithelium lining the infundibulum and the ampulla is ciliated columnar, that of the isthmus pseudostratified and low columnar with occasional ciliated cells, and that of the intramural portion simple columnar. The wall is formed by smooth muscle and adventitia.

Pancreas (Figure 9.44)

The pancreas is located in the mesentery of the duodenal loop and the transverse colon and in the greater omentum close to the stomach and spleen. In other species, the organ is subdivided in left lobe (tail), body, and right lobe (head), but such subdivision is not apparent in mice. The exocrine pancreas is a compound acinar gland. The acinar cells are pyramidal-shaped and in hematoxylin and eosin stained sections have basophilic cytoplasm and large nuclei in the basal portion and acidophilic zymogen granules in the apical portion. The cell shape and granule content depend on the secretory activity. The acini are connected to intercalated ducts leading to intralobular and then to interlobular ducts which open to the main excretory ducts. One or more main excretory pancreatic ducts lead to the duodenal papilla.

Pancreatic islets (insulae pancreaticae; Figure 9.44)

The pancreatic islets (islets of Langerhans) form the endocrine pancreas. The islets consist of pale staining polygonal cells and are well capillarized. The cell types include the glucagon-producing alpha cells, the somatostatin producing delta cells, and the pancreatic polypeptide-producing PP cells, all three kinds being located at the insular periphery. The insulin producing beta cells are located in the center of the islets.

Parathyroid gland (glandula parathyroidea; Figure 9.63)

The paired parathyroid glands are located bilaterally at the surface of the thyroid gland. Occasionally, they may lie deep within the thyroid, or be more than two. They consist of cords of polygonal cells which are either active, dark chief cells, or inactive light cells. Between the cells are numerous blood capillaries and sinusoids. The product of the parathyroid gland is parathormone (PTH).

Penis (Figure 9.45)

The penis consists of the root, body and glans. It contains the distal part of the urethra, cavernous bodies

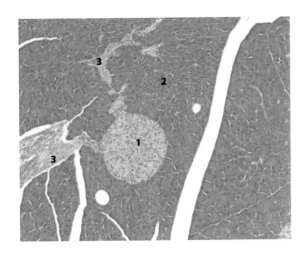

Figure 9.44 Pancreas.
1. pancreatic islet
2. exocrine acini
3. intralobular duct

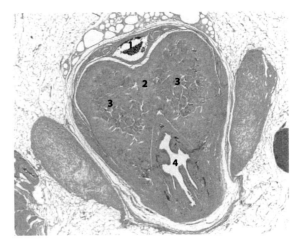

Figure 9.45 Penis.
1. dorsal vein and nerves
2. os penis (cartilaginous part)
3. corpora cavernosa
4. urethra

and a small os penis. The glans is enclosed in a fold of modified skin, the prepuce.

Peripheral nerve

As in other species, the peripheral nerves consist of unmyelinated and myelinated nerve fibers and connective tissue sheaths. In contrast to the central nervous system where myelin is formed by the oligodendrocytes, in the peripheral nervous system myelin is formed by the Schwann cells. A variety of mouse mutant strains serve as models of genetic diseases of human peripheral nerves (Krinke, 1996; Notterpek and Tolwani, 1999).

Pharynx

The pharynx is the site behind the nasal and oral cavities where the respiratory and digestive passages cross. It is lined by stratified squamous epithelium. It's dorsal part (oropharynx) receives the openings of the Eustachian tubes.

Pineal body (glandula pinealis, epiphysis cerebri)

The pineal body lies on the dorsal surface of the brain, in the midline, between the cerebral hemispheres and the cerebellum.

Pituitary gland (glandula pituitaria, hypophysis cerebri; Figure 9.46)

The gland is located on the ventral surface of the brain, in the midline, attached to the hypothalamus. In the

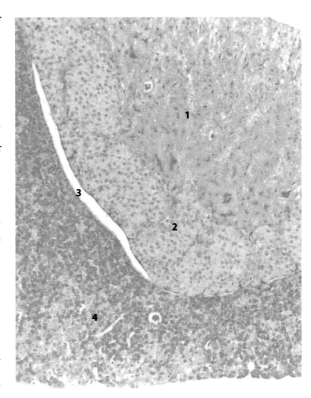

Figure 9.46 Pituitary gland.

1. neurohypophysis	3. hypophyseal cleft
2. pars intermedia	4. pars distalis

mouse, the gland is relatively larger in females than in males. It consists of the adenohypophysis and the neurohypophysis. The adenohypophysis is made up of the pars tuberalis, pars distalis and pars intermedia. Between the pars distalis and the pars intermedia there is a hypophyseal cleft lined by a layer of epithelial cells. Traditional classification of the secretory cells of the adenohypophysis is based on their appearance in hematoxylin and eosin stained sections. In this stain the cells are acidophils, basophils or chromophobes. Immunohistochemical markers enable more detailed classification to be made according to secretory products. The acidophils include growth hormone (GH) producing somatotrophs, and prolactin (PRL) producing mammotrophs. The basophilic gonadotrophs produce the FSH and LH. Also the thyrotrophs producing thyroid stimulating hormone (TSH) are basophils. The pars intermedia contains chromophobes, especially corticotrophs producing adrenocorticotropic hormone (ACTH) and melanotrophs, producing melanocyte stimulating hormone (MSH). The neurohypophysis contains terminal axons of the hypothalamic neurosecretory neurons producing vasopressin, oxytocin and antidiuretic hormone (ADH), and modified astroglia, so called 'pituicytes'.

Preputial gland (glandula preputialis; Figure 9.47)

The paired preputial glands are modified sebaceous glands, located in the subcutaneous adipose tissue lateral to the penis (in females to the clitoris). The excretory ducts are lined by stratified squamous epithelium, and have wide lumina. The gland opens at the border of prepuce and skin (clitoral fossa in females).

Prostate gland (glandula prostatica; Figure 9.48)

The prostate gland is the male accessory genital gland. It consists of anterior, dorsal and ventral lobes. The anterior lobes are known as the 'coagulating gland' and are attached to the seminal vesicles. The dorsal lobes surround the urethra as a single body, the ventral lobes lie between the urethra and the urinary bladder. All three lobes open into the urethra: the anterior lobes to the dorsal wall cranially to the seminal collicle, the dorsal lobes to the lateral walls and the ventral lobes to the ventral walls. The gland is tubuloalveolar, lined by epithelium

which may be flattened to columnar, depending on the secretory activity.

Rectum

See intestine

Salivary glands (glandulae salivales; Figures 9.49, 9.50, 9.80 and 9.81)

The mouse has three pairs of major salivary glands: the parotid gland, the submandibular (submaxillar, mandibular) gland and the sublingual gland. There are also minor glands within the tongue, palate, pharynx and larynx. The major salivary glands are located ventrally in the subcutaneous tissue of the neck. The parotid gland extends laterally to the base of the ear and lies adjacent to the exorbital lacrimal gland. All three major salivary glands are compound tubuloalveolar, the parotid is serous and the sublingual mucous. The submandibular gland is mostly described as mixed, serous and mucous. However, some authors describe it as

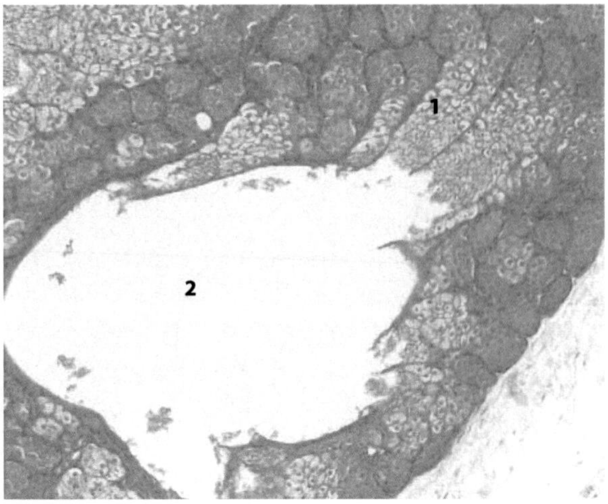

Figure 9.47 Preputial gland.
1. sebacious cells   2. excretory duct

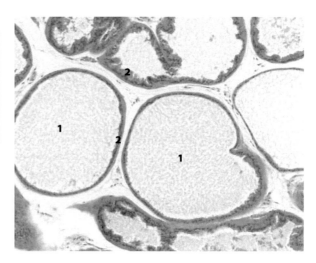

Figure 9.48 Prostate gland (ventral lobe).
1. alveolus containing secret
2. alveolar wall lined by epithelium of variable height

serous (Hummel *et al.*, 1975); indeed, the microscopic appearance of the acinar cells is usually serous and distinctly different from the mucous acini of the sublingual gland. However, occasionally glands with mucous acini can be encountered, so that the appearance of acini may depend on their physiological state. The excretory ducts of the submandibular and sublingual glands open caudally to the level of incisor teeth, and those of the parotid opposite the lower molars.

The submandibular gland exhibits prominent sexual dimorphism. In male mice the acinar cells, and especially the cells of convoluted (granular) ducts are larger than in females. The convoluted (granular) ducts occur only in the submandibular gland and produce biologically active polypeptides including nerve growth factor and epidermal growth factor. In all three glands there are intercalated, intralobular and interlobular excretory ducts. The serous acinar cells have cytoplasm which is basophilic at the base and granular eosinophilic in the apical portion. The mucous acinar cells have basally located nuclei and pale, slightly basophilic cytoplasm. The acinar cells are surrounded by myoepithelial cells. In the mouse parotid gland, foci of basophilic hypertrophic acinar cells may occur spontaneously (Chiu and Chen, 1986).

Seminal vesicle (glandula vesiculosa, vesicula seminalis; Figures 9.51 and 9.52)

The paired seminal vesicles are male accessory genital glands. They are relatively large, located dorsolaterally to the urinary bladder and attached to the anterior prostate (coagulating gland). The ducts of the seminal vesicles open at the seminal collicle, together with those of the prostatic lobes. The combined secretions of the seminal vesicles, the prostate and the bulbourethral glands form a copulatory plug which prevents outflow of semen from the vagina after ejaculation. The seminal vesicle has a wall composed of smooth muscle and tall columnar epithelium forming branching mucosal folds. When the gland is distended owing to a large content of secretory material, the folds stretch and become short.

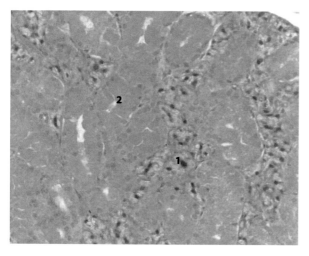

Figure 9.49 Salivary gland (male submandibular).
1. acini   2. convoluted ducts

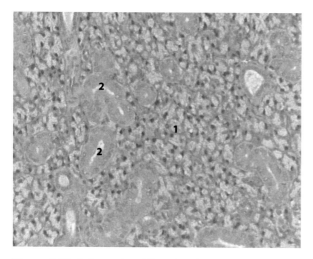

Figure 9.50 Salivary gland (female submandibular).
1. acini   2. convoluted ducts

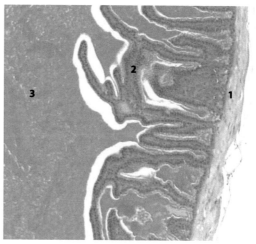

Figure 9.51 Seminal vesicle.
1. smooth muscle wall
2. tall columnar epithelium lining branching mucosal folds
3. secretory material (brightly eosinophilic)

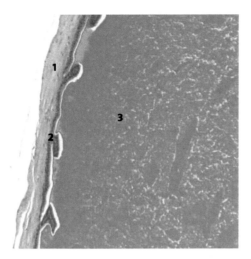

Figure 9.52 Seminal vesicle.
1. smooth muscle wall
2. epithelium forming only short mucosal folds, owing to its distention by large amount of secretory material
3. secretory material (brightly eosinophilic)

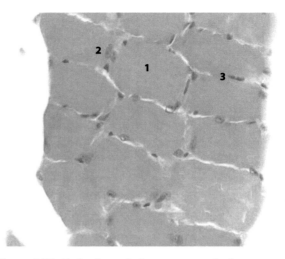

Figure 9.54 Skeletal muscle (transverse section).
1. cytoplasm
2. nuclei of muscle fibers (pale, oval)
3. endomysial nuclei (dark, elongated)

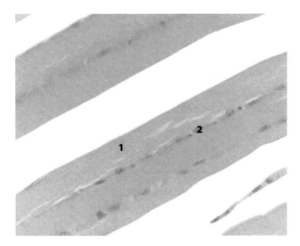

Figure 9.53 Skeletal muscle (longitudinal section).
1. cross striated cytoplasm of muscle fibers
2. peripherally located nuclei

Skeletal muscle (Figures 9.53 and 9.54)

As in other species, the skeletal muscle consists of striated, extrafusal and intrafusal muscle fibers, connective tissue, blood vessels, nerve fibers and motor and sensory nerve endings. Each muscle fiber is a multinucleated cell with nuclei located at the periphery and cytoplasm containing contractile myofibrils.

Skin (cutis, derma, integumentum commune; Figure 9.55)

The skin is composed of epidermis, dermis and subcutis (hypodermis). 'Skin adnexa' are hair follicles and sebaceous glands. The epidermis consists of four layers: the basal cell layer (stratum basale), the prickle cell layer (stratum spinosum), the granular cell layer (stratum granulosum) and a horny layer (stratum corneum). In pigmented (non-albino) strains, melanin pigment occurs in the cells of the basal layer, the hair follicles and hairs. It is produced by melanocytes. The dermis (corium) consists of fibrous connective tissue and has subepidermal papillary layer and a deeper reticular layer. The subcutis is formed by loose connective tissue with a moderate amount of fat tissue. A sheet of striated skeletal muscle lies between the dermis and the subcutaneous tissue and is prominent especially in the regions of neck, thorax and abdomen. The haircoat of the mouse (pelage) is formed by short hairs. Among the pelage hairs are scattered longer 'guard hairs', which have tactile function. Very large tactile hairs, the 'vibrissae' occur on the nose. The hairs are keratohyaline products of epithelial hair follicles which protrude from the epidermis into the dermis, these follicles being associated with sebaceous

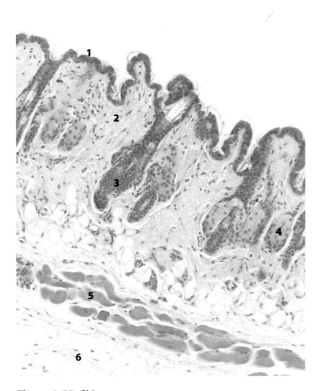

Figure 9.55 Skin.
1. epidermis
2. dermis
3. hair follicle
4. sebaceous gland
5. striated muscle
   (panniculus carnosus)
6. subcutis

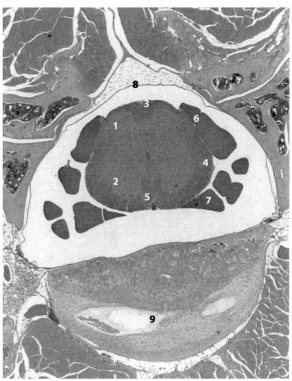

Figure 9.56 Spinal cord (lumbar segment).
1. gray matter (dorsal horn)
2. gray matter (ventral horn)
3. white matter (dorsal
   columns)
4. white matter (lateral
   columns)
5. white matter (ventral
   columns)
6. dorsal spinal root
7. ventral spinal root
8. pachymeninx
9. intervertebral disk

glands. In the mouse the sweat glands occur only on the footpads.

Spinal cord (medulla spinalis; Figure 9.56)

The spinal cord consists of central gray matter surrounded by columns of white matter. The dorsal columns contain ascending, sensory nerve fibers, the ventral columns descending, motor nerve fibers, and the lateral columns both. The gray matter has sensory dorsal horns and ventral motor horns. In the middle is the central spinal canal. The spinal cord has segmental organization and it forms a pair of spinal nerve roots in each segment. In the mouse, the roots of the fourth to eighth cervical and first and second thoracic segment contribute to the brachial plexus supplying the forelimbs, and the roots of the third to sixth lumbar segment contribute to the lumbosacral plexus supplying the hindlimbs.

Spleen (lien; Figure 9.57)

The spleen lies in the left dorsocranial part of the abdominal cavity, along the greater curvature of the stomach. It has elongated shape and is triangular in transverse section. The spleen has a connective tissue capsule which spreads into the parenchyma forming splenic trabeculae. The parenchyma consists of white and red pulp. The white pulp is organized into periarteriolar lymphoid sheets (PALS, representing mainly T cells) and lymphatic follicles, which become prominent in response to stimulation. The periphery of the white pulp is formed by a less densely cellular marginal zone. The red pulp consists of reticular tissue and venous spaces (in view of some authors not true sinuses, Hummel *et al.*, 1975) and is the site of extramedullary hematopoiesis, which normally occurs in the mouse spleen. Occasional accessory splenic tissue may occur in the pancreas.

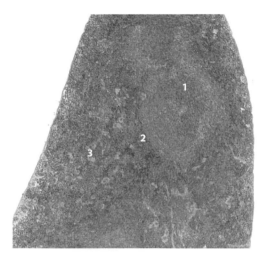

Figure 9.57 Spleen.
1. white pulp: PALS (periarteriolar lymphoid sheath)
2. white pulp: marginal zone
3. red pulp with prominent extremedullary hematopoietic activity

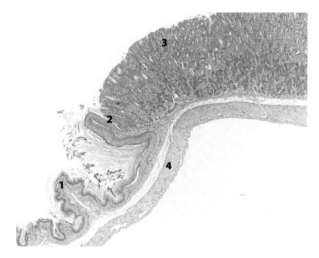

Figure 9.58 Stomach.
1. stratified squamous epithelium of the forestomach
2. limiting ridge (margo plicatus)
3. mucosa of the glandular stomach
4. muscularis

Stomach (gaster, ventriculus; Figures 9.58 and 9.59)

The stomach is located in the left cranial part of the abdominal cavity, partially covered by the left lateral hepatic lobe. Its left half is formed by the forestomach (pars cardiaca, saccus cecus), the right half by the glandular stomach (pars fundica et pars pylorica). The gastric wall consists of the mucous membrane, the smooth muscle muscularis and the serosa. The mucosa of the forestomach is lined by stratified squamous epithelium, that of the glandular stomach by epithelium forming gastric glands. The border of both kinds of epithelium is called the limiting ridge, or margo plicatus. The gastric glands are lined by single columnar epithelium and form 'gastric pits' or 'foveolae gastricae', which are perpendicular to the gastric wall. Close to the limiting ridge they contain mucous cells and are called 'cardiac glands'. The major part of the glandular stomach has the 'fundic glands' with granular eosinophilic parietal cells producing hydrochloric acid and, at the base of the glands, the basophilic chief cells, producing zymogen. The pyloric region of the glandular stomach contains mucous pyloric glands. The enteroendocrine, 'enterochromaffine-like' cells are scattered between the gastric glands. The muscularis has an inner oblique, a middle circular and an outer longitudinal layer.

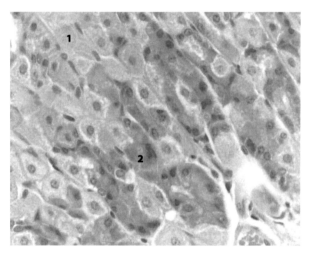

Figure 9.59 Stomach (glandular).
1. parietal cells   2. chief cells

Testis (testis, orchis; Figures 9.60 and 9.61)

The paired testes are located in the scrotum. The scrotal cavities communicate with the abdominal cavity through inguinal canals that remain open throughout life. When the testis is in the scrotum, the inguinal canal is occupied by the fat body attached to the epididymis. The testes are covered by tunica albuginea and tunica vaginalis, which also cover the inner surface of the scrotum. The testis contains convoluted seminiferous tubules. The germinal epithelium is arranged in layers, the early stages of maturation at the tubular basis, the most advanced stages at the lumen. The cell types, from the periphery towards the lumen are spermatogonia, spermatocytes, round and elongated spermatids. The Sertoli cells reach from the basal lamina to the tubular lumen. Maturing elongated spermatids are attached to their cell membrane. The mature spermatids become spermatozoa and are released. This process is called spermiation.

In the mouse the approximate time for spermatogonia to develop into spermatozoa is 35 days and 12 stages of the cycle of the seminiferous epithelium are recognized. The methodology of spermatogenic staging (determination of the stages of the spermatogenic cycle) was described by Russel *et al.* (1990) and reviewed by Creasy (1997). The seminiferous tubules end in tubuli recti which lead to the rete testis and then over a collecting chamber to efferent ducts and the head of epididymis. The rete testis is a small area under the tunica albuginea consisting of tubules lined by simple epithelium: they must not be mistaken for abnormal, atrophic seminiferous tubules. Among the seminiferous tubules are the endocrine Leydig cells, together with other peritubular cells, interstitial macrophages and interstitial vasculature.

Thymus (Figure 9.62)

The thymus consists of two lobes. It is located partly in the cervical and mostly the thoracic area, between the larynx cranially and the heart caudally. It is covered by a connective tissue capsule and, in the thoracic cavity, surrounded by mediastinal brown fat. The thymus consists of the cortex and the medulla and is divided into distinct lobules. In contrast to the lymph nodes, the thymus is epithelial in origin. The medulla is rich in epithelial cells occasionally arranged in Hassall's body-like formations. The cortex is densely filled with lymphocytes, especially the differentiating T-cells. The thymus retains its size until the young adult age and regresses thereafter by atrophy. Thymus does not develop in so called 'nude mouse', which is homozygous for *nu* gene. Such mice are hairless and lack T lymphocytes.

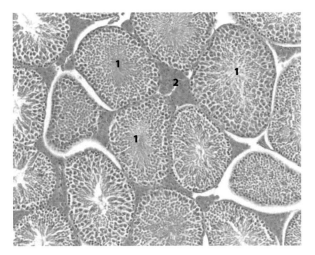

Figure 9.60 Testis.
1. seminiferous tubules   2. Leydig cells

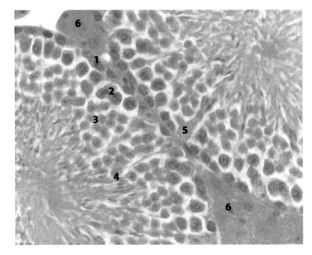

Figure 9.61 Testis.
1. spermatogonia   4. elongated spermatids
2. spermatocytes   5. Sertoli cells
3. round spermatids   6. Leydig cells

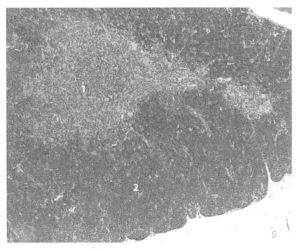

Figure 9.62 Thymus.
1. medulla   2. cortex

Thyroid gland (glandula thyroidea; Figures 9.63 and 9.64)

The thyroid gland consists of two lobes connected by a thin, ventral isthmus. The gland is located at the posterior part of the larynx and initial part of the trachea. In the mouse, there are usually two parathyroid glands located bilaterally at the surface of thyroid lobes. Portions of ectopic thymic tissue can be sometimes found at this position and can be mistaken for the parathyroid gland. The thyroid gland contains follicles filled with eosinophilic colloid and lined by epithelial cells which, depending on their secretory activity may be inactive, flattened, to highly active, high columnar. The products of follicular cells are thyroid hormones T_3 and T_4 (thyroxin). Between the thyroid follicles there are calcitonin-producing parafollicular 'C-cells'.

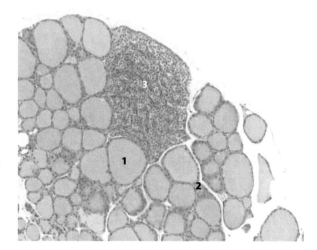

Figure 9.63 Thyroid gland (with parathyroid gland).
1. thyroid follicles   3. parathyroid gland
2. parafollicular C-cells

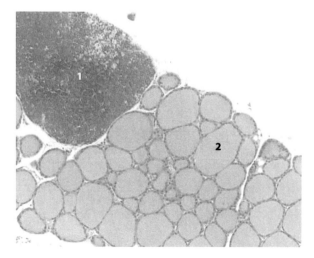

Figure 9.64 Thyroid gland (with ectopic thymus).
1. ectopic thymic tissue   2. thyroid follicles

Tongue (lingua; Figure 9.65)

The tongue is attached to the floor of oral cavity, and its portion anterior to the molars is free. The dorsal surface is rough, at the tip is a median dorsal groove and in the posterior part an elevated median intermolar eminence and the postmolar vallate papilla. The surface is covered by stratified squamous epithelium forming on the dorsal surface keratinized papillae. The tongue contains prominent striated muscles and connective tissue with minor salivary glands.

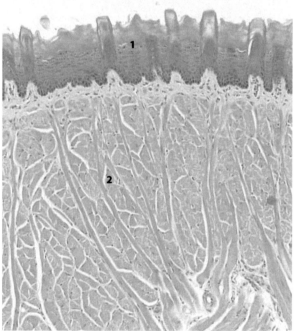

Figure 9.65 Tongue (dorsal surface).
1. keratinizing stratified squamous epithelium
2. striated muscle

Tooth (dens; Figure 9.66)

The dental formula in mice is: I 1/1, C 0/0, PM 0/0, M 3/3, i.e. an incisor and three molars on each side of the jaws, so that the total number of teeth is 16. Mice have only one set of teeth – there are no temporary decidual teeth. The histological layers of a tooth are the enamel, produced by the ameloblasts, the dentin, produced by the odontoblasts, and the cementum, produced by the cells of periodontal ligament. Inside the tooth is a cavity with dental pulp, consisting of the connective tissue, blood and lymphatic vessels and nerves. The incisors grow and are worn down continuously, so that their apical foramina remain open. Spontaneously occurring malformations of mouse maxillary incisors were described by Losco (1995).

Trachea (Figure 9.67)

The trachea connects the larynx to the left and right principal bronchi. It is located ventrally to the esophagus in the cervical area. It is formed by 15–18 C-shaped hyaline cartilages, with smooth muscle joining the ends. The mucosa is lined by pseudostratified columnar epithelium consisting of Clara cells, goblet cells, ciliated cells and basal cells.

Ureter

The paired ureters connect the kidneys to the urinary bladder. The uretral wall consists of transitional epithelium, muscularis with inner circular and outer longitudinal layer of smooth muscle fibers, and adventitia.

Urethra (Figure 9.45)

The urethra connects the urinary bladder to the body surface. In males the urethra is divided into the membraneous and the penile parts. The membranous urethra contains the colliculus seminalis and receives there the openings of vas deferens, prostate, seminal vesicles and ampullary glands. The seminal collicle (colliculus seminalis) is located on the dorsal wall of urethra, near the neck of urinary bladder. The area of seminal collicle is lined by columnar epithelium, the remaining area by transitional epithelium. Small mucous urethral glands open into the lumen of membraneous urethra (glands of Littré). Before its transition to the penile part the urethra forms a diverticulum and receives the openings of bulbourethral glands. The lumen of the penile urethra is lined by transitional epithelium and the external opening (orificium) by stratified squamous epithelium. In females, the urethra opens independently of the vagina – it empties into the clitoral fossa, cranially to the vaginal opening.

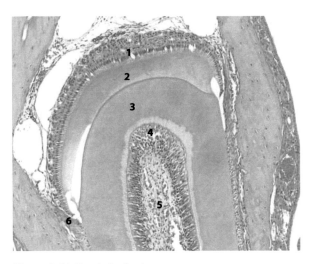

Figure 9.66 Tooth (incisor).

1. ameloblasts
2. enamel
3. dentin
4. odontoblasts
5. pulp
6. periodontal ligament

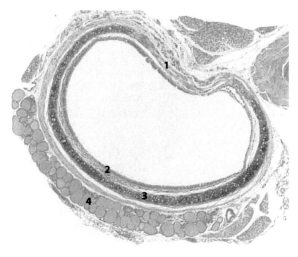

Figure 9.67 Trachea.
1. smooth muscle connecting the ends of cartilage
2. mucous membrane
3. cartilage
4. isthmus of the thyroid gland

Urinary bladder (vesica urinaria; Figures 9.68 and 9.69)

The bladder is located in the dorsocaudal abdominal cavity, ventrally to the colon. The wall is formed by transitional epithelium with well vascularized lamina propria, smooth muscle muscularis and adventitia. In the lamina propria nodules of lymphoid tissue may occur, especially in aging mice: they must not be mistaken for an inflammatory or neoplastic lesion. The thickness of bladder wall depends on the degree of distention by the content. Distended transitional epithelium is about two to three cell layers thick. Empty bladder has thick folds of transitional epithelium and lamina propria. The superficial cells of the transitional epithelium (also called 'umbrella cells') are large, may be binucleated and polyploid. The muscularis consists of smooth muscle bundles of irregular size and direction.

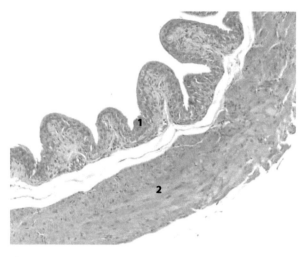

Figure 9.68 Urinary bladder.
1. transitional epithelium   2. muscularis

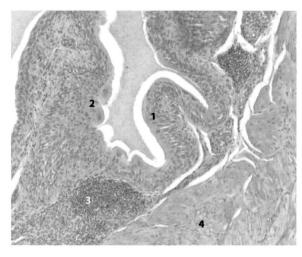

Figure 9.69 Urinary bladder.
1. transitional epithelium
2. superficial cells of transitional epithelium are large and may be polyploid
3. nodules of lymphoid tissue may occur in the lamina propria
4. muscularis

Uterus (Figures 9.70–9.73)

The mouse uterus consists of two long horns which join together in a single body which is connected by the cervix to the vagina. The cranial portion of the body contains two cavities separated by a medial septum, the caudal cervix is undivided. The uterine horns lie in the dorsal abdominal cavity, beginning at the oviducts, and the body and vagina lie ventrally to the rectum and dorsally to the urinary bladder. The wall is composed of mucosa (endometrium), an inner circular and an outer longitudinal smooth muscle layer (myometrium), and the adventitia. The endometrial mucosa is formed by simple columnar epithelium which extends tubular endometrial glands into the endometrial stroma (lamina propria). The

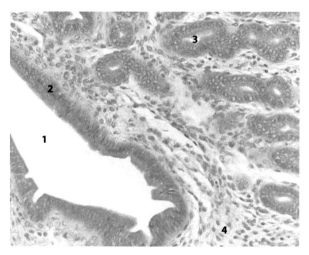

Figure 9.70 Uterus (proestrus).
1. lumen   3. endometrial gland
2. endometrial epithelium   4. endometrial stroma

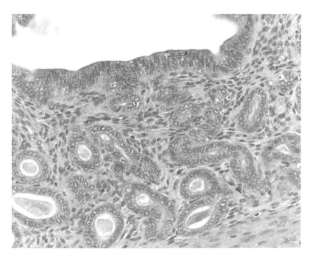

Figure 9.71 Uterus (estrus).

morphology of the endometrium is influenced by the estrous cycle. During proestrus and estrus the lumen is distended, and the stroma hyperemic. In metestrus the epithelium shows vacuolar degeneration (with apoptotic bodies), and during diestrus the epithelium regenerates. The reactivity of the uterus to estrogenic stimuli is used in the 'mouse uterotropic assay', a procedure designed for detecting potential estrogenic effects of synthetic chemicals. The stimulated uterus increases the weight and the height of endometrial epithelium. (Markey *et al.*, 2001). The uterus of aging mice frequently develops spontaneous adenomyosis (growth of endometrium into and beyond the myometrium). This has been demonstrated to result from an increased plasma level of PRL (Mori *et al.*, 1991).

Vagina (Figures 9.74–9.77)

The wall of the vagina has mucous membrane formed by prominent stratified squamous epithelium, and thin muscularis. During the estrous cycle the vaginal epithelium undergoes characteristic changes. In proestrus the superficial layers show mucous change and a layer of cornified cells develops underneath. The stratified squamous epithelium is thick. Estrus is characterized by thick stratified squamous epithelium with a distinct layer of cornified cells at the surface. In metestrus the

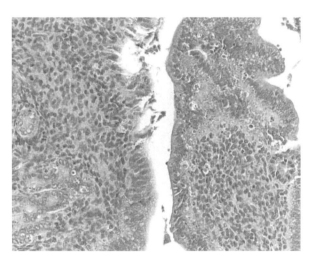

Figure 9.72 Uterus (metestrus).

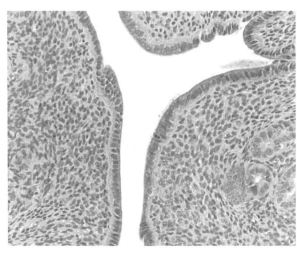

Figure 9.73 Uterus (diestrus).

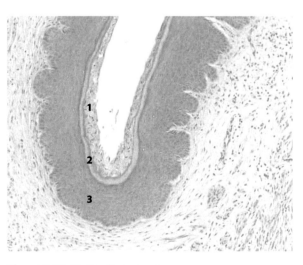

Figure 9.74 Vagina (proestrus).
1. mucous change of the superficial layer
2. layer of cornified cells
3. thick stratified squamous epithelium

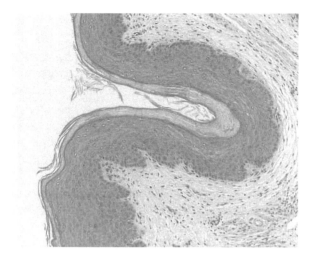

Figure 9.75 Vagina (estrus).

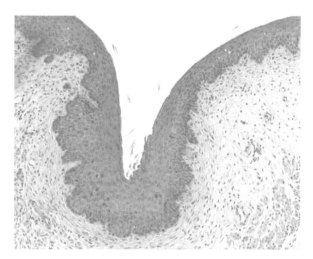

Figure 9.76 Vagina (metestrus).

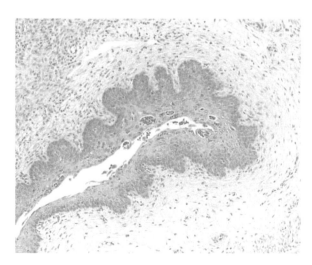

Figure 9.77 Vagina (diestrus).

cornified cells become detached and may still be present in the lumen, and in diestrus the stratified squamous epithelium becomes thinner and is infiltrated by poly-morphonuclear leucocytes. The muscularis is formed by inner circular and outer longitudinal smooth muscle lay-ers with considerable proportion of connective tissue. The adventitia is continuous with the connective tissue surrounding the urethra and rectum.

Vas deferens (ductus deferens; Figure 9.78)

The paired vas deferens begins at the tail of epididymis and ends at the seminal collicle. The terminal portion forms a widened ampulla. The wall is formed by ciliated columnar epithelium, thick middle circular and inner and outer longitudinal smooth muscle layers, and adventitia. The vas deferens together with associated blood vessels and nerves, forms the spermatic cord.

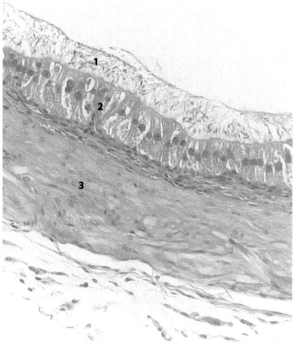

Figure 9.78 Vas deferens.
1. a layer of sperm   2. epithelium   3. smooth muscle

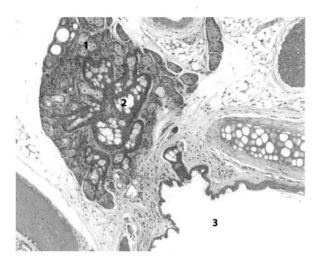

Figure 9.79 Zymbal's gland.
1. sebaceous cells   3. external ear canal
2. excretory ducts

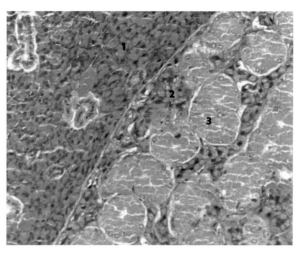

Figure 9.81
1. Parotid gland serous acini
2. acini of submandibular gland
3. convoluted ducts of submandibular gland

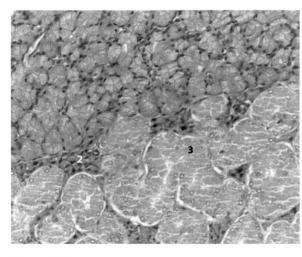

Figure 9.80
1. Sublingual gland, mucous acini
2. acini of submandibular gland
3. convoluted ducts of submandibular gland

Zymbal's gland (Figure 9.79)

The paired Zymbal's glands are auditory sebaceous glands. The gland consists of acinar sebaceous cells and excretory ducts lined by stratified squamous epithelium. It opens into the external ear canal.

References

Bannasch, P. and Gössner, W. (1994, 1997). *Pathology of Neoplasia and Preneoplasia in Rodents*, Vol. 1 and 2. Schattauer, Stuttgart.

Bolon, B., Bucci, T.J., Warbritton, A.R., Chen, J.J., Mattison, D.R. and Heindel, J.J. (1997). *Fund. Appl. Toxicol.* **39**, 1–10.

Bucci, T.J., Bolon, B., Warbritton, A.R., Chen, J.J. and Heindel, J.J. (1997). *Reprod. Toxicol.* **11**, 689–696.

Cardiff, R.D. (2002). *Comparative Med.* **52**, 12–31.

Chiu, T. and Chen, H.C. (1986). *Vet. Pathol.* **23**, 606–609.

Cotchin, E. and Roe F.J.C. (1967). *Pathology of Laboratory Rats and Mice*. Blackwell Scientific Publications, Oxford, Edinburgh.

Creasy, D.M. (1997). *Toxicol. Pathol.* **25**, 119–131.

Faccini, J.M., Abbott, D.P. and Paulus, G.J.J. (1990). *Mouse Histopathology*. Elsevier, Amsterdam, New York, Oxford.

Gude, W.D., Cosgrove, G.E. and Hirsch, G.P. (1982). *Histological Atlas of the Laboratory Mouse*. Plenum Press, London, New York.

Hummel, K.P., Richardson, F.L. and Fekete, E. (1975). Anatomy. In *Biology of the Laboratory Mouse, 2nd edn* (eds E.L. Green and E.U. Fahey), pp. 247–307. Dover Publications, New York.

Iwaki, T., Yamashita, H. and Hayakawa, T. (2001). *A Color Atlas of Sectional Anatomy of the Mouse*. Adthree, Tokyo.

Johnson, R.C., Dovey-Hartman, B.J., Syed, J., Leach, M.W., Frank, D.W., Sinha, D.P., Mirro, E.J., Little, J.M. and Halliwell, W.H. (1998). *Lab. Animal Pathol.* **26**, 789–792.

166

Kovac, W. and Denk, H. (1968). *Der Hirnstamm der Maus.* Springer Verlag, Wien.

Krinke, G.J., Schaetti, Ph.R. and Krinke, A.L. (1996). In *Pathobiology of the Aging Mouse*, Vol. 2 (eds U. Mohr, D.L. Dungworth, C.C. Capen, W.W. Carlton, J.P. Sundberg and J.M. Ward), pp. 139–152. ILSI Press, Washington DC.

Krinke, G.J. (1996). In *Pathobiology of the Aging Mouse*, Vol. 2 (eds U. Mohr, D.L. Dungworth, C.C. Capen, W.W. Carlton, J.P. Sundberg and J.M. Ward), pp. 83–103. ILSI Press, Washington DC.

Krueger, G. (1971). *Lab. Animal Sci.* **21**, 91–105.

Losco, P.E. (1995). *Toxicol. Pathol.* **23**, 677–688.

Markey, C.M., Michaelson, C.L., Veson, E.C., Sonnenschein, C. and Soto, A.M. (2001). *Environ. Health Perspect.* **109**, 55–60.

Maronpot, R.R., Boorman, G.A. and Gaul, B.W. (1999). *Pathology of the Mouse.* Cache River Press, Vienna, IL.

Mery, A., Gross, E.A., Joyner, D.R., Godo, M. and Morgan, K.T. (1994). *Toxicol. Pathol.* **22**, 353–372.

Mohr, U., Dungworth, D.L., Capen, C.C., Carlton, W.W., Sundberg, J.P. and Ward, J.M. (1996). *Pathobiology of the Aging Mouse, Volume 1 and 2.* ILSI Press, Washington DC.

Mohr, U. (ed.) (2001). *WHO—IARC International Classification of Rodent Tumors: The Mouse.* Springer, Berlin, Heidelberg.

Montemurro, D.G. and Dukelow, R.H. (1972). *A Stereotaxic Atlas of the Diencephalon and Related Structures of the Mouse.* Futura Publishing Co., Mount Kisco, NY.

Mori, T., Singtripop, T. and Kawashima, S. (1991). *Am. J. Obstet. Gynecol.* **165**, 232–234.

Notterpek, L. and Tolwani, R.J. (1999). *Lab. Animal Sci.* **49**, 588–599.

Paxinos, G. and Franklin, K.B.J. (2000). *The Mouse Brain in Stereotaxic Coordinates* (2nd edn with CD ROM). Academic Press, London.

Pedersen, T. and Peters, H. (1968). *J. Reprod. Fert.* **17**, 555–557.

Renne, R.A., Gideon, K.M., Miller, R.A., Mellick, P.W. and Grumbein, S.L. (1992). *Toxicol. Pathol.* **20**, 44–51.

Russel, L.D., Ettlin, R.A., Sinha Hikim, A.P. and Clegg, E.D. (1990). *Histological and Histopathological Evaluation of the Testis.* Cache River Press, Clearwater, FL.

Sagartz, J.W., Madarasz, A.J., Forsell, M.A., Burger, G.T., Ayres, P.H. and Coggins, C.R.E. (1992). *Toxicol. Pathol.* **20**, 118–121.

Sidman, R.L., Angevine, J.B. and Pierce, E.T. (1971). *Atlas of the Mouse Brain and Spinal Cord.* Harvard University Press, Cambridge, MA.

Stejskal, S.M. (1996). In *Pathobiology of the Aging Mouse, Volume 2* (eds U. Mohr, D.L. Dungworth, C.C. Capen, W.W. Carlton, J.P. Sundberg and J.M. Ward), pp. 155–177. ILSI Press, Washington DC.

Turusov, V. and Mohr, U. (eds) (1994). *Pathology of Tumours in Laboratory Animals, Volume 2—Tumours of the Mouse.* IARC Scientific Publications No 111, Lyon.

Vargas, K.J., Stephens, L.C., Clifford, C.B., Gray, K.N. and Price, R.E. (1996). *Lab. Animal Sci.* **46**, 572–575.

Internet addresses

EULEP – European Late Effects Project, Necropsy of the Mouse – Bibliography: www.eulep.org/Necropsy_of_the_Mouse/index.php?file=Bibliography.html

Mouse Atlas Project: www.loni.ucla.edu/MAP/index.html

Mouse Brain Atlases: http://mbl.org/mbl_main/atlas.html; www.hms.harvard.edu/research/brain/3D_atlasvDemo.html

MRC HGU, The Mouse Atlas and Gene Expression Database Project: http://genex.hgu.mrc.ac.uk/

The Virtual Mouse Necropsy: www.geocities.com/virtualbiology/

UC Davis Mouse Biology Program, The Visible Mouse: http://pathology.usdavis.edu/tgmice/visiblemouse/visiblemouse/web/Main.html

Whole Mouse Catalog – Organism: www.paperglyphs.com/wmc/domain_mouse.html

Yale Animal Resources Center, Mouse Phenotyping Service: www.med.yale.edu/yarc/mousephenotype.htm

Imaging

Roger E. Price
Small Animal Cancer Imaging Research Facility,
U.T. MD Anderson Cancer Center, Houston, TX, USA

Introduction

Advances in imaging technology now allow high resolution noninvasive and nondestructive imaging of mice with body weights of less than 30 g by magnetic resonance (MR) at 50 μm resolution *in vivo* and 10 μm *in vitro*, by computed tomography (CT) at 25 μm in plane resolution, and positron emission tomography (PET) at 1–2 mm resolutions (Johnson *et al.*, 1993; Budinger *et al.*, 1999; Paulus *et al.*, 2000, 2001; Balaban and Hampshire 2001). Similar technological advances are also being made in optical, ultrasound, and radiographic imaging. Imaging modalities also uniquely allow functional information to be obtained and viewed in the context of anatomic information. Judicious exploitation of the unique benefits of each imaging modality and the combination of information from multiple imaging modalities will permit expeditious economical phenotyping of animal models for mutations, pathologic changes, disease progression, and response to therapy.

Monitoring tissue characteristics using nondestructive imaging methods makes it feasible to conduct longitudinal studies on laboratory animals. The ability to acquire multiple data points and monitor disease evolution within an individual animal can reduce the number of animals required to compensate for biological and inter-individual variability in order to obtain significantly relevant data in experimental animal investigations. Noninvasive imaging methods can also be used for accurate determination of optimal times for interventional treatments or for tissue harvest and pathology, thus further reducing the number of animals required in animal research. In addition to the moral implications of reducing the number of animals required in research studies, the financial saving resulting from the reduced animal purchase and maintenance costs and the reduced space and personnel required to house and care for research animals, further mitigates the cost of performing the imaging studies.

Magnetic resonance imaging is a versatile, fast, non-destructive imaging technique that is capable of providing high resolution images with excellent soft tissue contrast and detailed information on tissue characteristics, anatomy, and function. The inherent isotropic three-dimensional nature of MR imaging allows detailed analysis of individual organs and retrospective studies through any arbitrary plane (Johnson *et al.*, 1993). Recent availability of increasing magnetic field and gradient strengths are producing dramatic improvements in signal-to-noise ratios and spatial and temporal resolution (Allport and Weissleder 2001).

The images in this chapter were acquired using a 4.7 T, 40 cm bore Bruker *Biospec* Avance MR scanner (Bruker Instruments, Karlsruhe, Germany) in the Small Animal Cancer Imaging Research Facility at The University of Texas M.D. Anderson Cancer Center Houston, Texas, USA. All images were acquired using a 6 cm actively shielded gradient system (950 mT/m, 110 μs rise time, and 8636 T/m/s slew rate) and a 25 mm linear volume resonator. Three regions of the animals were imaged (head, thorax, and abdomen). Axial, T_1-, T_2-, and proton density-weighted images were acquired from each region. The axial head images covered a 2.5 cm square field-of-view (FOV), while the sagittal head and axial thoracic, abdominal, and pelvic images had an FOV of 3 cm. Data was acquired at 512×256 data matrices then zero-filled to 512×512 before Fourier Transform, resulting in an isotropic in plane resolution of 59 μm for all axial images except those of the head, which were 49 μm resolution. All slices were 1 mm thick and separated by 0.25 mm gaps. The T_1-weighted images (TE/TR = 10.3/905 ms) were acquired in 1 h, while the T_2-weighted images (TE/TR = 55/2200 ms) required 2.5 h. The proton density weighted images (TE/TR = 10.3/5000 ms) were acquired in 1.5 h. Magnetic resonance images were displayed with the Image Analysis software Analyze® (CN Software, Southwater, UK) on a Silicon Graphics O^2 computer for presentation.

Images of the head and thoracic regions were acquired using adult female Hsd:ICR (CD-1®) mice (Charles Rivers Laboratories, Inc., Wilmington, Massachusetts, USA) and images of the abdomen and pelvic region of male mice were acquired from adult NOD.CB17-Prkdc scid/J mice (The Jackson Laboratory, Bar Harbor, Maine, USA). T_2-weighted images of the head were acquired on an unconscious animal which was immobilized using isoflurane inhalation anesthesia. For the remainder of the images of the head and other regions of the body, the animals were euthanized with CO_2 and imaged dead (to eliminate motion artifacts). Following imaging, the mice were left in the sleds in which they were imaged and quick frozen in a $-70°C$ freezer until rigid. Once frozen the imaged regions were cut in to transverse blocks of tissue for the axial images and into a sagittal block for the sagittal images of the head on a band saw and embedded in OCT embedding

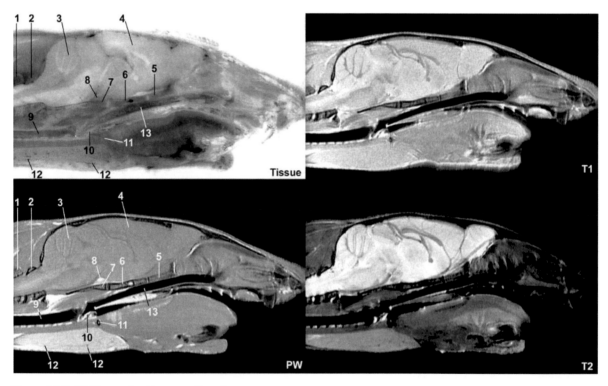

Figure 10.1 Midline sagittal images of head.

1. Cervical Vertebra II (Axis)	6. Basisphenoidal Bone	11. Basihyoideum
2. Cervical Vertebra I (Atlas)	7. Basilar Portion of Occipital Bone	12. Salivary Glands
3. Cerebellum	8. Pituitary	13. Nasal Cavity/Nasal Pharynx
4. Cerebrum	9. Trachea	
5. Presphenoid Bone	10. Epiglottis	

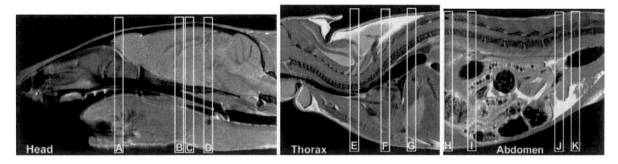

Figure 10.2 Midline sagittal T_1-weighted MR images with graphic slice locations for each imaged region.

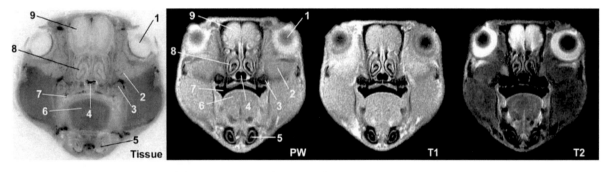

Figure 10.3 Axial images of head (Slice location A).

1. Eye
2. Lacrimal Gland
3. Maxilla
4. Nasopharynx
5. Mandible
6. Tongue
7. Oral Cavity
8. Endoturbinate
9. Olfactory Bulb of Brain

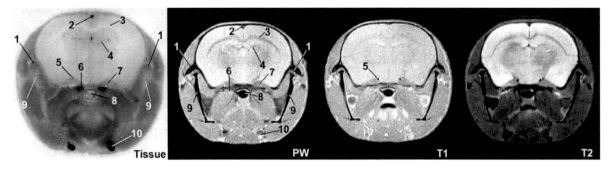

Figure 10.4 Axial images of head (Slice location B).

1. Temporomandibular Joint
2. Dorsal Sagittal Sinus
3. External Capsule
4. Lateral Ventricle
5. Caudal Cerebral Artery
6. Maxillary Vein
7. Trigeminal Nerve
8. Nasopharynx
9. Vertical Ramus of Mandible
10. Linguofacial Vein

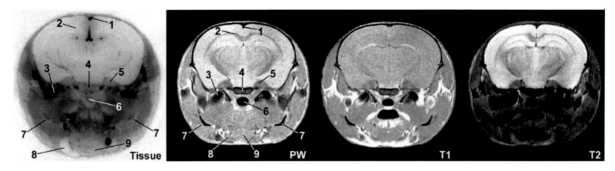

Figure 10.5 Axial images of head (Slice location C).

1. Dorsal Sagittal Sinus
2. Corpus Collosum and Internal Capsule
3. Temporal Vein
4. Basisphenoidal Bone
5. Trigeminal Nerve
6. Nasopharynx
7. Mandible
8. Mandibular Lymph Node
9. Mandibular Salivary Gland

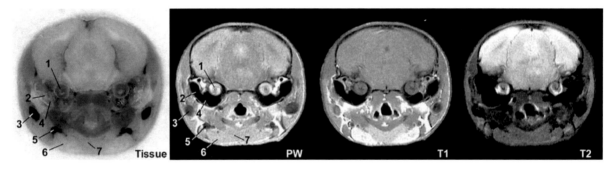

Figure 10.6 Axial images of head (Slice location D).

1. Inner Ear Portion of Petros Temporal Bone
2. External Ear Canal
3. Superficial Temporal Vein
4. Tympanic Membrane
5. Maxillary Vein
6. Sublingual Salivary Gland
7. Submandibular Salivary Gland

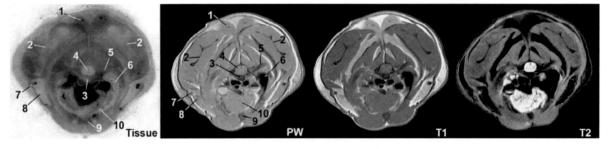

Figure 10.7 Axial images of thorax (Slice location E).

1. Brown Fat
2. Scapula
3. Thoracic Vertebra
4. Thoracic Spinal Cord
5. Rib
6. Lung
7. External Jugular Vein
8. Cervical Mammary Gland
9. Sternebra
10. Thoracic Lobe of Thymus

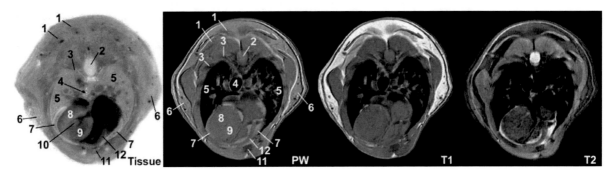

Figure 10.8 Axial images of thorax (Slice location F).

1. Brown Fat
2. Thoracic Spinal Cord
3. Rib
4. Trachea
5. Lung
6. Thoracic Mammary Gland
7. Thymus
8. Heart, Left Ventricle
9. Heart, Interventricular Septum
10. Heart, Papillary Muscle Left Ventricle
11. Sternebra
12. Heart, Right Ventricle

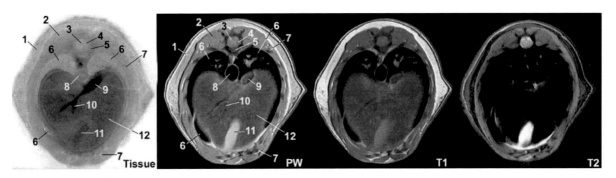

Figure 10.9 Axial images of thorax (Slice location G).

1. Mammary Gland
2. Epaxial Muscle
3. Spinal Cord (Gray Matter)
4. Spinal Cord (White Matter)
5. Vertebra
6. Lung
7. Rib
8. Esophagus
9. Posterior Vena Cava
10. Hepatic Vein
11. Gall Bladder
12. Liver

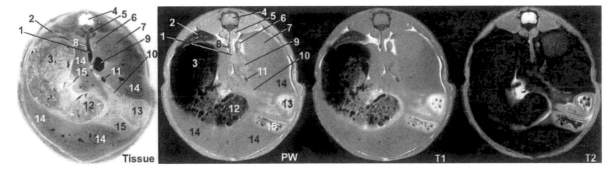

Figure 10.10 Axial images of abdomen (Slice location H).

1. Right Crus of Diaphragm
2. Spleen
3. Stomach, Fundus
4. Spinal Cord
5. Right Psoas Major Muscle
6. Right Adrenal Gland
7. Right Kidney
8. Aorta
9. Caudal Vena Cava
10. Pancreas
11. Portal Vein
12. Stomach, Pylorus
13. Duodenum
14. Liver
15. Jejunum

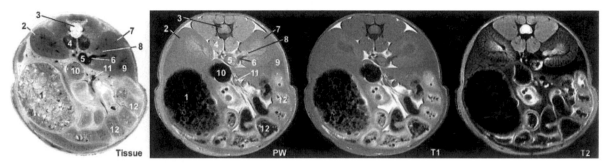

Figure 10.11 Axial images of abdomen (Slice location I).

1. Stomach, Fundus
2. Left Kidney
3. Spinal Cord
4. Left Psoas Major

5. Caudal Vena Cava
6. Right Renal Vein
7. Right Kidney, Cortex
8. Right Kidney, Papilla

9. Liver
10. Colon
11. Portal Vein
12. Jejunum

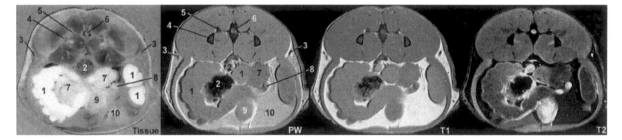

Figure 10.12 Axial images of abdomen (Slice location J). Slight differences in the positions and sizes of organs between tissue block face image and MR images due to movement of caudal abdominal organs during freezing process.

1. Vesicular Gland
2. Descending Colon
3. Superficial Circumflex Iliac Artery and Vein

4. Left Ilium, Body
5. Transverse Process, Lumbar Vertebra
6. Spinal Cord
7. Coagulation Gland

8. Right Deferent Duct
9. Urinary Bladder
10. Abdominal Fat

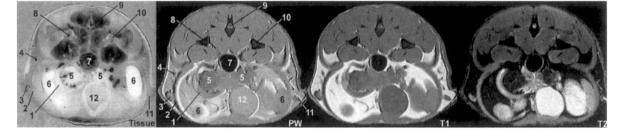

Figure 10.13 Axial images of abdomen (Slice location K).

1. Left Deferent Duct
2. Left External Oblique Muscle of Abdomen
3. Left Subiliac Lymph Node
4. Duct of Vesicular Gland

5. Coagulation Gland
6. Vesicular Gland
7. Descending Colon
8. Left Ilium, Body
9. Spinal Cord

10. Right Ilium, Body
11. Right Subiliac Lymph Node
12. Urinary Bladder

medium for frozen tissue specimens (Tissue-Tek, Sakura, Torrance, CA). The embedded blocks of tissue were cryosectioned at $-20°C$ through the axial plane of all three regions of the animal and the sagittal plane of the head in 250–500 μm increments. At each level, a high-resolution block-face digital image of the surface of the frozen block of tissue was captured using an Olympus 4-megapixel Camedia E-10 digital camera with a macro lens. In all images, Tissue = block-face image of frozen tissue block, PW = proton-weighted MR image, $T_1 = T_1$-weighted MR image, $T_2 = T_2$-weighted MR image (Figures 10.1–10.13).

Acknowledgments

The author would like to thank Belinda Rivera, B.S., R.V.T. for assistance in animal support and tissue processing, Jim Bankson, Ph.D. for acquiring the MR images, and Ms Sara Amra for cutting the frozen blocks of tissue for the tissue images. The work described herein was supported by The University of Texas M.D. Anderson Cancer Center Support Grant (CA16672) from the National Institutes of Health, National Cancer Institute, Department of Health and Human Services.

References

Allport, J.R. and Weissleder, R. (2001). *Exp. Hematol.* **29**, 1237–1246.

Balaban, R.S. and Hampshire, V.A. (2001). *ILAR J.* **42**, 248–262.

Budinger, T.F., Benaron, D.A. and Koretsky, A.P. (1999). *Annu. Rev. Biomed. Eng.* **1**, 611–648.

Johnson, G.A., Benveniste, H., Black, R.D., Hedlund, L.W., Maronpot, R.R. and Smith, B.R. (1993). *Magn. Reson. Q.* **9**, 1–30.

Paulus, M.J., Gleason, S.S., Kennel, S.J., Hunsicker, P.R. and Johnson, D.K. (2000). *Neoplasia* **2**, 62–70.

Paulus, M.J., Gleason, S.S., Easterly, M.E. and Foltz, C.J. (2001). *Lab. Anim. (NY)* **30**, 36–45.

Popesko, P., Rajtova, V. and Horak, J. (1992). *A Color Atlas of the Anatomy of Small Laboratory Animals 2. Mouse.* Wolfe Publishing Ltd., London.

Takamasa, I., Hiroshi, Y. and Toshiyuki, H. (2001). *A Color Atlas of the Sectional Anatomy of the Mouse.* Adthree, Inc., Japan.

CHAPTER 11

Introduction to Early Mouse Development

Andreas Kispert and Achim Gossler
Institut für Molekularbiologie,
Medizinische Hochschule, Hannover, Germany

Mouse embryonic development has been studied for a long time, and over the past two decades enormous progress has been made in the analysis of cell fates, developmental potential of cells in the early embryo and the understanding of molecular mechanisms regulating patterning and differentiation. Mouse embryonic development takes 18–20 days depending on the strain (Figure 11.1). 3.5 days after fertilization the first two differentiated cell types are present in the blastocyst and implantation takes place one day later. Development proceeds rapidly after implantation with the formation of the egg cylinder. Subsequently the formation of the three embryonic germ layers during gastrulation beginning around embryonic day 6.5 leads to dramatic morphological changes. During gastrulation, the basic body plan with its major axes is established, organogenesis commences, and around day 10.5 of embryonic development most organ primordia are present.

The aim of this chapter is to provide a basic introduction to mouse development until the early **postgastrulation** period. The intention is to give an overview rather than being comprehensive to help newcomers working with mice to understand better more detailed and specialized descriptions and experiments on various

aspects of mouse development. Detailed descriptions and illustrations of mouse embryonic development are found in Theiler (1989), Rugh (1990), Kaufman and Bard (1999) and Rossant and Tam (2001), instructions for various kinds of manipulations of pre- and postimplantation embryos as well as embryonic stem cells are given in Hogan *et al.* (1994), Monk (1997) and Copp and Cockroft (1990).

Fertilization and preimplantation development

Fertilization

Fertilization, the fusion of the female (oocyte) and male (sperm) gametes, activates the egg to commence embryonic development. The specialized structures of oocyte and sperm and the unique genetic complement

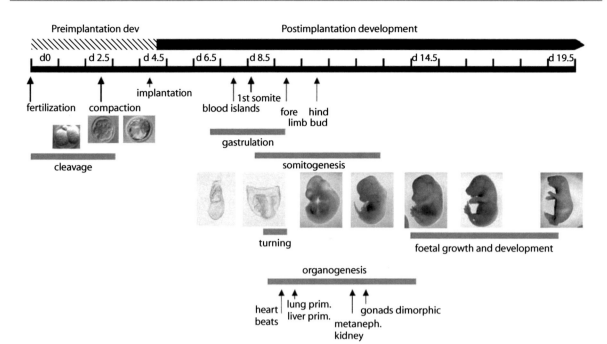

Figure 11.1 Overview of mouse development. Mouse embryonic development takes approximately 19 days. The pre- and postimplantation phases are indicated above the time line, critical events and processes are indicated below the time line. For details see text.

carried by each gamete as a consequence of meiotic recombination are generated during **gametogenesis**. In mature sperm the genetic material is reduced to a haploid set of chromosomes, whereas in the oocyte reduction to haploidy is achieved only after fertilization (see below). A fully grown mouse oocyte ready to undergo the final maturation steps measures about 85 μm in diameter and is surrounded by a thick extracellular envelope, the zona pellucida, consisting of three major glycoproteins (Bleil and Wassarman, 1980b), which is embedded in multiple layers of follicle or granulosa cells. The large nucleus, called the germinal vesicle, contains the chromosomes in the prophase of the first meiotic division. In each hormonal cycle, which takes about 4–5 days in the mouse, only a few follicles respond to an increase in the level of follicle stimulating hormone (FSH), mature, accumulate fluid and move towards the periphery of the ovary.

Shortly before ovulation the level of luteinizing hormone (LH) surges and the nuclear maturation of the egg commences with the disintegration of the nuclear membrane (germinal vesicle breakdown) and resumption of meiosis. One set of chromosomes is removed from the egg together with some cytoplasmic material as the first polar body. Nuclear maturation is arrested in the metaphase stage of the second meiotic division and only proceeds after fertilization. Upon ovulation eggs are transported into the ampulla, the anteriormost part of the oviduct by the movements of

the cilia of the epithelial cells lining the opening of the oviduct facing the ovary (**infundibulum**). Each egg is surrounded by its zona pellucida and follicle (cumulus) cells that are embedded in a matrix of proteins and hyaluronic acid.

After penetrating the viscous matrix of hyaluronic acid surrounding the cumulus cells and the egg, sperm associates with the surface of the zona pellucida and then binds more tightly in a species specific manner. This binding between egg binding proteins of the sperm head and the zona component ZP3 (Bleil and Wassarman, 1980a; Shur and Hall, 1982a,b) elicits the **acrosome** reaction and releases proteolytic and glycolytic enzymes that allow the sperm to penetrate the zona pellucida after limited proteolysis. After penetrating the zona pellucida the sperm head fuses with the egg membrane and triggers a cascade of events which prevent polyspermy and ultimately lead to the formation of the diploid zygote. (For a more detailed discussion of the processes of fertilization and molecules involved, see Snell and Stevens, 1991; Wassarman, 1995, 1999.)

Fertilization activates the egg and triggers the completion of meiosis. This results in the extrusion of the second polar body and leaves behind a haploid set of maternal chromosomes in the egg's female pronucleus. The nuclear membrane of the sperm nucleus breaks down, the chromatin decondenses, is reorganized and a new nuclear membrane is formed around the male pronucleus. Then the two pronuclei move towards each other and DNA replication takes place. Upon meeting, the two pronuclei do not fuse but their nuclear membranes break down, the chromosomes assemble on the metaphase plate and cleavage commences with the first cell division. Mature oocytes are transcriptionally silent. Significant transcriptional activity only resumes after the first cell division at the two cell stage. Thus, the final steps of egg maturation, early post-fertilization events and the first cell division are controlled by stored maternal proteins and mRNA. In accordance with this, transcriptional inhibitors like α-amanitin do not block development of fertilized eggs to the two-cell stage (Braude *et al.*, 1979). Thereafter transcription from the embryonic genome is required for normal development to proceed: all classes of new RNA are synthesized (Clegg and Piko, 1977, 1983) and further development is blocked by inhibition of transcription with α-amanitin (Warner and Versteegh, 1974). The mechanisms that control the transcriptional activity in the mature egg and after fertilization are not entirely clear and appear to be multifactorial. Changes in RNA polymerase II activity as well as chromatin structure appear to contribute to transcriptional silencing (Latham, 1999).

Cleavage divisions and compaction

Cleavage divisions in the mouse, like in other mammals, are slow. The first cell division occurs about 20 h after fertilization (Figure 11.2). The next divisions

follow at approximately 12 h intervals but are not truly synchronous between the different cells, which are called blastomeres. Up to the eight cell stage, blastomeres are spherical cells which are loosely attached at their sites of contact. At the eight cell stage blastomeres alter their adhesive behaviour and the embryo, now called the morula, undergoes a dramatic morphological change. Blastomeres flatten towards each other, maximize cell-to-cell contacts and the former grape-like structure is transformed into a compact aggregate of cells. This phenomenon is called compaction and is the prerequisite for the formation of the blastocyst. One major component of the compaction process is uvomorulin, a Ca^{2+} dependent cell adhesion protein, which is also known as E-cadherin (Hyafil *et al.*, 1980; Vestweber and Kemler, 1984; Yoshida-Noro *et al.*, 1984), is also required for cell adhesion in many epithelial cells later during development and in adult tissues. During compaction, tight junctions start to develop between blastomeres (Ducibella and Anderson, 1975), and in the periphery of the compact morula junctional complexes eventually constitute the zona occludens (Ducibella *et al.*, 1975), a tight permeability barrier which is required for the formation of the blastocoel, the fluid filled cavity of the blastocyst. After compaction, fluid accumulates between intercellular spaces and around the 32-cell stage the blastocoel becomes evident. Outer cells pump fluid into the nascent cavity which rapidly expands. The timing of cavitation seems to depend on the nucleo-cytoplasmic ratio or DNA or chromosomal replication, but does

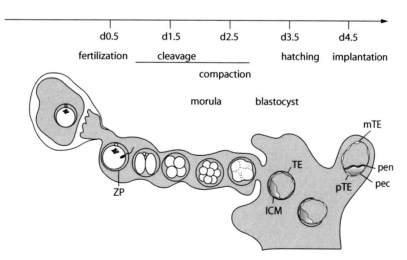

Figure 11.2 Schematic presentation of mouse preimplantation development. After fertilization, which occurs in the proximal part of the oviduct, the zygotes undergo cleavage while migrating through the oviduct towards the uterus. Compaction occurs at the eight cell stage as a prerequisite for blastocyst formation which occurs around day 3.5 in the uterus. Hatching frees the blastocyst from its zona pellucida and is required for implantation. ICM: inner cell mass; pec: primitive ectoderm; pen: primitive endoderm; TE: trophectoderm; pTE: polar trophectoderm; mTE mural trophectoderm; ZP: zona pellucida.

not depend on the absolute number of cells or cell divisions in the zygote. When the number of cells in the embryo was experimentally reduced or enlarged or cell divisions were suppressed with cytochalasin-B (which does not affect DNA replication) neither manipulation substantially affected the time of blastocyst formation (Smith and McLaren, 1977).

Two distinct cell populations are present in the blastocyst: an outer layer of trophectodermal (TE) cells which represents a true epithelium surrounding the blastocoel, and the inner cell mass (ICM) cells, a group of cells which is attached to one side of the inner surface of the TE. TE and ICM cells seem to remain totally distinct lineages from the onset of cavitation (Dyce et al., 1987). The TE cells give rise exclusively to extraembryonic tissue (Figure 11.3; see below). Shortly before implantation some of the ICM cells differentiate into a second epithelial cell type, the primitive endoderm, which arises on the free surface of the ICM facing the blastocoel. The remaining ICM cells will give rise to the embryo proper and to the extraembryonic mesoderm. The primitive endoderm will give rise to the embryonic membranes, i.e. the endodermal component of the visceral yolk sac and parietal yolk sac (see below).

Up to the eight cell stage blastomeres have a remarkable regulative ability. Single blastomeres of two- and four-cell embryos can form blastocysts in vitro

(Tarkowski and Wroblewska, 1967), and blastocysts formed in vitro from single blastomeres of two cell embryos can develop into normal mice after transfer into foster mothers (Tarkowski, 1959). In contrast, individual blastomeres from four- and eight-cell embryos cannot generate a mouse by themselves (Rossant, 1976b), which is probably due to the small number of cells in the resulting experimental blastocysts. However, this does not reflect a lack of developmental potency, since when single, isolated blastomeres are combined with (genetically) marked eight-cell embryos they can form normal chimaeric embryos and contribute to a broad range of embryonic and extraembryonic tissues (Kelly, 1977). The position of the 'single' blastomere during the aggregation seems to strongly influence its developmental fate: labelled blastomeres which were placed on the outside of aggregates of other blastomeres developed predominantly into TE cells in resulting blastocysts and were mainly found in trophoblast tissue at day 10 of development. When labelled blastomeres were surrounded by other blastomeres they contributed large numbers of daughter cells to the ICM and formed parts of the embryo rather than extraembryonic tissues at later stages of development (Hillman et al., 1972). The remarkable regulative capacity of individual blastomeres was long interpreted as an indication that the polarity of the oocyte and zygote (along the animal vegetal axis, the animal pole

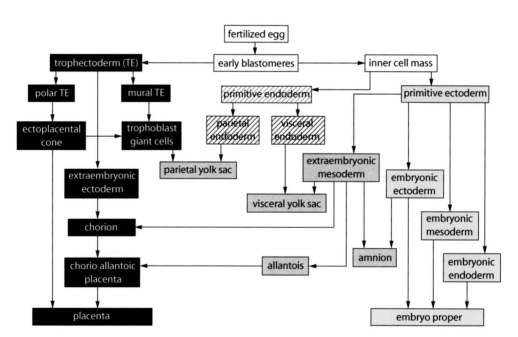

Figure 11.3 Cell lineages in the mouse embryo. The developmental potential of cells in the early embryo becomes progressively restricted. The first two distinct cell lineages, the TE and ICM, are present in the blastocyst. The primitive endoderm and primitive ectoderm lineages are established from the ICM. Both TE and primitive endoderm contribute to extraembryonic tissues, the primitive ectoderm gives rise to the germ layers of the embryo and to extraembryonic mesoderm which contributes to the extraembryonic tissues (amnion, yolk sac and allantois).

being defined by the location of the polar body) or localized maternal components are unlikely to play a role for patterning the embryo. However, there is a growing number of data strongly suggesting a correlation between the animal-vegetal axis of the zygote and an axis of bilateral symmetry present in the blastocyst, raising the possibility that the polarity of the egg might contribute to patterning of the embryo at least in normal undisturbed development (Gardner, 1997, 2001b; Ciemerych *et al.*, 2000; Piotrowska and Zernicka-Goetz, 2001; Zernicka-Goetz, 2002).

Shortly before compaction, individual blastomeres develop polarity with distinct apical and basolateral surfaces (Reeve and Ziomek, 1981). Concomitant with changes in adhesiveness, alterations in cytoskeletal architecture, lectin binding properties and distribution of membrane and cytoplasmic components occur (Johnson and Maro, 1984; Fleming and Pickering, 1985; Maro *et al.*, 1985; Pratt, 1985 and references therein). Once differences in the distribution of cellular components and structures (i.e. membrane proteins, cytoskeletal proteins and cell organelles) have been established within one cell, they can lead to differences in daughter cells after the next cell division: divisions parallel to the basal–apical axis result in two daughter cells both of which have inner (basolateral) and outer (apical) surfaces, while divisions perpendicular to this axis generates 'inner' and 'outer' cells which inherit different molecules from their mother cell (Johnson and Ziomek, 1981; Johnson, 1986). There is evidence that the 'inner' cells generated during cell divisions in the late morula give rise to the ICM cells of the blastocyst while outer cells predominantly form TE cells (Pedersen *et al.*, 1986; Fleming, 1987). However, this allocation of cells reflects developmental fate rather than developmental potential, since when cells isolated from the 'inside' of late morulae are reaggregated and cultured *in vitro*, they can form normal blastocysts with ICM and TE as is also true for 'outer' cells (Rossant and Lis, 1979; Rossant and Vijh, 1980), and the significance of polarization for the differentiation of ICM vs. TE remains unclear.

Implantation

Implantation occurs on the 5th day of development. Before the blastocyst can implant it has to shed its zona pellucida. This process is called hatching and is brought about by localized proteolysis of the zona (Perona and Wassarman, 1986) and contractions and expansion of the blastocyst. Once freed from its zona the blastocyst attaches to the epithelium of one of the lateral uterine walls with the murine TE (the TE opposite to and not facing the ICM). The uterine wall attached to the blastocyst responds by bulging into the lumen orienting the ICM either to the anterior or posterior end of the uterine horn. This and the following reorganization results in an invariable orientation of the early embryo. The axis through the ICM towards the opposite pole of the blastocyst parallels the dorsoventral (DV) axis of the mother, the ICM always facing the dorsal side. The future anterior–posterior (AP) axis of the embryo, which becomes evident around day 6.5 of development with the onset of **gastrulation**, is more or less perpendicular to the AP (longitudinal) axis of the uterine horn (Smith, 1980, 1985).

The significance of this invariant orientation of the embryo with respect to the uterus in the determination of the embryonic axes is not clear, however, since embryos can also develop normally *in vitro* from preimplantation stages up to the limb bud stage (Chen and Hsu, 1982). A detailed analysis of the orientation of mouse embryos during implantation and a discussion of how this might be achieved and might be related to embryonic axis formation is given by Smith (Smith, 1980, 1985). After attachment to the uterine wall the TE cells invade the degenerating uterine epithelium and penetrate into the endometrium (stroma) of the uterus. The mesenchymal stromal cells respond with increased proliferation resulting in the formation of a thick layer of mesenchymal tissue, the decidua, which encloses the embryo. The implantation sites are readily visible within 1 day after implantation by the **decidual** swellings of the uterus.

Early postimplantation development

Formation of the egg cylinder and cell type diversification

Preimplantation development results in the formation of the blastocyst which contains approximately 200 cells around the time of implantation. Three distinct cell types are present in the blastocyst at implantation, the TE, the ICM, now called primitive or embryonic ectoderm, and the primitive endoderm. These cell types will rapidly diverge further. Cells of the primitive ectoderm and overlying TE proliferate and form an elongated structure, the egg cylinder, projecting into the blastocoel. The egg cylinder shows a distinct junction between the distally located embryonic ectoderm and the proximal, TE-derived extraembryonic ectoderm. Proximal to the extraembryonic ectoderm some TE cells stop proliferating, undergo endoreduplication and form trophoblastic giant cells. Other TE cells proliferate and form the ectoplacental cone, which together with the extraembryonic ectoderm will form most of the foetal part of the placenta. The primitive endoderm cells will give rise to the parietal and visceral yolk sac endoderm. The primitive or embryonic ectoderm will give rise to the three definitive germ layers of the embryo during gastrulation, and will also contribute to extraembryonic tissues. Below, the development during early postimplantation stages of the three cell types present in the embryo at implantation will be described.

The epithelial layer of TE cells present in the late preimplantation blastocyst does not consist of a homogeneous cell population. The cells overlaying the ICM constitute the polar TE, the cells without contact to the ICM the mural TE. Mural TE cells stop proliferating and become large polyploid cells, the primary trophoblastic giant cells. In contrast, polar TE cells remain diploid, continue to proliferate and give rise to the ectoplacental cone and the extraembryonic ectoderm (Gardner and Johnson, 1973; Gardner et al., 1973; Rossant, 1976a;

Rossant et al., 1978; Papaioannou, 1982). Polar TE cells that move away from the embryonic pole differentiate into mural TE. Contact or proximity to the ICM or its derivatives control whether TE cells continue to proliferate or cease cell divisions and become polyploid giant cells. When trophoblast cells (these are the TE cells after implantation) were isolated from contact with ICM derivatives, they ceased proliferation and transformed into giant cells (Gardner and Johnson, 1972; Gardner, 1975) and contact with ICM cells appears to prevent endoreduplication of TE cells (Rossant and Ofer, 1977).

Recent studies suggest that the ICM stimulation of polar TE proliferation is mediated by FGF4 (Chai et al., 1998; Nichols et al., 1998; Tanaka et al., 1998). The polar TE gives rise to both the extraembryonic ectoderm and ectoplacental cone. The extraembryonic ectoderm projects into the blastocystic cavity while the ectoplacental cone extends in the opposite direction (Figure 11.4). Cells from the periphery of the ectoplacental cone form additional (secondary) trophoblastic giant cells. Ectoplacental cone and extraembryonic ectoderm give rise to the majority of cells in the foetal part of the placenta. The extraembryonic ectoderm becomes epithelial, moves back towards the ectoplacental cone and together with extraembryonic mesoderm cells constitutes the chorion. The chorion together with the allantois, another mesodermal tissue which gives rise to the umbilical cord (see below), forms the chorioallantoic placenta, or labyrinthine region of the placenta where exchange of metabolites and gases occurs between foetal and maternal blood.

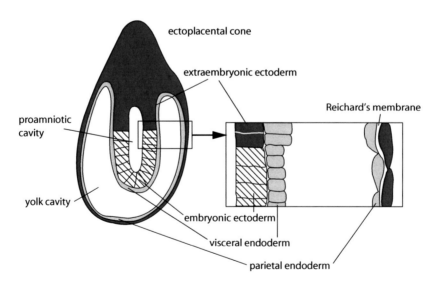

Figure 11.4 Schematic presentation of an early egg cylinder embryo. The early egg cylinder consists of the TE-derived ectoplacental cone and extraembryonic ectoderm, the embryonic ectoderm, and the visceral and parietal endoderm. The visceral endoderm covers the embryonic and extraembryonic ectoderm, the parietal endoderm the inner surface of the TE. For details see text.

Primitive endoderm and its derivatives

Prior to implantation primitive endoderm cells differentiate on the surface of the ICM facing the blastocystic cavity. The remaining cells of the ICM constitute the primitive or embryonic ectoderm (Figure 11.4). The primitive endoderm does not contribute to the definitive endoderm of the embryo (Gardner and Rossant, 1979). As the egg cylinder forms, primitive endoderm cells undergo further differentiation into two morphologically and biochemically distinct cell types, the visceral and parietal endoderm. The cells remaining in contact with and covering the egg cylinder constitute the visceral endoderm, cells that grow out and migrate on to the inner surface of the mural trophectoderm constitute the parietal endoderm (Snell and Stevens, 1991) (Figure 11.4). Visceral and parietal endoderm cells are part of the extraembryonic membranes, the visceral and parietal yolk sac respectively (see below).

Parietal endoderm cells start to grow and migrate onto the inner surface of the TE shortly after implantation and from day 6 on they cover the inner surface of the TE as a lawn of evenly spaced individual cells (Enders et al., 1978; Snell and Stevens, 1991). These cells produce and secrete large amounts of extracellular matrix material such as laminin, entactin, type IV collagen and heparan sulfate proteoglycan (Hogan et al., 1980, 1982; Carlin et al., 1981; Smith and Strickland, 1981) and lay down a thick basement membrane, known as Reichert's membrane, between the parietal endoderm and the underlying trophoblastic giant cells. Parietal endoderm, Reichert's membrane and trophoblastic giant cells together constitute the parietal (outer) yolk sac of the embryo. Until it starts to break down around day 16 of gestation, Reichert's membrane may serve as a major barrier and coarse filter between maternal and foetal environments. Nutrients from the mother can pass through this barrier, while penetration of maternal cells is prevented (Smith and Strickland, 1981).

The visceral endoderm cells become organized into a distinct epithelium the apical surface being covered by microvilli facing the former blastocoel which is now called the yolk cavity (Hogan and Tilly, 1981). On the basal side visceral endoderm cells are separated from the underlying embryonic and extraembryonic ectoderm cells by a thin basement membrane (Clark et al., 1982). Together with extraembryonic mesoderm cells the visceral endoderm constitutes the visceral yolk sac.

Besides its important absorptive and secretory functions signals emanating from the visceral endoderm are essential for AP patterning of the embryo (see below).

Embryonic ectoderm

The cell types described thus far only give rise or contribute to extraembryonic tissues which serve essential supportive functions for the developing embryo. The embryo proper is formed exclusively from descendants of primitive ectoderm cells (Figure 11.3) which in addition give rise to the germ line and extraembryonic mesoderm (Gardner and Rossant, 1979; for review see Beddington, 1983b). Around implantation, primitive ectoderm cells form a core of about 30–40 cells surrounded by primitive endoderm cells on the side facing the blastocoel, and juxtaposed to extraembryonic ectoderm cells on the other side. Shortly after implantation primitive ectoderm cells start to proliferate rapidly (Snow, 1977) and form the distal part (tip) of the egg cylinder. Between day 5.5 and 6 of embryonic development a small lumen called the proamniotic cavity forms in the centre of the primitive ectoderm and the cells form a columnar epithelium. The apical side faces the lumen of the cavity and cells are joined apically by junctional complexes. The basal surface is attached to the thin basal membrane separating primitive ectoderm from visceral embryonic endoderm. Around day 6 p.c. (post coitum) the central cavity extends more dorsally into the extraembryonic ectoderm resulting in a small lumen throughout the egg cylinder.

Gastrulation and development of the germ layers

At 6.5 d p.c. the stage is set for a morphogenetic process which will completely reshape the embryo: gastrulation. As a result a three germ-layered embryo composed of mesoderm, endoderm and ectoderm is generated. The rotational symmetry of the egg-cylinder is transformed into a bilaterally symmetrical organization with a distinct AP and a dorsal–ventral axis. The left–right axis is determined after onset of gastrulation and becomes morphologically obvious with the looping of the heart tube at 8.5 d p.c. (for recent reviews on mouse gastrulation see Tam and Behringer, 1997; Lu *et al.*, 2001; Ang and Behringer, 2002).

Morphogenetic movements and the generation of the germ layers

At the initiation of gastrulation, epiblast cells at the future posterior end of the embryonic portion of the egg cylinder undergo an epithelial–mesenchymal transition and leave the epithelial continuity of the primitive ectoderm. This region is called the primitive streak and extends soon after its appearance distally towards the tip of the egg cylinder. The epiblast derived mesenchymal cells move as a new tissue layer between the visceral endoderm, the epiblast and the extraembryonic ectoderm, and differentiate into mesodermal cells (Figure 11.5).

The first mesenchymal cells to emerge from the primitive streak migrate towards the extraembryonic ectoderm and will give rise to the extraembryonic mesoderm. At the margin of the embryonic part of the egg cylinder ectodermal cells bulge into the lumen of the egg cylinder together with the underlying extraembryonic mesoderm and form the amniotic folds. Formation of the amniotic folds progresses from posterior to anterior leading to a continuous constriction of the central cavity that is most advanced at the posterior end. The amniotic folds grow towards each other, meet finally and fuse. Concomitantly the mesoderm within the folds develops a central cavity, the exocoelom, pushes the extraembryonic ectoderm towards the ectoplacental cone and separates it from the embryonic ectoderm. On day 7.5 p.c., after the amniotic folds have fused, the proamniotic cavity of the egg cylinder has been divided into the amniotic, exocoelomic and ectoplacental cavities (Figure 11.5a,b,c,d,g,h,j,k). Extraembryonic ectoderm with the underlying extraembryonic mesoderm constitutes the chorion, visceral endoderm with the attached extraembryonic mesoderm forms the visceral yolk sac. The amnion consists of an ectodermal cell layer covered by extraembryonic mesoderm. At the posterior end of the embryo, extraembryonic mesoderm cells give rise to a finger-like structure which grows through the exocoelom towards the chorion. This tissue is called the allantois and will later fuse with the chorion, linking the embryo with the ectoplacental cone. The allantois will form the umbilical cord and together with the chorion will give rise to the chorioallantoic placenta (Figure 11.5c,f,i,k;

for reviews see Boucher and Pedersen, 1996; Watson and Tam, 2001; Rossant and Cross, 2002).

During 6.5 and 7.5 d p.c. the primitive streak extends from the posterior end of the embryo proper to the distalmost part of the egg cylinder (Figure 11.5a,b,j,k). During all that time mesodermal cells continuously form and move laterally and anteriorly away from the primitive streak (Poelmann, 1981a; Snell and Stevens, 1991). In addition, cells originating from the anterior region of the primitive streak displace the visceral embryonic endoderm cells into the yolk sac to form the definitive endoderm. They will colonize the midline region of the embryo and will eventually form the midgut (Lawson *et al.*, 1986; Lawson and Pedersen, 1987).

Around day 7 p.c. an ectodermal thickening emerges at the anterior end of the streak representing the node (Hensen's node as this structure is called in birds). Cells migrating through this area move anteriorly to form a transient embryonic structure lying in the midline of the embryo: the notochord (Figure 11.5b and k; Jurand, 1974; Beddington, 1981; Lawson *et al.*, 1986; Snell and Stevens, 1991). Endodermal cells from this region contribute to trunk endoderm (Beddington, 1981; Poelmann, 1981b; Lawson *et al.*, 1986; Snell and Stevens, 1991). The embryonic ectoderm cells overlying the notochord and its anterior extension, the prechordal plate, form the neural plate that folds in the midline to form the neural groove.

From day 7.5 p.c. onwards, extensive anterior growth and regression of the primitive streak extend the neural plate posteriorly. Concomitantly the primitive node moves back and cells migrating through the regressing node form the more posterior parts of the notochord. Ingression of cells through the primitive streak persists up to and through day 10 p.c. (midgestation) leading to posterior elongation of the embryo. Between 9.5 and 10.5 d p.c. the primitive streak gradually loses its identity and is then referred to as the tail bud instead. Gastrulation continues in the tail bud at the posterior end of the embryo until 13.5 d p.c. generating posterior trunk and tail tissue (Figure 11.5c and i).

Cells ingressing at different positions along the primitive streak (and the tail bud) have distinct developmental fates and give rise to different prospective mesodermal and endodermal tissues. Cells emerging from the posterior part of the streak move mainly into the extraembryonic mesoderm. Cells from the middle region of the streak give rise to lateral mesoderm (mesoderm located laterally to the paraxial mesoderm). Cells emerging anteriorly to the streak form paraxial mesoderm (giving rise to somites and head

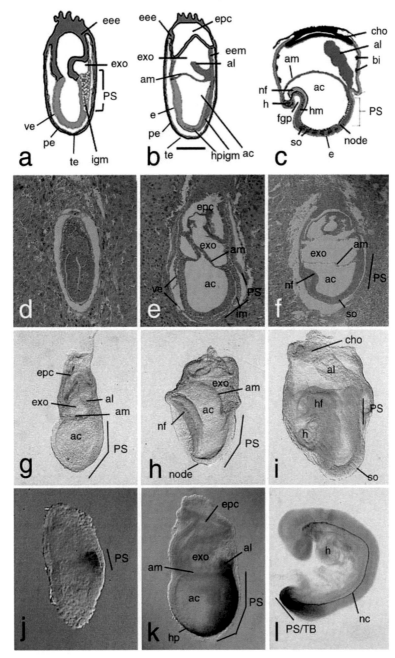

Figure 11.5 Mouse development from early primitive streak to early organogenesis stages. (a–c) Schematic representation of early postimplantation development. (a) 6.75 d p.c. embryo; (b) 7.5 d p.c. embryo; (c) 8.25 d p.c. embryo. Colour code. Yellow – (visceral, parietal, definitive) endoderm, red – mesoderm, grey – extraembryonic ectoderm, blue – embryonic ectoderm/epiblast, black – chorion. (d–f) Histological sections of early gastrulation stage mouse embryos with surrounding uterine tissue. 10 μm paraffin sections stained with hematoxylin and eosin. (d) 6.25 d p.c. embryo; (e) 7.0 d p.c. embryo; (f) 8.0–8.25 d p.c. embryo; (g–h) whole gastrulation stage embryos; (g) 7.5 d p.c. embryo; (h) 7.75 d p.c. embryo; (i) 8.5 d p.c. embryo. The black bar represents 200 μm. (j–l) Gastrulation stage embryos marked for the primitive streak/tail bud region. Immunohistological detection of Brachyury protein expression in whole embryos (Kispert and Herrmann, 1994). (j) 6.5 d p.c. embryo. Brachyury expression marks the primitive streak which has formed at the future posterior end of the embryo proper. (k) 7.5 d p.c. embryo. The primitive streak has extended to the distalmost tip of the egg cylinder. Brachyury protein expression also indicates the extraembryonic mesoderm (allantois) and the head process. (l) 9.0 d p.c. embryo. The primitive streak or tail bud marks the posterior pole of the embryo. Brachyury protein is detected all along the notochord. Anterior is to the left, posterior to the right, proximal is to the top, distal is to the bottom. ac – amniotic cavity, al – allantois, am – amnion, bi – blood islands, cho – chorion, e – (definitive) endoderm, eee – extraembryonic ectoderm, eem – extraembryonic mesoderm, epc – ectoplacental cavity, exo – exocoelom, fgp – foregut pocket, h – heart, hp – head process, igm – ingressing mesoderm, nc – notochord, nf – neural folds, pe – parietal endoderm, PS – primitive streak, so – somites, TB – tail bud, te – trophectoderm, ve – visceral endoderm. **(See also Colour Plate 5.)**

mesenchyme), and cells emerging from the anterior part of the primitive streak mainly contribute to notochord and gut (Tam and Beddington, 1987; Lawson, 1999; Watson and Tam, 2001). In addition to the position-dependent allocation of cells to different mesodermal tissues, there is a stage dependent potential of the streak to form different mesodermal cell types. While the early primitive streak (day 6.5–8) produces both embryonic and extraembryonic mesoderm, the older primitive streak (from day 8.5 onwards) continues to produce embryonic mesoderm but ceases to contribute to extraembryonic mesoderm (Tam and Beddington, 1987). Hence, there is time and space dependent translation of AP positional values in the primitive streak into an axial–lateral, i.e. DV patterning of the mesoderm.

Similarly, the fate of embryonic ectoderm cells in the day 7.5 embryo seems to depend on the position along the AP axis. Cells from the anterior regions give rise to neuroectoderm of the prosencephalon and mesencephalon, cells flanking the anterior end of the streak give rise to neuroectoderm of the rhombencephalon, and cells flanking the anterior and middle region of the streak give rise to the spinal cord. The future DV orientation of neuroectodermal cells in the neural tube seems to be already established at this stage of gastrulation. Cells closer to the midline end up in more ventral positions than cells which are located more laterally. Ectoderm from the posteriormost regions give rise to surface ectoderm and cells from positions most lateral to the midline are the presumptive neural crest cell precursors (Beddington, 1981, 1982; Tam, 1989; Lawson, 1999; Watson and Tam, 2001; Joyner, 2002). These cell fates, however, do not imply that cells are committed to specific lineages prior to gastrulation, since there is little regional restriction in the developmental potency of embryonic ectoderm cells (Beddington, 1983a).

Anterior–posterior patterning and early organizing centres

The first morphological indication of breaking the rotational symmetry of the egg cylinder is the emergence of the primitive streak. The position of the primitive streak defines the posterior pole of the AP axis of the emerging bilaterally symmetrical embryo. The pregastrulation epiblast is characterized by extensive cell mixing making it very unlikely that positional information for AP specification can be maintained within the epiblast (Gardner and Cockroft, 1998). However, extraembryonic tissues like the primitive endoderm grow coherently and could instruct the underlying epiblast

with positional information (Gardner, 1984; Gardner and Cockroft, 1998).

Evidence for molecular asymmetries inherent in the visceral endoderm before appearance of the primitive streak were recently provided by a number of genetic and embryological studies (reviewed in Bielinska et al., 1999; Perea-Gomez et al., 2001). Lineage tracing studies suggest a movement of visceral endoderm cells from the distal tip of the egg cylinder to the future anterior side of the proximal region (Thomas et al., 1998). This subregion of the visceral endoderm has been termed anterior visceral endoderm (AVE) and seems to play some role as a signalling centre for the underlying (future anterior neural plate) embryonic ectoderm. At the moment there are conflicting views on its precise role in AP patterning but it seems likely that the AVE is involved in maintaining or stabilizing anterior neural fates rather than in inducing them (Thomas and Beddington, 1996; Tam and Steiner, 1999). At present, the molecular events leading to specification of the AP polarity are unresolved, but it is likely that extraembryonic tissues like the visceral endoderm and the extraembryonic ectoderm play important roles in this process.

Molecular asymmetries are already present in the preimplantation as well as in the pregastrulation embryo. Although no experimental proof exists, it is conceivable that there is a flow of information originating in the zygote or even the oocyte. This information may be translated into AP polarity via the blastocyst to the egg cylinder embryo (for reviews see Gardner, 2001a, 2002; Lu et al., 2001; Zernicka-Goetz, 2002).

While relatively little is known about the initiation of the AP polarity of the gastrulation stage embryo, a great deal of evidence has accumulated to highlight the role of the node as an embryonic tissue to function as an organizer of AP-polarity within the embryo itself (for reviews see Smith and Schoenwolf, 1998; Ang and Behringer, 2002). The node only becomes morphologically visible as an indentation at the distal tip of the 7.75 d p.c. embryo (Figure 11.5h). However, lineage studies have shown that the precursors of the node can be traced back to the anterior end of the primitive streak in the 6.5 d p.c. embryo (Lawson et al., 1991; Tam et al., 1997; Kinder et al., 2001). Cells of this early node as well as the late node contribute to axial mesendoderm, the prechordal plate or head process, the notochord and the gut endoderm. Lengthening of the AP-axis of the embryo results in a posterior displacement of the node from 8.0 d p.c. on, leaving behind the mesodermal cells that undergo a convergent extension movement to form the notochord (Figure 11.5).

Transplantation experiments have shown that the anterior streak of the 6.5 d p.c. embryo and the node of the 7.5 d p.c embryo, respectively, not only give rise to axial mesendoderm but provide a patterning system that induces and organizes neighbouring tissues into an AP axis (Beddington, 1994; Tam et al., 1997). The late node and the early anterior streak behave as late and early gastrula organizers (EGO), respectively. They are functional homologues of the dorsal blastopore lip region of the amphibian gastrula, the paradigm for such an organizing centre (Smith and Schoenwolf, 1998). The gastrula organizer progressively induces more posterior cell fates. At present it seems that it never induces very anterior neural fate suggesting the existence of a separate head organizing activity. Alternatively, EGO and AVE activities may synergize in inducing anterior cell fates (Beddington and Robertson, 1998; Tam and Steiner, 1999).

Molecular analysis of organizer gene expression and function and the mouse embryo, have shown that EGO and node, rather than working by secretion of inducing factors, are a rich source of inhibitory molecules of Wnt and BMP signalling families. Locally restricted inhibition of BMP and Wnt pathways generates gradients of these signals important in patterning the AP and DV axis (for review see Beddington and Robertson, 1999; Ang and Behringer, 2002).

Left–right asymmetry

The external appearance of the mouse body like every mammalian body is overtly bilaterally symmetrical. However, most internal organs are asymmetric in shape or in position. Different mechanisms generate the asymmetry of the different organs. In some cases a differential looping, or turning programme is involved (e.g. heart, guts), in other cases differential growth is the driving force (e.g. lung), and last, differential remodelling of originally identical sides may lead to asymmetry as seen in the vascular system. Left–right asymmetry is established during gastrulation in embryogenesis, the first morphological sign being asymmetric looping of the heart tube at 8.5 d p.c. (for review see Mercola and Levin, 2001; Hamada et al., 2002; Hamada, 2002).

Genetic studies in the mouse as well as embryological manipulations in the chick have established a series of events that mediate left–right asymmetry. A major player in establishing left–right asymmetry is the node, the structure which also organizes DV and AP patterning of the embryo. Ventral node cells harbour monocilia

which all rotate in a clockwise fashion. Disruption of nodal cilia or perturbation of ciliary rotation leads to randomization of left–right asymmetry of internal organs suggesting that the directed rotation of nodal cilia creates a fluid flow from right to left (Brueckner, 2001; Nonaka et al., 2002). In turn, this might result in the differential transport of a signalling molecule across the nodal epithelium between 7.0 and 7.5 d p.c (Essner et al., 2002; Nonaka et al., 2002). The signal is then transferred from the left nodal side to the left lateral plate mesoderm where new molecular signals become asymmetrically expressed. They finally induce the asymmetric morphogenesis of the visceral organs starting from 8.5 d p.c. on.

Embryonic turning

The arrangement of ectoderm inside and endoderm on the outside of the embryo, which is found prior to and during early gastrulation, is known as inversion of the germ layers and common to mouse, rat, rabbit, guinea-pig and other closely related rodents. In sagittal sections the embryo has the appearance of a 'U', the midgut endoderm lines the outer curvature of the 'U' and fore- and hindgut follow at either end. At 8.25–8.5 d p.c., when the first 6–8 somites have formed (see below) the inversion of the germ layers is reverted by a process known as turning. During turning the embryo rotates anticlockwise around its AP axis. As a consequence, the curvature of the 'U' is reversed, the ectoderm comes to lie at the outer aspect of the embryo. The embryo becomes surrounded by the extraembryonic membranes, the amnion and the visceral yolk sac, since they are attached to the embryo along the boundary of the body wall and the future site of attachment of the umbilical cord. For a more detailed and illustrated description of turning, see Kaufman and Webb (1990).

Differentiation of mesoderm and early development of mesodermally-derived organs

Various types of mesoderm, which contribute to different tissues of the embryo are generated during gastrulation (for reviews see Tam and Behringer, 1997; Munoz-Sanjuan and Hemmati-Brivanlou, 2001). Mesoderm along the midline of the embryo (axial mesoderm) forms the prechordal plate and notochord, which, in turn, induce the overlying ectoderm to form the neural tube. Paraxial mesoderm flanks the noto-chord and the neural tube laterally on both sides as thick bands of mesoderm. Beginning at around day 7.75 p.c. balls of mesenchymal cells, somites, condense from paraxial mesoderm. The first somites form in the posterior head fold region of the embryo. Somite condensation progresses posteriorly, while new mesoderm continues to be generated from the primitive streak at the caudal end of the embryo. The region posterior to the first condensed somites is called presomitic or unsegmented mesoderm. New somites condense from presomitic mesoderm caudally to the first somites at regular intervals as the primitive streak regresses. The total number of about 65 somite pairs is formed at around day 13 of development. The metameric somites can be considered as segmental subdivisions along the AP axis.

Shortly after their condensation somites differentiate further. A central cavity forms within each somite and cells become epithelial, the apical sides facing the lumen of the cavity. Cells from the ventral part of the somites detach and migrate towards the notochord, which they enclose. These cells, called sclerotome, give rise to the axial skeleton (vertebrae, ribs). The cells left behind form a bilayered structure called dermamyotome. The dorsal dermatome cells give rise to the connective tissue of the skin, the dorsal myotome will form the skeletal muscles of the back (hypaxial musculature), whereas ventral myotomal cells form the muscles of the body wall and limbs (epaxial musculature; for reviews on somitogenesis see Pourquie, 2001; Gossler and Tam, 2002).

Mesodermal cells immediately flanking the somites (intermediate mesoderm) form the urogenital system comprising the gonads, the sex ducts and the kidneys. Further laterally lies the lateral plate mesoderm. Lateral plate mesoderm splits into the dorsal (or somatic) mesoderm underlying the ectoderm and the ventral (splanchic) mesoderm underlying the endoderm. Between these layers the coelom forms which will later be subdivided into the separate pleural, pericardial and peritoneal cavities. Lateral plate mesoderm cells form tissues such as the heart, connective tissues of the viscera and cartilage and bone of the limbs. The mesenchyme and bones of the head are derived from the cephalic neural crest cells (see below) on one side and from unsegmented cranial paraxial mesoderm on the other side.

Extraembryonic mesoderm of the visceral yolk sac is the first site of haematopoiesis in the developing embryo (Snell and Stevens, 1991; Palis and Yoder, 2001). From the 7th day of gestation onwards blood islands appear on the inner side of the visceral yolk sac. These are condensations of mesenchymal cells which form an irregular girdle around the exocoelom. The inner cells of these condensations become embryonic red blood cells (which are nucleated cells in contrast to the adult erythrocyte), the peripheral cells differentiate and form the endothelium of blood vessels of the yolk sac. Between embryonic day 9 and 10 haematopoiesis shifts into a region derived from the intermediate and lateral plate mesoderm referred to as AGM which contributes to the formation of the aorta, the gonads and the mesonephros (Godin et al., 1993; Medvinsky et al., 1993). Around day 12 of development the foetal liver takes over this function (for review see Speck et al., 2002).

Differentiation of ectoderm and early organogenesis of the nervous system

The nervous system develops from neural plate ectoderm which gives rise to the neural tube and the neural crest which in turn form all parts of the central and peripheral nervous system (for review see Joyner, 2002). Starting around day 7.5 the neural groove begins to form along the midline of the neural plate. While the primitive streak is regressing and the neural plate extending posteriorly, the neural groove deepens and the neural folds develop. In the cranial third of the embryo the head folds emerge rapidly and bulge deeply into the amniotic cavity, due to rapid growth and the indentation of the foregut pocket, which pushes the overlying neuroectoderm ahead of itself. As the folds become higher, the edges start to approach each other and finally meet and fuse to form the neural tube which underlies the surface ectoderm. Closure of the neural tube starts around day 8.25 at the position of the 4th to 5th somite and progresses anteriorly and posteriorly. The open ends of the neural tube are called the anterior and posterior neuropore. Development of the neural tube progresses more rapidly in the cranial region. The anterior neuropore is closed around day 9 while closure of the posterior neuropore is not complete until day 10 p.c. Cells from the edge of the neural folds between neuroectoderm and surface ectoderm give rise to the neural crest. The neural crest is a transient structure which is present only shortly after closure of the neural tube. The neural crest cells disperse rapidly and migrate through the embryo. They give rise to a variety of cell types depending from which part along the AP axis they originated and where they finally settle (Le Douarin, 1982). Among those are the neurons and glial cells of the spinal ganglia, the peripheral nervous system and the adrenal medulla, the melanocytes of the epidermis and most of the mesenchymal cells of the head (skeletal and connective tissue; for reviews in neural crest development see Nieto, 2001; Knecht and Bronner-Fraser, 2002).

Differentiation of endoderm and organogenesis of gut and its derivatives

The definitive endoderm together with the mesoderm derive from epiblast cells ingressing through the node and the anterior primitive streak during early gastrulation. While the mesodermal cell layer moves anteriorly and laterally, cells fated to become definitive endoderm migrate ventrally, intercalate in and finally displace the visceral endoderm cell layer. By the head fold stage (7.5–8 d p.c.), the definitive endoderm consists of some 500 cells which are organized in a single cup-like sheet. At this time, deep invaginations occur at the anterior and posterior end of the embryo to form the fore gut and hind gut pockets which will later make contact with the definitive endoderm of the midgut region. Concomitantly with embryonic turning (8.25–9.0 d p.c.), the lateral walls of the endoderm sheets are brought in juxtaposition and fuse, generating a continuous gut tube. During midgestation stages, the growth of the gut tube exceeds that of the body cavity leaving parts of the midgut herniating outside the ventral body wall (for reviews on endoderm development see Roberts, 2000; Hogan and Zaret, 2002).

A broad AP-patterning of the gut has already occurred at the late primitive streak stage, possibly by node and streak derived signals. Extensive morphogenetic movements between the late streak stage and midgestation juxtaposes the gut endoderm with various mesodermal cell types. Epithelial–mesenchymal interactions between gut endoderm and the surrounding splanchnic mesoderm result in a refined anterior/posterior patterning of the gut tube and induction of gut organ appendages. The gut tube differentiates into oesophagus, stomach, small intestine (duodenum, jejunum, ileum) and large intestine (caecum, colon, rectum), with highly specialized endodermal cell types and distinct radial patterns of the splanchnic mesoderm (Roberts, 2000). Along the AP axis of the gut, endodermal organs form by budding from the primitive gut: the thyroid at 8.5 d p.c. (Damante et al., 2001), the lungs at 9.5 d p.c. (Hogan and Kolodziej,

2002), the liver from 9 d on (Zaret, 2002) and the pancreas from a ventral and a dorsal bud of the foregut endoderm from embryonic day 9.5 p.c. (Edlund, 2002). Lung and pancreas development can be considered as typical for branching morphogenesis. The foregut endoderm buds out, proliferates and undergoes extensive branching while interacting with the surrounding mesenchymal cell layer.

Primordial germ cells

Primordial germ cells (PGCs), the ancestors of the gametes, originate in the mouse at least as early as on day 7 of development (for review see Wylie and Anderson, 2002). They arise from a population of pluripotent cells in the proximal epiblast close to the extraembryonic ectoderm (Lawson and Hage, 1994). They pass through the posterior primitive streak and are found first in the posterior part of the embryo at the base of the allantois (Copp et al., 1986; Ginsburg et al., 1990). They are large, round cells, which contain a high level of alkaline phosphatase activity. This enzymatic activity can be used to trace PGCs in the early embryo (Chiquoine, 1954). More recently, the POU transcription factor OCT4 was found as PGC marker (Schöler et al., 1990). A truncated Oct4 promotor was used to drive expression of green fluorescent protein in transgenic animals. Thus, migration of PGCs was visualized in a living embryo (Anderson et al., 1999, 2000). From day 8.5 onwards PGCs migrate through the hindgut and mesentery wall and colonize the genital ridges. The genital ridges, which give rise to the gonads, are a paired mesodermal tissue which lies beneath the dorsal mesentery of the body. By day 12.5 PGCs are largely confined to the developing gonads. In vitro studies suggest that colonization of the genital ridges is brought about by active, invasive movement of the PGCs and that PGCs lose their invasive motility after entering the gonad anlagen (Donovan et al., 1986). During their migration the PGCs proliferate and the population of about 10–100 PGCs present around day 7–8 increases to more than 20,000 in the colonized genital ridges around day 14 of development (Tam and Snow, 1981; Wylie et al., 1985). Genital ridges seem to release intrinsic factors which stimulate the proliferation of PGCs and which act as chemoattractants for PGCs in vitro (Godin et al., 1990).

At day 12.5 differences between male and female genital ridges become apparent and male and female germ cells embark on their specific developmental programmes. Male PGCs enter mitotic arrest around day

13 p. c. and continue development only after birth. In contrast, female mouse PGCs enter meiosis from day 13 of development onwards and already about 3–5 days after birth all germ cells have undergone oogonial development and are in the diplotene stage of meiosis (for review see Swain and Lovell-Badge, 2002; Wylie and Anderson, 2002).

Late embryonic development: completion of organogenesis and foetal growth

By midgestation (11th day of development) the basic body plan has essentially been established. The three embryonic axes have been laid down and patterning and cellular differentiation along these axes has progressed considerably. Whereas at the anterior end of the embryo tissue differentiation and organogenesis has already progressed and accelerated considerably, the axial elongation is only gradually coming to an end in the tail bud region. In the nervous system separation of the four brain vesicles proceeds and the major divisions of the brain are now clearly visible. The cellular differentiation of the nervous system, which begins around day 9 p.c., continues. Proliferating neuroblasts are found in the walls of the entire central nervous system and the spinal ganglia are well formed. The major elements of the circulatory system have developed and are functional to supply the growing embryo with nutrients and oxygen enriched blood. Fore- and hindlimb buds are present, the anterior limbs because they arise first, being more developed than the posterior ones. In the trunk region the development of the vertebrae commences and in this region somites start becoming less discernable. All major organ anlagen are present or emerge within the next few days.

Generally organs functional in the embryo are laid down more anteriorly and mature more quickly. Organs dispensable for the embryo are established more posteriorly and mature more slowly. Hence, organs like the kidneys and the lung only become functional with or after birth. Development of all organs is characterized by a highly coordinated programme of cell and tissue interactions, cell and tissue movements (morphogenesis) and locally controlled cellular differentiation pathways. Detailed discussions of these developmental programs are beyond the scope of this introductory chapter and can be found in Rossant and Tam (2002).

Besides locally controlled proliferation rates leading to directional outgrowth of organs and appendages like limbs, jaws or external genitalia, a massive increase in size occurs between 6.5 and birth. Global growth control occurs concomitantly with gastrulation and organogenesis but can be genetically uncoupled leading to small newborn pups (Yamaguchi et al., 1999). Global growth control is mediated by a cocktail of systemic factors which at some point during development are under hormonal control.

References

Anderson, R., Copeland, T.K., Scholer, H., Heasman, J. and Wylie, C. (2000). *Mech. Dev.* **91**, 61–68.

Anderson, R., Fassler, R., Georges-Labouesse, E., Hynes, R.O., Bader, B.L., Kreidberg, J.A., Schaible, K., Heasman, J. and Wylie, C. (1999). *Development* **126**, 1655–1664.

Ang, S.-L. and Behringer, R.R. (2002). In *Mouse Development* (eds J. Rossant and P.P.L. Tam), pp. 37–53. Academic Press, San Diego.

Beddington, R.S. (1982). *J. Embryol. Exp. Morphol.* **69**, 265–285.

Beddington, R.S. (1983a). *J. Embryol. Exp. Morphol.* **75**, 189–204.

Beddington, R.S. (1994). *Development,* **120**, 613–620.

Beddington, R.S. and Robertson, E.J. (1998). *Trends Genet.* **14**, 277–284.

Beddington, R.S. and Robertson, E.J. (1999). *Cell* **96**, 195–209.

Beddington, R.S.P. (1983b). In *Development in Mammals*, Vol. 5 (ed. M.H. Johnson), pp. 1–32. Elsevier, Amsterdam.

Beddington, S.P. (1981). *J. Embryol. Exp. Morphol.* **64**, 87–104.

Bielinska, M., Narita, N. and Wilson, D.B. (1999). *Int. J. Dev. Biol.* **43**, 183–205.

Bleil, J.D. and Wassarman, P.M. (1980a). *Cell* **20**, 873–882.

Bleil, J.D. and Wassarman, P.M. (1980b). *Proc. Natl. Acad. Sci. USA* **77**, 1029–1033.

Boucher, D.M. and Pedersen, R.A. (1996). *Reprod. Fertil. Dev.* **8**, 765–777.

Braude, P., Pelham, H., Flach, G. and Lobatto, R. (1979). *Nature* **282**, 102–105.

Brueckner, M. (2001). *Am. J. Med. Genet.* **101**, 339–344.

Carlin, B., Jaffe, R., Bender, B. and Chung, A.E. (1981). *J. Biol. Chem.* **256**, 5209–5214.

Chai, N., Patel, Y., Jacobson, K., McMahon, J., McMahon, A. and Rappolee, D.A. (1998). *Dev. Biol.* **198**, 105–115.

Chen, L.T. and Hsu, Y.C. (1982). *Science,* **218**, 66–68.

Chiquoine, A.D. (1954). *Anat. Rec.* **118**, 135–146.

Ciemerych, M.A., Mesnard, D. and Zernicka-Goetz, M. (2000). *Development* **127**, 3467–3474.

Clark, C.C., Crossland, J., Kaplan, G. and Martinez-Hernandez, A. (1982). *J. Cell Biol.* **93**, 251–260.

Clegg, K.B. and Piko, L. (1977). *Dev. Biol.* **58**, 76–95.

Clegg, K.B. and Piko, L. (1983). *J. Embryol. Exp. Morphol.* **74**, 169–182.

Copp, A.J. and Cockroft, D.L. (eds) (1990). *Postimplantation Mammalian Embryos. A Practical Approach.* IRL Press/Oxford University Press, New York.

Copp, A.J., Roberts, H.M. and Polani, P.E. (1986). *J. Embryol. Exp. Morphol.* **95**, 95–115.

Damante, G., Tell, G. and Di Lauro, R. (2001). *Prog. Nucleic Acid Res. Mol. Biol.* **66**, 307–356.

Donovan, P.J., Stott, D., Cairns, L.A., Heasman, J. and Wylie, C.C. (1986). *Cell* **44**, 831–838.

Ducibella, T., Albertini, D.F., Anderson, E. and Biggers, J.D. (1975). *Dev. Biol.* **45**, 231–250.

Ducibella, T. and Anderson, E. (1975). *Dev. Biol.* **47**, 45–58.

Dyce, J., George, M., Goodall, H. and Fleming, T.P. (1987). *Development* **100**, 685–698.

Edlund, H. (2002). *Nat. Rev. Genet.* **3**, 524–532.

Enders, A.C., Given, R.L. and Schlafke, S. (1978). *Anat. Rec.* **190**, 65–77.

Essner, J.J., Vogan, K.J., Wagner, M.K., Tabin, C.J., Yost, H.J. and Brueckner, M. (2002). *Nature* **418**, 37–38.

Fleming, T.P. (1987). *Dev. Biol.* **119**, 520–531.

Fleming, T.P. and Pickering, S. (1985). *J. Embryol. Exp. Morphol.* **89**, 175–208.

Gardner, R.L. (1975). In *The Developmental Biology of Reproduction.* 33rd Symposium of the Society for Developmental Biology (ed. C.L. Markert), pp. 207–238. Academic Press, New York.

Gardner, R.L. (1984). *J. Embryol. Exp. Morphol.* **80**, 251–288.

Gardner, R.L. (1997). *Development* **124**, 289–301.

Gardner, R.L. (2001a). *Int. Rev. Cytol.* **203**, 233–290.

Gardner, R.L. (2001b). *Development* **128**, 839–847.

Gardner, R.L. (2002). In *Mouse Development* (eds J. Rossant and P.P.L. Tam), pp. 21–35. Academic Press, San Diego.

Gardner, R.L. and Cockroft, D.L. (1998). *Development* **125**, 2397–2402.

Gardner, R.L. and Johnson, M.H. (1972). *J. Embryol. Exp. Morphol.* **28**, 279–312.

Gardner, R.L. and Johnson, M.H. (1973). *Nat. New. Biol.* **246**, 86–89.

Gardner, R.L., Papaioannou, V.E. and Barton, S.C. (1973). *J. Embryol. Exp. Morphol.* **30**, 561–572.

Gardner, R.L. and Rossant, J. (1979). *J. Embryol. Exp. Morphol.* **52**, 141–152.

Ginsburg, M., Snow, M.H. and McLaren, A. (1990). *Development* **110**, 521–528.

Godin, I., Wylie, C. and Heasman, J. (1990). *Development* **108**, 357–363.

Godin, I.E., Garcia Porrero, J.A., Coutinho, A., Dieterlen Lievre, F. and Marcos, M.A. (1993). *Nature* **364**, 67–70.

Gossler, A. and Tam, P.P.L. (2002). In *Mouse Development* (eds J. Rossant and P.P.L. Tam), pp. 127–153. Academic Press, San Diego.

Hamada, H. (2002). In *Mouse Development* (eds J. Rossant and P.P.L. Tam), pp. 55–73. Academic Press, San Diego.

Hamada, H., Meno, C., Watanabe, D. and Saijoh, Y. (2002). *Nat. Rev. Genet.* **3**, 103–113.

Hillman, N., Sherman, M.I. and Graham, C. (1972). *J. Embryol. Exp. Morphol.* **28**, 263–278.

Hogan, B., Beddington, R., Constantini, F. and Lacy, E. (1994). *Manipulating the Mouse Embryo – A Laboratory Manual.* Cold Spring Harbor Laboratory Press, Cold Spring Harbor.

Hogan, B.L. and Kolodziej, P.A. (2002). *Nat. Rev. Genet.* **3**, 513–523.

Hogan, B.L. and Tilly, R. (1981). *J. Embryol. Exp. Morphol.* **62**, 379–394.

Hogan, B.L. and Zaret, K.S. (2002). In *Mouse Development* (eds J. Rossant and P.P.L. Tam), pp. 301–330. Academic Press, San Diego.

Hogan, B.L.M., Cooper, A.R. and Kurkinen, M. (1980). *Dev. Biol.* **80**, 289–300.

Hogan, B.L.M., Taylor, A. and Cooper, A.R. (1982). *Dev. Biol.* **90**, 210–214.

Hyafil, F., Morello, D., Babinet, C. and Jacob, F. (1980). *Cell* **21**, 927–934.

Johnson, M. and Ziomek, C.A. (1981). *Cell,* **24**, 71–80.

Johnson, M.H. (1986). In *Developmental Biology: A Comprehensive Synthesis* (ed. R.B.L. Gwatkin), pp. 279–296. Plenum Press, New York.

Johnson, M.H. and Maro, B. (1984). *J. Embryol. Exp. Morphol.* **82**, 97–117.

Joyner, A.L. (2002). In *Mouse Development* (eds J. Rossant and P.P.L. Tam), pp. 107–126. Academic Press, San Diego.

Jurand, A. (1974). *J. Embryol. Exp. Morph.* **32**, 1–33.

Kaufman, M.H. and Bard, J.B.L. (1999). *The Anatomical Basis of Mouse Development.* Academic Press, London.

Kaufman, M.H. and Webb, S. (1990). *Development* **110**, 1121–1132.

Kelly, S.J. (1977). *J. Exp. Zool.* **200**, 365–376.

Kinder, S.J., Tsang, T.E., Wakamiya, M., Sasaki, H., Behringer, R.R., Nagy, A. and Tam, P.P. (2001). *Development* **128**, 3623–3634.

Kispert, A. and Herrmann, B.G. (1994). *Dev. Biol.* **161**, 179–193.

Knecht, A.K. and Bronner-Fraser, M. (2002). *Nat. Rev. Genet.* **3**, 453–461.

Latham, K.E. (1999). *Int. Rev. Cytol.* **193**, 71–124.

Lawson, K.A. (1999). *Int. J. Dev. Biol.* **43**, 773–775.

Lawson, K.A. and Hage, W.J. (1994). *Ciba Found. Symp.* **182**, 68–84; discussion 84–91.

Lawson, K.A., Meneses, J.J. and Pedersen, R.A. (1986). *Dev. Biol.* **115**, 325–339.

Lawson, K.A., Meneses, J.J. and Pedersen, R.A. (1991). *Development* **113**, 891–911.

Lawson, K.A. and Pedersen, R.A. (1987). *Development* **101**, 627–652.

Le Douarin, N.M. (1982). *The Neural Crest.* Cambridge University Press, Cambridge.

Lu, C.C., Brennan, J. and Robertson, E.J. (2001). *Curr. Opin. Genet. Dev.* **11**, 384–392.

Maro, B., Johnson, M.H. and Pickering, S.J. (1985). *J. Embryol. Exp. Morphol.* **90**, 287–309.

Medvinsky, A.L., Samoylina, N.L., Müller, A.M. and Dzierzak, E.A. (1993). *Nature* **364**, 64–67.

Mercola, M. and Levin, M. (2001). *Annu. Rev. Cell. Dev. Biol.* **17**, 779–805.

Monk, M. (ed.) (1997). *Mammalian Development. A Practical Approach.* IRL Press Limited, Oxford.

Munoz-Sanjuan, I. and Hemmati-Brivanlou, A. (2001). *Dev. Biol.* **237**, 1–17.

Nichols, J., Zevnik, B., Anastassiadis, K., Niwa, H., Klewe-Nebenius, D., Chambers, I., Scholer, H. and Smith, A. (1998). *Cell* **95**, 379–391.

Nieto, M.A. (2001). *Mech. Dev.* **105**, 27–35.

Nonaka, S., Shiratori, H., Saijoh, Y. and Hamada, H. (2002). *Nature* **418**, 96–99.

Palis, J. and Yoder, M.C. (2001). *Exp. Hematol.* **29**, 927–936.

Papaioannou, V.E. (1982). *J. Embryol. Exp. Morphol.* **68**, 199–209.

Pedersen, R.A., Wu, K. and Balakier, H. (1986). *Dev. Biol.* **117**, 581–595.

Perea-Gomez, A., Rhinn, M. and Ang, S.L. (2001). *Int. J. Dev. Biol.* **45**, 311–320.

Perona, R.M. and Wassarman, P.M. (1986). *Dev. Biol.* **114**, 42–52.

Piotrowska, K. and Zernicka-Goetz, M. (2001). *Nature* **409**, 517–521.

Poelmann, R.E. (1981a). *Anat. Embryol. Berl.* **162**, 29–40.

Poelmann, R.E. (1981b). *Anat. Embryol. Berl.* **162**, 41–49.

Pourquie, O. (2001). *Annu. Rev. Cell Dev. Biol.* **17**, 311–350.

Pratt, H.P.M. (1985). *J. Embryol. Exp. Morphol.* **90**, 101–121.

Reeve, W.J. and Ziomek, C.A. (1981). *J. Embryol. Exp. Morphol.* **62**, 339–350.

Roberts, D.J. (2000). *Dev. Dyn.* **219**, 109–120.

Rossant, J. (1976a). *J. Embryol. Exp. Morphol.* **36**, 163–174.

Rossant, J. (1976b). *J. Embryol. Exp. Morphol.* **36**, 283–290.

Rossant, J. and Cross, J.C. (2002). In *Mouse Development* (eds J. Rossant and P.P.L. Tam), pp. 155–180. Academic Press, San Diego.

Rossant, J., Gardner, R.L. and Alexandre, H.L. (1978). *J. Embryol. Exp. Morphol.* **48**, 239–247.

Rossant, J. and Lis, W.T. (1979). *Dev. Biol.* **70**, 249–254.

Rossant, J. and Ofer, L. (1977). *J. Embryol. Exp. Morphol.* **39**, 183–194.

Rossant, J. and Tam, P. (2001). *Mouse Development: Patterning, Morphogenesis, and Organogenesis.* Academic Press, San Diego.

Rossant, J. and Tam, P.P.L. (eds) (2002). *Mouse Development.* Academic Press, San Diego.

Rossant, J. and Vijh, K.M. (1980). *Dev. Biol.* **76**, 475–482.

Rugh, R. (1990). *The Mouse. Its Reproduction and Development.* Oxford University Press, New York.

Schöler, H.R., Ruppert, S., Suzuki, N., Chowdhury, K. and Gruss, P. (1990). *Nature* **344**, 435–439.

Shur, B.D. and Hall, N.G. (1982a). *J. Cell. Biol.* **95**, 574–579.

Shur, B.D. and Hall, N.G. (1982b). *J. Cell. Biol.* **95**, 567–573.

Smith, J.L. and Schoenwolf, G.C. (1998). *Curr. Top. Dev. Biol.* **40**, 79–110.

Smith, K.K. and Strickland, S. (1981). *J. Biol. Chem.* **256**, 4654–4661.

Smith, L.J. (1980). *J. Embryol. Exp. Morphol.* **55**, 257–277.

Smith, L.J. (1985). *J. Embryol. Exp. Morphol.* **89**, 15–35.

Smith, R. and McLaren, A. (1977). *J. Embryol. Exp. Morph.* **41**, 79–92.

Snell, G.D. and Stevens, L.C. (1991). In *Biology of the Laboratory Mouse* (ed. E.L. Green), pp. 205–245. Dover Publications, New York.

Snow, M.H.L. (1977). *J. Embryol. Exp. Morphol.* **42**, 293–303.

Speck, N., Peeters, M. and Dzierzak, E. (2002). In *Mouse Development* (eds J. Rossant and P.P.L. Tam), pp. 191–210. Academic Press, San Diego.

Swain, A. and Lovell-Badge, R. (2002). In *Mouse Development* (eds J. Rossant and P.P.L. Tam), pp. 371–393. Academic Press, San Diego.

Tam, P.P. (1989). *Development* **107**, 55–67.

Tam, P.P. and Behringer, R.R. (1997). *Mech. Dev.* **68**, 3–25.

Tam, P.P. and Snow, M.H. (1981). *J. Embryol. Exp. Morphol.* **64**, 133–147.

Tam, P.P., Steiner, K.A., Zhou, S.X. and Quinlan, G.A. (1997). *Cold Spring Harb. Symp. Quant. Biol.* **62**, 135–144.

Tam, P.P.L. and Beddington, R.S.P. (1987). *Development* **99**, 109–126.

Tam, P.P.L. and Steiner, K.A. (1999). *Development* **126**, 5171–5179.

Tanaka, S., Kunath, T., Hadjantonakis, A.K., Nagy, A. and Rossant, J. (1998). *Science* **282**, 2072–2075.

Tarkowski, A.K. (1959). *Nature* **184**, 1286–1287.

Tarkowski, A.K. and Wroblewska, J. (1967). *J. Embryol. Exp. Morphol.* **18**, 155–180.

Theiler, K. (1989). *The House Mouse. Atlas of Embryonic Development.* Springer Verlag, New York.

Thomas, P. and Beddington, R. (1996). *Curr. Biol.* **6**, 1487–1496.

Thomas, P.Q., Brown, A. and Beddington, R.S. (1998). *Development* **125**, 85–94.

Vestweber, D. and Kemler, R. (1984). *Exp. Cell Res.* **152**, 169–178.

Warner, C.M. and Versteegh, L.R. (1974). *Nature* **248**, 678–680.

Wassarman, P.M. (1995). *Curr. Opin. Cell Biol.* **7**, 658–664.

Wassarman, P.M. (1999). *Cell* **96**, 175–183.

Watson, C.M. and Tam, P.P. (2001). *Cell Struct. Funct.* **26**, 123–129.

Wylie, C. and Anderson, R. (2002). In *Mouse Development* (eds J. Rossant and P.P.L. Tam), pp. 181–190. Academic Press, San Diego.

Wylie, C.C., Stott, D. and Donovan, P.J. (1985). In *Developmental Biology* Vol. 2 (ed. L.W. Browder), pp. 433–448. Plenum Press, New York.

Yamaguchi, T.P., Bradley, A., McMahon, A.P. and Jones, S. (1999). *Development* **126**, 1211–1223.

Yoshida-Noro, C., Suzuki, N. and Takeichi, M. (1984). *Dev. Biol.* **101**, 19–27.

Zaret, K.S. (2002). *Nat. Rev. Genet.* **3**, 499–512.

Zernicka-Goetz, M. (2002). *Development* **129**, 815–829.

P A R T 3

Pathophysiology (Including Non-infectious Diseases)

Contents

Skin and Adnexa of the Laboratory Mouse

John P Sundberg

The Jackson Laboratory, Bar Harbor, ME, USA

Introduction

The skin is the largest of the intermediate-sized organs (Goldsmith, 1990). Dermatology, anatomy, and histology textbooks assign simple functions to this organ system which, in reality, is as complicated as any organ in the body. More importantly, it is integrated with every organ of the body, not simply a wrapping to hold things together. The list of functions of the skin are constantly expanding. Table 12.1 is a summary from a recent published debate on this topic (Chuong *et al.*, 2002).

    Spontaneous and genetically engineered mutations in laboratory mice have changed the basis of our knowledge of the function of the skin and how gene expression in the skin may be a reflection of similar expression in different organs (Sundberg and King, 2000). For example, for a long time there was a general thought that mice without hair (alopecia) have some form of immunodeficiency. This was largely based on the nude mouse. These mutant mice appear to lack hair at the gross level and lack a cell mediated immune system due to failure of the thymus to develop normally.

TABLE 12.1: Functions of the skin and adnexa
Protection from the environment
Defense
Weapons
Communication with outside species
Communication with internal organs
Respiration (especially in lower species)
Chemical reactions (activation of compounds by light)
Locomotion (especially in lower species)
Thermoregulation
Progeny support (lactation in mammals)
Source: Summarized from Chuong *et al.*, 2002.

In fact, these mice have hair follicles and produce hair fibers but the structures are defective due to the role of the nude gene (*Foxn1^{nu}*) that acts as a transcription factor to down regulate hard keratin production (Mecklenburg *et al.*, 2001). This gene also plays a role in terminal differentiation of keratinocytes at various anatomic sites (Baxter and Brissette, 2002). Hairless (*hr*), another mutant mouse that has been available for over half a century (Gaskoin, 1856) also has a minor abnormality in its immune system (Sprecher *et al.*, 1990). The advent of the severe combined immuno-deficiency mutant mice (*Prkdc*^{scid}) with normal **pelage** and hair cycle changed that limited correlation (Sundberg and Shultz, 1994). We now know that each skin defect can be unique and may or may not be associated with visceral lesions (Sundberg and King, 2000).

Numerous mutations have occurred in laboratory mice spontaneously, induced by radiation or various chemical mutagens, or created using transgenesis or targeted mutagenesis (Sundberg, 1994a; Sundberg *et al.*, 1995; Sundberg and King, 1996a, b; 2000; Nakamura *et al.*, 2002a, b; Randall *et al.*, 2003). It is beyond the scope of this chapter to cover these mutant mice. In addition to the reviews provided, much information is available on the Mouse Genome Informatics web site (www.informatics.jax.org).

This chapter will provide an overview and references as sources for more specific information on normal anatomy, development, and cycling of the skin and its adnexa. It will also provide information on routine methods to prepare specimens for analysis. General, systematic descriptions of necropsy procedures evaluating all organ systems can be found in another chapter in this book.

Clinical evaluation, tissue collection, and preservation of the skin

Clinical evaluation

The normal mouse is covered with hair. Careful examination reveals that at least two hair types are obvious, as is the case with most domestic mammals. A fine short hair coat covers most of the body (truncal hairs) while long hairs are evident around the head (vibrissae, incorrectly called whiskers by many investigators since there is no anatomical similarity to androgen induced facial hair in humans). If the hair is carefully studied there are many hair types present. Within the pelage hairs there are classically four types: guard hairs are long, straight, thick, and protrude above the level of most hairs; auchene hairs are nearly as long as guard hairs with a gradual bend at the distal end; awl are also straight with a bend

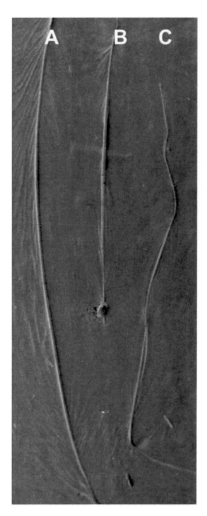

Figure 12.1 Scanning electron micrograph of telogen stage plucked guard (A), awl (B), and zigzag (C) hairs from the dorsal truncal skin of an adult mouse.

at the distal end but are short and thin; lastly zigzag hairs are the underhairs that have two bends giving in a 'Z' shape (Figure 12.1). These hair types are best differentiated in plucked samples examined with magnification. Historically many people stuck hairs on double sticky tape to a glass microscope slide. An easier method is to place a few hairs on a glass slide, add a drop of mounting medium, and drop on a cover slip. This forces the hairs to lay in one plane. The hairs can be examined with a microscope, photographed, and a variety of light sources used that can provide diagnostic information (Sundberg and Hogan, 1994; Sundberg *et al.*, 1998).

In addition to the pelage or truncal hairs there are many other specialized hair types in the mouse. The tail is covered with very short, broad fibers. Ears have a variety of very short fine fibers (Figure 12.2). Eyes have vibrissae above the eyelids and a network of long hairs protruding from the lid margins called cilia.

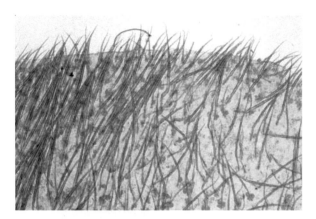

Figure 12.2 Subgross photograph of pilosebaceous units (hair follicles with the sebaceous gland at its base) in cleared skin from the ear of an adult mouse.

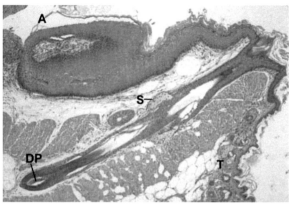

Figure 12.4 Anagen stage perianal hair. Anus (A), dermal or follicular papilla (DP) within the bulb, sebaceous gland (S), telogen stage truncal hair follicles (T).

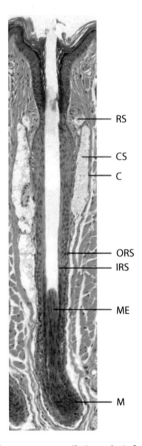

Figure 12.3 Anagen stage vibrissae hair follicle from the muzzle of an adult mouse. Ring sinus (RS), cavernous sinous (CS), capsule (C), outer root sheath (ORS), inner root sheath (IRS), medulla (ME), matrix (M).

Vibrissae found around the mouth, eyelids, and near the foot pads on the lower legs (Figure 12.3). Perianal hairs are large thin structures that form a network above the opening of the anus (Figure 12.4). Hairs change around nipples and the base of the ear. These differences often can only be seen using a hand lens,

dissection microscope, scanning electron microscopy, or some other means of magnification.

Hairs are usually thin and straight with a uniform distribution pattern within a strain. Variations, especially hair loss, can suggest that the mice have a mutant phenotype but only after simple diagnostic methods rule out infectious causes or infestations. Ectoparasites remain common in many mouse rooms and will result in alopecia often mistaken the novice for a mutant phenotype (Figure 12.5). Mites are easily diagnosed by placing a piece of hairy skin in a closed petri dish into a refrigerator then examining it after an hour or so with a hand lens. Mites migrate to the tips looking for another host. They can also be easily identified by histologically if hairs are not shaved during preparation of the skin (Figure 12.5). Other infectious diseases require the assistance of a trained veterinary pathologist for correct diagnosis.

Tissue collection and preservation for histologic evaluation of the skin

Every pathologist has his or her own preference for fixation of tissues. It is always best to work with the pathologist who will be evaluating the tissues before proceeding. Neutral buffered formalin solution is the most universal fixative used. Tissues are often left in formalin for long periods. Due to cross linking of amino groups by the aldehydes, many epitopes are modified making immunohistochemistry difficult or impossible. Fekete's acid alcohol formalin minimizes this problem, especially when tissues are transferred to 70% ethanol after overnight fixation. Commercially available zinc based preservatives are being promoted to

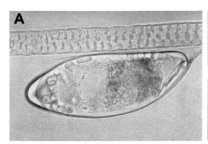

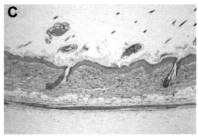

Figure 12.5 Photomicrograph of an egg (nit) anchored onto a hair fiber (A). Scanning electron microscopy reveals the mite (*Myocoptes musculinus*) holding onto a hair fiber (B). Ectoparasites (mites) above the epidermis in a histologic section (C).

maintain epitopes, optimize immunohistochemical results while maintaining some degree of the histologic quality that pathologists are used to with paraffin sections. Bouin's solution is popular as a general fixative but it hyalinizes collagen fibers so fine detail of the skin can be difficult to interpret. Use of Bouin's solution requires washing in tap water and transfer to ethanol. Failure to do so results in major artifacts often making the tissue unusable. These and other fixatives are discussed in Chapter 30 on Necropsy Methods which includes formulations for their preparation.

Since skin, and especially hair follicles, vary dramatically by location, several locations should be sampled in order to evaluate potential changes. Collection of tissue consistently throughout a study will make specimens comparable. Dorsal skin can be collected over the thorax making sure to label cranial and caudal orientations so the tissue will be trimmed correctly. Ventral skin covering the thorax is also taken. Both dorsal and ventral skin are very similar histologically so they can be placed in separate cassettes when trimmed together with other skin that has distinct histologic features. Vibrissae on the head are collected by removing all the skin on the head as a complete unit. Vibrissae on the muzzle are trimmed as one piece. Eyelids are sectioned from this piece of skin as well to include upper and lower lids. Ear and tail are removed from the body and fixed by immersion. Tail skin can be removed from the bone and muscle or collected together. If the latter is done the bone must be decalcified. Footpads are also collected. Details are provided in Chapter 30 on Necropsy Methods and elsewhere (Sundberg *et al.*, 1998; Relyea *et al.*, 2000a). Nails are collected attached to the feet and digits. Distal limbs can be disarticulated and fixed *in toto*. If the paw is to be examined, it can be fixed under weight to lay it flat then sectioned horizontally to include all the joints after decalcification. Sagittal sections are the most useful. Digits are processed *in toto* and serially sectioned lengthwise after decalcification.

Scanning electron microscopy of the skin and hair fibers

Scanning electron microscopy provides a detailed three dimensional view of structures at various magnifications. X-ray microanalysis can determine the relative elemental content of a specimen which may be useful for evaluation of some mutant mice. Hairs are made up of what used to be called the high sulfur keratins or hard keratins, now called the hair keratins. Changes in sulfur levels can be detected that suggest abnormalities in these hairs. Such is the case with the ichthyosis mutant mouse that has a form of trichothiodystrophy, low sulfur levels in the hair fibers (Figure 12.6) (Itin *et al.*, 1990). Toxic agents, especially heavy metals, can also be identified using this method (Chatt *et al.*, 1990; Takeuchi *et al.*, 1990; Bache *et al.*, 1991).

Whole mounts of skin or nails can be easily made by removing tissues at the time of necropsy, spreading soft tissues out on a firm nylon membrane to fix them flat, and placing them in buffered glutaraldehyde using standard methods. Electron microscopists will critically point dry the specimen, coat it with gold, and then examine it with the investigator (Bechtold, 2000).

Hairs can be examined in whole mounts or manually removed and examined individually. Adult mouse hair follicles are in telogen for prolonged periods so the hairs can be easily removed manually from lightly anesthetized animals without causing pain but more importantly, damage to fibers is rare since they come out easily. Hairs are placed in a dry vial and processed routinely. Shipments we receive from collaborating laboratories for evaluation are routinely disinfected on the outside surface and are then filled with 70% ethanol and stored for a week or longer before processing due to the common infestation of mites in many research colonies. This approach kills the mites thus avoiding introduction into your colonies.

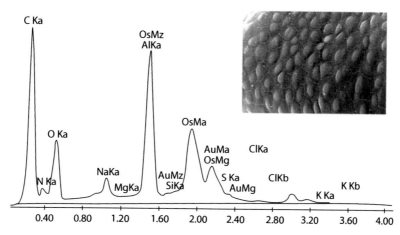

Figure 12.6 X-ray microanalysis of skin of the tip of a digit from a mouse embryo. The surface of the skin (keratinocytes) is evident. High element peaks represent the specimen preparation materials. Sulfur levels (S) can be quantitated.

Transmission electron microscopy of the skin and hair fibers

Transmission electron microscopy can provide a great deal of information but is technically difficult and labor intensive. Tissue is removed during necropsy but should be finely minced into 1 mm³ pieces since glutaraldehyde fixatives do not penetrate tissues deeply. Cacodylate or phosphate buffered glutaraldehyde are commonly used but others are available and described elsewhere. Tissues should be stored refrigerated and embedded soon after collection to minimize artifacts (Bechtold, 2000).

Other methods

Many different methods have been developed to evaluate skin. We have tested a thermal imaging device that measures infrared radiated from mice under general anesthesia (Thermogenic Imaging, Billerica, MA). It appeared to be a useful device for determining response to treatment for mutant mice with thick, scaly, neovascularized skin or those with various forms of alopecia (Figure 12.7). Longitudinal studies revealed that thermal changes over time reflected the hair cycle in both mutant and control mice since the hypodermal fat layer, and therefore the insulation value of the skin, varied dramatically through the hair cycle.

Transepidermal water loss is an important measurement in mice with abnormalities in the cutaneous water barrier. Mice are first sedated with 100 mg/kg ketamine HCl plus 0.5 mg/kg xylazine intraperitoneally. Dorsal hair is removed with electric clippers and then depilated for 5 min with a chemical agent such as Neet

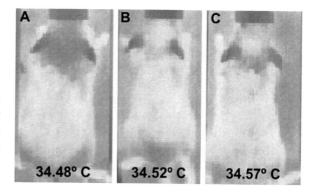

Figure 12.7 Thermal images of a normal C3H/HeJ mouse (A), one with focal alopecia areata (B), and one with diffuse alopecia areata (C). There is a quantifiable increase in heat loss associated with increased hair loss.

(Reckitt and Coleman, Wayne, NJ). Transepidermal water loss is measured 24 h later by placing a Servo Med Evaporimeter EPI probe (Servomed AB, Stockholm, Sweden) on the bald area (Serup and Jemec, 1995; Sundberg *et al.*, 2000).

Surface lipids can be collected by dipping euthanized mice into 40 ml of acetone 10 times and drying the acetone under argon gas. The residue is dissolved in toluene and plated in separate lanes on silica gel G chromatographic plates (Merck, Rahway, NJ). The plates are developed to 19 cm in hexane–ether–acetic acid (80 : 20 : 1). Following drying of the plate it is sprayed with 50% sulfuric acid (Downing and Stranieri, 1980; Sundberg *et al.*, 2000).

Kinetic studies can be easily done if considered at the time of necropsy. Mice can be injected with bromodeoxyuridine (50 μg/gm body weight) 1 h before necropsy (Smith *et al.*, 2000). A consistent time interval between injection and necropsy is critical since it will determine the rates at which this compound is incorporated into DNA currently being synthesized. Unstained sections are processed routinely for immunohistochemistry (Relyea *et al.*, 2000b) and an anti-bromodeoxyuridine antibody used. Positive cells in 'S' or the DNA synthesis phase of the cell cycle will have nuclei that are brown or red depending upon the chromogen and enzyme system used (Relyea *et al.*, 2000b; Smith *et al.*, 2000). An alternative is to use tritiated thymidine. This radionuclide requires special safety precautions, takes 3–6 weeks for development, and can be difficult to interpret so it is less commonly used today. Interpretation is complicated and can depend upon what types of proliferation rates are needed for evaluation of a particular mutant. Standard approaches are described for interfollicular skin such as counting the number of positive nuclei per 1000 basal cell nuclei or per linear millimeter of skin, if it lays flat (Leblond *et al.*, 1964; Skerrow and Skerrow, 1985; Kwochka and Rademakers, 1989; Kwochka, 1990; Sundberg *et al.*, 1994a). Some mutant mice have marked proliferation of the infundibulum which requires modifications and special adaptation of counting criteria (Sundberg *et al.*, 1997a).

Gene arrays are a relatively new technology that is becoming more widely accessible to investigators. A variety of methods are now available. The critical starting material is high quality RNA. What tissue to select and how to prepare it are developing and controversial topics as the technology evolves. We have used the entire skin of mice that develop a generalized cutaneous phenotype. The advantage is that an adequate volume can be obtained to provide enough RNA for many experiments. The disadvantage is that hair follicles in various stages are obtained, anatomically discreet areas are mixed and not all areas are affected. Assuming similar anatomical effects are found in age and gender matched controls, the differences in gene expression profiles should represent those related to the disease under investigation. More specific sites can be chosen as the disease is better understood (Carroll *et al.*, 2002). The main advantage of the gene arrays is that complex pathways that can take a great deal of time to analyze using traditional methods can be screened with a small group of animals in a matter of days to generate the data but weeks to months to analyze it. For example, the mouse flaky skin (*fsn*) mutation develops a psoriasiform dermatitis (Sundberg *et al.*, 1994a; Sundberg *et al.*, 1997b; Pelsue *et al.*, 1998). One run with an Affymetrix Gene Chip® determined that this mutant mouse had primarily a Th2 immune response in contrast to the Th1 response found in human psoriasis patients (Figure 12.8). This explained variations in

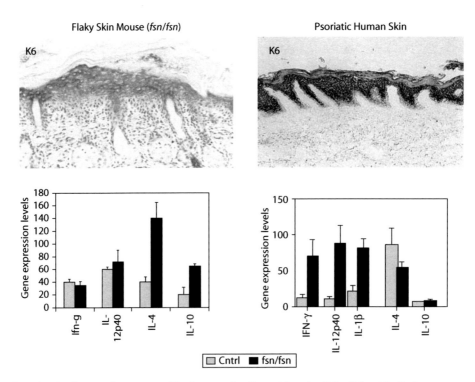

Figure 12.8 Keratin 6 is abnormally expressed in the suprabasilar epidermis of the flaky skin (*fsn*) mutant mouse in a pattern similar to that of a human psoriasis biopsy. Gene expression profiles from these tissues revealed a different immunological mechanism underlying clinically similar phenotypes (black bars represent *fsn/fsn* mice or psoriatic human skin, gray bars represent +/+ normal littermate mice or normal human skin RNA).

therapeutic responses between the species (Carroll and Sundberg, unpublished data).

Tissue arrays are the next technology with direct application to many research projects. Tissue arrays are build on traditional histology methods whereby paraffin blocks are systematically punched at prescribed sizes and the cores placed into predrilled holes in a new paraffin block (Moch *et al.*, 2001). Large numbers of tissues from many different organs or different case materials of similar lesions from the same organs can be used. This is providing a tool to specifically evaluate the cells producing proteins from the up and down regulated transcripts detected using gene arrays. Currently custom or predesigned arrays are commercially available. As the technology becomes more accepted, many institutions will probably be adopting it.

Development of the normal skin and adnexa

Each hair follicle type starts to develop at different time points during embryogenesis. Therefore, it is not surprising to find clusters of hair follicles in a section at different stages of development. The large vibrissae develop earliest and are nearly fully developed by birth. In spite of this, all hair follicles develop in a similar anatomic fashion. This developmental scheme is detailed both historically and anatomically elsewhere (Paus *et al.*, 1999) and serves as a guide for the summary below.

The sequential stages of hair follicle development begin with the pregerm stage which is hard to recognize histologically but can be defined with various immunohistochemical markers. It consists of a sharply demarcated plaque of basal and suprabasal epidermal keratinocytes. In stage 1, the pregerm develops into an histologically evident epidermal thickening where the keratinocytes display a vertically polarized orientation compared with the more cuboidal appearance of adjacent basal cells. Concurrently, dermal fibroblasts increase in number immediately below this structure forming what will become the dermal (follicular) papilla. Stages 2–4 produce a column of epidermal keratinocytes that develop a cap, invagination of the dermal papilla, and formation of the basic hair follicle

structure. The root sheaths begin to form and differentiate. Stage 5 is the bulbus peg stage with elongation of the inner root sheath and development of the bulge and first **sebocytes**. Melanin begins to form at this stage in pigmented mice. At stage 6 the follicle begins to extend below the level of the dermis into the hypodermal fat layer. The hair canal can now be identified. In stage 7 the tip of the hair fiber leaves the inner root sheath and enters the hair canal at the level of the infundibulum of the forming sebaceous gland. Stage 8 is the maximum length of the hair follicle where it extends down to the panniculus carnosus muscle and the hair fiber emerges through the epidermis. This process begins *in utero* and is completed for all follicle types by 5 days post partum when the hair is evident on the skin of most strains of normal mice.

The epidermis develops from a single layer into a multilayered structure. In newborn mice it is thick at all anatomic sites and keratinocytes follow a classical differentiation scheme for stratified squamous epithelium (Figure 12.9). Cuboidal basal cells (keratins 5 and 14 positive) are located on the basal lamina (HogenEsch *et al.*, 1999). Above this layer the cells differentiate into the statum spinosum or prickle-cell layer. Here the cells begin to elongate along the axis of the skin and have prominent intercellular bridges (desmosomes) that are evident under high magnification. These spine-like structures are due to artifactual shrinkage of the tissues during preparation. This layer can be identified by the presence of keratins 1 and 10. The next layer, the stratum granulosum, has cells that are flattened along the axis of the skin and contain prominent basophilic granules (keratohyalin granules). Two types of granules are present in the mouse, profilaggrin (P) and loricrin (L) granules. The larger profilaggrin granules are the blue structures seen by light microscopy (Presland *et al.*, 2000). The most superficial layer, the stratum corneum, is brightly eosinophilic and consists of compacted, flattened keratinocytes. This is the critical portion of the skin that provides a strong aqueous barrier due to the presence of lamellar bodies, small lipid based structures only detectable by special staining and transmission electron microscopy (Elias, 1988).

The epidermis of a newborn mouse is thick. As the mouse ages, within two weeks, the truncal epidermis thins to only about two cell layers with the stratum granulosum and corneum becoming very thin and often hard to visualize by light microscopy. Other anatomic sites do not change. The tail skin remains thick throughout the mouse's life. The muzzle skin is thinner than at birth but thicker than truncal skin. Foot pads remain thick.

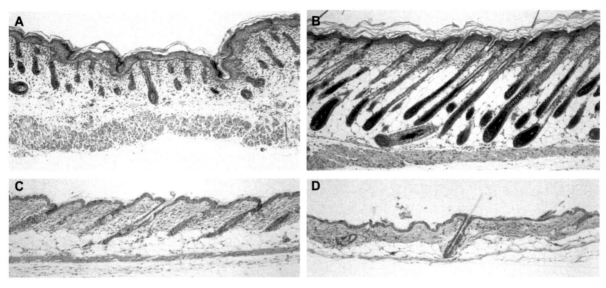

Figure 12.9 Hair cycle of the mouse. Newborn mouse skin (A) has a thick epidermis with incomplete development of hair follicles. By 1 week of age (B) hair follicles are fully developed in anagen and producing hair fibers that emerged at 5 days of age. At 14 days of age (C) the follicles enter catagen and begin to regress undergoing apoptosis. With 3–5 days the follicles are in the resting (telogen) phase (D). Note that the epidermis thins and remains thin for life under normal circumstances by 2 weeks of age (C).

Normal anatomy of the skin and the hair cycle

Histology of the normal skin

This anatomy of the skin and hair follicles are illustrated in Figures 12.4 and 12.9. The top layer of epithelial cells is called the epidermis. This layer differentiates from the cuboidal basal cells in the stratum basale into the polygonal cells of the stratum spinosum, then more flattened cells with fine blue granules in the stratum granulosum, and ultimately into the flat cells that lack a nucleus and become very eosinophilic at the surface in the stratum corneum. The outermost layer of cells separating from the surface are sometimes called the stratum dysjunctum. The hair follicle is a very complicated structure that invaginates into the dermis and hypodermal fat undergoing major changes on a regular basis with the hair cycle (see below). A large gland protrudes from its side that consists of swollen pale cells with fine uniform vacuoles. These are sebaceous glands that produce oils to coat the surface of the skin and hair fiber. The oils can be visualized in frozen sections stained with oil red O, sudan black, or other histologic means to follow how the lipids spread out over the

surface of the skin in normal compared with mutant mice (Sundberg *et al.*, 1997a). The dermis consists of dense irregular collagenous connective tissue, elastic connective tissue, blood vessels, nerves, smooth muscle (arrector pili muscles that lift hair follicles and fibers; Figure 12.10), and a variety of individual cell types including fibroblasts, mast cells, and small numbers of cells from the immune system. One important feature of skin that is characteristically found in many rodents, especially laboratory mice, is that apocrine sweat glands are *not* present. Modified apocrine glands, mammary glands are abundant because of the large litter size most mice have (Sundberg *et al.*, 1996). The mouse does not normally have **rete ridges** where the lower aspect of the epidermis forms ridges of cells that extend into the dermis. The dermis between such ridges is commonly called the dermal papillae, a term also used by hair biologists for the specialized fibroblasts that populate the base of an anagen hair follicle called the bulb. Because of this, the fibroblasts within the bulb are also called the follicular papilla. Rete ridge-like structures become prominent when mouse skin heals following ulceration. These changes resemble those found in neoplasms of the epidermis such that the changes are referred to as pseudoepitheliomatous or pseudocarcinomatous hyperplasia. Below the dermis is a layer of fat, the hypodermal fat layer. The thickness of this fat layer changes with the hair cycle being thickest during anagen when follicles need a great deal of energy to produce a hair

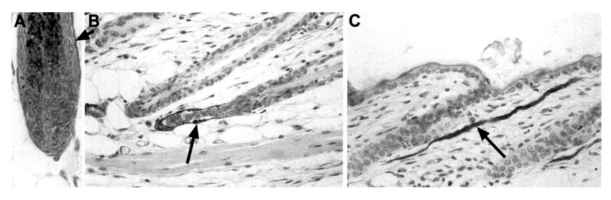

Figure 12.10 Smooth muscle actin expression (gray, arrows) is located around the outer root sheath of anagen and catagen follicles (A, B). It is also present in the arrector pili muscles (C).

fiber. The panniculus muscle separates the hypodermal fat layer from the adventitia, loose collagenous connective tissue that attaches the skin to the underlying musculature and fat. Mammary glands are found in the fat below this skeletal muscle layer.

Hair cycle in the mouse

The skin undergoes major changes during the first 2 weeks of life (Figure 12.9). Hair follicles continue to develop and enter late stage anagen 5 days postpartum when hair fibers emerge through the epidermis. The truncal epidermis is relatively thick at birth and thins to normal by 2–3 weeks of age. Hair follicles produce fibers until around 14 days of age over the thorax at which time the lower portion undergoes apoptosis. The dermal papilla is retracted by actin filaments (Figure 12.10) that reside below the isthmus in the resting or telogen stage until the hair cycle is reinitiated. This usually lasts about 3 days in young mice. The follicle develops into a new anagen stage follicle pushing the old follicle laterally. The new fiber emerges adjacent to the old one. At some point the old fiber is lost in what is now called the exogen stage (Milner *et al.*, 2002). The general features of the different stages of the hair cycle are illustrated in Figure 12.9. Development of the hair follicle embryologically and progression through the hair cycle regularly throughout life have been dissected anatomically and with molecular and immunological markers to differentiate numerous stages within each major portion of the hair cycle that have been detailed and reviewed elsewhere (Paus *et al.*, 1999; Muller-Rover *et al.*, 2001; Millar, 2002). What is commonly called the second hair cycle, or the first real hair cycle after embryogenesis, has a short anagen stage and prolonged telogen stage. The hair cycles in a cranial to caudal pattern that can be easily visualized in pigmented mice. Unlike humans, pigment in the mouse skin is limited to the bulb of anagen follicles and hair fibers. Interfollicular epidermis rarely contains pigment and when it does, it is usually only in mutant mice (Sundberg *et al.*, 1994b). If the mice are shaved, irregular pigmented areas will be seen. These are areas containing anagen follicles. If mice are followed daily, these pigmented patches will migrate caudally. This feature is dramatic in mutant mice such as hairless. These mice have normal hair 5 days after birth but not subsequent hair cycles. Beginning at 2 weeks of age their hair is shed from head to tail (Sundberg, 1994a). Other hair follicle types have different hair cycles. This is why hairless mice appear to retain vibrissae while being completely bald. The length of the hair cycle determines the length of the hair fiber. This is consistent with why hairless mice have long, persistent vibrissae while short pelage hairs are lost. This feature was demonstrated with angora (*Fgf5go*) mutant mice that have a three day prolongation of their truncal hair cycle and as a result often have long, shaggy hair compared with normal littermates (Sundberg *et al.*, 1997c).

Numerous genes regulate development and cycling of the hair follicles (Hardy, 1992; Paus, 1996, 1998; Millar, 1997; Chuong and Noveen, 1999; Stenn and Paus, 2001; Millar, 2002). Classic work done half a century ago detailed changes in the skin and hair follicles as they cycle, not just the changes in the follicles but also changes in sebaceous gland size and shape as well as the thickness of the hypodermal fat layer (Chase *et al.*, 1951, 1953; Chase, 1954, 1955; Chase and Eaton, 1959; Straile, 1960, 1965, 1969; Straile *et al.*, 1961). Furthermore, hormones cause changes as well (Deplewski and Rosenfield, 2000). These are important to understand when comparing differences between wild type, normal mice, and mutant mice. Not only should the mice be age and gender matched in such studies but it is critical to match the stage of the hair cycle as well.

Nails

The mouse has nails or claws on each digit, just like most other mammals including humans. The name claw suggests these structures are different than human nails which may be why little attention has been paid to them. In fact, anatomically they are very similar at the gross and histologic levels. Human nails are dorsoventrally flattened to form a plate while rodent nails are laterally flattened. These changes, not restricted to humans since similar refinements are found in many nonhuman primates, are associated with the function of the nails in primates as a refined tool associated with manual dexterity rather than as a weapon or digging tool. Sagital sections illustrate that mice have a nail matrix, nail plate, nail bed, hyponychium, and other structures (Figure 12.11) identical to but smaller than the human nail. To veterinarians this is not at all surprising since all mammals have nails or claws that are variations on this general theme. Nails can be extremely difficult to prepare and interpret histologically. However, changes in mutant mice can be dramatic when these structures are magnified with a dissection microscope or by scanning electron microscopy (Sundberg and King, 2000).

Other specialized glands

Mammary glands are specialized forms of apocrine sweat glands with a complex developmental and lactation cycle that will not be discussed here. Other glands found at specific anatomic sites are modified sebaceous glands, a type of holocrine gland. These include preputial and clitoral glands around the genitals, meibomian glands in the eyelid, and cerruminous glands within the outer ear. All are large glands with a structure similar to that found in the sebaceous glands associated with hair follicles. The major difference is that each has a stratified squamous epithelial lined duct that empties directly onto the structure where it is located (Sundberg and King, 2000). Hair follicles are specialized and have sebaceous glands associated with them that vary in size. The most notable are the perianal hairs which have large sebaceous glands. Salivary, lacrimal, and harderian glands are very different and are described in chapters dealing with the organs they are associated with.

Skin and adnexal mutant phenotypes

It is beyond the scope of this chapter to describe or even list all mutant mice with skin and/or hair/nail phenotypes. As a general starting point we have grouped phenotypes into 10 classes (Table 12.2). Detailed lists, descriptions, references, and illustrations are published elsewhere (Sundberg, 1994a; Sundberg and King, 1996a, b, 2000; Nakamura et al., 2002a, b; Randall et al., 2003).

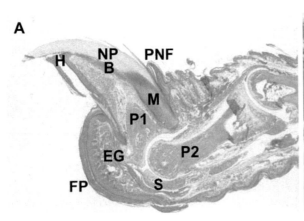
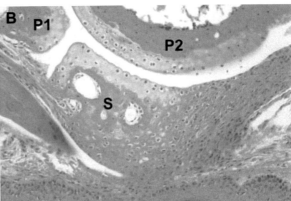

Figure 12.11 Normal sagital section of the nail from an adult mouse (A). Hyponechium (H), nail plate (NP), nail bed (B), proximal nail fold (PNF), matrix (M), phalanx 1 (P1), phalanx 2 (P2), sesamoid bone (S), eccrine gland (EG), foot pad (FP). High magnification of a sesamoid bone under P2.

TABLE 12.2: General categories of mutant mouse cutaneous phenotypes

Hair and skin color (pigmentation)

Eccrine gland defects

Sebaceous gland defects

Primary scarring disorders

Hair follicle cycling disorders

Structural defects of hair fibers

Hair texture abnormalities

Missing hair fiber and follicle types

Noninflammatory (ichthyosiform and keratodermas) skin diseases

Inflammatory (psoriasiform and proliferative) skin diseases

Papillomatous skin diseases

Cutaneous carcinogenesis

Bullous and acantholytic skin diseases

Structural and growth defects of the nails

References

Bache, C.A., Lisk, D.J., Scarlett, J.M. and Carbone, L.G. (1991). *J. Toxicol. Environ. Health*, **34**, 423–431.

Baxter, R.M. and Brissette, J.L. (2002). *J. Invest. Dermatol.* **118**, 303–309.

Bechtold, L.S. (2000). In *Systematic Characterization of Mouse Mutations* (eds J.P. Sundberg and D. Boggess), pp. 121–129. CRC Press, Boca Raton FL.

Carroll, J., McElwee, K.J., King, L.E.J., Byrne, M.C. and Sundberg, J.P. (2002). *J. Invest. Dermatol.* **119**, 392–402.

Chase, H.B. (1954). *Physiol. Rev.* **34**, 113–126.

Chase, H.B. (1955). *J. Soc. Cosmetic Chem.* **6**, 9–14.

Chase, H.B. and Eaton, G.J. (1959). *Ann. NY Acad. Sci.* **83**, 365–368.

Chase, H.B., Montagna, W. and Malone, J.D. (1953). *Anat. Rec.* **116**, 75–82.

Chase, H.B., Rauch, H. and Smith, V.W. (1951). *Physiol. Zool.* **24**, 1–10.

Chatt, A., Holzbecher, J. and Katz, S.A. (1990). *Biol. Trace Elem. Res.* **26–27**, 513–519.

Chuong, C.-M. and Noveen, A. (1999). *J. Invest. Dermatol. Symp. Proc.* **4**, 307–311.

Chuong, C.M., Nickoloff, B.J., Elias, P.M., Goldsmith, L.A., Macher, E., Maderson, P.A., Sundberg, J.P., Tagami, H., Plonka, P.M., Thestrup-Pederson, K., Bernard, B.A., Schroder, J.M., Dotto, P., Chang, C.M., Williams, M.L., Feingold, K.R., King, L.E. Kligman, A.M., Rees, J.L. and Christophers, E. (2002). *Exp. Dermatol.* **11**, 159–187.

Deplewski, D. and Rosenfield, R.L. (2000). *Endocrine Rev.* **21**, 363–392.

Downing, D.T. and Stranieri, A.M. (1980). *J. Chromatogr.* **192**, 208–211.

Elias, P.M. (1988). *Drug Develop. Res.* **13**, 97–105.

Gaskoin, J.S. (1856). *Proc. Zoo. Soc. London* **24**, 38–40.

Goldsmith, L.A. (1990). *Arch. Dermatol.*, **126**, 301–302.

Hardy, M.H. (1992). *Trends Genet.* **8**, 55–61.

HogenEsch, H., Boggess, D. and Sundberg, J.P. (1999). *Pathobiology* **67**, 45–50.

Itin, P.H., Sundberg, J.P., Dunstan, R.W. and Pittelkow, M.R. (1990). *J. Invest. Dermatol.* **94**, 537.

Kwochka, K.W. (1990). *Am. J. Vet. Res.* **51**, 1570–1573.

Kwochka, K.W. and Rademakers, A.M. (1989). *Am. J. Vet. Res.* **50**, 587–591.

Leblond, C.P., Greulich, R.C. and Pereira, J.P.M. (1964). In *Advances in Biology of the Skin* (eds W. Montagna and R.E. Billingham), pp. 39–67. Pergamon Press, New York.

Mecklenburg, L., Nakamura, M., Sundberg, J.P. and Paus, R. (2001). *Exp. Mol. Pathol.* **71**, 171–178.

Millar, S. (1997). In *Cytoskeletal-membrane Interactions and Signal Transduction* (eds P. Cowin and M.W. Klymkowsky), pp. 87–102. Landes Bioscience, Austin, TX.

Millar, S.E. (2002). *J. Invest. Dermatol.* **118**, 216–225.

Milner, Y., Sudnik, J., Filippi, M., Kizoulis, M., Kashgarian, M. and Stenn, K. (2002). *J. Invest. Dermatol.* **119**, 639–644.

Moch, H., Kononen, J., Kallioniemi, O.-P. and Sauter, G. (2001). *Adv. Anat. Pathol.* **8**, 14–20.

Muller-Rover, S., Handjiski, B., vanderVeen, C., Eichmuller, S., Foitzik, K., McKay, I.A., Stenn, K.S. and Paus, R. (2001). *J. Invest. Dermatol.* **117**, 3–15.

Nakamura, M., Sundberg, J.P. and Paus, R. (2002a). *Exp. Dermatol.* **10**, 369–390.

Nakamura, M., Tobin, D.J., Richards-Smith, B., Sundberg, J.P. and Paus, R. (2002b). *J. Dermatol. Sci.* **28**, 1–33.

Paus, R. (1996). *Curr. Opin. Dermatol.* **3**, 248–258.

Paus, R. (1998). *J. Dermatol.* **25**, 793–802.

Paus, R., Muller-Rover, S., vanderVeen, C., Maurer, M., Eichmiller, S., Ling, G., Hofmann, U., Foitzik, K., Mecklenburg, L. and Handjiski, B. (1999). *J. Invest. Dermatol.* **113**, 523–532.

Pelsue, S.C., Schweitzer, P.A., Schweitzer, I.B., Christianson, S.W., Gott, B., Sundberg, J.P., Beamer, W.G. and Shultz, L.D. (1998). *Eur. J. Immunol.* **28**, 1379–1388.

Presland, R.B., Boggess, D., Lewis, S.P., Hull, C., Fleckman, P. and Sundberg, J.P. (2000). *J. Invest. Dermatol.*, **115**, 1072–1081.

Randall, V.A., Sundberg, J.P. and Philpott, M.P. (2003) *J. Invest. Dermatol. Sym. Proc.* **8**, 39–45.

Relyea, M.J., Miller, J., Boggess, D. and Sundberg, J.P. (2000a). In *Systematic approach to evaluation of mouse mutations* (eds J.P. Sundberg and D. Boggess), pp. 57–90. CRC Press, Boca Raton, FL.

Relyea, M.J., Sundberg, J.P. and Ward, J.M. (2000b). In *Systematic Approach to Evaluation of Mouse Mutations* (eds J.P. Sundberg and D. Boggess), pp. 131–144. CRC Press, Boca Raton, FL.

Serup, J. and Jemec, G.B.E. (1995). *Handbook on Non-invasive Methods and the Skin*, CRC Press, Boca Raton, FL.

Skerrow, D. and Skerrow, C.J. (eds) (1985). *Methods in Skin Research*. John Wiley & Sons Ltd, Chinchester.

Smith, R.S., Martin, G. and Boggess, D. (2000). In *Systematic Approach to Evaluation of Mouse Mutations* (eds J.P. Sundberg and D. Boggess), pp. 111–119. CRC Press, Boca Raton, FL.

Sprecher, E., Becker, Y., Kraal, G., Hall, E. and Shultz, L.D. (1990). *Arch. Dermatol.*, **282**, 188–193.

Stenn, K.S. and Paus, R. (2001). *Physiol. Rev.*, **81**, 449–494.

Straile, W.E. (1960). *Am. J. Anat.*, **106**, 133–148.

Straile, W.E. (1965). In *Biology of Skin and Hair Growth* (eds A.G. Lyne and B.F. Short), pp. 35–37. Angus & Robertson, Sydney.

Straile, W.E. (1969). *Adv. Biol. Skin*, **9**, 369–390.

Straile, W.E., Chase, H.B. and Arsenault, C. (1961). *J. Exp. Zool.* **148**, 205–216.

Sundberg, J., X, M. and Boggess, D. (1998). In *Cutaneous Appendages* (ed. M. Chuong), pp. 421–435. Molecular Biology Intelligence Unit I, Landes Company, Austin, TX.

Sundberg, J.P. (1994a). In *Handbook of Mouse Mutations with Skin and Hair Abnormalities: Animal Models and Biomedical Tools* (ed. J.P. Sundberg), pp. 291–312. CRC Press, Boca Raton.

Sundberg, J.P. (1994b). *Handbook of Mouse Mutations with Skin and Hair Abnormalities. Animal Models and Biomedical Tools.* CRC Press, Inc, Boca Raton, FL.

Sundberg, J.P. and Hogan, M.E. (1994). In *Handbook of Mouse Mutations with Skin and Hair Abnormalities: Animal Models and Biomedical Tools* (ed. J.P. Sundberg), CRC Press, Boca Raton, FL.

Sundberg, J.P. and King, L.E. (1996a). *J. Invest. Dermatol.* **106**, 368–379.

Sundberg, J.P. and King, L.E.J. (1996b). *Dermatol. Clin.* **14**, 619–632.

Sundberg, J.P. and King, L.E. (2000). In *Pathology of Genetically Engineered Mice* (eds J.M. Ward, J.F. Mahler, R.R. Maronpot and J.P. Sundberg), pp. 181–213. Iowa State University Press, Ames.

Sundberg, J.P. and Shultz, L.D. (1994). In *Handbook of Mouse Mutations with Skin and Hair Abnormalities* (ed. J.P. Sundberg), pp. 423–429. CRC Press, Boca Raton, FL.

Sundberg, J.P., Dunstan, R.W., Roop, D.R. and Beamer, W.G. (1994a). *J. Invest. Dermatol.* **102**, 781–788.

Sundberg, J.P., Dunstan, R.W., Roop, D.R. and Beamer, W.G. (1994b). *J. Invest. Dermatol.* **102**, 781–788.

Sundberg, J.P., Orlow, S.J., Sweet, H.O. and Beamer, W.G. (1994c). In *Handbook of Mouse Mutations with Skin and Hair Abnormalities: Animal Models and Biomedical Tools* (ed. J.P. Sundberg), pp. 159–164. CRC Press, Boca Raton, FL.

Sundberg, J.P., HogenEsch, H. and King, L.E. (1995). *Mouse Models for Scaly Skin Diseases.* CRC Press, Inc, Boca Raton.

Sundberg, J.P., Hogan, M.E. and King, L.E. (1996). In *Pathobiology of the Aging Mouse*, Vol. 2 (eds U. Mohr, D.L. Dungworth, C.C. Capen, W.W. Carlton, J.P. Sundberg, and J.M. Ward), pp. 303–323. ILSI Press, Washington, DC.

Sundberg, J.P., Boggess, D., Hogan, M.E., Sundberg, B.A., Rourk, M.H., Harris, B., Johnson, K., Dunstan, R.W. and Davisson, M.T. (1997a). *Am. J. Pathol.*, **151**, 293–310.

Sundberg, J.P., France, M., Boggess, D., Sundberg, B.A., Jenson, A.B., Beamer, W.G. and Shultz, L.D. (1997b). *Pathobiology* **65**, 271–286.

Sundberg, J.P., Rourk, M., Boggess, D., Hogan, M.E., Sundberg, B.A. and Bertolino, A. (1997c). *Vet. Pathol.* **34**, 171–179.

Sundberg, J.P., Boggess, D., Sundberg, B.A., Eilersten, K., Parimoo, S., Filippi, M. and Stenn, K. (2000). *Am. J. Pathol.* **156**, 2067–2075.

Takeuchi, T., Nakano, Y., Aoki, A., Ohmori, S. and Kasuya, M. (1990). *Biol. Trace Elem. Res.* **26–27**, 263–268.

CHAPTER 13

The Cardiovascular System

Lloyd H Michael, George E Taffet,
Nikolaos G Frangogiannis, Mark L Entman
and Craig J Hartley
Baylor College of Medicine, Houston, Texas, USA

Introduction and background

During the past decade, tremendous strides have been made in assessing the murine cardiovascular system. This has occurred because of the logarithmic increase in opportunities to use transgenic mice and embryonic stem cell and homologous recombination approaches to target gene expression and to ablate/overexpress specific gene products. These gene mutations and targeted genes produce animals that have altered cardiovascular phenotypes. Hence, the need to accurately assess cardiovascular function is vitally important. This chapter considers the known normal state of the cardiovascular system of mice as well as the changes measured when there is heart pathophysiology such as in myocardial hypertrophy and myocardial ischemia and reperfusion. No attempt will be made to describe the altered hemodynamic states of the hundreds of relevant transgenic/knockout mice used to study the heart and cardiovascular system; rather, select examples will be cited for emphasis. The surgical technique to induce these altered states: restriction of aortic blood flow and occlusion of a coronary artery, respectively, will be outlined. A variety of methods currently used to measure the mouse cardiovascular anatomy and function will be summarized. A discussion will also focus on the age-related differences in response to surgically induced pathophysiological states in the mouse.

Compared with the dog, the mouse has an approximate 1000 fold body weight difference as seen in a 0.025 kg mouse and a 25 kg dog. This 1000 fold difference extends to heart weight between the dog and the mouse and a 100-fold difference in area, and a 10-fold difference in linear dimensions. This is illustrated in Figure 13.1, a photograph of a dog, a rat, and a mouse heart. In order to measure cardiovascular function in such a small system, many development projects have been undertaken to miniaturize sensors and transducers needed to measure these functions. Besides the size, mice have a very fast resting heart rate of 500–600 bpm, requiring instrumentation with higher signal fidelity and better temporal and spatial resolution.

The Laboratory Mouse
Copyright 2004 Elsevier
ISBN 0-1233-6425-6

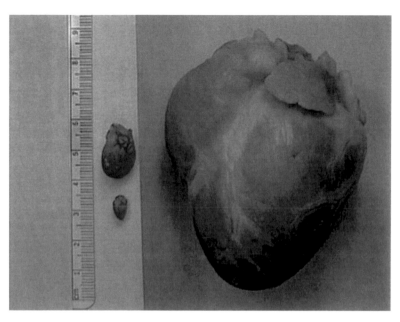

Figure 13.1 Dog, Rat, Mouse heart. Comparison of heart sizes from a 25,000 g dog, a 250 g rat, and a 25 g mouse.

Anatomical considerations

The most complete description of the anatomy of the cardiovascular system of the mouse may be found in an article by Cook (1965), detailing all aspects of the system anatomy. Gross anatomy of the venous and arterial supply to all regions of the body including limbs is diagrammatically illustrated. A brief description of the heart and major blood vessels is also given by Green *et al.* (1966). The major branches of the mouse aorta are similar to that seen in humans. The right innominate artery leaves the aortic arch and divides into the right common carotid artery and the right subclavian artery. Approximately 1 mm distal on the aortic arch is the left common carotid artery followed laterally adjacent by the left subclavian artery. This anatomical arrangement is different to that of an animal species such as the dog where the right innominate artery leads to both right and left common carotid arteries. The striking differences in the mouse heart and vessels appear to be in the arrangement of the venous system of the heart. Cardiac veins are the most prominent structures on the epicardial surface of the left ventricle, far exceeding the visibility of the coronary arteries. Small cardiac veins abut at right angles to the largest coronary vein, the left cardiac vein, which proceeds from the ventral surface of the left ventricle and the apex of the heart toward the dorsum of the heart to drain into the left anterior vena cava at its

junction with the right anterior and posterior vena cava connection with the right atrium. This vessel distribution appears similar to that seen in rat heart by Halpern (1953). In addition, Halpern describes two major veins which drain the conal region of the right ventricle and the ventrocephalic region of the left ventricle. These were called extra coronary cardiac veins because they originated on the heart and terminated in vessels not otherwise associated with the coronary circulation, i.e. in this case, the anterior vena cava. In summary, one is struck by the great abundance of veins as well as size of veins on the surface of the heart. At first glance, the venous architecture may be confused with the underlying and more subtle coronary arterial system which is embedded in the musculature of the heart. Observation of the coronary artery system of the mouse heart is much more difficult and requires a source of intense light and magnification to visualize these deeper and more hidden vessels. The coronary artery anatomy of the mouse is shown in Figure 13.2, which illustrates that the coronary **ostia** originate within 2 mm of the aortic valve. The right coronary artery usually divides into two major branches, one supplying the right ventricle and the second the septal region. The left coronary artery divides generally into a major septal branch and a left anterior descending coronary artery (LAD) supplying the free wall of the left ventricle, part of the septum, and the apical region of the left ventricle. The pattern of coronary artery distribution is shown in Figure 13.3 and schematically illustrates the various patterns as seen in six separate

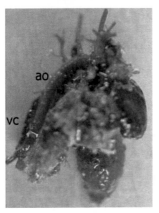

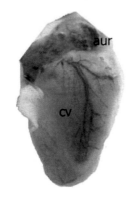

Figure 13.2 Top Panel: Photograph of latex cast of mouse arterial system including coronary arteries. ca = carotid artery; lad = left anterior descending coronary artery; ao = aortic arch; rca = right coronary artery. (Reprinted with permission from *Am. J. Physiol. Heart Circ. Physiol.* **269**, H2147–H2154). Middle Panel: Latex cast of major arteries and veins in and around heart and lungs; ao = aortic arch; vc = posterior vena cava. Bottom Panel: Latex cast of major cardiac veins (cv) on the lateral wall of left ventricle; aur = auricle . **(See also Colour Plate 6.)**

mice. In all of these animals, the left circumflex coronary artery, which is a major branch of the left coronary artery in other animal species and humans, is not clearly a major vessel in these mice, appearing rudimentary. Therefore, the variability in epicardial coronary architecture is a very important consideration even in the same genetic stock.

Histopathological characteristics of the normal mouse heart

The development of transgenic and knockout animals led to the widespread use of mice in studies investigating the pathophysiology of cardiovascular disease. Mouse models have provided insight into cardiac development (Kwon *et al.*, 2002) and have contributed to our understanding of the pathogenesis of myocardial infarction (MI; Michael *et al.*, 1995; Heymanns *et al.*, 1999; Briaud *et al.*, 2001), myocardial hypertrophy (MacLellan and Schneider, 2000; Zhang *et al.*, 2000), myocarditis (Opavsky *et al.*, 2002), and atrial fibrillation (Hong *et al.*, 2002). Although much smaller in size, the adult mouse heart shares many common pathological features with the hearts of higher mammals.

The murine atria have a very small mass, creating significant problems for investigations studying the generation of atrial fibrillation (Wakimoto *et al.*, 2001). The leaflet tissue of the two murine atrioventricular valves is a continuous veil showing no commissures or clefts. The right atrioventricular valve of the mouse is not morphologically tricuspid. The mitral valve is served by two papillary muscles, which do not become independent from the ventricular wall, resembling trabeculae carnae rather than true papillary muscles (Icardo and Colvee, 1995).

Cardiomyocytes in the murine ventricles can be divided into three layers: **myofibers** in the middle layer run mainly circumferentially (Figure 13.4a), whereas those in the inner and outer layers run parallel or oblique to the apical–basal axis (McLean and Prothero, 1991). The murine myocardium has a rich vascular supply, composed of relatively thin walled arterioles, venules and a well-organized capillary network (Figure 13.4b). Vascular pericytes of the murine myocardium are extensively branched cells that form an incomplete layer around the capillary endothelium and postcapillary venules (Forbes *et al.*, 1977). Arterioles are easily identified (Figure 13.4c). Resident inflammatory cells, such as macrophages and mast cells (Figure 13.4d) are rare in normal adult mouse hearts compared with other species (Gersch *et al.*, 2002), found occasionally in the pericardium and in close proximity to vascular structures. In contrast, murine arterial trunks exhibit a large population of adventitial mast cells that may be involved in regulation of vascular tone.

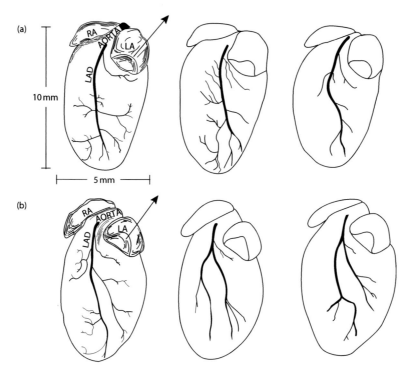

Figure 13.3 Schematic of the major coronary arteries of the mouse heart. Line drawings of coronary artery anatomy of six individual mouse hearts showing patterns of bifurcation. (a) Major singular LAD pattern; (b) major bifurcation pattern (Reprinted with permission from *Am. J. Physiol. Heart Circ. Physiol.* **269**, H2147–H2154.)

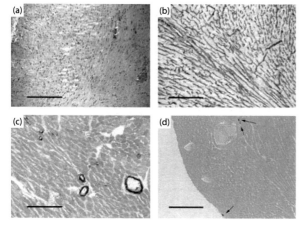

Figure 13.4 (a) Hematoxylin/eosin staining of a normal murine heart. Cardiomyocytes in the murine left ventricle can be divided into three layers: myofibers in the middle layer run mainly circumferentially, whereas those in the inner and outer layers run parallel or oblique to the apical–basal axis. (b) Normal mouse heart stained with an antibody to CD31/PECAM-1 identifying the vascular endothelium. The murine myocardium exhibits a rich vascular supply with a dense capillary network. (c) α-smooth muscle actin staining identifies arterioles in the normal mouse heart. (d) The murine heart shows a small number of resident inflammatory cells. Toluidine blue staining identifies occasional mast cells located around vessels and in close proximity to the pericardium (arrows). Black bars in figures are 75 μm in length.

Mouse electrocardiography

A very comprehensive review of mouse electrocardiography (ECG) was recently published by Wehrens *et al.* (2000) detailing the current state of measuring and quantifying the mouse ECG. Goldbarg *et al.* (1968) investigated the ECG characteristic of normal strains of mice and useful definition of Q-T segments. While the first mouse ECG was recorded on a string galvonometer in 1929, the current efforts have produced a system which records a 12-lead ECG Electrophysiological (EP) techniques have been used in a mouse model of **hypertrophic** cardiomyopathy (HCM; Berul *et al.*, 1997; Spirito *et al.*, 1997). Given the fact that Richards *et al.* (1953) were unable to show a clear T-wave following the QRS complex, he directed experiments to investigate whether the T-wave could be made distinct from the main QRS complex. He was able to show that the notch on the QRS wave, which potentially represented the T-wave, was separated from the main QRS wave on cooling and was accentuated with increases in potassium. Methods have been developed to determine the various components of the ECG during myocardial ischemia, long QT syndrome mutations, and atrial fib-rillation. Within any individual study in a longitudinal manner, the electrocardiographic data are substantially reproducible. However, variation in the data between studies is many times illustrative of differences in strain, anesthesia, or other factors. Genetically modified mice are continually produced to study cardiac depolar-ization and repolarization phenomena and to produce the long QT syndrome and atrial fibrillation. An implantable telemetry system should be helpful in monitoring changes in Q-T interval in transgenic mice with ion channel defects (Mitchell *et al.*, 1998). Mutations of the underlying genes controlling the vari-ous ion channels and gap junctions offer promise for understanding the origin of various parts of the ECG and abnormalities produced by disease. Our laboratories continuously monitor the ECG of all surgically operated or hemodynamically monitored mice and this has been especially important in studies of myocardial ischemia and reperfusion. While the T-wave is only a 'notch' in the ECG on the down-slope of the QRS wave com-plex, this site on the ECG shows striking changes in voltage when the animal is challenged with myocardial ischemia by coronary artery occlusion. The notch is increased in height with a plateau (generally named 'S-T segment'), which becomes 50–90% of the QRS volt-age peak height. When the measured peak voltage in the S-T segment produced at the end of a 1-h coronary occlusion is plotted against the infarct damage mea-sured as a percent of area at risk at 24 h after the ischemic insult, there is a direct relationship between the peak ST segment and the infarct size. This is shown in Figure 13.5 where 25 min into a 60-min coronary artery occlusion created an ST segment change that was plotted against infarct damage measured at autopsy 24 h later. Clearly, the ECG ST segment is reflective of

the damage that occurs in the myocardium with ischemia. It is also clear that mice do not die of ventric-ular fibrillation post MI or reperfusion of an occluded vessel. However, reentry and fibrillation in the mouse heart was produced by sustained burst pacing (Vaidya *et al.*, 1999). In our hands, any ECG tracing inter-preted as ventricular fibrillation reverts to normal rhythm when the artificial voltage disturbance is dis-continued.

Functional cardiovascular measurements in the mouse – invasive vs. noninvasive

Intense efforts have been made to develop instruments which will be able to monitor flow, pressure, and dimension in the mouse heart. When developing instrumentation to monitor the cardiovascular system in mice, it is important to realize the challenges inher-ent in both invasive and noninvasive measurement methods. Several reviews have focused on the methods to evaluate cardiovascular physiology in mice (Doevendans *et al.*, 1998; Hoit and Walsh, 1998; James *et al.*, 1998; Kass *et al.*, 1998; Hartley *et al.*, 2002). Invasive experiments generally involve placing sensors and/or catheters in a vessel or the heart in an anesthetized open or closed chest state. Generally, the experiment is completed within hours and the animal cannot be used for repeat measurements (Lorenz and Robbins, 1996; Georgakopoulos *et al.*, 1998; Kubota *et al.*, 1998; Feldman *et al.*, 2000; Wang *et al.*, 2000). Conscious-mouse methods, such as telemetry, require great care in surgical dissection to implant the sensors for chronic studies (Kramer *et al.*, 1998; Mitchell *et al.*, 1998). There are several noninvasive methods reported in this chapter, such as magnetic resonance imaging (MRI) and echocardiography. Even these methods rely on anesthetizing the animal on the day of the experiment, except in less common conscious-mouse experiments where some degree of stress still may be present. Some noninvasive protocols are minimally invasive i.e., nuclear ventriculography in which a jugu-lar vein is cannulated for injection of the radioisotope

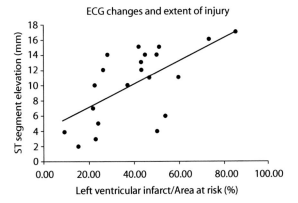

Figure 13.5 ECG vs. infarct size. The S-T segment elevation at 25 min during a 60-min coronary artery occlusion plotted against the infarct weight as % of the area at risk measured 24 h later at autopsy with triphenyltetrazolium stain.

(Hartley *et al.*, 1999). Other studies attempted to look at conscious noninvasive animals as exemplified by telemetry studies and use of implanted sensors (Mitchell *et al.*, 1998). In experiments to measure cardiac output, regional blood flow, and intravascular volumes in the 'conscious' mouse, injections were made through the femoral artery in animals that had just recovered from anesthesia but were still restrained. Heart rates in this case were lower than normal. Resting heart rate normally is between 500 and 600 bpm and values lower than 300 might be considered nonphysiological whereas rates above 750 are rare.

Doppler ultrasound system

In order to detect aortic blood flow velocities, and blood flow velocities across the mitral valve as seen in human studies, a murine pulsed **Doppler** system was designed for use in mice to measure serially the function of the cardiovascular system including pulse wave velocity. (Hartley *et al.*, 1995, 1997, 1999; Taffet *et al.*, 1996). The Doppler system uses probes consisting of 1 mm diameter 10 or 20 MHz ultrasonic crystals mounted at the end of 2 mm diameter stainless steel tubes small enough to be oriented parallel to direction of flow from a site at the border of the sternum. In this system the best velocity resolution is 5 mm/s and the best temporal resolution is 0.1 ms. The maximum measurable velocity is approximately 9 m/s. This system also monitors the ECG simultaneously throughout the experiment and relates the temporal features of the velocity wave to the ECG.

Doppler aortic flow velocities reflect cardiac systolic function and flow velocities across the mitral valve reflect diastolic function as blood flows into the relaxed left ventricle. Both aortic and peak mitral flow velocities were measured in myocardial ischemia and reperfusion experiments longitudinally in time for months (Michael *et al.*, 1999). Figure 13.6 illustrates the peak aortic velocity (a) and peak early filling velocity (b) in three types of animals – sham, permanent occlusion, and occlusion followed by reperfusion. In these experiments Doppler velocities indicated a return to normal values in reperfused animals but not those permanently occluded.

Pulsed Doppler blood velocity signals can be obtained noninvasively from many peripheral vessels as well as from the heart. Figure 13.7 shows 20 MHz spectral Doppler signals representing blood velocity vs. time from the left and right carotid arteries, the left and right renal arteries, the celiac artery, and several sites along the aorta in one anesthetized mouse. The renal

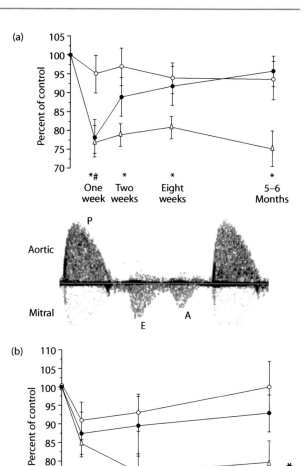

Figure 13.6 (a) Peak aortic flow velocity followed for 5–6 months in mice subjected to sham operation (○), 2-h occlusion followed by reperfusion (●), and permanent occlusion (△). Data are % of preoperative values and are expressed means ± SE. Preoperative values: sham ($n = 15$), 104 ± 20 cm/s; permanent occlusion ($n = 24$), 111 ± 17 cm/s; reperfusion ($n = 13$), 102 ± 10 cm/s. *$P < 0.05$, permanent occlusion vs. sham; *$P < 0.05$, reperfusion vs. sham. (b) Peak early filling velocity followed for 5–6 months in mice subjected to sham operation (○), 2-h ischemia followed by reperfusion (●), and permanent occlusion (△). Data are % of preoperative values and are expressed means ± SE. Preoperative values: sham ($n = 15$), 69 ± 3.1 cm/s; permanent occlusion ($n = 24$), 71 ± 1.9 cm/s; reperfusion ($n = 13$), 66 ± 2.1 cm/s. *$P < 0.05$ vs. sham. The central figure shows representative aortic and mitral Doppler signals taken from the ascending aorta (aortic) and from the mitral valve (mitral). Peak aortic flow velocity (P) was measured during systole; peak flow velocities across the mitral value were measured during diastole: E = peak flow velocity; A = peak flow velocity resulting from atrial contraction. (Modified from *Am. J. Physiol. Heart Circ. Physiol.* **277**, H660–H668.)

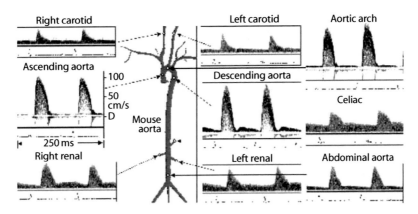

Figure 13.7 Doppler flow velocity profiles at various sites on the mouse arterial system: right and left carotid arteries, transverse aorta, ascending and descending aorta, abdominal aorta with both right and left renal arteries. (Reprinted with permission from *ILAR J.* **43**(3), Institute for Laboratory Animal Research, National Academy of Sciences, 500 Fifth Street NW, Washington, DC, 2001 (*www.national-academies.org/ilar*).)

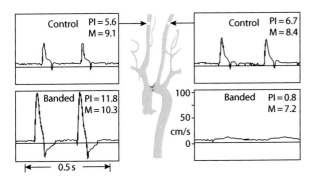

Figure 13.8 Schematic diagram of carotid arteries, ascending and descending aorta, and aortic arch with indication of banded site. Doppler signals are shown for right and left carotid arteries. After banding (lower panels) flow velocity is increased in the right carotid artery and decreased in the left carotid artery distal to stenosis. PI = pulsatility index; M = mean; PI = (max−min)/mean. The ratio of pulsatility indices (right/left) is a useful indicator of degree of banding.

artery signals were obtained with the mouse in the prone position and the others were obtained with the mouse supine. All have high signal quality with magnitudes and waveforms similar to those recorded from the same vessels in man and larger animals. The variations in magnitude and waveforms are due to differences in peripheral vascular impedance in the arterial bed distal to the measurement site. Many of the murine models and interventions produce alterations in cardiac performance and regional **vascular impedance** which change these signals in characteristic ways. In the aortic banding model described later, which produces pressure overload, there are dramatic alterations in the blood flow waveforms in the right and left carotid arteries with only small changes in the average or mean velocity. Figure 13.8 shows right and left carotid artery

velocity signals before and immediately after placement of the aortic constriction. The signals are comparable in magnitude and waveform before banding, but after banding the maximum velocity and **pulsatility index (PI)** are increased in the right and decreased in the left carotid arteries (PI = (max − min)/mean, M = mean). Some of the observed changes in velocity are due to differences in driving pressure, but most are due to adaptive changes in peripheral impedance (resistance and compliance) in order to maintain cerebral perfusion and to minimize the load seen by the heart during systole. Pressure decrease across the aortic band ranges from 30 to 50 mmHg.

The Apo-E knockout mouse may be considered as a model of atherosclerosis similar to that seen in humans (Osada *et al.*, 2000). ApoE knockout mice have atherosclerotic lesions which cover approximately 50% of the aorta by one year of age with elevated cardiac output, elevated pulse wave velocity, and cardiac hypertrophy (Hartley *et al.*, 2000). A unique alteration in aortic arch acceleration was discovered in these mice. In wild-type normal mice, velocity in the aortic root and arch during acceleration is smooth and continuous with the maximum slope or peak acceleration occurring early as shown in Figure 13.9. In ApoE knockout mice, acceleration in the aortic root appears normal, but at the aortic arch, acceleration occurs in two distinct phases with a slower initial rise followed by a later peak. This unusual waveform is probably caused by alterations in aortic impedance and a strong and early-arriving reflected wave from the stenosed carotid bifurcation (Hartley *et al.*, 2000). This study also illustrated that pulse wave velocity is an indication of changes in vascular stiffness. Differences in pulse arrival times in the aortic arch and abdominal aorta allowed

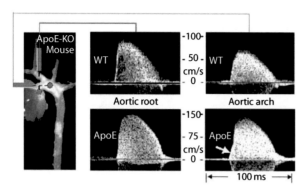

Figure 13.9 Schematic diagram of carotid arteries, aortic root, aorta, and aortic arch with representative signals from aortic root and arch in a wild-type and an ApoE knockout mouse.

(a)

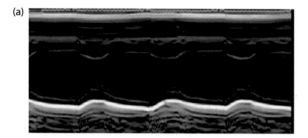

(b)

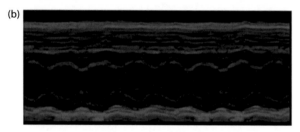

(c)

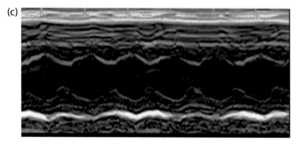

Figure 13.10 Representative echocardiogram of (a) dilated, (b) normal, and (c) hypertrophied mouse hearts.

calculation of pulse wave velocity and demonstrated that ApoE knockout mice had substantially elevated values reflecting a less compliant arterial system.

Echocardiography

Cardiac mass and function are now being evaluated routinely using high frequency cardiac ultrasound. Figure 13.10 shows echocardiographic tracings from normal, hypertrophic and dilated mouse hearts.

Overcoming the temporal and spatial resolution has been a challenge in using a mouse with such a fast heart rate and small size (Hoit and Walsh, 1997; Williams *et al.*, 1998). Left ventricular imaging by transthoracic **echocardiography** allowed assessment of left ventricular mass and systolic function (Gardin *et al.*, 1995; Pollick *et al.*, 1995; Scherrer-Crosbie *et al.*, 1998; Youn *et al.*, 1999). A two-dimensionally directed M-mode approach allowed good visualization but should be used cautiously in LV segment wall motion abnormalities or in instances where the left ventricle hypertrophies in a nonhomogenous fashion. Myocardial shortening in these mice was found to be approximately 57% (Gardin *et al.*, 1995). Tanaka *et al.* (1996) used echocardiography to estimate LV function in normal and transgenic mice including mice that had overexpression of the β_2 adrenergic receptor and the *H-ras* gene. Good correlation has been shown between m-mode echocardiography and left ventricular mass determined at autopsy (Manning *et al.*, 1994; Hoit and Walsh, 1997). Transoesophageal echocardiography was developed and applied to mice to evaluate murine right ventricular size and function as it can be used to image left ventricular function (Scherrer-Crosbie *et al.*, 1998). This type of study is particularly important in the pathophysiology of pulmonary hypertension.

Compared with transthoracic echocardiography, transesophageal echocardiography requires that the mouse trachea be intubated so that there is no collapse of the airway. Doppler echocardiography has been pushed to new levels by studies showing that embryonic mouse hearts may be analyzed. This method was particularly useful to monitor normal mouse embryos and compare them with embryos which might die during gestation and/or have cardiac failure (Gui *et al.*, 1996). An additional feature in echocardiography has been the use of contrast medium to monitor myocardial perfusion in humans. This has now been applied to a mouse model of myocardial ischemia illustrating that perfusion defects could be imaged in mice to quantitate the non-perfused myocardium (Scherrer-Crosbie *et al.*, 1999).

Magnetic resonance imaging

MRI may be very useful in mouse imaging. See Chapter 10 on Imaging. The procedure is relatively non-invasive and suitable for longitudinal studies. There are significant advantages compared with M-mode or two-dimensional echocardiography murine studies. For example, MRI may be used for three-D reconstruction of all chambers and for twisting movement to measure torsion with the cardiac cycle (Henson et al., 2000). Magnetic resonance imaging provides data about the right heart which is difficult to image otherwise. In some studies, an 'average volume' is provided which is not likely to reflect end-diastolic or end systolic measurements (Sze et al., 2001), but as frame rates increase, more time-resolved information may be obtainable. Chacko et al. (2000) combined P31 MR Spectroscopy with MRI allowing simultaneous determination of anatomical/functional data and high energy phosphate status. Using a similar approach and specific probes, simultaneous measurement of function and intracellular hypoxia, pH, or other biochemical parameters appear feasible. Others have visualized small structures such as coronary arteries and valves in vivo with MRI that have not been visualized by other techniques (Ruff et al., 2000). Magnetic resonance imaging can provide reasonable serial estimates of LV and heart mass for developmental studies (Wiesmann et al., 2000). Determination of LV ejection fraction does not require many of the assumptions used by the echocardiographers. However, of necessity, the MR image acquisition was gated to both cardiac and respiratory cycles by acquiring ECG and respiratory signals in studies to determine hypertrophic changes in mouse heart mass. (Slawson et al., 1998).

While MRI equipment is very expensive to purchase, maintain, and operate, the quality of images continue to improve. Most investigators use dedicated coils for mouse cardiac imaging. At best, the frame acquisition is still slow, more than 8 ms (which converts to 12.5 frames per 100 ms cardiac cycle). This factor should continue to decrease as frame rates have increased significantly in the past years. Because of the dependence on numerous acquisitions to generate one image, gating using the ECG for timing must be used and another sensor for respiration may be necessary. This will provide challenges in mice with irregular heart rhythms, a property of some phenotypes. Additionally, the animals must remain absolutely immobile for the

repeated acquisitions, though algorithms do exist to handle minimal subject movements. Magnetic resonance imaging provides no real assay of diastolic function.

Magnetic resonance imaging has been used to document the systolic dysfunction and cardiac dilation induced by tumour necrosis factor (TNF)-α expression (Franco et al., 1999). It has also been used to show the hypertrophy with relatively maintained systolic function in the GLUT4 null mouse (Stenbit et al., 2000).

The scaling issues of MRI are similar to those faced by the echo. The mouse will fit nicely into a relatively small bore magnet for studies under anesthesia (see Table 13.1).

Ventriculography

Radiographic imaging the mouse heart to obtain ventricular volumes and ejection fractions used constant fluoroscopy and a nonionic contrast medium (Rockman et al., 1994). Both right and left ventricular function was evaluated, however temporal resolution was limited to the equivalent of 60 frames/s.

Radionuclide ventriculography using a multiwire gamma camera allowed quantitation of left and right ventricular ejection fractions in normal and infarct damaged mouse hearts (Hartley et al., 1999). The technique utilized a short half-life isotope Tantallum-178 injected into the jugular vein and imaging with a nuclear camera fitted with a pinhole lens to achieve a 2 mm resolution with frame rate of 160/s. The average value for the normal mouse was 58%; decreasing to 15–25% for hearts damaged by ischemia.

Values based on scaling equations

A variety of mouse cardiovascular parameters were calculated by using scaling laws for animals developed by Dawson (1991) which relate a measured value to animal

Investigator	LVEDV (μl)	LVESV (μl)	LVEF (%)	Cardiac output (ml/min)	LV mass (mg)	Stroke volume (μl)
Franco (1998)	51.7 ± 12.8	13.4 ± 6.7	75 ± 8		156 ± 14	38.3 ± 8.8
Franco (1999)	60 ± 9.8	21 ± 2.6	66 ± 2	15.6 ± 0.7		32 ± 2
Ruff (1998)	45.2 ± 9.3	14.6 ± 5.5	68.6 ± 6.6			
Weismann (2000)	63.6 ± 6.6	23.5 ± 4.4	65 ± 3.5	14.3 ± 0.5	101.3 ± 8	40.2 ± 2.7
Chacko (2000)			65 ± 7			

TABLE 13.1: WT parameters (3–4 month old mice) measured with MRI

LVEDV = left ventricuar end-diastolic volume; LVESV = left ventricular end-systolic volume; LVEF = left ventricular ejection fraction.

TABLE 13.2: Relationship of measured parameters to animal body weight. General scaling equation: Y = a BW^b

Parameter	Relationship to BW (kg)	Value (BW = 25 g)
Heart rate	α BW$^{-1/4}$ 230 BW$^{-1/4}$	578 bpm
Heart weight	α BW^1 4.3 BW	112 mg
LV volume	α BW^1 2.25 BW	56 µl
Stroke volume	α BW^1 0.95 BW	24 µl
Cardiac output	α BW$^{3/4}$ 224 BW$^{3/4}$	14 ml/min
Aortic diameter	α BW$^{3/8}$ 3.6 BW$^{3/8}$	0.9 mm
Arterial pressure	α BW^0 100	100 mmHg
Peak aortic velocity	α BW^0 100	100 cm/s
Mitral E velocity	α BW^0 60	60 cm/s
Pulse-wave velocity	α BW^0 500	500 cm/s

body weight. The power law form utilizes best-fit methods of analysis of the measured parameters vs. body weight wherein a coefficient alpha and exponent are derived (see Table 13.2). These derived values for normal mice are usually very close to the actual measured values for mouse parameters. Blood flow velocities and mean blood pressure are independent of body weight, hence, the exponent is 0 requiring no correction of measured value. These latter values for mice are strikingly similar to those seen in other normal animals and human.

Mouse hemodynamic values vary somewhat as result of strain differences and, more so, as a result of experimental conditions such as anesthetic regimen, invasive vs. noninvasive procedures and genetic alteration as seen with transgenic or knockout maneuvers. Similarities of wave forms of pressure and velocity across species indicate that arterial time constants scale with cardiac period (Westerhoff, 2002).

Experimental models

Myocardial ischemia and reperfusion

Anesthesia and surgical technique

In order to study myocardial ischemic injury, a coronary vessel is selected for occlusion. In the mouse, this vessel is the LAD. Other arterial vessels on the left ventricular free wall and septal region are not prominent. The right coronary artery is a major supplier of the septal region and also the right ventricle. It is important to stress that a consistent coronary artery occlusion site must be chosen so that the area at risk is the same in each heart.

The first method to be described is one in which the thorax is opened and the LAD is occluded and/or reperfused acutely (Michael *et al.*, 1995, 1999). The second method requires placement of an occluding device and the occlusion activation does not occur for several days, a chronic model. (Nossuli *et al.*, 2000). In both methods, surgical manipulation requires that the animal be placed on a respirator; this may involve anesthesia using 2% isofluorane or anesthetizing the animal with an agent, such as sodium pentobarbital, prior to placement on the respirator. There are several important points which allow the models to be successfully utilized. First, the animal needs tracheal insertion of a tube and this is generally more easily accomplished by a slight incision in the skin of the ventral neck over the laryngeal region; careful reflection of the muscles of the neck allows direct visualization of the trachea. At this point, the tongue is extended and polyethylene (PE)-90 tubing, with a beveled point is slid in through the mouth, past the pharyngeal region, and into the trachea just past the laryngeal cords. Second, the animal is ventilated very carefully, so that the lungs are not over-expanded or under-expanded. In some systems, this is allowed by having a loose fitting connection between the respirator and the endotracheal tube giving the animal sufficient respiratory volume but not over-expansion of the lungs.

Proper lighting and magnification of the field of interest is essential in order to identify the LAD and to place an 8-0 sterile suture underneath and around the vessel in order to secure the occluding device. In the first method a 2 mm piece of PE-10 tubing is placed on the surface of the heart over the LAD and then the suture is brought up alongside the vessel and tied around the PE tubing firmly in order to compress the coronary artery against the under surface of the PE tubing. This creates a coronary artery occlusion and results in myocardial ischemia. For reperfusion studies, the ligature on top of the PE tubing is cut and allows the release of the PE tubing and hence restores flow into the coronary vessel. If a permanent coronary artery occlusion is desired, the ligature placed around the coronary artery is simply tied with a double knot without placing an intermediate piece of PE tubing. Ligatures used in closing the chest wall are generally either a 7-0 or 6-0 size suture. As soon as the chest is closed, the animal is allowed to recover by removal from the respirator and, within a few minutes, begins spontaneous breathing.

The second major way to promote a LAD coronary occlusion is the chronic model of coronary occlusion, specifically for ischemia followed by reperfusion experiments in a closed-chest mouse several days post surgery. This allows a time period, after implanting the occlusion device, to allow the dissipation of inflammation and the trauma associated with the surgical manipulation. This model is especially helpful in studying cytokines, chemokines, and other inflammatory events which are promoted by the myocardial ischemic state. The surgical manipulation and resulting release of a variety of these elements are separated in time from the ischemic event. In this model, a median thoracotomy is not performed as for the acute open chest occlusion model. The chest is opened with a left lateral cut with fine scissors along the sternum cutting through ribs to approximately mid sternum. The chest wall is then retracted and the pericardium gently dissected to see the coronary artery. After passing the needle of the 8-0 suture underneath the LAD, the needle is cut from the suture and the two ends of the 8-0 suture are then placed through a 0.5 mm piece of PE-10 tubing. This forms a loose snare around the LAD. The sutures are then exteriorized through each side of the chest wall and the chest closed. The ends of the 8-0 suture are then placed underneath the skin and the skin closed with 6-0 suture. At this point, the animal is removed from the respirator, allowed to recover consciousness, and placed in the **intensive care unit** (ICU) for recovery.

At various later times, days to weeks, the animal is then re-anesthetized and the appropriate occlusion and/or reperfusion protocol performed without opening the chest. This occurs by ventilating the animal with isofluorane anesthesia and proceeding to extract carefully the two ends of the suture from underneath the skin; coronary occlusion occurs by pulling on the ends of the suture laterally which then creates a compression on the coronary artery as the small piece of PE tubing is forced downward. Reperfusion results when the lateral tension is stopped. A variety of experiments may be designed using this chronic model such as experiments where brief periods of ischemia, 5 or 15 min, followed by reperfusion of minutes to hours or experiments where there are longer occlusion or reperfusion periods. We have investigated the role of several chemokines and cytokines and their release during numerous experimental paradigms using this model (Nossuli *et al.*, 2001).

Myocardial hypertrophy

Pressure overload hypertrophy is induced by transaortic banding in mice (Rockman *et al.*, 1993; Zhang *et al.*, 2000). The technique involves opening the anterior chest wall sufficiently to expose the transverse aortic arch. This entails dissecting the fatty material immediately juxtaposed to the arch after placing a ligature loosely around the aortic arch at that site and then applying a 3 mm section of a 27-gauge needle. The ligature is tightened around the needle and aorta sufficient to occlude the vessel completely. Two knots are then applied to tie it in place. Immediately, the small section of steel tubing is removed and, hence, the vessel blood flow will be similar to the diameter of that removed piece of tubing. After this, the animal's chest wall is closed and the skin is sutured. With practice, this technique allows 80–90% occlusion of the blood flow velocity in the aortic arch. This allows left ventricular hypertrophy and heart weight/body weight ratio increase within 10–14 days. This is similar to the time course of myocardial hypertrophy seen in similar experiments in rats.

Pathology of MI in mice

Murine models of MI, patterned after the models previously described, have contributed to our understanding of the pathogenetic mechanisms operative in the ischemic myocardium and are now being performed in numerous laboratories. Occlusion of the LAD coronary artery in the mouse generates an extensive infarction involving the anterior and lateral left ventricular wall. Evidence of cardiomyocyte injury, such as contraction band necrosis and wavy fibers is found in the early post-infarction stages (Michael *et al.*, 1995). Murine MI is associated with an intense local inflammatory response, which (similar to higher mammalian species) is significantly accentuated with reperfusion of the myocardium (Frangogiannis *et al.*, 1998b, 2002). Reperfused murine infarcts exhibit intense leukocyte infiltration, leading to accumulation of myofibroblasts and deposition of collagen, and thence to the rapid formation of thinned relatively acellular scars (Figure 13.11). Reperfusion of the murine myocardium appears to reduce the degree of infarct expansion, even under circumstances in which infarct size is not altered, inducing more effective ventricular repair and preservation of ventricular function (Michael *et al.*, 1999) (Figure 13.12).

Figure 13.13 illustrates the collagen matrix stained with sirius red in cross sections of a mouse heart which

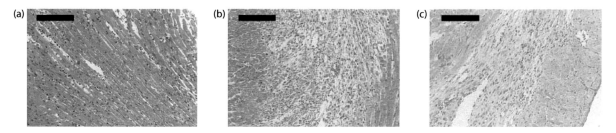

Figure 13.11 Hematoxylin/eosin staining of infarcted murine myocardium. (a) After 1 h of coronary occlusion and 24 h of reperfusion the mouse myocardium exhibits extensive leukocyte infiltration. (b) After 72 h of reperfusion myocyte replacement with granulation tissue is noted. (c) After 7 days of reperfusion the healing infarct has thinned and demonstrates a relatively low cellular content. Black bars are 75 μm in length.

Figure 13.12 Transverse sections of representative mouse hearts: sham, permanent LAD coronary artery occlusion for 8 weeks, and LAD coronary artery occlusion for 60 min followed by reperfusion for 8 weeks. (Reprinted with permission from *Am. J. Physiol. Heart Circ. Physiol.* **277**, H660–H668.)

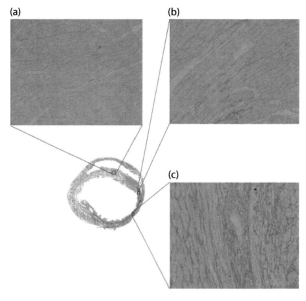

Figure 13.13 Cross section of mouse heart at autopsy after a 2-h coronary artery occlusion followed by 2 weeks of reperfusion. The collagen stain picro sirius red of (a) thinned left ventricular free wall; (b) border zone; (c) septum.

had been subjected to 24 h of LAD occlusion followed by 2 weeks of reperfusion. Use of the murine model of experimental myocardial ischemia and reperfusion has elucidated important aspects of the pathophysiology of MI. Studies in large mammals have suggested that a cytokine cascade is triggered by the release of TNF-α in the ischemic heart ultimately leading to healing and scar formation (Frangogiannis *et al.*, 1998a). A recent investigation in a murine model of experimental infarction indicated that TNF-α may exacerbate myocardial ischemic injury at an early stage of reperfusion by activating NF-kappaB, thereby inducing chemokines and adhesion molecules and facilitating

leukocyte infiltration (Maekawa *et al.*, 2002). Other studies, however, indicated that TNF signaling gives rise to one or more cytoprotective signals that prevent and/or delay the development of cardiac myocyte apoptosis after acute ischemic injury (Kurrelmeyer *et al.*, 2000), emphasizing the **pleiotropic** effects of cytokines in inflammatory reactions. The importance of adhesion molecules in MI was illustrated using animals deficient in both ICAM-1 and P-selectin and demonstrating impaired neutrophil trafficking without a difference in infarct size (Briaud *et al.*, 2001).

Mouse models of experimental MI have been used to investigate the importance of specific mediators in infarct healing and post-infarction remodeling. Recent investigations suggested a critical role for proteases regulating extracellular matrix remodeling in infarct healing: deficiency of urokinase-type plasminogen activator (u-PA) protected against cardiac rupture, whereas lack of gelatinase-B protected against rupture (Heymans *et al.*, 1999). However, u-PA−/− mice showed impaired scar formation and infarct revascularization, even after treatment with vascular endothelial growth factor, and died of cardiac failure due to depressed contractility (Heymans *et al.*, 1999). In addition, targeted deletion of the Matrix Metalloproteinase (MMP)-9 gene attenuated left ventricular dilation after experimental MI in mice. The decrease in collagen accumulation and the enhanced expression of other MMPs suggested that MMP-9 plays a prominent role in post-infarction extracellular matrix remodeling (Ducharme *et al.*, 2000). Investigations using genetically altered animals to study the pathological basis of MI will improve our understanding of the cellular and molecular steps involved in myocardial injury and repair. It should be emphasized, however, that significant species differences may exist and should be considered when extrapolating findings derived from murine studies to the pathogenesis of the human disease process.

Stem cells and cardiac regeneration

Myocardial cell death is inevitable when blood flow ceases to a region of the heart. As adult cardiac myocytes do not proliferate, an exciting new area of cardiovascular research using the mouse relates to use of a variety of stem cells to promote repair of the irreversibly injured myocardium. The adult mouse heart contains approximately 20–30% myocardial cells with the remainder composed of endothelial cells, vascular smooth muscle cells, and fibroblasts (Soonpaa *et al.*, 1996). The potential plasticity of stem cells to form other cells is found in bone marrow, skeletal muscle, cardiac muscle, liver bile ducts, vascular endothelium, and other sources. One source of stem cells was an enriched ROSA bone marrow hematopoietic side population of stem cells which were injected into irradiated mice. After 10–12 weeks, the mice were subjected to 1 h of coronary artery occlusion followed by 2–4 weeks of reperfusion. Donor-derived cardiomyocytes as well as endothelial cells were identified in the periinfarct region (Jackson *et al.*, 2001). Differentiation of various kinds of stem cells to myocardial cells in infarcted mouse heart has now been reported by several laboratories (Condorelli *et al.*, 2001; Orlic *et al.*, 2001; Toma

et al., 2002). While a variety of stem cells appear to offer some promise of partial repair by formation of new myocardial cells, this method of repair may not be the most efficient. Difficulties in cell orientation, integration, and electrical syncytium formation remain as major concerns. The existence of resident cardiac progenitor cells within the heart appears to be a new and promising pathway to form new heart cells. In a recent study, progenitor cells differentiate into cardiac muscle when injected intravenously after 1 h of ischemia followed by 6 h of reperfusion. In this study, undifferentiated cells from the adult heart therefore targeted injured myocardium after a systemic injection (Oh, 2002).

Pathology of myocardial hypertrophy

Cardiac muscle hypertrophy and cardiac enlargement include increased myocyte size, sarcomeric formation, reactivation of a fetal gene program, including upregulation of genes such as beta myosin heavy chain (MHC), atrial natriuretic factor (ANF), and skeletal alpha actin (SkA) (Mulvagh *et al.*, 1987; Izumo *et al.*, 1988; Schiaffino *et al.*, 1989; Chassagne *et al.*, 1993; Vikstrom *et al.*, 1998; Zhang *et al.*, 2000). A large number of manipulations of the mouse genome have resulted in murine myocardial hypertrophy. The majority of the models have employed transgenic overexpression of the candidate gene using the α-MHC promoter; fewer studies have attempted to ablate a gene of interest. Although in many cases the level of overexpression of the transgene vastly exceeds that observed in naturally occurring pathological states, useful mechanistic information has been derived from generation and analysis of these genetically engineered mice.

It appears that hypertrophy is not always associated with increased ventricular expression of ANF induction which has quite often been considered the best general indicator of the transcriptional response (Vikstrom *et al.*, 1998). This was deduced using a transgenic mouse model of HCM where hypertrophy occurred in the absence of increased ventricular levels of ANF message and levels of mRNA were absent where cardiac hypertrophy was detected. Localized changes in gene expression, however, did correlate with areas of tissue pathology. This is counter to studies where much greater increases in ANF gene expression occurred in models of acute pressure or volume overload (Mercadiar et al., 1989; Feldman et al., 1993; Calderone et al., 1995), which, however, may represent a different pathogenic response. In other transgenic models of HCM a mouse with *in vivo* expression of the mutant cardiac sarcomeric protein, troponin T-Q[92], led to impaired local cardiac systolic function and increased interstitial collagen (Lim et al., 2000).

Genetic screening

In studies where no particular gene was implicated, differential screening was attempted in order to identify any and all genes regulated during hypertrophy (Johnatty et al., 2000). Experiments were performed using subtractive hybridization between cDNA from the hearts of aortic-banded compared with sham-operated mice. In these experiments, more than 50 genes were identified as being upregulated following the mechanical challenge of pressure overload in mouse hearts. The results revealed similarities between the genetic programs of the neonatal and pressure overloaded hearts.

Cardiac growth induced by mechanical load requires coupling of extracellular stimuli to gene transcription. Transforming growth factor beta-activated kinase (TAK1) is a member of the **mitogen** activated protein kinase (MAPK) family which is involved in this coupling. TAK1 kinase activity is upregulated after aortic banding in mice and induces hypertrophy and the expression of the transforming growth factor beta (Zhang et al., 2000). Similar studies, where TAK1 is expressed in the myocardium of transgenic mice, provided sufficient stimulus to produce a P38 MAPK phosphorylation leading to myocardial hypertrophy with various disruptive elements presaging heart failure including interstitial fibrosis and severe myocardial dysfunction.

One of the potential mediators of hypertrophy is a broad spectrum G protein called Gq. Hence G protein-coupled receptors initiate complex MAPKs (Clerk and Sugden, 2000; Molkentin and Dorn, 2001). One of these MAPKs, called MEKK1 (mitogen activated protein kinase/extracellular signal-regulated protein kinase), was activated by specific overexpression of Gq in wild type mouse cardiac muscle (Minamino et al., 2002). When MEKK1 was absent, most of the features of hypertrophy induced by Gq (i.e. cardiac mass and myocyte enlargement) were eliminated. Importantly, in the absence of MEKK1, there was protection from negative effects of Gq on heart function.

One of the basic features of myocardial hypertrophy is the increase in RNA and protein per cell. Recently, it was shown that hypertrophy, which may be triggered by signaling proteins such as Gq, calcineurin or chronic mechanical stress, activates cyclin-dependent, kinases such as Cdk9 which are required for RNA increase. It appears that the kinases Cdk9 and Cdk7 are down regulated as the heart matures, but both are activated by Gq, calcineurin, or chronic mechanical stress to promote cardiac hypertrophic growth. Figure 13.14 illustrates that when Cdk9 is increased several fold in transgenic mice, there is concomitant concentric hypertrophy and substantially increased myocyte size (Sano et al., 2002).

Inflammatory gene expression

Stress on the mouse heart by banding the aorta and creating mechanical load or acute hemodynamic pressure

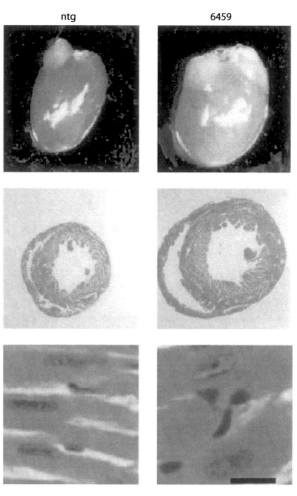

ntg   6459

Figure 13.14 Transgenic and hypertrophied mouse hearts. Activation of Cdk9 by cyclin T1 in transgenic mice produce increase in heart weight/body weight ratio with 50% increase in myocardial cell size and concentric hypertrophy. (Reprinted with permission from *Nat. Med.* **8**, 1310–1317.)

overload promotes inflammatory cytokine gene expression and several proinflammatory agents such as TNF, interleukin (IL)-1 beta, and IL6 RNA levels within hours after the mechanical stress (Mann, 1996). Interestingly it was noted that this was a transient effect and, in fact, after several hours to 3 days, without any change in loading conditions, the proinflammatory cytokines in the heart decreased (Kapadia et al., 1997; Baumgarten et al., 2002). This suggests that there may be both load dependent and independent mechanisms operative in this period. It should be noted that cytokines may be released as a result of acute surgical intervention and this must always be considered when examining inflammatory mediators (Nossuli et al., 2000).

Congestive heart failure

Congestive heart failure (CHF) is a very important condition to model in the mouse. In humans, most CHF is the end result of MI, longstanding hypertension, or other process. Significant MIs were produced in young mice (8–12 weeks) by occluding the LAD anticipating that CHF would develop after infarction. While large infarcts and low perioperative mortality resulted after recovery from the procedure, no clinical manifestations of CHF were noted in these young post-MI mice. (Michael et al., 1999). There was no mortality beyond the perioperative period associated with respiratory difficulty, weight gain or loss, edema, or other manifestation of decreased cardiac output or increased filling pressures. There was frequent aneurysmal dilation of the left ventricular free wall, remodeling and hypertrophy of the non-infarcted ventricle, and depressed fractional shortening on echocardiography, but no evidence of CHF. After adding deoxycorticosterone acetate (DOCA by subcutaneous pellet and 8% saline in lieu of water, no evidence of the CHF syndrome presented and there were no increases in lung weight or lung water in the young mice at necropsy. Therefore, the young mouse, perhaps by using compensatory mechanisms, did not routinely or reproducibly develop CHF post-infarction even if allowed to live to 16 months.

In contrast, older mice (12–14 months of age) develop CHF post-infarct after LAD occlusion and late mortality associated with respiratory difficulty, decreased grooming and weight loss, and increases in lung weight at necropsy (Gould et al., 2001). Treatment with the angiotensin converting enzyme inhibitor, captopril,

was associated with improved survival in these older mice. This is one of the most robust findings in trials of CHF in humans and obviously could not be tested in the younger mice, which experienced no mortality. Using a modification of human criteria, we consider heart failure to be present in the mouse when at least one major criterion is present (**rales**, cardiomegaly, pulmonary edema determined by elevated lung weight or wet/dry ratio, and decreased survival) and, additionally, the presence of several minor criteria (hepatomegaly as determined by elevated liver weight or wet/dry ratio, pleural effusion, weight loss or weight gain, and decreased grooming).

Murine models in aortic disease

Aneurysmal aortic disease is a significant cause of morbidity and mortality in western societies. Despite intensive research, the pathogenesis of aortic aneurysms remains unclear. Recently, murine studies using genetically altered animals have provided us with valuable insight into the mechanisms involved in aortic aneurysm formation. Periarterial application of calcium chloride and elastase perfusion have been established as convenient and reliable models for creating abdominal aortic aneurysms in mice. Elastase-induced aneurysmal degeneration was suppressed by treatment with a nonselective MMP inhibitor (doxycycline) and by targeted gene disruption of MMP-9, but not by isolated deficiency of MMP-12 (Pyo et al., 2000). In addition, in a model of abdominal aortic aneurysm induction by abluminal application of calcium chloride, no aneurysm formation was observed after treatment in either the MMP-9 deficient or the MMP-2 knockout mice (Longo et al., 2002). Experimental murine aortic aneurysm generation may contribute to our understanding of aortic wall remodeling; however it should be emphasized that the human disease process is pathogenetically complex and may not be adequately simulated by existing experimental models.

Acknowledgements

Grateful appreciation is given to Annabelle Lozano, Roy Hendley, Jennifer Pocius, Thuy Pham, and Pawel Zymek for expert assistance in mouse experiments;

Justin Zacchariah for computer analyses and figure preparation; Sharon Malinowski for excellent editing and manuscript preparation.

References

Baumgarten, G., Knuefermann, P., Kalra, D., Gao, F., Taffet, G.E., Michael, L., Blackshear, P.J., Carballo, E., Sivasubramanian, N. and Mann, D.L. (2002). *Circulation* **105**, 2192–2197.

Berul, C., Christe, M., Aronovitz, M.J., Seidman, C., Seidman, J. and Mendelsohn, M. (1997). *J. Clin. Invest.* **99**, 570–576.

Briaud, S.A., Ding, Z.M., Michael, L.H., Entman, M.L., Daniel, S. and Ballantyne, C.M. (2001). *Am. J. Physiol. Heart Circ. Physiol.* **280**, H60–H67.

Calderone, A., Takahashi, N., Izzo, N., Thaik, C.M. and Colucci, W.S. (1995). *Circulation* **92**, 2385–2390.

Chacko, V.P., Aresta, F., Chacko, S.M. and Weiss, R.G. (2000). *Am. J. Physiol.* **279**(5), H2218–H2224.

Chassagne, C., Wisnewsky, C. and Schwartz, K. (1993). *Circ. Res.* **72**, 857–864.

Clerk, A. and Sugden, P. (2000). *Cir. Res.* **86**, 1019–1023.

Condorelli, G., Borello, U., De Angelis, L., Latronico, M., Sirabella, D., Coletta, M., Galli, R., Balconi, G., Follenzi, A., Frati, G., Cusella De Angelis, M.G., Gioglio, L., Amuchastegui, S., Adorini, L., Naldini, L., Vescovi, A., Dejana, E. and Cossu, G. (2001). *Proc. Natl. Acad. Sci. USA* **98**, 10733–10738.

Cook, M.J. (1965). *The Anatomy of the Laboratory Mouse.* Academic Press, New York.

Dawson, T.H. (1991). *Engineering Design of the Cardiovascular System of Mammals*, 179 pp. Prentice Hall, Englewood Cliffs.

Doevendans, P.A., Daemen, M.J., DeMuinck, E.D. and Smits, J.F. (1998). *Cardiovasc. Res.* **39**, 34–49.

Ducharme, A., Frantz, S., Aikawa, M., Rabkin, E., Lindsey, M., Rohde, L.E., Schoen, F.J., Kelly, R.A., Werb, Z., Libby, P. and Lee, R.T. (2000). *J. Clin. Invest.* **106**, 55–62.

Feldman, A.M., Weinberg, E.O., Ray, P.E. and Lorell, B.H. (1993). *Circ. Res.* **73**, 184–192.

Feldman, M., Erikson, J., Korcarz, C., Lang, R. and Freeman, G. (2000). *Am. J. Physiol.* **179**, H1698–H1707.

Forbes, M.S., Rennels, M.L. and Nelson, E. (1977). *Am. J. Anat.* **149**, 47–70.

Franco, F., Dubois, S., Peshock, R.M., and Shohet, R.V. (1998). *Am J Physiol* **274**, H679–H683.

Franco, F., Thomas, G.D., Giroir, B., Bryant, D., Bullock, M.C., Chwialkowski, M.C., Victor, R.G. and Peshok, R.M. (1999). *Circulation* **99**(3), 448–454.

Frangogiannis, N.G., Lindsey, M.L., Michael, L.H., Youker, K.A., Bressler, R.B., Mendoza, L.H., Spengler, R.N., Smith, C.W. and Entman, M.L. (1998a) *Circulation* **98**, 699–710.

Frangogiannis, N.G., Youker, K.A., Rossen, R.D., Gwechenberger, M., Lindsey, M.H., Mendoza, L.H., Michael, L.H., Ballantyne, C.M., Smith, C.W. and Entman, M.L. (1998b). *J. Mol. Cell. Cardiol.* **30**, 2567–2576.

Frangogiannis, N.G., Smith, C.W. and Entman, M.L. (2002). *Cardiovasc. Res.* **53**, 31–47.

Gardin, J., Siri, F., Kitsis, R., Edwards, J. and Leinwand, L. (1995). *Circ. Res.* **76**, 907–914.

Georgakopoulos, D., Mitzner, W.A., Chen, C.H., Byrne, B.J., Millar, H.D., Hare, J.M. and Kass, D.A. (1998). *Am. J. Physiol. Heart Circ. Physiol.* **274**, H1414–H1422.

Gersch, C., Dewald, O., Zoerlein, M., Michael, L.H., Entman, M.L. and Frangogiannis, N.G. (2002). *Histochem. Cell Biol.* **118**, 41–49.

Goldbarg, A., Hellerstein, H., Bruell, J. and Daroczy, A. (1968). *Cardiovasc. Res.* **2**, 93–99.

Gould, K.E., Taffet, G.E., Michael, L.H., Christie, R.M., Konkol, D.L., Pocius, J.S., Zachariah, J.P. Chaupin, D.F., Daniel, S.L., Sandusky, G.E., Hartley, C.J. and Entman, M.L. (2001). *Am. J. Physiol.* **282**, H615–H621.

Green, E.L. (ed.) and Staff of Jackson Lab. (1966). In *Biology of the Laboratory Mouse*, 2nd edn, pp. 249–251, 574–576. McGraw Hill, New York.

Gui, Y.H., Linask, K.K., Khowsathit, P. and Huhta, J.C. (1996). *Pediat. Res.* **40**, 633–642.

Halpern, M. (1953). *Am. J. Anat.* **101**, 307–327.

Hartley, C.J., Michael, L.H. and Entman, M.L. (1995). *Am. J. Physiol. Heart Circ. Physiol.* **268**, H499–H505.

Hartley, C.J., Taffet, G.E., Michael, L.H., Pham, T.T. and Entman, M.L. (1997). *Am. J. Physiol. Heart Circ. Physiol.* **273**, H494–H500.

Hartley, C.J., Lacy, J.L., Dai, D., Nayak, N., Taffet, G.E., Entman, M.L. and Michael, L.H. (1999). *Nat. Med.* **5**, 237–239.

Hartley, C.J., Reddy, A.K., Madala, S., Martin-McNulty, B., Vergona, R., Sullivan, M.E., Halks-Miller, M., Taffet, G.E., Michael, L.H., Entman, M.L. and Wang, Y.X. (2000). *Am. J. Physiol. Heart Circ. Physiol.* **279**, H2326–H2334.

Hartley, C.J., Taffet, G.E., Reddy, A.K., Entman, M.L. and Michael, L.H. (2002). *ILAR J.* **43**, 147–158.

Henson, R.E., Song, S.K., Pastorek, J.S., Ackerman, J.J. and Lorenz, C.H. (2000). *Am. J. Physiol. Heart Circ. Physiol.* **278**(4), H1117–1123.

Heymans, S., Luttun, A., Nuyens, D., Theilmeier, G., Creemers, E., Moons, L., Dyspersin, G.D., Cleutjens, J.P., Shipley, M., Angellilo, A., Levi, M., Nube, O., Baker, A., Keshet, E., Lupu, F., Herbert, J.M., Smits, J.F., Shapiro, S.D., Baes, M., Borgers, M., Collen, D., Daemen, M.J. and Carmeliet, P. (1999). *Nat. Med.* **5**, 1135–1142.

Hoit, B.D. and Walsh, R. (1997). *Trends Cardiovasc. Med.* **7**, 129–134.

Hoit, B.D. and Walsh, R.A. (eds) (1998). *Cardiovascular Physiology in the* Genetically Engineered Mouse, pp. 424. Kluwer Academic Publishers, Norwell, MA.

Hong, C.S., Cho, M.C., Kwak, Y.G., Song, C.H., Lee, Y.H., Lim, J.S., Kwon, Y.K., Chae, S.W. and Kim do, H. (2002). *FASEB J.* **16**, 1310–1312.

Icardo, J.M. and Colvee, E. (1995). *Anat. Rec.* **241**, 391–400.

Izumo, S., Nadal-Ginard, B. and Mahdavi, V. (1988). *Proc. Natl. Acad. Sci. USA* **85**, 339–343.

Jackson, K., Mojka, S., Wang, H., Pocius, J., Hartley, C., Majesky, M., Entman, M., Michael, L., Hirschi, K. and Goodell, M. (2001). *J. Clin. Invest.* **107**, 1395–1402.

James, J.F., Hewett, T.E. and Robbins, J. (1998). *Circ. Res.* **82**(4), 407–415.

Johnatty, S., Dyck, J., Michael, L., Olson, E. and Ardellatie, M. (2000). *J. Mol. Cell. Cardiol.* **32**, 805–815.

Kapadia, S., Oral, H., Lee, J., Nakemo, M., Taffet, G. and Mann, D. (1997). *Circ Res.* **81**, 187–195.

Kass, D.A., Hare, J.M. and Georgakopoulos, D. (1998). *Circ. Res.* **82**, 519–522.

Kramer, K., Voss, J., Grimbergen, J. and Bast, A. (1998). *Lab. Animal* **27**, 23–26.

Kubota, T., Mahler, C., McTiernan, C.F., Wu, C., Feldman, M., Feldman A., and Frye (1998). *J. Mol. Cell. Cardiol.* **30**, 357–363.

Kurrelmeyer, K.M., Michael, L.H., Baumgarten, G., Taffet, G.E., Peschon, J.J., Sivasubramanian, N., Entman, M.L. and Mann, D.L. (2000). *Proc. Natl. Acad. Sci. USA* **97**, 5456–5461.

Kwon, Y.T., Kashina, A.S., Davydov, I.V., Hu, R.G., An, J.Y., Seo, J.W., Du, F. and Varshavsky, A. (2002). *Science* **297**, 96–99

Lim, D-S., Oberst, L., McCluggage, M., Youker, K., Lacy, J., DeMayo, F., Entman, M.L., Roberts, R., Michael, L.H. and Marian, A.J. (2000). *J. Mol. Cell. Cardiol.* **32**, 365–374.

Longo, G.M., Xiong, W., Greiner, T.C., Zhao, Y., Fiotti, N. and Baxter, B.T. (2002). *J. Clin. Invest.* **110**, 625–632.

Lorenz, J. and Robbins, J. (1996). *Am. J. Physiol.* **272**, H1137–H1146.

MacLellan, W.R. and Schneider, M.D. (2000). *Annu. Rev. Physiol.* **62**, 289–319.

Maekawa, N., Wada, H., Kanda, T., Niwa, T., Yamada, Y., Saito, K., Fujiwara, H., Sekikawa, K. and Seishima, M. (2002). *J. Am. Coll. Cardiol.* **39**, 1229–1235.

Mann, D. (1996). *Cytokine Growth Factor Rev.* 7, 341–354.

Manning, W., Wei, J., Katz, S., Litwin, S. and Douglas, P. (1994). *Am. J. Physiol.* **266**, H1672–H1675.

McLean, M. and Prothero, J. (1991). *Am. J. Anat.* **192**, 425–441.

Mercadier, J., Samuel, J., Michel, J., Zongazo, M., de la Bastie, D., Lompré, A., Wisnewsky, C., Rappaport, L., Levy, B. and Schwartz, K. (1989). *Am. J. Physiol.* **257**, H979–H987.

Michael, L.H., Entman, M.L., Hartley, C.J., Youker, K.A., Zhu, J., Hall, S.R., Hawkins, H.K. and Ballantyne, C.M.

(1995). *Am. J. Physiol. Heart Circ. Physiol.* **269**, H2147–H2154.

Michael, L.H., Ballantyne, C.M., Zachariah, J.P., Gould, K.E., Pocius, J.S., Taffet, G.E., Hartley, C.J., Pham, T.T., Daniel, S.L., Funk, E. and Entman, M.L. (1999). *Am. J. Physiol. Heart. Circ. Physiol.* **277**, H660–H668.

Minamino, T., Yujiri, T., Terada, N., Taffet, G., Michael, L., Johnson, G. and Schneider, M. (2002). *Proc. Natl. Acad. Sci. USA* **99**, 3866–3871.

Mitchell, G., Jeron, A. and Koren, G. (1998). *Am. J. Physiol.* **274**, H747–H751.

Molkentin, J. and Dorn, I. (2001). *Annu. Rev. Physiol.* **63**, 391–426.

Mulvagh, S., Michael, L., Perryman, B., Roberts, R. and Schneider, M. (1987). *Biochem. Biophys. Res. Commun.* **147**, 627–636.

Nossuli, T.O., Lakshminarayanan, V., Baumgarten, G., Taffet, G.E., Ballantyne, C.M., Michael, L.H. and Entman, M.L. (2000). *Am. J. Physiol. Heart Circ. Physiol.* **278**, H1049–1055.

Nossuli, T.O., Frangogiannis, N.G., Knuefermann, P., Lakshminarayanan, V., Dewald, O., Evans, A.J., Peschon, J., Mann, D.L., Michael, L.H. and Entman, M.L. (2001). *Am. J. Physiol.* **281**, H2549–H2558.

Oh, H., Bradfute, S.B., Gallardo, T.D., Nakamura, T., Gaussin, V., Mishina, Y., Pocius, J., Michael, L.H., Behringer, R.R., Garry, D.J., Enman, M.L., Schneider, M.D. (2003). *Proc. Natl. Acad. Sci. USA* **100**, 12313–12318.

Opavsky, M.A., Martino, T., Rabinovitch, M., Penninger, J., Richardson, C., Petric, M., Trinidad, C., Butcher, L., Chan, J. and Liu, P.P. (2002). *J. Clin. Invest.* **109**, 1561–1569.

Orlic, D., Kajstura, J., Chimenti, S., Jakoniuk, I., Anderson, S.M., Li, B., Pickel, J., McKay, R., Bernardo, N.-G., Bodine, D.M., Leri, A. and Anversa, P. (2001). *Nature* **410**, 701–705.

Osada, J., Joven, J. and Maeda, N. (2000). *Curr. Opin. Lipidol.* **11**, 25–29.

Pollick, C., Hale, S. and Kloner, R. (1995). *J. Am. Soc. Echocardiogr.* **8**, 602–610.

Pyo, R., Lee, J.K., Shipley, J.M., Curci, J.A., Mao, D., Ziporin, S.J., Ennis, T.L., Shapiro, S.D., Senior, R.M. and Thompson, R.W. (2000). *J. Clin. Invest.* **105**, 1641–1649.

Richards, A.G., Simonson, E. and Visscher, M.B. (1953). *Am. J. Physiol.* **174**, 293–298.

Rockman, H.A., Ono, S., Ross, R.S., Jones, L.R., Karimi, M., Bhargava, V., Ross, J., Jr. and Chen, K.R. (1994). *Proc. Natl. Acad. Sci. USA* **91**, 2694–2698.

Ruff, J., Wiesmann, F., Hiller, K.H., Voll, S., vonKienlin, M., Bauer, W.R., Rommel, E., Neubauer, S., Haase, A. (1998). *Magn. Reson. Med.* **40**, 43–48.

Ruff, J., Wiesmann, F., Lanz, T. and Haase, A. (2000). *J. Magn. Reson.* **146**(2), 290–296.

Sano, M., Abdellatif, M., Oh, H., Xie, M., Bagella, L., Giordano, A., Michael, L., DeMayo, F. and Schneider, M. (2002). *Nat. Med.* **8**, 1310–1317.

Scherrer-Crosbie, M., Steudel, W., Hunziker, P., Foster, G., Garrido, L., Liel-Cohen, N., Zapol, W. and Picard, M. (1998). *Circulation* **98**, 1015–1021.

Scherrer-Crosbie, M., Steudel, W., Ullrich, R., Hunziker, P., Liel-Cohen, N., Newell, J., Zroff, J., Zapol, W. and Picard, M. (1999). *Am. J. Physiol.* **277**, H986–H992.

Schiaffino, S., Samuel, J.L., Sassoon, D., Lompré, A.M., Garner, I., Marotte, F., Buckingham, M., Rappaport, L. and Schwartz, K. (1989). *Circ. Res.* **64**, 937–948.

Slawson, S., Roman, B., Williams, D. and Koretsky, A. (1998). *MRM* **39**, 980–987.

Soonpaa, M., Kim, K., Pajak, L., Franklin, M. and Field, L. (1996). *Am. J. Physiol.* **271**, H2183–H2189.

Spirito, P., Serdman, C.E., McKenna, W. and Mason, B.J. (1997). *N. Engl. J. Med.* **336**, 775–785.

Stenbit, A.E., Katz, E.B., Chatham, J.C., Geenen, D.L., Factor, S.M., Weiss, R.G., Tsao, T.S., Malhotra, A., Chacko, V.P., Ocampo, C., Jelicks, L.A. and Charron, M.J. (2000). *Am. J. Physiol. Heart Circ. Physiol.* **279**, H313–H318.

Sze, R.W., Chan, C.B., Dardzinski, B.J., Dunn, S., Sanbe, A., Schmithorst V., Robbins, J., Holland, S.K. and Strife, J.L. (2001). *Pediatr. Radiol.* **3192**, 55–61.

Taffet, G.E., Hartley, C.J., Wen, X., Pham, T.T., Michael, L.H. and Entman, M.L. (1996). *Am. J. Physiol. Heart Circ. Physiol.* **270**, H2204–H2209.

Takeishi, Y. and Walsh, R. (2001). *Acta Physiol. Scand.* **173**, 103–111.

Tanaka, N., Dalton, N., Mao, L., Rockman, H.A., Peterson, K.L., Gottshall, K.R., Hunter, J.J., Chien, K.R. and Ross, J. (1996). *Circulation* **94**, 1109–1117.

Toma, C., Pittenger, M., Cahill, K., Byrne, B. and Kessler, P. (2002). *Circulation* **105**, 93–98.

Vaidya, D., Morley, G., Samie, F. and Jalife, J. (1999). *Circ. Res.* **85**, 174–181.

Vikstrom, K., Bohlmeyer, T., Factor, S. and Leinwand, L. (1998). *Circ. Res.* **82**, 773–778.

Wakimoto, H., Maguire, C.T., Kovoor, P., Hammer, P.E., Gehrmann, J., Triedman, J.K. and Berul, C.I. (2001). *Cardiovasc. Res.* **50**, 463–473.

Wang, Y.X., Halks-Miller, M., Vergona, R., Sullivan, M.E., Fitch, R., Mallari, C., Martin-McNulty, B., Da Cunha, V., Freay, A., Rubanyi, G.M. and Kauser, K. (2000). *Am. J. Physiol. Heart Circ. Physiol.* **278**, H428–H434.

Wehrens, X.H., Kirchhoff, S. and Doevendans, P.A. (2000). *Cardiovasc. Res.* **45**, 231–237.

Westerhoff, N. (2002). In *Recent Progress in Cardiovascular Mechanics* (eds S. Hosoda, M. Yaginuma, M. Sugawara, M. Taylor and C. Caro), pp. 115–127. Harwood Academic Publishers, New York.

Wiesmann, F., Ruff, J., Hiller, K.H., Rommel, E., Haase, A. and Neubauer, S. (2000). *Am. J. Physiol. Heart Circ. Physiol.* **278**, H652–H657.

Williams, R.V., Lorenz, J.N., Witt, S.A., Hellard, D.T., Khoury, S.F. and Kimball, T.R. (1998). *Am. J. Physiol. Heart Circ. Physiol.* **274**, H1828–H1835.

Youn, H.J., Rokosh, G., Lester, S.J., Simpson, P., Schiller, N.B. and Foster, E. (1999). *J. Am. Soc. Echocardiogr.* **12**, 70–75.

Zhang, D., Gaussin, V., Taffet, G.E., Belaguli, N.S., Yamada, M., Schwartz, R.J., Michael, L.H., Overbeek, P.A. and Schneider, M.D. (2000). *Nat. Med.* **6**, 556–563.

CHAPTER 14

Respiratory Tract

Armin Braun, Heinrich Ernst, Heinz G Hoymann, and Susanne Rittinghausen
Fraunhofer Institute of Toxicology and Experimental Medicine
Hannover, Germany

Introduction

The laboratory mouse has developed into the preferred model system for biomedical research. Many investigators have turned to murine models of lung disease for several reasons. First, dense genetic and physical maps of the murine genome have been constructed and are available to the public. Because of similar **synteny** in murine and human genomes, knowledge of a defect in a murine gene leading to a disease phenotype may elucidate corresponding human genes responsible for that genotype. Second, the understanding of murine immunology has dramatically increased and inflammatory reactions have been analyzed in detail. Third, technology has been developed that allows the programmed overexpression of target genes or the functional ablation of their protein products (Drazen *et al.*, 1999). Therefore the murine lung has become the focus of basic research, toxicology and drug development. New measurement techniques for physiological parameters e.g. lung function measurements have been developed. Although the mouse is genetically very closely related

to man, lung anatomy and function differs significantly (Table 14.1). In addition there are striking differences in the physiology of the lung between mice and men. For example, murine airway mast cells respond to a

TABLE 14.1: Comparison between murine and human lung

	Mouse	Human
anatomy	right: 4 lobes	right: 3 lobes
	left: 1 lobe	left: 2 lobes
diameter main bronchus	1 mm	10–15 mm
diameter bronchioli	0.01–0.05 mm	<1 mm
diameter terminal bronchioli	0.01 mm	0.6 mm
diameter respiratory bronchioli	not existent	0.5 mm
diameter alveoli	0.00039–0.0069 mm	0.2–0.4 mm

The Laboratory Mouse
Copyright 2004 Elsevier
ISBN 0-1233-6425-6

variety of stimuli like substance P and compound 48/80, these having no effect on human airways, with the release of serotonin rather than histamine.

In spite of these differences, the mouse is widely used for asthma, tumour, and chronic obstructive pulmonary disease (COPD) research. For these models the standard histological, molecular biology and immunological tools that are available for the mouse were adapted to the lung. In contrast, the lung function measurement is extremely expensive due to the very small dimensions and the high breathing frequency. In this chapter we will describe common pathological lesions of the respiratory tract and the state of the art methods to analyze the physiology of the mouse lung and the main applications in biomedical research.

Physiological measurements

Determination of inflammation

Much of our current understanding of the immune reaction in the lung has been derived from studies carried out in mice. Analysis of cytokines, chemokines, growth factors and also cellular differentiation can easily be performed in the broncho-alveolar lavage fluid (BAL). After sacrificing the animals by cervical dislocation or an overdose of anesthetic, the trachea is canulated and airways are lavaged two times with 0.8 ml ice-cold saline containing proteinase inhibitor. BAL fluids from each mouse are pooled and the recovered volume and total cell number are determined. The cells can be analyzed by flow cytometry and/or by analyzing cytospin preparations using hematoxylin and eosin (H&E) staining. They can then be classified by light microscopy according to common morphological criteria. The cell-free supernatants can be stored and analyzed for cytokine content. Broncho-alveolor lavage fluid is a very powerful tool for analyzing acute inflammatory reactions in infectious diseases and allergic immune reactions. It mainly represents the processes in the airways so that to acquire the processes in the lung mucosa other methods like measurements in lysates, enzymatic lung digestion followed by cellular analysis and histological staining need to be performed (Braun et al., 1999).

Generation and deposition of aerosol and particles in the murine lung

For toxicological and pharmacological studies, the deposition of aerosols in the lung is very critical. Since the murine airway architecture is very different in comparison to the human, aerosol deposition is different. Humans are able to inhale all particles in an aerosol with a diameter of up to $7 \mu m$ (total respiratory tract deposition). Small laboratory animals would only inhale about 55% of an aerosol of $7 \mu m$ aerodynamic diameter. Inhalability in the animals is predicted to be 95% or greater for particles up to approximately $0.7 \mu m$ diameter (Menache et al., 1995). In conscious mice, for particles with 1 and $6 \mu m$ the total respiratory tract deposition is 28% and 54% but the total lung deposition due to the filter function of the nose is only 16% and 1.8% (Raabe et al., 1988). In contrast, in intubated animals lung deposition can be markedly increased (to more than 60%).

Therefore, aerosol generation is very critical for lung function measurements with aerosols. They can be generated by various dispersing and condensing processes. Particle generation starts from bulk material which is either liquid or powder. For dispersion, different physical mechanisms for breaking adhesive forces are employed. Nebulization of liquids can be achieved by interaction with a pressurized gas or by focusing ultrasonic energy on to the surface of the liquid (Koch, 1998).

Methods for measuring lung function

The measurement of pulmonary function in mice is very sophisticated due to the small dimensions and the resulting technical challenge. Table 14.2 gives an idea of the difference in respiratory parameters between mice and humans. For COPD and asthma research, the broncho-obstruction is a frequently used target parameter. In asthma models, airway hyperreactivity (AHR), defined as the unspecific broncho-obstruction in response to pharmacological stimuli like methacholine, histamine or serotonine, is widely used (Drazen et al., 1999). The techniques available for the measurement of airway functions are given in Table 14.3.

Depending on the method chosen for AHR-measurement, different pathways can be distinguished: (a) altered neuronal regulation of airway tone,

(b) increases in muscle content or function, and (c) increased epithelial mucus production and airway oedema (Table 14.4; Wills-Karp, 1999).

In vitro *EFS*

It has been demonstrated that *in vitro* EFS of tracheal segments specifically reflects neuronal airway obstruction. Administration of both atropin (disruption of cholinergic pathways) and capsaicin (depletion of sensory neurons) completely blocks responsiveness of tracheal segments to EFS (Andersson and Grundstrom, 1983; Ellis and Undem, 1992).

Airway smooth muscle responsiveness can be assessed by EFS as described in Figure 14.1; (Larsen *et al.*, 1992;

Braun *et al.*, 1998). Tracheal smooth muscle segments (~0.5 cm) are removed and hung between stainless steel wire triangular supports. The contraction in response to EFS stimulus (12 V, 200 mA, 0.5–30 Hz) is measured via an isometric force transducer. The frequency that causes 50% of the maximal contraction is calculated from logarithmic plots of the contractile response versus the frequency of EFS, and expressed as the ES_{50}.

Ex vivo *lung function measurement in the isolated perfused lung*

Perfusion of murine lungs has only rarely been reported and was restricted mostly to toxicological investigations. Recently the group of Uhlig adapted the technique of

TABLE 14.2: Respiratory parameters in mice and humans

Parameter	Mouse	Human	Unit
Tidal volume	0.16–0.20	500	ml
Respiratory rate	215–230	12	min^{-1}
Minute ventilation	33.5–47.5	6 000	ml/min
Total lung capacity	0.9–1.44	6 000	ml
Residual volume	0.11–0.14	1 500	ml
Lung compliance	0.053–0.13	200	ml/cmH_2O
Airway resistance	1.5	0.0016	$cmH_2O/ml^{-1}*s^{-1}$
Pa_{CO_2}	34–35	40	mmHg
Pa_{O_2}	78–84	80–100	mmHg
pH	7.37	7.4	

Source: Adapted from Rao and Verkman, 2000.

TABLE 14.4 : Target cells of different frequently used stimuli in lung function measurements

Stimulus	Effector cells	Major pathways postulated
EFS	sensory neurons	unspecific
	motor neurons	depolarization
Methacholine	smooth muscle cells	M_3 receptors
Histamine	smooth muscle cells	H_1 receptors
	sensory neurons	
	motor neurons	
Serotonin	sensory neurons	$5 HT_1$ receptors
	motor neurons	$5 HT_3$ receptors
Capsaicin	sensory neurons	Vanilloid receptor
hypotonic H_2O	sensory neurons	unspecific

TABLE 14.3: Frequently used methods available for the measurement of airway mechanics in the mouse

In vitro/ex vivo	Invasive/in vivo	Non-invasive/in vivo
isolated airway segments e.g. electrical field stimulation (EFS)	airway pressure measurements during mechanical ventilation	dual chamber plethysmography
isolated lung	flow and pressure measurements during spontaneous respiration, pulmonary resistance, and compliance	head-out plethysmography (EF_{50})
Precision cut lung slice (PCLS) e.g. video analysis		barometric plethysmography (PenH)

Source: Adapted from Drazen *et al.*, 1999.

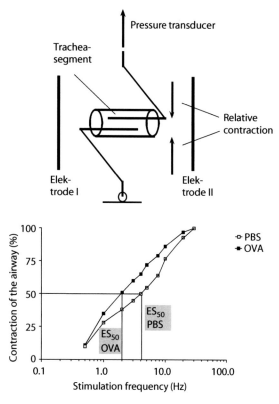

Figure 14.1 Schematic drawing of lung function measurement by EFS.

the isolated perfused lung to the species mouse as an expansion of the one that has previously described in detail for the isolated perfused rat lung (Held *et al.*, 1999, 2001; Held and Uhlig, 2000a,b). The experimental set up for the isolated perfused mouse lung is given in detail by von Bethmann *et al.* (1998). Use of the perfused murine lung allows study of several important features of lung physiology: (1) respiratory mechanics including pulmonary resistance and compliance (therefore measurements of bronchoconstriction and AHR can be made; Held and Uhlig, 2000); (2) vascular responsiveness (Held *et al.*, 1999); and (3) mediator release e.g. cytokines such as Tumor necrosis factor-alpha (TNF-a) and interleukin (IL)-6 (von Bethmann *et al.*, 1998;) (Held *et al.*, 2001).

Ex vivo *PCLS in combination with video microscopy*

A recently developed method for studying pulmonary responses is the use of PCLS, allowing microscopic investigation of the constriction (Martin *et al.*, 1996; Held *et al.*, 1999). This method was validated against lung function measurements using the isolated perfused lung and can be used for the characterization of

airway and vascular responses in the murine lung (Held *et al.*, 1999). A major advantage of this method is reduced use of animals since up to 30 slices can be obtained from one lung. In addition, functional pharmacological responses of small as well as large airways under cell culture conditions can be made.

In vivo *invasive lung function measurement*

The invasive lung function measurement is the golden standard for exact determination of lung function as pulmonary resistance and dynamic compliance can be measured (Figure 14.2). They are derived by calculating breath by breath continuous data for tidal airflow, tidal volume and transpulmonary pressure (Martin *et al.*, 1988). Measurements in spontaneously breathing animals after endotracheal intubation via the oral route are closer to physiological conditions than in ventilated animals and allow repeated studies. This technique has been extensively published for rats (Likens and Mauderly, 1982; Costa and Tepper, 1988; Hoymann and Heinrich, 1998) and has been adapted to the mouse quite recently (Brown *et al.*, 1999). Well defined aerosol treatments or challenges can be performed. The major advantages are: (1) spontaneously breathing animals, (2) diagnostic precision, (3) high deposition and precise dosage of aerosols for drug treatment or challenges, (4) exclusion of the nasal passages when the lung is the target of interest, and (5) repeated measurements are possible. One disadvantage is that extensive measurements of transpulmonary pressure and tidal airflow require extensive technical equipment. In addition, the animals must be anesthetized. Anesthetic agents may alter the lung function due to changes in neuronal function (Drazen *et al.*, 1999). Airway responses in the intact animal depend on airway smooth muscle contractility, chest wall compliance, bronchiolar mucus plugging, airway fibrosis and other factors (Leong and Huston, 2001). Airway responses e.g. AHR are quantitated by measuring the amount of a contractile stimulus e.g. methacholine or histamine required to achieve a certain contractile response.

In vivo *non-invasive lung function head out plethysmography*

An alternative way to measure lung function in the mouse is the head-out body **plethysmography** (Figure 14.3; Neuhaus-Steinmetz *et al.*, 2000; Glaab *et al.*, 2001). Originally developed for toxicological studies the system was adapted for asthma research (Vijayaraghavan

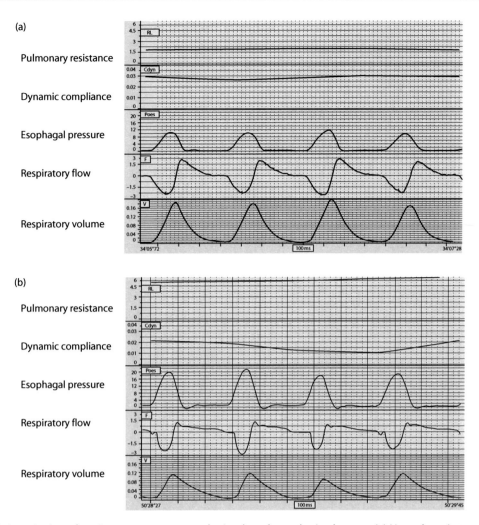

Figure 14.2 Invasive lung function measurement on the intubated anesthetized mouse. (a) Normal respiratory pattern before challenge: base values of resistance, dynamic compliance, esophagus pressure, respiratory flow and volume (from top to bottom), the x axis shows the time (1 mark corresponds to 1/10 s); (b) signals immediately after inhalation challenge with ovalbumin: increase of resistance and esophagus pressure as well as decrease of respiratory flow and volume (figure by H.G. Hoymann, Fraunhofer Institut, Hannover, Germany).

et al., 1993, 1994; Braun *et al.*, 1999). Up to four mice can be placed in four body plethysmographs attached to an exposure chamber (Figure 14.3). Airflow is measured by means of a differential pressure transducer (Table 14.5). For the determination of bronchoconstriction; mid-expiratory airflow (EF_{50}), i.e. the expiratory airflow when 50% of the tidal volume is exhaled, is calculated (Figure 14.4). Changes of EF_{50} in response to bronchoconstricting agonists e.g. methacholine or to allergen delivered by an aerosol generator can be measured during aerosol exposure. The major advantages of this system are: (1) spontaneously breathing animals; (2) simultaneous analysis of 4 animals; (3) no anesthesia required; (4) lung function measurement during aerosol challenge; and (5) repeated measurements are possible. Disadvantages: The classical lung function

parameters e.g. lung resistance and compliance are not available and nasal passage of the inhaled air filters a major part of the delivered aerosols.

In vivo *non-invasive lung function whole body plethysmography*

For barometric plethysmography, unrestrained animals are placed in a chamber and the pressure fluctuations that occur due to breathing are recorded (Figure 14.5; Hamelmann *et al.*, 1997). The pressure difference between the main chamber containing the animal and a reference chamber is measured. These box pressure changes are considered to be caused by volume and resultant pressure changes in the main chamber during the respiratory cycle of the animal. From the resulting

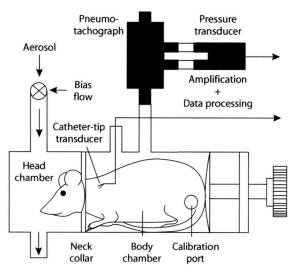

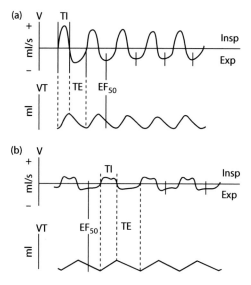

Figure 14.3 Schematic drawing of the head-out body plethysmograph illustrating the exposure system and the equipment used for the measurement of tidal midexpiratory flow (EF50), tidal volume (VT), expiratory time (TE), inspiratory time (TI), and breathing frequency (f) and for monitoring of dynamic lung compliance (Cdyn) and pulmonary conductance (GL). The system was operated with the addition of a catheter-tip differential pressure transducer to measure transpulmonary pressure (Ptp). The conscious animal was placed in a glass plethysmograph that was attached to a head exposure chamber. The output of a jet nebulizer was directed to the inlet of the head exposure chamber that was continuously ventilated with a bias flow of 0.2 l/min. Before data collection, mice were allowed to acclimatize for 15 min in the body plethysmograph (adapted from Glaab et al., 2001).

Figure 14.4 Head-out plethysmography: Characteristic modifications to the normal breathing pattern in unanesthetized BALB/c mice. (a) Normal breathing pattern of BALB/c mice while breathing room air; (b) characteristic pattern of airway obstruction during aerosol challenge with methacholine, illustrating the decline in EF50. (a) and (b), top tracings: pneumotachograph airflow signals recorded at 40 mm/s. (a) and (b), bottom tracings: corresponding integrated VT signal as calculated by the computer program from the collected voltage digitalizations. A horizontal line at 0 flow separates inspiratory (Insp; upward; 1) from expiratory (Exp; downward; 2) airflow. V, tidal flow (adapted from Glaab et al., 2001).

TABLE 14.5: Lung function measurement by head out body plethysmography of frequently used mice strains at 10 weeks of age						
	EF50 (ml/min)		Tidal volume (ml)		Frequency (1/min)	
	Mean	Range (±)	Mean	Range (±)	Mean	Range (±)
CBAJ	1.89	0.91	0.19	0.09	226.40	107.92
C57BL/6J	1.97	0.90	0.17	0.08	258.74	122.08
BALB/C	1.92	0.90	0.14	0.06	267.63	127.00
Source: R. Bälder, Fraunhofer Institut, Hannover, Germany.						

signal several parameters such as expiration time or relaxation time can be calculated. For the determination of broncho-obstruction the enhanced pause (Penh), a dimensionless parameter, is calculated (Hamelmann et al., 1997). The method is widely used for the determination of hyperreactivity in asthma models (Figure 14.6; Drazen et al., 1999; Finotto et al., 2001), but the physiological meaning of Penh is considered controversial

(Mitzner and Tankersley, 1998). The advantages of the method are: (1) measurement in conscious, spontaneously breathing animals; (2) no anesthesia required; (3) lung function measurement during aerosol challenge, (4) repeated measurements are possible; and (5) measurements are easy to perform with commercially available equipment (e.g. Buxco). Disadvantages: Classic lung function parameters e.g. lung resistance

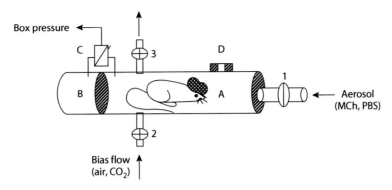

Figure 14.5 Schematic diagram of the whole-body plethysmograph. (A) Main chamber containing the mouse; (B) reference chamber; (C) pressure transducer connected to analyzer; (D) pneumotacho-graph; (1) main inlet for aerosol closed by valve; (2) inlet for bias flow with four-way stopcock; (3) outlet for aerosol with four-way stopcock (adapted from Hamelmann *et al.*, 1997).

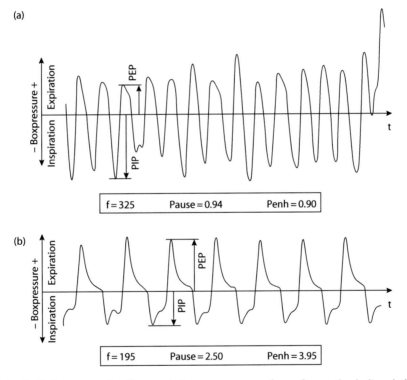

Figure 14.6 Whole-body Plethysmography: Changes in box pressure waveform after methacholine challenge. Waveform of the box pressure signal derived from a normal mouse after 3 min of nebulization with (a) aerosolized PBS or (b) aerosolized methacholine (50 mg/ml in PBS). f 5 respiratory rate (breaths/min); Pause, Penh (enhanced pause), PIP 5 peak inspiratory pressure (ml/s), maximal negative box pressure occurring in one breath; PEP 5 peak expiratory pressure (ml/s), maximal positive box pressure occurring in one breath; f 5 frequency (breaths/min) (adapted from Hamelmann *et al.*,1997).

and compliance are not available and the physiological interpretation of the data is not exactly defined.

Lung diseases and pathology

Murine models of asthma

Human asthma is characterized by variable airflow obstruction in response to allergen (early- and late phase response) and airway hyperreactivity. Structurally, the airways of asthmatics are characterized by the presence of chronic allergic inflammation with intense infiltration of the bronchial mucosa by lymphocytes (especially T-cells type Th2) and eosinophils, accompanied by epithelial desquamation, goblet cell hyperplasia and thickening of the submucosa (airway remodeling) (Wills-Karp, 1999). An animal model of asthma must reproduce these features of the human disease to be credible.

Several murine models that show particular features of asthma have been developed. In standard protocols, inbred mice are first sensitized to allergen systemically and then challenged by aerosol. The following are critical for the induction of an asthma-like phenotype with allergic airway inflammation and AHR: (1) Selection of the antigen: frequently used allergens are proteins (e.g. ovalbumin and Aspergillus fumigatus extract) or microorganisms (e.g. Aspergillus fumigatus, Schistosoma mansoni egg); (2) Selection of the mouse strain: the choice of the mouse strain is very dependent on the allergen used. Balb/c mice are often used for ovalbumin protocols; and (3) Selection of the protocol: the protocol should ascertain a reproducible sensitisation against the allergen (e.g. repeated systemic injection of the allergen associated with an adjuvant such as alum) and the induction of a long lasting allergic airway inflammation (e.g. repeated aerosol challenges over weeks). An excellent overview of the most commonly used protocols is given in Lloyd et al., (2001). The acute models are very effective in inducing acute airway inflammation – mainly consisting of Th2 cells and eosinophils – and AHR (Braun et al., 1998; Hogan et al., 1998; Hansen et al., 1999). In order to induce a chronic inflammation, the protocols were developed further. Multiple aerosol challenges and/or live Aspergillus fumigatus conidia were used to induce a long lasting (chronic) inflammation in the airways that was associated with airway remodeling and AHR (Braun et al., 1999; Kumar and Foster, 2001; Blease et al., 2002; Kumar et al., 2002; Schuh et al., 2002). Recent progress in developing these models and improvement in lung function measurements has made it possible to acquire early- as well as late-phase responses due to allergen provocation in mice (Cieslewicz et al., 1999 546; Neuhaus-Steinmetz et al., 2000). Most of those new therapeutical strategies for asthma, that are designed to neutralize central mediators in asthma pathology, were developed and tested in murine asthma models. Prominent examples are anti-IgE, anti-interleukin(IL)-5, anti-IL-13 and anti-eotaxin (for a review see Lloyd et al., 2001).

Murine models for COPD

Chronic obstructive pulmonary disease may be defined as chronic airflow obstruction due to a mixture of emphysema and peripheral airway obstruction from chronic bronchitis. The main causes of COPD are cigarette smoking and air pollution (e.g. sulfur oxide and particulates). The disease is associated with inflammatory changes and the infiltration of neutrophils and macrophages into the lung (Barnes, 2002). Animal models for COPD are extremely difficult to establish since the human disease develops over decades. Exposure to tobacco smoke may induce emphysema. Other possibilities are the use of genetically altered animals (see also Alveolar about emphysema above). There are some natural mutations such as the Blotchy mouse that develops a connective tissue disorder, or knockout animals. Knockout models for emphysema include those for Platelet Derived Growth Factor A (PDGFA) and the double knockout for Fibroblast Growth Factors 3 and 4 (FGF3 and 4; for review see Dawkins and Stockley, 2001).

Pathology of the respiratory tract

Non-neoplastic lesions

In the following section some of the more common spontaneous non-neoplastic lesions occurring in the respiratory tract of mice are briefly addressed. Specific infectious diseases are not included in this chapter and dealt with elsewhere in this book.

Nasal cavity, larynx, trachea

Epithelial hyperplasia   Hyperplastic changes are usually observed as adaptive response to irritants and may involve all the cell types that are normally found in the upper respiratory tract, i.e. cells of the squamous, transitional, respiratory and olfactory epithelia as well as basal cells, neuroendocrine cells and cells of the subepithelial glands (Herbert and Leininger, 1999; Dungworth et al., 2001). Epithelial hyperplasia is often associated with degeneration and inflammation, but is frequently reversible. Hyperplasia is characterized by a thickening of the epithelium due to an increase in the number of the respective cell type and may result in an undulating rugose surface. In mucous (goblet) cell hyperplasia, mucosal invagination with formation of intraepithelial 'pseudoglands' may occur. Epithelial hyperplasia can also include proliferation of atypical or pleomorphic basal or undifferentiated cells, but without disruption of the underlying basal lamina (Dungworth et al., 2001).

Squamous cell metaplasia   Adaptive squamous cell metaplasia (Figure 14.7) can occur within areas of epithelial regeneration or hyperplasia and characterizes the replacement of the more susceptible respiratory or

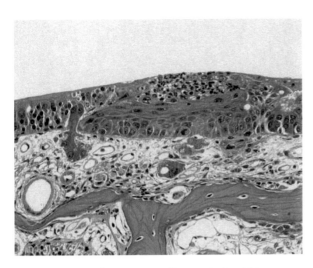

Figure 14.7 Focal squamous cell metaplasia within olfactory epithelium of nasal cavity.

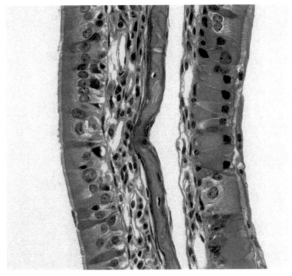

Figure 14.8 Nasal septum with normal respiratory epithelium (left side) and cells showing hyaline eosinophilic cytoplasmic inclusions (eosinophilic globules) and condensed peripheral nuclei (right side).

olfactory epithelium by the more resistant squamous epithelium (Monticello *et al.*, 1990). In the larynx, squamous cell metaplasia affects mainly the base of the epiglottis, the area of transition from stratified squamous to respiratory epithelium (Renne *et al.*, 1992).

Respiratory epithelial metaplasia After damage, the olfactory epithelium may be replaced by a columnar epithelium, with or without cilia, that resembles respiratory epithelium. In the area of respiratory epithelial metaplasia, subepithelial glands are also usually replaced by respiratory epithelium (Dungworth *et al.*, 2001).

Eosinophilic globules (hyalinosis) Eosinophilic proteinaceous inclusions are commonly seen in all types of hyperplastic and non-hyperplastic nasal epithelial cells (Figure 14.8). They were initially described as 'eosinophilic globules' (Buckley *et al.*, 1985), are age-related in occurrence and are thought to represent either a non-specific adaptive response (Herbert and Leininger, 1999) or a degenerative change (Dungworth *et al.*, 2001). In advanced cases, these eosinophilic inclusions may be associated with intra- and/or extra-cellular crystal formation and are also found in other organs including the lung (see below). Recent studies have identified the eosinophilic proteins as Ym1 and Ym2 (members of the chitinase family; Guo *et al.*, 2000; Ward *et al.*, 2001).

Lung

Chronic passive congestion Cardiac failure or age-related thrombogenic changes in blood constituents may lead to spontaneous left atrial thrombosis, which is generally accompanied by pathological lung alterations

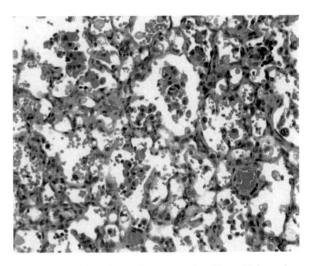

Figure 14.9 Chronic passive congestion. Note thickened edematous and slightly collagenized interalveolar septa and alveoli containing red blood cells and macrophages with phagocytosed erythrocytes.

referred to as 'chronic passive congestion' (Figure 14.9). The pathological spectrum of small and acute atrial thrombi includes interstitial oedema, intra-alveolar serofibrinous exudate, hemorrhage and **hemosiderin**-laden macrophages that stain positive for iron (so-called 'heart failure cells'). Larger and chronic thrombi may also cause mixed interstitial and alveolar inflammatory cell infiltration and interstitial fibrosis with a similar appearance to the end stage of chronic fibrosing alveolitis or interstitial pneumonia (Faccini *et al.*, 1990).

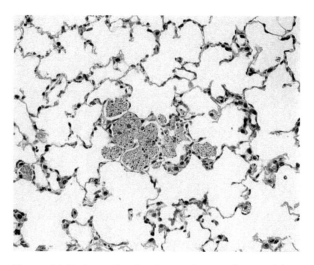

Figure 14.10 Intra-alveolar cluster of macrophages with numerous cytoplasmic pigment granules. Note the absence of a significant septal response.

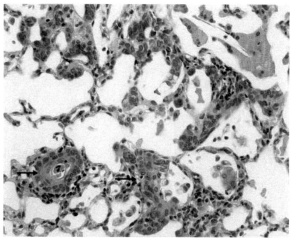

Figure 14.11 Area of bronchiolo-alveolar hyperplasia (alveolar type). Note focal squamous cell metaplasia (arrows) and foamy macrophages surrounding a cholesterol crystal cleft (asterisk).

Pigmentation   Small intra-alveolar aggregates of macrophages containing dark brown granules (Figure 14.10) are sometimes observed in the lungs of old mice. These cells stain intensely for iron and less reliably for lipofuscin and are not usually accompanied by a proliferative or inflammatory response. Pathogenetically, there seems to be an association with phagocytosed erythrocytes (hemosiderin) from preceding minor hemorrhages. However, in aged mice, accumulation of lipofuscin resulting from peroxidation of phagocytosed surfactant lipid may also contribute to the cytoplasmic pigmentation.

Alveolar histiocytosis   Alveolar histiocytosis describes focal accumulations of alveolar macrophages (histiocytes) containing foamy lipid material (Ernst *et al.*, 1996). This lesion occurs mainly in peripheral regions of the lung and may be associated with mixed inflammatory cell infiltration, focal fibrosis and sometimes in association with cholesterol clefts (Figure 14.11). If the lesion consists mainly of cholesterol crystals surrounded by epithelioid macrophages and fibrosis, the term 'cholesterol granuloma' may be used.

Alveolar lipoproteinosis   Alveolar lipoproteinosis is a rare finding in mice and is characterized by the presence of intra-alveolar phospholipid derived from alveolar surfactant together with smaller amounts of other proteins from dying foamy macrophages or plasma proteins derived from leaky vessels, as has been shown in studies with the rat (Heppleston *et al.*, 1972; Adamson and Bowden, 1984). Lipoproteinosis usually involves accumulation of intra-alveolar foamy macrophages (alveolar histiocytosis) as well, especially if the pathogenesis is a defective or decreased clearance of the lipoproteinaceous

material by alveolar macrophages. In the case of excessive production of surfactant, hypertrophic/hyperplastic alveolar type II cells are observed. Probably due to a defect in surfactant homeostasis, immunodeficient SCID mice (C.B-1-*Prkdc*^{scid}) and SCID-beige (*Prkdc*^{scid} *Lyst*^{bg}) mice develop spontaneous alveolar lipoproteinosis with marked increases in phosphatidylcholine and surfactant proteins A and B in the alveolar lipoproteinaceous material (Jennings *et al.*, 1995; Warner and Balish, 1995). Surfactant proteins and phospholipids also accumulate in the alveolar spaces of mice deficient in granulocyte-macrophage colony-stimulating factor (GM-CSF; Lieschke *et al.*, 1994; Hallman and Merritt, 1996).

Crystalline inclusions and deposits (acidophilic macrophage pneumonia)   Endogenously formed eosinophilic crystalline inclusions within the cytoplasm of macrophages and epithelial cells or extracellular crystals have been reported in the respiratory system of several strains of mice. Ultrastructurally, crystals within the cytoplasm of alveolar macrophages were needle-shaped, showed a lattice pattern and appeared to lie in membrane-bound vacuoles (Murray and Luz, 1990). The crystals stain strongly eosinophilic with H & E, are black-blue with toluidine blue and Alcian blue- and periodic acid-schiff (PAS)-negative (Rehm *et al.*, 1985). Crystal-laden macrophages or free-lying intra-alveolar crystals are often associated with bronchopneumonias or pulmonary neoplasms. In aged mice, crystalline deposits may also be seen within gland-like infoldings of the tracheobronchial epithelium. The so-called 'acidophilic macrophage pneumonia' (Figure 14.12) is characterized

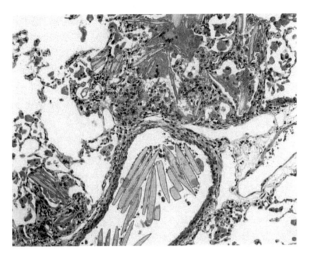

Figure 14.12 Accumulation of macrophages with crystalline inclusions in 'acidophilic macrophage pneumonia'. Some needle-like and rectangular crystals are lying free within airspaces.

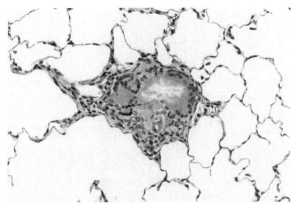

Figure 14.13 Well-demarcated foreign-body granuloma showing plant fragments surrounded by macrophages and multinucleated giant cells.

by the simultaneous presence of crystal-laden macrophages, alveolar crystalline deposits and other inflammatory cells (Rehm *et al.*, 1985; Murray and Luz, 1990).

The composition of the crystalline material may be very heterogeneous. When related to pulmonary hemorrhage, the crystals stain strongly positive for hemoglobin (Frith *et al.*, 1983) or are morphologically indistinguishable from hematoidin, which results from hemoglobin breakdown following uptake of erythrocytes by alveolar macrophages (Shultz *et al.*, 1984). Other crystalline inclusions contained various amounts of alpha-1-antitrypsin, IgG and IgA (Rijhsinghani *et al.,* 1988). It has been suggested that crystalloid inclusions of alveolar macrophages in acidophilic macrophage pneumonia may have been derived from breakdown products of granulocytes, especially eosinophils (Murray and Luz, 1990).

In homozygous C57BL/6J viable motheaten (*Hcph^{me}*) mice, pulmonary crystal formation represents a major feature of the fatal lung inflammation induced by macrophage dysregulation (Ward, 1978; Khaled *et al.*, 1999; Guo *et al.*, 2000). The alveolar macrophages of these animals, which have a mean lifespan of about 10 weeks, produce significantly increased levels of TNF-a (Thrall *et al.*, 1997). Mass spectrometric analysis of proteins from BAL of moth-eaten mice has identified the crystals as Ym1 (T-lymphocyte-derived eosinophil chemotactic factor) which is homologous with chitinase (Guo *et al.*, 2000). These authors suggested that crystal formation contributes to pulmonary inflammation by mechanical damage and enzymatic degradation.

Aspiration pneumonia/foreign body granuloma Aspiration of foreign material, usually of plant origin and derived from food or bedding material, may produce an acute suppurative bronchitis/bronchiolitis or bronchopneumonia, sometimes with formation of lung abscesses. Frequently associated lesions are erosion or ulceration of airway epithelium bordering the foreign bodies and mucous (goblet) cell hyperplasia with production of intraluminal mucus plugs.

A more common response to inhaled particles is formation of well circumscribed foreign body granulomas, in which the particles are centrally lodged and surrounded by a peripheral rim of macrophages and multinucleate giant cells of the foreign body type (Figure 14.13).

Enlargement of airspaces/alveolar emphysema As in humans, destruction of alveolar walls in mice should also be considered as an essential criterion for the diagnosis of emphysema. Alveolar dilatation without destruction of alveolar walls is called 'enlargement of airspaces' or 'hyperinflation' (Ernst *et al.*, 1996).

Enlargement of airspaces characterized by dilated alveolar ducts and flattened alveoli, with corresponding changes in morphometric parameters (increased lung volume, mean linear intercept and total alveolar duct and alveolar volume; decreased internal surface area and elastic fiber length per unit lung volume), but without evidence of alveolar wall destruction has been reported in the senescence-accelerated mouse (SAM; Kurozumi *et al.*, 1994; Teramoto *et al.*, 1994).

Enlargement of alveolar airspaces may also develop secondarily to obstructive and inflammatory lesions in the upper airways. Occasionally, these primary conditions may produce well circumscribed compensatory emphysematous lesions in the peripheral lung characterized

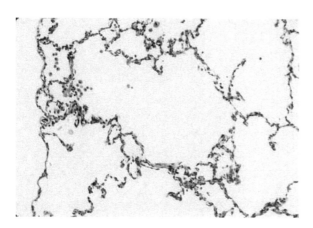

Figure 14.14 Alveolar emphysema secondary to obstruction showing distorted enlarged alveoli, blunt-ending ruptured septa and some inflammatory cell infiltration.

by severely enlarged alveoli, distorted and separated by septa which are partially ruptured and show chronic inflammatory changes and fibrosis (Figure 14.14).

Several genetic models of spontaneous pulmonary emphysema in mice are available, e.g. the blotchy ($Atp7a^{Mo-blo}$) mouse (Fisk and Kuhn, 1976), the tight-skin ($Fbn1^{Tsk}$) mouse (Szapiel et al., 1981; Rossi et al., 1984), the pallid ($Pldn^{pa}$) mouse (Martorana et al., 1993), the surfactant protein D ($Sftpa^{tm1Kor}$) mouse (Wert et al., 2000), the osteopetrotic ($Csf1^{op}$) mouse (Shibata et al., 2001) and the klotho (Kl^{tm1Yin}) mouse (Suga, 2002) and correlates well with the prevalence of bronchioloalveolar tumours (Frith, 1987; Faccini et al., 1990).

The mutant blotchy mouse develops a progressive panlobular emphysema due to an alteration in copper transport. It was this model that demonstrated that injury to the elastic fiber of the lung plays a key role in the development of emphysema (Snider et al., 1986).

Elastolytic processes, with breakdown and loss of interstitial elastic fibers due to a deficiency in serum elastase inhibitory capacity, have been shown to be responsible for the development of emphysema in the tight-skin (tsk) mouse model (Gardi et al., 1990; O'Donnel et al., 1999). In the pallid mouse, emphysema develops as a result of an elastolytic process due to a severe inborn deficiency of serum alpha 1-antitrypsin (Martorana et al., 1993).

Ablation of the SP-D gene caused chronic inflammation, emphysema and fibrosis in the lungs of $Sftpa^{tm1Kor}$ mice. These lesions were associated with increased activity of matrix metalloproteinases and increased hydrogen peroxide production of alveolar macrophages (Wert et al., 2000). Increased secretion of matrix metalloproteinases by alveolar macrophages

and abnormal elastin deposition could be related to spontaneous emphysema development in osteopetrotic ($Csf1^{op}$) mice, which have no detectable macrophage colony-stimulating factor (M-CSF) and show macrophage abnormalities in various other tissues as well (Shibata et al., 2001). In the homozygous mutant klotho (Kl^{tm1Yin}) mouse, which is deficient in klotho gene expression, a defect in matrix synthesis and/or in type II pneumocyte function is considered to be involved in development of pulmonary emphysema (Suga, 2002). Spontaneous emphysema and pulmonary fibrosis has also been observed in transgenic mice overexpressing human transforming growth factor-alpha (TGF-alpha) during the period of postnatal alveolarization (Hardie et al., 1997, 2001).

Bronchiolo-alveolar hyperplasia Bronchiolo-alveolar hyperplasia originates from secretory bronchiolar (Clara) cells, alveolar type II cells or a combination of both. It is frequently present in chronically inflamed lungs. In untreated mice, the incidence increases with age and correlates well with the prevalence of bronchiolo-alveolar tumours (Frith, 1987; Faccini et al., 1990). There are three main histological types of bronchiolo-alveolar hyperplasia according to the type of epithelium involved, but the boundaries between the different types are not distinct. For the first type, i.e. 'bronchiolar type', bronchiolar cells (ciliated cells and Clara cells) either extend peripherally on to adjacent alveolar walls (= alveolar bronchiolization) as suggested by Nettesheim and Szakal (1972) or alveolar type II cells transdifferentiate into bronchiolar cells (Adamson and Bowden, 1977). In the second type, i.e. 'alveolar type', alveoli are lined by a single layer of cuboidal, sometimes vacuolated alveolar type II cells (Figure 14.11). It is this type of hyperplasia where formation of papillary projections or solid cell clusters marks the transition toward neoplasia. The third type, i.e. 'mixed type' of bronchiolo-alveolar hyperplasia, is rarely observed in clean control lungs and shows a combination of bronchiolar and alveolar type II cells.

Squamous cell metaplasia Squamous cell metaplasia may arise by transdifferentiation of Clara cells (Rehm and Kelloff, 1991) and/or alveolar type II cells (Figure 14.11) or their poorly differentiated precursor cells. Persistent squamous cell metaplasia of bronchiolar and alveolar epithelial cells has been described in mice recovering from infection, especially with influenza and Sendai viruses (Semkow et al., 1979; Castleman, 1983).

Mucous cell hyperplasia Mucous (goblet) cell hyperplasia in the bronchioles (Figure 14.15) is usually

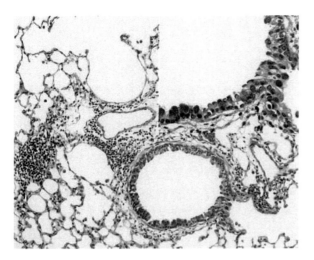

Figure 14.15 Allergic inflammation with prominent bronchiolar mucous (goblet) cell hyperplasia, interstitial inflammatory cell infiltration and slight peribronchiolar fibrosis. Numerous hypertrophic PAS/Alcianblue-positive mucous cells are the predominant cell type in the bronchiolar epithelium (inset).

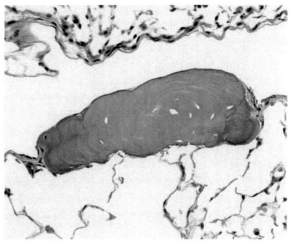

Figure 14.16 Subpleural focal osseous metaplasia.

related to foci of chronic inflammation. Increased production of PAS/Alcian blue-positive mucus and occasionally formation of mucus-filled bronchioles may be associated changes (Faccini *et al.*, 1990). Prominent mucous (goblet) cell hyperplasia has also been induced in CBA/J mice after sensitization and challenge with *Aspergillus fumigatus* conidiae (Hogaboam *et al.*, 2000) or by ovalbumin sensitization in BALB/c and C57BL/6 mice to mimic late asthmatic responses (Hayashi *et al.*, 2001).

Psammoma bodies (corpora amylacea)/alveolar microlithiasis Concentrically laminated **psammoma bodies** are frequently found in the lungs of old mice where they seem to occur within alveolar wall capillaries (Rehm *et al.*, 1985). In the lungs of adult mutant *nackt* mice, which develop alopecia and CD4+ T-cell deficiency, the occurrence of multiple round, PAS-positive and mineralized concretions with a diameter between 30 and 100 μm has recently been reported and termed 'alveolar microlithiasis' (Starost *et al.*, 2002).

Osseous metaplasia Foci of osseous metaplasia consisting of osteoid or bone with viable osteocytes are occasionally seen in subpleural or centriacinar regions of the lung (Figure 14.16) and must be differentiated from metastases of osteosarcoma (Faccini *et al.*, 1990; Dixon *et al.*, 1999).

Amyloidosis Amyloid deposits in the interalveolar septa or in the wall of pulmonary arteries may be observed in mice as part of a systemic amyloidosis (Burek *et al.*, 1982; Shimizu *et al.*, 1993).

Tumors of the respiratory system

Naturally occurring tumors of the upper respiratory tract are extremely rare in laboratory mice. However, primary neoplasms of the lung are common in aged animals of many strains of laboratory mice. The most important respiratory tract tumors in this species are bronchiolo-alveolar tumors which arise in distal lung parenchyma. The great majority of lung tumors in mice is observed in animals older than 18 months. Generally, the tumor rates in male mice are higher than in females.

The classification and descriptions of epithelial respiratory tract tumors are based on that published by Rittinghausen *et al.* (1996) and for World Health Organization (WHO)/International Agency for Research on Cancer (IARC) by Dungworth *et al.* (2001). The descriptions of respiratory tract tumors include the histogenesis and the main diagnostic features of the different tumor types.

Nasal cavity, paranasal sinus, nasopharynx

Adenoma Adenomas usually arise in the most anterior part of the nasal cavity, originating from the transitional, respiratory or glandular epithelial cells of the mucosa of the naso- or maxilloturbinates or from the lateral wall of the anterior nasal cavity. They are characterized by an expansive growth with occasional protrusion into the nasal- or paranasal cavities, but can show endophytic growth as well (Renne *et al.*, 1986 (Renne *et al.*, 1986, Brown, 1990).

Location of the tumor usually determines the type of adenoma (transitional or respiratory epithelial type). Transitional tumors are mostly found in the lateral meatus of the anterior aspect of the nasal cavity.

Adenocarcinoma   Adenocarcinomas are localized in the anterior nasal cavity arising in subepithelial glands or in the posterior nasal cavity originating in the mucosa of the ethmoturbinates. They arise from transitional, respiratory or subepithelial glandular epithelium, or from olfactory Bowman's glands or by malignant change occurring in an adenoma. The neoplasms consist of solid, pseudoglandular, papillary or tubular formations. Lumina within formations of large cuboidal to columnar, or anaplastic cells may be filled with mucosubstances (Brown, 1990; Maronpot, 1990).

Squamous cell papilloma   Squamous cell papillomas can originate from transitional, respiratory or olfactory epithelium, or from the squamous epithelium of the nasal vestibule. Usually they consist of an exophytic mass of uniform, regularly arranged more or less keratin producing squamous cells forming papillary or filiform structures. A vascularized stalk of connective tissue is covered by epithelial cells. Occasionally they grow beneath the mucosal surface (Brown, 1990; Brown et al., 1991; Reznik et al., 1994).

Squamous cell carcinoma   Squamous cell carcinomas arise from squamous differentiation of transitional, respiratory or olfactory epithelial cells, epithelial cells of the subepithelial glands, or from the squamous epithelium of the nasal vestibule. They are seen most often in the anterior nasal cavity, arising in the lateral walls, septum, naso- and maxilloturbinates, and in the ethmoturbinates. They are composed of solid, often branching, cords or masses of cells with various degrees of anaplasia. Shape and size of cells is irregular, large and polygonal, or flattened and stratified. Neoplasms show frequent mitoses, cellular and/or nuclear atypia, and/or invasion into surrounding tissues (Brown 1990; Brown et al., 1991; Reznik et al., 1994).

Adenosquamous carcinoma   Adenosquamous carcinomas originate from respiratory epithelial cells, ductal cells of submucosal glands or basal and sustentacular cells of the olfactory epithelium, and from metaplastic areas of these epithelial cells. Portions of the tumor consist of malignant squamous cells, other portions reveal a malignant glandular part. Squamous components of the tumors can show typical keratin pearl formation. Undifferentiated epithelial cells may occur

(Brown, 1990; Rittinghausen et al., 1996; Dungworth et al., 2001).

Neuroepithelial carcinoma   Neuroepithelial carcinomas arise from **sustentacular** cells, basal cells, immature sensory cells and possibly ductal cells of Bowman's glands of the olfactory epithelium, which covers a small dorsocranial portion of the septum and much of the posterior nasal cavity. Sheets of neoplastic cells frequently show compartmentalization into lobules by fibrovascular septa. Rosettes and pseudorosettes can be present. The frequency and morphology of rosette structures are highly variable. Tumor cells are small round or columnar with poorly defined, pale-staining cytoplasm. Neuroepithelial carcinomas often invade the ethmoid bone and the brain (Brown et al., 1991; Reznik et al., 1994).

Larynx, trachea, bronchus, bronchiole

Papilloma   Papillomas arise from respiratory epithelium which may have undergone squamous metaplasia. The airways are expanded or distorted by growth of branching papillary structures with a central connective tissue stalk and peripheral lining of cuboidal respiratory epithelial cells. The branching connective tissue stalks are usually covered by varying proportions of cuboidal or columnar respiratory epithelium, or occasionally by squamous cells (Figure 14.17).

If a papilloma originates in the terminal bronchioles, there may be growth by expansion into alveolar parenchyma. There is no clear evidence of invasion of adjacent structures. Proliferating cells can migrate out into the alveolar parenchyma either at the end of a

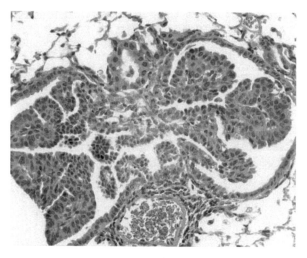

Figure 14.17 Papilloma of a bronchiole in a female mouse (H&E).

terminal bronchiole or through pores in its wall (Rehm *et al.*, 1994).

Adenocarcinoma Adenocarcinomas originate from respiratory epithelium of a conducting airway. There is evidence of invasion of basement membranes or adjacent pulmonary structures. Foci of mucinous cell differentiation can be present. In early stages, neoplasms show papillary growth with a central connective tissue stalk lined by cuboidal to columnar or pleomorphic epithelium. (Rehm *et al.*, 1994; Dungworth *et al.*, 2001). Irregular tubular or glandular structures and cytological features of anaplasia can be present (Dixon and Maronpot, 1991).

Squamous cell carcinoma Squamous cell carcinomas are rarely reported to be experimentally-induced. They arise from respiratory epithelium that has undergone squamous metaplasia. Cells are arranged in clusters or irregular structures with central keratinization, or without overt keratinization but forming distinct intercellular bridges. Dysplasia and anaplasia are common cytological features. Cells can frequently be quite pleomorphic, including the formation of very large cells (Rehm and Kelloff, 1991).

Lung

Bronchiolo-alveolar adenoma Bronchiolo-alveolar adenomas with a solid growth pattern are generally believed to be composed of cells expressing alveolar type II cell features. Papillary tumors, however, are considered either to be less well-differentiated type II cell tumors progressing toward a malignant phenotype or to be of Clara cell origin (Thaete and Malkinson, 1991 Belinsky *et al.*, 1992 Dungworth *et al.*, 2001).

Bronchiolo-alveolar adenomas are frequently located at the lung periphery and usually small in size. Solid adenomas show alveolar spaces obliterated by proliferating round to oval cells (Figure 14.18). Papillary adenomas are composed predominantly of papillary structures lined by cuboidal to columnar cells. Tumors can show prominent tubular formations. Mitoses are rare or absent (Dungworth *et al.*, 2001).

Bronchiolo-alveolar carcinoma Bronchiolo-alveolar carcinomas can originate from either alveolar type II cells (Ward and Rehm, 1990; Belinsky *et al.*, 1992; Rehm *et al.*, 1991b; 1994) or Clara cells (Thaete and Malkinson, 1991; Dixon and Maronpot, 1991; Gunning *et al.*, 1992). Carcinomas containing a proportion of cells with neuroendocrine immunohistochemical features were reported in CC10-hASH transgenic mice (Linnoila *et al.*, 2000). Depending on the experimental design and/or carcinogen, tumors can be induced selectively or different types can be observed in a single mouse.

Neoplasms show irregular nodular growth and are moderately well to poorly circumscribed, and can occupy an entire pulmonary lobe. Most frequently they are composed of papillary connective tissue lined by cuboidal to columnar or pleomorphic cells (Figure 14.19). Late stages of differentiation may be associated with formation of spindle cells, round atypical cells, scirrhous changes, increased mitotic rate, hemorrhages and necrosis.

The histological growth pattern, number and size of induced tumors depend also on the design of a study, e.g. on the age of the mice, when a carcinogen is

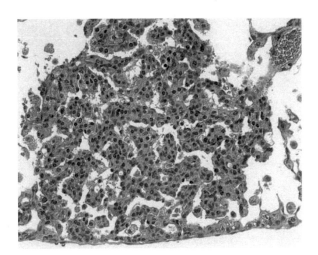

Figure 14.18 Bronchiolo-alveolar adenoma in the lung of a female mouse (H&E).

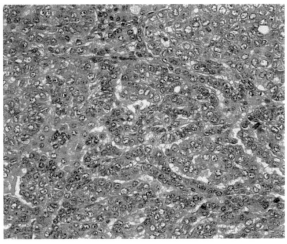

Figure 14.19 Bronchiolo-alveolar carcinoma in the lung of a female mouse (H&E).

administered (Branstetter and Moseley 1991; Rehm *et al.*, 1991b), or on the strain of mouse (Malkinson and Thaete, 1986).

Acinar carcinoma Acinar carcinomas are extremely rare naturally-occurring tumors, but are reported to be induced by a variety of chemicals. They are believed to originate from bronchiolar Clara cells or to arise directly from respiratory epithelium of airways. They are diffusely expansive with irregular margins or more circumscribed nodules. The growth pattern is predominantly glandular or acinar. They are composed of cuboidal to columnar or pleomorphic cells without distinguishing features or more commonly, mixed with ciliated cells or mucous cells (Figure 14.20). The neoplasms show clear indication of malignancy such as penetration of basement membranes and tissue destruction (Rehm and Kelloff, 1991).

Adenosquamous carcinoma Adenosquamous carcinomas are believed to derive directly from acinar carcinomas, or possibly bronchiolo-alveolar carcinomas, with clonal shifts to malignant squamous cell phenotype. They are nodular, or diffusely expansive with irregular margins. They are composed of both adenocarcinomatous and malignant squamous cell components that may show keratinization. Neoplasms usually show clear indication of malignancy such as penetration of basement membranes and tissue destruction (Dixon and Maronpot, 1991; Rehm and Kelloff, 1991).

Squamous cell carcinoma Squamous cell carcinomas are extremely rare spontaneous tumors in lungs of mice. The neoplasms arise from squamous cell metaplasia of alveolar epithelium or Clara cells. Cells grow in clusters or irregular nests with central or overt keratinization (Figure 14.21). Cells have cytological features of malignancy such as atypia, disorganization and increased mitotic rate. Cells can frequently be quite pleomorphic, including the formation of very large cells. Tumors can invade the adjacent lung parenchyma, pleura, vessels or bronchi (Dixon and Maronpot, 1991; Rehm and Kelloff, 1991; Rehm *et al.*, 1991b).

Mesenchymal and nerve tissue tumors

Any of the vascular, connective, supporting and nerve tissues of the respiratory tract can conceivably give rise to primary tumors. Mesenchymal and nerve tissue tumors are classified according to the apparent cell of origin (Ernst *et al.*, 2001).

Tumor metastases in the lung

Besides primary tumors, metastases may occur in the lung and may be found in the lung more frequently than primary pulmonary tumors, depending on the strain of mouse and its susceptibility to primary lung tumors. Microscopically, metastatic tumors usually resemble the primary lesions, although they may be either better or less differentiated. Therefore, metastases can easily mimic primary pulmonary tumors.

In many strains of mice, malignant lymphomas and histiocytic sarcomas may also affect the lungs relatively frequently. This was reported for BALB/C mice by Sheldon and Greenman (1979), and is a common feature in the majority of mice bearing systemic tumors.

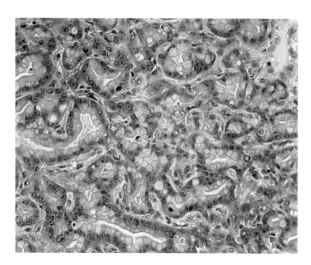

Figure 14.20 Acinar carcinoma in the lung of a female mouse (H&E).

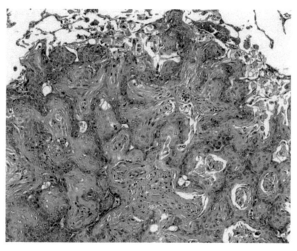

Figure 14.21 Squamous cell carcinoma in the lung of a female mouse (H&E).

Metastases of osteosarcomas have been described to be frequent in BALB/C mice (Frith *et al.*, 1981)). In several strains of mice (e.g. A, BALB/cfC3H, C3H, RIII), metastases found in the lung are frequently from mammary gland carcinomas (Sass and Liebelt, 1985). In CD-1 mice, metastatic deposits from primary liver tumors have been seen quite often (Faccini *et al.*, 1990).

Conclusion

Due to the relevance of mice for biomedical research the mouse lung has become the focus of lung physiologists, toxicologists and pharmacologists. Although the murine lung significantly differs in anatomical and physiological details from the human lung it is widely used for modelling human lung diseases.

References

Adamson, I.Y.R. and Bowden, D.H. (1977). *Am. J. Pathol.* **87**, 569–580.

Adamson, I.Y.R. and Bowden, D.H. (1984). *Am. J. Pathol.* **117**, 37–43.

Andersson, R.G. and Grundstrom, N. (1983). *Eur. J. Respir. Dis. Suppl.* **131**, 141–157.

Barnes, P.J. (2002). *Nat. Rev. Drug Discov.* **1**, 437–446.

Belinsky, S.A., Devereux, T.R., Foley, J.F., Maronpot, R.R. and Anderson, M.W. (1992). *Cancer Res.* **52**, 3164–3173.

Blease, K., Schuh, J.M., Jakubzick, C., Lukacs, N.W., Kunkel, S.L., Joshi, B.H., Puri, R.K., Kaplan, M.H. and Hogaboam, C.M. (2002). *Am. J. Pathol.* **160**, 481–490.

Branstetter, D.G. and Moseley, P.P. (1991). *Exp. Lung Res.* **17**, 169–179.

Braun, A., Appel, E., Baruch, R., Herz, U., Botchkarev, V., Paus, R., Brodie, C. and Renz, H. (1998). *Eur. J. Immunol.* **28**, 3240–3251.

Braun, A., Lommatzsch, M., Mannsfeldt, A., Neuhaus-Steinmetz, U., Fischer, A., Schnoy, N., Lewin, G.R. and Renz, H. (1999). *Am. J. Respir. Cell Mol. Biol.* **21**, 537–546.

Brown, H.R. (1990). *Environ. Health Perspect.* **85**, 291–304.

Brown, H.R., Monticello, T.M., Maronpot, R.R., Randall, H.W., Hotchkiss, J.R. and Morgan, K.T. (1991). *Toxicol. Pathol.* **19**, 358–372.

Brown, R.H., Walters, D.M., Greenberg, R.S. and Mitzner, W. (1999). *J. Appl. Physiol.* **87**(6), 2362–2365.

Buckley, L.A., Morgan, K.T., Swenberg, J.A., James, R.A., Hamm, T.E. Jr. and Barrows, C.S. (1985). *Fundam. Appl. Toxicol.* **5**, 341–352.

Burek, J.D., Molello, J.A. and Warner, S.D. (1982). In *The Mouse in Biomedical Research, Vol. II, Diseases* (eds H.L. Foster, J.D. Small and J.G. Fox), pp. 425–440. Academic Press, San Diego.

Castleman, W.L. (1983). *Am. Rev. Respir. Dis.* **128**, S83–S87.

Cieslewicz, G., Tomkinson, A., Adler, A., Duez, C., Schwarze, J., Takeda, K., Larson, K.A., Lee, J.J., Irvin, C.G. and Gelfand, E.W. (1999), *J. Clin. Invest.* **104**, 301–308.

Costa, D.L. and Tepper, J.S. (1988). In *Toxicology of the Lung.* (eds D.E. Gardner, J.D. Crapo, E.J. Massaro), p. 147. Raven Press, New York.

Dawkins, P.A. and Stockley, R.A. (2001). *Thorax* **56**, 972–977.

Dixon, D. and Maronpot, R.R. (1991). *Toxicol. Pathol.* **19**, 540–556.

Dixon, D., Herbert, R.A., Sills, R.C. and Boorman, G.A. (1999). In *Pathology of the Mouse* (ed. R.R. Maronpot), pp. 293–332. Cache River Press, Vienna.

Drazen, J.M., Finn, P.W., and De Sanctis, G.T. (1999). *Annu. Rev. Physiol.* **61**, 593–625.

Dungworth, D.L., Rittinghausen, S., Schwartz, L., Harkema, J.R., Hayashi, Y., Kittel, B., Lewis, D., Miller, R.A., Mohr, U., Morgan, K.T., Rehm, S. and Slayter, M.V. (2001). In (eds U. Mohr, C.C. Capen, D.L. Dungworth, P. Greaves, J.F. Hardisty, Y. Hayashi, N. Ito, P.H. Long and G. Krinke *International Classification of Rodent Tumors. The Mouse* pp. 87–137. Springer, Berlin, Heidelberg, New York.

Ellis, J.L. and Undem, B.J. (1992). *J. Pharmacol. Exp. Ther.* **262**, 646–653.

Ernst, H., Dungworth, D.L., Kamino, K., Rittinghausen, S. and Mohr, U. (1996). In *Pathobiology of the Aging Mouse* (eds U. Mohr, D.L. Dungworth, C.C. Capen, W.W. Carlton, J.P. Sundberg and J.M. Ward), pp. 281–300. ILSI Press, Washington, DC.

Ernst, H., Carlton, W.W., Courtney, C., Rinke, M., Greaves, P., Isaacs, K.R., Krinke, G., Konishi, Y., Mesfin, G.M. and Sandusky, G. (2001). In (eds U. Mohr, C.C. Capen, D.L. Dungworth, P. Greaves, J.F. Hardisty, Y. Hayashi, N. Ito, P.H. Long and G. Krinke), *International Classification of Rodent Tumours. Part II: The mouse* pp. 361–415. Springer, WHO/IARC, Berlin, Heidelberg, New York.

Faccini, J.M., Abbott, D.P. and Paulus, G.J.J. (1990). *Mouse Histopathology*, pp. 48–63. Elsevier, Amsterdam, The Netherlands.

Finotto, S., De Sanctis, G.T., Lehr, H.A., Herz, U., Buerke, M., Schipp, M., Bartsch, B., Atreya, R., Schmitt, E., Galle, P.R., Renz, H. and Neurath, M.F. (2001): *J. Exp. Med.* **193**, 1247–1260.

Fisk, D.E. and Kuhn, C. (1976). *Am. Rev. Resp. Dis.* **113**, 787–797.

Frith, C.H. (1987). *Incidence of Neoplastic and Nonneoplastic Lesions in Several Strains of Mice*, pp. 1–57. Toxicology Pathology Associates, Little Rock, USA.

Frith, C.H., Zuna, R.E. and Morgan, K.T. (1981). *J. Natl. Cancer Inst.* **67**, 693–699.

Frith, C.H., Highman, B., Burger, G. and Sheldon, W.D. (1983). *Lab. Anim. Sci.* **33**, 273–286.

Gardi, C., Martorana, P.A., van Even, P., de Santi, M.M. and Lungarella, G. (1990). *Exp. Mol. Pathol.* **52**, 46–53.

Glaab, T., Daser, A., Braun, A., Neuhaus-Steinmetz, U., Fabel, H., Alarie, Y. and Renz, H. (2001). *Am. J. Physiol. Lung Cell Mol. Physiol.* **280**, L565–L573.

Gunning, W.T., Goldblatt, P.J. and Stoner, G.D. (1992). *Am. J. Pathol.* **140**, 109–118.

Guo, L., Johnson, R.S. and Schuh, J.C. (2000). *J. Biol. Chem.* **17**, 8032–8037.

Hallman, M. and Merritt, T.A. (1996). *J. Clin. Invest.* **97**, 589–590.

Hamelmann, E., Schwarze, J., Takeda, K., Oshiba, A., Larsen, G.L., Irvin, C.G., and Gelfand, E.W. (1997). *Am. J. Respir. Crit. Care Med.* **156**, 766–775.

Hansen, G., Berry, G., DeKruyff, R.H. and Umetsu, D.T. (1999). *J. Clin. Invest.* **103**, 175–183.

Hardie, W.D., Bruno, M.D., Huelsman, K.M., Iwamoto, H.S., Carrigan, P.E., Leikauf, G.D., Whitsett, J.A. and Korfhagen, T.R. (1997). *Am. J. Pathol.* **151**, 1075–1083.

Hardie, W.D., Piljan-Gentle, A., Dunlavy, M.R., Ikegami, M. and Korfhagen, T.R. (2001). *Am. J. Physiol. Lung Cell Mol. Physiol.* **281**, L1088–L1094.

Hayashi, T., Hasegawa, K., Nakai, S., Hamachi, T., Adachi, Y., Yamauchi, Y. and Maeda, K. (2001). *J. Comp. Pathol.* **125**, 208–213.

Held, H.D. and Uhlig, S. (2000a). *J. Appl. Physiol.* **88**, 2192–2198.

Held, H.D. and Uhlig, S. (2000b). *Am. J. Respir. Crit. Care Med.* 162:1547–1552.

Held, H.D., Martin, C. and Uhlig, S. (1999). *Br. J. Pharmacol.* **126**, 1191–1199.

Held, H.D., Boettcher, S., Hamann, L. and Uhlig, S. (2001). *Am. J. Respir. Crit. Care Med.* **163**, 711–716.

Heppleston, A.G., Fletcher, K. and Wyatt, I. (1972). *Experientia* **28**, 938–939.

Herbert, R.A. and Leininger, J.R. (1999). In *Pathology of the Mouse* (ed. R.R. Maronpot), pp. 259–292. Cache River Press, Vienna.

Hogaboam, C.M., Blease, K., Mehrad, B., Steinhauser, M.L., Standiford, T.J., Kunkel, S.L. and Lukacs, N.W. (2000). *Am. J. Pathol.* **156**, 723–732.

Hogan, S.P., Matthaei, K.I., Young, J.M., Koskinen, A., Young, I.G. and Foster, P.S. (1998). *J. Immunol.* **161**, 1501–1509.

Hoymann, H.G. and Heinrich, U. (1998). In *Methods in Pulmonary Research* (eds S. Uhlig, and A.E. Taylor) pp. 1–28. Birkhäuser Verlag, Basel.

Jennings, V.M., Dillehay, D.L., Webb, S.K. and Brown, L.A. (1995). *Am. J. Respir. Cell Mol. Biol.* **13**, 297–306.

Khaled, A.R., Butfiloski, E.J., Sobel, E.S. and Schiffenbauer, J. (1999). *Autoimmunity* **30**, 115–128.

Koch, W. (1998). In *Methods in Pulmonary Research* (eds S. Uhlig *et al.*), pp. 485–507.

Kumar, R.K. and Foster, P.S. (2001). *Immunol. Cell Biol.* **79**, 141–144.

Kumar, R.K., Herbert, C., Yang, M., Koskinen, A.M., McKenzie, A.N. and Foster, P.S. (2002). *Clin. Exp. Allergy* **32**, 1104–1111.

Kurozumi, M., Matsushita, T., Hosokawa, M. and Takeda, T. (1994). *Am. J. Respir. Crit. Care Med.* **149**, 776–782.

Larsen, G.L., Renz, H., Loader, J.E., Bradley, K.L. and Gelfand, E.W. (1992). *J. Clin. Invest.* **89**, 747–752.

Leong, K.P. and Huston, D.P. (2001). *Ann. Allergy Asthma Immunol.* **87**, 96–109.

Lieschke, G.J., Stanley, E., Grail, D., Hodgson, G., Sinickas, V., Gall, J.A., Sinclair, R.A. and Dunn, A.R. (1994). *Blood* **84**, 27–35.

Likens S.A. and Mauderly J.L. (1982). *J. Appl. Physiol.: Respir. Environ. Exercise Physiol.* **52**(1), 141.

Linnoila, R.I., Zhao, B., DeMayo, J.L., Nelkin, B.D., Baylin, S.B., DeMayo, F.J. and Ball, D.W. (2000). *Cancer Res. JID – 2984705R* **60**, 4005–4009.

Lloyd, C.M., Gonzalo, J.-A., Coyle, A.J. and Gutierrez-Ramos, J.-C. (2001). *Adv. Immunol.* **77**, 263–295.

Malkinson, A.M. and Thaete, L.G. (1986). *Cancer Res.* **46**, 1694–1697.

Maronpot, R.R. (1990). *Environ. Health Perspect.* **85**, 331–352.

Martin, T.R., Gerard, N.P., Galli, S.J. and Drazen, J.M. (1988). *J. Appl. Physiol.* **64**, 2318–2323.

Martin, C., Uhlig, S. and Ullrich, V. (1996). *Eur. Respir. J.* **9**, 2479–2487.

Martorana, P.A., van Even, P., Gardi, C. and Lungarella, G. (1989). *Am. Rev. Respir. Dis.* **139**, 226–232.

Martorana, P.A., Brand, T., Gardi, C., van Even, P., de Santi, M.M., Calzoni, P., Marcolongo, P. and Lungarella, G. (1993). *Lab. Invest.* **68**, 233–241.

Menache, M.G., Miller, F.J. and Raabe, O.G. (1995). *Ann. Occup. Hyg.* **39**, 317–328.

Mitzner, W. and Tankersley, C. (1998). *Am. J. Respir. Crit. Care Med.* **158**, 340–341.

Monticello, T.M., Morgan, K.T. and Uraih, L. (1990). *Environ. Health Perspect.* **85**, 249–274.

Murray, A.B. and Luz, A. (1990). *Vet. Pathol.* **27**, 274–281.

Nettesheim, D.E. and Szakal, A.K. (1972). *Lab. Invest.* **26**, 210–219.

Neuhaus-Steinmetz, U., Glaab, T., Daser, A., Braun, A., Lommatzsch, M., Herz, U., Kips, J., Alarie, Y. and Renz, H. (2000). *Int. Arch. Allergy Immunol.* **121**, 57–67.

O'Donnell, M.D., O'Connor, C.M., FitzGerald, M.X., Lungarella, G., Cavarra, E. and Martorana, P.A. (1999). *Matrix Biol.* **18**, 357–360.

Raabe, O.G., Al-Bayati, M.A. and Teague, S.V. (1988). *Ann. Occup. Hyg.* **32**, 53–63.

Rao, S. and Verkman, A.S. (2000). *Am. J. Physiol. Cell Physiol.* **279**, C1–C18.

Rehm, S. and Kelloff, G.J. (1991). *Exp. Lung Res.* **17**, 229–244.

Rehm, S., Wcislo, A. and Deerberg, F. (1985). *Lab. Anim.* **19**, 224–235.

Rehm, S., Devor, D.E., Henneman, J.R. and Ward, J.M. (1991a). *Exp. Lung Res.* **17**, 181–195.

Rehm, S., Lijinsky, W., Singh, G. and Katyal, S.L. (1991b). *Am. J. Pathol.* **139**, 413–422.

Rehm, S., Ward, J.M. and Sass, B. (1994). In *Pathology of Tumours in Laboratory Animals. Vol. II: Tumours of the Mouse* (eds U. Mohr, and V.S. Turusov), pp. 325–355. IARC Sci. Publ. No. 111, Lyon.

Renne, R.A., Giddens, W.E., Boorman, G.A., Kovatch, R., Haseman, J.E. and Clarke, W.J. (1986). *J. Natl. Cancer Inst.* **77**, 573–582.

Renne, R.A., Gideon, K.M., Miller, R.A., Mellick, P.W. and Grumbein, S.L. (1992). *Toxic. Pathol.* **20**, 44–51.

Reznik, G.K., Schüller, H.M. and Stinson, S.F. (1994). In *Pathology of Tumours in Laboratory Animals*. Vol. II: Tumours of the Mouse (eds U. Mohr, and V.S. Turusov), pp. 305–324. IARC Sci. Publ. No. 111, Lyon.

Rijhsinghani, K., Abrahams, C., Swerdlow, M.A. and Ghose, T. (1988). *Cancer Detect. Prevent.* **11**, 279–286.

Rittinghausen, S., Dungworth, D.L., Ernst, H. and Mohr, U. (1996). In *Pathobiology of the Aging Mouse* (eds U. Mohr, D.L. Dungworth, C.C. Capen, W.W. Carlton and J.M.Ward), pp. 301–314. ILSI Press, Washington, DC.

Rossi, G.A., Hunninghake, G.W., Gadek, J.E., Szapiel, S.V., Kawanami, O., Ferrans, V.J. and Crystal, R.G. (1984). *Am. Rev. Respir. Dis.* **129**, 850–855.

Sass, B. and Liebelt, A.G. (1985). In *ILSI Monographs on Pathology of Laboratory Animals. Respiratory System* (eds T.C. Jones, U. Mohr and R.D. Hunt), pp. 138–159. Springer, Berlin, Heidelberg, New York, Tokyo.

Schuh, J.M., Blease, K., Kunkel, S.L. and Hogaboam, C.M. (2002). *Am. J. Physiol. Lung Cell Mol. Physiol.* **283**, L198–L204.

Semkow, R., Wilczynski, J., Krus, H. and Bogacka-Zatorska, E. (1979). *Acta Virol.* **23**, 479–499.

Sheldon, W.G. and Greenman, D.L. (1979). *J. Environ. Pathol. Toxicol. Oncol.* **3**, 155–167.

Shibata, Y., Zsengeller, Z., Otake, K., Palaniyar, N. and Trapnell, B.C. (2001). *Blood* **98**, 2845–2852.

Shimizu, K., Morita, H., Niwa, T., Maeda, K., Shibata, M., Higuchi, K. and Takeda, T. (1993). *Acta Pathol. Jpn.* **43**, 215–221.

Shultz, L.D., Coman, D.R., Bailey, C.L., Beamer, W.G. and Sidman, C.L. (1984). *Am. J. Pathol.* **116**, 179–192.

Snider, G.L., Lucey, E.C. and Stone, P.J. (1986). *Am. Rev. Resp. Dis.* **133**, 149–169.

Starost, M.F., Benavides, F. and Conti, C.J. (2002). *Vet. Pathol.* **39**, 390–392.

Suga, T. (2002). *Nihon Kokyuki Gakkai Zasshi* **40**, 203–209.

Szapiel, S.V., Fulmer, J.D., Hunninghake, G.W., Elson, N.A., Kawanami, O., Ferrans, V.J. and Crystal, R.G. (1981). *Am. Rev. Respir. Dis.* **123**, 680–685.

Teramoto, S., Fukuchi, Y., Uejima, Y., Teramoto, K., Oka, T. and Orimo, H. (1994). *Am. J. Respir. Crit. Care Med.* **150**, 238–244.

Thaete, L.G. and Malkinson, A.M. (1991). *Exp. Lung Res.* **17**, 219–228.

Thrall, R.S., Vogel, S.N., Evans, R. and Shultz, L.D. (1997). *Am. J. Pathol.* **151**, 1303–1310.

Vijayaraghavan, R., Schaper, M., Thompson, R., Stock, M.F. and Alarie, Y. (1993). *Arch. Toxicol.* **67**, 478–90.

Vijayaraghavan, R., Schaper, M., Thompson, R., Stock, M.F., Boylstein, L.A., Luo, J.E., Alarie, Y., Vijayaraghavan, R., Schaper, M., Thompson, R., Stock, M.F. and Alarie, Y. (1994). *Arch. Toxicol.* **68**, 490–499.

von Bethmann, A.N., Brasch, F., Nusing, R., Vogt, K., Volk, H.D., Muller, K.M., Wendel, A. and Uhlig, S. (1998). *Am. J. Respir. Crit. Care Med.* **157**, 263–272.

Ward, J.M. (1978). *Vet. Pathol.* **15**, 170–178.

Ward, J.M. and Rehm, S. (1990). *Exp. Pathol.* 40, 301–312.

Ward, J.M., Yoon, M., Anver, M.R., Haines, D.C., Kudo, G., Gonzalez, F.J. and Kimura, S. (2001). *Am. J. Pathol.* **158**, 323–332.

Warner, T. and Balish, E. (1995). *Am. J. Pathol.* **146**, 1017–1024.

Wert, S.E., Yoshida, M., LeVine, A.M., Ikegami, M., Jones, T., Ross, G.F., Fisher, J.H., Korfhagen, T.R. and Whitsett, J.A. (2000). *Proc. Natl. Acad. Sci.* **97**, 5972–5977.

Wills-Karp, M. (1999). *Annu. Rev. Immunol.* **17**, 255–281.

Gastrointestinal System and Metabolism

Carolyn D Berdanier

Nutrition and Cell Biology, University of Georgia,
Athens, Georgia, USA

Introduction

The gastrointestinal (GI) system of the mouse is quite similar to that of other species in the rodentia family. The mouse, unlike the rat, is a day feeder. The normal mouse will consume the majority of its food during the light period (Bailey *et al.*, 1975; Meier and Cincotta, 1996). Exceptions to this feeding pattern have been reported in strains of mice having an obesity phenotype (Bailey *et al.*, 1975). **Hyperphagia** is a common feature of mice that have a genetic mutation in the neuroendocrine system that signals hunger and satiety (Roberts and Greenberg, 1996; Chua and Leibel, 1997; Wolff, 1997). Not all of these mutations are known. In mice with aberrant signals for satiety, the increased food intake has effects on both the GI system and metabolism. These will be reviewed in this chapter.

The digestive system

Anatomy and physiology

The digestive system begins with the mouth and ends with the anus (Figure 15.1). In the mouth there are two pairs of incisors, two on the top and two on the bottom plus cheek teeth. There is a space called the diastema that separates the incisors and cheek teeth. The incisors grow continually. When the diet provided is a powder or unpelleted dry mixture, these incisor teeth must be periodically clipped. The frequency needed for this clipping varies depending on the nature of the diet and its nutrient content. In the wild, these teeth are ground down by the hard seed coats and other rough textured edibles in the environment. In some instances, mice

The Laboratory Mouse
Copyright 2004 Elsevier
ISBN 0-1233-6425-6

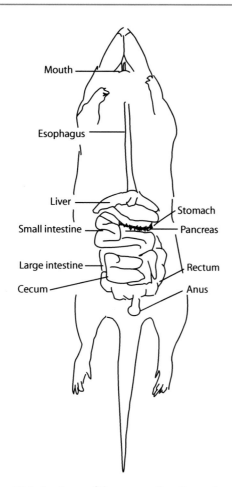

Figure 15.1 Anatomy of the mouse digestive system.

will grind their teeth against caging materials or food/water dispensers thereby shortening them.

Located near the mouth and emptying into it are the salivary glands: the parotid gland lies just behind the ear and extends over the ventro-lateral surface of the neck to the shoulder; the mandibular glands are ventral to the parotids and the sublingual glands are under the base of the tongue. These glands contain ascinar cells that drain by way of an interlobular duct. The parotid glands drain into the parotid duct while the mandibular and sublingual glands drain into the mouth via Wharton's duct. This duct terminates in small papillae near the incisors. All three glands produce and release saliva into the mouth. The sublingual secretion is almost entirely mucus while the secretions of the other two glands vary in the amount of mucus produced. The parotid saliva contains salivary amylase as well as salivary lipase. These two enzymes initiate the digestion of carbohydrate and lipid respectively. However, the amount of digestion that occurs in the mouth and esophagus is minimal. These two enzymes are inactivated by the low pH of the stomach.

The tongue is long and flexible. It is attached to the floor of the mouth at the rear of the mouth. On the surface of the tongue can be found taste buds. These buds (circumvallate and fungiform papilla) consist of elongated cells arranged around a central lumen that opens to the papillary furrow by a taste pore. It is assumed that salt, sweet and bitter tastes can be detected by these taste buds. However, because there are far fewer of these buds in the mouse than in the human, it is also assumed that the sensory perception of taste is less acute in this species than in the human.

At the back of the mouth is the epiglottis, a raised flap of tissue caudal to the tongue that guards the glottis. Behind this is the soft palate. The soft palate is merely a continuation of the hard palate (roof of the mouth) that in turn gives way to the opening to the pharynx or entryway to the esophagus.

The esophagus connects the pharynx to the stomach. It is lined by a thick layer of squamous epithelia cells covered by an acellular layer of cornified tissue. Just internal to the squamous layer is a very thin layer of smooth muscle called the *muscularis mucosa* and a slightly wider band of connective tissue, the *laminar propia*.

The stomach consists of the forstomach or entry from the esophagus, the fundus, and the pyloric region. The fundus is the first true portion of the stomach. It is lined by columnar epithelial cells arranged in deep gastric pits. In these pits are found two types of cells: the chief cells that produce the enzyme pepsin in its precursor form (prepepsin) and the mucin-secreting non-chief cells. On the outer border of the gastric pits are parietal cells. These cells produce a precursor to hydrochloric acid. The pyloric region of the stomach also has an epithelial lining of columnar cells but the cells lining the pits produce only mucin. In the course of digestion, the function of the stomach is to mix the ingested food with both acid and enzyme. Pepsin, secreted in a precursor form, is activated by the reduced pH that occurs when the stomach contents are mixed with the hydrochloric acid. The muscles of the stomach are arranged such that their contractions and relaxations result in a churning and mixing of the stomach contents now known as chyme. The stomach is connected to the small intestine by the pyloric valve. This valve functions to regulate the flow of the stomach contents into the duodenum of the small intestine.

The duodenum is the first portion of the small intestine. It is here that the exocrine secretions of the pancreas and Brunner's glands as well as the bile produced by the liver are mixed with the chyme. The digestive enzymes found in these secretions act on the proteins, fats and carbohydrates of the food producing small easily absorbed molecules (monosaccharides, amino acids and some fatty acids). The digestive enzymes and their functions are listed in Table 15.1. Although the mouse has an active lactase at birth, like the rat and some humans, this activity disappears as the neonate develops. By weaning (~21 days) very little lactase activity can be found.

Most maintenance mouse diets have very little fat (less than 5% by weight) thus a very low lipase activity will be found in the duodenal contents. Low fat mouse diets have a longer shelf life than diets having a higher % fat. Since the mouse can grow satisfactorily on this level of dietary fat, the typical mouse pelleted ration supplied to mouse production facilities will contain 4–5% fat. Higher fat diets have been prepared for specific purposes and these will induce a more active lipid digestive system (Surwit *et al.*, 1988). In addition, those mice produced through **transgenics** where the mutant gene for one or more of the lipid carrying proteins has been inserted will become lipemic when fed high fat or cholesterol enriched diets (Paigen *et al.*, 1985).

In general, the mouse maintenance diet contains varying amounts of protein (10–15% by weight) and the remaining part of the diet as a mixture of simple and complex carbohydrates. Mice in the wild may exist on seeds, nuts or scavenged food widely varying in composition. The composition of the diet and frequency of feeding will determine the amount and activity of the digestive enzymes released into the duodenum as well as the rate of passage of food from the mouth to the anus. In addition to the release of digestive fluids by both the intestinal enterocytes and the exocrine pancreas, there are also mucus producing goblet cells intermittently located between the columnar epithelial cells.

The epithelial layer of the small intestine has numerous finger-like projections called villi. These are the absorptive units of the intestine. Amino acids, simple sugars and short chain fatty acids are absorbed by the villous absorptive cells called enterocytes. The villi are very densely placed in the duodenum. Their density becomes less as the intestinal tract proceeds towards the large intestine. The need for absorptive cells also becomes less as the simple nutrients from the diet are absorbed. Simple sugars and amino acids are the first to disappear from the chyme. Electrolytes disappear rapidly from both stomach and duodenum. Calcium, magnesium, iron and other essential minerals are absorbed slowly all along the intestinal tract. Water soluble vitamins are absorbed in the duodenum while the fat soluble vitamins

TABLE 15.1: Digestive enzymes and their substrates		
Enzyme	**Location**	**Target or substrate**
Pepsin	Stomach	Peptide bonds involving aromatic amino acids
Trypsin	Small intestine	Peptide bonds involving arginine, lysine
Chymotrypsin	Small intestine	Peptide bonds involving tyrosine, tryptophane, phenylalanine, methionine, leucine
Elastase	Small intestine	Peptide bonds involving alanine, serine, glycine
Carboxypeptidase A	Small intestine	Peptide bonds involving valine, leucine, isoleucine, alanine
Carboxypeptidase B	Small intestine	Peptide bonds involving lysine, arginine
Endopeptidases	Enterocytes	Di or tripeptides that enter the enterocytes
αAmylase	Mouth, small intestine	Starch, amylopectin, glycogen
αGlucosidase	Small intestine	αLimit dextrin
Lactase	Small intestine	Lactose
Maltase	Small intestine	Maltose
Sucrase	Small intestine	Sucrose
Lingual lipase	Mouth	Triglycerides
Duodenal lipases	Small intestine	Triglycerides
Esterase	Small intestine	Cholesterol esters

follow the pattern of the long chain fatty acids and cholesterol. This occurs in the ilium and to some extent, the jejunum. In general, the essential nutrients (glucose, essential amino acids, vitamins) are absorbed via an active transport system. Some of the essential minerals require dedicated mineral transport proteins while others can share transporters or even compete for these transporters.

The epithelial cells as described above comprise the mucosal layer of the intestinal tract. Beneath the mucosal layer is a relatively thin smooth muscle layer, the muscularis mucosa. These muscles contract and relax rhythmically so as to propel the contents through to the large intestine and anus. Both longitudinal and horizontal contractions occur.

The jejunum is the middle segment of the small intestine between the duodenum and the ileum. Digestion begun in the duodenum continues here as does absorption. Generally, the simple sugars and readily available amino acids have already been absorbed but the larger macromolecules are still in need of digestion so as to release more absorbable nutrients. This is also the case for the ileum, the final portion of the small intestine. This portion is characterized by lymph nodules (*Peyer's patches*) in the submucosa. The villi of the ileum contain thin walled lymph vessels (*lacteals*) in the lamina propria that function in the absorption of dietary fat. Dietary fat and vitamins A, D, E and K are absorbed here. The triacylglycerides are hydrolyzed and then resynthesized in the process of absorption. With resynthesis they are joined to protein carriers for transport to the peripheral tissues for storage or to the liver for use. Cholesterol likewise is processed. If esterified, the fatty acid is removed and the free cholesterol is joined to a protein carrier and carried via the lymph to the thoracic duct.

The lipid–protein complexes so formed at the site of the intestine are called chylomicrons. Lipid–protein transport complexes differ in the ratios of lipid and protein and thus differ in density. The chylomicrons are the least dense. Table 15.2 shows these protein carriers and indicates their function. The chylomicrons utilize proteins labeled apoA-I, A-IV, B and CII. At the fat cell, the triglycerides are released through the action of the interstitial lipoprotein lipase. These lipids pass into the fat cell and are stored. The loss of the glycerides results in a less dense lipid–protein complex containing primarily cholesterol. The lipid carrying protein Apo E is added and the complex moves to the liver. Here the cholesterol is taken up and the lipid carrying proteins released for reuse. Lipids synthesized in the liver are transported to the fat cells using the proteins labeled Apo E, B, B-100, CI, II and III. These same proteins are used for lipid recycling. In the mouse with an active *de novo* lipogenic system, endogenous lipid transport proteins are more common than those used for the transport of food lipid. Because mice are routinely fed low fat–high carbohydrate diets, *de novo* lipogenesis is quite active. If the carbohydrate in the diet is a simple sugar then lipogenesis becomes very active and indeed, liver lipid, normally 3–4 % of the organ by weight, can rise to 5–6%. The flux through the lipogenic pathway in both liver and adipocyte increases as does the activity of the rate limiting enzyme, acetyl CoA carboxylase and enzymes involved in the production of reducing equivalents needed to support lipogenesis. With time the normal mouse will adapt to this high sugar diet and develop more efficient lipid packaging and export systems. At this point the liver lipid level will return to its normal level.

The remaining non absorbed food plus the desquamated epithelial cells and residual digestive enzymes

TABLE 15.2: Lipid carrying proteins	
Protein	Function
Apo AII	Protein in the high density lipoprotein
Apo B-48	Protein in the chylomicrons
Apo A-I	Protein in chylomicrons
Apo C-III	Protein in very low density lipoproteins
Apo A-IV	Protein in chylomicrons
CETP	Cholesterol transport protein originating in peripheral cells
LCAT	Protein in high density protein
Apo E	Mediates binding of low density lipoproteins to receptor
Apo C-I	Protein in low density lipoprotein
Apo B-100	Protein in the very low density protein

and mucins now leave the ileum entering the colon or large intestine. At this juncture is the cecum. In rodents, the cecum is a rather large appendage projecting caudally from the ileum–colon juncture. In rodents this sac functions as a fermentation vat. In the mouse it can be up to a third of the length of the large intestine. Its size is dependent on the composition of the diet. Diets containing large amounts of complex carbohydrates will result in larger cecums than diets containing very little complex carbohydrate. In this fermentation vat, bacteria will act on undigested fibers (as well as other undigested materials) and produce metabolically useful products. Short chain fatty acids are produced here and these can be readily absorbed and used by the body. Some vitamin synthesis occurs here as well as in the colon and this synthesis has a benefit to the animal. In particular, vitamin K is synthesized and will be absorbed by the epithelial cells of the colon. Some of the B vitamins are also produced and absorbed or excreted in the feces. If the animal is maintained on a wire mesh floor that prevents **coprophagy** (consumption of feces), some of this synthesized vitamin will be lost to the animal.

Finally, at the end of the colon is the rectum. Here, the last remnants of the intestinal contents are 'stored' prior to defecation. Mucus is produced to lubricate and enhance defecation. The longitudinal layer is thin compared to that of the colon except for two distinct longitudinal bundles called *taenia coli*. The opening at the end of the rectum is called the anus and consists of keratinized, stratified squamous cells. The lamina propria is very thick and contains *circumanal* glands that open by short broad ducts to the anal canal. The striated muscle fibers are specialized into internal and external sphincters needed to expel the feces.

Regulation of food intake

The cells of the GI tract have both an endocrine and an exocrine function. The release of the digestive enzymes has already been described. As such, these cells have a function with respect to nutrient digestion and absorption. They also have an endocrine function in that they can release substances that have hormone activity. The definition of a hormone is a substance that is released by one cell type into the blood and then has an effect on a distant cell. Cholesystokinin and gastrin meet this definition as does somatostatin (the long form, SS-28). SS-28 serves as a paracrine to inhibit gastrin release and hydrochloric acid release. The GI tract also releases neuromedin B. This neuropeptide counteracts SS-28. Galanin, vasoactive peptide (VIP), gastrointestinal peptide (GIP), motilin, calcitonin gene-regulated peptide (CGRP), endothelin, neurotensin, met-enkephalin, leu-enkephalin, enteroglucagon and bombesin are all 'hormones' produced by cells in the GI tract. Some of these have a role in the generation of a hunger signal while others serve to signal satiety. These substances and their functions are listed in Table 15.3. Most of these signals are of very short duration (Weigle, 1994) and, although they can stimulate eating or satiety, their effects are not long lasting. Longer lasting signals to the brain are thought to control the duration of feeding and

TABLE 15.3: Hormones released by the gastrointestinal tract	
Hormone	**Function**
Cholecystokinin (CCK)	Stimulates exocrine pancreas to release digestive enzymes
	Found in the brain and thought to signal satiety
Gastrin	Stimulates gastric acid and pepsin release; found in the brain
Secretin	Stimulates bicarbonate release
Enkephalins	Inhibit gut motility, 'may' signal satiety
Enteroglucagon	Trophic factor for intestinal mucosa
GIP	Inhibits gastric acid secretion
Motilin	Stimulates movement of chyme through intestine
Neurotensin	Inhibits insulin release
Pancreatic polypeptides	One of these, neuropeptide Y, stimulates food intake
Somatostatin	Inhibits growth hormone release
Bombesin	Signals satiety
Tachkinin	Signals satiety?
Endothelin	Vasoconstrictor

its cessation. These signals include leptin (Liebel, 1997), circulating glucose (Langhans,1996) and fatty acids (Harris and Martin, 1984). Some amino acids and proteins may also play a role.

Endocrine aspects of digestion and absorption

In addition to those satiety signals generated by cells of the GI system, there are those released by adipose tissue (leptin) and the central nervous system (CNS) that affect food intake as well as metabolism. Hormones (insulin, somatostatin, glucagon) generated by the endocrine pancreas and the signals (mainly metabolites) generated by the liver affect GI function as well as metabolism. For this reason, the pancreas and the liver are included in the description of the digestive system. The exocrine pancreas produces digestive enzymes as described above and the liver produces the bile needed for fat absorption. In addition, the liver is the first tissue to receive, via the portal vein, the products of digestion and absorption. As such, it is the central integrator of metabolism as it pertains to the use of nutrients derived from food.

The pancreas is both an exocrine and an endocrine gland. As an endocrine gland it produces and releases three hormones that have major regulatory authority over the use of food components by the body. These hormones include: insulin, produced by the β cells of the islets of Langerhans, glucagon, produced by the α cells and somatostatin (the short form, SS-14) produced by the d cells of the islets. These three hormones determine the fate of intermediary metabolism whether it be catabolic or anabolic. That is, whether macromolecules are degraded or synthesized. Somatostatin serves as a regulator of the balance of insulin and glucagon. It inhibits the release of either of these hormones and thus serves to prevent excess levels of either. Insulin acts primarily in the fed animal to facilitate the use of glucose and amino acids while glucagon acts primarily in the starving animal to facilitate the mobilization of energy stores. Glucose from the ingested food is absorbed by the enterocytes and transferred to the blood. Rising blood glucose signals the pancreas to release insulin. Insulin, in turn, stimulates the uptake and metabolism of this glucose by the rest of the body. When the animal ceases to eat and the circulating glucose falls, glucagon is released that in turn facilitates or stimulates the release of glucose from glycogen and also stimulates the synthesis of glucose (gluconeogenesis) from metabolites such as pyruvate.

Metabolism

Pathways of intermediary metabolism

The term intermediary metabolism covers all those reactions in the body concerning the conversion of the products of digestion to useful molecules. It includes the synthesis of macromolecules as well as macromolecule breakdown. As such, metabolism is divided into two parts: catabolism, macromolecule breakdown and anabolism, macromolecule synthesis. Catabolic processes include glycolysis, pentose shunt, glycogenolysis, fatty acid oxidation, lipolysis and amino acid catabolism. Anabolic pathways include glycogenesis, protein synthesis, lipogenesis and cholesterogenesis. Each of these pathways functions continuously yet their activity is determined by the nutritional and hormonal state of the animal. Starving animals are primarily catabolic. Glycogen stores are raided, fat stores are used and some amino acids are oxidized. Glucose is synthesized (gluconeogenesis) to provide this essential fuel to the CNS. Some anabolism does take place during starvation but it is minimal compared with that which takes place in the non-starving animal. Similarly, the fully fed, resting animal is primarily anabolic. Some catabolism is taking place to provide the substrates needed to support anabolism but major macromolecule destruction is minimal. In between starvation and full feeding, there is a balance between anabolism and catabolism that occurs such that the physiological state of the animal is optimized. In this respect there is a constant and steady state of metabolism that is maintained until this state is perturbed. Perturbations can be the result of changes in the environment (food supply, temperature, changes in lighting schedule) or internal processes (chronic disease, pathogens, reproductive activity, age). In any event, normal animals adjust to such perturbations and a new steady state is established to ensure survival. If such perturbations are overwhelming, the animal will not be able to adjust and death will ensue.

Metabolic control

Metabolic processing of carbohydrate, lipid and protein is carefully regulated. Glycolysis, shown in Figure 15.2 has three steps that regulate the flux from glucose to pyruvate. The reverse of glycolysis, gluconeogenesis, is shown in part in this figure and again in Figure 15.3. Also shown in this figure is the pentose phosphate

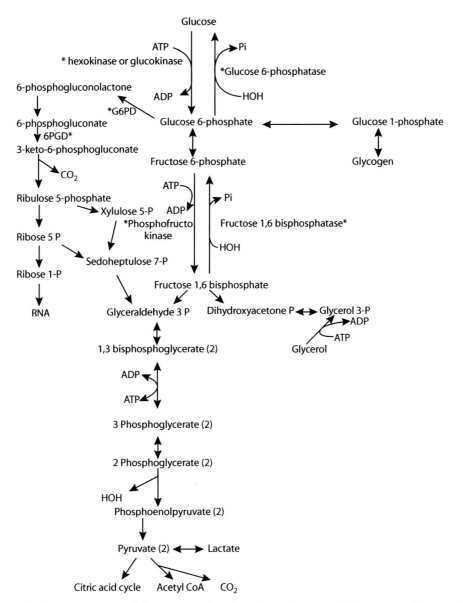

Figure 15.2 The glycolytic sequence and the pentose phosphate shunt. Key rate limiting steps are indicated*. Part of the gluconeogenic sequences is also shown through the reverse arrows in the glycolytic sequence. The relationship of the glycolytic sequences to glycogen is indicated.

shunt that provides reducing equivalents for the support of lipogenesis. When lipogenesis is very active, as in a mouse fed a sugar rich diet, pentose shunt activity rises. The shunt also produces phosphorylated ribose that is used for DNA and RNA synthesis. The rate limiting steps for glycolysis, pentose shunt and gluconeogenesis are listed in Table 15.4. The end product of glycolysis, pyruvate, enters the citric acid cycle (Figure 15.4) for oxidation and produces reducing equivalents that in turn are transferred to the respiratory chain (Figure 15.5). The energy developed by the respiratory chain is captured in the high energy bond of ATP and this high energy material transfers its energy to those reactions

requiring it. ATP synthesis and the respiratory chain are coupled in the process called oxidative phosphorylation or OXPHOS.

Once glucose is phosphorylated it can go into several pathways: glycolysis, the pentose phosphate shunt (Figure 15.2) or into glycogenesis. Glycogenesis is minimally shown in Figure 15.2. Glucose 6-phosphate is isomerized to glucose 1-phosphate and then converted to UDP glucose. Through the action of active glycogen synthetase, glucose units are added stepwise to preexisting glycogen molecules. The animal never uses all of its glycogen even when starving. It always reserves a small amount to serve as a primer for subsequent glycogen

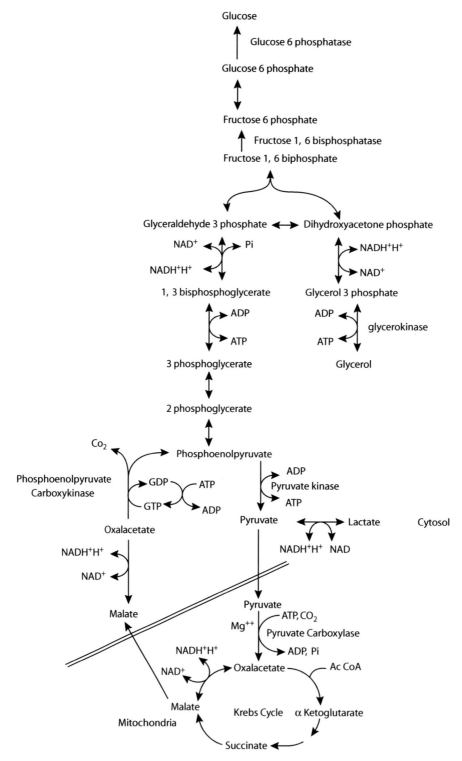

Figure 15.3 Gluconeogenesis. This pathway is very expensive with respect to its energy need and is used primarily in the starving animal. Its primary regulatory step is that catalyzed by phosphoenolpyruvate carboxykinase.

synthesis. It might be unmeasurable in quantity but it is always there. Glycogen breakdown (Figure 15.6) provides a quick source of energy to working muscle. The release of glucose from glycogen is carefully regulated via a cascade of events initiated by one or more of the catabolic hormones. Epinephrine for example stimulates glycogenolysis. This catecholamine stimulates the activation of adenyl cyclase that catalyzes the conversion of ATP to cAMP. Cyclic AMP stimulates the activation of cAMP dependent protein kinase that in turn

TABLE 15.4: Rate limiting enzymes in metabolic pathways	
Pathway	**Rate limiting enzymes**
Glycolysis	Hexokinase(glucokinase), phosphofructokinase, pyruvate kinase
Pentose shunt	Glucose 6 phosphate dehydrogenase, 6-phosphogluconate dehydrogenase
Glycogen synthesis	Glycogen synthase
Glycogenolysis	Glycogen phosphorylase (a reaction cascade)
Gluconeogenesis	Phosphoenolpyruvate carboxykinase, glucose 6-phosphatase, glucose 1,6 phosphatatse
Fatty acid synthesis	Acetyl CoA carboxylase, fatty acid synthetase
Citric acid cycle	Pyruvate dehydrogenase, pyruvate kinase, ATP-citrate lyase
Lipolysis	Lipoprotein lipase
Fatty acid oxidation	Carnitine palmitoyl transferase
Cholesterol synthesis	HMG CoA reductase

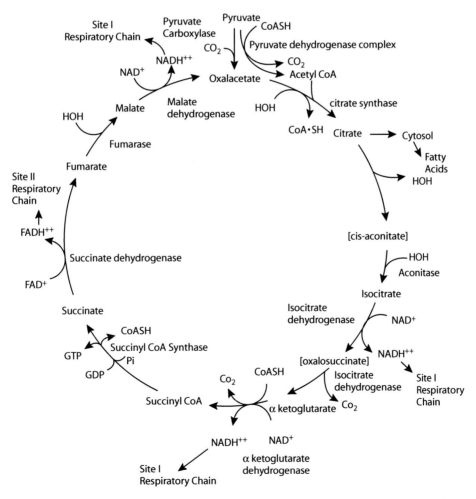

Figure 15.4 The Krebs Citric Acid Cycle in the mitochondria. This cycle is sometimes referred to as the tricarboxylate cycle.

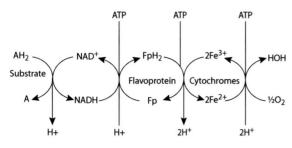

Figure 15.5 The respiratory chain. Products of glycolysis and fatty acid oxidation contribute their reducing equivalents to this chain. Pyruvate and α ketoglutarate contribute their reducing equivalents via lipoate at the first entry site of the chain. Proline, 3-hydroxyacyl-CoA, 3-hydroxybutyrate, glutamate, malate and isocitrate contribute reducing equivalents via NAD linked dehydrogenase, and succinate, choline, glycerol 3-phosphate, acyl CoA, sarcosine and dimethylglycine all contribute via FAD linked flavoproteins. Each time a proton gradient is developed ATP is synthesized. A total of 3 ATPs can be synthesized when reducing equivalents enter via NAD. Two ATPs are synthesized when reducing equivalents are contributed via FAD.

activates phosphorylase b that in turn activates glycogen phosphorylase a. Once glycogen phosphorylase a is activated it catalyzes the phosphorylation of glycogen and glucose 1-phosphate is released. After isomerization to glucose 6 -phosphate the glucose is available for use. Gluconeogenesis is shown in detail in Figure 15.5. The main rate controlling step is the conversion of malate to phosphoenolpyruvate through the action of the enzyme PEPCK, phosphoenopyruvate carboxykinase. Malate availability is determined by the outward flow of malate from the mitochondrial compartment into the cytosolic compartment. This flux known as the malate aspartate shuttle, is in turn influenced by the phosphorylation state of the compartment that is in turn influenced by OXPHOS. PEPCK is very active in starvation and is controlled by hormonal state (Phillips and Berry, 1970). Lipogenesis (Figure 15.7) and lipolysis plus fatty acid oxidation (Figures 15.8 and 15.9) complete the energy storing and releasing processes. Again, the rate limiting steps are shown in Table 15.4. Cholesterol synthesis uses the same starting substrate as

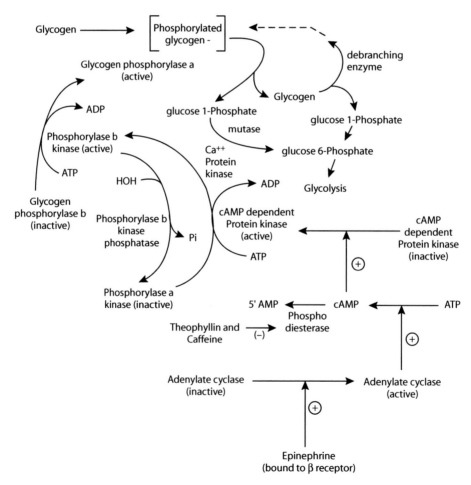

Figure 15.6 Glycogen breakdown catalyzed by a cascade of reactions initiated by one or more catabolic hormonal signals such as epinephrine.

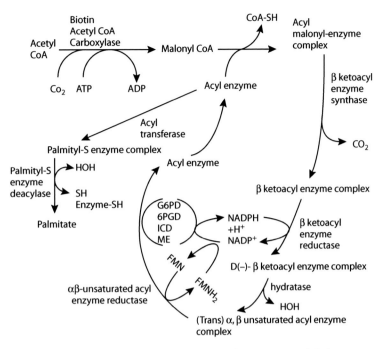

Figure 15.7 Lipogenesis. The synthesis of fatty acids from the two carbon unit, acetyl CoA.

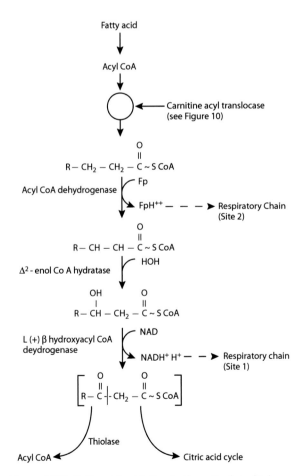

Figure 15.8 Fatty acid oxidation in the mitochondrial compartment. The key limitation of fatty acid oxidation rests with the entry of the fatty acid into the mitochondrial compartment.

fatty acid synthesis, acetyl CoA. Acetyl CoA is joined to acetoacyl CoA to form HMG CoA. This product is converted to melvalonate and it is this step that is rate limiting. Melvalonate is converted to farnesyl phosphate then to squalene then to lanosterol and finally to cholesterol. This cholesterol can be used for the synthesis of the sex hormones, for vitamin D, for glucocorticoids, mineral corticoids and can be esterified to produce cholesterol esters. Protein degradation and amino acid recycling into new protein as well as amino acid oxidation are sometimes included in the pathways of intermediary metabolism. However, in terms of energy balance, the contribution of amino acids as an energy source is minor compared with that of the lipids and carbohydrates. Protein synthesis and degradation are heat producing processes and cannot be ignored. These processes will not be discussed in this section.

Heat production (**thermogenesis**) is a feature of all metabolic processing. The energy lost and gained by metabolism contributes to the energy balance of the animal. The animal is by no means energetically efficient. Most of the energy lost from metabolic processing is lost as heat. This heat is needed to sustain body temperature and optimize enzyme activity. Some energy is used for neural transmissions and vision (chemical energy or electrical energy) and some is used for the work of the body (mechanical energy). Above and beyond the normal amount of heat energy lost through metabolic processing, additional heat can be generated on demand should environmental conditions so warrant.

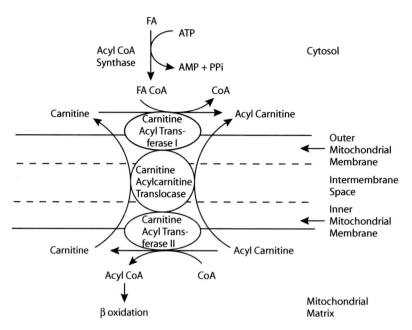

Figure 15.9 Scheme for the entry of fatty acids into the mitochondrial compartment.

Thermogenesis by brown fat pads is an important contributor to this extra energy release. A mouse suddenly thrust into a 4°C environment will begin adapting to this cold by stimulating brown fat thermogenesis. The catecholamines and the thyroid hormones play an important role in this thermogenic process. Upon exposure to cold the normal mouse will immediately release norepinephrine that in turn induces the production of an uncoupling protein that in turn dissipates the proton gradient generated by the mitochondrial respiratory chain. This results in an uncoupling of oxidative phosphorylation in the brown fat cells and the energy usually trapped in the high-energy bond of ATP is released as heat instead. Cold exposed genetically obese mice (Lep^{ob}/Lep^{ob} and others) cannot rapidly respond to this cold environment and die from hypothermia. When the temperature is gradually lowered they are able to adapt so this feature of genetic obesity is characteristic of the phenotype not the cause of its phenotype.

With respect to metabolism in general, the controls of metabolism are similar to those of other mammals. Glycolysis, pentose phosphate shunt, citric acid cycle, oxidative phosphorylation, lipogenesis, lipolysis, ketone formation, glycogen synthesis, gluconeogenesis, glycogenolysis, protein synthesis and degradation are all controlled not only by the availability of the starting substrates as determined by nutritional state but also by certain of the enzymes in the metabolic pathway as well as by the hormonal state of the animal.

Hormonal status can be genetically determined. The mice with mutations that phenotype as diabetes for example will have their metabolic pathways perturbed by their developing diabetic state. For example, Roesler and Khandelwal (1985) followed $Lepr^{db}$ mice and their controls for 16 weeks. Initially, there were no differences in glycogen metabolism between the two groups of mice. However, with age, glycogen synthetase activity rose in the $Lepr^{db}$ mouse reaching a peak around 8–9 weeks. Glycogen synthesis also peaked at this time. Subsequently, synthesis fell as glycogenolysis began to rise in these mice compared with their normal controls. The rise in glycogen breakdown was measured through the activity of phosphorylase a. These changes in glycogen metabolism coincided with the development of the diabetic state in the $Lepr^{db}$ mouse.

Other hormones (in addition to insulin) are also active in the control of intermediary metabolism. The adrenal cortical hormones, the adrenal medullary hormones, the pituitary hormones, the thyroid hormones all have an impact on the flux of substrates through metabolism. Some of these hormonal effects are direct while others are indirect. Nonetheless, all are needed for the integration of metabolism such that nutrients from the food are absorbed, oxidized, or used for the synthesis of macromolecules in the body. The details of the actions of these hormones will be covered in Chapters 19 and 20.

Differences in metabolism

The regulation of food intake is important to the understanding of metabolism. Where hyperphagia occurs due

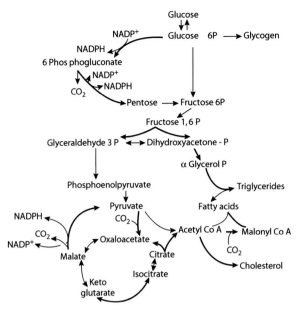

Figure 15.10 Intermediary metabolism. Heavy arrows show steps that are more active in lipogenic animals.

TABLE 15.5: Glucokinase activities in various mouse strains	
Strain	Enzyme activity (nmoles substrate used/min/mg protein)
CBA/J	19.2 ± 1.25
SM/J	18.2 ± 0.53
C3H/HeJ	16.8 ± 0.63
C57Bl/6J	13.9 ± 0.55
AKR/J	13.1 ± 0.36
RF/J	7.41 ± 0.38
C58/J	7.33 ± 0.37
Transgenic FBP/m[a]	14.0 ± 2.0

[a]Mice expressin uman hepatic glucokinase.

to an error in the production/release/response to satiety factors, the mouse will have an enlarged GI tract due to its increased food consumption and will be more anabolic than catabolic. As hyperphagia means the consumption of more than the needed amount of food to sustain health, this surplus food will provide additional substrates for the anabolic reactions that are part of metabolism. Figure 15.10 is an abbreviated outline of metabolic processing of carbohydrate and its conversion to fat. The rate limiting steps are shown as heavy arrows and indicate that the hyperphagic mouse with the obese phenotype will have a more active reaction sequence than a normal mouse. However, obesity aside, there are variations in the activities of key regulatory steps in these metabolic pathways. As an example, Table 15.5 provides a strain comparison in the activity of hepatic glucokinase (Coleman, 1977; Hariharan et al., 1997). The phosphorylation of glucose via the enzyme glucokinase, is the first rate limiting step in hepatic glycolysis (Ferre et al., 1996). In the pancreas the glucokinase reaction is the glucose sensor of the endocrine pancreatic β cell and signals insulin release. Note that considerable variation in hepatic glucokinase activity exists. In part, this variation may be related to genetic variation in the expression of the gene for glucokinase regulatory protein (Schaftingen et al., 1994). Glucokinase activity is also regulated by the amount of substrate (glucose) available (Towle et al., 1996) and the amount of long-chain acyl-CoAs (Dawson and Hales, 1969; Tippett and Neet, 1982). The latter inhibits enzyme activity while the former stimulates it. However, one should also realize that the amount of enzyme or its activity measured *in vitro* is not a true measure of the activity of the pathway in question. Enzymes rarely work at saturation *in vivo*. The measurement of such activity *in vitro* is an optimized measurement. That is, all the needed substrates and cofactors are provided in optimal amounts. The best measurement of a metabolic pathway is through a dynamic measurement of the flux through the pathway. This is rarely done because of the difficulty in assessing back flow as well as disposal rates of the end products of the pathway. In some instances, corrections for recycling must be applied and these are very difficult indeed to calculate and measure. An example of flux measurement reported by Smith et al. (1986) is shown in Table 15.6. These workers compared obese-diabetic C57BL/6-Lep^{ob}/Lep^{ob} and normal (C57BL/6) mice. As expected, the obese hyperglycemic mice were less glycolytic than the normal controls as shown by the reduction in the fractional glucose use. Blood glucose levels were higher in the fed state in both genotypes and the mutant genotype had higher blood glucose levels regardless of feeding status. Note also that the mutant genotype had higher glucose synthesis rates and glucose recycling was higher when the mice were in the fed state. Liver glycogen reflected the abnormal glucose metabolic flux: the mutant mice had far higher levels of glycogen in both fed and fasted state and this liver glycogen may reflect a defense of the animal against elevated blood glucose levels. By synthesizing and storing glucose as glycogen the blood glucose level can be reduced.

Similar shifts in fatty acid synthesis, deposition and lipolysis are observed in mice that have the obesity phenotype. The hyperphagia of these genotypes increases

TABLE 15.6: Glucose flux in fed and fasted lean and obese mice				
Measurement	Lean, fed	Lean, fasted	Obese, fed	Obese, fasted
Blood glucose, (μmol/ml)	10.4 ± 0.5	6.0 ± 0.3	15.4 ± 0.9	9.9 ± 0.7
Frac. glucose use (%/min)	3.22 ± 0.60	3.52 ± 0.14	2.83 ± 0.17	3.18 ± 0.20
Glucose synthesis (μmol/min)	5.6 ± 0.6	4.4 ± 0.2	8.7 ± 0.8	6.4 ± 0.8
Cori cycle (%) (glucose recycling)	16.8 ± 0.8	24.1 ± 1.6	23.9 ± 2.4	25.4 ± 2.1
Liver glycogen (μmol/liver)	316 ± 28	34 ± 5	773 ± 33	356 ± 47

the intake of carbohydrate that must be metabolized. The carbohydrate can be oxidized, recycled as shown above or converted to fatty acids and stored as triglycerides. The increase in the amount of substrate that must be converted to fat induces an increase in the activities of those enzymes needed for such synthesis, as well as those proteins needed for transport and those needed to facilitate storage. Numerous reports exist that document the increase in de novo lipogenesis that occurs in obese strains of mice. Typically these strains have a fatty liver as well as enlarged adipose tissue storage depots.

In animals having mutations in genes relating to glucose use, the phenotype is that of diabetes (Leiter, 1989). In some, the phenotype can also include obesity but in others, obesity is absent. The NOD mouse typifies the latter while the Lep^{ob}/Lep^{ob} and $Lepr^{dp}/Lepr^{dp}$ typify the former. The genetic defect thus determines the metabolic defect. In the obese Lep^{ob}/Lep^{ob} mouse, the mutant gene encodes leptin, the hormone produced by the adipocyte that signals satiety to the brain. In the $Lepr^{dp}/Lepr^{dp}$ mouse the defective gene encodes the leptin receptor. In both instances the satiety signal is aberrant and in both hyperphagia is part of the phenotype and can explain the metabolic patterns in these mice.

The defect in the NOD mouse involves a defect in one or more components of the immune system. These mice are not obese. Their pancreases have signs of autoimmune disease. The insulin production of these mice decline with age and diabetes is apparent. The metabolic patterns are those of insulin deficit. Glucose is not used appropriately and fatty acid mobilization occurs. This means an increase in peripheral lipolysis, and increase in fatty acid oxidation and because of the insulin deficit, an increase in ketone production. When these mice develop these signs of abnormal intermediary metabolism they must receive exogenous insulin or they will die.

There are other genetically determined variants in metabolism in the mouse and these have been exploited by researchers interested in nutrition. Specific dietary components can influence the phenotypic expression of a specific genotype as illustrated by Surwit et al. (1988). These investigators used two strains of mice, C57BL/6J and A/J fed either a stock diet or a high fat–simple carbohydrate diet. The latter diet induced obesity but only the C57BL/6J mice became glycemic. Actually, the composition of the diet can determine when a particular phenotype will be observed. Leiter et al. (1983) fed diets differing in carbohydrate content to db/db mice. They found that a 60% simple sugar diet elicited the diabetic state more rapidly than did 8% or 24% sugar diet. Further, when these mice were fed a starch diet, the phenotypic expression of the diabetes genotype was delayed and lifespan was extended.

Diurnal variation in metabolism occurs in normal mice just as it occurs in other species. The synchronizer of these rhythms is not fully known but the lighting schedule (hours of dark and light) influences the feeding pattern. In turn the feeding pattern influences metabolic flux. When glycolysis is high, gluconeogenesis is low; when lipogenesis is high, lipolysis is low; when glycogen synthesis is high, glycogenolysis is low. These highs and lows follow a daily pattern that appears to be cued by light (Bailey et al., 1975; Hems et al., 1975; Cornish and Cawthorne, 1978; Meier and Cincotta, 1996; Bartness et al., 2002). In contrast, genetically obese mice seem to lose this diurnal rhythm and eat fairly constantly throughout the day–night cycle. As a result, they also lose the above described rhythm in metabolism. Changes in food availability will also affect metabolic rhythm. If mice are forced to consume all of their food in a single 2-h meal instead of feeding ad libitum, the rhythm of anabolic processing of this food will be cued by the timing of food availability rather than by the lighting cycle (Cornish and Cawthorne, 1978).

Food restriction can not only affect metabolic rhythms but can also affect lifespan. In turn, lifespan can be genetically determined. Variation in the responses of different strains of mice to restricted feeding have been published (Harrison and Archer, 1987). Food restriction to two thirds that of ad libitum fed mice extended the lifespan of several different strains of mice. Those that were normally long lived (B6CBAF-1 hybrids) had substantially longer lifespans than the shorter-lived ob/ob mice. Yet, food restriction benefited the ob/ob mice as well. A third mouse strain(C57BL/6J) when food restricted had the reverse response. That is, their lifespans were reduced when food restricted. No overt differences in metabolism were observed in these three strains yet differences in response were found.

Finally, it must be noted that the literature on metabolism is enormous. Mice have been studied extensively both as they respond to experimental manipulation of their diets and as they respond to genetic manipulation through the use of transgenic technology. Mice are excellent models for humans because of the similarity in metabolic patterns and the similarity of digestive systems. Two outstanding differences do exist however. One, the mouse lacks the gall bladder found in the human and the mouse has a relatively large cecum compared to the human. These differences aside, the small size and short lifespan of the mouse makes it an ideal research tool for studies of metabolism and nutrient-gene interactions. What can be learned from the mouse can be applied to some degree to the human.

References

Bailey, C.J., Atkins, T.W., Conner, M.J., Mantley, C.G. and Matty, A.J. (1975). *Hormone Res.* **6**, 380–386.

Bartness, T.J., Demas, G.E. and Song, C.K. (2002). *Exp. Biol. Med.* **227**, 363–376.

Cornish, S. and Cawthorne, M. (1978). *Metab. Res.* **10**, 286–290.

Chua, S. and Leibel, R.L. (1997). *Diabetes Rev.* **5**, 2–7.

Coleman, D.L. (1977). *Biochem. Genet.* **15**, 297–305.

Dawson, C.M. and Hales, C.N. (1969). *Biochim. Biophys. Acta* **176**, 657–659.

Ferre, T., Riu, E., Bosch, F. and Valera, A. (1996). *FASEB J.* **10**, 1213–1218.

Hariharan, N., Farrelly, D., Hagan, D., Hillyer, D., Arbeeny, C., Sabrah, T., Treloar, A., Brown, K., Kalinowski, S. and Mookhtiar, K. (1997). *Diabetes* **46**, 11–16.

Harris, R.B.S. and Martin, R.J.(1984). *Nutr. Behav.* **1**, 253–275.

Harrison, D.E. and Archer, J.R.(1987). *J. Nutr.* **117**, 376–382.

Hems, D.A., Roth, E.A. and Verrinder, T.R. (1975). *Biochem. J.* **150**, 167–173.

Langhans, W. (1996). *Proc. Nutr. Soc.* **55**, 497–515.

Liebel, R.L. (1997). *J. Nutr.* **127**, 1908S.

Leiter, E.H., Coleman, D.L., Ingram, D.K. and Reynolds, M.A. (1983). *J. Nutr.* **113**, 184–195.

Leiter, E.H. (1989). *FASEB J.* **3**, 2231–2241.

Meier, A.H. and Cincotta, A.H. (1996). *Diabetes Rev.* **4**, 464–487.

Paigen, B., Morrow, A., Brandon, C., Mitchell, D. and Holmes, P. (1985). *Atherosclerosis* **57**, 65–73.

Roberts, S.B. and Greenberg, A.S. (1996). *Nutr. Rev.* **54**, 41–49.

Phillips, L.J. and Berry, L.J. (1970). *Am. J. Physiol.* **219**, 797–701.

Roesler, W.J. and Khandelwal, R.L. (1985). *Diabetes* **34**, 395–402.

Schaftingen, E.V., Detheux, M. and Da Cunha, M.V. (1994). *FASEB J.* **8**, 414–419.

Smith, S.A., Cawthorne, M.A. and Simson, D.L. (1986). *Diabetes Res.* **3**, 83–86.

Surwit, R.S., Kuhn, C.M., Cochrane, C., McCubbin, J.A. and Feinglos, M.N. (1988). *Diabetes* **37**, 1163–1167.

Tippett, P.S. and Neet, K.E. (1982). *J. Biol. Chem.* **257**, 12839–12845.

Towle, H.C., Kaytor, E.N. and Shih, H.M. (1996). *Biochem. Soc. Trans.* **24**, 364–368.

Weigle, D.S. (1994). *FASEB J.* **8**, 302–310.

Wolff, G.L. (1997). *J. Nutr.* **127**, 1897S–1901S.

CHAPTER 16

Inbred Mouse Models for Autoimmune Disease

Rikard Holmdahl

Medical Inflammation Research, Lund University,
Lund, Sweden

Abstract

Traditionally mouse models for autoimmune disease have largely been used to analyse hypothetical functions of newly discovered genes or proteins thereby achieving a link to a relevant disease. This approach successfully addressed specific molecular interactions in pathophysiological pathways but has been less productive for the understanding of specific autoimmune disease. A specific autoimmune disease is the result of a complex interaction between many genes and between genes and environmental factors. In addition, the crucial events for understanding of pathologic development is the chronicity of these diseases. Both complexity and chronicity are best approached by developing a better understanding of the animal model diseases themselves. These diseases can subsequently be dissected through the identification of the relevant genes and the important environmental factors. This approach is however in its infancy and this review mainly addresses the available animal models and the first steps used to reveal unique pathologic characteristics of the various diseases and to identify crucial genes. The material will be taken mostly from experience with arthritis models.

Introduction

Autoimmune disease is an undefined entity and represents a collection of highly diverse diseases. In a strict sense we will classify autoimmune diseases as being caused and driven by an immune reaction towards self tissue. In reality, however, this is often a group of diseases in which we have not yet found a distinct environmental cause such as an infectious agent and in which pathogenic autoimmune recognition occurs. A typical example would be Grave's disease in which antibodies to the thyroid stimulatory hormone receptor play a pathogenic role. In its broadest sense we will group together diseases in which autoimmune responses occur at any stage for example in artherosclerosis where T cell responses to vascular antigens are seen or in rheumatoid arthritis (RA) in which autoimmune responses to joint antigens are often recorded. It is also clear that all of these diseases are more or less complex as they are dependent on several genes and several environmental factors. They are the result of a lack of balance between our genes and a changing environment and probably reflects our difficulty in coping with the rapid evolution of infectious agents, which has been dramatically changed depending not only on the infectious agents themselves but also changing life style. In other words, most common autoimmune diseases are induced diseases and a deeper understanding of their complexity is required. This is also highly relevant for autoimmune animal models, which should be created to act as a mirror for the human diseases.

Because autoimmune diseases and their respective animal models are highly diverse, the focus here will be on one condition that includes the relevant considerations for using autoimmune animal models: this condition is arthritis.

Induced arthritis models

There are many agents that induce arthritis in mice or rats (Table 16.1). This includes live bacteria (Bremell *et al.*, 1990) or bacterial components (Pearson, 1956), pure adjuvant oils (Vingsbo *et al.*, 1996), ubiquitous antigens (Kouskoff *et al.*, 1996) and cartilage specific proteins (Courtenay *et al.*, 1980; Glant *et al.*, 1987;

TABLE 16.1: Overview of animal arthritis models

Model	Species	Genetics	Disease characteristics	References
Arthritis caused by infection				
Mycoplasma induced arthritis	Rats and mice	More pronounced in B cell deficient mice	Mild chronic arthritis	(Cole *et al.*, 1971; Berglöf *et al.*, 1997)
Borrelia induced Sarthritis	Mice	MHC, non-MHC loci identified	Severe and erosive arthritis with spirochetes in the joints	(Schaible, 1991; Yang *et al.*, 1994; Roper *et al.*, 2001)
Staphylococcus induced arthritis	Rats and mice	MHC	Severe arthritis	(Bremell *et al.*, 1991, 1994; Abdelnour *et al.*, 1997)
Yersinia induced arthritis	Rats and mice	LEW and SHR but not DA and BN rats	Severe arthritis with bacteria in the joints	(Hill and Yu, 1987; Gripenberg-Lerche *et al.*, 1994)
Arthritis caused by fragments from bacteria persisting in joints				
Adjuvant (mycobacterium cell wall) induced arthritis (AA)	Rats	MHC, non-MHC genes (LEW > F344)	Acute and generalised inflammatory disease including erosive arthritis	(Pearson and Wood, 1959; Kawahito *et al.*, 1998; Joe *et al.*, 2002)
Streptococcal cell wall induced arthritis	Mice and rats	Non-MHC genes (LEW > F344), (DBA/1 = BALB/c > B10)	Severe and erosive arthritis	(Dalldorf *et al.*, 1980; Wilder *et al.*, 1982; Koga *et al.*, 1985)
Arthritis induced by adjuvant injection				
Avridine induced arthritis (AvIA)	Rats	MHC (*f*)	Very severe, erosive and chronic arthritis	(Chang *et al.*, 1980; Vingsbo *et al.*, 1995)
Oil (mineral oil) induced arthritis (OIA)	DA rats	Non-MHC loci on RNO 4, 10	Acute and self-limited inflammation of peripheral joints	(Holmdahl and Kvick, 1992; Holmdahl *et al.*, 1992a; Lorentzen *et al.*, 1998)

TABLE 16.1: (Continued)				
Model	**Species**	**Genetics**	**Disease characteristics**	**References**
Pristane induced arthritis (PIA)	Rats	MHC (f), non-MHC loci on RNO 4, 6, 12, 14	Chronic and erosive arthritis in peripheral joints	(Vingsbo et al., 1996; Vingsbo-Lundberg et al., 1998)
PIA	Mice	MHC (q, d)? BALB/c, DBA and C3H gene backgrounds	Generalized inflammatory disease predominately affecting joints. Not similar to PIA in rats.	(Potter and Wax, 1981; Wooley et al., 1989)
Arthritis induced by cartilage protein immunization				
CII (heterologous or homologous CII in mineral oil) induced arthritis (CIA)	Rats	MHC (a, l, f and u), non-MHC loci on RNO 1, 4, 7, 10	Chronic and erosive arthritis in peripheral joints.	(Trentham et al., 1977; Holmdahl et al., 1992b; Remmers et al., 1996)
CII (heterologous or homologous CII in CFA) induced arthritis (CIA)	Mice	MHC (q and r), non-MHC loci on MMU 3	Erosive arthritis in peripheral joints	(Courtenay et al., 1980; Wooley et al., 1981; Holmdahl et al., 1986a; Jirholt et al., 1998)
CXI (rat CXI in IFA) induced arthritis	Rats	MHC ($f > u > a$), DA > LEW rats	Chronic relapsing arthritis	(Lu et al., 2002)
Human proteoglycan (in CFA) induced arthritis	BALB/c c mice		Chronic arthritis	(Glant et al., 1987)
COMP (in mineral oil) induced arthritis	Rats	MHC (u)	Acute arthritis	(Carlsén et al., 1998)
'Spontaneous' arthritis models				
HLA-B27 transgenic animals	Mice and rat	B27 heavy chain transgene	Ankylosing spondylitis, colitis, balanitis, arthritis	(Hammer et al., 1990; Khare et al., 1996)
The MRL-*Tnfrsf6*^{lpr} mouse (mutation in the *Tnfrsf6* gene controlling apoptosis)	Mice	*Tnfrsf6*^{lpr}	Generalized inflammation as a part of lupus disease which also affect joints	(Hang et al., 1982)
Inter male aggressiveness stress induced arthritis (SIA)	DBA/1 mice	Non-MHC genes	Destruction of peripheral joints with no evidence for immune involvement	(Holmdahl et al., 1992c; Corthay et al., 2000)
TNF-α transgenic mice (overproduction of TNF-α)	Mice	TNF-α transgene	Erosive arthritis as well as generalized tissue inflammation	(Keffer et al., 1991; Butler et al., 1997)
GPI TCR transgenic mouse (T cell autoreactivity)	Mice	TCR transgene	Severe arthritis and immune complex disposition on internal organs	(Kouskoff et al., 1996)
HTLV transgenic mouse	Mice	HTLV transgene	Erosive arthritis	(Iwakura et al., 1991)
HTLV transgenic rat	Rats	HTLV transgene	Generalized tissue inflammation	(Yamazaki, et al., 1997)
IL-1Ra deleted mouse	Mice	Lacking IL-1Ra	Chronic arthritis	(Horai et al., 2000)

Guerassimov *et al.*, 1998). Although the nature of the autoantigens in RA is not fully understood, joint specific antigens such as collagen type II (CII) are candidates. Immunization with CII or the major cartilage proteoglycan, aggrecan, induces arthritis in a number of different inbred strains of mice (Wooley *et al.*, 1981; Otto *et al.*, 1999, 2000).

The most commonly used model for RA in mice and rats is collagen-induced arthritis (CIA) (Trentham *et al.*, 1977; Courtenay *et al.*, 1980). Collagen-induced arthritis is induced by intradermal injections of CII together with mycobacteria suspended in oil (Courtenay *et al.*, 1980) or with oil only (Trentham *et al.*, 1977; Svensson *et al.*, 2002) (Figure 16.1). Most commonly, heterologous CII of bovine or chick origin is used to induce the disease but severe and chronic arthritis can also be induced by homologous CII, mouse CII in mice (Holmdahl *et al.*, 1986a) and rat CII in rats (Larsson *et al.*, 1990). Collagen-induced arthritis is major histocompatibility complex (MHC) dependent and is characterized by erosive joint inflammation mediated by both T cells and B cells. However, the pathogenesis and disease course vary considerably depending on the genetic background of the mice and on the origin of the CII used for disease induction.

A bottleneck in the development of disease is the T cell recognition of an immunodominant CII-derived peptide bound to MHC class II with a certain affinity. The critical importance of a single MHC class II gene was shown in a transgenic mouse model for arthritis. By introducing a change in the *H2^p* resulting in four amino acids change of the beta chain of the MHC class II A molecule in transgenic mice, it mimics exactly the beta chain of the corresponding molecule in the *H2^q* haplotype (Table 16.2). The mice carrying this transgene were now susceptible for CIA (Brunsberg *et al.*, 1994). This is the first example where resistance to an autoimmune disease could be attributed to a single MHC class II molecule. It provides evidence for the importance of class II molecules but does not exclude additional polymorphic genes in the MHC region influencing the disease course of CIA and RA. It should be emphasized that the importance of the A^q molecules in CIA is quantitative and it has been clearly shown that other **MHC haplotypes** in the mouse also contribute with various degree of permissiveness for development of arthritis (Courtenay *et al.*, 1980; Holmdahl *et al.*, 1988; Campbell *et al.*, 2000). The identification of the MHC class II molecule in CIA opened up research to address autoimmune specificity. Full understanding of autoimmune specificity also needs the understanding of **tolerance** induction; self versus non-self or danger versus danger discrimination.

Type II collagen in cartilage is normally exposed to the immune system, and since CII reactive B cells and T cells are of major importance in the pathogenesis of CIA, a fundamental question for understanding CIA, and probably RA as well, is how the immune system is made tolerant to cartilage proteins. The CIA model provides a unique opportunity to address this question since

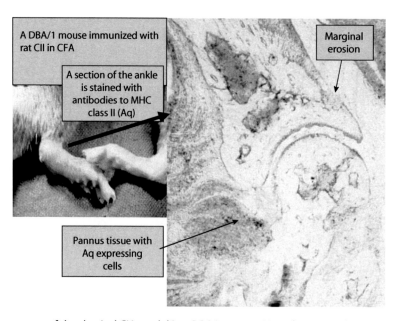

A DBA/1 mouse immunized with rat CII in CFA

A section of the ankle is stained with antibodies to MHC class II (Aq)

Marginal erosion

Pannus tissue with Aq expressing cells

Figure 16.1 The appearance of the classical CIA model in a DBA/1 mouse. Note the severe destruction of the joints with pannus formation and marginal erosions.

TABLE 16.2: MHC class II molecules associated with CIA in the mouse					
MHC class II	CII peptide	MHC binding	T cell activation	CIA	Reference
A_q	259–270	+ +	+ +	+	(Brunsberg, et al., 1994)
A_p	259–270	+	+	–	(Brunsberg, et al., 1994)
DR4*0401	260–270	+ +	+ +	+	(Andersson et al., 1998)
DR1*0101	260–270	+ +	+ +	+	(Rosloniec, et al., 1997)
DQ6	?	?	?	–	(Bradley, et al., 1997)
DQ8	?	?	+ +	+	(Nabozny, et al., 1996)

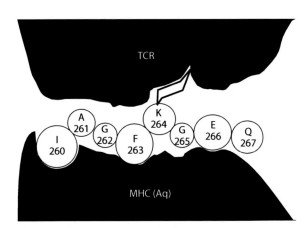

Figure 16.2 The immunodominant galactosylated CII260–267 peptide indicating interactions with the MHC class II Aq molecule and T cells.

the immunodominant collagen peptide (CII260–270) that binds to the Aq molecule (see Figure 16.2) contains an amino acid unique to mouse at position 266; aspartic acid (D) in the mouse and glutamic acid (E) in the human, rat, bovine and chick CII. This leads to a reduced affinity of the binding to the class II molecule and the lower incidence after induction with mouse CII in mouse could be explained by a partial ignorance of the immune system to recognize it but also to being physiologically tolerized against it. To address the question as to whether the availability of CII is important for tolerance and arthritis induction, transgenic mice expressing a mutated CII (at 266) were generated (Malmström *et al.*, 1996). In one transgenic line (TSC) this change was introduced into a collagen type I (CI) construct allowing the systemic expression of the rat CII **epitope**, thereby available to the immune system. In a second transgenic line (MMC), a CII construct was generated with an exchange at position 266 allowing a joint restricted expression of the rat epitope. The systemic expression of the immunodominant epitope conferred total resistance to arthritis in the mice

while a more restricted expression in the cartilage still allowed arthritis development but with a lower susceptibility. This clearly demonstrates that CII induce immune tolerance but that even in a partially tolerant state arthritis may develop. Having a closer look at the T cell response reveals that the T cells no longer proliferate in response to CII but are still able to fulfil effector functions such as the production of the cytokine IFN-γ and B cell support for the anti-CII antibody response. These experiments show that T cell tolerance mechanisms are operating in vivo and to a degree dependent on the availability of the antigen. Interestingly, the B cell compartment is less skewed due to tolerance and autoreactive B cells are easily activated if T cell tolerance is broken. In this situation the B cells produce anti-CII antibodies mediating one important pathway leading to arthritis. B cell deficient mice are resistant to CIA (Jansson and Holmdahl, 1993) and arthritis is inducible by transfer of collagen specific sera (Stuart and Dixon, 1983) and monoclonal antibodies against CII (Holmdahl *et al.*, 1986b; Terato *et al.*, 1992). This model enables investigation of the genes and pathways controlling the later phase of arthritis, the effector phase, after the initiation of the disease process. The transfer of CII specific antibodies, on the other hand, induces arthritis in several inbred mouse strains. However, the penetrance varies depending on the background genome of different mouse strains, indicating a genetic influence on the antibody effector pathway (Watson and Townes, 1985). The importance of antibody-mediated pathways in arthritis development was investigated further by studying the resistance of CIA induction in mice deficient for complement (Watson and Townes, 1985; Hietala *et al.*, 2002), FcγIII receptors (Kleinau *et al.*, 2000) or T cell receptor (TCR)-α chain (Corthay *et al.*, 1999).

Arthritis is also inducible with various types of adjuvant. Rats but not mice are, for unknown reasons, highly susceptible to this type of arthritis induction.

The classical model is the induction of arthritis with complete Freund's adjuvant, which is a mixture between mycobacterial cell walls and mineral oil (Pearson, 1956). It has now been shown that these components trigger separate types of arthritis development (Holmdahl *et al.*, 2001). Only mineral oil, without mycobacteria, and pure adjuvants, such as pristane and squalene, induce severe arthritis in many rat strains. Models, like the PIA in rats, are optimal models for RA since they fulfil the RA criteria including a chronic relapsing disease course (see Figure 16.3). Arthritogenic adjuvants like pristane, avridine, squalene and mineral oil, are not immunogenic since they do not contain MHC binding peptides. Nevertheless, the diseases are MHC-associated and dependent on the activation of abTCR expressing T cells. However, it has not been possible to link the immune response to joint antigens or other endogenous components although immunization with various cartilage proteins induce arthritis (Horai *et al.*, 2000) but with different pathogeneses.

Spontaneous models of arthritis

There are several examples of genetically manipulated mouse strains, which spontaneously develop arthritis (Table 16.1). Various pro-inflammatory cytokines, such as tumour necrosis factor-alpha (TNF-α), interleukin

(IL)-1 and IL-6, are overexpressed in joints of RA patients which is why their importance in the development of arthritis have been suggested. Evidence for a critical role of TNF-α was demonstrated by the spontaneous development of arthritis in mice overexpressing the human TNF-α transgene (Keffer *et al.*, 1991). The disease in transgenic mice was prevented by treatment with antibodies against TNF-α or by blocking IL-1. In addition, mice deficient in IL-1 receptor antagonist (IL-1ra), which is the natural occurring inhibitor of IL-1 and competes for the receptor, also spontaneously developed arthritis (Horai *et al.*, 2000). Similar to TNF-α transgenic mice, the character of the disease in these mice is dependent on the genetic background of different mouse strains. The numerous studies using these and similar transgenic mice provided evidence for the role of the pro-inflammatory cytokines for the development of arthritis in mice and potentially in human RA as well.

Transgenic mice have also been used as models to evaluate the potential role of environmental factors in the aetiology of RA. Human T cell leukaemia virus type-I (HTLV-I) is a retrovirus known to be an oncogenic virus for humans, initially found as a causative agent for adult T cell leukaemia. Transgenic mice carrying the HTLV-1 genome develop spontaneous chronic arthritis resembling that of human RA, accompanied by hyperproduction of IgG immunoglobulins (Iwakura *et al.*, 1991). The incidence and severity of arthritis were dependent on the genetic background of the mice used. The HTLV-1 virus encodes a transcriptional **transactivator**, Tax, in the env-pX region. The Tax transactivates transcription from the cognate viral promoter as well as

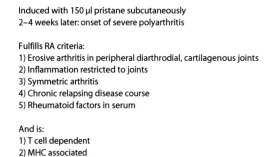

PIA

Induced with 150 μl pristane subcutaneously
2–4 weeks later: onset of severe polyarthritis

Fulfills RA criteria:
1) Erosive arthritis in peripheral diarthrodial, cartilagenous joints
2) Inflammation restricted to joints
3) Symmetric arthritis
4) Chronic relapsing disease course
5) Rheumatoid factors in serum

And is:
1) T cell dependent
2) MHC associated
3) Polygenically controlled
4) Acute phase protein response – orosomucoid
5) Elevation of circulating cartilage proteins (COMP)

Pristane:

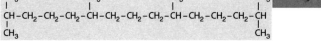

Figure 16.3 Characteristics of the PIA model in DA rats. Hind paws from a rat with PIA is compared with paws from a resistant rat.

many host cell promoters, including cytokines, cytokine receptors and immediate early transcriptional factor genes, which may explain the observed chronic arthritis of the transgenic mice and serve as a model for the involvement of HTLV in RA disease mechanisms.

Cartilage specific proteins, such as CII, are candidate auto-antigens in RA. However, it has also been suggested that ubiquitous antigens, such as glucose 6-phosphate isomerase (GPI) may also be targets for an autoimmune and arthritogenic response. Arthritis spontaneously developed in a cross between the autoimmune-prone non-obese diabetic (NOD) mice and mice transgenic for a TCR specifically recognizing bovine pancreas ribonuclease (RNase). Surprisingly the T cells in this cross also recognized the self-antigen GPI (Matsumoto et al., 1999). T cells, B cells and the specific MHC haplotype of the NOD mice were found to be essential for the development of arthritis. Transfer of serum or purified IgG from the arthritic animals induced severe arthritis and thus provided a new tool to study the genetics of the antibody effector pathway as has been recently demonstrated by identifying the mast cell as crucial player in the development of arthritis (Lee et al., 2002). Interestingly, mapping of genes of crosses between NOD and C57BL strains revealed the same major loci associated with both anti-GPI serum transferred arthritis as well as CIA, indicating a partly shared effector pathway (Ji et al., 2001).

Pathological pathway diversity in arthritis

It is obvious from the wide panorama of arthritis models, and from the broad definition of RA, that there are several different pathways that may eventually lead to arthritis. The main use of animal models will delineate each of them and there are therefore not one optimal model but several, each demonstrating unique pathways. At the cellular and molecular level, both the innate and the acquired immune system are operating in concert in the different pathological pathways. Different arthritis models can be grouped into three different mechanistic principles. Firstly, the cell-mediated pathology will, by definition, be based on T helper 1 (TH1)-like responses leading to macrophage activation and granuloma formation in the joints. Secondly, joint specific antibodies can directly bind to cartilage or

bone and attract, after complement activation, mast cells, neutrophils and macrophages causing oedematous inflammation and may also cause erosions later on. And thirdly, an aggressive response mediated by mesenchymal synovial cells could lead to both destruction and pathologic deformation of joints.

The classical CIA model is probably dependent on both TH1-like responses involving IFN-γ producing T cells initiating a more cell mediated pathology and TH2-like mediated inflammatory mechanisms facilitating humoral factors like antibodies and complement (Svensson et al., 2002). The complexity of arthritis models, and the wide diversity of the different pathways, provide the possibility of defending widely divergent hypotheses if the proper model or experimental situation is selected. Therefore, we need to know how nature herself has selected pathways leading to arthritis.

A disease oriented approach based on genetics

Beside the influence of the MHC region to the development of arthritis, there is a wide variety of unknown non-MHC susceptibility genes among inbred mouse strains. In crosses between susceptible and resistant strains, a number of gene regions have been identified that control the disease (Figure 16.3). Several of these loci probably include clusters of regulating genes. This situation clearly illustrates the complexity of genetic susceptibility to arthritis in mice just as in human RA. Notably, co-localization of susceptibility loci for arthritis in mice and rats as well as suggestive loci for human RA, illustrate the evolutionary conservation of arthritogenic pathways. One illustrative example has so far been achieved in rat models, using PIA. The PIA model is a highly relevant model for RA as it fulfils the criteria for RA. It is also inducible with adjuvant leaving the rat to decide how to respond immunologically, and is very reproducible. This makes it useful for genetic analysis. Pristane is not immunogenic and does not contain MHC binding peptides. Nevertheless, the disease is MHC-associated and dependent on the activation of abTCR-expressing T cells. However, it has not been possible to link the immune response to joint antigens or other endogenous components although immunization with various cartilage proteins induce arthritis but with different pathogeneses. To identify the genes controlling this disease we have made crosses between susceptible and resistant rat strains. Several loci that control the onset of arthritis as well as the severity and chronicity of the disease have been identified and been confirmed in **congenic** strains (Vingsbo-Lundberg et al., 1998; Holmdahl et al., 2001). One gene, Ncf1, has been identified as controlling the severity of arthritis (Olofsson et al., 2003). The Ncf1 gene unexpectedly controlled T cell activation through release of reactive oxygen species. Thus, a new pathway of importance in arthritis was revealed. Interestingly, the pathogenesis of PIA is different from antibody-induced arthritis, as described for antibodies against glucose phosphoisomerase or against CII (Watson et al., 1987; Ji et al., 2001, 2002; Johansson et al., 2001a,b; Burkhardt et al., 2002; Lee et al., 2002; Matsumoto et al., 2002; Svensson et al., 2002). It also opens up the possibility of directly addressing the chronic phase of the disease (Holmdahl et al., 1994; Vingsbo et al., 1996; Svensson et al., 2002), which is clearly more relevant for comparison with human RA. The possibility of identifying the naturally selected genes controlling a complex disease like arthritis, clearly points to a new way of using animal models and to delineate relevant disease mechanisms and pathways.

Humanization of animal models

Another recent twist for the use of animal models is the possibility of humanizing them, i.e. to introduce relevant human genes. This has been demonstrated in models for both MS and RA. In MS a mouse was created, which developed an MS-like disease by the introduction of the relevant MHC class II genes along with genes coding for a TCR, rescued from an MS patient (Madsen et al., 1999). It demonstrated directly that molecular interactions in humans can be studied in mice. However, it also highlighted the complexity as the disease course was found to be highly diverse probably due to the variable genetic background in the mice (Madsen et al., 1999).

This approach has also been successful in arthritis (Table 16.2). In RA a set of MHC class II molecules, sharing the same peptide binding pocket, has been suggested as being associated with disease (Gregersen et al., 1987). Interestingly, this particular set of human class II

molecules bind the same immunodominant peptide as described above in relation to the murine Aq molecule, although shifted slightly in its MHC binding properties. In addition, a lysine side chain at a specific position of the CII peptide (in position 264) is recognized by both DR4 restricted and A^q restricted T cells (Andersson et al., 1998). Interestingly, as shown in the mouse, the T cell recognition predominantly involves various posttranslational modified forms of lysine side chains (Bäcklund et al., 2002). Transgenic expression of human class II molecules with this RA associated binding pocket, like DR1 and DR4, made the mice as susceptible as if they expressed the murine Aq molecule (Rosloniec et al., 1997, 1998; Andersson et al., 1998; Holmdahl et al., 1999). This provides a direct molecular link between RA and CIA and clearly increases the usefulness of animal models.

Concluding remark

Autoimmune disease is not only a highly diverse set of disorders but so is also each specific autoimmune disease. Certain tissues are more commonly subjected to an inflammatory attack than others, an important example being arthritis which is a common disease in humans. The diversity of the different arthritis models has been summarized and the importance of addressing this, in the same way as nature has selected them, has been highlighted.

References

Abdelnour, A., Zhao, Y.-X., Holmdahl, R. and Tarkowski, A. (1997). Scand. J. Immunol. **45**, 301.

Andersson, E.C., Hansen, B.E., Jacobsen, H., Madsen, L.S., Andersen, C.B., Engberg, J., Rothbard, J.B., Sönderstrup-McDevitt, G., Malmström, V., Holmdahl, R., Svejgaard, A. and Fugger, L. (1998). Proc. Natl. Acad. Sci. USA **95**, 7574.

Bäcklund, J., Carlsen, S., Höger, T., Holm, B., Fugger, L., Kihlberg, J., Burkhardt, H. and Holmdahl, R. (2002). Proc. Natl. Acad. Sci. USA **99**, 9960.

Berglöf, A., Sandstedt, K., Feinstein, R., Bölske, G. and Smith, C.I. (1997). Eur. J. Immunol. **27**, 2118.

Bradley, D.S., Nabozny, G.H., Cheng, S., Zhou, P., Griffiths, M.M., Luthra, H.S. and David, C.S. (1997). J. Clin. Invest. **100**, 2227.

Bremell, T., Lange, S., Svensson, L., Jennische, E., Gröndahl, K., Carlsten, H. and Tarkowski, A. (1990). *Arthritis Rheum.* **33**, 1739.

Bremell, T., Lange, S., Yacoub, A., Ryden, C. and Tarkowski, A. (1991). *Infect Immun.* **59**, 2615.

Bremell, T., Lange, S., Holmdahl, R., Ryden, C., Hansson, G.K. and Tarkowski, A. (1994). *Infect Immun.* **62**, 2334.

Brunsberg, U., Gustafsson, K., Jansson, L., Michaëlsson, E., Ährlund-Richter, L., Pettersson, S., Mattsson, R. and Holmdahl, R. (1994). *Eur. J. Immunol.* **24**, 1698.

Burkhardt, H., Koller, T., Engstrom, A., Nandakumar, K.S., Turnay, J., Kraetsch, H.G., Kalden, J.R. and Holmdahl, R. (2002). *Arthritis Rheum.* **46**, 2339.

Butler, D.M., Malfait, A.-M., Mason, L.J., Warden, P.J., Kollias, G., Maini, R.N., Feldmann, M. and Brennan, F.M. (1997). *Biochem. J.* **326**, 763.

Campbell, I.K., Hamilton, J.A. and Wicks, I.P. (2000). *Eur. J. Immunol.* **30**, 1568.

Carlsén, S., Hansson, A.S., Olsson, H., Heinegård, D. and Holmdahl, R. (1998). *Clin. Exp. Immunol.* **114**, 477.

Chang, Y.H., Pearson, C.M. and Abe, C. (1980). *Arthritis Rheum.* **23**, 62.

Cole, B.C., Ward, J.R., Jones, R.S. and Cahill, J.F. (1971). *Infect. Immun.* **4**, 344.

Corthay, A., Johansson, Å., Vestberg, M. and Holmdahl, R. (1999). *Int. Immunol.* **11**, 1065.

Corthay, A., Hansson, A.S. and Holmdahl, R. (2000). *Arthritis Rheum.* **43**, 844.

Courtenay, J.S., Dallman, M.J., Dayan, A.D., Martin, A. and Mosedal, B. (1980). *Nature* **283**, 666.

Dalldorf, F.G., Cromartie, W.J., Anderle, S.K., Clark, R.L. and Schwab, J.H. (1980). *Am. J. Pathol.* **100**, 383.

Glant, T.T., Mikecz, K., Arzoumanian, A. and Poole, A.R. (1987). *Arthritis Rheum.* **30**, 201.

Gregersen, P.K., Silver, J. and Winchester, R.J. (1987). *Arthritis Rheum.* **30**, 1205.

Gripenberg-Lerche, C., Skurnik, M., Zhang, L., Söderström, K.O. and Toivanen, P. (1994). *Infect. Immun.* **62**, 5568.

Guerassimov, A., Zhang, Y., Banerjee, S., Cartman, A., Leroux, J.Y., Rosenberg, L.C., Esdaile, J., Fitzcharles, M.A. and Poole, A.R. (1998). *Arthritis Rheum.* **41**, 1019.

Hammer, R.E., Maika, S.D., Richardson, J.A., Tang, J.P. and Taurog, J.D. (1990). *Cell* **63**, 1099.

Hang, L., Theofilopoulos, A.N. and Dixon, F.J. (1982). *J. Exp. Med.* **155**, 1690.

Hietala, M.A., Jonsson, I.M., Tarkowski, A., Kleinau, S. and Pekna, M. (2002). *J. Immunol.* **169**, 454.

Hill, J.L. and Yu, D.T. (1987). *Infect. Immun.* **55**, 721.

Holmdahl, R., Jansson, L., Larsson, E., Rubin, K. and Klareskog, L. (1986a). *Arthritis Rheum.* **29**, 106.

Holmdahl, R., Rubin, K., Klareskog, L., Larsson, E. and Wigzell, H. (1986b). *Arthritis Rheum.* **29**, 400.

Holmdahl, R., Jansson, L., Andersson, M. and Larsson, E. (1988). *Immunology* **65**, 305.

Holmdahl, R. and Kvick, C. (1992). *Clin. Exp. Immunol.* **88**, 96.

Holmdahl, R., Goldschmidt, T.J., Kleinau, S., Kvick, C. and Jonsson, R. (1992a). *Immunol.* **76**, 197.

Holmdahl, R., Vingsbo, C., Hedrich, H., Karlsson, M., Kvick, C., Goldschmidt, T.J. and Gustafsson, K. (1992b). *Eur. J. Immunol.* **22**, 419.

Holmdahl, R., Jansson, L., Andersson, M. and Jonsson, R. (1992c). *Clin. Exp. Immunol.* **88**, 467.

Holmdahl, R., Vingsbo, C., Malmström, V., Jansson, L. and Holmdahl, M. (1994). *J. Autoimmun.* **7**, 739.

Holmdahl, R., Andersson, E.C., Andersen, C.B., Svejgaard, A. and Fugger, L. (1999). *Immunol. Rev.* **169**, 161.

Holmdahl, R., Lorentzen, J.C., Lu, S., Olofsson, P., Wester, L., Holmberg, J. and Pettersson, U. (2001). *Immunol. Rev.* **184**, 184.

Horai, R., Saijo, S., Tanioka, H., Nakae, S., Sudo, K., Okahara, A., Ikuse, T., Asano, M. and Iwakura, Y. (2000). *J. Exp. Med.* **191**, 313.

Iwakura, Y., Tosu, M., Yoshida, E., Takiguchi, M., Sato, K., Kitajima, I., Nishioka, K., Yamamoto, K., Takeda, T., Hatanaka, M., Yamamoto, H. and Sekiguchi, T. (1991). *Science* **253**, 1026.

Jansson, L. and Holmdahl, R. (1993). *Clin. Exp. Immunol.* **94**, 459.

Ji, H., Gauguier, D., Ohmura, K., Gonzalez, A., Duchatelle, V., Danoy, P., Garchon, H.J., Degott, C., Lathrop, M., Benoist, C. and Mathis, D. (2001). *J. Exp. Med.* **194**, 321.

Ji, H., Ohmura, K., Mahmood, U., Lee, D.M., Hofhuis, F.M., Boackle, S.A., Takahashi, K., Holers, V.M., Walport, M., Gerard, C., Ezekowitz, A., Carroll, M.C., Brenner, M., Weissleder, R., Verbeek, J.S., Duchatelle, V., Degott, C., Benoist, C. and Mathis, D. (2002). *Immunity* **16**, 157.

Jirholt, J., Cook, A., Emahazion, T., Sundvall, M., Jansson, L., Nordquist, N., Pettersson, U. and Holmdahl, R. (1998). *Eur. J. Immunol.* **28**, 3321.

Joe, B., Cannon, G.W., Griffiths, M.M., Dobbins, D.E., Gulko, P.S., Wilder, R.L. and Remmers, E.F. (2002). *Arthritis Rheum.* **46**, 1075.

Johansson, Å.C.M., Hansson, A.-S., Nandakumar, K.S., Bäcklund, J. and Holmdahl, R. (2001a). *J. Immunol.* **167**, 3505.

Johansson, Å.C.M., Sundler, M., Kjellen, P., Johannesson, M., Cook, A., Lindqvist, A.K., Nakken, B., Bolstad, A.I., Jonsson, R., Alarcon-Riquelme, M. and Holmdahl, R. (2001b). *Eur. J. Immunol.* **31**, 1847.

Kawahito, Y., Cannon, G.W., Gulko, P.S., Remmers, E.F., Longman, R.E., Reese, V.R., Wang, J., Griffiths, M.M. and Wilder, R.L. (1998). *J. Immunol.* **161**, 4411.

Keffer, J., Probert, L., Cazlaris, H., Georgopoulos, S., Kaslaris, E., Kioussis, D. and Kollias, G. (1991). *EMBO J.* **10**, 4025.

Khare, S.D., Hansen, J., Luthra, H.S. and David, C.S. (1996). *J. Clin. Invest.* **98**, 2746.

Kleinau, S., Martinsson, P. and Heyman, B. (2000). *J. Exp. Med.* **191**, 1611.

Koga, T., Kakimoto, K., Hirofuji, T., Kotani, S., Ohkuni, H., Watanabe, K., Okada, N., Okada, H., Sumiyoshi, A. and Saisho, K. (1985). *Infection Immun.* **50**, 27.

Kouskoff, V., Korganow, A.S., Duchatelle, V., Degott, C., Benoist, C. and Mathis, D. (1996). *Cell* **87**, 811.

Larsson, P., Kleinau, S., Holmdahl, R. and Klareskog, L. (1990). *Arthritis Rheum.* **33**, 693.

Lee, D.M., Friend, D.S., Gurish, M.F., Benoist, C., Mathis, D. and Brenner, M.B. (2002). *Science* **297**, 1689.

Lorentzen, J.C., Glaser, A., Jacobsson, L., Galli, J., Fakhrai-Rad, H., Klareskog, L. and Luthman, H. (1998). *Proc. Natl. Acad. Sci. USA* **95**, 6383.

Lu, S., Carlsen, S., Hansson, A.-S. and Holmdahl, R. (2002). *J. Autoimmun.* **18**, 199.

Madsen, L.S., Andersson, E.C., Jansson, L., Krogsgaard, M., Andersen, C.B., Engberg, J., Strominger, J.L., Svejgaard, A., Hjorth, J.P., Holmdahl, R., Wucherpfennig, K.W. and Fugger, L. (1999). *Nat. Genet.* **23**, 343.

Malmström, V., Michaëlsson, E., Burkhardt, H., Mattsson, R., Vuorio, E. and Holmdahl, R. (1996). *Proc. Natl. Acad. Sci. USA* **93**, 4480.

Matsumoto, I., Staub, A., Benoist, C. and Mathis, D. (1999). *Science* **286**, 1732.

Matsumoto, I., Maccioni, M., Lee, D.M., Maurice, M., Simmons, B., Brenner, M., Mathis, D. and Benoist, C. (2002). *Nat. Immunol.* **3**, 360.

Nabozny, G.H., Baisch, J.M., Cheng, S., Cosgrove, D., Griffiths, M.M., Luthra, H.S. and David, C.S. (1996). *J. Exp. Med.* **183**, 27.

Olofsson, P., Holmberg, J., Tordsson, J., Lu, S., Åkerström, B. and Holmdahl, R. (2003). *Nat. Genet.* **33**, 25.

Otto, J.M., Cs-Szabo, G., Gallagher, J., Velins, S., Mikecz, K., Buzas, E.I., Enders, J.T., Li, Y., Olsen, B.R. and Glant, T.T. (1999). *Arthritis Rheum.* **42**, 2524.

Otto, J.M., Chandrasekeran, R., Vermes, C., Mikecz, K., Finnegan, A., Rickert, S.E., Enders, J.T. and Glant, T.T. (2000). *J. Immunol.* **165**, 5278.

Pearson, C.M. (1956). *Proc. Soc. Exp. Biol. Med.* **91**, 95.

Pearson, C.M. and Wood, F.D. (1959). *Arthritis Rheum.* **2**, 440.

Potter, M. and Wax, J.S. (1981). *J. Immunol.* **127**, 1591.

Remmers, E.F., Longman, R.E., Du, Y., O'Hare, A., Cannon, G.W., Griffiths, M.M. and Wilder, R.L. (1996). *Nat. Genet.* **14**, 82.

Roper, R.J., Weis, J.J., McCracken, B.A., Green, C.B., Ma, Y., Weber, K.S., Fairbairn, D., Butterfield, R.J., Potter, M.R., Zachary, J.F., Doerge, R.W. and Teuscher, C. (2001). *Genes Immun.* **2**, 388.

Rosloniec, E.F., Brand, D.D., Myers, L.K., Whittington, K.B., Gumanovskaya, M., Zaller, D.M., Woods, A., Altmann, D.M., Stuart, J.M. and Kang, A.H. (1997). *J. Exp. Med.* **185**, 1113.

Rosloniec, E.F., Brand, D.D., Myers, L.K., Esaki, Y., Whittington, K.B., Zaller, D.M., Woods, A., Stuart, J.M. and Kang, A.H. (1998). *J. Immunol.* **160**, 2573.

Schaible, U.E., Kramer, M.D., Wallich, R., Tran, T. and Simon, M.M. (1991). *Eur. J. Immunol.* **21**, 2397.

Stuart, J.M. and Dixon, F.J. (1983). *J. Exp. Med.* **158**, 378.

Svensson, L., Nandakumar, K.S., Johansson, A., Jansson, L. and Holmdahl, R. (2002). *Eur. J. Immunol.* **32**, 2944.

Terato, K., Hasty, K.A., Reife, R.A., Cremer, M.A., Kang, A.H. and Stuart, J.M. (1992). *J. Immunol.* **148**, 2103.

Trentham, D.E., Townes, A.S. and Kang, A.H. (1977). *J. Exp. Med.* **146**, 857.

Vingsbo, C., Jonsson, R. and Holmdahl, R. (1995). *Clin. Exp. Immunol.* **99**, 359.

Vingsbo, C., Sahlstrand, P., Brun, J.G., Jonsson, R., Saxne, T. and Holmdahl, R. (1996). *Am. J. Pathol.* **149**, 1675.

Vingsbo-Lundberg, C., Nordquist, N., Olofsson, P., Sundvall, M., Saxne, T., Pettersson, U. and Holmdahl, R. (1998). *Nat. Genet.* **20**, 401.

Watson, W.C. and Townes, A.S. (1985). *J. Exp. Med.* **162**, 1878.

Watson, W.C., Brown, P.S., Pitcock, J.A. and Townes, A.S. (1987). *Arthritis Rheum.* **30**, 460.

Wilder, R.L., Calandra, G.B., Garvin, A.J., Wright, K.D. and Hansen, C.T. (1982). *Arthritis Rheum.* **25**, 1064.

Wooley, P.H., Luthra, H.S., Stuart, J.M. and David, C.S. (1981). *J. Exp. Med.* **154**, 688.

Wooley, P.H., Seibold, J.R., Whalen, J.D. and Chapdelaine, J.M. (1989). *Arthritis Rheum.* **32**, 1022.

Yamazaki, H., Ikeda, H., Ishizu, A., Nakamaru, Y., Sugaya, T., Kikuchi, K., Yamada, S., Wakisaka, A., Kasai, N., Koike, T., Hatanaka, M. and Yoshiki, T. (1997). *Int. Immunol.* **9**, 339.

Yang, L., Weis, J.H., Eichwald, E., Kolbert, C.P., Persing, D.H. and Weis, J.J. (1994). *Infect. Immun.* **62**, 492.

Hematology of the Mouse

Nancy Everds
Haskell Laboratory for Health and Environmental
Sciences, Newark, Delaware, USA

Introduction

This chapter is intended to provide useful and practical advice to those involved in assaying and interpreting hematologic results from mice. The interpretation of hematologic changes is similar whether the changes are due to an infectious disease, a toxin, or a mutation. Identification of underlying causes is dependant on knowing which hematologic tests should be used and how to interpret the test results.

A complete review of hematologic malignancies of the mouse is beyond the scope of this chapter. However, the investigator who is studying murine hematology must be aware of the common hematologic malignancies of mice, and is referred to a review by Frith *et al.* (1993).

Much of the information in this chapter is not published, but has been gleaned from practical experience and from discussions with colleagues. When applicable, exact references have been specifically cited.

Terminology

Hematology is the study of the physiology and pathology of the cellular elements of blood. The three major cellular components of blood are red blood cells (erythrocytes), white blood cells (leukocytes), and platelets (thrombocytes). Basic hematologic concepts are similar across most mammalian species, and can be found in several human and veterinary textbooks. The reader may consult any of the excellent hematology reference books listed at the end of this chapter for guidance about general hematologic principles.

Clinical terms (e.g. anemia, leukopenia, lymphocytosis, pancytopenia) have specific medical connotations. These terms generally refer to conditions in which parameters fall outside of an appropriate reference range for a widely inclusive population (i.e. male humans, female dogs, dairy cows). In research, changes in hematologic parameters for mice are usually compared to a control group or a very narrowly defined reference interval.

The Laboratory Mouse
Copyright 2004 Elsevier
ISBN 0-1233-6425-6

Therefore, the use of these clinical terms is generally not appropriate in mouse research, and should not be used. Small changes can be referred to as 'increases' or 'decreases' in the affected parameter. For example, instead of stating that an experimental treatment caused anemia, one can state that mean hemoglobin concentration decreased 1.5 g/dl, to 90% of the control group mean.

Blood collection and handling

Blood collection techniques

Several references describe methods for blood collection from mice (Hoff, 2000; Schnell *et al.*, 2002). For hematologic testing, it is important to collect blood quickly with a minimum of tissue trauma. Thus, methods that collect blood directly from a vessel or plexus are preferred to those that may cause more tissue trauma. The following sites are most commonly used for blood collection for hematology in mice: orbital sinus, tail vein, heart, aorta, and vena cava. Depending on the site used, blood collection must be a terminal procedure (heart, aorta, vena cava) or may be a survival procedure (orbital sinus, tail). Blood collected from the saphenous vein is inappropriate for hematology due to hair and tissue contamination.

Blood collected from a mouse should be immediately placed in a tube containing an anticoagulant. The preferred anticoagulant for routine hematologic testing is ethylenediamine tetra-acetic acid (EDTA). If the blood is collected for smears only, blood without anticoagulant can be used, provided that smears are made immediately after seconds of blood collection. Heparin should not be used as an anticoagulant because it tends to cause clumping of platelets, especially in mice, and negatively impacts the **tinctorial** quality of Romanowsky staining. It is important to use plain rather than heparin-coated capillary tubes for blood collection.

Preparation of blood smear

Blood smear preparation is a skill that is easily learned, and is covered in standard hematologic texts referenced at the end of this chapter. A blood smear can be prepared with a minimum of blood (less than 50 μl); it can easily be conducted as a survival procedure in mice.

Basics of hematologic evaluation

Methods for evaluation

The scope of hematologic evaluation may vary greatly depending on the needs of the investigator and the capabilities of the clinical pathology laboratory. The methods used will determine the type and number of parameters measured and the accuracy of the results. The hematologic evaluation can include some or all of the following tests; these are listed below in the order of most simple and least expensive to those that are most complicated and requiring significant investment in instrumentation.

Blood smear evaluation

The blood smear is examined microscopically. The density of white and red blood cells and platelets is estimated. A differential white blood cell count is performed. The blood smear is reviewed for morphologic changes. For this examination, only a microscope is required. The validity of results is highly dependent on the skill and experience level of the examiner.

Spun hematocrit, hemacytometer cell counts, and blood smear examination

A microcapillary tube is centrifuged to determine the spun hematocrit (also called packed cell volume). White and red blood cells and platelets are counted microscopically, using a hemacytometer. Hemoglobin concentration may be measured, and some or all red blood cell indices are calculated. The blood smear is examined microscopically, and a differential white blood cell count is performed. Absolute white blood cell differential counts are calculated. The blood smear is reviewed for morphologic changes. This method requires a minimum of equipment (microcapillary hematocrit centrifuge, hemacytometer, microscope, +/− method of determining hemoglobin concentration). The results are moderately accurate.

Instrument cell counts and blood smear examination

Whole blood is analyzed on a hematology cell counter. The cell counter determines red and white blood cell counts, platelet counts, and hemoglobin concentration, and measures or calculates red blood cell indices. The blood smear is examined microscopically, and a differential white blood cell count is performed. Absolute white blood cell differential counts are calculated. The blood smear is reviewed for morphologic changes. This method requires a dedicated hematology analyzer and trained personnel. The results are very accurate if the instrument has been shown to be valid for the determination of mouse hematologic tests, and if the operators have sufficient training to operate the instrument.

Instrument cell counts and differential with blood smear review

Whole blood is analyzed on a flow cytometer-based hematology instrument. The instrument determines red and white blood cell counts, platelet counts, and hemoglobin concentration, and measures or calculates red blood cell indices. The instrument also estimates the differential white blood cell counts. The blood smear is examined microscopically to confirm the automated differential, and is also reviewed for morphologic changes. This method requires a sophisticated hematology analyzer and well-trained personnel. The results are very accurate if the instrument has been shown to be valid for the determination of mouse hematologic tests, and if the operators have sufficient training to operate the instrument. The Bayer Advia 120 and the Abbott Cell-Dyn 3500 are the two most commonly used flow cytometry-based hematology analyzers with software that can determine mouse white blood cell differential counts (Figures 17.1 and 17.2).

Almost all hematology cell counters and flow cytometry-based analyzers were developed to determine hematologic parameters for humans. The instruments have specialized software to discriminate and count the different cellular constituents of blood. Some hematology analyzers have been adapted for animal species by using species-specific software or settings. Analyzers adapted for animal species vary markedly in their ability to accurately count and differentiate animal blood cells. This is especially true with blood from rodents. Therefore, for accurate results, it is essential to use a laboratory with instrumentation validated for the analysis of mouse blood, and whose employees are skilled at mouse blood evaluation.

The complete blood count

The report of results from a standard hematologic evaluation is called a complete blood count (CBC). The CBC generally includes most or all of the parameters listed below. As discussed briefly above, these parameters are determined using one or all of the following methods: automated hematology analyzers, hematology analyzers plus microscopic examination, or by a combination of manual and automated methods. In most countries, hematology results are expressed in SI units, as described by Laposata (1992).

Red blood cell mass parameters

The functional red blood cell mass is measured by three parameters: red blood cell count (RBC), hematocrit (HCT), and hemoglobin concentration (HGB).

Red blood cell count (RBC)

Red blood cell count is the number of red blood cells in a given volume of whole blood. It is usually determined using an automated counter. In this method, red blood cells are counted while they flow through an aperture in single file, using either impedance or optical technology (Figure 17.3). For manual hemacytometer counts, a commercially available diluent system (Unopette©, Becton-Dickinsen, Test 5850) is used to dilute blood prior to counting red blood cells. A qualitative estimation of the red blood cell count can be determined by microscopic evaluation of well-prepared blood smears. Mouse red blood cell counts are higher than most other species, because of the small size of their red blood cells. Counts range from approximately 7 to 11×10^{12}/L (Bannerman, 1983; Hall, 1992). Mouse red blood cells have smaller mean cell volumes (MCVs) than other species, but since their red cell counts are higher, mouse hematocrits are similar to other species.

The primary function of red blood cells is to carry oxygen from the lungs to tissues, and to carry carbon dioxide back to the lungs. In the mouse, red blood cells are normally produced in bone marrow and in the spleen. This is in contrast to other species in which splenic hematopoiesis does not occur in health. In the mouse, the amount of extramedullary hematopoiesis in the spleen and liver increases greatly when there is increased demand for red blood cells.

Compared to other mammalian species, the mouse red blood cell has a fairly short lifespan of other species. Red blood cell lifespan has been estimated at between 30 and 40 days (Bannerman, 1983; Car and Eng, 2001). In contrast, the red blood cell lifespan other species is much longer (rats 45–50 days, dogs 110 days, and humans 120 days). Because the lifespan of murine red cells is so short, there are a higher percentage of circulating

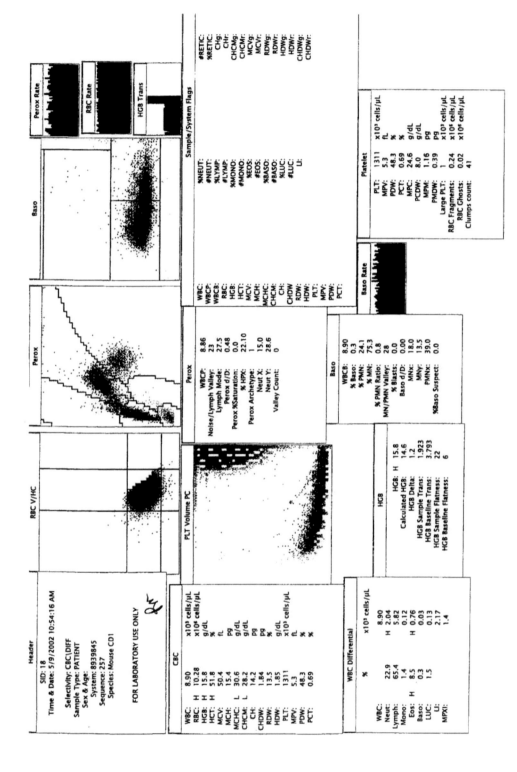

Figure 17.1 Example of analysis of mouse blood on a Bayer Advia 120 Automated Hematology Analyzer. This instrument uses flow cytometry-based analysis to measure individual white blood cell types (automated differential). Notice the scatterplots used to classify the various cell types.

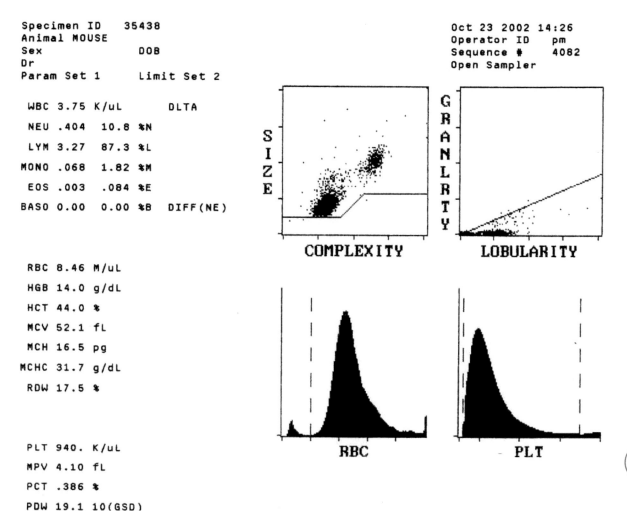

```
Specimen ID     35438                          Oct 23 2002 14:26
Animal MOUSE                                   Operator ID    pm
Sex                DOB                          Sequence #     4082
Dr                                             Open Sampler
Param Set 1        Limit Set 2

  WBC  3.75 K/uL         DLTA

  NEU  .404   10.8 %N

  LYM  3.27   87.3 %L

 MONO  .068   1.82 %M

  EOS  .003   .084 %E

 BASO  0.00   0.00 %B     DIFF(NE)


  RBC  8.46 M/uL

  HGB  14.0 g/dL

  HCT  44.0 %

  MCV  52.1 fL

  MCH  16.5 pg

 MCHC  31.7 g/dL

  RDW  17.5 %


  PLT  940. K/uL

  MPV  4.10 fL

  PCT  .386 %

  PDW  19.1 10(GSD)
```

Figure 17.2 Example of analysis of mouse blood on a cell–Dyn 3500. This instrument uses flow cytometry and impedence-based analysis to measure individual white blood cell types (automated differential). Notice the scatterplots used to classify the various cell types.

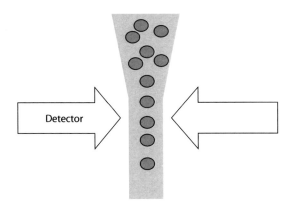

Figure 17.3 Automated cell counting: Automated cell counters use either optical or electronic detection methods to enumerate and measure the size of individual cells as they pass single file through the detector.

immature red blood cells at any given time. Immature red blood cells are larger than mature erythrocytes and stain with a blue tint. Therefore, polychromasia (blue-staining cells) and anisocytosis (variably sized red blood cells) occur to a greater extent in normal healthy mice, compared to humans and many other animals. In addition, Howell–Jolly bodies (remnants of nuclear DNA) are also more common in circulating red blood cells in mice.

Hemoglobin (HGB)

Hemoglobin concentration is the measurement of total hemoglobin per volume of whole blood. It is determined spectrophotometrically after lysis of red blood cells. All forms of hemoglobin, whether functional or not, are included in the measurement of hemoglobin concentration. Mouse hemoglobin ranges from 130–180 g/L.

Hematocrit (HCT)

Hematocrit is a measurement of the volume of red blood cells as a percent of whole blood. For automated procedures, the hematocrit is the product of the red blood cell count and the MCV (see below). For manual determinations, the hematocrit is measured after centrifugation of a microcapillary tube filled with whole blood. The percent of blood composed of red blood cells is the hematocrit (sometimes called packed cell volume). Manual or 'spun' hematocrits tend to be a few percentage points higher than calculated hematocrits, because trapped plasma is included in the apparent red blood cell volume. Hematocrit is expressed as a number without units between 0.00 and 1.00. Hematocrit values for mice are generally between 0.40 and 0.50, but may range up to 0.60 depending on sampling site and fasting status.

Red blood cell indices and other red blood cell parameters

MCV

Mean cell volume is the average size of red blood cells. When red blood cell counts are determined by instrument, the MCV is measured. When cell counts are determined by hemacytometer, MCV is calculated. For instrument-generated cell counts, red blood cell volume is measured during the cell count described above. A histogram is generated from the red blood cell count and size (Figure 17.4). The MCV is determined from this histogram. If cells are counted by hemacytometer, the MCV is determined by dividing the hematocrit by the red blood cell count. Mean cell volumes of mice are approximately 40–55 fl (Bannerman, 1983; Hall, 1992).

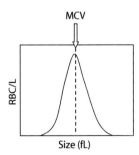

Figure 17.4 Red blood cell histogram: As each red blood cell is counted, its size is measured. The results of these measurements are plotted on a histogram. The histogram is used to determine the red blood cell count, the mean cell volume, and the red blood cell distribution width.

Mean cell hemoglobin (MCH)

Mean cell hemoglobin is the average amount of hemoglobin found in each individual red blood cell. It is determined by dividing the hemoglobin concentration by the red blood cell count. In general, MCH is the least useful hematology parameter, because it is insensitive to change and provides no additional information than other red blood cell parameters. MCH for mice ranges from 13 to 17 pg. MCH is sometimes expressed as femtomoles (fmol).

Mean cell hemoglobin concentration (MCHC)

Mean cell hemoglobin concentration is the hemoglobin concentration divided by hematocrit, and thus is the average concentration of hemoglobin in all red blood cells. MCHC is much more relevant than the MCH, discussed above, because it is more sensitive to changes affecting red blood cells. Mouse MCHC generally ranges between 260 and 320 g/L.

Red blood cell distribution width (RDW)

Red blood cell distribution width is the coefficient of variation of the red blood cell volume, and is calculated from the red blood cell histogram described above and in Figure 17.3. Red blood cell distribution width is therefore a quantitative indicator of the variation in red blood cell size (**anisocytosis**). Mouse RDW generally falls between 11% and 15%.

Reticulocyte count (RETIC)

Reticulocytes are immature red blood cells containing residual RNA. Reticulocytes can be counted by instrumentation or by hemacytometer. Automated reticulocyte counts are performed similarly to automated red blood cell counts. Before counting, red blood cells are stained with a dye that stains for nucleic acid (such as acridine orange) to differentiate reticulocytes from mature red blood cells. For manual reticulocyte counts, whole blood is mixed with a supravital dye such as new methylene blue, and blood smears are prepared. The dye causes clumping and staining of residual nucleic acid present in immature cells. The stained cells (reticulocytes) are counted as a percentage of total red blood cells. The absolute reticulocyte count is determined by multiplying the total red blood cell count by the percent of reticulocytes. The number of circulating reticulocytes is higher in mice ($200–500 \times 10^9$/L) than in most other species, due to the short lifespan of the mouse red blood cell. Units for reticulocyte count vary among laboratories, but generally are reported in the same units as red blood cells or in the same units as platelets (cells $\times 10^9$/L).

Platelet parameters

Platelet count (PLT)

Platelets are essential for primary hemostasis, and form a temporary hemostatic plug prior to activation of the clotting cascades. Platelet count is the number of platelets in a given volume of whole blood. It is usually determined using an automated counter, using either impedance or optical technology. Platelets are counted as they flow through an aperture in single file, similar to red blood cells. Platelets can also be counted microscopically with a hemacytometer. A qualitative estimation of the platelet count can be determined by microscopic evaluation of well-prepared blood smears. Counting platelets in mouse blood is problematic because mouse platelets tend to form clumps. The presence of clumped platelets can interfere with the accuracy of both platelet and white blood cell counts, and also invalidates the platelet count. Depending on the analyzer, clumped platelets can either be counted as white blood cells, or specifically as eosinophils. Therefore, regardless of the method used to count platelets, the smear must be evaluated for platelet clumps prior to acceptance of the platelet count.

Mice have the highest circulating platelet count of any laboratory animal species (approximately 1000–2000 $\times 10^9$/L) (Hall, 1992). Mouse platelet half-life is approximately 5 days (Ault and Knowles, 1995; Manning and McDonald, 1997). Platelets are variable in morphology, but range from 1 to 3 μm in diameter and 4 to 7 fl in volume. The primary growth factor controlling platelet production is thrombopoietin, but erythropoietin and iron status also affect platelet production. Under conditions of increased demand for platelets, platelet production is increased, and giant platelets are sometimes observed.

Mean platelet volume (MPV)

Mean platelet volume is an estimation of the average size of platelets, and is analogous to the MCV of red blood cells. The MPV is only available on automated cell counters. As platelets are counted, their size is measured. A histogram is generated from the platelet counts and platelet sizes. The MPV is determined from this histogram. Units for MPV are femtoliters (fl).

White blood cell parameters

White blood cell count (WBC)

White blood cells participate in immune and inflammatory processes. The white blood cell count may be determined quantitatively (automated analyzer counts or manual hemacytometer counts) or qualitatively (slide review). Prior to quantitative white blood cell determinations, red blood cells are lysed using a hypotonic solution. The resulting preparation contains only white blood cells and platelets. Automated counters measure white blood cells as they flow single-file past a detector. For manual hemacytometer counts, a commercially available diluent system (Unopette©, Becton-Dickinsen, Test 5855) is used to dilute blood and lyse red blood cells prior to counting white blood cells. A qualitative estimation of the white blood cell count can be determined by microscopic evaluation of well-prepared blood smears by a trained observer. White blood cell counts of mice range from 2 to 10 $\times 10^9$/L.

Differential white blood cell count (DIFF)

The differential white blood cell count enumerates the various individual white blood cell types found in peripheral blood. The predominant circulating leukocyte in mice is the lymphocyte, followed by neutrophils, monocytes, eosinophils, and lastly basophils.

The differential white blood cell count can be performed on an automated instrument or by microscopy. Automated analyzers with species-specific software are flow cytometers dedicated to differentiating peripheral white blood cells, and use nuclear and cytoplasmic characteristics of white blood cells for classification (Figures 17.1 and 17.2). Differential white blood cell counts can also be determined microscopically by examining and categorizing 100 white blood cells on a peripheral smear. The percent of each cell type is multiplied by the total white blood cell count to arrive at absolute differential counts for the various cell types. Interpretation of leukocyte changes should be based on absolute numbers (number of a given cell type per unit volume of blood) rather than relative numbers (percent of a given cell type). Units for absolute differential white blood cell counts are the same as for total white blood cell counts (cells $\times 10^9$/L).

The main categories of white blood cells counted during a differential count are lymphocytes, neutrophils, monocytes, eosinophils, and basophils. Additional cell types may be observed; these might be subcategories of the five major cell types, or perhaps other cells not normally observed in peripheral blood. Interpretation of white blood cell changes is covered in a later section of this chapter.

- Lymphocytes are responsible for immune surveillance, production of antibodies, production of cytokines, and antigenic memory. Unlike other circulating leukocytes, lymphocytes are still capable of division. Mouse lymphocytes are generally similar in appearance to those of other species. Both small and large lymphocytes can be observed on peripheral smears. Lymphocytes are approximately 7–12 µm in diameter, and have round to oval dark blue nuclei with pale blue scant cytoplasm on Romanowsky-stained blood smears. In the mouse, approximately 60–90% of circulating leukocytes are lymphocytes ($2–8 \times 10^9$/L).
- Neutrophils are circulating phagocytes and modulators of the immune response. In the mouse, neutrophils comprise approximately 5–20% of the circulating leukocyte population ($0.5–3.0 \times 10^9$/L). The morphologic appearance of the mouse neutrophil is similar in most respects to those of other species. Mouse neutrophils are 10–25 µm in diameter (Bannerman, 1983) and have pale granules. Mouse neutrophils sometimes exhibit circular doughnut-shaped nuclei; similar morphology is sometimes observed in rat neutrophils as well.
- Monocytes and eosinophils are minor cell types, and range from about 0.05 to 0.10×10^9/L. Generally there are more monocytes than eosinophils in peripheral blood. Basophils are very rarely observed in the peripheral blood of mice. Some authors have questioned the presence of basophils in mouse blood, however microscopic and ultrastructural characteristics of murine basophils have been described in the literature (Dvorak *et al.*, 1982; Dvorak, 2000).

Morphologic evaluation of the blood smear

The blood smear is evaluated microscopically for alterations in appearance (morphology) of white blood cells, red blood cells, and platelets. At the same time, any other unusual findings are noted. Examples of appropriate comments would be observations of variation in size, shape, and coloration of cells, cellular organization (agglutination, **rouleaux**, platelet clumps), parasites, and other relevant findings.

Bone marrow evaluation

Bone marrow smears are analyzed to determine the underlying pathophysiology of peripheral blood changes.

Among the common methods used to prepare bone marrow cells are cytospin preparations, paint brush smears, and squash preparations. Paintbrush smears are prepared by spreading cells with a fine natural bristle paintbrush (No. 0) slightly moistened with either autologous serum or 5% bovine serum albumin in saline. The femur (or other appropriate long bone) is cracked and the moistened paintbrush is introduced into the marrow cavity to pick up a small amount of marrow (Figures 17.5 and 17.6). The brush is then drawn several times down the length of one or two clean glass slides, taking care to spread cells out fully. After the slides are air-dried, they are stained using a Romanowsky-type stain (Figure 17.7). Generally, bone marrow slides are stained twice on a platen stainer, or twice as long as blood smears if using a dip-type stain. The bone marrow smears should be prepared so that individual cells are easily identifiable.

Both qualitative and quantitative evaluations of bone marrow cells can be useful in determining the cause of peripheral blood changes. Bone marrow smears can be qualitatively evaluated for relative proportions, maturity of precursor cells, storage pool, and other changes. Quantitative evaluations of various cell types are generally not necessary, but may be done on occasion if extremely precise data is required. When performing quantitative evaluations, careful formulation of the hypothesis will allow for targeted quantitative evaluation; this is more efficient and generally produces appropriate information.

Bone marrow may also be evaluated by histology or flow cytometry. For histology, a section of the femur is placed in formalin, decalcified, embedded, stained, and evaluated. Alternatively, a portion of bone marrow may be removed from the femur and preserved without decalcification. Histologic bone marrow slides are excellent for evaluating bone marrow cellularity, but bone marrow smears are better to evaluate for cellular morphology. Flow cytometry may be used to determine various subpopulations of cells in the bone marrow; however, a discussion of flow cytometric techniques is beyond the scope of this chapter.

Variables affecting hematologic results

The results of hematology tests are not only affected by pathologic processes, but are also affected by the status

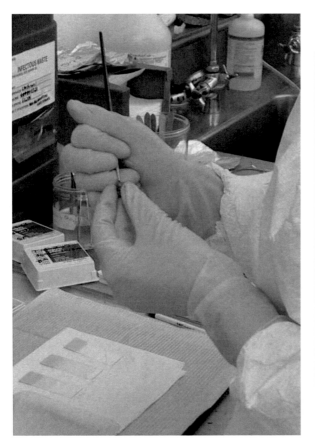

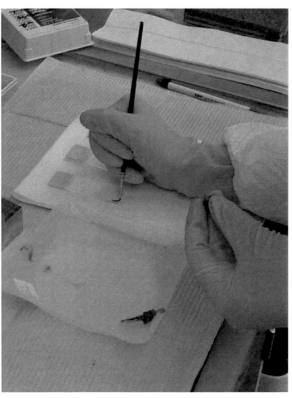

Figures 17.5 Collection of bone marrow: The moistened paintbrush is introduced into the marrow cavity to pick up a small amount of marrow.

Figures 17.6 Preparation of paintbrush bone marrow smears: The brush is drawn several times down the length of one or two clean glass slides, taking care to spread cells evenly.

of the mouse at the time of blood collection, collection techniques, and handling of blood prior to and during analysis. Therefore, it is essential to keep variables as consistent as possible so that results are comparable across experiments.

Hematologic parameters vary with the status of the animal being tested. Fasting status, hydration status, time of collection, prior experimental manipulations, prior and concurrent anesthesia, sex, and a myriad of other factors will influence the results of hematology tests. In general, mice are not fasted prior to hematologic sample collection, because fasted mice tend to drink less water and may become dehydrated.

Age has very significant effect on hematologic parameters. For example, the erythroid parameters for newborn mice are very different from those of mice aged 6 weeks or more. Newborn mice have much higher reticulocyte counts with lower red blood cell mass parameters. During the first few weeks of life, red blood cell mass actually decreases, but then rebounds steadily and is near adult levels at weaning.

Order of blood collection can affect hematologic results. For this reason, it is important to collect blood from control alternately with treated (experimentally manipulated) mice. In general, blood collected from central arteries or heart tends to have lower white and red blood cell counts than blood collected from distal sites such as ocular plexus or tail (Schermer, 1967). Under-filling of tubes results in excess anticoagulant for the amount of blood, and can result in artifacts in hematologic parameters. It is important to fill blood tubes to the volume recommended by the manufacturer.

Once samples are collected, it is important that all samples are stored appropriately (generally room temperature or refrigerated), and are analyzed at approximately the same amount of time post-collection.

Analytical variables that need to be controlled include operator error, bias due to order of analysis, instrument malfunction, outdated reagents or control material, dilution errors, calculation errors, and calibration errors.

Figure 17.7 Air-dried slides are stained using a Romanowsky-type stain. The bone marrow smears should be prepared so that individual cells are easily identifiable in many areas of the smear.

Numerous factors can interfere with accurate analysis of hematology parameters. Potential interfering factors for analysis of mouse blood are listed on Table 17.1. Some of these interfering factors only occur with mouse samples, while others can occur with samples from other species.

Pathophysiology and interpretation of results

Red blood cells

The most important effects on red blood cells are those that result in changes in red blood cell mass. These effects can be divided into those that increase red blood cell mass, and those that decrease red blood cell mass. Red blood cell mass changes are summarized in Table 17.2.

Increased red blood cell mass

Increases in red blood cell mass can be classified as either relative or absolute changes. Relative increases in red blood cell mass are the result of dehydration, which can occur very rapidly in mice. Dehydration may be associated with other signs of poor health, including

TABLE 17.1: Selected interferences in mouse hematology measurements			
Cause	Change in analyte	Reason	Instruments affected
Turbidity caused by hemolysis, lipemia, or Heinz bodies	Falsely increased hemoglobin concentration	Spectrophotometer	Several instruments
Platelet clumps	Falsely increased white cell count	Platelet clumps counted as WBC by impedance counters	Coulter instruments, others
Platelet clumps	Falsely decreased platelet count	Only individual platelets counted	All instruments
Platelet clumps (mouse only)	Falsely increased eosinophil count	PEROX channel classifies mouse platelet clumps as eosinophils	Bayer Advia 120
High percentage of immature red cells	Falsely increased white cell count	Immature red cells are lyse-resistant and are counted as white cells	Impedance counters
Nucleated red blood cells	Falsely increased white cell count	Nucleated red cells are counted as white cells	Hemacytometer and most automated counters

TABLE 17.2: Alterations in red blood cell mass parameters in mice

Causes	Characteristics
INCREASED RED CELL MASS	Increased hemoglobin, hematocrit, red blood cell count
Relative increased red cell mass	
Dehydration	Recognized by clinical signs
	May see increased total protein or albumin
Absolute increased red cell mass	
Increased erythropoietin activity	Increased extramedullary erythropoiesis
Myeloproliferative disease	Increased reticulocytes
DECREASED RED CELL MASS	Decreased hemoglobin, hematocrit, red blood cell count
Relative decrease in red cell mass	
Pregnant dams	Plasma volume expansion
Neonates	Rare occurrence
Absolute decreases in red cell mass	
Red cell loss (hemorrhage)	
Internal or external (overt, gastrointestinal, genitourinary)	Recognize by clinical signs
Red cell destruction	
Toxins, immune-mediated	↑ reticulocytes
Cell membrane alterations	↑ MCV
Biochemical alterations	↑ RDW
Vascular injury/turbulence	↓ MCHC
	↑ extramedullary erythropoiesis
	↑ splenic weights
Decreased red cell production	
Decreased erythropoietin	↓ reticulocytes
Renal disease	↓ MCV
Endocrine disease	↓ RDW
Chronic inflammatory disease	no change or ≠ MCHC
Bone marrow toxicity	
Abnormal maturation	
Abnormal heme or nucleic acid synthesis	
Hematopoietic neoplasia	

inactivity, hunched appearance, decreased appetite, and poor skin **turgor**.

Absolute increases in red blood cell mass result from increased red blood cell production, and may occur whenever there is increased production of growth factors or cytokines that stimulate erythropoiesis. The primary cytokine driving erythropoiesis is erythropoietin (EPO), but other cytokines, especially thrombopoietin, can also affect red blood cell production.

Increased red blood cell mass can be secondary and appropriate (occurring in response to a disorder with decreased oxygen delivery to tissues) or may be primary and inappropriate (occurring in the absence of any need for increased oxygen delivery). Causes of appropriately increased red blood cell mass include cardiovascular disorders, pulmonary disorders, or abnormal oxygen carrying capacity of hemoglobin. Causes of inappropriately increased red blood cell mass include conditions resulting in excess erythropoietin (autonomous production) or excess stimulation of the erythropoietin receptor (activating mutations). Polycythemia, a myeloproliferative disease, also results in increased red blood cell mass.

In mice, increased red blood cell production, regardless of the cause, is associated with extramedullary erythropoiesis in the spleen and liver, hypercellularity of bone marrow, and increased reticulocytes in the peripheral blood.

Decreased red blood cell mass

Decreases in circulating red blood cell mass are indicated by decreases in red blood cell counts, hemoglobin concentration, and hematocrit. Decreased red blood cell mass can be either relative (expansion of plasma volume) or absolute.

Relative decreases in red blood cell mass are rare, but may occur in the pregnant dam and in neonates, or as a result of other perturbations in plasma volume homeostasis.

Absolute decreases in red blood cell mass are much more common, and result from loss (hemorrhage), increased destruction (hemolysis), or decreased production of red blood cells. Each of these causes can be recognized by particular changes in hematologic parameters.

Hemorrhage

Loss of red blood cells is called hemorrhage, and can be either overt or occult. Overt hemorrhage can result from disorders of hemostasis, ulcerated masses, surgical procedures, or other trauma. Overt hemorrhage is a clinical diagnosis, and is best identified by observing the physical condition of the mouse. Occult hemorrhage is generally due to loss of red blood cells into the alimentary tract or urogenital tract, and can be detected by tests for occult hemorrhage such as HemoCult© for feces or dipstick tests for urine blood. In mice, loss of red blood cells is generally accompanied by marked regeneration (increased reticulocytes), while very chronic loss of blood may result in decreased red blood cell production due to lack of iron.

Increased destruction

Increased destruction of red blood cells, or hemolysis, results in decreased circulating half-life of red blood cells. In most cases, increased destruction of red blood cells occurs by premature removal from circulation (extravascular hemolysis) by the reticuloendothelial system of the liver and spleen, rather than by rupture of red blood cells within the vasculature (intravascular hemolysis). Normally, red blood cell lifespan is 30–40 days in the mouse. With hemolysis, the lifespan of red blood cells can be decreased markedly. Destruction of red blood cells may be due to metabolic, immune-mediated, or physical causes.

Hemolytic processes result in compensatory increased red blood cell production (regeneration).

Regeneration is the result of increased erythropoietin in response to decreased oxygen tension. Erythropoietin recruits stem cells to differentiate into red blood cell precursors, and promotes survival of committed red blood cell precursors, resulting in increased circulating immature red blood cells in peripheral blood, and increased absolute reticulocyte count. The immature red blood cells (which include reticulocytes) are larger than mature red blood cells, and contain less hemoglobin. Therefore, in regenerative hemolytic anemia, the altered parameters are associated with cell size (increased MCV, RDW, macrocytosis) and MCHC (decreased MCHC).

In mice, even minimal hemolysis results in compensatory hypercellularity of the bone marrow and splenic extramedullary erythropoiesis. This is in contrast to other species, in which extramedullary erythropoiesis only occurs under more severe hematologic stress. In mice, splenic weights can be used as sensitive and objective measures of red blood cell regeneration. The presence of extramedullary erythropoiesis should be confirmed histologically.

Decreased production

Any process that has a deleterious effect on red blood cell precursors in the bone marrow can cause decreased red blood cell production. The effects can be grouped into effects on stem cells, growth factors, or synthesis of hemoglobin or nucleic acids.

Mice with decreased red blood cell mass due to decreased red blood cell production will have inappropriately low reticulocyte counts in peripheral blood, with respect to the change in red blood cell mass. The decrease in red blood cell mass due to decreased production is referred to as a non-regenerative or poorly regenerative process.

Reticulocytes are larger and have less hemoglobin than mature red blood cells. Mice normally have a larger percentage of reticulocytes than other species. Therefore, in mice, decreased reticulocytes may result in decreased polychromatophilic cells on the blood smear, decreased MCV, and increased MCHC. The morphology of peripheral blood and/or bone marrow may help to elucidate the mechanism for decreased red blood cell production. When decreased red blood cell production is suspected, bone marrow should be evaluated to determine an underlying cause.

Anemia of chronic disease

The most common anemia in humans, and probably the most common anemia in animals, is called anemia

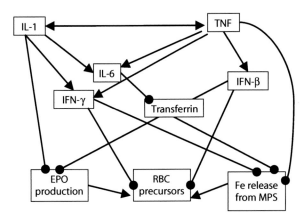

Figure 17.8 Anemia of chronic disease. This diagram outlines factors that contribute to the pathogenesis of the anemia of chronic disease. Lines ending in circles indicate inhibition, while those ending in arrows indicate stimulatory effects (adapted from Means and Krantz (1992). *Blood* **80**, 1639–1647).

Decreased lymphocytes can be due to stress (Drozdowicz *et al.*, 1990), altered trafficking, or genetic manipulation resulting in immunodeficiency syndromes. Decreased lymphocytes due to stress occurs secondary to the effects of increased corticosteroids. This stress response in lymphocytes occurs within hours of the stressful event, and can persist for days to weeks. Often, decreased lymphocytes are incorrectly attributed to stress without correlative findings supporting the diagnosis. Unless supporting evidence is present (such as corticosteroid concentration or thymic atrophy), this diagnosis should be made only with great caution.

Total peripheral blood lymphocyte counts are a crude measurement of lymphoid status. Unexplained effects on lymphocytes may be further explored using cytochemical or flow cytometric techniques. In addition, histopathology of other lymphoid tissues, such as lymph nodes and thymus, might be useful in the understanding of peripheral lymphocyte changes.

of chronic disease (Means and Krantz, 1992). This condition is usually secondary to hormonal or inflammatory conditions. It results from decreased red blood cell production and increased red blood cell destruction. Although the term 'chronic' implies that this anemia only occurs after a long period, it can actually be observed within days to weeks. Because mice have shorter red blood cell half-lives, effects of disease on red blood cell mass are observed more quickly than in other species. Figure 17.8 outlines the pathophysiology of anemia of chronic disease.

Leukocytes

Interpretation of changes in total white blood cell count should be made on the basis of changes in the absolute differential leukocyte counts. This is necessary because different leukocyte types have different reactions to the same stimulus. Leukocytes are summarized in Table 17.3.

Lymphocytes

In response to inflammation, mice usually have increased lymphocyte counts; this response is in contrast to that of most other laboratory species, but similar to that of rats. Increased circulating lymphocytes are also observed secondary to catecholamines (excitement), altered trafficking, or lymphoid malignancy. Lymphocyte increases due to catecholamines are generally transitory (minutes to hours), whereas inflammatory-related increases are more persistent (days to weeks).

Neutrophils

Increases in circulating neutrophils can be due to increased neutrophil production or decreased neutrophil egress. Increased production is generally caused by increases in colony stimulating factors (G-CSF or GM-CSF), as a result of inflammation. Increased production can rarely be recognized by the presence of immature neutrophils in the peripheral blood. These neutrophils may have the morphologic appearance of band neutrophils, or may show signs consistent with toxic change (increased granulation, basophilic cytoplasm, and/or foamy cytoplasm), indicating accelerated neutrophil production. Decreased egress can be caused by lack of functional adhesion molecules (P-selectin or E-selectin deficiency) or absence of chemotactic factors to recruit neutrophils into tissues. Under conditions of decreased egress, increased nuclear segmentation may be observed in a subset of circulating neutrophils.

Decreased neutrophils can result from decreased bone marrow production, increased egress into tissue, or destruction of neutrophils. Because neutrophils circulating lifespan is short (7–14 h) (Jain, 1993), decreased neutrophil count is often the first peripheral result of bone marrow toxicity, and occurs prior to platelet or red blood cell decreases. Increased egress can be due to peracute inflammation. In the initial stages of inflammation, many circulating neutrophils may be called to the site of inflammation by inflammatory cytokines. This may leave a temporary decrease in circulating neutrophils that is ameliorated when the bone marrow increases production of neutrophils to meet

TABLE 17.3: Alterations in peripheral white blood cell counts in mice	
Alteration	**Cause**
Increased lymphocyte counts	Inflammation
	Catecholamines (exogenous or endogenous)
	Altered trafficking
	Lymphoproliferative disease
Decreased lymphocyte counts	Corticosteroids (exogenous or endogenous)
	Altered trafficking
	Disorders of immunity
	Immunosuppressive treatments
Increased neutrophil counts	Catecholamines (endogenous or exogenous)
	Corticosteroids (endogenous or exogenous)
	Inflammation
	Increased colony stimulating factors
	Decreased egress into tissues
	Hematopoietic neoplasia
Decreased neutrophil counts	Decreased bone marrow production
	Decreased bone marrow release
	Increased egress into tissues
	Destruction of circulating neutrophils
Increased monocyte counts	Inflammation
	Corticosteroids (endogenous or exogenous)
	Hematopoietic neoplasia
Decreased monocyte counts	Corticosteroids (endogenous or exogenous)
Increased eosinophil counts	Increased production
	Decreased egress
	Hypersensitivity/allergy
	Hematopoietic neoplasia
Decreased eosinophil counts	Corticosteroids (endogenous or exogenous)
Increased basophil counts or decreased monocyte, eosinophil, or basophil counts	Not generally recognized in mice

the peripheral demand. Decreased neutrophils can also result from their destruction, either by immunological or non-immunological mechanisms.

Monocytes, eosinophils, and basophils

Increased circulating monocytes generally indicate an increased demand for tissue macrophages, and thus is an indicator of inflammation. A decrease in circulating monocytes is uncommon, but may indicate increased egress into tissues, or may be a result of increased corticosteroid activity (Hall, 1992).

Increased circulating eosinophils result from allergic or inflammatory conditions. Generally, increases in circulating eosinophils are accompanied by increased tissue eosinophils as well. Decreased eosinophils may occur in conjunction with decreased lymphocytes, as a result of increased circulating corticosteroids. Increased basophil counts are uncommon in mice.

Platelets

Effects on platelets may alter their number, size, or function. The most common reason for increase in platelets is as a secondary response to accelerated production of red blood cells (increased erythropoietin). Other reasons for increased platelets include iron deficiency and increased thrombopoietin.

Decreased platelets can result from decreased production or increased destruction. Decreased production occurs as a result of injury to megakaryocytes, the bone marrow cell responsible for production of platelets. If decreased production is suspected, megakaryocyte number and morphology may help to elucidate the cause of decreased platelets. Decreased numbers or abnormal morphology of megakaryocytes support the hypothesis that decreased platelet production has occurred. Often, processes that affect bone marrow production of one cell line can have ramifications in one of the other two cell lines produced in the bone marrow. Because platelets and leukocytes have shorter half-lives than red blood cells, effects on bone marrow will often first manifest as peripheral blood abnormalities on these two cell lines.

Increased destruction/removal of platelets occurs if there is accelerated activation of platelets or immune recognition of platelets. If platelet half-life is shortened, megakaryocytes release large immature platelets containing rRNA. These 'reticulated platelets' can be measured using nucleic acid-binding dyes and flow cytometry. Alternatively, these large platelets can be detected by measuring MPV. Bone marrow generally shows increased numbers of megakaryocytes, with or without an increase in younger megakaryocytes.

Platelet function can also be studied in mice. Techniques such as aggregometry, bleeding time assays, and flow cytometry (Ware *et al.*, 2000; Car and Eng, 2001) can be used to elucidate functional changes.

Bone marrow evaluation

Murine hematopoiesis (production of granulocytes, monocytes, platelets, and erythrocytes) generally occurs in bone marrow, although a small amount of hematopoiesis also occurs in spleen in healthy mice. Therefore, alterations in peripheral blood counts may be investigated by examination of the bone marrow. Mouse bone marrow includes stromal cells, macrophages, mast cells, megakaryocytes, and megakaryocyte precursors; erythrocytes and erythroid precursors; granulocytes, monocytes, and their precursors; and lymphocytes, as well as non-hematopoietic cells such as osteoblasts and osteoclasts.

Granulocytes (neutrophils, eosinophils, and basophils) and red blood cells arise from precursor cells. The precursor cells of both granulocytes and red blood cells can be divided into proliferating cells

TABLE 17.4: Maturation stages of granulocyte and erythrocyte precursors

	Granulocyte	Erythrocyte
Proliferating	Myeloblast	Rubriblast
	Promyleocyte	Prorubricyte
	Myelocyte	Rubricyte
Maturing	Metamyelocyte	Metarubricyte
	Band	Reticulocyte
	Segmented	Mature red cell

(cells that are capable of undergoing division), and maturing cells (end-stage cells in the process of maturation) (Table 17.4). During a complete bone marrow differential cell count, these cells are counted, along with those cells mentioned above.

The following is a short list of common terms used to describe possible changes in bone marrow during qualitative bone marrow evaluation:

Cellularity. A qualitative evaluation of the number of cells on a bone marrow smear. Altered cellularity can be a function of bone marrow changes or collection technique.

Increased or decreased myeloid : erythroid ratio. A qualitative estimate of the relative proportions of all myeloid cells to all erythroid cells.

Left shift. A relative increase in proliferating stages of a particular cell type. For megakaryocytes, a left shift refers to an increase in immature megakaryocytes.

Hypoplasia. Numbers of a particular cell type are decreased.

Hyperplasia. Numbers of a particular cell type are increased.

Erythrophagocytosis. Presence of intracellular erythrocytes in phagocytic cells (usually macrophages).

Acknowledgments

Acknowledge Bruce Car, Bob Hall, Denise Hoban, and Beth Wilkinson.

References

Ault, K.A. and Knowles, C. (1995). *Exp. Hematol.* **23**, 996–1001.

Bannerman, R. (1983). In *The Mouse in Biomedical Research* (eds H. Foster, D. Small and J.G. Fox), pp. 293–312. Academic Press, New York.

Car, B.D. and Eng, V.M. (2001). *Vet. Pathol.* **38**, 20–30.

Drozdowicz, C.K., Bowman, T.A., Webb, M.L. and Lang, C.M. (1990). *Am. J. Vet. Res.* **51**(11), 1841–1846.

Dvorak, A.M., Nabel, G., Pyne, K., Cantor, H., Dvorak, H.F. and Galli, S.J. (1982). *Blood* **59**, 1279–1285.

Dvorak, A.M. (2000). *Blood* **96**, 1616–1617.

Frith, C.H., Ward, J.M. and Chandra, M. (1993). *Toxicol. Pathol.* **21**(2), 206–218.

Hall, R.L. (1992). In *Animal Models in Toxicology* (ed. Gad Schengeles), pp. 765–811. Marcel Dekker Inc., NY.

Hoff, J. (2000). *Lab. Anim.* **29**(10), 47–53.

Jain, N.C. (1993). *Essentials of Veterinary Hematology*, Lea and Febiger, Philadelphia, PA.

Jones, T.C., Ward, J.M., Mohr, U. and Hunt, R.D. (eds) (1990). *Hematopoietic System* (Monographs on pathology of laboratory animals). Springer-Verlag, Berlin.

Knoll, J.S. (2000). In *Schalm's Veterinary Hematology* (eds B.F. Feldman, J.G. Zinkl and N.C. Jain). Lippincott Williams & Wilkins, Boston.

Laposata, M. (1992). *SI Conversion Guide*. Massachusetts Medical Society/*New England Journal of Medicine*, Boston, MA.

Manning, K.L. and McDonald, T.P. (1997). *Exp. Hematol.* **25**(10), 1019–1024.

Means, R.T. and Krantz, S.B. (1992). *Blood* **80**, 1639–1647.

Moore, D.M. (2000). In *Schalm's Veterinary Hematology*, 5th edn (eds B.F. Feldman, J.G. Zinkl and N.C. Jain). Lippincott Williams & Wilkins, Boston.

Rebar, A.H. (1993). *Toxicol. Pathol.* **21**(2), 118–129.

Schermer (1967). *The Blood Morphology of Laboratory Animals*. FA Davis, Philadelphia, PA.

Schnell, M.A., Hardy, C., Hawley, M., Propert, K.J. and Wilson, J.M. (2002). *Hum. Gene Ther.* **13**, 155–162.

Vacha, J. (1983). In *Red Blood Cells of Domestic Mammals* (eds N.S. Agar and P.G. Board). Elsevier, New York.

Ware, H., Russell, S. and Ruggeri, Z.M. (2000). *Proc. Natl. Acad. Sci.* **97**, 2803–2808.

Foucar, Kathryn. (2001). *Bone Marrow Pathology*, 2nd edn. American Society Clin Pathologists, Chicago.

Hoffman, R., Benz, E.J., Shattil, S.J., Furie, F., Cohen, H.J. and Silberstein, L.E. (eds) (1999). *Hematology*, 3rd edn. Churchill Livingstone, New York.

Lee, G. Richard (1999). *Wintrobe's Clinical Hematology*, 10th edn. Lippincott Williams & Wilkins, Baltimore, MD.

Animal hematology

Feldman, B.F., Zinkl, J.G. and Jain, N.C. (2000). *Lippincott Schalm's Veterinary Hematology*, 5th edn. Williams & Wilkins, Boston.

Jain, N.C. (1993). *Essentials of Veterinary Hematology*. Lea and Febiger, Philadelphia, PA.

Latimer, K.S., Mahaffey, E.A., Prasse, K.W., Duncan, J. and Proctor, N.W. (2003). *Duncan and Prasse's Veterinary Laboratory Medicine: Clinical Pathology*, 4th edn. Iowa State Press, Ames, IA.

Hawkey, C.M. and Dennett, T.B. (1989). *Color Atlas of Comparative Veterinary Hematology*. Iowa State University Press, Ames, IA.

Smith, C.A., Andrews, C.M., Collard, J.K., Hall, D.E. and Walker, A.K. (1994). *A Color Atlas of Comparative, Diagnostic, and Experimental Hematology*. Mosby-Yearbook, London.

Hematologic methods

Stiene-Martin, A.E., Lotspeich-Steininger, C.A. and Koepke, J.A. (1998). *Clinical Hematology: Principles, Procedures, Correlations*, 2nd edn. Lippincott Williams & Wilkins, Baltimore, MD.

O'Connor, B.H. (1984). *A Color Atlas and Instruction Manual of Peripheral Blood Cell Morphology*. Williams & Wilkins, Baltimore, MD.

Bain, B.J. (1989). *Blood Cells: A Practical Guide*. JB Lippincott, Philadelphia, PA.

Suggested reading

General hematologic texts

Beutler, E., Lichtman, M.A., Collier, B.S., Kipps, T.J. and Siligsohn, U. (eds) (2001). *Williams' Hematology*, 6th edn. Mcgraw-Hill Book Company, New York.

CHAPTER 18

The Social Behaviour of Mice and its Sensory Control

Arnold Keith Dixon

Ethology Research Unit, Neuenegg, Switzerland

Introduction

The behaviour described in this article refers to that of unadulterated laboratory strains of Mus musculus L and the wild Mus musculus domesticus. It is often argued that domestication has reduced and almost eradicated the 'natural' behaviour of rodents. This is not true. Laboratory mice are captive animals and, because captivity deprives animals of many natural means of response, the full gamut of their behaviour may not be observed (Dixon, 1978). In fact, rearing mice in enclosures will, if supplying enough food, water and cover, allow the 'wild' mouse in many laboratory strains of Mus to emerge. As with any captive strain of animal, it is necessary to understand the behavioural and social organization of their wild counterparts in order to make sense of what is observed in the laboratory. It is also necessary to discuss some aspects of the behaviour of other rodents, particularly the rat in order to highlight the commonalities but also the differences between their behaviour and that of the mouse.

Mice are the commonest rodents used in the laboratory. Often they are chosen because of their size and

ease of breeding and husbandry and, more recently, because they lend themselves to genetic manipulations more readily than do rats. Unfortunately, there is a tendency to regard mice as small rats! However the mouse and rat differ in important respects particularly with respect to their social behaviour. Even though behavioural studies often employ mice and rats as single animals it is important to realize that they are social animals and that much of their behaviour has evolved to mediate intraspecific interactions. Consequently, this article is mainly concerned with the **ethology** of Mus musculus and the ways in which social behaviour of mice and other rodents is assessed in a laboratory setting. Mice are regularly used for behavioural experiments in the laboratory, particularly in industrial laboratories. The most common behavioural procedures in use involve tests of learning and memory e.g. passive avoidance, tests of spatial memory e.g. Morris Maze, and tests of putative models of psychiatric disorders e.g. behavioural despair. Such procedures are not the subject of this review since they are covered in many standard texts. Rather, this article is focussed on the social behaviour of the mouse and the way this is mediated through sensory mechanisms.

Measuring social behaviour

Social behaviour in mice is usually determined using the methods pioneered by Grant (1963) and Grant and Mackintosh (1963). Close inspection has shown that the behaviour of mice consists of individual acts and postures i.e. behavioural elements many of which have signal value i.e. they convey information to another animal about intent, state of excitement, etc. Once identified, the elements are described in an unambiguous manner, named and coded in a shorthand notation to assist quantification. The analysis involves counting how often and when in a social encounter between a resident male mouse and a group-housed intruder, specific behavioural elements are displayed by the participants. This is done by making video recordings of the social interaction under reversed day–light conditions (mice are primarily nocturnal animals) from which, trained observers then enter into a computer the incidence of the elements whenever they occur during the social encounter. This gives, in real-time, a permanent record of the duration, frequency and sequence of all non-verbal elements. The elements do not occur at random but occur in sequences whereby certain elements almost always precede or follow each other whereas others do not. In order to rule out fortuitous occurrences between elements powerful statistical techniques have been developed which have allowed cohesive clusters or categories of elements to be identified based upon their probabilities of occurrence. The elements in a given cluster share strong mutual associations with each and appear to be functionally related. The statistical power of the associations give a measure of 'tendency' which is akin to the strength of a drive but really describes the likelihood that the element (or category) will be expressed. The product of such an analysis is the *ethogram* which is a catalogue of elements classified into categories each having functional significance.

The mouse social ethogram

The ethogram is used to construct social profiles of individuals and, because it is a measurement of what animals actually do, provides a behavioural 'fingerprint' which can be used to compare individual profiles or even profiles across species. The ethogram mostly used today stems from Grant and Mackintosh (1963) although it has been revised and modified to suit laboratory use (Dixon and Mackintosh, 1971; Dixon *et al.*, 1990). Table 18.1 depicts some 60 behavioural acts and postures of the male mouse which have been classified according to their function into several distinct categories of behaviour based on both sequence analysis (Mackintosh, 1981), sensory deprivation studies (Strasser, 1987) and drug studies (e.g. Dixon *et al.*, 1990).

Non-Social elements refer to all elements shown by a solitary mouse and which do not require a second animal for their expression. The remaining social categories include elements that arise when two or more animals interact. Social Attention in which the animal detects and orients itself towards a conspecific and Social Investigation that involves mutual bodily inspection and the exchange of olfactory, tactile and visual information, are closely linked. Related too, is the category of Sexual Activity, which in this case describes elements of the male mating pattern. Nevertheless, all three categories can, on occasion be separated. Offence refers to all elements leading to attack and is the most extreme form of aggressive behaviour whereas Escape consists of high intensity flight elements leading to separation of the mice. Between these two extremes lie the categories of Offensive-, Defensive- and Distance-Ambivalence in which two (or more) incompatible tendencies, in these cases aggression and flight, compete with each other for expression. They are denoted by opposing vectors, which distort the body or movement of an animal according to the relative strength of the opposing tendencies. Offensive Ambivalence denotes postures in which aggression and flight are both aroused but the former predominates. Distance Ambivalence is similarly weighted towards aggression but consists of acts in which approach and avoidance components alternate e.g. zigzag walking. When a strong tendency to escape competes with a weaker tendency to attack this results in Defensive Ambivalence, which includes postures having predominantly negatively-oriented vectors e.g. upright stance with averted head. This category is closely associated with Escape. Arrested Flight consists of static elements e.g. freeze and crouch which arise when escape routes are restricted or blocked (Dixon, 1998). Although crouch is more apparent in rats, mice make frequent use of freeze when under attack especially in cages (Dixon and Kaesermann, 1987). The ethogram just described is peculiar to male–male encounters. Slightly varied forms occur for male–females and female–female encounters as well as for parental care.

TABLE 18.1: A social ethogram of the laboratory mouse Mus musculus	
Categories	**Elements**
Non-Social	Explore, scan, wash, self-groom, scratch, dig, push-dig, kick-dig, shake, jump, eat, drink, sit, flop, leave, stretched-attend-posture, substrate, leave, off-bars
Social Attention	attend, stretched-attend, approach, turn
Social Investigation	investigate, nose, groom, sniff
Sexual activity	follow, attempted-mount, mount, genital-groom, push-under, crawl over, push-past, intromission, elevated crouch
Offence	threat, attack, bite, chase, aggressive-groom
Offensive Ambivalence	offensive-sideways posture, offensive-upright posture, sideways-posture
Distance Ambivalence	zig-zag, circle, walk-round, rattle
Defensive Ambivalence	oblique, upright posture, defensive-upright posture (DUP), defensive-sideways posture
Arrested Flight	crouch, freeze
Escape	Flag, evade, retreat, flee, on-bars

The Table shows the names of individual elements i.e. acts and postures displayed by mice during a dyadic social encounter with another mouse. The categories on the left represent functional clusters of elements and are based on the results of sequence analysis and upon the effects of sensory deprivation on the behaviour of mice. Of the ten categories, Non-Social refers to all activities which are displayed by solitary mice. The other nine categories are social categories which are shown towards another mouse. Social Attention, Investigation and Sexual activity are closely related and comprise Social Contact. Offence, Offensive Ambivalence and Distance Ambivalence are associated with Aggression whereas the remaining categories are closely related to Escape. For further details see the text.

Defensive behaviour of the mouse and rat

One of the first findings to emerge from the ethograms was that mice and rats display different flight strategies. Grant (1963) showed that rats respond to attacks via two pathways. One pathway leads to crouching. It occurs in cage encounters but is really a blocked form of escape and tends to be replaced by retreat or flee when more space and therefore escape routes become available (as in enclosures). The second pathway, which occurs in cages and enclosures involves submission. When attacked rats commonly flip over onto their backs and hold this position for several seconds or even minutes (Figure 18.1).

Such on-back postures both hide aggression-provoking cues (Blanchard et al., 1986, 1989; the rump and back are the main sites of attack) and attenuate the attacks of aggressive opponents, which usually desist. Grant (1963) called this a submissive posture and it was later shown to well predict the structure of hierarchies in caged rats (Grant and Chance, 1958; Baenninger, 1966), the lowest ranking rat showing the

most submissive postures and the fewest aggressive acts. In laboratory rats rank orders are readily established in makrolon cages Type IV of groups of 5–7 and, given sufficient space, in larger groups.

Mice, in contrast to rats, do not possess a submissive posture (Mackintosh, 1981) and hence must escape to resolve a fight. In cages, where escape is limited, mice suddenly freeze their movement and commonly display an upright posture which Grant and Mackintosh (1963) named the 'defensive upright posture' or DUP (Figure 18.2). This posture, adopted by subordinate animals, is the commonest element of defensive-ambivalence and shows the two vectoral components of ambivalence in which the ventral surface is directed towards the attacker whilst the head is turned up and away.

A mouse adopting this posture generally keeps its eyes open and its ears erect whilst slowly raising its head backwards. This posture is often accompanied by 'offensive sideways posture' in the opponent (Figure 18.2) but generally a pause in the aggression ensues. Instead, the attacker walks around the opponent to try and plant bites on the rump. The victim responds by rotating

Figure 18.1 The submissive or On-Back posture of the rat. The rat showing submission adopts an On-Back posture, the head tends to move up and back and eye-closure (cut-offs) may occur. The figure shows the opponent in an Aggressive-Posture in which cut-offs can also occur. (after Grant, 1963).

Figure 18.2 Defensive upright and offensive sideways postures in mice. This mouse on the left is showing a DUP and is the commonest element of defensive-ambivalence shown by subordinate mice. The ventral surface is directed towards the attacker whilst the head is turned up and away. A mouse adopting this posture generally keeps its eyes open and its ears erect. This posture is often accompanied by an 'offensive sideways posture' in the opponent in which the nearside vibrissae are brought into contact with the subordinate mouse and the nearside eye is closed (cut-off). For further details see the text.

the body such that its abdomen is still directed towards the attacker and, if circumstances permit, prolongs the pause long enough to win respite and escape.

Agonistic behaviour in mice

Aggression in mice

It will be evident from the above ethogram that a considerable portion of the behaviour of male mice is devoted to fighting or agonistic behaviour. Aggression is a set of offensive behaviours usually employed by an animal to effect the retreat or removal of a threatening conspecific e.g. a rival male. Whilst often regarded as a nuisance in laboratory-stocks of mice, under natural conditions, aggression serves to space animals out over the terrain so that local food supplies are not over-exploited and is used to drive off rivals as, for example in territorial conflicts (Crowcroft and Rowe, 1963; Mackintosh, 1970). Indeed a general observation among researchers is the readiness with which mice are rendered aggressive by individual housing, a phenomenon, which probably reflects its territorial propensity (Brain, 1974). Rats are much less prone to becoming aggressive when caged-individually than are mice. Presumably, this was one of the reasons why recourse to foot shocks and other aversive means were used to study aggression in earlier laboratory studies of rats. Aggression develops in laboratory stocks of male mice from about 7 weeks. Data on wild stocks is more difficult

to come by but scars of trapped individuals suggest anything from 7 to 11 weeks of age.

It should be noted that aggression is not confined to male mice. Females too show aggression towards conspecifics under certain conditions (Parmigiani *et al.*, 1988). Outside the nest site and at times of low reproductive activity they are relatively benign in their behaviour. However, female mice or rats attack males vigorously when they defend their nest and young. Lactating female mice are particularly aggressive towards males (Parmigiani *et al.*, 1989) and occasionally even other females. These observations show that offensive behaviour in mice is displayed by both sexes and is under the influence of hormones. Furthermore the pattern of aggression appears to be gender-specific. Thus, examination of wound sites on the bodies of attacked animals show that most bites delivered by males to other males are confined to the rump and back of the opponent, sometimes on the flanks and head but rarely on the ventral surfaces i.e. their aggression is ritualised (Brain, 1981). These observations accord with the bite-deflecting function of the DUP discussed previously. In contrast, male opponents attacked by females display bites at all accessible sites including the ventral surface and face, indicating that attacks by female mice are indiscriminate (Brain 1981; Parmigiani *et al.*, 1988). This observation suggests that the DUP has primarily evolved as a defensive measure against males rather than females and that the brain of females is 'programmed' differently. Females rarely attack other females and males rarely attack females although exceptions do occur. When population densities are high, infants are sometimes killed by adults including parents (Brown, 1953; Southwick, 1955) and infanticide has been regarded as a form of aggression (Paul, 1986; Parmigiani, 1994). In that infanticide can serve to alleviate demands upon local resources as well as favour future mating strategies, it can be regarded as an adaptive form of defence.

Social structure of mice

Mice are one of the most successful mammals. They live and breed in many different environments e.g. corn ricks, houses, open fields and even carcasses held in cold stores (cited by Mackintosh, 1978). In view of their adaptability to different environmental demands and given the highly varied topographical features, which they inhabit, a comprehensive description of the social structure of mice living in the wild is difficult. Stable rank-orders such as that found in rats are not seen in mice. Although early studies (e.g. Uhrich, 1938; Scott and Frederickson, 1951) claimed to observe hierarchies in caged male mice, it is now recognized that they are not of the type observed in rats (Grant and Chance, 1958; Baenninger, 1966) but take the form of one or two mice exerting despotic dominance over several subordinates (Dixon, 1978). Rudimentary rank-orders consisting of a despot, a sub-dominant and subordinates, occur in artificial enclosures and may occur in the wild when high population densities are associated with local food supplies e.g. corn ricks (Southwick, 1958; Anderson, 1961). Also, further differentiation within groups of subordinates may occur i.e. some are attacked more than others, and some subdominants may even receive more immediate attacks from the dominant than some subordinates (Mackintosh, 1978). Nevertheless, these phenomena have yet to be fully investigated and are subtler compared with the overt despotic dominance so readily observed in male mice.

Territoriality in mouse populations

Experiments with artificial colonies of wild mice (Crowcroft, 1955, 1966; Crowcroft and Rowe, 1963;

Reimer and Petras, 1967) and laboratory mice (Mackintosh, 1970, 1973) together with direct observations of feral populations (Anderson, 1961; Mackintosh, 1978; Gray and Hurst, 1996) and trapping results (Berry, 1968, 1970) all indicate that male mice frequently establish individual territories from which male rivals are expelled. A territory is defined here as an area within a defined boundary, which is defended by the territory holder against other rivals, usually males and territoriality, is one of the most widely studied behaviours (Crowcroft and Rowe, 1963; Eibl-Eibesfeldt, 1970; Hinde, 1970; Mackintosh, 1970). There are various functions of territoriality including the procurement of a mate, establishing a home base and ensuring access to a localized food supply. Territories are also an important mechanism for dispersing the population over a **biotope**. Those animals that gain preferred habitats presumably have a greater chance of survival than those forced to compete for with poorer resources (Berry, 1981).

Territory formation in laboratory mice

In laboratory strains of mice, most notably with CFW, C57BL, C57BR and DBA, territoriality can be demonstrated using artificial enclosures. The studies of Crowcroft and Rowe (1963) and Mackintosh (1970, 1973) exemplify best the general conditions for establishing territories in semi-natural conditions. Whereas Crowcroft allowed mice to occupy disused aircraft hangers, Mackintosh made use of sheet aluminium enclosures measuring $2 \times 2 \times 1$ m separated in two halves by a metal partition 1 m high (Figure 18.3). The enclosure contained sawdust, food pellets and water bottles as well as small boxes for providing cover.

Adult male mice were placed 4–6 on each side of the partition. Within a few hours fighting broke out and one (occasionally two) males emerged in each half which attacked all other mice. When the partition was removed a day later, these 'dominants' were found to control their respective halves of the enclosure and vigorously expelled the rival dominant from the other half. By plotting the movements of each mouse on scale maps of the enclosure, it could be shown that each dominant mouse enjoyed complete freedom of movement within its own half of the enclosure but rarely ventured into its rival's half (Figure 18.4).

The junction between the two exclusive zones marked the territory boundaries. Within its boundary, the territory holder attacked and chased any foreign intruder. If its chase took it near to its own boundary its

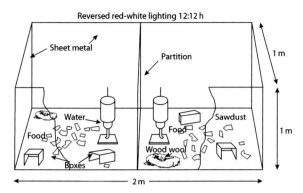

Figure 18.3 Enclosure for establishing territories in male mice. The enclosure is made of aluminium but may be constructed of wood or any durable synthetic material. It is important that sufficient cover in the form of wood-wool, paper or opaque boxes is supplied as this greatly assists the formation of territories. Usually four male mice are placed on either side of the partition and the latter is raised 24–48 h later. Depending upon the size, water bottles can be doubled and food can be distributed freely or localized. Modern video-imaging techniques now replace the older paper methods of recording the tracks of movements of the mice within the enclosure.

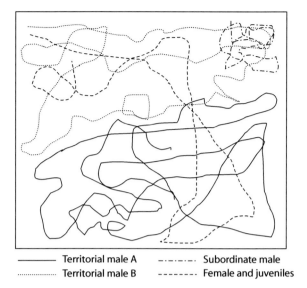

—————— Territorial male A	–·—·—·– Subordinate male
················ Territorial male B	- - - - - - - Female and juveniles

Figure 18.4 Movements of mice within a territory. The traces are those typical of adult male and female mice and of juveniles inhabiting an enclosure in which two males have established territories. Adult females and immature juveniles of both sexes enjoy complete freedom of movements and can cross from one territory to another. Within the confines of its own territory, a dominant male also has complete freedom of movement. However, subordinate mice are generally restricted in their movements unless the territory holder is sleeping.

behaviour became ambivalent with both offensive and defensive components becoming more apparent (Dixon, 1978). Outside the boundary the territory-holder exhibited a more furtive defensive mode and was quick to retreat back into the safety of its own territory when challenged. These types of observations emphasize an important feature of the territorial boundary, namely, that despite the incorporation of landmarks, a territory boundary, like individual space, is essentially a behavioural entity (Dixon, 1998). Mackintosh (1973) provided evidence that territory boundaries are recognized both by odour and by the use of visual cues. Thus, removing territory holders from their respective halves of the enclosure and exchanging the soiled sawdust of each half with the one opposite prompted the mice to switch sides when they were reintroduced into the enclosure (Mackintosh, 1973). Since the objects i.e. drinking bottles, boxes, etc., were not altered olfactory cues appear to override these visual cues. Nevertheless, moving artificial black patches that had been hung over the existing boundary at a height of roughly 70 cm led to the boundary being moved in the same direction and the mice also appeared to orient themselves by using the lamps on the ceiling of the experimental room. Whilst these observations support the use of visual objects in fixing boundaries they presumably were recognized during the white-light phase and, in the wild, probably operate above ground during the day when mice occasionally patrol their territories.

Movements of mice within territories

It has already been mentioned that the territory holder enjoyed complete freedom of movement within its defended area but was restricted from entering a foreign territory. Within a given territory, subordinate mice i.e. mice without own territories, were restricted in their movements and only moved around when the territory holder was asleep or preoccupied. In some cases mice that had been vigorously attacked were so restricted that they lived on top of one of the wooden houses or even on top of water bottles. Our own observations show that within enclosures, territory holders tend to dictate the activity rhythms of the mice. (Figure 18.5). Round the clock observations showed that under conditions of a reversed 12:12 h dark–night cycle (red light of 5–7 lx) from 08:00 to 20:00 h and white light during the night, dominant mice were active in patrolling their territories and showed most aggression during the dark phase although they occasionally patrolled at the end of the light phase.

Subordinates were much less active during the dark phase than were the dominants and were frequently active during the day since this was the only time the territory-holders were sleeping. Subdominants i.e. mice which were attacked by the territory-holder but directed its own attacks towards the subordinates proved to have an intermediate cycle which approximated to a 6 h cycle and appeared to be relatively active during day and night. Indeed the subdominants appeared to be very active and 'stressed'. The available evidence shows that if territory holders are removed (either deliberately) or die, the subdominants usually take over the vacated territory. There is also some laboratory evidence that when given the chance to escape from an enclosure containing established territories, it is the subdominants rather than the subordinates which escape and form territories of their own (Dixon, 1973; Mackintosh, 1978).

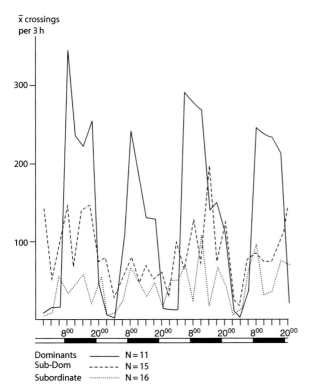

Figure 18.5 Circadian characteristics of general motor activity in male LAC mice: territory situation. The plots of activity for each class of mouse were obtained from scale maps of the enclosure, consisting of a grid of 10 cm² squares on which lines traced the track of each mouse over a 10 min observation period performed every 3 h. By counting the number of lines crossing each square a measure of the activity of each animal was obtained over four nights and days. The day–night cycle was reversed by illuminating the enclosure with red light for 12 h during the day and white light during the night. The white–black bar denotes day and night periods respectively, together with the appropriate clock time. Note the periods of activity by subordinate and sub-dominant mice during the dominant's rest periods. Data from five enclosure experiments (see text for more details).

In contrast to subordinates and territorial males, females or immature juveniles of both sexes are rarely attacked although the juvenile males are slightly more susceptible than are females (Dixon and Mackintosh, 1976) and enjoy complete freedom of movement and even wander unmolested from one territory to another (Mackintosh, 1970). These observations raise questions as to the mechanisms underlying the apparent immunity seen in these mice.

Sensory mediation of mouse social behaviour

Close observation of mice show that at the start of a social encounter they point their heads such that their sense organs (eyes, ears, nose, vibrissae) are directed towards the conspecific. They then stretch their bodies take a few tentative steps and approach. The animals first touch noses and then sniff the head, body and ano-genital region of the partner. At some stage the sensory information gained from this period of mutual inspection leads to one of several responses that may be aggressive, defensive, sexual or disengagement. Clearly, there is a cascade of actions taking place that will demand the interplay of one or more sensory modalities. Mice lead an essentially nocturnal existence although they are occasionally trapped during daylight hours and have been observed to patrol above ground. Nevertheless, vision would be expected to play a less important role than sound and olfaction. Indeed mice are generally considered to be **macrosmatic** and studies have shown that olfaction plays a major role in coordinating their behaviour. Whilst they possess several sources of odours (breath, saliva, skin secretions, anal and preputial glands) the one most investigated is mouse urine.

Behavioural releasing effects of mouse odours

Male urinary odours

Mackintosh and Grant (1966) showed that aggression was increased between pairs of familiar male mice when one partner was marked with urine from a foreign male donor but was attenuated when familiar male urine was swabbed onto an introduced stranger. It was concluded that the urine of male mice contained substances, which conveyed information as to the degree of familiarity of the donors. About the same time Ropartz (1966a) showed that airborne odours of foreign mice evoked locomotor excitement in a recipient group of mice and attributed the effect to a plantar factor of the feet (Ropartz, 1966b). It is now known that urine is the source of a number of odours or 'factors' which exert behavioural-releasing effects in conspecifics. The aggression-releasing effect of male mouse urine is dependent upon male hormones and castration removes the effect (Mugford and Nowell, 1970; Dixon, 1973) moreover, immature males do not appear to possess the aggression releasing factor in their urine (Dixon and Mackintosh, 1976). The substances involved are yet to be unequivocally identified but are referred to here as the male urinary factors. They appear to indicate both adult male gender as well as social status. Thus, urine from territory holders evokes intense aggression in caged mice, that have won encounters but little aggression in mice that have lost encounters (losers). However, mice from subordinates evoked little interest from winners but high levels of aggression from losers (Dixon, 1973). This pattern of effects is entirely consistent with the social structure of the mice. Thus dominants will display most aggression to rival territory holders and less towards subordinates. However, subordinates that are forced to migrate in search of their own territories would be expected to be in direct competition with each other since established territory-holders are extremely hard to usurp.

Female urinary odours

The urinary odours of female mice also exert powerful effects on males. When urine from adult female mice was swabbed onto one of a pair of aggressive males, the aggression between them was radically reduced the greatest effect being in the unmarked mice (Mugford and Nowell, 1970; Dixon and Mackintosh, 1971). Urine from juvenile females but not immature males exerted the same effect (Dixon and Mackintosh, 1976) thus showing that the protection of the immature males was passive and that of the females active. A comparison of the effects of urine taken from female donors in oestrous and in dioestrous showed that both of them inhibited aggression in fighting males (Mugford and Nowell,

1971a,b; Dixon and Mackintosh, 1975). In addition, however, male mice exposed to urine from oestrous females displayed sexual activity towards their marked male partners, an effect not evoked by urine from dioestrous females or from immature female donors (Dixon and Mackintosh, 1975, 1976). These studies showed that female mouse urine contained at least two factors one releasing sexual behaviour and perhaps serving as a sex-attractant (Dixon, 1973; Hayashi and Kimura, 1974) and the other reducing aggression. The former proved to be under ovarian control as ovariectomy removed it, (Dixon and Mackintosh, 1975) and oestrogen treatment restored it (Mugford and Nowell, 1971b). This factor denotes maturity and reproductive state. In the case of the aggression-inhibitor this was independent of ovarian control and presumably provided olfactory information of female gender. Table 18.2 summarizes the behavioural effects of murine urinary odours and their putative function as deduced from the available data.

Priming effects of mouse odours

The effects of odours are not just confined to behavioural releasing effects. Adult male urine is also known to terminate pregnancy in freshly mated female mice provided it stems from a male other than the stud. This is the so-called Bruce effect (Bruce 1960, 1970; Dominic, 1964, 1965). Moreover, the long periods of anoestrous occurring in long-term caged groups of female mice (Lee and Boot, 1955, 1956; Whitten, 1957, 1959) can be overcome not just by the presence of an adult male (Whitten, 1956) but also by exposure to his urine (Marsden and Bronson, 1964). Indeed, exposure to male odours will accelerate and synchronize the oestrous cycles of female mice such that the highest proportion of subsequent matings occur on the third night (Whitten, 1956). These so-called *priming* effects of mouse odours differ from the *releasing* effects in that the odours do not need to be specifically attached to the donor but can be left deposited on the ground. Obviously, such physiological effects can ultimately alter behaviour, particularly reproductive behaviour and also aggression.

Many species mark their territorial boundaries with scent marks e.g. urine, or include physical landmarks, e.g. stones or bushes, when establishing the boundaries and mice certainly make use of odours and landmarks to recognize their boundaries. Such traces may have related but not necessarily identical effects on behaviour as those stemming from the mouse itself.

TABLE 18.2: Summary of main urinary odours in mice			
Possessed by	**Effect**	**Function**	**Characteristics**
Adult Female Juvenile Female	Aggression reduced in Adult Males	Signals female gender Territory-penetration	Independent of ovariesand of oestrous cycle
Adult Female	Releases Sexual activity Reduces Aggression	Reproductive state	Oestrous, ovarian andoestrogen- dependent
Adult Males	Releases Aggression in Foreign Males	Signals male gender and territoriality	Gonadal and androgen dependent
Adult Males	Different aggressive response in doms. and subs.	Denotes social status	Gonadal and androgen dependent
Adult Males	Releases sex in females	Female attractant	Gonadal and androgen dependent
Juvenile Males	No Aggression-releaser No Aggression-inhibitor	Gender denoted by other cues?	Small size may afford protection

The Table shows the effects of the urinary odours possessed by adult and immature male and female mice together with their characteristics and presumed function as determined on the basis of experiments reviewed in the text. The chemical nature of these various urinary factors are still being elucidated.

Physico-chemical properties of urinary odours

In mice, airborne odours act via two pathways, one that stimulates the olfactory epithelium where information is transmitted to the olfactory cortex in the brain and the other which involves a bundle of nerve cells in the nose, the vomeronasal organ which sends messages to the hypothalamus, a structure intimately associated with reproduction and homeostatic control of autonomic functions. The vomeronasal organ is suspected of being responsible for many of the so-called pheromonal effects of odours including those affecting mating and aggression. Identification of a gene, which encodes a protein TRP2, showed that it sits only on the vomeronasal organ (Stowers et al., 2002). Mice lacking this gene tried to mount foreign males and even castrated males whereas intact mice attacked them. This implies that the TRP2 gene, and hence the vomeronasal organ may be important for receipt and central processing of the urinary odours described above. The exact nature of the urinary factors still remain unknown although substances in male urine such as thiazole and brevicomin are known to have signalling function with respect to their aggressive status (Novotny et al., 1985; Hurst et al., 1998; Humphries et al., 1999). These volatile substances bind to lipocalins,

a group of proteins known as major urinary proteins which appear to effect the slow release of volatile chemicals from urinary traces (Hurst et al., 1998; Darwish et al., 2001). There is some evidence that the unbound volatile constituents exert an attractant effect towards a competitor's scent marks whereas the bound protein–ligand complexes induce countermarking. The dissociation of the effects of odours left as traces is complex as the information they convey will change with decay of the trace. In addition, the effects of freshly deposited odours exert social releasing effects only when they are directly associated with specific animals (Dixon, 1973). Thus, context and accessibility as well as the association with a specific mouse may evoke different effects even with the same odour.

An interesting pharmacological aspect is that production of aggression-releasing substances in male mouse urine is stimulated by treating the donors with diazepam, a benzodiazepine derivative, usually used to treat anxiety disorders in humans (Dixon, 1982). This effect of diazepam on aggression has been duplicated in both LAC and C57BL mice, and is removed if the donors are castrated (Dixon et al., 1984; Ryzhova et al., 1989). It remains to be shown that the urinary substances induced by diazepam treatment are the same as those occurring in intact, untreated male mice.

Sensory deprivation studies

Reversible techniques

The studies with urine all involved the presentation of additional olfactory stimuli. However, insight into the sensory control of behaviour can be obtained by depriving animals of specific sensory modalities. Although several authors have deprived mice of specific senses these have almost all involved techniques hazardous to the mice. For example zinc sulphate (ZnSO$_4$) solution (Alberts and Galef, 1971) instilled into the nose of the mice will destroy the nasal mucosa but also induce gastric toxicity and even death in some mice (Sieck and Baumbach, 1974). Visual deprivation has also been applied to rodents but here too drastic techniques such as surgical enucleation have rendered the animals permanently blind. In a series of elegant experiments Strasser (1987) investigated the relative roles of the senses in mediating territorial defence by male LAC mice as well as a small population of wild mice caught from the local feral population. He devised a series of reversible deprivation techniques that allowed specific sensory modalities to be temporarily disenabled during behavioural encounters both in enclosures and in cages. Visual deprivation was achieved by fitting the mice bilaterally with specially designed opaque contact lenses constructed out of heat moulded cellulose acetate foil and bilateral auditory deprivation was achieved using silicone ear plugs, both deprivation utensils being removed after the experiments (Strasser and Dixon, 1986). For sensory deprivation the mystacial vibrissae on the muzzle of the mouse were removed with scissors. Olfactory deprivation employed irrigation of zinc sulphate solution using a reversed siphon (liquid entered orally but exited nasally) so as to avoid gastric irritation and toxicity noted earlier (Strasser, 1987). The aim of these studies was to determine what major sensory modalities guided the detection and tracking of mice in enclosures and what modalities guided agonistic behaviour at close quarters as in cages.

Sensory deprivation of territorial mice

Detection phase

Intact territorial mice first detect intruders by *attending*, a specific movement of the head that aligns the face and eyes, nose and ears towards another mouse. If an intruder is stationary this occurs at a mean distance of about 10 cm (for wild and CFW mice) distance whereas moving intruders are detected as far away as 70 cm, i.e. moving animals are detected at greater distances and more readily than when they are stationary. When visually deprived, the mice could detect stationary intruders except when they are within touching distance, as one might expect if vision was important. Unexpectedly, however, the ability of the visually-deprived mice to detect moving intruders remained unimpaired (mean distance measured for attends occurred at 70.1 cm and for intact mice 71.4 cm (Strasser and Dixon, 1986) thus indicating that vision was not important for distant (early) detection of moving conspecifics. Mice fitted with earplugs resembled controls in that they first detected stationary intruders at a distance of 10.5 cm (vs. controls at 8.9 cm). However they appeared unable to detect moving intruders until the latter were as close as 18 cm (vs. 71.4 for intact mice). When mice were both deafened and visually impaired then attends only occurred within touching distance of an intruder irrespective of whether it was moving or not. These findings strongly suggest that initial detection of intruders by a territorial male first occurs at a distance if the intruder is moving and is primarily mediated by auditory cues, visual cues being mainly employed at close quarters!

Attack and chase phase

Sensory deprivation also exerted dramatic effects on aggressive behaviour of the territory holder. Attacking and chasing the intruder appeared unaffected by visual deprivation although the outcome of the attacks switched from successful to unsuccessful i.e. the territory-holder lost the intruder at close quarters. In contrast, mice fitted with earplugs showed far fewer chases and attacks on intruders than did intact mice and when they chased, they lost the intruder in 83% of the cases. Both deprivation techniques combined enhanced these deficits but, like acoustic deprivation, resulted in a net increase in locomotion as if the mice were searching for auditory cues.

When a dominant chases an intruder, the two mice weave and turn in unison as if held together by an invisible thread and the paths taken characteristically describing a series of large arcs. Close analysis revealed that during the pursuit, intact territory-holders kept a mean distance from the intruder of 29 cm but accelerated to attack whenever they came within 18–20 cm proximity. Mice fitted with earplugs kept a much closer distance of 19 cm i.e. within visual detection limits but were unable to track and chase mice at longer distances and usually lost them even though they could see them. Mice fitted with opaque contact lenses were generally able to track and chase intruders effectively but lost them at close quarters. These findings clearly show that the detection, chasing and tracking of foreign intruders is predominantly guided by auditory cues and the attacks themselves required visual and perhaps tactile cues to be successful. Since, measurements did not detect any audible or ultrasonic calls from the intruders, Strasser concluded that mice track and chase their rivals by means of a 'Passive Acoustic Tracking' mechanism based on the audible but non-vocal sounds of the escaping mice (Strasser, 1987).

Sensory deprivation of caged mice

Equally dramatic effects of sensory deprivation have been found with caged mice during close-quarter social encounters. **Anosmia** produced a complete attenuation of all aggressive activities of isolated resident mice. The effects were detected within 24 h of instilling the $ZnSO_4$ and lasted about 4 days. Recovery was complete by 21 days by which time the olfactory mucosal began to regenerate. This latter finding fits well with previous studies (Matulionis, 1976) of the effects of $ZnSO_4$ on the olfactory mucosa,. The behavioural findings are also in line with the major role of olfaction in mediating agonistic behaviour, reviewed earlier.

Resident mice fitted with earplugs showed a reduction in the element attend just as they had shown this in enclosures. However, in contrast to the findings in enclosures they did not show any other major effects and their aggression remained intact. Thus except for attend, audition does not seem to operate in any major way at during close quarter agonistic encounters between mice.

When residents were fitted with opaque contact lenses they, like intact mice, also vigorously attacked male intruders. However, closer examination showed that the 'blinded mice' showed little ambivalent behaviour. Moreover, in contrast to intact mice, the attacks were delivered straight ahead including the face and head i.e. their behaviour resembled to some extent the indiscriminate attacks of lactating females (Strasser, 1987). Although vision seems to be implicated in guiding the bites, the fact that their vigour was not impaired by visual deprivation suggests vision plays a subsidiary role in biting *per se*. Close inspection of intact mice provided further information relative to biting.

Immediately prior to delivering a bite, an attacking mouse often adopts an offensive sideways posture in which the head is brought towards the opponent but held slightly lateral to it. Inspection shows that the aggressive animal closes its nearside eye whereas the outside eye remains open (see Figure 18.2). In addition, the mystacial vibrissae are brought into close contact with the opponent before placing a bite and the mouse may even lift the nearside paws and touch its opponent. These observations prompted the idea that tactile contact might be substituting for visual control of biting and some evidence for this was obtained (Strasser, 1987). Thus, when the residents were bilaterally vibrissectomized, bites were massively reduced. At the same time, a marked increase in offensive ambivalence occurred together with 'parry' in which the resident used its paws to tap its opponent. Those bites that were delivered, like those of the visually deprived animals, were inaccurate. This result indicates that the ultimate direction of bites is guided by tactile cues obtained through the use of the paws and vibrissae and assisted by visual cues. Elsewhere it has been argued that mice close their nearside eye to offset or cut off flight-evoking stimuli emanating from the opponent who, by inducing ambivalence in the attacker could offset a final successful attack (Dixon, 1998). Since visual deprivation reduced all forms of ambivalence yet left offence this is evidence that flight-inducing stimuli from the opponent were being effectively cut-off. Indeed, cut-off acts, which reduce the input of flight-evoking stimuli, are widespread in animals and man (Chance, 1962; Dixon, 1998) and have been described in a number of rodent species. The more popular view that eyes are closed during close quarter combat to protect the eyes is not invalidated by the cut-off concept. Table 18.3 summarizes the major effects of sensory deprivation on mice.

Overall, the data indicate that whilst overlap between the modalities occurs during agonistic encounters between mice, the various phases of detection, tracking, chasing, attacking and biting each appear to be guided by a dominant sensory modality with delicate interplay of sight, tactile senses and olfaction as the mice attain close proximity.

Ultrasounds

In this article attention has been given to the role of the main senses in mediating agonistic behaviour and the role of olfactory stimuli in sexual encounters. However, the actual exchange of signals between mice is extremely complex and the use of calls, particularly ultrasonic calls have not been dealt with here. In fact, young mice emit ultrasonic cries of about 70 kHz and adult males emit ultrasonic cries in response to stimuli emanating from females. These appear to be

TABLE 18.3: Effects of sensory deprivation on the behaviour of laboratory mice

Modality deprived	Behaviour altered	Changes observed
Olfaction ($ZnSO_4$)	Attention	Reduced
	Aggression	Reduced
	Sexual	Reduced
Audition (earplugs)	Proximal Attention	Reduced
	Distance detection	Impaired
	Chasing and Tracking	Impaired
Vision (opaque contact lens)	Proximal Attention	Reduced
	Distance detection	Intact
	Ambivalence	Reduced
	Attacks in enclosures	Unsuccessful
	Biting (inaccurate)	Intensified
Tactile (Vibrissectomy)	Pawing of opponent	Increased
	Bites	Reduced

Summarized here are the main behavioural consequences of selective deprivation of sensory modalities in male LAC mice using the reversible techniques described by Strasser (1973). The changes observed cover behaviour shown at a distance (as in enclosures) as well as at close-quarters (in cages) and were determined on territory-holders or on dominant caged mice (see text for details).

odour-induced (Nyby *et al.*, 1977) and mediate sexual activity between males and female mice (Sales, 1972). A comprehensive and informative account of the senses including ultrasonic calls, in mediating behaviour can be found in Smith (1981). Whether ultrasonic calls augment the passive protection of juvenile males from attacks by adult males remains to be assessed.

Conclusions

Although this article is intended as a contribution to a handbook of the laboratory mouse much of its contents has been concerned with a description of the putative social structure and social behaviour of mice living under natural conditions. As mentioned before this is a risky undertaking as it is extremely difficult to observe mice in the wild and hence, the picture needs to be augmented by controlled studies of mice in captivity, albeit under semi-natural conditions and the results of such studies have been presented here. Hopefully, it will be evident from the foregoing, that many of the problems encountered in animal laboratories e.g. biting of laboratory stock by rogue males, excessive hyperirritability and aggression of individually-housed mice and the occasional cannibalism or fighting in females, all have their origins in the way precursor mice survive and breed in their natural habitats. The way mice use different sensory modalities to mediate their agonistic and social activities, though not exhaustively reviewed here, nevertheless shows that communication between mice is exceedingly complex and still not fully investigated. Although regarded as macrosmatic animals it is clear that other senses, particular audition plays a very important role in their social activities. Finally, although mice are generally maintained in rather impoverished conditions (e.g. standard cages) for reasons of practicality, the use of more mouse-friendly cages such as the Cambridge Mouse Cage or Wallace Cage (Wallace, 1982) which allows the mice cover, nest site, separate from defecation sites and even allows them to be transferred from one cage to another without actually handling them physically can do much to maintain some of their natural patterns of behaviour which may otherwise remain wanting in laboratory strains of mice.

Acknowledgements

I am indebted to Mrs C. Huber, Mrs M. Szilagyi, Mrs F. Kaesermann, Mr K. Rotach and Mr H. Asper for their practical involvement in many of the studies reviewed here. This article is dedicated to the memory of the late Stephan Strasser who showed us, yet again, of what extraordinary feats mice are capable.

References

Alberts, J.R. and Galef, B.G. (1971). *Physiol. Behav.* **6**, 619–621.

Anderson, P.K. (1961). *Trans. NY Acad. Sci. Ser II* **23**, 447–451.

Amir, S., Brown, Z.W., Amit, Z. and Ornstein, K. (1981). *Life Sci.* **10**, 1189–1194.

Baenninger, L.P. (1966). *Anim. Behav.* **14**, 367–371.

Berry, R.J. (1968). *J. Anim. Ecol.* **37**, 445–470.

Berry, R.J. (1970). *Field Studies* **3**(2), 219–262.

Berry, R.J. (1981). *Symp. Zool. Soc. Lond.* **47**, 395–425.

Blanchard, R.J., Flannelly K.J. and Blanchard, D.C. (1986). *J. Comp. Psychol.* **100**, 101–107.

Blanchard, D.C., Blanchard, R.J., Cholvanich, P., Scott, I. and Mayer, I.S. (1989). In *Ethoexperimental Approaches to the Study of Behaviour* (eds R.J. Blanchard, P. Brain, St Parmiagani, and J. Rosenblatt), pp. 338–360. Kluwer Academic Publishers, Dordrecht, The Netherlands.

Brain, P.F. (1974). *Life Sci.* **16**, 187–200.

Brain, P.F. (1981). In *Multidisciplinary Approaches to Aggression Research* (eds P.F. Brain and D. Benton), pp. 53–78. Elsevier, Amsterdam.

Brown, R.Z. (1953). *Ecol. Monogr.* **23**, 217–240.

Bruce, H.M. (1960). *J. Reprod. Fertil.* **1**, 96–103.

Bruce, H.L. (1970). *Brit. Med. Bull.* **26**, 10–13.

Calhoun, J.B. (1948). *J. Wildl. Manage.* **12**, 167–171.

Chance, M.R.A. (1962). *Symp. Zool. Soc. Lond.* **8**, 71–89.

Chance, M.R.A. and Russell, W.M.S. (1959). *Proc. Zool. Soc. Lond.* **132**, 65–70.

Crowcroft, P (1955). *J. Mamm.* **36**, 189–301.

Crowcroft, P. (1966) In *Mice all over*. Foulis & Co. Ltd. London.

Crowcroft, P and Rowe, F.P. (1963). *Proc. Zool. Soc. Lond.* **140**, 517–531.

Darwish, M.A., Veggerby, C., Robertson, D.H., Gaskell, S.J., Hubbard, S.J., Martinsen, L., Hurst, J.L and Beynon, R.J. (2001). *Protein Sci.* **10**(2), 411–417.

Dixon, A.K. (1973). The effect of olfactory stimuli upon the social behaviour of laboratory mice (*Mus musculus* L). Ph.D. Thesis, University of Birmingham.

Dixon, A.K. (1978). In *Das Tier im Experiment* (ed. W.H. Weihe), pp. 128–146. Hans Huber Verlag, Berne.

Dixon, A.K. (1982). *Psychopharmacology* **77**, 246–252.

Dixon, A.K. (1986). *Arch. Schw. Neurol. Psychiatry* **137**, 151–163.

Dixon, A.K. (1998). *Brit. J. Med. Psychol.* **71**, 417–445.

Dixon, A.K. and Kaesermann, H.P. (1987). In *Ethopharmacology of Agonistic Behaviour in Animals and Man* (eds B. Olivier, J. Mos, and P.F. Brain), pp. 46–79. Nijhoff Publishers, Dordrecht, Martines.

Dixon, A.K. and Mackintosh, J.H. (1971). *Anim. Behav.* **19**, 138–140.

Dixon, A.K. and Mackintosh, J.H. (1975). *Anim. Behav.* **23**, 513–520.

Dixon, A.K. and Mackintosh, J.H. (1976). *Z. Tierpsychol.* **41**, 225–234.

Dixon, A.K., Huber, C. and Kaesermann, F. (1984). In *Ethopharmacological Aggression Research* (eds K.A. Miczek, M.R. Kruk, and B. Olivier), pp. 81–91. Alan R Liss Inc., New York.

Dixon, A.K., Fisch, H.U., Huber, C. and Walser, A. (1989). *Pharmacopsychiatry* **22**(Suppl.), 44–50.

Dixon, A.K., Fisch, H.U. and McAllister, K.H. (1990). *Adv. Study Behav.* **19**, 171–204.

Dominic, C.J. (1964). *J. Reprod. Fertility.* **8**, 266–267.

Dominic, C.J. (1965). *J. Reprod. Fert.* **10**, 469–472.

Eibl-Eibesfeldt, I. (1970). *Ethology: The Biology of Behaviour*. Holt, Reihardt and Winston, New York.

Grant, E.C. and Chance, M.R.A. (1958). *Anim. Behav.* **6**, 183–193.

Grant, E.C. (1963). *Behaviour* **21**, 260–281.

Grant, E.C. and Mackintosh, J.H. (1963). *Behaviour* **21**, 246–259.

Gray, S.J and Hurst, J.L. (1996). *Anim. Behav.* **53**(3), 511–524.

Hayashi, S. and Kimura, T. (1974). *Physiol. Behav.*, **13**(4), 563–567.

Hinde, R.A. (1970). *Animal Behaviour: A Synthesis of Ethology and Comparative Psychology*, 2nd edn. McGraw-Hill, New York.

Humphries, R.E., Robertson, D.H., Beynon, R.J. and Hurst, J.L. (1999). *Anim. Behav.* **58**(6), 1177–1190.

Hurst, J.L., Robertson, D.H.L, Tolladay, U. and Beynon, R.J. (1998). *Anim. Behav.* **55**(5), 1289–1297.

Lee, van der. S. and Boot, L.M. (1955). *Acta Physiol. Pharmacol. Neerl.* **4**, 442–443.

Lee, van der. S and Boot L.M. (1956). *Acta Physiol. Pharmacol. Neerl.* **5**, 213–214.

Mackintosh, J.H. (1970). *Anim. Behav.* **18**, 177–183.

Mackintosh, J.H. (1973). *Anim. Behav.* **21**, 464–470.

Mackintosh, J.H. (1978). In *Population Control of Social Behaviour* (eds I.J. Ebling and D.M. Stoddart), pp. 157–180. Inst. Biol., London.

Mackintosh, J.H. (1981). In *Biology of the House Mouse* (ed. R.J. Berry). *Symp. Zool. Soc. Lond.* **47**, 337–365.

Mackintosh, J.H. and Grant, E.C. (1966). *Z. Tierpsychol.* **23**, 584–58.

Marsden, H.M. and Bronson, F.H. (1964). *Science* **144**, 1469.

Matulionis, D.H. (1976). *Am. J. Anat.* **142**, 67–90.

Mugford, R.A. and Nowell, N.W. (1970). *Nature (Lond.)* **226**, 967–968.

Mugford, R.A. and Nowell, N.W. (1971a). *Anim. Behav.* **19**, 153–155.

Mugford, R.A. and Nowell, N.W. (1971b). *J. Endocrinol.* **49**, 225–232.

Novotny, M. (1985). *Proc. Natl. Acad. Sci.* **82**, 2059–2061.

Nyby, J., Wysocki, C.J., Whitney, G. and Dizinno, G. (1977). *Anim. Behav.* **25**, 333–341.

Parmigiani, S. (1994). In *Infanticide and Parental Care* (eds S. Parmigiani and F. vom Saal), pp. 341–363. Harwood Academic Publishers, Chur.

Parmigiani, S., Brain, P.F., Mainardi, D. and Brunoni, W. (1988). *J. Comp. Psychol.* **102**, 287–293.

Parmigiani, S., Brain, P.F. and Palanza, P. (1989). In *Ethoexperimental Approaches to the Study of Behaviour* (eds R.J. Blanchard *et al.*), pp. 338–360. Kluwer Academic Publishers, Dordrecht, The Netherland.

Paul, L. (1986). *Aggr. Behav.* **12**, 1–11.

Reimer, J.D. and Petras, M.J. (1967). *J. Mammalogy* **48**, 88–99.

Ropartz, P. (1966a). *C.R. Acad. Sci. Paris* **262**, 507–510.

Ropartz, P. (1966b). *C.R. Acad. Sci. Paris* **263**, 525–528.

Ryzhova, L. Yu. and Novikov, S.N. (1989). *Dokl. Akad. Nauk. SSSR* **306**(4), 1005–1007.

Sales, G.D. (1972). *J. Zool. Soc. Lond.* **168**, 149–164.

Scott, J.P. and Frederickson, E. (1951). *Physiol. Zool.* **24**, 273–309.

Sieck, M.H. and Baumbach, H.D. (1974). *Physiol. Behav.* **13**, 407–425.

Smith, C.J. (1981). In *Biology of the House Mouse* (ed. R.J. Berry). *Symp. Zool. Soc. Lond.* **47**, 367–393.

Southwick, C.H. (1955). *Ecology* **36**, 627–634.

Southwick C.H. (1958). *Proc. Zool. Soc. Lond.* **131**, 163–175.

Stowers, L. Holy, T.E., Meister, M., Dulac, C. and Koentges, G. (2002). *Science* **295**, 1493–1500.

Strasser, S. (1987). Neuron-ethology of mouse agonistic behaviour. Ph.D. Thesis, University of Zurich, Zbinden Druck und Verlag AG, Basel.

Strasser, S. and Dixon, A.K. (1986). *Physiol. Behav.* **36**(4), 773–778.

Uhrich, J. (1938). *J. Comp. Physiol. Psychol.* **25**, 373–413.

Whitten, W.K. (1956). *J. Endocrinol.* **13**, 399–404.

Whitten, W.K. (1957). *Nature (Lond.)* **180**, 1436.

Whitten, W.K. (1959). *J. Endocrinol.* **18**, 102–107.

Wallace, M.E. (1982). *Int. J. Stud. Anim. Prob.* **3**(3), 234–242.

CHAPTER 19

Neuroendocrine–Immune Network in Stress

Yukiko Kannan

Graduate School of Agriculture and Biological
Sciences, Osaka Prefecture University, Japan

Stress

As Hans Selye (who defined the current meaning of 'stress') has stated that 'stress is life and life is stress,' stress is an unavoidable consequence of life. The survival of an organism and its species depends on its ability to adapt to a changing environment.

The scientific investigation of stress stems from Claude Bernard's (1878) recognition that maintaining a relatively constant internal environment is important for survival. In 1932, Walter Cannon elucidated the mechanisms required to maintain the internal environment within appropriate tolerable limits and developed the term 'homeostasis' to describe the stable state maintained by these mechanisms. Selye (1946) extended Cannon's notions of homeostasis and proposed that in addition to the specific homeostatic and local responses (such as inflammation) a single stereotypic stress response or general adaptation syndrome (GAS) is elicited by any demands, including psychosocial demands, upon the body. He described GAS as triphasic, beginning with an alarm reaction followed by a resistance stage and ending with the stage of exhaustion. Agents that induce stress are termed stressors, and

Selye (1975) recognized the existence of both good (eustress) and bad (distress) stressors. Stressors may be physical/metabolic (e.g. surgery, cold, starvation) or psychological (e.g. fear, maternal separation, restraint) and differ in controllability, intensity, duration and frequency. Selye's concept of a single stereotypic syndrome induced by any demand must be modified to reflect the differences in response patterns to various stressors.

The stress system

Mammalian organisms, including mice, have developed a complex system to coordinately adapt to various stressors (Mason, 1968a, b). Collectively called the stress system, the main components are the hypothalamic–pituitary–adrenocortical (HPA) and the sympathetic–adrenomedullary (SA) systems (Figure 19.1). The stress system functions primarily to maintain homeostasis during rest and stress states and its activation leads to behavioral changes, such as regulation of arousal and attention and the generation of fear. In addition, the stress system influences many peripheral functions such as gastrointestinal (GI) and immune/inflammatory activities.

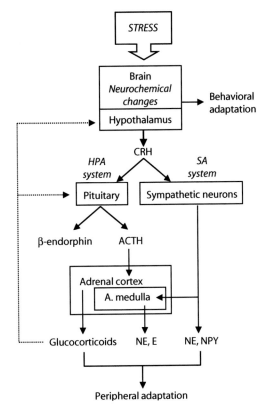

Figure 19.1 Schematic diagram of the stress system. Dotted lines indicate the negative feedback modulation of the HPA system by glucocorticoids. Abbreviations: ACTH, adrenocorticotropin; CRH, corticotropin-releasing hormone; E, epinephrine; HPA, hypothalamic–pituitary–adrenocortical; NE, norepinephrine; NPY, neuropeptide Y; SA, sympathetic–adrenomedullary.

Cannon (1914) first demonstrated increased plasma concentrations of catecholamines in cats exposed to barking dogs and identified the catecholamine sources as the sympathetic nervous system and the adrenal medulla. The postganglionic sympathetic neurons release norepinephrine (NE) both locally and into the general circulation. Activation of the adrenal medulla by preganglionic sympathetic neurons results in secretion of epinephrine and, in most species, NE, into the general circulation. Cannon suggested that peripheral catecholamine levels redistribute blood flow and reorient metabolism toward the production of glucose as energy for the musculature to enable 'fighting or fleeing'. On the other hand, Seyle (1936a) demonstrated the importance in the final HPA activation stage of the adrenal cortex and glucocorticoid secretion. The HPA system is initiated by secretion of corticotropin-releasing hormone (CRH) from hypothalamic neurons with terminals in the median eminence (Vale *et al.*, 1981). CRH is then transported by portal blood flow to the anterior pituitary, where it stimulates adrenocorticotropin (ACTH)

secretion. Circulating ACTH stimulates the secretion of glucocorticoids (cortisol and/or corticosterone, depending on the species) from the adrenal cortex. The anterior pituitary secretes β-endorphin concomitantly with ACTH in response to CRH (Conte-Devolx *et al.*, 1983). Glucocorticoids are the major effectors of the HPA system and provide negative feedback at their upper streams (Keller-Wood *et al.*, 1984). But CRH, ACTH and β-endorphin also have independent effects during stress.

Cerebral mechanisms in stress

A large number of neurochemical changes occur in the brain during or following stress. Stressful treatments frequently decrease cerebral NE content, especially in the hypothalamus, due to the increased release of NE coupled with the inability of synthetic mechanisms to keep pace during intense or prolonged stress. Further, the increased synthesis and catabolism of catecholamines during stress and the stress-related increased extracellular NE concentrations suggest that stress increases the synaptic release of NE (Dunn and Berridge, 1987). Cerebral dopaminergic and adrenergic neurons are also activated in response to physical and behavioral stressors (De Souza and Van Loon, 1986; Dunn and Berridge, 1987). Thus, brain neurons containing all three catecholamines may contribute to stress responses.

Cerebral serotonin (5-hydroxytryptamine, 5-HT) systems also appear to be important in the stress response. Numerous studies have demonstrated alteration in 5-HT content or increased metabolism and release during stress (De Souza and Van Loon, 1986).

Corticotropin-releasing hormone acts as a central coordinator in stress. CRH is found in neurons in a variety of brain regions, not only in the paraventricular hypothalamic neurons important in activation of the HPA system (Frim *et al.*, 1990). Intracerebral administration of CRH produces a wide variety of effects, induced by activation of the HPA system (De Souza and Van Loon, 1984), that resemble those observed in stress. Intracerebral CRH is also a potent activator of both central and peripheral catecholaminergic neurons (Dunn and Berridge, 1987) and induces many behavioral effects, such as increased grooming (Dunn *et al.*,

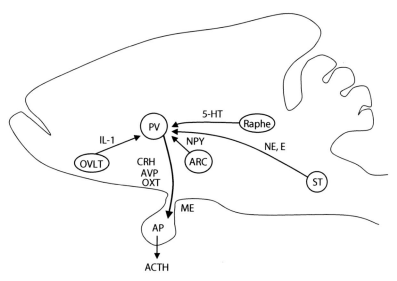

Figure 19.2 Schematic model of the stimulatory regulation of CRH and ACTH release by various neurons in the rodent brain. Abbreviations: ACTH, adrenocorticotropin; AP, anterior pituitary; ARC, arcuate nucleus; AVP, arginine vasopressin; CRH, corticotropin-releasing hormone; E, epinephrine; IL-1, interleukin-1; ME, median eminence; NE, norepinephrine; 5-HT, serotonin; NPY, neuropeptide Y; OVLT, organum vasculosum of lamina terminalis; OXT, Oxytocin; PV, paraventricular nucleus; ST, solitary tract nucleus.

1987) and decreased exploratory behavior (Berridge and Dunn, 1986).

The neuropeptides arginine vasopressin (AVP) and oxytocin (OXT), which differ in only two amino acids, are synthesized mainly in the magnocellular neurons of the hypothalamic paraventricular and supraoptic nuclei that project to the posterior pituitary. Parvicellular neurons of the paraventricular nucleus (PVN) co-express AVP and CRH (which are released into the portal blood), coordinate HPA activity and project to the external layer of the median eminence. Arginine vasopressin synergizes strongly with CRH to stimulate ACTH release during stress (Rivier and Vale, 1984a). Oxytocin is less effective than AVP but potentiates CRH-induced ACTH release (Gibbs, 1986).

Neuropeptide Y (NPY), a known co-transmitter of NE in the sympathetic nervous system (Stjarne *et al.*, 1986), is one of the most abundant and widely distributed neuropeptides within the central nervous system (CNS), with particularly high concentrations in the hypothalamus along with other neurotransmitters like somatostatin, galanin, GABA and the catecholamines. Neuropeptide Y is involved in a variety of neuroendocrine functions such as stress responses, circadian rhythms, central autonomic functions and eating and drinking behavior (Leibowitz, 1991).

The first cytokines discovered were mediators of communication between various types of immune cells but genes encoding various cytokines and their receptors have since been found to be expressed in vascular, glial

and neuronal structures of the CNS. Cytokines exert profound effects on physiological processes related to food intake, fever, neuroendocrine regulation, long-term potentiation and behavior (Saper and Breder, 1992).

Figure 19.2 schematically diagrams various neuronal projections into the PVN where CRH neurons are activated during stress.

Stress and growth/ reproduction

Growth and reproduction, linked directly to the stress system, are profoundly influenced by the HPA system (Figure 19.3).

The CRH-induced increase in somatostatinergic tone has been implicated as a potential mechanism of the stress-induced suppression of growth hormone (GH), thyrotropin-releasing hormone (TRH) and thyrotropin secretion (Terry *et al.*, 1976; Rivier and Vale, 1985), resulting in growth and thyroid function suppression.

The hypothalamic–pituitary–gonadal axis is maintained by the activity of hypothalamic neurons secreting gonadotropin-releasing hormone (GnRH). The nerve terminals of these neurons project to the median eminence where GnRH is released into the hypophysial portal circulation. The pulsatile secretion and surge of

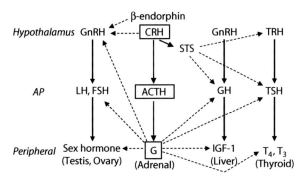

Figure 19.3 Schematic diagram of interactions between the HPA and growth/reproduction axes. Solid lines indicate activation, and dashed lines indicate inhibition. Abbreviations: ACTH, adrenocorticotropin; AP, anterior pituitary; CRH, corticotropin-releasing hormone; FSH, follicle-stimulating hormone; G, glucocorticoid; GH, growth hormone; GnRH, gonadotropin-releasing hormone; IGF-1, insulin-like growth factor-1; LH, luteinizing hormone; STS, somatostatin; T_3, triiodothyroxine; T_4, thyroxine; TRH, thyrotropin-releasing hormone; TSH, thyrotropin.

luteinizing hormone (LH) depends on the activity of GnRH neurons. Restraint stress has been shown to reduce the pregnancy rate and average litter size of mice (Wiebold *et al.*, 1986) and to inhibit the proestrus LH surge and ovulation in intact cycling rats (Roozendaal *et al.*, 1995). The reproductive axis is inhibited at all levels by various components of the stress system. Corticotropin-releasing hormone may negatively regulate GnRH secretion (Rivier and Vale, 1984b). Central factors such as endogenous opioids are also reported to restrain LH secretion (Petraglia *et al.*, 1986).

Glucocorticoids exert broad inhibitory effects on growth and reproduction at the hypothalamus, pituitary and peripheral hormone producing sites to render target tissues resistant to the corresponding hormones (Chrousos, 2000).

Stress and gastrointestinal function

Activation of the stress system influences gastrointestinal (GI) function. A CRH-mediated decrease in gastric acid secretion, gastric emptying time, small bowel transit time and, alternatively, increase in large bowel transit time have been observed in restrained rats (Lenz *et al.*, 1988).

These GI responses may be mediated by activation of sympathetic neurons and/or suppression of vagus nerves innervating the GI system (Yamamoto *et al.*, 1998). Central AVP and dopamine (Bueno *et al.*, 1992) as well as central and peripheral cholecystokinin (Bueno, 1993) may also be involved in stress-induced GI changes. Severe stress (such as cold-restraint stress) often induces gastric ulcers, caused at least in part by vagal hyperactivity (Cho *et al.*, 1996; Matsushima *et al.*, 1999).

Stress and immune function

A substantial body of data has recently revealed a bidirectional interaction between the brain and immune system (Figure 19.4). The HPA and the sympathetic nervous system represent major pathways by which the brain regulates the immune process. Selye (1936b) noted that rats exposed to diverse stressors simultaneously developed enlarged adrenal glands and shrunken thymuses and lymph nodes and suggested that a cause and effect relationship might be present between the two, with the former secreting a hormone that suppressed the latter. Immune cells have receptors that bind a number of hormones, neuropeptides and neurotransmitters. Changes in neuroendocrine and/or autonomic activities may therefore lead to modulation of immune responses. As communication mediators from the immune system to the brain, inflammatory cytokines such as tumor necrosis factor-α (TNFα), interleukin (IL)-1 and IL-6 may act upon the HPA and the sympathetic system, locally or via the CNS.

Although stress is widely regarded as impairing immune system function, the relationship appears to be more complicated. Considerable evidence suggests that stress not only suppresses but also enhances immune functions.

Effect of the HPA system on immune response

Corticotropin-releasing hormone has a variety of effects on immune cells (Elenkov *et al.*, 1999). As the

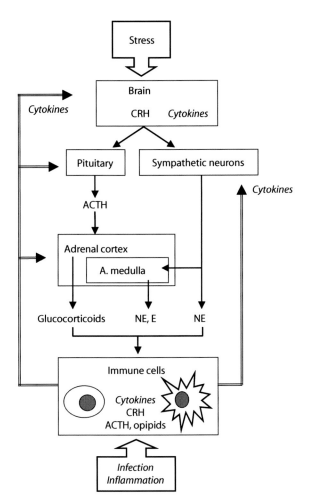

Figure 19.4 Schematic diagram of the bidirectional relationship between the brain and immune system. The brain modulates the immune functions via the HPA and the SA systems, and the immune cell-derived cytokines affect neuroendocrine functions via the CNS and peripheral organs. Corticotropin-releasing hormone and cytokines are also produced by immune cells and within the brain, respectively.

glucocorticoids cause stimulatory effects within the normal physiological range and inhibitory effects at higher levels found in chronically stressed animals.

The immunomodulative effects of central CRH are mediated at least partly via the central stimulation of sympathetic outflow, too. Anatomical studies have demonstrated that sympathetic noradrenergic fibers innervate the vasculature, other smooth muscle compartments and the parenchyma in the bone marrow, thymus, spleen and lymph nodes in numerous species (Felten *et al.*, 1985). The nerve endings are in close proximity to lymphocytes (mainly T cells) and more commonly macrophages. Thus, these cells may be exposed to high local concentrations of catecholamines. Norepinephrine stimulates α- and β-adrenergic receptors expressed by immune cells. α-adrenergic receptors are generally believed to be stimulatory while β-adrenergic receptors are believed to inhibit immune responses, although the detailed responses are very complex and critically dependent upon timing, dose and stimulated cell types. The inhibitory effects of stress on the lytic activity of natural killer (NK) cells and cytotoxic T lymphocytes are mediated by the activation of the SA system, as the decline in the cytotoxic activity is reversed by a ganglionic blocker and β-adrenergic receptor antagonists (Sheridan *et al.*, 1998; Ben-Eliyahu *et al.*, 2000).

Corticotropin-releasing hormone and its receptors are located not only throughout the CNS, but also in the immune system. 'Immune CRH' appears to have proinflammatory actions and is associated with initiating, propagating and/or regulating inflammatory responses (Mastorakos *et al.*, 1995; Elenkov *et al.*, 1999).

Both ACTH and opioid peptides (α, β and γ-endorphin) have also been found to influence many aspects of immune responses and act directly on immune cells by interacting with cellular receptors (Brooks, 1990; Gaveriaux *et al.*, 1995).

Effect of the inflammatory cytokines on the HPA and SA systems

TNF-α, IL-1, IL-6 and other cytokines produced in response to inflammation activate the HPA system. TNF-α, which has tumoricidal activity and causes

primary regulator of the HPA system, CRH modulates immune/inflammatory reactions through receptor-mediated actions of glucocorticoids on target immune cells. Glucocorticoids prevent the migration of leukocytes from the circulation into extravascular fluid spaces, reduce the accumulation of monocytes and granulocytes at inflammatory sites and suppress the production and/or action of many cytokines and inflammatory mediators (Hermann *et al.*, 1995; Dobbs *et al.*, 1996; Sheridan *et al.*, 1998; Zhang *et al.*, 1998). Of note, glucocorticoid effects are not exclusively immunosuppressive. Rather, glucocorticoids assist in maintaining normal immunity. Adrenally compromised animals are more susceptible to infection and the administration of physiological doses of glucocorticoids is essential for normal recovery from infections in adrenalectomized animals (Ruzek *et al.*, 1999). Thus, cachexia, is first to appear in the cascade of inflammatory events. TNF-α peaks approximately 1 h after administration of lipopolysaccharide, a component of Gram-negative bacterial cell walls, and then rapidly declines, whereas IL-1 and IL-6 levels peak somewhat later (2–4 h) and are sustained longer (Shalaby et al., 1989). The inflammatory cytokine route of access across the blood–brain barrier (BBB) to the CNS has not yet been clearly elucidated. At least two routes are possible and evidence exists for both: (1) active transport across the BBB by cytokine-specific carriers, and (2) message transfer where the BBB is 'leaky' (i.e. in the 'sensory' circumventricular organs, particularly the organum vasculosum laminae terminalis (OVLT)) on the midline of the preoptic area (POA), by an unknown group of neurons projecting into the OVLT from the brain. But alternative pathways are also possible and support for some exists: (1) cytokine-induced generation of BBB-permeable prostaglandin E_2 (PGE$_2$; the most proximal, putative mediator of fever) by endothelial cells of the cerebral microvasculature or perivascular microglia and meningeal macrophages, and (2) direct transmission to the POA of pyrogenic messages via peripheral (largely vagal) afferent nerves activated by cytokines (Blatteis, 2000).

In addition to their hypothalamic effects, cytokines directly stimulate pituitary ACTH and adrenal glucocorticoid secretion (Angeli, et al., 1999; Arzt et al., 1999). There is also evidence that several inflammatory cytokines stimulate sympathetic nerve growth (Kannan et al., 1996, 2000).

Stress and host defense

Several examples of model stressors on host defenses in mice are introduced below. Restraint is a relatively mild stressor, and is often used as an experimental model stressor of emotional stress in humans. Usually, an animal is introduced into an adequately ventilated container (e.g. a cylinder or a conical tube) with adjustable head and tail gates.

Repeated, prolonged periods of restraint (16 h/ overnight for four nights) have been shown to depress anti-viral cellular immunity in C57BL/6 mice (Hermann et al., 1994a). The restrained mice infected with influenza virus showed a pattern of reduced mononuclear cell infiltration and lung consolidation which coincided with elevated plasma corticosterone levels; the cellular infiltration was reversed by blocking corticosterone activity by administration of the glucocorticoid receptor antagonist RU486. Moreover, viral cellular immunity (e.g. IL-2 secretion) was significantly depressed, while the magnitude of the viral humoral immune response (e.g. anti-viral IgG antibody levels) was unaffected. Interestingly, restraint stress enhanced the probability of survival in another strain (DBA/2) of inbred mice (Hermann et al., 1994b). In this mouse, the interplay between glucocorticoids and catecholamines appears to be coordinated to optimize the immune response to viral infection. Repeated, prolonged periods of restraint also suppressed NK cell activity and the primary development of herpes simplex virus (HSV)-specific cytotoxic T lymphocytes (Bonneau et al., 1998). Higher titers of infectious HSV were detected at the site of the viral injection in restrained mice compared with unrestrained mice. Both corticosterone and catecholamine-mediated mechanisms have been demonstrated to operate in the stress-induced suppression of antiviral cellular immunity (Sheridan et al., 1998).

Contrary to the immune-suppressive effects of chronic restraint stress, acute restraint appears to enhance immune responses. Dhabhar (2000) demonstrated that acute restraint (2–5 h) immediately before the introduction of an antigenic challenge enhances a cutaneous delayed-type hypersensitivity (DTH) response in rats, caused at least partially by elevated corticosterone levels during restraint. Increasing the severity of acute stress, by combining immobilization with shaking, enhanced the DTH response. In contrast, lengthening the stress significantly suppressed the DTH response. A similar enhancement of cutaneous hypersensitivity was shown in BALB/c mice, but not in C57BL/6 mice (Flint and Tinkle, 2001). C57BL/6 mice are reportedly relatively stress resistant and although their basal corticosterone levels are similar to the levels in BALB/c mice, they produce lower concentrations of ACTH and corticosterone in response to acute restraint.

Forced swimming in cold water is another factor that has been used as a stressor in the study of the effect of stress on host defense mechanisms. Exposure of mice to cold water reduces the number of thymocytes and splenocytes, decreases T-cell blastogenesis in response to mitogen stimulation, reduces NK activity and alters macrophage function. Macrophages were activated in an apparently unregulated manner by cold-water stress (Chancellor-Freeland et al., 1995). Some functions such as cytokine and PGE$_2$ production were enhanced, while other functions, such as the ability to induce immune regulation associated antigen expression by interferon-γ were suppressed. This is interesting since cold stress elevates glucocorticoid levels and it blocks the production of PGs and cytokines by macrophages. Substance P (SP), an 11-amino acid neuropeptide widely distributed throughout the central and peripheral nervous system, regulates a wide variety of inflammatory responses and may activate macrophages. Substance P stimulates phagocytic and chemotactic capacity as well as macrophage cytokine and PG production. Cold stress induced SP in the peritoneal fluid that bathes peritoneal macrophages and capsaicin depleted SP, thus reducing cytokine production in response to stress.

Laboratory animals may encounter a number of non-experimental stress events including psychosocial stress (e.g. isolation, overcrowding and social rank interactions), handling or transportation.

Mice exposed to social isolation stress developed a suppressed immune response resulting in enhanced metastasis of carcinoma cells (Wu et al., 2001). On the other hand, overcrowding induces dramatic body weight reduction and significantly increased adrenal gland weight. Simultaneously, it induced dry skin and impaired barrier function (Aioi et al., 2001).

In the wild, adult male mice exhibit high levels of aggressive behavior toward conspecific males. In laboratory conditions, fighting behavior in male mice can be elicited after several weeks of social isolation. This psychosocial stress not only activates the HPA axis, but also stimulates the release of nerve growth factor (NGF) from salivary glands into the blood in male mice. On the other hand, restraint or cold water swimming does not alter serum NGF levels (Alleva and Santucci, 2001). Moreover, NGF levels in fighting mice are correlated with their social status, that is, serum NGF levels in mice that repeatedly experience defeat and submission are two-fold higher than NGF levels in dominant attacking mice. Nerve growth factor is well known to be important in the survival, differentiation and maintenance of sensory and sympathetic neurons and some cholinergic neurons in the CNS, especially the hippocampus (Levi-Montalcini et al., 1995). In addition to its neurotrophic effects, NGF exerts various biological actions, including regulation of immune/inflammatory responses (Kannan et al., 1991; Kawamoto et al., 1995). Exogenous administration of NGF results in adrenal gland hypertrophy and glucocorticoids have been shown to enhance submissiveness in male mice (Bigi et al., 1992). Thus, the higher NGF levels and hypertrophic adrenals in subordinates suggest that a regulatory loop involving NGF-promoted adrenal hormone activation enhances a subordination

profile. Nerve growth factor may also accelerate cutaneous wound healing in subordinates (Matsuda et al., 1998).

Mice are generally affected more than rats by environmental stress such as handling and transportation (Tabata et al., 1998). Physiological responses to handling may occur very rapidly (Lapin, 1995); when groups of mice are removed from their home shelf and randomly captured and bled over a period of time, 10-fold increases in plasma corticosterone levels can occur in 30 min (Riley, 1981). Therefore, reliable baseline measurements of plasma corticosterone can be obtained only within several minutes following an initial handling disturbance. However, mild, long-term intermittent handling during the early postnatal period attenuates the stress-induced behavioral and neurochemical reactivity in adulthood correlated with low corticosterone secretion (Vallee et al., 1997). In addition, the systemic effects of handling are modulated by genetic background. Expression of the immediate early gene-related proteins was examined in DBA/2 and C57BL/6 mice in response to acute or repeated handling and injection stress (Ryabinin et al., 1999). A strong induction of the proteins in discrete areas of the hypothalamus and other brain regions occurred 2 h after an acute intraperitoneal injection of saline in both strains. Repeated handling and injections for two weeks led to complete habituation in C57BL/6 mice, but the elevated early-gene was still induced in many brain regions of the DBA/2 mice.

Transportation is another commonplace stressor for laboratory animals. Simply moving mice from one room to another significantly elevated corticosterone levels and caused weight loss (Wallace, 1976; Tuli et al., 1995). Shipping is highly stressful; corticosterone levels in mice were markedly increased at arrival and remained high for a 48-h period, regardless of the method of shipment. Immune-functions were also significantly depressed at arrival, but returned to baseline within 48 h, indicating that a minimum few-day stabilization period is extremely important for new arrivals of mice (Landi et al., 1982).

Conclusions

Organisms live in dynamic equilibrium, or homeostasis. When homeostasis is challenged by extrinsic or intrinsic forces, nonspecific and specific behavioral and physiological responses occur from activation of the stress system. The stress system is driven by the neuroendocrine–immune network to coordinate adaptive responses to stressors. If the stress system is overly taxed or impaired, the organism may be vulnerable to pathology. In other words, normal development and survival are dependent on the functioning stress system.

Acknowledgment

The author wishes to thank Prof. Hiroshi Matsuda for the opportunity to write this chapter.

References

Aioi, A., Okuda, M., Matsui, M., Tonogaito, H. and Hamada, K. (2001). *J. Dermatol. Sci.* **25**, 189–197.

Alleva, E. and Santucci, D. (2001). *Physiol. Behav.* **73**, 313–320.

Angeli, A., Masera, R.G., Sartori, M.L., Fortunati, N., Racca, S., Dovio, A., Staurenghi, A. and Frairia, R. (1999). *Ann. NY Acad. Sci.* **876**, 210–220.

Arzt, E., Pereda, M.P., Castro, C.P., Pagotto, U., Renner, U. and Stalla, G.K. (1999). *Front. Neuroendocrinol.* **20**, 71–95.

Bernard, C. (1878). *Les phénomènes de la vie*, Vol. 1, p. 879. Librairie J.B. Baillière et Fils, Paris, France.

Ben-Eliyahu, S., Shakhar, G., Page, G.G., Stefanski, V. and Shakhar, K. (2000). *Neuroimmunomodulation* **8**, 154–164.

Berridge, C.W. and Dunn, A.J. (1986). *Regul. Pept.* **16**, 83–93.

Bigi, S., Maestripieri, D., Aloe, L. and Alleva, E. (1992). *Physiol. Behav.* **51**, 337–343.

Blatteis, C.M. (2000). *J. Physiol.* **526**, 470.

Bonneau, R.H., Zimmerman, K.M., Ikeda, S.C. and Jones, B.C. (1998). *J. Neuroimmunol.* **82**, 191–199.

Brooks, K.H. (1990). *J. Mol. Cell. Immunol.* **4**, 327–337.

Bueno, L. (1993). *J. Physiol.* (*Paris*) **87**, 301–306.

Bueno, L., Gue, M. and Delrio, C. (1992). *Am. J. Physiol.* **262**, G427–G431.

Cannon, W.B. (1914). *Am. J. Physiol.* **33**, 356–372.

Cannon, W.B. (1932). *The Wisdom of the Body*, p. 24. Norton, WW, NY.

Chancellor-Freeland, C., Zhu, G.F., Kage, R., Beller, D.I., Leeman, S.E. and Black, P.H. (1995). *Ann. NY Acad. Sci.* **771**, 472–484.

Cho, C.H., Qui, B.S. and Bruce, I.C. (1996). *J. Gastroenterol. Hepatol.* **11**, 125–128.

Chrousos, G.P. (2000). *Int. J. Obes. Relat. Metab. Disord.* **24**, S50–S55.

Conte-Devolx, B., Rey, M., Boudouresque, F., Giraud, P., Castanas, E., Millet, Y., Codaccioni. J.L. and Oliver, C. (1983). *Peptides* **4**, 301–304.

De Souza, E. and Van Loon, G.R. (1984). *Experientia* **40**, 1004–1006.

De Souza, E. and Van Loon, G.R. (1986). *Brain Res.* **367**, 77–86.

Dhabhar, F.S. (2000). *Ann. NY Acad. Sci.* **917**, 876–893.

Dobbs, C.M., Feng, N., Beck, F.M. and Sheridan, J.F. (1996). *J. Immunol.* **157**, 1870–1877.

Dunn, A.J. and Berridge, C.W. (1987). *Pharmacol. Biochem. Behav.* **27**, 685–691.

Dunn, A.J., Berridge, C.W., Lai, Y.I. and Yachabach, T.L. (1987). *Peptides* **8**, 841–844.

Elenkov, I.J., Webster, E.L., Torpy, D.J. and Chrousos, G.P. (1999). *Ann. NY Acad. Sci.* **876**, 1–13.

Felten, D.L., Felten, S.Y., Carlson, S.L., Olschowka, J.A. and Livnat, S. (1985). *J. Immunol.* **135**, 755s–765s.

Flint, M.S. and Tinkle, S.S. (2001). *Toxicol. Sci.* **62**, 250–256.

Frim, D.M., Robinson, B.G., Pasieka, K.B. and Majzoub, J.A. (1990). *Am. J. Physiol.* **258**, E686–E692.

Gaveriaux, C., Peluso, J., Simonin, F., Laforet, J. and Kieffer, B. (1995). *FEBS Lett.* **369**, 272–276.

Gibbs, D.M. (1986). *Neuroendocrinology* **42**, 456–458.

Hermann, G., Tovar, C.A., Beck, F.M. and Sheridan, J.F. (1994a). *J. Neuroimmunol.* **49**, 25–33.

Hermann, G., Beck, F.M., Tovar, C.A., Malarkey, W.B., Allen, C. and Sheridan, J.F. (1994b). *J. Neuroimmunol.* **53**, 173–180.

Hermann, G., Beck, F.M. and Sheridan, J.F. (1995). *J. Neuroimmunol.* **56**, 179–186.

Kannan, Y., Ushio, H., Koyama, H., Okada, M., Oikawa, M., Yoshihara, T., Kaneko, M. and Matsuda, H. (1991). *Blood* **77**, 1320–1325.

Kannan, Y., Bienenstock, J., Ohta, M., Stanisz, A.M. and Stead, R.H. (1996). *J. Immunol.* **157**, 313–320.

Kannan, Y., Moriyama, M., Sugano, T., Yamate, J., Kuwamura, M., Kagaya, A. and Kiso, Y. (2000). *Neuroimmunomodulation* **8**, 132–141.

Kawamoto, K., Okada, T., Kannan, Y., Ushio, H., Matsumoto, M. and Matsuda, H. (1995). *Blood* **86**, 4638–4644.

Keller-Wood, M., Shinsako, J. and Dallman, M.F. (1984). *Am. J. Physiol.* **247**, E489–E494.

Landi, M.S., Kreider, J.W., Lang, C.M. and Bullock, L.P. (1982). *Am. J. Vet. Res.* **43**, 1654–1657.

Lapin, I.P. (1995). *J. Pharmacol. Toxicol. Methods* **34**, 73–77.

Leibowitz, S.F. (1991). *Brain Res. Bull.* **27**, 333–337.

Lenz, H.J., Raedler, A., Greten, H., Vale, W.W. and Rivier, J.E. (1988). *Gastroenterol.* **95**, 1510–1517.

Levi-Montalcini, R., Dal Toso, R., della Valle, F., Skaper, S.D. and Leon, A. (1995). *J. Neurol. Sci.* **130**, 119–127.

Mason, J.W. (1968a). *Psychosom. Med.* **30**, 576–607.

Mason, J.W. (1968b). *Psychosom. Med.* **30**, 666–681.

Mastorakos, G., Bouzas, E.A., Silver, P.B., Sartani, G., Friedman, T.C., Chan, C.C., Caspi, R.R. and Chrousos, G.P. (1995). *Endocrinology* **136**, 4650–4658.

Matsuda, H., Koyama, H., Sato, H., Sawada, J., Itakura, A., Tanaka, A., Matsumoto, M., Konno, K., Ushio, H. and Matsuda, K. (1998). *J. Exp. Med.* **187**, 297–306.

Matsushima, Y., Kinoshita, Y., Watanabe, M., Hassan, S., Fukui, H., Maekawa, T., Okada, A., Kawanami, C., Kishi, K., Watanabe, N., Nakao, M. and Chiba, T. (1999). *Digestion* **60**, 34–40.

Petraglia, F., Vale, W. and Rivier, C. (1986). *Endocrinology* **119**, 2445–2450.

Riley, V. (1981). *Science* **212**, 1100–1109.

Rivier, C. and Vale, W. (1984a). *Endocrinology* **114**, 914–921.

Rivier, C. and Vale, W. (1984b). *Endocrinology* **114**, 2409–2411.

Rivier, C. and Vale, W. (1985). *Endocrinology* **117**, 2478–2482.

Roozendaal, M.M., Swarts, H.J., Wiegant, V.M. and Mattheij, J.A. (1995). *Eur. J. Endocrinol.* **133**, 347–353.

Ruzek, M.C., Pearce, B.D., Miller, A.H. and Biron, C.A. (1999). *J. Immunol.* **162**, 3527–3533.

Ryabinin, A.E., Wang, Y.M. and Finn, D.A. (1999). *Pharmacol. Biochem. Behav.* **63**, 143–151.

Saper, C.B. and Breder, C.D. (1992). *Prog. Brain Res.* **93**, 419–429.

Selye, H. (1936a). *Br. J. Exp. Pathol.* **17**, 234–248.

Selye, H. (1936b). *Nature* **138**, 32.

Selye, H. (1946). *J. Clin. Endocrinol.* **6**, 117–230.

Selye, H. (1975). *J. Hum. Stress* **1**, 37–44.

Shalaby, M.R., Waage, A., Aarden, L. and Espevik, T. (1989). *Clin. Immunol. Immunopathol.* **53**, 488–498.

Sheridan, J.F., Dobbs, C., Jung, J., Chu, X., Konstantinos, A., Padgett, D. and Glaser, R. (1998). *Ann. NY Acad. Sci.* **840**, 803–808.

Stjarne, L., Lundberg, J.M. and Astrand, P. (1986). *Neuroscience* **18**, 151–166.

Tabata, H., Kitamura, T. and Nagamatsu, N. (1998). *Lab. Anim.* **32**, 143–148.

Terry, L.C., Willoughby, J.O., Braseau, P., Martin, J.B. and Patel, Y. (1976). *Science* **192**, 565–567.

Tuli, J.S., Smith, J.A. and Morton, D.B. (1995). *Lab. Anim.* **29**, 132–138.

Vale, W., Spiess, J., Rivier, C. and Rivier, J. (1981). *Science* **213**, 1394–1397.

Vallee, M., Mayo, W., Dellu, F., Le Moal, M., Simon, H. and Maccari, S. (1997). *J. Neurosci.* **17**, 2626–2636.

Wallace, M.E. (1976). *Lab. Anim.* **10**, 335–347.

Wiebold, J.L., Stanfield, P.H., Becker, W.C. and Hillers. J,K. (1986). *J. Reprod. Fertil.* **78**, 185–192.

Wu, W., Murata, J., Hayashi, K., Yamaura, T., Mitani, N. and Saiki, I. (2001). *Biol. Pharm. Bull.* **24**, 772–776.

Yamamoto, O., Niida, H., Tajima, K., Shirouchi, Y., Kyotani, Y., Ueda, F., Kise, M. and Kimura, K. (1998). *J. Pharmacol. Exp. Ther.* **287**, 691–696.

Zhang, D., Kishihara, K., Wang, B., Mizobe, K., Kubo, C. and Nomoto, K. (1998). *J. Neuroimmunology* **92**, 139–151.

Circadian Rhythms of the Mouse

Burghart Jilge and Emanuela Kunz
Laboratory Animal Research Unit, University of Ulm,
Germany

Chronobiology is a discipline of 'life sciences' dealing with temporal fluctuations of organisms. Rhythms of about 24 h are most evident in all of our life, and the laboratory animal's life as well as in for example the alternation of sleep and wakefulness, the rhythms of body temperature, corticosterone, liver glycogen, etc. Perhaps, due to their ubiquity, it is widely thought that they are a mere passive reflection of the external light : dark (LD) cycle. In addition, since some rhythms, such as body temperature, gonadotropic hormone release or the mitotic rate in some compartments of the small intestine have a low amplitude of <30–50% only, they often are ignored completely. Hence, it is not uncommon to hear that 24-h rhythms need not be considered since they are '…minimal in extent and trivial in mechanism'. We will show here that both of these statements are wrong: the 'mechanism' of physiological 24-h rhythms is quite intricate and the ranges of many cell-, organ- or whole-organism functions within the boundaries of **homeostasis** may amount up to several hundred percent. In fact there is almost no function of an organism which does not exhibit a 24-h fluctuation. Thus, in many if not most cases it would

be an artifact not to consider 24-h rhythms in an animal experiment using mice.

While there are some other physiological low-frequency rhythms of significance, such as ultradian and circannual rhythms, those exhibiting a 24-h period are unequivocally of greatest relevance to the laboratory animal scientist.

Genesis of circadian rhythms

A nocturnal animal species like the mouse in a regular 12 : 12 LD cycle (LD 12 : 12) expresses most of locomotor activity, food and water consumption etc. during the hours of darkness. The activity pattern during a persistent LD cycle is highly strain specific (Figure 20.1a). When the LD regimen as the primary external **zeitgeber** is withdrawn for an extended time, the former 24-h alternation of activity and rest, feeding and fasting etc.

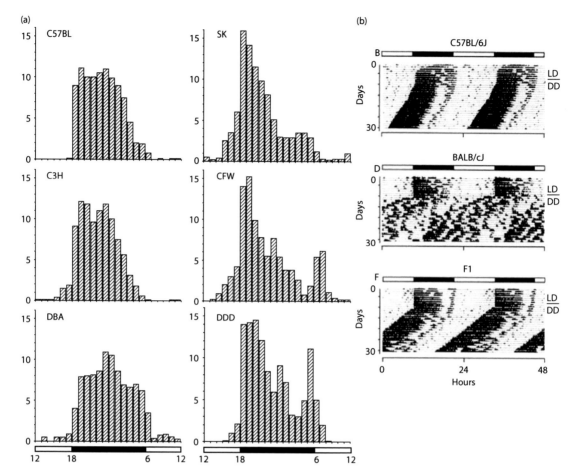

Figure 20.1 (a) Pattern of wheel-running activity of six different strains of mice living in a regular 12 h LD zeitgeber schedule (LD 12:12; redrawn from data of Ono *et al.*, 1991). (b) Original plots of wheel-running activity of a C57BL/6J and a BALB/cJ male mouse and their F1 male offspring. The data were recorded during 7 days in LD 12:12 and then in continuous darkness. According to chronobiological standards the data were double-plotted, i.e. the data of 48 h are plotted on one line, the data of the second day being repeated on the left of the line below followed by the data of the next day etc. Activity is represented by vertical black marks. Note the marked differences of the activity pattern between both inbred strains: while the B6 mouse has a very stabile rhythm with an average τ of 23,61 ± 0.068 h, the activity rhythm of the BALB/c mouse is rather scattered and has a significantly shorter period length (average: 22,88 ± 0,602 h). The activity rhythm of the F1 progeny closely resembles that of the B6 parental strain indicating the presence of B6 alleles that regulate this trait in a dominant manner. For further details of genetic analysis see Shimomura *et al.*, 2001 (plots from Shimomura *et al.*, 2001).

during DD or LL will continue but subsequently desynchronize from the phase obtained during the preceding schedule. Usually the period length of nocturnal animals in DD conditions is less than 24 h and expands during LL. Period length may vary conspicuously between individuals even of an inbred strain (Pittendrigh, 1976; Pittendrigh and Daan, 1976a; Shimomura *et al.*, 2001). Within an individual animal the rhythm in most strains and stocks is very precise, i.e. the day-to-day-variability can amount to few minutes per day only (Pittendrigh and Daan, 1976a,b; Welsh *et al.*, 1986). With increasing age, precision and robustness of the rhythm decrease – but some may continue, such as that of body temperature, until death (Weinert *et al.*, 2002) – whereas the length of the

free-running period of the mouse increases (Possidente *et al.*, 1995; Mayeda *et al.*, 1997; Valentinuzzi *et al.*, 1997; Aujard *et al.*, 2001; Weinert *et al.*, 2001). The autonomous rhythm of an organism in constant conditions with a period length (τ) of about – but significantly different from – 24 h is called a 'free-running circadian rhythm' (from latin 'circa' = about and 'dies' = day).

The genetic set-up of the mouse determines the length of the circadian period as well as the pattern, precision and robustness of the free-running rhythm (Figure 20.1b; Ebihara *et al.*, 1978; Schwartz and Zimmerman, 1990; Beau, 1991; Ono *et al.*, 1991; Hofstetter *et al.*, 1995; Possidente and Stephan, 1998; Suzuki *et al.*, 2000; Tankersley *et al.*, 2002). Table 20.1 lists the period length of locomotor activity obtained in

TABLE 20.1: Average length of the free-running period (τ) of wheel-running activity in 12 different inbred strains of mice living in the absence of any external zeitgeber (continuous dark (DD) conditions)

Strain	n	Mean τ_{DD} ± SEM
129/J[a]	6	23.93 ± 0.07
RF/J[a]	6	23.92 ± 0.06
C57BL/6J	10	23.77 ± 0.02
SWR/J[a]	3	23.70 ± 0.02
SEC/ 1ReJ	8	23.59 ± 0.04
AKR/J[a]	5	23.52 ± 0.04
DBA/2J	3	23.46 ± 0.05
C57BL/10J	5	23.43 ± 0.01
C57L/J	4	23.42 ± 0.13
A/J[a]	10	23.37 ± 0.07
B10.D2(58N)/Sn	4	23.34 ± 0.15
BALB/cByJ[a]	8	22.94 ± 0.06

Source: Data from Schwartz and Zimmerman (1990).
[a] Albino strains.

12 inbred strains of mice: since experimental variables such as the way of monitoring activity (by photodiodes, spring suspended cage, running wheel of different diameter etc.) may have an impact on τ, Table 20.1 represents data obtained under identical experimental conditions.

In all animal species and in man the adjustment of $\tau \neq 24$ with the external T = 24.0 h LD schedule occurs via a phasically varying response to light. When a light pulse meets the first half of the subjective night – when the nocturnal animal is awake and active – then the rhythm is abruptly delay-shifted. A light pulse falling into the second half of the subjective night is responded to by an advance of the rhythm, as a rule via a couple of transients. Light during the subjective day does not shift the circadian rhythm at all (even when animals are kept in DD). The quantitative shift of a circadian rhythm following light stimulation is portrayed by a phase response curve (PRC). The general shape of the PRC is about the same in nocturnal and in diurnal animal species and in humans (Figure 20.2; Pittendrigh, 1976; Daan and Pittendrigh, 1976; Aschoff, 1981; Moore-Ede et al., 1982; Schwartz and Zimmerman, 1990).

The time of re-entrainment of a free-running rhythm following reinstatement of the LD regimen depends on the phase of the rhythm relative to the zeitgeber. As a rule of thumb in most cases 4 weeks will be sufficient for shifting body functions after reinstatement or a shift of the LD zeitgeber; longlasting 'after-effects', however, have been demonstrated for ~65 days (d) (Pittendrigh and Daan, 1976a). During the time of reentrainment a jet-lag like internal desynchronization of different body functions can occur.

The existence of a persisting free running circadian rhythm in constant conditions and its entrainment by a LD schedule implies that the rhythm is generated endogenously and is synchronized by an external zeitgeber. It is generally accepted that the master-generator of the 24-h rhythms in mammals resides in the suprachiasmatic nuclei (SCN), a bilaterally symmetric accumulation of 16,000–20,000 neurons dorsal to the optic chiasm and located at either side of the third ventricle (Figure 20.3; Ralph et al., 1990; Moore, 1995; Viswanathan and Davis, 1995; Weaver, 1998; Ibata et al., 1999; Reppert and Weaver, 2000; Abrahamson and Moore, 2001; Harmer et al., 2001). Light enters the SCN via the monosynaptic retino-hypothalamic tract (RHT) connecting the retina with the SCN and the intergeniculate leaflets (Speh and Moore, 1993; Moore, 1995). Experiments with retinally deficient ($Pde6b^{rd}$/$Pde6b^{rd}$) mice on various inbred strain backgrounds have given evidence for the fact that neither rods nor cones transmit the light message to the master clock. The receivers are in fact additional photopigments (including mCRY1 and mCRY2) residing in the retina (Foster et al., 1991; Lucas and Foster, 1999; Lucas et al., 1999; Mrosovsky et al., 1999; Vitaterna et al., 1999; Lowrey and Takahashi, 2000; Sancar, 2000; Foster and Helfrich-Forster, 2001; VanGelder, 2001; Froy et al., 2002).

The efferent transduction of the 'zeitgeber' message from the SCN is apparently far more complex: while some behavioral functions like locomotor activity evidently do not need neural efferents, the latter are apparently obligatory for controlling endocrine rhythms (Lehman et al., 1995; Abrahamson and Moore, 2001; Bartness et al., 2001; Nelms et al., 2002). In this context the pineal organ (epiphysis cerebri) is playing an important role: while it is getting the message from the SCN via neural sympathetic afferents its major product, melatonin is synthesized and secreted during the dark time exclusively. Thus, the external LD cycle is generally reflected and transduced to each cell of the body by the endocrine melatonin level (see Figure 20.7a and b). In mammalian species the pineal organ, however, is completely dependent on neural afferents and, thus, without afferents cannot produce one single oscillation of its own.

At the molecular level at least nine different genes have been identified which are generating and controlling

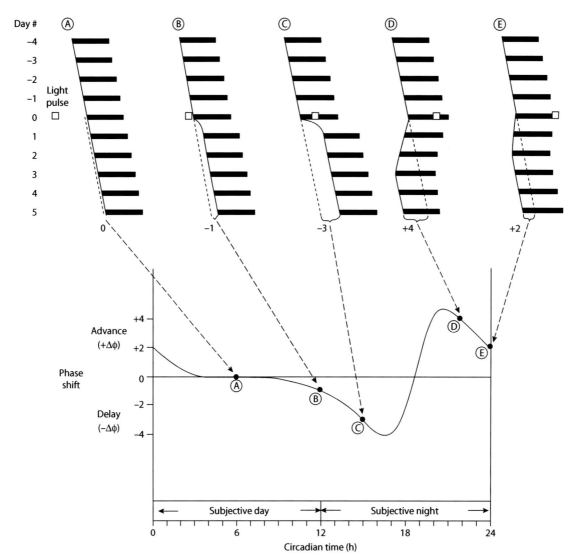

Figure 20.2 Schematic representation of the principle of the phase dependent response of the free-running rhythm to a light stimulus (PRC). For details see text (Moore-Ede *et al.*, 1982).

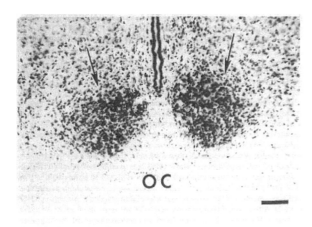

Figure 20.3 Photomicrograph of a Nissl-stained coronal section from rat hypothalamus showing the SCN (arrows) as two dense accumulations of cells lying above the optic chiasm (OC) and lateral to the third ventricle. Scale bar 100 μm (Moore, 1995).

circadian rhythms inside the cell (bmal1, clock, ckI€, cry1, cry2, per1, per2, per3, tim). Thus, while a couple of different cells and organs may be able to generate circadian rhythms – which is important to consider in *in vitro* studies – the central coordination within the whole organism is done by the SCN (Kolker and Turek, 1999; Lowrey and Takahashi, 2000; Reppert and Weaver, 2000).

While the LD cycle is the zeitgeber used by most eukaryotic organisms, a periodic schedule of restricted feeding (RF) can be used as zeitgeber too, at least by some mouse strains. As compared with LD, RF as a zeitgeber appears to be weaker. However, it can set the phase of many gastrointestinal, endocrine and metabolic functions. Food anticipatory activity is expressed even in SCN-lesioned animals and the peak of mPer1- and mPer2 genes are expressed around the time of food access in the liver, the cerebral cortex and hippocampus but are unaffected in the SCN. Thus it appears that RF entrains additional circadian clocks outside the SCN. Hence, an RF-induced phase setting for example at a light-time feeding schedule will affect a great many of physiological and pathophysiological functions and so far the consequences for homeostasis of a permanent temporal displacement of functions driven by the LD and those controlled by an RF regimen are widely unknown (Pauly *et al.*, 1975; Abe *et al.*, 1989; Ohdo *et al.*, 1996b; Damiola *et al.*, 2000; Holmes and Mistlberger, 2000; Hara *et al.*, 2001; Stokkan *et al.*, 2001; Wakamatsu *et al.*, 2001; Turek and Allada, 2002).

Circadian functions of the mouse

Within the boundaries of homeostasis almost every cell, tissue and organ of the mouse shows a 24-h rhythm. During the last 35 years a plethora of facts and data have been explored for different functions and in different mouse strains. Hence, this chapter will focus on some general and representative examples of common interest.

Behavior

The nocturnal mouse in an environment with minimal human interference and a persistent, regular LD cycle expresses most locomotor activity, exploratory and aggressive behavior, climbing on lids, seizure susceptibility, food and water intake (though there is not total agreement on this point as seen in Chapter 15 on Gastrointestinal System and Metabolism), feces and urine excretion and grooming during the hours of darkness. As has been stated above, with respect to behavioral parameters conspicuous differences exist between mouse strains and stocks (Ziesenis *et al.*, 1975; Poirel, 1988; Schwartz and Zimmerman, 1990; Ono *et al.*, 1991; Harri *et al.*, 1999). In general the aggressive behavior against intruders during D is about twice that seen in L (Paterson and Vickers, 1984). A more recent investigation in C57BL/6J mice reports that altruistic behavior may be elevated in D as compared with the hours of light (Nejdi *et al.*, 1996). Naturally the NREM and REM sleep states are significantly elevated during the 12 h of light time (Meerlo and Turek, 2001).

The strain specific patterns of locomotor activity are reflected, too, in core body temperature and heart rate (HR). The differences between strains and reciprocal first generation offsprings in most cases are significant and the range of telemetrically monitored core body temperature within an individual animal amounts to 1.6–2.0°C from a light-time minimum to the dark time maximum. The HR was found to have a range of 116–191 beats/min (minimum: begin of L, maximum: begin of D; see below; Tankersley *et al.*, 2002).

Reproduction

The sexual cycle of the female mouse has a much lower frequency lasting in most strains and stocks 4 or 5, in some even 6 days to complete itself. There is, however, a circadian gate for a couple of overt and underlying parameters. During a continuous regular LD regimen ovulation occurs on the day of estrus in the middle of the dark period, between 24 and 3 a.m. and this is the time when the females show maximal **lordosis** behavior and when in the presence of a male copulatory activity is maximal (Bingel and Schwartz, 1969; Chan *et al.*, 2001). If ovulation does not occur at that time, e.g. due to experimental interference etc., it will occur about 24 h or 48 h later around mid D (Kaiser, 1967; Champlin *et al.*, 1980). Correspondingly the endocrine regulatory and feedback loops controlling the overt sexual rhythm are cyclic, starting from the gonadotropin-releasing hormone (GnrH) pulse generator in the mediobasal hypothalamus and the pituitary gonadotropins down to the sexual steroid hormones. The number of ovarian luteinizing hormone (LH) receptors are maximal during the second half of the light phase of the day of proestrus (DeLeon *et al.*, 1990). Therefore, when administrating exogenous gonadotropins, in order to superovulate animals, for example, the best response is achieved with injections given during the second part of L. This holds true, too, following complete reentrainment with an inverted LD cycle. Thus, the latter unequivocally is the zeitgeber for circadian endocrine sexual functions (Bindon and Lamond, 1966). During proestrus a surge of LH occurs and attains its maximum just before the end of the light period and triggers ovulation about 6–9 h later. During diestrus the 24-h rhythm of progesterone culminates shortly after the onset of darkness (peripheral plasma progesterone >24 nmol/L). The progesterone level during diestrus is much lower as compared with proestrus. During the latter the peak of ~130 nmol/L also occurs after the onset of darkness (Bailey, 1987; DeLeon *et al.*, 1990).

While in the mouse, as in most mammalian species, parturition mainly occurs during the first hours of resting time in L, some strains are reported to deliver during D (Kaiser, 1967; Ghiraldi *et al.*, 1993). During the whole course of lactation 63–65% of nursing time was found to lie during the hours of quiescence in light while non-social behavior was elevated during the hours of darkness (Chapman and Cutler, 1987).

Gastrointestinal tract and metabolism

As a direct consequence of nocturnally increased food intake follow-up functions like the contents of the stomach, serum gastrin, the activity of gastrointestinal enzymes, bile and exocrine pancreas secretion, absorptive functions in the intestines and the blood glucose level are significantly elevated during the hours of darkness. In the same way glycogen stores in liver and muscle tissues are on their minimum at the beginning of the dark period and attain peak levels during the first hours of L, the range between the evening trough and the morning peak values of liver glycogen amounting to about 200% (Halberg, 1959; Halberg *et al.*, 1960b; Weinert *et al.*, 1994) (Figure 20.4a). It is self evident that the physiological periodicity of food intake and fasting results in periodic absorption of metabolites and storing and depletion of energy. Hence, it appears important to point out that the rhythm of storage and depletion of glycogen in liver cells of BALB/c mice was found to persist even during the fasting of the animals (Haus and Halberg, 1966) and that the 24-h muscle glycogen rhythm is also reported to continue *in vitro* (Saffe de Vilchez *et al.*, 1972; Pessaq and Gagliardino, 1975).

Similarly, serum gastrin and plasma cholecystokinin are reported to exhibit a significant 24-h periodicity with a large span from their minimum during L to a maximum at the end of D (range ~150%). The 24-h rhythm of either function also persisted in fasted mice, though at a lower level (Pasley *et al.*, 1987).

As a rule the mitotic index of liver, epidermal and corneal cells attains maximum values during the hours of light (ranging up to ~300%) whereas that of parenchymal and striatal cells of the adrenal cortex is counterphasic (range 160–130% of 24-h mean; Halberg, 1959; Haus and Halberg, 1966; Scheving and Pauly, 1973). Scheving *et al.* (1983, 1989) found significant 24-h rhythms of cell proliferation in various parts of the gastrointestinal tract, from the tongue down to the rectum and including spleen and thymus: in general the peaks were seen during the light time and

the percentual range of change extended from 28% (tongue) up to 945% (cornea; Figure 20.4b; Scheving *et al.*, 1983).

Cardiovascular parameters and blood constituents

With miniaturized microelectronic devices unbiased telemetric data of cardiovascular parameters even of a small animal like the mouse have been acquired during recent years. Characteristic strain differences are also found in cardiovascular parameters. In general, HR and mean arterial pressure (MAP) are significantly higher during D as compared with L, the HR ranging from ~450 (L) to ~650 bpm (D) and the MAP from ~90 (L) to ~130 mmHg (D) (Figure 20.5; Li *et al.*, 1999; Carlsson and Wyss, 2000; Kramer *et al.*, 2000; Mills *et al.*, 2000; Butz and Davisson, 2001; Tankersley *et al.*, 2002).

In a comprehensive survey on 24-h rhythms of hematologic reference values of 3 different strains of mice Swoyer *et al.* (1987) demonstrated that red cell parameters, lymphocytes, neutrophils, monocytes and eosinophils are on a conspicuously and in most cases statistically significant higher level during the first hours of the light phase in male and female animals as well. The eosinophilic rhythm was the very first blood parameter of the mouse to be shown to have a significant 24-h rhythm (Figure 20.7c; Halberg, 1953; Halberg *et al.*, 1958). The eosinophils and lymphocytes exhibit the highest variability ranging for >100% (depending on strain) within 24 h while red blood cell number, hemoglobin and hematocrit have a range of ~10% (Figure 20.6). The administration of granulocyte colony-stimulating factor (G-CSF) was found to elevate the leucocyte level and to exaggerate the 24-h rhythm: the leucocyte counts during L were significantly higher than during the first hours of D, a reflection of the rhythm which was also found in the myeloid colonies of the bone marrow *in vitro* (Ohdo *et al.*, 1998). It is well known that the proportion of bone marrow cells undergoing DNA synthesis varies by 50% or more during 24 h in mice. Knowledge of the circadian organization of cell proliferation may be crucial for carrying out experiments on the effectiveness of cytostatic drugs in mice (see below). Summarizing results of measurements of cell proliferation of hemopoietic stem cells of the bone marrow, Smaaland (1996, 1997) reported that DNA synthesis is highest during D and that the highest mitotic index is generally found during L. The circadian rhythm in femoral bone marrow cells

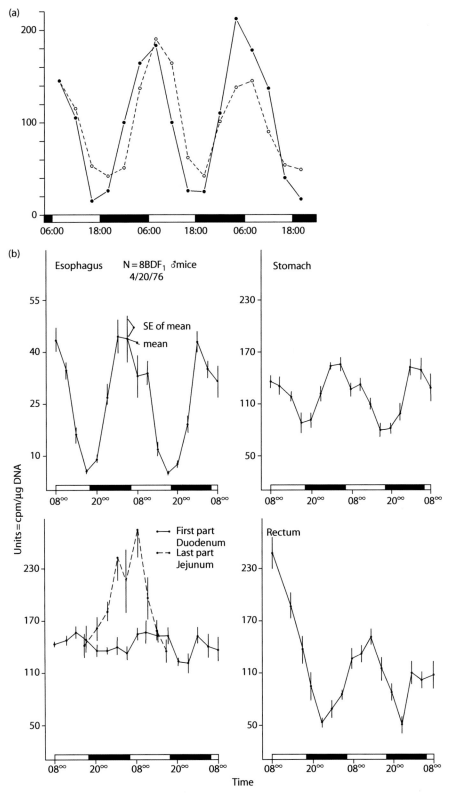

Figure 20.4 (a) 24-h rhythm of hepatic glycogen of the mouse (female BALB/c) fed *ad libitum* (solid) and deprived of food 16 h prior to sampling of first data (dashed). Data are plotted as deflections from the 24 h mean equivalent to 100%. Note that the liver glycogen rhythm persisted even in the food-deprived animals, though with a lower amplitude (Data replotted from Haus and Halberg, 1966). (b) 24-h rhythms of [³H]TdR incorporation into DNA over a 48 h span in five different segments of the intestinal tract of BDF₁ mice. The rhythm with the lowest amplitude was found in the duodenum (reproduced from Scheving *et al.*, 1983).

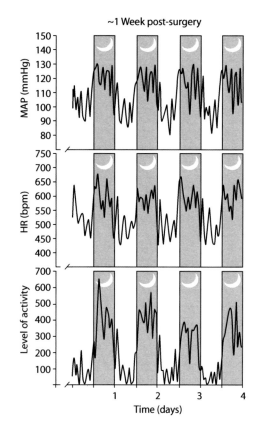

~1 Week post-surgery

Figure 20.5 24-h rhythms of MAP (top), HR (middle) and locomotor activity (bottom) of a representative female C57BL/6 mouse during four consecutive days. Data were obtained about 1 week after implantation of the telemetry transmitter. The plots represent mean values over each hour of a regular LD 12 : 12 (reproduced from Butz and Davisson, 2001).

from mice continues *in vitro* for up to 4 days (Smaaland, 1997; Kolaczkowska *et al.*, 2001; Bourin *et al.*, 2002; Yellon and Tran, 2002).

Endocrinology

Melatonin is considered to be the main endocrine mediator of dark time. A conspicuous strain specificity of melatonin concentration exists. A few mouse strains (C57BL/6J, AKR/J, BALB/c, NZB/BlNJ) due to a deficit either of *N*-acetyltransferase (NAT) or of hydroxiindol-O-methyltransferase (HIOMT) synthesis are reported not to transform serotonin to melatonin within the pinealozyte. While in these strains almost no melatonin is found at all (Ebihara *et al.*, 1987; Goto *et al.*, 1989), the melatonin of C3H/He or CBA strains ranges between nocturnal 500–1800 fmol/pineal gland and 30–180 pmol in the plasma of peripheral blood (Figure 20.7a). Due to reduced neuronal activity of the SCN cells during the light period almost no melatonin is produced within the pinealocyte. It is important to know that even a short light pulse during D immediately truncates melatonin synthesis (Figure 20.7b; Tamarkin *et al.*, 1985; Reiter, 1991, 1993; Kennaway *et al.*, 2002). Interestingly those strains which are almost devoid of melatonin reentrain significantly faster than those with a normal melatonin synthesis rate (Kopp *et al.*, 1998), which is in accordance with findings obtained in rat and hamster. Recent investigations using much more

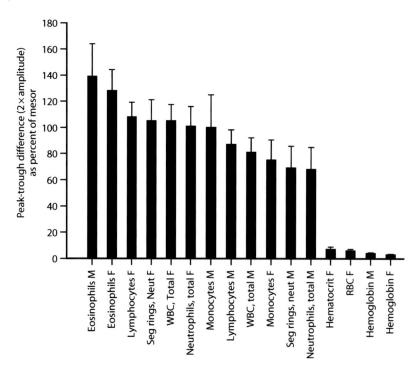

Figure 20.6 Range of some hematologic 24-h rhythms in the BDF₁ mouse (replotted from Swoyer *et al.* 1987).

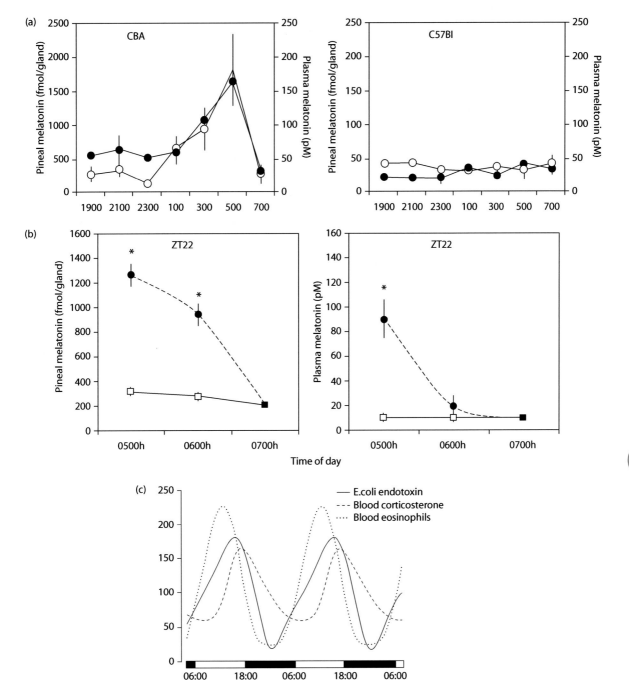

Figure 20.7 (a) Pineal gland melatonin content (black dot) and plasma melatonin concentration (open circles) in two inbred strains of mice (reproduced from Kennaway *et al.*, 2002). (b) Effect of a light pulse (200 lx, 15 min) switched 2 h before the end of the dark period on pineal and plasma melatonin of male CBA mice (data of untreated control animals are represented by black dots, those exposed to the light pulse by open squares (reproduced from Kennaway *et al.*, 2002). (c) 24-h rhythms of plasma corticosterone, blood eosinophils and lethality-rate following E. coli endotoxin injection in the mouse (strain and gender not reported; replotted from data of Halberg, 1959); data are plotted as deflections from the 24-h mean equivalent to 100%.

sensitive assays and at a higher sampling frequency have demonstrated, however, that even BALB/c mice synthesise melatonin although at a rather low level (<100 pg/ml serum; Vivien-Roels *et al.*, 1998; Barriga *et al.*, 2001). To the best of our knowledge so far, no comparative data have been published on the strain

dependency of (cytosolic or membrane-bound) melatonin receptors in the SCN and other organs of mice.

Corticosterone has a rhythmic pattern as seen in other nocturnally active rodent species: several hours prior to the onset of activity around the beginning of D the serum corticosterone level peaks at about

200–300 ng/ml and declines thereafter to trough levels of <50 ng/ml in the second half of D and the first part of L (Figure 20.7c). Correspondingly the response of corticosterone to an injection of ACTH at different times of the day varies significantly. As in many other functions, age, gender and genome are shown to modify the shape of the rhythm (Halberg et al., 1958; Haus et al., 1974; Fitzpatrick et al., 1992; Weinert et al., 1994; Barriga et al., 2001).

Pharmacology/toxicology/irradiation

As stated above precise 24-h rhythms are present in cells, organs and the whole organism. As a consequence the response to external interference, whether it is a chemical (drug) or physical (light or ionic radiation) or even a social (human or companion animal interference) factor in most cases is responded to differently at different phases of the 24-h day. Thus, the outcome of a test on acute toxicity, the pharmacokinetics of a substance or the therapeutic effect of a drug as a rule depends significantly on the time of day when the substance is given, absorbed, distributed, metabolized and excreted. Chronobiologists are aware of that fact and many papers have been published in this field. The monographs of Lemmer (1989) and Redfern and Lemmer (1997) summarize many aspects of this area. In a very recent volume of Chronobiology International (Vol. 19, 2002) circadian aspects of cancer therapy are compiled. In the following some representative examples are presented in order to focus the attention of experimenters on the phase dependency for example during tests on the toxicity or efficacy of substances or therapeutic interventions.

It is most important to consider that there are no general and common phases of reduced and enhanced susceptibility or tolerance versus external interference by substances. Rather each individual drug can have a specific phase of increased or reduced susceptibility. The range between highest and lowest susceptibility following administration of a toxic substance can amount to several hundred percent. As a consequence large differences in the tolerance of a substance can be present within few hours. In many cases the 'one time only'-regimen, will reach a very limited conclusion.

One of the classical examples of 'chronotolerance' is the response of mice to high doses of alcohol and E. coli endotoxin: The lethality following intraperitoneal injection of about 2 g of alcohol was found to be maximal around the beginning of dark time (death rate 50–64% 4 h post administration) and significantly

lowest in mid D and during the first part of L (death rate 8–33%; Haus and Halberg, 1959). Following intraperitoneal E. coli endotoxin application mice were found to be most susceptible at the end of L and least susceptible around mid D (Figure 20.7c). When equal doses of the toxin were injected into different groups of mice more than 80% of animals died when it was administered at the end of L and <10% when injected at mid D (Halberg et al., 1960a).

The teratogenic effects following alcohol administration to pregnant mice were also found to be maximal during D: the fetal weight was lowest and the rate of dead and abnormal fetuses greatest in mid-D. The teratogenic response to dexamethasone was found to be quite similar (Sauerbier, 1986, 1987a,b) whereas that following sodium valproate injection was significantly phase-advanced (highest effect end of L, lowest rate of malformations mid D; Ohdo et al., 1995, 1996a). Searching for an alternative for chloramphenicol which is known to cause ablastic anemia, the toxicity of florfenicol was tested and found to have a significant phase dependence when the LD_{50} of 1958 mg/kg was administered: while 87.5% of the animals died when injected in mid D, 40% of animals died when injected at mid L (Picco et al., 2001).

A rather counterphasic susceptibility rhythm has been found in the cytostatic drugs 5-fluorouracil (5-FU) and oxaliplatin: while 5-FU in mice is best tolerated during (the middle of) light time (survival rate ~75%) and worst during mid D (s.r. < 40%), oxaliplatinum was best tolerated during mid D (s.r. ~75%) and worst during mid L (s.r. ~35%; Figure 20.8), the fastest plasma elimination occurring in mid D (Burns and Beland, 1984; Boughattas et al., 1994; Lévi, 1997). The antiphasic combination of both drugs resulted in a significant improvement of chemotherapy in human patients suffering from colorectal cancer (Adlard et al., 2002; Fischel et al., 2002). Many other examples of phase dependent toxicity of cytostatic drugs have been compiled quite recently by Granda and Lévi (2002) and Granda et al. (2001, 2002). One other striking example of phase dependency is the effect following administration of the cyclooxigenase inhibitor celecoxib which exhibits a cytotoxic effect in different tumor cells to nu/nu mice bearing mammary tumor xenografts. The lowest tolerance was found to lie in D (survival rate ~10%) and the highest during mid L (100% of survivors). The regressive effect on tumor cells was found to be maximal during the first half of L (Blumenthal et al., 2001). Other examples of phase dependent effects of cytostatic drugs in mice are listed in Table 20.2 which is adapted from Granda and Lévi (2002).

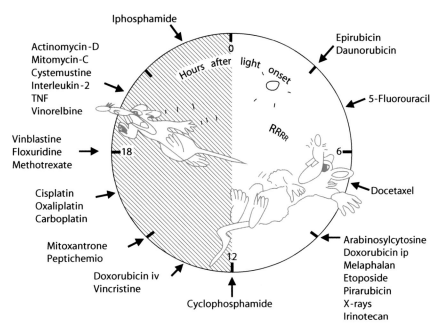

Figure 20.8 Circadian chronotolerance for some cytostatic drugs in rats and mice living in a regular LD 12:12. The time of highest tolerance (lowest mortality) is indicated by an arrow (reproduced from Lévi, 2002).

TABLE 20.2: Relationship between the circadian rhythm of mice in tolerance and efficacy of cytostatic drugs in experimental tumors (table taken from Granda and Lévi, 2002)

Experimental tumor	Drug	Least toxicity	Best efficacy
P03 pancreatic adenocarcinoma	Docetaxel	3, 7, 11	7, 11
MA13/C mammary adenocarcinoma	Docetaxel	7	7
Lewis lung cancer	Pirarubicin	15	15
CO26, CO38 colon carcinoma	5-Fluorouracil	0	0–2.5
Glasgow osteosarcoma	IRI	7	7
	OXA	15	15
	IRI + OXA		7 + 15
Sarcoma	DOX	16	16
T10-MCA induced	CPA + CPA	N.R.	2 + 6
	CPA + Vinblastine		2 + 18
Meth-A sarcoma	Interleukin-2	N.R.	~11
L1210 leukemia	Ara C	5, 8	5, 8
	CPA + DOX		13
	CPA + Ara C	11–14	11–14
Ehrlich's ascites	CPA	N.R.	Light span
B16 melanoma	Interferon-β	2	2

Abbreviations: N.R., not reported; Ara C, arabinosylcytosine; CPA, cyclophosphamide; DOX, doxorubicin; IRI, irinotecan; L-PAM, L-Phenylalanine mustard (melphalan); OXA, oxaliplatin.

It is well known that the damaging effect of ionizing radiation significantly depends on time. As a rule of thumb, whole body irradiation of mice around the end of L is tolerated much better than during D and the beginning of L. Following whole body X-Irradiation of BALB/c mice the peak of radioresistance was found to be around the end of the light period. While 100% of mice died within 8 days after irradiation (5.5 Gy)

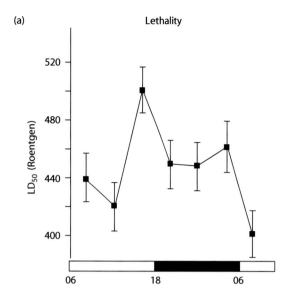

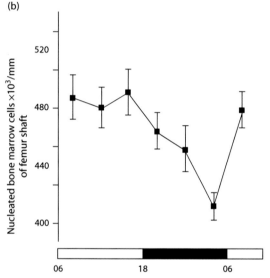

Figure 20.9 Circadian rhythm of radiosensitivity in adult male BALB/c mice kept in a regular LD 12 : 12 zeitgeber schedule: whole body X-irradiation at different circadian phases. (a) Circadian variation of LD_{50}; (b) depression of bone marrow cells (number of remaining nucleated cells per mm of femur shaft 4 days after exposure to a single dose of 3.5 Gy; redrawn from Haus, 2002).

at mid D (mean survival time ~5 d), no mouse died within this time when irradiated at the end of L (mean survival time ~12 d; Figure 20.9a). The greatest depression of the rapidly proliferating cells of the bone marrow following 3.5 Gy, was found to be at the time of lowest radioresistance of the animals during the last hours of D and the first hours of L (Figure 20.9b; Haus, 2002; Haus *et al.*, 1974; Scheving *et al.*, 1983). Interestingly the toxicity of camptothecin which is used for radiation-sensitization exhibits a 24-h rhythm which almost parallels that of radiosensitivity: the lowest survival rate of mice was found at mid D and a significantly

elevated survival (up to 100%) during the light period (Rich and Kirichenko, 2000). *In vitro* the highest radiosensitivity of dividing cells was usually found during G2 phase, the least toward the end of the S phase. Cell division in different mammalian tissues does not occur at random but is in fact strictly periodical but not in synchrony in various organs. Thus, it appears to be important to consider the specific phase of the rhythm of the organ of interest while carrying out respective experiments (Scheving *et al.*, 1991).

Conclusion

Within the boundaries of homeostatic regulation almost all functions of the mouse exhibit 24-h fluctuations which can have rather large amplitudes. 24-h rhythms are generated endogenously and their phases are controlled by the external environment. The LD regimen is the most important zeitgeber for the master oscillator of the mouse, the SCN. A periodic feeding schedule can act as a secondary zeitgeber for many functions. Thus, a consistent and standardized, periodic environment results in predictable 24-h rhythms of the mouse. Since circadian fluctuations are ubiquitous and can have large amplitudes it would appear wise to consider them when carrying out any experiment. Otherwise only very limited information will be obtained and this could have rather dramatic consequences for example when testing the efficacy of a drug or the toxicity of a substance.

References

Abe, H., Kida, M., Tsuli, K. and Mano, T. (1989). *Physiol. Behav.* **45**, 397–402.

Abrahamson, E.E. and Moore, R.Y. (2001). *Brain Res.* **916**, 172–191.

Adlard, J.W., Richman, S.D., Seymour, M.T. and Quirke, P. (2002). *Lancet Oncol.* **3**, 75–82.

Aschoff, J. (1981). In *Handbook of Behavioral Neurobiology, Vol. 4: Biological Rhythms* (ed. J. Aschoff), pp. 81–93. Plenum Press, New York and London.

Aujard, F., Herzog, E.D. and Block, G.D. (2001). *Neuroscience* **106**, 255–261.

Bailey, K.J. (1987). *J. Endocr.* **112**, 15–21.

Barriga, C., Martin, M.I., Tabla, R., Ortega, E. and Rodriguez, A.B. (2001). *J. Pineal Res.* **30**, 180–187.

Bartness, T.J., Song, C.K. and Demas, G.E. (2001). *J. Biol. Rhythms* **16**, 196–204.

Beau, J. (1991). *Behav. Gen.* **21**, 117–129.

Bindon, B.M. and Lamond, D.R. (1966). *J. Reprod. Fertil.* **12**, 249–261.

Bingel, A.S. and Schwartz, N.B. (1969). *J. Reprod. Fertil.* **19**, 223–229.

Blumenthal, R.D., Waskewich, C., Goldenberg, D.M., Lew, W., Flefleh, C. and Burton, J. (2001). *Clin. Cancer Res.* **7**, 3178–3185.

Boughattas, N.A., Hecquet, B., Fournier, C., Bruguerolle, B., Trabelsi, H., Bouzouita, K., Omrane, B. and Lévi, F. (1994). *Biopharm. Drug Dispos.* **15**, 761–773.

Bourin, P., Ledain, A.F., Beau, J., Mille, D. and Lévi, F. (2002). *Chronobiol. Int.* **19**, 57–67.

Burns, E.R. and Beland, S.S. (1984). *Pharmacology* **28**, 296–300.

Butz, G.M. and Davisson, R.L. (2001). *Physiol. Genom.* **5**, 89–97.

Champlin, A.K., Beamer, W.G., Carter, S.C., Shire, J.G. and Whitten, W.K. (1980). *Biol. Reprod.* **22**, 164–172.

Chan, J., Ogawa, S. and Pfaff, D.W. (2001). *Proc. Natl. Acad. Sci. USA* **98**, 700–704.

Chapman, J.B. and Cutler, M.G. (1987). *Neuropharmacology* **26**, 607–612.

Daan, S. and Pittendrigh, C.S. (1976). *J. Comp. Physiol.* **106**, 253–266.

Damiola, F., Le Minh, N., Preitner, N., Kornmann, B., Fleury-Olela, F. and Schibler, U. (2000). *Genes Dev.* **14**, 2950–2961.

DeLeon, D.D., Zelinski-Wooten, M.B. and Barkley, M.S. (1990). *J. Reprod. Fertil.* **89**, 117–126.

Ebihara, S., Tsuji, K. and Kondo, K. (1978). *Physiol. Behav.* **20**, 795–799.

Ebihara, S., Hudson, D.J., Marks, T. and Menaker, M. (1987). *Brain Res.* **416**, 136–140.

Fischel, J.L., Formento, P., Ciccolini, J., Rostagno, P., Etienne, M.C., Catalin, J. and Milano, G. (2002). *Br. J. Cancer.* **86**, 1162–1168.

Fitzpatrick, F., Christeff, N., Durant, S., Dardenne, M., Nunez, E.A. and Homo-Delarche, F. (1992). *Life Sci.* **50**, 1063–1069.

Foster, R.G., Provencio, I., Hudson, D., Fiske, S., DeGrip, W. and Menaker, M. (1991). *J. Comp. Physiol.* A **169**, 39–50.

Foster, R.G. and Helfrich-Forster, C. (2001). *Philos. Trans. R. Soc. Lond.* B *Biol. Sci.* **356**, 1779–1789.

Froy, O., Chang, D.C. and Reppert, S.M. (2002). *Curr. Biol.* **12**, 147–152.

Goto, M., Oshima, I., Tomita, T. and Ebihara, S. (1989). *J. Pineal Res.* **7**, 195–204.

Ghiraldi, L.L., Plonsky, M. and Svare, B.B. (1993). *Horm. Behav.* **27**, 251–268.

Granda, T.G. and Lévi, F. (2002). *Chronobiol. Int.* **19**, 21–41.

Granda, T.G., Filipski, E., D'Attino, R.M., Vrignaud, P., Anjo, A., Bissery, M.C. and Lévi, F. (2001). *Cancer Res.* **61**, 1996–2001.

Granda, T.G., D'Attino, R.M., Filipski, E., Vrignaud, P., Garufi, C., Terzoli, E., Bissery, M.C. and Lévi, F. (2002). *Br. J. Cancer* **86**, 999–1005.

Halberg, F. (1953). *J. Lancet* **73**, 20–32.

Halberg, F. (1959). *Z. Vitamin Hormon. Fermentforsch.* **10**, 225–296.

Halberg, F., Peterson, R.E. and Silber, R.H. (1958). *Endocrinology* **64**, 222–230.

Halberg, F., Johnson, E.A., Brown, B.W. and Bittner, J.J. (1960a). *Proc. Soc. Exp. Biol. Med.* **103**, 142–144.

Halberg, F., Albrecht, P.G. and Barnum, jr. C.P. (1960b). *Am. J. Physiol.* **199**, 400–402.

Hara, R., Wan, K., Wakamatsu, H., Aida, R., Moriya, T., Akiyama, M. and Shibata, S. (2001). *Genes Cells* **6**, 269–278.

Harri, M., Lindblom, J., Malinen, H., Hyttinen, M., Lapvetelainen, T., Eskola, S. and Helminen, H.J. (1999). *Lab. Anim. Sci.* **49**, 401–405.

Haus, E. (2002). *Chronobiol. Int.* **19**, 77–100.

Haus, E. and Halberg, F. (1959). *J. Appl. Physiol.* **14**, 878–880.

Haus, E. and Halberg, F. (1966). *Experientia* **22**, 113–114.

Haus, E., Halberg, F., Kühl, J.F.W. and Lakatua, D.J. (1974). In *Chronobiological Aspects of Endocrinology* (eds J. Aschoff, F. Ceresa and F. Halberg), pp. 269–304. Schattauer Verlag, Stuttgart-New York.

Hofstetter, J.R., Mayeda, A.R., Possidente, B. and Nurnberger, J.I. (1995). *Behav. Genet.* **25**, 545–556.

Holmes, M.M. and Mistlberger, R.E. (2000). *Physiol. Behav.* **68**, 655–666.

Ibata, Y., Okamura, H., Tanaka, M., Tamada, Y., Hayashi, S., Iijima, N., Matsuda, T., Munekawa, K., Takamatsu, T., Hisa, Y., Shigeyoshi, Y. and Amaya, F. (1999). *Front. Neuroendocrinol.* **20**, 241–268.

Kaiser, I.H. (1967). *Am. J. Obstet. Gynecol.* **99**, 772–784.

Kennaway, D.J., Voultsios, A., Varcoe, T.J. and Moyer, R.W. (2002). *Am. J. Physiol. Regul. Intergr. Comp. Physiol.* **282**, R358–R365.

Kolker, D.E. and Turek, F.W. (1999). *J. Psychopharmacol.* **13**, S5–S9.

Kopp, C., Vogel, E., Rettori, M.-C., Delagrange, P., Guardiola-Lemaitre, B. and Misslin, R. (1998). *Physiol. Behav.* **63**, 577–585.

Kolaczkowska, E., Chadzinska, M., Seljelid, R. and Plytycz, B. (2001). *Lab. Anim.* **35**, 91–100.

Kramer, K., Voss, H.P., Grimbergen, J.A., Mills, P.A., Huetteman, D., Zwiers, L. and Brockway, B. (2000). *Lab. Anim.* **34**, 272–280.

Lehman, M.N., LeSauter, J., Kim, C., Berriman, S.J., Tresco, P. and Silver, R. (1995). *Cell Transplant.* **4**, 75–81.

Lemmer, B. (1989). *Chronopharmacology.* Marcel Dekker Inc., New York and Basel.

Lévi, F. (1997). In *Physiology and Pharmacology of Biological Rhythms* (eds P.H. Redfern and B. Lemmer), pp. 299–331. Springer Verlag, Berlin, Heidelberg.

Lévi, F. (2002). *Chronobiology International* **19**, 1–19.

Lowrey, P.L. and Takahashi, J.S. (2000). *Ann. Rev. Genet.* **34**, 533–562.

Lucas, R.J. and Foster, R.G. (1999). *J. Biol. Rhythms* **14**, 4–10.

Lucas, R.J., Freedman, M.S., Munoz, M., Garcia-Fernandez, J.M. and Foster, R.G. (1999). *Science* **284**, 505–507.

Mayeda, A.R., Hofstetter, J.R. and Possidente, R. (1997). *Chronobiol. Int.* **14**, 19–23.

Meerlo, P. and Turek, F.W. (2001). *Brain Res.* **907**, 84–92.

Mills, P.A., Huetteman, D.A., Brockway, B.P., Zwiers, L.M., Gelsema, A.J.M., Schwartz, R.S. and Kramer, K. (2000). *J. Appl. Physiol.* **88**, 1537–1544.

Moore, R.Y. (1995). *Ciba Found. Symp.* **1831**, 88–99.

Moore-Ede, M.C., Sulzman, F.M. and Fuller, C.A. (1982). *The Clocks that Time us.* Harvard University Press, Cambridge.

Mrosovsky, N., Foster, R.G. and Salmon, P.A. (1999). *J. Comp. Physiol.* A **184**, 423–428.

Nejdi, A., Gustavino, J.M. and Lalonde, R. (1996). *Physiol. Behav.* **59**, 45–47.

Nelms, J.L., LeSauter, J., Silver, R. and Lehman, M.N. (2002). *Exp. Neurol.* **174**, 72–80.

Ohdo, S., Watanabe, H., Ogawa, N., Yoshiyama, Y. and Sugiyama, T. (1995). *Eur. J. Pharmacol.* **293**, 281–285.

Ohdo, S., Watanabe, H., Ogawa, N., Yoshiyama, Y. and Sugiyama, T. (1996a). *Jpn. J. Pharmacol.* **70**, 253–258.

Ohdo, S., Ogawa, N., Nakano, S. and Higuchi, S. (1996b). *J. Pharmacol. Exp. Ther.* **278**, 74–81.

Ohdo, S., Arata, N., Furukubo, T., Yukawa, E., Higuchi, S., Nakano, S. and Ogawa, N. (1998). *J. Pharmacol. Exp. Ther.* **285**, 242–246.

Ohdo, S., Wang, D.S., Koyanagi, S., Takane, H., Inoue, K., Aramaki, H., Yukawa, E. and Higuchi, S. (2000). *J. Pharmacol. Exp. Ther.* **294**, 488–493.

Ono, T., Shimizu, T. and Yoshida, M. (1991). *J. Mamm. Soc. Japan* **16**, 47–58.

Pasley, J.N., Barnes, C.L. and Rayford, P.L. (1987). *Prog. Clin. Biol. Res.* **227**, A371–A378.

Paterson, A.T. and Vickers, C. (1984). *Pharmacol. Biochem. Behav.* **21**, 495–499.

Pauly, J.E., Burns, E.R., Halberg, F., Tsai, S., Betterton, H.O. and Scheving, L.E. (1975). *Acta Anat.* **93**, 60–68.

Pauly, J.E., Scheving, L.E., Burns, E.R. and Tsai, T.H. (1976). *Anat. Rec.* **184**, 275–284.

Pessaq, M.T. and Gagliardino, J.J. (1975). *Chronobiologia* **2**, 205–209.

Picco, E.J., Diaz, D.C., Valtorta, S.E. and Boggi, J.C. (2001). *Chronobiol. Int.* **18**, 567–572.

Pittendrigh, C.S. (1976). In *The Molecular Basis of Circadian Rhythms* (eds J.W. Hastings and H.G. Schweiger), pp.11–48. Dahlem Konferenzen, Abakon Verlagsgesellschaft, Berlin.

Pittendrigh, C.S. and Daan, S. (1976a). *J. Comp. Physiol.* **106**, 223–252.

Pittendrigh, C.S. and Daan, S. (1976b). *J. Comp. Physiol.* **106**, 333–355.

Poirel, C. (1988). *J. Gen. Psychol.* **115**, 187–201.

Possidente, B. and Stephan, F.K. (1988). *Behav. Genet.* **18**, 109–117.

Possidente, B., McEldowney, S. and Pabon, A. (1995). *Physiol. Behav.* **57**, 575–579.

Ralph, M.R., Foster, R.G., Davis, F.C. and Menaker, M. (1990). *Science* **247**, 975–978.

Redfern, P.H. and Lemmer, B. (1997). *Physiology and Pharmacology of Biological Rhythms.* Springer Verlag, Berlin, Heidelberg.

Reiter, R.J. (1991). *Endocr. Rev.* **12**, 151–180.

Reiter, R.J. (1993). *Experientia* **49**, 654–664.

Reppert, S. and Weaver, D.R. (2000). *J. Biol. Rhythms* **15**, 357–364.

Rich, T.A. and Kirichenko, A. (2000). *Ann. NY Acad. Sci.* **922**, 334–339.

Saffe de Vilchez, I., Lobarbo, S. and Vilchez, C.A. (1972). *Enzyme* **13**, 240–245.

Sancar, A. (2000). *Annu. Rev. Biochem.* **2000**, 31–67.

Sauerbier, I. (1986). *Drug Chem. Toxicol.* **9**, 25–31.

Sauerbier, I. (1987a). *Am. J. Anat.* **178**, 170–174.

Sauerbier, I. (1987b). In *Chronobiology and Chronomedicine* (eds G. Hildebrandt, R. Moog and F. Raschke), pp. 265–269. P. Lang Verlag, Frankfurt, Bern, New York, Paris.

Schanz, M. von, Lucas, R.J. and Foster, R.G. (1999). *Mol. Brain Res.* **72**, 108–114.

Scheving, L.E. and Pauly, J.E. (1973). *Int. J. Chronobiol.* **1**, 269–286.

Scheving, L.E. and Pauly, J.E. (1977). In *Nova Acta Leopoldina: Die Zeit und das Leben*, pp. 237–258. Deutsche Akad. Naturf. Leopoldina, Halle.

Scheving, L.E., Tsai, T.H. and Scheving, L.A. (1983). *Am. J. Anat.* **168**, 433–465.

Scheving, L.E., Tsai, T.H., Pauly, J.E., Scheving, L.A., Feuers, R.J., Kanabrocki, E.L. and Lucas, E.A. (1989). *Chronobiologia* **16**, 307–329.

Scheving, L.E., Tsai, T.H., Scheving, L.A. and Feuers, R.J. (1991). *Ann. NY Acad. Sci.* **618**, 182–227.

Schwartz, W.J. and Zimmerman, P. (1990). *J. Neurosci.* **10**, 3685–3694.

Shimomura, K., Low-Zeddies, S.S., King, D.P., Steeves, T.D., Whiteley, A., Kushla, J., Zemenides, P.D., Lin, A., Vitaterna, M.H., Churchhill, G.A. and Takahashi, J.S. (2001). *Genome Res.* **11**, 959–980.

Smaaland, R. (1996). *Prog. Cell Cycle Res.* **2**, 241–266.

Smaaland, R. (1997). In *Physiology and Pharmacology of Biological Rhythms* (eds P.H. Redfern and B. Lemmer), pp. 487–532. Springer, Berlin, Heidelberg, New York.

Speh, J.C. and Moore, R.Y. (1993). *Dev. Brain Res.* **76**, 171–181.

Stokkan, K.A., Yamazi, S., Tei, H., Sakaki, Y. and Menaker, M. (2001). *Science* **291**, 490–493.

Swoyer, J., Haus, E. and Sackett-Lundeen, L. (1987). In *Advances in Chronobiology, Part A* (eds J.E. Pauling and L.E. Scheving), pp. 281–296. Alan R. Liss Inc., New York.

Tamarkin, L., Baird, C.J. and Almeida, O.F.X. (1985). *Science* **227**, 714–720.

Tankersley, C.G., Irizarry, R., Flanders, S. and Rabold, R. (2002). *J. Appl. Physiol.* **92**, 870–877.

Turek, F.W. and Allada, R. (2002). *Hepatology* **35**, 743–745.

Valentinuzzi, V.S., Scarbrough, K., Takahashi, J.S. and Turek, F.W. (1997). *Am. J. Physiol.* **273**, 1957–1964.

VanGelder, R.N. (2001). *Ophthalmic Genet.* **22**, 195–205.

Viswanathan, N. and Davis, F.C. (1995). *Brain Res.* **686**, 10–16.

Vitaterna, M.H., Selby, C.P., Todo, T., Niwa, H., Thompson, C., Fruechte, E.M., Hitomi, K., Thresher, R.J., Ishikawa, T., Miyazaki, J., Takahashi, J.S. and Sancar, A. (1999). *Proc. Natl. Acad. Sci. USA* **96**, 12114–12119.

Vivien-Roels, B., Malan, A., Rettori, M.C., Delagrange, P., Jeanniot, J.P. and Pevet, P. (1998). *J. Biol. Rhythms* **13**, 403–409.

Wakamatsu, H., Yoshinobu, Y., Aida, R., Moriya, T., Akiyama, M. and Shibata, S. (2001). *Eur. J. Neurosci.* **13**, 1190–1196.

Weaver, D.R. (1998). *J. Biol. Rhythms* **13**, 100–112.

Weinert, D., Eimert, H., Erkert, H.G. and Schneyer, U. (1994). *Chronobiol. Int.* **11**, 222–231.

Weinert, H., Weinert, D., Schurov, I., Maywood, E.S. and Hastings, M.H. (2001). *Chronobiol. Int.* **18**, 359–365.

Weinert, H., Weinert, D. and Waterhouse, J. (2002). *Biol. Rhythms Res.* **33**, 199–212.

Welsh, D., Engle, E.M.R.A., Richardson, G.S. and Dement, W.C. (1986). *J. Comp. Physiol.* A **158**, 827–834.

Yellon, S.M. and Tran, L.T. (2002). *J. Biol. Rhythms* **17**, 65–75.

Ziesenis, J.S., Davis, D.E. and Smith, D.E. (1975). *Anim. Behav.* **23**, 941–948.

Gerontology

Miriam R Anver and Diana C Haines

SAIC Frederick, Inc., National Cancer Institute,
Frederick, Maryland, USA

Introduction

Mice older than 20 months may be considered senescent. The geriatric mouse, however, is an **iatrogenic** phenomenon since the feral mouse rarely lives beyond 10 months of age due to predation, disease, or climatic extremes (Austad, 1997). Several strains, including GEM (genetically engineered mice which include transgenics and targeted mutations) have foreshortened life expectancy due to development of neoplasms or other lesions that cause death at less than a year of age. Other GEM and inbred strains, such as the senescence-accelerated mouse (SAM), exhibit premature aging or extended life spans and serve as animal models for human gerontology.

Longevity in the laboratory mouse varies by strain. Average longevity (age at which 50% are dead) for several mouse strains has been reported: IDR Swiss, 22–24 months; BALB/c, less than 24 months; CBA, 29 months; RFM, 19.8 months; NZB, 14–18 months; C57BL/6 and C57BL/10, 25–30 months; B6CF1, 32 months; B6C3F1, 29–31 months; B6D2F1, 27 months; DBA/2, 25 months. Individual mice from some of these and other strains may live beyond 44 months of age (Zurcher et al., 1982; Sheldon et al.,

1996; Rao, 1999; Forster et al., 2003). A specific cause of death cannot always be determined for the aged mouse, since some die without neoplasia or other lesions of sufficient severity to ascribe as cause of death.

Mechanisms of cellular aging

Cellular aging, in both humans and mice, involves generalized progressive decline in mitochondrial oxidative phosphorylation as well as decreases in synthesis of nucleic acids, enzymes, cell receptors and transcription factors which negatively impacts the cell's ability to absorb and utilize nutrient and repair defects in DNA. By-products of glycosylation and oxidation accumulate with resultant protein cross-linkage and damage to proteins, lipids, and nucleic acid (Cotran et al., 1999). Several mechanistic hypotheses have been proposed to explain aging including free radical damage, telomere shortening, DNA mutations, and protein misfolding. These pathways are not mutually exclusive. Indeed, they probably interact in various configurations depending on tissue, individual organism, and environmental factors. All of the pathways, with the possible exception of telomeres, appear to function in both humans and mice. A brief discussion of mechanistic hypotheses follows.

Free radicals

The free radical hypothesis of aging, first proposed by Harman (1956), considers the accumulation of molecular damage resulting from free radicals to be the primary cause of senescence and aging. Free radicals, due to presence of an unpaired electron, are unstable chemical species that react with adjacent molecules and can propagate formation of additional free radicals (free radical cascade). Free radicals can be produced by the action of ionizing radiation, enzymatic metabolism of various chemicals and drug, intracellular reactions involving iron or copper, reactions involving nitric oxide, and the reduction–oxidation reaction of normal aerobic metabolic activity (Cotran et al., 1999). The latter is probably the most important producer of intracellular free radicals. During normal aerobic mitochondrial and peroxisomal metabolic activities, reactive oxygen species (ROS) are produced that can cause molecular damage to vital cellular components, including proteins, lipids, and DNA, thus negatively impacting cellular integrity and physiological functioning. Protective mechanisms including antioxidants, metal storage and transport proteins, and enzymes including superoxide dismutase (SOD), catalase, and glutathione peroxidase counterbalance production of ROS and other free radicals. When production of free radicals increases and/or protective mechanisms decrease, increased oxidative stress (OS) ensues with resulting peroxidation of lipids; cross-linkage and fragmentation of proteins; and single strand breaks and mutations in both nuclear and mitochondrial DNA (Wei and Lee, 2002). However, production of ROS is not always detrimental to the organism. Reactive oxygen species participate in the immune response to pathogens, help regulate gene expression by signal transduction pathways, and can even aid in the termination of free radical chain cascades induced by very low levels of oxidants (Sohal et al., 2002).

Telomeres and telomerase

Hayflick (1965) first observed that human fibroblasts grown in culture had a fixed number of replications before entering a post-mitotic state (Hayflick phenomenon). This preset replicative senescence could

have profound effects on the physiological function of an organ as the percentage of cells entering non-replicative stage increases. Decrease in length of the telomeres is considered to be a major cause of replicative senescence in humans. Telomeres are specialized nucleoproteins that cap the ends of chromosomes thereby preventing abnormal recombination or degradation. This prevents structural instability and loss of genetic information that could ultimately lead to neoplastic transformation or cell death through apoptosis (Kim et al., 2002). Shortening of the telomeres is due to the inability of DNA polymerase to replicate completely the ends of linear chromosomes. With each replication cycle of the cell, telomeres progressively shrink by 100–200 bp. Oxidative damage, as suggested by Von Zglinicki (2000), may also play a role in regulating telomere length. Telomerase, a DNA polymerase responsible for synthesis of telomeric DNA, is active in mammalian germline, lymphoid subsets, early embryonic cells, and stem cells, but is quiescent in the majority of tissues of adult humans.

The telomerase complex consists of a catalytic subunit telomerase reverse transcriptase (TERT), a telomerase-associated protein 1 (TEP1), and a telomerase RNA subunit. Telomerase activity can be down-regulated by dephosphorylation induced by protein phosphatase 2A (PP2A), or alternatively, upregulated by protein phosphorylation induced by protein kinase C. Additionally, human TERT (hTERT) gene expression, which is the rate-limiting factor in telomerase activity, can be inhibited by a number of transcription repressors including p53 protein. Binding sites for a number of transcription factors, including the oncogene c-myc, are present in the hTERT promoter, indicating additional mechanisms for activation of telomerase (Liu, 2001). Although telomere shortening is seen in progeria in humans, and telomerase activity is present in most neoplastic cells, there is not a direct correlation between telomere length and maximal life span between species. Indeed, mice have extremely long telomeres and high telomerase activity (see section on Mouse Models of Aging). Furthermore, fibroblasts taken from centenarians are still capable of 20–50 replications in cell culture, suggesting that telomere and telomerase may play roles but are not the cause per se of aging (Goyns, 2002).

Genes and mutations

Several endogenous repair mechanisms exist to correct defects in the structure of DNA resulting from oxidative damage or other mechanisms. Nevertheless, due to increased damage and/or decreased repair, DNA mutations accumulate with age with resultant alteration in gene activity and function. Additionally, methylation of DNA, which can inhibit gene expression, may be randomly altered by both exogenous (UV light, dietary folate) and endogenous (Dnmt1 gene expression, methyltransferases, demethylase) factors. Resultant changes in expression of age-associated genes could result in delayed or accelerated senescence (Richardson, 2002). A comprehensive and ever growing list of genes associated with the aging process is accessible on the aging gene/interventions database (AGEID) on the world wide web: http://sageke.sciencemag.org/cgi/genesdb. This database, whose aim is to catalog all published research where life span is measured, includes information on species, strain, gene function, phenotype, homologs, references, and related links. Description of the web site and means of accessing the data (subscription is required) is given by Kaeberlein et al. (2002).

Of the genes associated with senescence, the p53 tumor suppressor gene is of particular interest because of its pivotal role in the cell cycle. When DNA defects occur, p53 becomes activated and induces cell cycle arrest by means of the CDK inhibitor p21 that prevents phosphorylation of pRb and the entry of the cell into the S phase. If DNA repair is successful during this pause in cell cycling, p53 activates mdm2 that in turn down-regulates p53 expression and cell cycling continues. If DNA repair is unsuccessful, p53 activates bax (which antagonizes the apoptosis-inhibiting protein bcl-2) and IGF-BP3 (which blocks IGF-mediated intracellular signaling) and the cell undergoes apoptosis (Cotran et al., 1999). By this mechanism, cells that might otherwise exhibit abnormal function, including possible neoplastic transformation, are removed from the cellular pool. Telomere shortening can induce p53 activation and p53 can inhibit telomerase (see above). Increased p53 activity is associated with senescence in cultured cells and may play a role in the reduction of stem cell function seen with aging (Donehower, 2002). It has been further proposed that if not p53 itself, then genes working upstream (p66[shc]) or downstream (Sir2) might be gene therapy targets for delaying senescence (Sharpless and DePinho, 2002). However, it should be pointed out the p53 is mutated or lost in the vast majority of human neoplasms, and mice defi-cient in p53 die early due to neoplasms. Thus, control mechanisms, such as p53 (and also telomere shortening), maintain genome stability at the expense of longevity.

Folding and turnover of proteins

Improper functioning and reduced turnover of proteins have been associated with aging. In the process of protein formation, correct folding (configuration) of the polypeptide chains alpha helices or beta sheets is essential for proper transportation and function of the resulting protein. **Molecular chaperones**, both constitutive and stress-induced, e.g. heat shock protein, assist in the normal folding process, prevent stress-induced abnormal folding or, in cases of defective folding or other damage, label abnormal protein for degradation through the proteasome or lysosome systems (Carrard et al., 2002). Accumulation of abnormally folded and other defective proteins occurs during the process of aging and has been associated with some forms of amyloidosis as well as Parkinson and Alzheimer diseases (Cotran et al., 1999). Several mechanisms may be involved. Free radicals (see above) can cause oxidative damage to proteins, proteases and subunits of the proteasome that results in degradation-resistant proteins or decrease in functional capacity of the proteolytic system (Szweda et al., 2002). Non-enzymatic reactions between sugars and protein result in advanced glycation end products (AGEs) that resist degradation and induce free radical formation (Baynes, 2001). Molecular misreading due to a dinucleotide deletion frameshift in RNA transcript can result in mutated defective protein (+1 protein).

Ubiquitin, a molecular chaperone that tags defective proteins for degradation, when present as ubiquitin[+1], is ineffective as a chaperone and functions to block the proteasome (Van Leeuwen et al., 2002). Beta amyloid precursor protein[+1] (APP[+1]) as well as ubiquitin[+1] are present in the neuritic plaques and neurofibrillar tangles associated with advanced age, Alzheimer's disease, and Down's syndrome (Leeuwen et al., 2002). Lipofuscin is a nondegradable, electron-dense, brownish-yellow, autofluorescent, fluorochrome that consists primarily of cross-linked protein residues formed by iron-catalyzed oxidative process. The pigment accumulates in lysosomes of post-mitotic cells in a number of organs. Presence of lipofuscin may have a negative effect on the autophagocytosis function of lysosomes by monopolizing lysosomal enzymes. Furthermore, iron associated with the lipofuscin may promote additional ROS formation resulting in lysosome instability, rupture, and ensuing apoptosis of the cell (Brunk and Terman, 2002).

In conclusion, cellular aging is a complex tapestry of interacting, and often opposing, factors; a balance between controlled growth and maintenance and uncontrolled growth (neoplasia) and death.

Aging in the mouse

The website of the National Institute on Aging (NIA), which has descriptions of intramural and extramural research programs can be accessed at http://www.nih.gov/about/almanac/organization/NIA.htm.

The cells of higher organisms, including man and mouse, must contend with external influences that affect their internal molecular orchestration. Research resulting from molecular studies must be extrapolated, both through hypothesis and testing, to the organism as a whole. Effects of general nutrition, hormones, growth factors, cytokines, immune status, and intercurrent disease may become important when research moves out of the tissue culture flask and into a mouse model or human therapy.

Longevity, as well as neoplastic and non-neoplastic lesions, have been shown to vary in genetically similar mice depending on breeding source (Engelhardt et al., 1993). Although genetic drift may be partially responsible, environmental conditions such as intercurrent disease and animal facility husbandry are undoubtedly contributing factors.

Caloric restriction

The single most consistent iatrogenic manipulation, which can significantly delay senescence in a number of species, including both man and mouse, is caloric restriction (CR; Weindruch and Walford, 1988). Caloric restriction involves decreasing energy intake 25–60% while still providing all essential nutrients. Caloric restriction has been shown to reduce age-associated neoplastic and non-neoplastic lesions in mice (Bronson and Lipman, 1991). Decrease in inflammation in a skin paint study was related to CR (Perkins et al., 1998). Delay in development of autoimmune disease, renal disease, and diabetes also has been associated with CR. Caloric restriction has been shown to decrease age-associated neuronal loss, and prevent age-associated decline in psychomotor and spatial memory tasks in mice (Koubova and Guarente, 2003). In a mouse study reported by Sheldon et al. (1996), longevity (age at which 50% of the cohort are dead) was increased

19–36% depending on sex and strain (B6C3F1, B6D2F1 and C57BL6). Maximal life span (average age of death of the oldest 10% of the cohort) was increased 6–32%.

Mechanisms by which CR exerts its effect are still being elucidated. Through use of animal modeling, CR has been shown to cause a metabolic shift toward increased protein turnover (Weindruch et al., 2001), reduce the production of AGEs (Baynes, 2001), maintain optimal functioning of proteasomes (Carrard et al., 2002), enhance heat-shock protein response (Masoro, 1998), and retard lipofuscin accumulation (Brunk and Terman, 2002). Caloric restriction has also been shown to delay the onset of changes in gene expression seen in aging mouse muscle and brain (Lee et al., 1999). Since the relative effects of CR on survival curves and tumor development have been shown to be similar in p53−/− and p53+/+ mice, delay in tumorigenesis may work through a p53-independent pathway. (Hursting et al., 1997). Caloric restriction's ability to reduce oxidative stress (Sohal and Weindruch, 1996) has been interpreted as a possible primary mechanism of action in increasing longevity. Merry (2002) describes how CR might decrease ROS production by effectively lowering the mitochondrial membrane potential. Decrease in growth hormone (GH) and insulin-like growth factor-1 (IGF-1) associated with CR may be another important mechanism (Dunn et al., 1997). Growth hormone is produced in the anterior pituitary and activates IGF-1 production (primarily from the liver). Insulin-like growth factor-1 promotes growth and reproduction and is a potential mitogen also decreases thyroid stimulating hormone (TSH) plasma glucose, and insulin. Concurrently there is an increase in insulin sensitivity, glucocorticoids, catecholamines, and glucagons.

Koubova and Guarente (2003) have recently described a hypothetical pathway for how CR works. Although involving some extrapolation from lower organisms, the pathway integrates several mechanisms mentioned previously. They suggest that CR-induced initial decrease in blood glucose and reduction in rate of conversion of NAD to NADH is sensed (possibly through Sir2) by pancreatic islet cells and glucose-monitoring neurons in the hypothalamus. This in turn reduces the production of insulin and GH, respectively. Decreased GH lowers IGF-1. Reduction in levels of insulin and IGF-1 activate forkhead transcription factors, which in turn increases resistance to OS. They further postulate that altered metabolism might upregulate Sir 2 resulting in decreased apoptosis.

Mouse models of aging

Mouse models of **progeria**/accelerated senescence as well as models of increased longevity are valuable for gerontology research. Mutant models of various species for prolonged lifespan are reviewed by Mahler (2001). Murine genetic engineering technology and examples of GEM used in gerontology research are reviewed by Andersen (2001). Mouse models also can be used to study the relationship between aging and carcinogenesis. Anisimov (2001) emphasizes that evaluation of newly generated mouse models should involve the study of both cancer and aging. He has reviewed the incidences of neoplasms in GEM and mutant mouse models of longevity or accelerated aging, as well as spontaneous tumor incidence in 20 additional mouse strains.

Oxidative stress models

Animal models of OS include the SOD2 mutant mouse. Superoxide dismutase 2 is a mitochondrial form of SOD. Mice null for SOD2 ($SOD2^{-/-}$) die as neonates with dilated cardiomyopathy, hepatic lipidosis, ketosis, metabolic acidosis, reduced mitochondrial enzymatic activity and oxidatively damaged DNA. Of interest, mice null for cytosolic or extracellular SOD are phenotypically normal, suggesting that free radical damage is of most significance when it occurs within the mitochondria (Melov, 2002). Conversely, transgenic mice over expressing SOD do not exhibit increased longevity (Huang et al., 2000).

A targeted mutation in the $Shc1$ (Syn. $p66^{shc}$) gene ($Shc1^{tm1Pgp}$) extends the lifespan of mice by 30% over wild type. This mutation makes the mouse resistant to the effects of ROS (Lithgow and Andersen, 2000).

Telomere/telomerase models

The mouse, when compared to man, has extremely long telomeres as well as an abundance of telomerase. To overcome this difference, telomerase null mice ($Terc^{tm1Rdp}$) have been produced. After several generations, during which telomere length progressively shortens, these mice show decreased longevity and reduced capacity to respond to stresses such as wound healing and hematopoietic loss (Rudolph et al., 1999). After 4–6

generations the mice become infertile. There is partial embryonic mortality due to a defect in neural tube closure and increased neuronal apoptosis. Intestinal and splenic atrophy occur. Mitogenic stimulation produces reduced proliferation of B- and T-lymphocytes. Germinal center (GC) function (T-lymphocyte-dependent clonal expansion of B-lymphocytes for production of antibodies and memory cells) after immunization is impaired. This results in a decrease in circulating antigen-specific IgM and IgG antibodies. This impairment in GC function is not seen in first generation $Terc^{tm1Rdp}$ mice, suggesting that the impairment is due to the shortening of the telomere rather than to the absence of telomerase. Incidence of neoplasms is reduced unless the mutant is derived from a p53-deficient background strain (Blasco, 2002).

Mice deficient or lacking Ku86 or ATM, proteins associated with telomere functioning, also are reported to exhibit early aging (Goytisolo and Blasco, 2002).

Gene/mutation models

p53 models

The $p53^{+/m}$ mouse developed by Tyner et al. (2002) overexpresses p53. These mice show a decrease in spontaneous neoplasms, but have reduced longevity. The mice exhibit early onset of bone atrophy and generalized organ atrophy as well as a decreased tolerance for stress such as general anesthesia, wound healing, and recoverability of hematopoietic precursor cells after ablation by 5-fluorouracil treatment. Similarly pL53 transgenic mice, that have a temperature-sensitive mutant allele, display a phenotype of early aging including reduction in wound healing.

GH/IGF-1 models

The Snell dwarf and Ames dwarf mutant mice are homozygous (null) for $Pit1^{dw}$ and $Prop-1$, on chromosome 16 and 11, respectively. $Pit1^{dw}$ is responsible for the cellular differentiation of somatotrophs, lactotrophs and thyrotrophs in the anterior pituitary. $Prop-1$ is responsible for the development of $Pit1^{dw}$. Both mutants have an underdeveloped anterior pituitary and lack GH, prolactin (PRL), and TSH. They are phenotypically similar. Although of normal size at birth, they have reduced growth rate and are distinguishable from their normal littermates by 10–14 days of age due to smaller size, delayed eye opening and altered shape of the head. The adult weight is only one third that of normal littermates and the females are usually infertile. Mean longevity is increased over controls by 40–65% depending on background strain and gender. The survival curve is similar to the controls suggesting that delayed aging is the cause of the extended life span. Increases have been reported for catalase (liver, kidney), SOD (hypothalamus), and insulin sensitivity. Decreases have been noted in core body temperature ($-1.5°C$), metabolic rate, plasma glucose, plasma insulin, and IGF-1. Deficiency in GH/IGF1 has been interpreted to be the major reason for enhanced longevity in these dwarfs (Bartke *et al.*, 2001).

Growth hormone receptor knockout (*Ghr^tm1*) mice are resistant to GH signal. These mice have 50% reduction in adult body size, are fertile, have increased prolactin levels, and mild hypothyroidism. Insulin-like growth factor-1 plasma levels are decreased and longevity is increased. Transgenic mice which overexpress GH have shortened life spans and exhibit premature aging (Bartke *et al.*, 2001).

Mice mutant for IGF-1 receptor (*Igf1r^tm*) show minimal growth reduction, normal age of sexual maturation, and are fertile. Longevity in females, but not males, is increased 33% over the wild type. The increase in life span has been associated with increased tolerance for oxidative stress and reduced phosphorylation of the Shc gene (Tatar *et al.*, 2003).

Senescence-accelerated mouse

The SAM has been presented as a model for aging research as reviewed by Takeda (1999). There are 14 senescence-prone strains (SAMP) and four senescence-resistant strains (SAMR). They were derived in a conventional facility from several pairs of AKR/J mice; there are also some specific-pathogen-free SAM strains. The average life span of the SAMP is 9.7 months, 40% shorter than the SAMR strains (16.3 months). The various SAMP strains develop a plethora of pathobiological phenotypes tabulated by Takeda (1999). Oxidative damage due to mitochondrial dysfunction may be one of the pathogenic mechanisms for accelerated senescence (Mori *et al.*, 1998; Butterfield *et al.*, 2001). Some SAMP are described in this chapter under the various organ systems.

A complication of the SAM model, particularly as it involves brain research, is the presence of high levels of endogenous ecotropic murine leukemia in brain and other tissues from a number of the SAMP strains: SAMP 1, SAMP 2, SAMP 7, SAMP 8, SAMP 9, SAMP 10, and SAMP 11 (Carp *et al.*, 2000).

Kotho mouse

The kotho mouse has an extremely short life span (9 weeks). The mice are infertile, have growth retardation, decreased spontaneous activity, thymic cortical atrophy, metastatic calcification, skin atrophy, arteriosclerosis, lower bone density, and pulmonary emphysema. This model, which is more appropriately a model for progeria, and other mouse models of aging are reviewed by Kuro-o (2001).

Organ systems

Clinical presentation of an aged mouse may include decreased activity, dull haircoat, stiff gait, and decreased ability in learning and agility tests.

Only select organ-specific, age-associated lesions will be discussed in this chapter. Neoplasms, discussed elsewhere in this book, will not be emphasized. By necessity, due to the variabilities between strains and the number of strains available, discussion of organ systems is generalized.

Nervous system and behavior

Changes in the brain during normal aging of mice have been described by Sturrock (1996). Aged neurons have cytoplasmic lipofuscin pigment, loss of Nissl substance, and changes in nuclear size and shape. Decreases in neuronal numbers occur in various regions of the brain. Oligodendroglial cells have little structural evidence of age, although vacuolated myelin sheaths increase with age. Microglia have increased lipofuscin but no apparent quantitative changes. Astrocytes have increased lipofuscin and microfilaments. Scott and Mandybur (1996) found that hypertrophied (gemistocytic) astrocytes were a more common finding of aging than increased numbers. Astrocytes had stronger immunoreactivity for glial fibrillary acidic protein (GFAP) as animals aged. They noted that numerical changes in glial cells with age are difficult to analyze between different studies that used a variety of techniques and mouse strains and ages.

Neuroaxonal dystrophy occurs in the central nervous system of a number of specific neurological mutants but is also present at lower frequency in aging mice (Bronson *et al.*, 1992), particularly in the lumbosacral spinal cord. Dystrophic axons appear as rounded eosinophilic globules of varying sizes by light microscopy and are composed ultrastructurally of collections of organelles and fibrillary and amorphous material sometimes surrounded by myelin.

Krinke (1996) found spinal cords have 15–20% decrease in motor neurons in BALB/c mice at 110 weeks of age compared with young animals (age unspecified). Spinal neurons of aged mice also accumulate lipofuscin and lose postsynaptic dendritic specializations; loss of dendrites precedes cell death. Aging C57BL6/J mice develop inflammatory demyelinating polyneuropathy with lesions most extensive in the lumbar cord. This lesion is not prominent in aging C57BL6/NCr mice (D.C. Haines, pers. comm.).

Neurobehavioral changes with age have been noted in T-maze activity (spontaneous alteration) in several strains of mice (Lalonde, 2002). These changes may be due to age-related losses in memory or due to reduced willingness to explore specific environmental stimuli (behavioral testing methods are covered in more detail elsewhere in this book). An age-related decline in exploratory activity occurs among various strains of mice (Ingram, 2000). This decline is postulated to be related to changes in the dopaminergic neurotransmitter system, especially decreased concentration of the D_2 dopamine receptor. There is a strong correlation between increasing body weight and decreasing activity. Tou and Wade (2002) hypothesize that decreased activity is a response rather than a contributor to weight gain and that decline in activity may be related to decreased leptin levels.

The proto-oncogene c-fos is presumed to have an important role in long term regulation of neuronal function (D'Costa *et al.*, 1991). Following a single convulsive electroshock in B6C3F1 mice, there was an overall decrease in brain c-fos immunoreactivity in 13- and 28-month old compared to 6-month old mice.

Human senile dementia (Alzheimer's disease, AD) and other age-related neurodegenerative diseases, e.g. Huntington's disease, are major medical and social problems as the mean age of the populations of affluent Western countries increases. Mouse models of these age-related conditions are important for understanding pathogenetic mechanisms and interventions. There are numerous transgenic models of AD as reviewed by van Leuven (2000) and Ashe (2001). A comprehensive transgenic mouse directory can be accessed on the website of the Alzheimer Research Forum at http://staging.alzforum.org/res/com/tra/default.asp.

It must be noted that any brain/behavioral studies using transgenic mice on the FVB background can be complicated by this strain's spontaneous neurologic problems (Ward *et al.*, 2000; Ashe, 2001) and

hereditary blindness due to retinal atrophy (Ward *et al.*, 2000).

Several strains of the SAM have been used as models of human aging/neurodegenerative disease. SAMP 10 mice develop age-dependent cerebral atrophy and altered behavior (Shimada, 1999). SAMP 8 have spongiform degeneration with micro- and astrogliosis in their brain stem, dystrophic axons in the hippocampus and impairment of the blood brain barrier (Hosokawa and Ueno, 1999). Neuronal degeneration, neuropil vacuolation, gliosis and effects on endothelium may be related to endogenous murine virus replication (Kawamata *et al.*, 1998). These SAMP 8 mice have age-related defects in learning and memory, emotional abnormality (reduced anxiety behavior) and disrupted circadian rhythms (Nomura and Okuma, 1999).

Oxidative stress of the brain results when sources of free radicals outpace the free radical scavenging systems. It is manifested by protein oxidation and lipid peroxidation. These findings are present in brains of SAMP mice, several AD models and a chemically induced (3-nitroproionic acid) model of Huntington's disease (Butterfield *et al.*, 2001).

In a number of models of AD, CR had a neuroprotective effect, increasing the resistance of neurons to oxidative, metabolic, and excitotoxic insults. However, in a model of amylotropic lateral sclerosis (transgenic mice expressing mutant Cu/Zn SOD), CR accelerated disease onset (Mattson, 2000).

Hematopoietic

In contrast to humans, both the long bones and the red pulp of the spleen remain active in hematopoiesis throughout the life span of the mouse. Fatty infiltrate of the marrow tends to be mild and involves primarily the metaphyseal, and to a lesser extent epiphyseal, marrow. Fibro-osseous lesion, more commonly seen in females than males, may affect hematopoiesis by physically diminishing the marrow space (Wijnands *et al.*, 1996).

Schlessinger and Van Zant (2001) studied bone marrow stem cell function in mouse chimera made by aggregating C57BL/6 (a long-lived strain) and DBA/2 (a short-lived strain) embryos. They found that DBA/2 stem cell contributions declined with age. After 2 years, blood cells were entirely of C57BL/6 origin. However, if bone marrow cells from 31-month old mice were transplanted into lethally irradiated recipients, DBA/2 stem cells were reactivated and contributed significantly to hematopoiesis. With time, the DBA/2 stem cells again became quiescent, only to be reactivated in a second transplantation. Thus, they concluded that aging was associated with declining mobilization of hematopoietic stem cells, i.e. functional rather than numerical depletion. They postulated that an age-related change in the stem cells' environment could render the cells either less accessible or less responsive to activation. While older mice of various strains do not develop severe anemias or leukopenias, Schlessinger and Van Zant (2001) found a strong negative correlation in an *in vitro* assay between hematopoietic progenitor cells in S-phase of the cell cycle and lifespan of eight inbred strains. These findings suggest that functional depletion of stem cells may in itself be a principal feature of aging.

Spleen, lymph nodes, and mucosa-associated lymphoid tissue (MALT) frequently exhibit decreased percentage of mature lymphocytes with concurrent hyperplasia of immature forms. Increased pigment occurs due to accumulation of hemosiderin in splenic and lymph node histiocytes. Sinusoidal histiocytosis (sometimes atypical), reticular cell hyperplasia, sinusoidal ectasia, and hilar/medullary fatty infiltrates are other lesions commonly seen in lymph nodes (Wijnands *et al.*, 1996).

Involution of the thymus results in decreased thickness of the cortex due to loss of T-cell component. B-cell hyperplasia, with formation of secondary follicles, within the medulla may occur to a notable degree in some mice (personal observation). Epithelial elements in the medulla may increase in prominence due to loss of cortical mass or due to actual hyperplasia.

Proper immune function involves a complex interaction among antigen-presenting cells (macrophages; dendritic cells), helper T-lymphocytes, cytotoxic T-lymphocytes, B-lymphocytes, natural killer (NK) cells, neutrophils, and cytokines (for review see Abbas *et al.*, 2000). Studies in aging mice have revealed alterations in several components of the immune system, many of which emulate changes seen in humans.

Differentiation of pro-B to pre-B cells is reduced. This reduction may relate to impaired Ig heavy chain VDJ gene segment rearrangement in pro-B cells and/or decreased production (or sensitivity to) interleukin-7 (IL-7) from bone marrow stromal cells (Weksler *et al.*, 2002). The number of peripheral B-lymphocytes does not decrease, despite the decrease in bone marrow production of new B-cells. This appears to be due to marked increase in longevity of the peripheral lymphocytes. A 6-fold increase in peripheral B-lymphocyte longevity has been reported for 23–25-month-old mice when compared to 1.5–3.5-month-old mice (Kline

et al., 1999). An increase in number and activity of B1a lymphocytes and concurrent decrease in B2 lymphocytes also occurs. This is significant because B1a cells are largely T-cell independent and primarily produce IgM autoantibodies. B2 cells are involved in T-cell dependent production of antibodies to exogenous antigens. Indeed although concentration of circulating Ig does not decrease (and may indeed increase), there is a shift from IgG to IgM and an increase in autoantibodies (Weksler *et al.*, 2002).

T-lymphocytes also show a decrease in naïve cells (possibly resulting from a noted decrease in IL-7) and increased longevity of memory cells. T-lymphocytes exhibit impaired expression of CD40 ligand, which can negatively impact T-cell-dependent B-cell activation. The ability of the T-lymphocyte to become activated and produce IL-2 decreases (Hodes, 2002). Cytotoxic CD8[+] T cells exhibit decreased response to telomerase activity that may represent one pathway to their senescence (Goyns, 2002).

Follicular dendritic cells (FDC) also show altered function. Production from the bone marrow decreases and the FDC have a reduced capacity to trap and transport antigens. This may play a major role in decreased number and functioning of germinal centers in lymphoid organs (Szakal *et al.*, 2002). An increase in circulating neutrophils has been reported in 27-month-old mice. This may represent a compensatory response to T/B-lymphocyte dysfunction. (Blasco, 2002).

The cumulative effect of these alterations is a dysfunctional immune system with an increased susceptibility of aged mice to infection and neoplasm formation.

Another consequence of immune dysfunction may be amyloid deposition. Senile amyloidosis in mice consists of extracellular accumulation of AapoAII amyloid protein. This is a systemic condition involving multiple organs including the intestines, adrenal gland, lung, heart, thyroid gland, reproductive organs, kidney, liver, and spleen. Amyloid deposition associated with inflammatory processes in aging mice is composed of AA amyloid protein deposition. Long-term corticosteroid-induced immunosuppression has been shown to decrease AapoAII and increase AA amyloid (HogenEsch *et al.*, 1996).

Respiratory

Studies have shown the lungs of aging mice to have an increase in number and size of interalveolar pores (Ranga and Kleinerman, 1980), increase in total alveolar surface area and lung volume (BALB/cNNia; Kawakami *et al.*, 1984), and a decrease in elastic fibers and elastin

(Ranga *et al.*, 1979). Decreased internal surface area with dilated alveolar ducts, flattened alveoli and increased total lung volume has been reported in the SAMP 1 mouse that may serve as an animal model for senile hyperinflation of humans (Kurozumi *et al.*, 1994). Similar to man, respiratory infections have more severe effects with slower recovery in older mice. This has been shown to be true with Sendai virus (BALB/c; Jacoby *et al.*, 1994) and influenza (Bender *et al.*, 1991). General immune senescence, particularly of cytotoxic T-lymphocytes, may play a role in this increased susceptibility to infection. Additionally, a 30% decrease in pulmonary glutathione concentration reported in aging mice (Chen *et al.*, 1990) indicates a possible mechanism for increased susceptibility to the toxic effects of inherent metabolic oxidation. Bronchioloalveolar hyperplasia and neoplasia increase in incidence in a number of strains. Activation of the proto-oncogene K-ras has been associated with a majority of the mouse lung tumors (for review see Rittinghausen *et al.*, 1996).

Increase in **hyalinosis** of nasal epithelium and acidophilic macrophage pneumonia may be noted in some strains. Hyalinosis presents as bright acidophilic intracytoplasmic accumulations. The protein isolate has been identified as Ym2, a member of the chitinase family (Ward *et al.*, 2001).

Cardiovascular

Idiopathic calcification of the heart is not uncommon, being more prominent in some strains (DBA/2, C3H, BALB/c) than others (C57BL). The lesion may present as large epicardial plaques, particularly on the right ventricle, but may also occur as scattered foci throughout the myocardium. The mineralization occurs earlier and more notably in multiparous females and in mice fed high-fat diets (Eaton *et al.*, 1978). Atrial thrombosis of varying severity is also a common finding. Chronic cardiomyopathy (myocardial degeneration with associated chronic inflammation and fibrosis) shows wide variation in incidence by strain (BALB/c, sensitive; C57BL, resistant) and may be influenced by diet, hormones, infection, and immune status (for review see Price and Papadimitriou, 1996). The heart may also be affected by systemic diseases: amyloidosis or periarteritis (coronary arteries and aorta in particular).

Although medial hypertrophy and periarteritis (polyarteritis nodosa associated with immune dysregulation) of arterioles are common in many strains, atherosclerosis is rare without diet intervention, and even then susceptibility varies by strain (Paigen *et al.*, 1985).

Endocrine

Pituitary

Atrophy of the pituitary gland is rarely given as a morphological diagnosis, although functional changes undoubtedly occur. More commonly, focal hyperplasia of the pars distalis and/or pars intermedia, are noted. Hyperplasia may be inherent to the gland or secondary to negative feedback due to aging changes in one or more of the other endocrine glands. Although hyperplastic lesions are common, transformation into neoplasms (adenoma/carcinoma) is much less common. Schecter *et al.* (1981) reported an 80% incidence of adenomas in the pars distalis of female C57BL/6J over 22 months of age; however, this appears to be the exception. In an ongoing study of aging C57BL/NCr by the author (Haines) the current incidence of pars distalis adenomas is 19% in females (8/42) and 0% in males (0/32). The incidence of hyperplasia in the pars distalis of the same cohort is 17% in the females and 0% in the males. The FVB/NCr female mice 18- to 23-month old have an incidence of 19% prolactin-secreting adenomas and 52% hyperplasia (Wakefield *et al.*, 2003). Spontaneous adenomas of the pars distalis are reported at even lower percentages in many other mouse strains: 2% adenomas in both sexes and incidence of hyperplasia 11% in female, 4% in male B6;129 mice (Haines *et al.*, 2001).

Thyroid

Although the total number of follicles in the thyroid remains the same, there is an increased percentage of 'cold' follicles which have decreased ability to bind iodine. These follicles, with enlarged colloid-filled lumen and flattened lining epithelium, result in the increased size and weight of the thyroid noted in aged mice. Circulating T3 and T4 may decrease due to defective colloid resorption resulting from mechanical interference due to over distention of these follicles or cytoskeletal defect affecting endocytosis (Gerber *et al.*, 1987). The remaining active follicles may undergo hyperplasia.

Parathyroid

The parathyroid (PTH) exhibits increased interstitial connective tissue with increased indentation of the nucleus. As seen by electron microscopy, there is a decrease in ribosomes and secretory granules with concurrent increase in dense bodies, lipid droplets and multivesicular bodies, suggesting a decrease in PTH (dd mouse; Capen *et al.*, 1996). However, in mouse strains with chronic renal failure, decreased synthesis of 1;25-dihydroxy-vitamin D with a subsequent decrease in blood calcium, may result in diffuse PTH hyperplasia with an elevation of PTH production.

Adrenal

In aging mice, the adrenal cortex response to adrenocorticotropic hormone (ACTH) declines and production of corticosterone decreases (DBA/2J and C57Bl/6J; Eleftheriou *et al.*, 1974). Morphological changes may include hyperplasia and/or atrophy. Hyperplasia of subcapsular cells is common and focal hypertrophy and focal hyperplasia of the zona fasiculata may alternate with areas of atrophy. Cystic dilatation of cortical sinusoids may be severe enough to cause macroscopic enlargement of the gland. PAS-positive, acid-fast-positive lipogenic pigment, which accumulates at the corticomedullary junction, may be increased by high-fat diets or vitamin E deficiencies (Davies *et al.*, 1987). In mice with systemic amyloidosis, the adrenal cortex is a common site with deposits usually more pronounced in the inner cortex.

Only a few mouse models, such as F1 heterozygous Nf1-knockout mutants (Jacks *et al.*, 1994) and c-mos transgenics (Schulz *et al.*, 1992), have significant incidences of adrenal medullary neoplasms. Focal or diffuse hyperplasia and neoplasms of the adrenal medulla, although reported, tend to be of low incidence in most mouse strains even with advanced age.

Endocrine pancreas (islets)

Pancreatic islet hypertrophy (increase in size of individual cells) and hyperplasia (increase in the number of cells) is a progressive aging change in many strains of mice. There is a resultant increase in insulin, increased efficiency of glucose clearance, as measured by glucose tolerance testing, and decrease in blood glucose (C57BL/6J; Leiter *et al.*, 1988). Perfetti *et al.* (1994) has reported insulin and glucose transporter-2 (Glut2) mRNA as well as mRNA for regenerating (*Reg*) gene product (expressed in both pancreatic endocrine and exocrine cells; may be important in beta-cell proliferation), to decrease in aging mice.

Gastrointestinal

Liver, gall bladder

A complex perturbation of hepatic enzymes occurs with aging, and there are alterations in carbohydrate and lipid metabolism. DNA and RNA synthesis decrease with age; DNA repair and methylation also are affected. The susceptibility of the liver to toxic effects generally increases with age but the susceptibility to chemical carcinogens may decrease except for certain nongenotoxic carcinogens such as phenobarbital (Harada *et al.*, 1996).

Foci of cellular alteration and hepatocellular neoplasms increase with age in some strains of mice. Nonneoplastic lesions present in aged mice are the foci of coagulative necrosis, subacute inflammation, fatty change, apoptotic hepatocytes (manifested as acidophilic globules), and pigment (lipofuscin, hemosiderin, and bile) deposition. Hepatocellular hypertrophy, hepatocytomegaly, and karyomegaly are common in aging mice. Karyomegaly is often related to hepatocyte nuclear polyploidy (Frith and Ward, 1988; Harada *et al.*, 1996).

The incidence of gall bladder nonneoplastic lesions in aged mice is low except for epithelial hyalinosis in some strains (Ward *et al.*, 2001).

Salivary gland

Of the murine salivary glands, aging changes are primarily seen in the submandibular gland. This gland exhibits sexual dimorphism: in males the gland is larger, with more zymogen granules present. Atrophy of the granular convoluted tubule (GCT) occurs in males, possibly related to declines in androgen and other endocrine factors. The aged female submandibular gland may have a more masculinized, hypertrophic morphology. Parotid salivary glands may have an increased incidence of basophilic hypertrophic foci (Seely, 1996; Botts *et al.*, 1999). Perivascular lymphocytic infiltrates in the submandibular and parotid glands increase in incidence and severity with age. Salivary gland secretion of mucin and production by the GCT of biologically active peptides such as epidermal growth factor decrease with age.

Teeth

Incisor teeth may have varying degrees of dental dysplasia (Cohen *et al.*, 1983; Maekawa *et al.*, 1996; Long and Leininger, 1999a). Dental dysplasia may range from small invaginations of the enamel organ into the dental pulp to large distorting lesions with proliferation of dentine, enamel and odontoblasts. Dental dysplasia increases in incidence and severity with aging and is more frequent in males than females. Many strains of mice are affected.

Esophagus

Hyperkeratosis has been reported in B6C3F1 mice (Maekawa *et al.*, 1996). Esophageal impaction/mega-esophagus is found in several of the background strains used to generate GEM. The mucosa of the dilated esophagus has hyperkeratosis with admixed bacterial colonies. The distended esophagus may be impacted with food material; secondary inhalation pneumonia can result (Ward *et al.*, 2000; Haines *et al.*, 2001). This lesion is a significant cause of mortality in 129S4/SvJae mice.

Stomach

Ulcers may occur in the nonglandular stomach although this is not a consistent finding among strains/laboratories and may be related to environmental factors. In the glandular stomach, gland dilatation, mucosal atrophy, and chronic inflammation increase with age; the volume of the gastric secretion and hydrochloric acid output may decrease (Maekawa *et al.*, 1996). Gastric plaques consisting of epithelial hyperplasia, hyalinosis, and chronic inflammation are a common age-associated finding in 129 mice and in some targeted mutations using this background strain (Ward *et al.*, 2000) but were infrequent in aged B6;129 mice (Haines *et al.*, 2001).

Intestines

There are no consistently reported age-related nonneoplastic lesions in the gut. Some strains and stocks have age-related amyloidosis in the small intestine while this finding is uncommon or absent in other strains/stocks (Maekawa *et al.*, 1996).

Significant work has been done by Potten *et al.* (2001) on mouse small intestine stem cells. Using tritiated thymidine labeling, they studied small intestinal (ileal) crypt cells, comparing young (5–7 months) to old (30 month) C57BL/IcrFa mice. They identified two populations of stem cells. The first is a single self-maintaining lineage ancestor stem cell (ASC) located in the crypt base immediately above the Paneth cells. An ACS may divide 1000 times in a 3-year lifespan of a laboratory mouse. The second are potentially **clonogenic** stem cells located immediately above the ASC. Cells higher in the crypt are committed to differentiation.

Following a gamma irradiation insult, intestinal stem cells from old mice had enhanced sensitivity to apoptosis and regenerated less efficiently. Old mice had a delay in onset of expression of p53 in the stem cell region as well as slightly delayed p21 expression. These data and those from other studies cited in the paper suggest that intestinal stem cell deterioration with age is probably due to alterations at the molecular level.

Exocrine pancreas

Common nonneoplastic lesions involving acini in aged mice include small foci of chronic inflammation and acinar atrophy, sometimes associated with ductule proliferation (Enomoto *et al.*, 1996; Boorman and Sills, 1999). Atrophy may be focal or lobular. The interstitium may be infiltrated by adipocytes. Lymphocytic infiltrates may also be present. Cystic ducts are occasionally seen. Foci of basophilic and eosinophilic cytoplasmic alterations, similar to those seen in the rat, can occur in older mice (Boorman and Sills, 1999).

Musculoskeletal

Muscle

Muscle strength and endurance as well as motor skills decrease in mice with age. Measured maximum tetanic forces in several muscle groups were approximately 25% less in 26–27-month-old compared with 9–10-month-old C57BL/6 mice. Weakness is due to functional loss of motor neurons with loss of myelinated axons. Histologically, muscle fibers were atrophic and varied in size. There was increased endomyseal and perimyseal connective tissue (Novilla and Smith, 1996).

Transgenic mouse models have been generated to study muscle aging: carriers of HAP transcription unit driven by myosin light chain 1 (MLC1) promoter and downstream MLC enhancer to mark specifically fast fibers; nuclear localizing LacZ gene under the control of the myosin light chain 3F isoform. Satellite cells are the major source of myogenic precursors. Using a transgenic mouse carrying the LacZ reporter gene under the desmin promoter, satellite cells from old muscle were shown to have less proliferative capability and/or less access to host trophic substances. (Musaró and Rosenthal, 1999).

Age-related stiffening of connective tissues occurs due to alterations in collagen and elastin. Intermolecular cross-linking occurs due to glycation reactions. Modification of side-chains alters normal cell–matrix interaction (Bailey, 2001).

Joints

Arthritic lesions in the apophyseal joints of the thoracic vertebra have been described in B6C3F1 mice

(Sheldon *et al.*, 1996). The precursor lesion was focal chondrocyte necrosis starting at 12 months of age. Cartilage damage with proliferation increased with age, progressing to full blown noninflammatory osteoarthritis by 24 months in *ad libitum* fed mice. Degenerative vertebral joint disease in dietary restricted mice developed later and was less severe. Degenerative vertebral joint disease has been reported in aging B6; 129 mice (Haines *et al.*, 2001). Degenerative joint disease can also occur in the temporal–mandibular joint, especially in SAMP 3.

Some of the SAM mice have systemic amyloidosis (Takeda *et al.*, 1996). A number of theses strains have microdeposition of amyloid in various synovial joints and intervertebral discs.

Bone

Long bone cortices become thin with age. Atrophy, rather than osteoporosis or osteopenia, is the recommended terminology for this murine lesion (Long and Leininger, 1999b). Osteoporosis has clinical connotations of hormone/age-related bone loss in humans while osteopenia denotes a non-specific bone mass decrease which could be caused by fibrous osteodystrophy or hyperosteoidosis as well as atrophy. The incidence of atrophy (loss of bone mass) is strain- and sex-dependent with females more commonly affected.

Fibro-osseous lesion is an age-related bone change occurring primarily in female mice, located most frequently in long bones, the sternum, and vertebrae. The lesion occurs in the marrow and consists of prominent fibrovascular stroma with osteoblasts and frequently multinucleated osteoclasts (Albassam and Courtney, 1996).

Urinary

From maturity through senescence, the kidney to body weight ratio, number of glomeruli, rate of renal protein and glucose synthesis and gluconeogenesis remain unchanged, while spontaneous DNA methylation increases. Tubule cell proliferation rate, glomerular filtration rate, renin messenger RNA, and glutathione and cysteine concentrations decrease (Wolf and Hard, 1996). Since the kidney is an important source for cysteine, which is a rate-limiting amino acid in protein synthesis, this decrease may negatively impact protein synthesis in non-renal tissues (Wolf and Hard, 1996). Chronic progressive nephropathy (chronic nephritis), which begins as tubule lesions, as well as membranoproliferative glomerulopathy, which begins as an immune-mediated glomerular lesion, increase in mice with age (discussed in more detail elsewhere in this book).

Reproductive

Cessation of reproduction in female mice tends to occur between 300 and 400 days of age. Time of onset of reproductive senescence in the male mouse is more variable (Loeb *et al.*, 1996).

Males

Degeneration of the seminiferous tubules of the testes proceeds from the more mature spermatids to the less mature spermatozoa with spermatogonia being the most resistant of the germinal cells to be lost due to aging. Multinucleated giant cells (spermatid symplasts), formed by fusion of defective round spermatids, are indicative of degeneration (similar cells, however, may occasionally be seen in juvenile testes; personal observation). Leydig (interstitial) cells are usually preserved and may appear to be more numerous due to reduction in testicular mass resulting from extensive loss of the germinal cells. Sertoli cells are quite hardy and may be the only cells left in severely atrophic testes (Radovsky *et al.*, 1999). Foci of mineralization are not uncommon, both within atrophic seminiferous tubules and perivascular. Secretory dilatation/constipation of accessory male sex glands (seminal vesicles, coagulating glands, prostate, bulbourethral glands, preputial gland) may occur. Alternatively, if testosterone influence is greatly reduced due to marked atrophy of the Leydig cells and/or decrease in number or function of hormone receptors, these organs may decrease in size with decreased height of the epithelium (atrophy). Preputial gland ducts may be **ectatic** with marked atrophy of the acinar epithelium.

Females

Atrophy of the ovary is a common finding in aged mice. The gland is decreased in size and weight with loss of follicles and corpus lutea. Interstitial cells become prominent and often have increased cytoplasmic volume due to accumulation of **ceroid**. The germinal epithelium on the surface of the gland is also prominent and frequently forms cords and tubules deep into the parenchyma of the ovary. Epithelial lined cysts, of variable size, may cause gross enlargement of the gland. Angiectasis, with occasional thrombosis, may impart a red coloration (Davis *et al.*, 1999).

Changes seen in the uterus include simple dilatation, cystic endometrial hyperplasia, adenomyosis, and endometrial stromal polyps (Maekawa and Maita, 1996). Angiectasis alone or in combination with vascular medial hypertrophy, periarteritis, and/or thrombosis may occur, and these vascular lesions need to be differentiated from neoplasia (hemangiomas/hemangiosarcomas). It must be noted that mice do not experience menses; therefore the terms 'menstruation' and 'menopause' are inappropriate for use with mouse models.

Special senses

Auditory

Hearing loss is an age-associated change in mice. Willott (1996) provides a detailed review of hearing loss in a number of inbred strains, F1 hybrids, an outbred stock, and some mutants. Some strains have minimal age-related loss while others have progressive loss with varying degrees of severity. C57BL/6 mice are the most extensively affected. Lesions in mice with hearing loss are in the cochlea and the neuropil.

Ocular

Glaucoma associated with aging has been reported in DBA2/J mice with ocular changes beginning at 6 months of age and progressing to severe damage by 22 months. Cataracts develop secondary to glaucoma (Smith et al., 1996).

Incidence of cataracts generally increases with age. For example, there are incidences of 25% in CD1 mice at 18 months of age (Frame and Slone, 1996), and 13% in aging 129S4/SvJae mice (Ward et al., 2000). SAMP 9 mice have age-related cataracts which become apparent at 10 weeks of age, with incidence up to 50–60% in males at 20 weeks of age and up to 90% in females at the end of their lifespan. Cataract development in SAMP is related to a persistent hyaloid vasculature (Hosokowa and Ashida, 1996).

Integumentary

Chronic ulcerative dermatitis is a major clinical problem in C57BL/6 mice (Sundberg et al., 1996). This idiopathic condition increases with age and often is severe enough to cause significant morbidity and mortality, complicating the use of black mice in gerontology research.

Mammary gland

Age-related changes in the mammary glands of mice are influenced by the mouse's genetic makeup, reproductive history, and other endocrine factors. For example, by 13 months of age, the mammary glands of 40% of virgin FVB/NCr mice have the morphologic appearance of glands during pregnancy. This lobuloalveolar hyperplasia is related to prolactin-secreting hyperplasias and adenomas of the pituitary pars distalis (Wakefield et al., 2003). Proliferative lesions and neoplasms increase with age and parity in some genotypes. They are reviewed by Rehm and Liebelt (1996), while the spectrum of neoplasms in GEM and a proposed classification are presented by Cardiff et al. (2000).

The mouse mammary gland is used as an experimental system to study stem cells and their role in growth and senescence (Kenney et al., 1996; Smith and Boulanger, 2002). Three distinct multipotent mammary epithelial cells have been identified by transplantation studies: a cell which gives rise to secretory lobules; a cell which gives rise to branching mammary ducts and a multipotent cell which gives rise to both structures. The stem cell niche exists in the mammary gland throughout the life span of the mouse. Growth senescence of transplants may be related to aging of these stem cells. The in vitro lifespan of normal, non-transformed mammary epithelial cells is limited, correlating with the lifespan of the species.

Clinical biochemistry

Although clinical biochemistry can vary by sex and strain, general trends reported for several strains during aging include decreases in serum glucose, triglycerides, corticosterone, testosterone, urine osmolarity and urine concentrating ability, and increases in aspartate aminotransferase (AST) and insulin. Minimal to no changes are reported for serum sodium, serum osmolarity, and TSH. Findings for parameters which vary by studies (increased, decreased, or no change) include alkaline phosphatase, alanine aminotransferase, total serum protein, and urea nitrogen (the latter largely dependent on presence or absence of renal disease; Loeb et al., 1996).

Conclusion

The process of aging is multifactorial and is affected by both genes and environment. Thus, the mouse is an excellent species for use in gerontology research. Its genome has been characterized, and genetic engineering technology is well established. It has a reasonably short lifespan and its environment (diet, bedding, light cycles, disease status, chemical exposure, etc.) can be readily controlled.

Study design is key to a successful outcome of a gerontology research project. Miller and Nadon (2000) provide some helpful guidelines for animal use. These include advice on choice of ages, numbers of groups and animal numbers; choice of strains or stocks; the necessity to use specific pathogen free mice; the need for pathology support and the pitfalls of pooling cells/tissues.

Many mouse models of aging have been generated. Longevity and age-related neoplasms and non-neoplastic lesions have been documented for a wide number of inbred strains, outbred stocks, F1 hybrids and GEM. When choosing a mouse model, the scientist needs to evaluate data available on longevity and lesions (including age of onset) not only for the specific model but also for its background strain(s). Even with this information available, use of appropriate sex/age matched controls is essential for valid interpretation of gerontology studies.

Acknowledgments

This project has been supported in part with Federal funds from the National Cancer Institute, National Institutes of Health under contract no. NO1-CO-12400. The content of this publication does not necessarily reflect the views or policies of the Department of Health and Human Services, nor does mention of trade names, commercial projects or organization imply endorsements by the U.S. Government. We thank Stephanie Bowers for assistance in manuscript preparation, and Dr C. Dahlem Smith for reviewing this work.

References

Abbas, A.K., Lichtman, A.H. and Pober, J.S. (2000). *Cellular and Molecular Immunology*. W.B.Saunders Co., Philadelphia.

Albassam, M.A. and Courtney, C.L. (1996). In *Pathobiology of the Aging Mouse*, Vol. 2 (eds U. Mohr, D.L. Dungworth, C.C. Capen, W.W. Carlton, J.P. Sundberg and J.M. Ward), pp. 425–437. ILSI Press, Washington, DC.

Andersen, J.K. (2001). *Mol. Biotechnol.* **19**, 45–57.

Anisimov, V.N. (2001). *Mech. Ageing Dev.* **122**, 1221–1255.

Ashe, K.H. (2001). *Learn. Mem.* **8**, 301–308.

Austad, S.N. (1997). *Exp. Gerontol.* **32**, 23–38.

Bailey, A.J. (2001). *Mech. Ageing Dev.* **122**, 735–755.

Bartke, A., Brown-Borg, H., Mattison, J., Kinney, B., Hauck, S. and Wright, C. (2001). *Exp. Gerontol.* **36**, 21–28.

Baynes, J.W. (2001). *Exp. Gerontol.* **36**, 1527–1537.

Bender, B.S., Johnson, M.P. and Small, P.H. (1991). *Immunology* **72**, 514–519.

Blasco, M.A. (2002). *Springer Semin. Immunopathol.* **24**, 75–85.

Boorman, G.A. and Sills, R.C. (1999). In *Pathology of the Mouse* (eds R.R. Maranpot, G.A. Boorman and B.W. Gaul), pp. 185–205. Cache River Press, Vienna, IL.

Botts, S., Jokinen, M., Gaillard, E.T., Elwell, M.R. and Mann, P.C. (1999). In *Pathology of the Mouse* (eds R.R. Maranpot, G.A. Boorman and B.W. Gaul), pp. 49–79. Cache River Press, Vienna, IL.

Bronson, R.T. and Lipman, R.D. (1991). *Growth Dev. Aging* **55**, 169–184.

Bronson, R.T., Sweet, H.O., Spencer, C.A. and Davisson, M.T. (1992). *J. Neurogenet.* **8**, 71–83.

Brunk, U.T. and Terman, A. (2002). *Free Radic. Biol. Med.* **33**, 611–619.

Butterfield, D.A., Howard, B.J. and LaFontaine, M.A. (2001). *Curr. Med. Chem.* **8**, 815–828.

Capen, C.C., Grone, A., Bucci, T.J. and Rosol, T.J. (1996). In *Pathobiology of the Aging Mouse*, Vol. 1 (eds U. Mohr, D.L. Dungworth, C.C. Capen, W.W. Carlton, J.P. Sundberg and J.M. Ward), pp. 109–110. ILSI Press, Washington.

Cardiff, R.D., Anver, M.R., Gusterson, B.A., Henninghausen, L., Jensen, R.A., Merino, M.J., Rehm, S., Russo, J., Tavassoli, F.A., Wakefield, L.M., Ward, J.W. and Green, J.E. (2000). *Oncogene* **19**, 968–988.

Carp, R.I., Meeker, H.C., Kozlowski, P. and Sersen, E.A. (2000). *Trends Microbiol.* **8**, 39–42.

Carrard, G., Bulteau, A.-L., Petropoulos, I. and Friguet, B. (2002). *Int. J. Biochem. Cell Biol.* **34**, 1461–1474.

Chen, T.S., Richie, F.P. Jr and Lang, C.A. (1990). *Drug Metab. Dispos.* **18**, 882–887.

Cohen, B.J., Crisp, C.E., McClatchey, K.D. and Anver, M.R. (1983). *Gerodontology* **2**, 43–50.

Cotran, R.S., Kumar, V. and Collins, T. (1999). *Robbins Pathologic Basis of Disease*, pp. 1–49, 276–296. W.B. Saunders Co., Philadelphia.

Davies, I., Davidson, Y. and Fotheringham, A.P. (1987). *Exp. Gerontol.* **22**, 127–137.

Davis, B.J, Dixon, D. and Herbert, R.A. (1999). In *Pathology of the Mouse* (eds R.R. Maronpot, G.A. Boorman and B.W. Gaul), pp. 409–440. Cache River Press, Vienna, IL.

D'Costa, A., Breese, C.R., Boyd, R.L., Booze, R.M. and Sonntag, W.E. (1991). *Brain Res.* **567**, 204–211.

Donehower, L.A. (2002). *J. Cell Physiol.* **192**, 23–33.

Dunn, S.E., Kari, F.W., French, J., Leninger, J.R., Travlos, G., Wilson, R. and Barrett, J.C. (1997). *Cancer Res.* **57**, 4667–4672.

Eaton, G.J., Curtis, R.P., Johnson, H.N. and Stabenow, K.T. (1978). *Am. J. Pathol.* **90**, 173–186.

Eleftheriou, B.E. (1974). *Gerontologia* **20**, 224–230.

Engelhardt, J.A., Gries, C.L. and Long, G.G. (1993). *Toxicol. Pathol.* **21**, 538–541.

Enomoto, M., Hirouchi, Y., Tsutsumi, M. and Konishi, Y. (1996). In *Pathobiology of the Aging Mouse*, Vol. 2 (eds U. Mohr, D.L. Dungworth, C.C. Capen, W.W. Carlton, J.P. Sundberg and J.M. Ward), pp. 251–259. ILSI Press, Washington, DC.

Forster, M.J., Morris, P. and Sohal, R.S. (2003). *FASEB J.* **17**, 690–692.

Frame, S.R. and Slone, T.W. (1996). In *Pathobiology of the Aging Mouse*, Vol. 2 (eds U. Mohr, D.L. Dungworth, C.C. Capen, W.W. Carlton, J.P. Sundberg and J.M. Ward), pp. 97–103. ILSI Press, Washington, DC.

Frith, C.H. and Ward, J.M. (1988). *Color Atlas of Neoplastic and Non-Neoplastic Lesions in Aging Mice*, pp. 10–14. Elsevier, New York.

Gerber, H., Peter, H.J. and Studer, H. (1987). *Endocrinology* **120**, 1758–1764.

Goyns, M.H. (2002). *Mech. Ageing Dev.* **123**, 791–799.

Goytisolo, F.A. and Blasco, M.A. (2002). *Oncogene* **21**, 584–591.

Haines, D.C., Chattopadhyay, S. and Ward, J.M. (2001). *Toxicol. Pathol.* **29**, 653–661.

Harada, T., Maronpot, R.R., Enomoto, A., Tamano, S. and Ward, J.M. (1996). In *Pathobiology of the Aging Mouse*, Vol. 2 (eds U. Mohr, D.L. Dungworth, C.C. Capen, W.W. Carlton, J.P. Sundberg and J.M. Ward), pp. 207–241. ILSI Press, Washington, DC.

Harman, D. (1956). *J. Gerontol.* **11**, 298–300.

Hayflick, L. (1965). *Exp. Cell Res.* **37**, 614–636.

Hodes, R.J. (2002). *Springer Semin. Immunopathol.* **24**, 1–5.

HogenEsch, H., Gruys, E. and Higuchi, K. (1996). In *Pathobiology of the Aging Mouse*, Vol. 1 (eds U. Mohr, D.L. Dungworth, C.C. Capen, W.W. Carlton, J.P. Sundberg and J.M. Ward), pp. 237–244. ILSI Press, Washington, DC.

Hosokawa, M. and Ashida, Y. (1996). In *Pathobiology of the Aging Mouse*, Vol. 2 (eds U. Mohr, D.L. Dungworth, C.C. Capen, W.W. Carlton, J.P. Sundberg and J.M. Ward), pp. 105–115. ILSI Press, Washington, DC.

Hosokawa, M. and Ueno, M. (1999). *Neurobiol. Aging* **20**, 117–123.

Huang, T.T., Carlson, E.J., Gillespie, A.M., Shi, Y. and Epstein, C.J. (2000). *J. Gerontol.* **55**, B5–B9.

Hursting, S.D., Perkins, S.N., Brown, C.C., Haines, D.C. and Phang, J.M. (1997). *Cancer Res.* **57**, 2843–2846.

Ingram, D.K. (2000). *Med. Sci. Sports Exerc.* **32**, 1623–1629.

Jacks, T., Shih, T.S., Schmitt, E.M., Bronson, R.T., Bernards, A. and Weinberg, R.A. (1994). *Nat. Genet.* **7**, 353–361.

Jacoby, R.O., Bhatt, P.N., Barthold, S.W. and Brownstein, D.G. (1994). *Exp. Gerontol.* **29**, 89–100.

Kaeberlein, M., Jegalian, B. and McVey, M. (2002). *Mech. Ageing Dev.* **123**, 1115–1119.

Kawakami, M., Paul, J.L. and Thurlbeck, W.M. (1984). *Am. J. Anat.* **170**, 1–21.

Kawamata, T., Akiguchi, I., Maeda, K., Tanaka, C., Higuchi, K., Hosokawa, M. and Takeda, T. (1998). *Microsc. Res. Tech.* **43**, 59–67.

Kenney, N.J., Hosick, H., Herrington, E. and Smith, G.H. (1996). In *Pathobiology of the Aging Mouse*, Vol. 2 (eds U. Mohr, D.L. Dungworth, C.C. Capen, W.W. Carlton, J.P. Sundberg and J.M. Ward), pp. 369–379. ILSI Press, Washington, DC.

Kim, S.-H., Kaminker, P. and Campisi, J. (2002). *Oncogene* **21**, 503–511.

Kline, G.H., Hayden, T.A. and Klinman, N.R. (1999). *J Immunol.* **162**, 3342–3349.

Koubova, J. and Guarente, L. (2003). *Genes Dev.* **17**, 313–321.

Krinke, G.J. (1996). In *Pathobiology of the Aging Mouse*, Vol. 2 (eds U. Mohr, D.L. Dungworth, C.C. Capen, W.W. Carlton, J.P. Sundberg and J.M. Ward), pp. 83–93. ILSI Press, Washington, DC.

Kuro-o, M. (2001). *Trends Mol. Med.* **7**, 179–181.

Kurozumi, M., Matsushita, T., Hosokawa, M. and Takeda, T. (1994). *Am. J. Respir. Crit. Care Med.* **149**, 776–782.

Lalonde, R. (2002). *Neurosci. Biobehav. Rev.* **26**, 91–104.

Lee, C.K., Klopp, R.G., Weindruch, R. and Prolla, T.A. (1999). *Science* **285**, 1390–1393.

Leiter, E.H., Premdas, F., Harrison, D.E. and Lipson, L. (1988). *FASEB J.* **2**, 2807–2811.

Lithgow, G.J. and Andersen, J.K. (2000). *BioEssays* **22**, 410–413.

Liu, J.-P. (2001). In *Telomerase, Aging and Disease* (eds M.P. Mattson and T. Pandita), pp. 33–59. Elsevier Science, Amsterdam.

Loeb, W.F., Das, S.R., Harbour, L.S., Turturro, A., Bucci, T.J. and Clifford, C.B. (1996). In *Pathobiology of the Aging Mouse*, Vol. 1 (eds U. Mohr, D.L. Dungworth, C.C. Capen, W.W. Carlton, J.P. Sundberg and J.M. Ward), pp. 3–19. ILSI Press, Washington, DC.

Long, P.H. and Leininger, J.R. (1999a). In *Pathology of the Mouse* (eds R.R. Maranpot, G.A. Boorman and B.W. Gaul), pp. 13–28. Cache River Press, Vienna, IL.

Long, P.H. and Leininger, J.R. (1999b). In *Pathology of the Mouse* (eds R.R. Maranpot, G.A. Boorman and B.W. Gaul), pp. 645–677. Cache River Press, Vienna, IL.

Maekawa, A. and Maita, K. (1996). In *Pathobiology of the Aging Mouse*, Vol. 1 (eds U. Mohr, D.L. Dungworth, C.C. Capen, W.W. Carlton, J.P. Sundberg and J.M. Ward), pp. 469–480. ILSI Press, Washington, DC.

Maekawa, A., Enomoto, M., Hirouchi, Y. and Yamakawa, S. (1996). In *Pathobiology of the Aging Mouse*, Vol. 2 (eds U. Mohr, D.L. Dungworth, C.C. Capen, W.W. Carlton, J.P. Sundberg and J.M. Ward), pp. 267–285. ILSI Press, Washington, DC.

Mahler, J.F. (2001). *Toxicol. Pathol.* **29**, 673–676.

Masoro, E.J. (1998). *J. Toxicol. Environ. Health B Crit. Rev.* **1**, 243–257.

Masoro, E.J. (2000). *Exp. Gerontol.* **35**, 299–305.

Mattson, M.P. (2000). *Brain Res.* **886**, 47–53.

Melov, S. (2002). *Int. J. Biochem. Cell Biol.* **34**, 1395–1400.

Merry, B.J. (2002). *Int. J. Biochem. Cell Biol.* **34**, 1340–1354.

Miller, R.A. and Nadon, N.L. (2000). *J. Gerontol. A Biol. Sci. Med. Sci.* **55**A, B117–B123.

Mori, A., Utsumi, K., Liu, J. and Hosokawa, M. (1998). *Ann. NY Acad. Sci.* **854**, 239–250.

Musaró, A. and Rosenthal, N. (1999). *Exp. Gerontol.* **34**, 147–156.

Nomura, Y. and Okuma, Y. (1999). *Neurobiol. Aging* **20**, 111–115.

Novilla, M.N. and Smith II, C.K. (1996). In *Pathobiology of the Aging Mouse*, Vol. 2 (eds U. Mohr, D.L. Dungworth, C.C. Capen, W.W. Carlton, J.P. Sundberg and J.M. Ward), pp. 401–413. ILSI Press, Washington, DC.

Paigen, B., Marrow, A., Brandon, C., Mitchell, D. and Holmes, P. (1985). *Atherosclerosis* **57**, 65–73.

Perfetti, R., Egan, J.M., Zenilman, M.E. and Shuldiner, A.R. (1994). *Transplant. Proc.* **26**, 733.

Perkins, S.N., Hursting, S.D., Phang, J.M. and Haines, D.C. (1998). *J. Invest. Dermatol.* **111**, 292–296.

Potten, C.S., Martin, K. and Kirkwood, T.B. (2001). *Novartis Found. Symp.* **235**, 66–79.

Price, P. and Papadimitriou, M. (1996). In *Pathobiology of the Aging Mouse*, Vol. 1 (eds U. Mohr, D.L. Dungworth, C.C. Capen, W.W. Carlton, J.P. Sundberg and J.M. Ward), pp. 373–383. ILSI Press, Washington, DC.

Radovsky, A., Mitsumori, K. and Chapin, R.C. (1999). In *Pathology of the Mouse* (eds R.R. Maronpot, G.A. Boorman and B.W. Gaul), pp. 381–405. Cache River Press, Vienna, IL.

Ranga, V. and Kleinerman, J. (1980). *Am. Rev. Respir. Dis.* **122**, 477–481.

Ranga, V., Kleinerman, J. and Sorensen, J. (1979). *Am. Rev. Respir. Dis.* **119**, 369–376.

Rao, G.N. (1999). In *Pathology of the Mouse* (eds R.R. Maronpot, G.A. Boorman and B.W. Gaul), pp. 7–11. Cache River Press, Vienna, IL.

Rehm, S. and Liebelt, A.G. (1996). In *Pathobiology of the Aging Mouse*, Vol. 2 (eds U. Mohr, D.L. Dungworth, C.C. Capen, W.W. Carlton, J.P. Sundberg and J.M. Ward), pp. 381–398. ILSI Press, Washington, DC.

Richardson, B.C. (2002). *J. Nutr.* **132**, 2401S–2405S.

Rittinghausen, S., Dungworth, D.L., Ernst, H. and Mohr, U. (1996). In *Pathobiology of the Aging Mouse*, Vol. 1 (eds U. Mohr, D.L. Dungworth, C.C. Capen, W.W. Carlton, J.P. Sundberg and J.M. Ward), pp. 301–314. ILSI Press, Washington.

Rudolph, K.L., Cang, S., Lee, H.W., Blasco, M., Gottlieb, G.J., Greider, C. and DePinho, R.A. (1999). *Cell* **96**, 701–712.

Schechter, J., Felicio, L.S., Nelson, F. and Finch, C.E. (1981). *Anat. Rec.* **199**, 423–432.

Schlessinger, D. and Van Zant, G. (2001). *Mech. Ageing Dev.* **122**, 1537–1553.

Schulz, N., Propst, F., Rosenberg, M.P., Linnoila, R.I., Paules, R.S., Kovatch, R., Ogiso, Y. and Vande Woude, G. (1992). *Cancer Res.* **52**, 450–455.

Scott, S.A. and Mandybur, T. I. (1996). In *Pathobiology of the Aging Mouse*, Vol. 2 (eds U. Mohr, D.L. Dungworth, C.C. Capen, W.W. Carlton, J.P. Sundberg and J.M. Ward), pp. 39–52. ILSI Press, Washington, DC.

Seely, J.C. (1996). In *Pathobiology of the Aging Mouse*, Vol. 2 (eds U. Mohr, D.L. Dungworth, C.C. Capen, W.W. Carlton, J.P. Sundberg and J.M. Ward), pp. 261–265. ILSI Press, Washington, DC.

Sharpless, N.E. and DePinho, R.A. (2002). *Cell* **110**, 9–12.

Sheldon, W.G., Bucci, T.J. and Turturro, A. (1996). In *Pathobiology of the Aging Mouse*, Vol. 2 (eds U. Mohr, D.L. Dungworth, C.C. Capen, W.W. Carlton, J.P. Sundberg and J.M. Ward), pp. 21–26, 445–453. ILSI Press, Washington, DC.

Shimada, A. (1999). *Neurobiol. Aging* **20**, 125–136.

Smith, G.H. and Boulanger, C.A. (2002). *Mech. Ageing Dev.* **123**, 1505–1519.

Smith, R.S., Chang, B., Nawes, N., Heckenlively, J.R., Roderick, T.H. and Sundberg, J.P. (1996). In *Pathobiology of the Aging Mouse*, Vol. 2 (eds U. Mohr, D.L. Dungworth, C.C. Capen, W.W. Carlton, J.P. Sundberg and J.M. Ward), pp. 125–130. ILSI Press, Washington, DC.

Sohal, R.S. and Weindruch, R. (1996). *Science* **273**, 59–63.

Sohal, R.S., Mockett, R.J. and Orr, W.C. (2002). *Free Radic. Biol. Med.* **33**, 575–586.

Sturrock, R.R. (1996). In *Pathobiology of the Aging Mouse*, Vol. 2 (eds U. Mohr, D.L. Dungworth, C.C. Capen, W.W. Carlton, J.P. Sundberg and J.M. Ward), pp. 3–37. ILSI Press, Washington, DC.

Sundberg, J.P., Sundberg, B.A. and King, Jr., L.E. (1996). In *Pathobiology of the Aging Mouse*, Vol. 2 (eds U. Mohr, D.L. Dungworth, C.C. Capen, W.W. Carlton, J.P. Sundberg and J.M. Ward), pp. 325–337. ILSI Press, Washington, DC.

Szakal, A.K., Aydar, Y., Balogh, P. and Tew, J.G. (2002). *Semin. Immunol.* **14**, 267–274.

Szweda, P.A., Friguet, B. and Szweda, L.I. (2002). *Free Rad. Biol. Med.* **33**, 29–36.

Takeda, T. (1999). *Neurobiol. Aging* **20**, 105–110.

Takeda, T., Shimizu, K. and Chen, W.-H. (1996). In *Pathobiology of the Aging Mouse*, Vol. 2 (eds U. Mohr, D.L. Dungworth, C.C. Capen, W.W. Carlton, J.P. Sundberg and J.M. Ward), pp. 455–471. ILSI Press, Washington, DC.

Tatar, M., Bartke, A. and Antebi, A. (2003). *Science* **299**, 1346–1351.

Tou, J.C.L. and Wade, C.E. (2002). *Exp. Biol. Med.* **227**, 587–600.

Tyner, S.D., Venkatachalam, S., Choi, J., Jones, S., Ghebranious, N., Igelmann, H., Lu, X., Soron, G., Cooper, B., Brayton, C., Park, S.H., Thompson, T., Karsenty, G., Bradley, A. and Donehower, L.A. (2002). *Nature* **415**, 45–53.

van Leuven, F. (2000). *Prog. Neurobiol.* **61**, 305–312.

Van Leeuwen, F.W., Gerez, L., Benne, R. and Hol, E.M. (2002). *Int. J. Biochem. Cell Biol.* **34**, 1502–0505.

Von Zglinicki, T. (2000). *Ann. NY Acad. Sci.* **908**, 99–110.

Wakefield, L.M., Thordarson, G., Nieto, A.I., Shyamala, G., Galvez, J.J., Anver, M.R. and Cardiff, R.D. (2003). *Comp. Med* **53**, 424–432.

Ward, J.R., Anver, M.R., Mahler, J.F. and Devor-Henneman, D.E. (2000). In *Pathology of Genetically Engineered Mice* (eds J.M. Ward, R.R. Maranpot and J.P. Sundberg), pp. 161–179. Iowa State University Press, Ames.

Ward, J.M., Yoon, M., Anver, M.R., Haines, D.C., Kudo, G., Gonzalez, F.J. and Kimura, S. (2001). *Am. J. Pathol.* **158**, 323–332.

Weksler, M.E., Goodhardt, M. and Szabo, P. (2002). *Springer Semin. Immunopathol.* **24**, 35–52.

Wei, Y.-H. and Lee, H.-C. (2002). *Exp. Biol. Med.* **227**, 671–682.

Weindruch, R. and Walford, R.L. (1988). *The Retardation of Aging and Disease by Dietary Restriction.* CC Thomas, Springfield, IL.

Wiendruch, R., Kayo, T., Lee, C.-K. and Prolla, T.A. (2001). *J. Nutr.* **131**, 918S–923S.

Wijnands, M.V.W., Kuper, C.F., Schuurman, H.-J. and Woutersen, R.A. (1996). In *Pathobiology of the Aging Mouse*, Vol. 1 (eds U. Mohr, D.L. Dungworth, C.C. Capen, W.W. Carlton, J.P. Sundberg and J.M. Ward), pp. 205–218. ILSI Press, Washington, DC.

Willott, J.F. (1996). In *Pathobiology of the Aging Mouse*, Vol. 2 (eds U. Mohr, D.L. Dungworth, C.C. Capen, W.W. Carlton, J.P. Sundberg and J.M. Ward), pp. 179–204. ILSI Press, Washington, DC.

Wolf, D.C. and Hard, G.C. (1996). In *Pathobiology of the Aging Mouse*, Vol. 1 (eds U. Mohr, D.L. Dungworth, C.C. Capen, W.W. Carlton, J.P. Sundberg and J.M. Ward), pp. 331–344. ILSI Press, Washington, DC.

Zurcher, C., van Zwieten, M.J., Solleveld, H.A. and Hollander, C.F. (1982). In *The Mouse in Biomedical Research, Volume IV, Experimental Biology and Oncology* (eds H.L. Foster, J.D. Small and J.G. Fox), pp. 11–33, Academic Press, New York, NY.

Diversity of Spontaneous Neoplasms in Commonly Used Inbred Strains and Stocks of Laboratory Mice

Igor Mikaelian
The Jackson Laboratory, Bar Harbor, Maine, USA

Tsutomu Ichiki
The Jackson Laboratory, Bar Harbor, Maine, USA
Dainippon Ink and Chemicals, Inc., Chiba, Japan

Jerrold M Ward
The National Cancer Institute, Fredrick, Maryland, USA

John P Sundberg
The Jackson Laboratory, Bar Harbor, Maine, USA

Introduction

Humans have maintained and domesticated many species over the centuries for food and labor. Animals have also been maintained as companion animals or, in the case of rodents, often as curiosities. Mice with spontaneous mutations that resulted in dramatic physical changes from normal, what we now call phenotypic deviants, were particularly prized. A notable example is the rhinoceros mouse that lost all hair with age and developed prominent wrinkling (Gaskoin, 1856). An extension of these observations was the understanding that these animals, particularly rodents, were useful as tools, biological models, for understanding similarities with human biology and especially disease. The Jackson Laboratory was founded in 1929 with the goal of using the laboratory mouse as a biomedical tool to

The Laboratory Mouse
Copyright 2004 Elsevier
ISBN 0-1233-6425-6

unlock the secrets of mammary cancer that led to the discovery that a filterable agent (the Bitner Agent), later determined to be a retrovirus (the mouse mammary tumor virus, MMTV), was the major cause of mammary cancer in some strains of mice (Staff of the Roscoe B Jackson Memorial Laboratory, 1933).

It was once thought to be impossible to inbreed animals. In spite of this dogma, Little and Tyzzer in the early 1900s succeeded and this began the process of developing a large variety of inbred strains in many laboratories around the world (Holstein, 1979). As these strains became large colonies, a variety of diseases appeared in some colonies but not in others. As husbandry conditions improved through the twentieth century, thereby eliminating most serious infectious diseases, background levels of cancer became more evident, particularly in aging studies. Background levels of cancer were important to understand because these lesions had to be differentiated from those caused by the experimental design of various studies (Mahler, 2000). The advent of genetic engineering has further emphasized the need to understand background diseases in strains used so as to be able to differentiate lesions induced by genetic manipulation from those that arise consistently in the commonly used strains (Booth and Sundberg, 1996; Mahler *et al.*, 1996; Husler *et al.*, 1998; Ward *et al.*, 2000). As mouse research has expanded exponentially in recent years, there is a great need to understand the spontaneous background diseases that occur in each strain that is commonly used so that experiments can be interpreted correctly.

Inbred mice are essentially identical except for sex because their genome is homogenous and stable. As we come to recognize that certain diseases only arise in some strains and not others, especially complex diseases such as cancer, the value of these mice as models to dissect the genetic bases of disease and their mechanisms becomes great. For example, rhabdomyosarcomas are rare malignant neoplasms of striated muscle that occur almost exclusively in BALB/cJ, BALB/cByJ, A/J, and C58/J strains (Sundberg *et al.*, 1991). It is now possible, with modern gene mapping tools, to dissect the complex polygenic nature of this neoplasm since genetic diversity between individuals is not an issue with such a model. Such studies are currently in progress. This approach has been done with the juvenile ovarian granulosa cell tumor model using recombinant inbred approaches (Beamer *et al.*, 1998).

This chapter reviews archival records of diagnostic work done on the most commonly used inbred strains of laboratory mice. The focus is on the most important

strains currently used in research today as defined by panels of experts who advise the Mouse Tumor Biology (http://tumor.informatics.jax.org/FMPro?-db = TumorInstance&-format=mtdp.html&-view) and Mouse Phenome (http://aretha.jax.org/pub-cgi/phenome/mpd-cgi?rtn=docs/home) databases. Annotated color images of specific mouse neoplasms or non neoplastic diseases with extensive bibliographies are available from these databases and others as provided in the list of web sites at the end of this chapter.

Source materials

The Jackson Laboratory is a private, nonprofit research institution that focuses on mammalian genetics using the laboratory mouse as a model. This institution was founded in 1929 and has evolved into one of the largest producers and distributors of inbred, **congenic**, recombinant inbred, and mutant (both spontaneous and induced by various methods) mice in the world. During 1986 we developed a retrievable relational database to store medical records (Sundberg and Sundberg, 1990, 2000). This system was developed to keep track of diagnostic necropsies in a disease surveillance program aimed at early detection of potential infectious disease outbreaks. Routine histopathology of neoplasms was recorded with history and individual data (signalment). Records from 1987 through 2000 were searched and compiled by organ system to generate Table 22.1.

Mice from this series were obtained from two primary sources. Production colonies were the largest source of mice with spontaneous neoplasms, and ranged in size from a few breeder pairs to thousands of trio matings (two females with one male) for the high demand colonies. Mice within these colonies are raised through the age of high reproductive performance (high fecundity period) which varies with each inbred strain (Green and Witham, 1997). Age of rotation (age at which fecundity drops for a particular strain) dictates the mean age of the colony, which is provided as the mean age for the mice diagnosed with neoplasia in this study (Green and Witham, 1997). It ranges from a low in the C3H/HeJ strain of 5 months due to high frequency of ovarian neoplasia and cysts to a high of 12 months in BALB strains. Mice were selected by animal care technicians using criteria established to **triage** mice into groups of potential infectious diseases or neoplasia (so-called 'sick mice'), spontaneous mutations, or those to be culled due to common changes such as overgrowth of

TABLE 22.1: Relative frequency of tumors in nine inbred strains of mice and one substrain

System	Organ	Tumor	C57BL6/J Fᵇ	C57BL6/J Mᵃ	129P3/J F	129P3/J M	A/J F	A/J M	BALB/cByJ F	BALB/cByJ M	BALB/cJ F	BALB/cJ M	C3H/HeJ F	C3H/HeJ M	CBA/J F	CBA/J M	DBA/2J F	DBA/2J M	HRS/J F	HRS/J M	SJL/J F	SJL/J M
			1644ᵇ	1108	100	138	281	236	516	256	899	349	765	580	395	247	184	168	202	236	557	423
			143ᶜ	140	148	113	161	136	119	142	179	157	191	152	119	114	112	137	183	210	159	148
Digestive	Liver	Hepatocellular adenoma											525	543	227							
		Hepatocellular carcinoma												332								
Endocrine	Adrenal	Adenoma	260ᵈ																			
	Pituitary gland	Hyperplasia, nodular											555									
		Adenoma, pars intermedia																				
Lymphopoietic		Lymphoma	218	201			187	330	168	168	157	139					156	168	238	249	197	235
		Lymphoproliferative disease								87											306	265
		Histiocytic sarcoma																			306	265
		Plasmacytoma																			240	259
Musculoskeletal	Skeletal muscles	Rhabdomyosarcoma					256	399	198	148	141	185										
	Skeleton	Osteosarcoma	238	265																		
Nervous	Choroid plexuses	Choroid plexus papilloma		354																		
Reproductive	Ovary	Adenoma											550									
		Cystadenoma											544									
		Luteoma											495									
	Testes	Leydig cell tumor										229										
		Teratoma				58																
Respiratory	Lung	Bronchoalveolar adenoma					215	256	284	258	241											
Skin and adnexae	Haired skin	Basal cell carcinoma								150												
		Fibropapilloma																				
		Papilloma							247	457	227	213										

TABLE 22.1: (Continued)

System	Organ	Tumor	C57BL6/J Fᵃ	C57BL6/J Mᵃ	129P3/J F	129P3/J M	A/J F	A/J M	BALB/cByJ F	BALB/cByJ M	BALB/cJ F	BALB/cJ M	C3H/HeJ F	C3H/HeJ M	CBA/J F	CBA/J M	DBA/2J F	DBA/2J M	HRS/J F	HRS/J M	SJL/J F	SJL/J M
			1644[b]	1108	100	138	281	236	516	256	899	349	765	580	395	247	184	168	202	236	557	423
			143[c]	140	148	113	161	136	119	142	179	157	191	152	119	114	112	137	183	210	159	148
		Pilomatrixoma							274													
		Squamous cell carcinoma					122		248										251			
		Trichoepithelioma							349													
	Mammary gland	Adenocarcinoma/ Carcinoma	238						253		247		214				224					218
	Subcutis	Hemangioma							218													
		Hemangiosarcoma									192	245										
		Lipoma					202															
		Lipomatous hamartoma	86	47									29	131								
		Myoepithelioma					191	196	224	225	216	246		170			216					
		Sarcoma		142																		
		Cumulative frequency per gender[e,f]	4.4	4.3	7.0	51.4	17.8	13.1	14.1	10.9	38.7	28.9	29.7	8.8	1.3	1.6	7.6	4.2	52.5	55.1	18.0	14.2
		Cumulative frequency per strain[e,f]	4.3		32.8		15.7		13.1		36.0		20.7		1.4		6.0		54.0		16.3	

The values were color-coded for easy comparison between groups. Frequencies of tumors of the reproductive tract and of the mammary gland were adjusted for the sex. Unstained boxes indicate a frequency of less than 0.5%. Pale gray ( ) indicates a frequency of 0.5–1%. Dark gray ( ) indicates a frequency of 1.01–10%. Black ( ) indicates a frequency >10%.

[a] F: female; M: male.

[b] Total number of mice examined per each group.

[c] Mean age of examined animals (with and without tumors) per each sex in each strain and substrain.

[d] Mean age for animal with specific types of cancer.

[e] In %; includes tumors not listed.

[f] Lipomatous hamartomas are not included.

incisors (T. Cunliffe-Beamer, unpublished criteria). The second source of mice in this study was from research colonies. Mice that developed untoward results in an experiment or spontaneous disease were submitted. All spontaneous mutant mice and genetically engineered mice were eliminated from this study.

Mice meeting the 'sick mouse' criteria were euthanized by CO_2 asphyxiation using AVMA approved methods for euthanasia and necropsied (Relyea et al., 2000).

Grossly evident neoplasms were removed, trimmed for adequate fixation, and fixed by immersion in Fekete's acid–alcohol–formalin solution. Lungs from all mice with suspected neoplasms were inflated with fixative and then immersed. After overnight fixation tissues were trimmed, processed routinely, embedded in paraffin, sectioned at 6 μm, and stained with hematoxylin and eosin (H&E) or other special stains as needed to arrive at a diagnosis. Many neoplasms that arose regularly in particular strains were also fixed in cacodylate or phosphate buffered glutaraldehyde for transmission electron microscopy, snap frozen, or prepared by other means for a variety of specialized diagnostic workups as described elsewhere (Relyea et al., 2000; Seymour et al., in press).

Data generated through necropsies performed at The Jackson Laboratory convey information on tumors developing during the first half of the life span of an inbred laboratory mouse. Data for tumors occurring in aging mice have been reported elsewhere. Summaries are available on line (http://tumor.informatics.jax.org/straingrid.html).

The diagnoses were abbreviated to reduce the size of the table and allow comparisons. For example, all lymphomas, regardless of the phenotypic subtype, were categorized as 'lymphoma'. All myoepitheliomas, regardless their precise anatomical origin, were clustered in a single diagnosis since the precise origin of many of these tumors could not be determined. All malignant mammary tumors were categorized as mammary adenocarcinoma/carcinoma.

Strains

The Mouse Phenome Project has defined the inbred strains of greatest importance in modern biomedical research (http://aretha.jax.org/pub-cgi/phenome/mpd-cgi?rtn=strains/list). These represent not only inbred strains used for a variety of basic research projects but also those that are currently used for **transgenesis** (primarily C57BL/6 and FVB/N) or targeted mutagenesis (primarily various substrains of 129 and C57BL/6) experiments. The Jackson Laboratory maintains most of these strains but many are very small colonies so spontaneous neoplasia frequency data are limited. Due to space limitations, we limited the data presented here to nine strains and one substrain (Table 22.1). Mice from these strains and this substrain account for 36.7% (9280/25264) necropsy accessions for the study period. These are six Castle's mouse-related strains (129P3/J, A/J, BALB/cJ, C3H/HeJ, CBA/J, and DBA/2J), one Castle's mouse-related substrain (BALB/cByJ), one C57 mouse-related strain (C57BL/6J), and two Swiss mouse-related strains (HRS/J and SJL/J) (Beck et al., 2000).

Information for some other strains (AKR/J, C57BLKS/J, C57L/J, C58/J, CE/J, FVB/NJ, LP/J, MOLF/Ei, NOD/LtJ, NON/LtJ, NZB/B1NJ, PL/J, RF/J, SM/J, STX/LeJ, SWL/J), substrains (129/SvJ, A/HeJ, BUB/BnJ, C3H/HeOuJ, C3HeB/FeJ, C57BL/10J, C57BL/10SnJ, C57BR/cdJ, CBA/CaJ, I/LnJ, NZW/LacJ, RBF/DnJ, SEA/GnJ), and mutants are available on the Mouse Tumor Biology Database (http://tumor.informatics.jax.org/pathweb/path.html). This database also provides a comprehensive selection of annotated photomicrographs of slides stained with hematoxylin eosin, special stains, or immunohistochemistry as well as some electron micrographs with links to extensive references.

Results/discussion

Due to space limitation, data in Table 22.1 only include those tumors diagnosed at least twice in a strain with a frequency of 0.50% or more. Rarer tumors are interpreted as non-strain related tumors. Frequency is defined as the proportion of mice with tumors in animals submitted for necropsy.

The frequency of tumors differed greatly among strains, ranging from 1.4% in CBA/J mice to 35.7% in BALB/cJ mice. Frequency represents the percentage of cases with a specific diagnosis of all mice submitted for necropsy of that strain between 1987 and 2000. Retroviral infection accounts for high frequencies of tumors such as mammary tumors and lymphomas in some strains, especially BALB/cJ and C3H/HeJ strains (Vaage et al., 1986). Interestingly, consistent with a higher ability of Mouse Mammary Tumor Virus

(MMTV) to cause *int-1* gene rearrangement in C3H mice than in BALB/c mice (Marchetti *et al.*, 1991), the frequency of mammary tumors was higher in C3H mice (24.3%) than in BALB/c mice (7.9%). Alternatively, infection by MMTV strains with a lower oncogenic potential may account for a lower frequency of mammary tumors in BALB/c mice compared with C3H mice (Drohan *et al.*, 1981).

The overall frequency of tumors in this study is lower than generally reported in the literature (Table 22.2; Ward *et al.*, 2000; Haines *et al.*, 2001). This low frequency partly reflects the young age of mice in this study. It also reflects the selection procedure because sampling was limited to mice with clinical signs of disease or with the presence of an easily detectable mass. Thus, some mice with internal malignancies at an early stage of development were missed.

Profound differences segregate with some strains based on their development. C57-related strains in this study have a low overall frequency of cancer. A high frequency of pituitary tumors has been reported in female C57BL/6 (Felicio *et al.*, 1980). However, pituitary tumors in C57BL/6 develop mostly in the second half of life, an age group that was not available for this study. A low to intermediate frequency of lymphopoietic malignancies has been reported in C57BL strains. The low frequency of these tumors in this study may reflect the young age of mice.

SJL/J strain showed a high frequency of tumors, especially tumors of the lymphohemopoietic system, including lymphoma and plasmacytoma. SJL/J mice also have a high frequency of lymphoproliferative disease, which is interpreted as a preneoplastic process of the lymphocytic lineage (Ponzio, 1983; Tang, Ho *et al.*, 1998a,b). These findings are consistent with those expected for strains derived from the Swiss mouse strain.

Tumor frequencies greatly differed among strains derived from Castle's mice. DBA/J and CBA/J strains had low tumor frequency. In contrast, C3H/HeJ and BALB/cJ strains had high tumor frequency. Genetic selection and drift may account for differences in tumor frequency among these strains because most of these strains have been created between 1910 and 1935. Interestingly, the overall tumor frequency greatly differed between the BALB/cByJ substrain and the parental strain BALB/cJ, although the former was derived in the late 1970s from the parental strain. This is particularly true for myoepitheliomas and mammary adenocarcinomas, with frequencies of 2.6% and 19.2%, and 0.6 and 5.3% for BALB/cByJ and BALB/cJ, respectively. These

differences are statistically significant ($P < 0.001$, Chi square test). Differences in cancer frequency between these substrains may be the result of different immune-responses of these strains to cancer cells because these substrains display different immune-responses to viral infection (Babu *et al.*, 1986; Nicholson *et al.*, 1994).

Data for Castle's-derived strains overall fitted those expected from the literature: AKR/J strain (data not shown) had a high frequency of lymphoma, with a lower frequency in BALB/cByJ, BALB/cJ, and DBA/2J. Mammary adenocarcinomas were most common in C3H/HeJ and BALB/cJ strains, and this high frequency is likely to be the result of **enzootic** infection by MMTV. Mouse mammary tumor virus has now been eradicated from our colonies.

Finally, high frequency of myoepithelioma was detected in BALB/cJ and A/J mice, with lower frequency in the BALB/cByJ substrain. BALB/c and A strains were developed from Castle's strains between the 1920s and the 1940s. Myoepitheliomas metastasize more frequently in A/J than in BALB/cJ mice suggesting minor genetic changes between closely related inbred strains may have major effects on the biology of common cancers (Sundberg *et al.*, 1991).

Mouse cancer websites

Several mouse tumor databases have been developed over the past few years (Table 22.3). At the present time, their contents reflect the interests of their curators, and thus their scope remains limited. However, they provide room for more in-depth illustration than any textbook or regular paper in a scientific journal. Researchers are encouraged to contribute to these databases to increase the impact of their work.

Some of these databases are already linked to literature databases (such as PubMed and The Jackson Laboratory's Mouse Genome Informatics Database). Some also provide information about immunohistochemical methods with links to antibodies manufacturers. With time, these databases are expected to host some of the supplementary illustrations that cannot be published in regular scientific journals. Also, as technology becomes available, links will be developed with tumor gene databases.

TABLE 22.2: Reported relative frequency to tumors in BALB/c, C57BL/6, and FVB/N strains of mice

System	Organ	Tumor	BALB/c (Frith et al., 1983) >18 mo[a]		C57BL/6N (Ward et al., 1975) 24 mo	C57BL/6J (Frith et al., 1983) >18 mo		C57BL/6 (Blackwell et al., 1995) <24 mo		C57BL/6 (Blackwell et al., 1995) >24 mo		FVB/N (Mahler et al., 1996) 14 mo		FVB/N (Mahler et al., 1996) 24 mo	
			M[b]	F[b]	M	M	F	M	F	M	F	M	F	M	F
Digestive	Liver	Hepatocellular adenoma or carcinoma	10/248 (4.0)[c]	4/353 (1.1)	5/136 (13.9)	13/281 (4.6)	5/331 (1.5)	10/83 (12.0)	3/75 (4.0)	10/78 (12.8)	1/42 (2.3)			6/29 (21)	
	Forestomach	Papilloma			4/36 (11.1)										
	Intestine	Polyp/tumor	6/273 (2.2)	4/347 (1.2)		2/247 (0.8)	5/337 (1.5)								
Endocrine	Adrenal	Adenoma	9/191 (4.7)	17/312 (5.4)		1/215 (0.5)									
		Pheochromocytoma													4/71 (6)
	Pituitary gland	Adenoma			1/36 (2.8)	2/205 (1.0)	30/305 (9.8)	3/83 (3.6)	17/75 (22.7)		32/42 (76.2)		5/98 (5)		10/71 (14)
	Thyroid	Tumor							3/75 (4.0)	1/78 (1.3)	7/42 (16.7)				
Lymphopoietic		Lymphoma	34/250 (13.6)	90/352 (25.6)	7/36 (19.4)	51/286 (17.8)	83/358 (23.2)	9/83 (3.6)	24/75 (32.0)	3/78 (3.8)	10/42 (23.8)				4/71 (6)
		Histiocytic sarcoma	1/250 (0.4)	1/352 (0.28)	3/36 (8.3)	6/288 (2.1)	14/358 (3.9)	34/83 (41.0)	19/75 (25.3)	55/78 (70.5)	18/42 (42.9)		1/98 (1)		4/71 (6)
Reproductive	Ovary	Tumors (combined)		8/328 (2.4)			8/318 (2.5)						2/98 (2)		5/71 (7)

TABLE 22.2: (Continued)

System	Organ	Tumor	BALB/c (Frith et al., 1983) >18 mo[a] M[b]	F[b]	C57BL/6N (Ward et al., 1975) 24 mo M	C57BL/6J (Frith et al., 1983) >18 mo M	F	C57BL/6 (Blackwell et al., 1995) <24 mo M	F	C57BL/6 (Blackwell et al., 1995) >24 mo M	F	FVB/N (Mahler et al., 1996) 14 mo M	F	FVB/N 24 mo M	F
	Testis	Leydig cell tumor	4/242 (1.7)												
	Uterus	Tumor							3/75 (4.0)		3/42 (7.1)				
Respiratory	Lung	Adenoma/carcinoma	104/249 (41.8)	100/346 (28.9)	6/36 (17)	14/280 (5.0)	6/352 (1.7)	3/83 (3.6)	2/75 (2.7)	4/78 (5.1)	3/42 (7.1)	6/45 (14)	14/98 (14)	12/29 (41)	26/71 (37)
Senses	Eye and adnexae	Harderian gland tumor	22/244 (9.0)	44/342 (12.9)		8/271 (2.9)	12/338 (3.6)	2/83 (2.4)	1/75 (1.3)	5/78 (6.4)				2/29 (7)	4/71 (6)
Skin and adnexae	Mammary gland	Carcinoma													
	Subcutis	Neural crest tumor											1/98 (1)	3/29 (10)	2/71 (3)
		Myoepithelioma		3/372 (0.8)											
Urinary	Kidney	Adenoma/carcinoma	4/247 (1.6)												
	Multiple	Hemangioma/hemangiosarcoma	5/250 (2.0)	19/352 (5.4)		15/288 (5.2)	7/220 (3.2)	4/83 (4.8)	3/75 (4.0)	6/78 (7.7)					

[a]Age in months.
[b]Gender: M: male; F: female.
[c]Affected/examined (percentage).

TABLE 22.3: Mouse tumor databases accessible on the Web	
Host	**Address**
National Cancer Institute	http://histology.nih.gov/
The Jackson Laboratory	http://tumor.informatics.jax.org/
Registry Nomenclature Information System (RENI; access restricted to subscribers)	http://www.ita.fhg.de/reni/i
European Late Effects Project	http://www.eulep.org/Pathbase/

Conclusions

Most reported tumor rates in mice are based on 2-year aging experiments. This study provides tumor frequencies for mice during their first year of life. Data for additional strains and substrains are available on the web site of the Mouse Tumor Biology Database. Knowledge of tumor rates in young mice is crucial to the interpretation of most studies since most of them involve mice in their first months of life.

Tumor frequency greatly differs among strains, possibly as a result of strain specific genetic differences. Scientists involved with mouse research need to be aware of these differences before initiating their research.

In addition to the strain, numerous factors affect the rate of cancer in mice. These factors include stress, temperature, bedding, caging density, altitude, diet, laboratory practices, and time period (Gopinath, 1994). Some of these factors may evolve with time as the diet and the husbandry practices evolve to reflect the advance of regulations. Hence, spontaneous tumor rates are expected to evolve with time. Genetic drift may also alter cancer rates. The Jackson Laboratory and other repositories play a key role in preventing the genetic drift of inbred strains of mice.

Acknowledgments

This work was supported in parts by grants from the National Institutes of Health (RR173, CA34196, CA89713) and Transgenics Inc.

The authors thank B.A. Sundberg, J. Miller, L. Davis, K.S. Brown, and M.J. Relyea for their technical assistance.

References

Babu, P.G., Huber, S.A. *et al.* (1986). *Am. J. Pathol.* **124**, 193–198.

Beamer, W.G., Shultz, K.L. *et al.* (1998). *Toxicol. Pathol.* **26**, 704–710.

Beck, J.A., Lloyd, S. *et al.* (2000). *Nat. Genet.* **24**(1), 23–25.

Blackwell, B.N., Bucci, T.J. *et al.* (1995). *Toxicol. Pathol.* **23**(5), 570–582.

Booth, C. and Sundberg, J.P. (1996). In *Pathobiology of the Aging Mouse*, Volume 1 (eds U. Mohr, D.L. Dungworth, C. C. Capen *et al.*) pp.51-65. 1ILSI Press, Washington, DC.

Drohan, W., Teramoto, Y.A. *et al.* (1981). *Virology* **114**, 175–186.

Felicio, L.S., Nelson, J.F. *et al.* (1980). *Exp. Gerontol.* **15**(2), 139–143.

Frith, C.H., Highman, B. *et al.* (1983). *Lab. Anim. Sci.* **33**, 273–286.

Gaskoin, J.S. (1856). *Proc. Zool. Soc. London* **24**, 38–40.

Gopinath, C. (1994). *Toxicol. Pathol.* **22**(2), 160–164.

Green, M.C. and Witham, B.A. (1997). *Handbook of Genetically Standardized Jax Mice*. The Jackson Laboratory, Bar Harbor, ME.

Haines, D.C., Chattopadhyay, S. *et al.* (2001). *Toxicol. Pathol.* **29**(6), 653–661.

Holstein, J. (1979). *The First Fifty Years at the Jackson Laboratory*. The Jackson Laboratory, Bar Harbor.

Husler, M.R., Beamer, W.G. *et al.* (1998). *J. Exp. Anim. Sci.* **38**, 165–180.

Mahler, J.F. (2000). In *Pathology of Genetically Engineered Mice*. (eds J.M. Ward, J.F. Mahler, R.R. Maronpot and J.P. Sundberg), pp. 137–144. Iowa State University Press, Ames.

Mahler, J.F., Stokes, W. *et al.* (1996). *Toxicol. Pathol.* **24**, 710–716.

Marchetti, A., Robbins, J. *et al.* (1991). *J. Virol.* **65**(8), 4550–4554.

Nicholson, S.M., Peterson, J.D. *et al.* (1994). *J. Neuroimmunol.* **52**, 19–24.

Ponzio, N.M. (1983). *Behring Inst. Mitt.* **72**, 28–36.

Relyea, M.J., Miller, J. *et al.* (2000). In *Systematic Approach to Evaluation of Mouse Mutations.* (eds J.P. Sundberg and D. Boggess), pp. 57–90. CRC Press, Boca Raton.

Seymour, R., Ichiki, T. *et al.* (2004). In *Laboratory Mouse* (ed. H.J. Hedrich), pp. 495–516. Elsevier Science, London.

Staff of the Roscoe B Jackson Memorial Laboratory (1933). *Science* **78**, 465–466.

Sundberg, B.A. and Sundberg, J.P. (1990). *Lab. Anim.* **19**, 55–58.

Sundberg, B.A. and Sundberg, J.P. (2000). In *Systematic Approach to Evaluation of Mouse Mutations* (eds J.P. Sundberg and D. Boggess), pp. 47–55. CRC Press, Boca Raton.

Sundberg, J.P., Adkison, D.L. *et al.* (1991). *Vet. Pathol.* **28**, 200–206.

Sundberg, J.P., Hanson, C.A. *et al.* (1991). *Vet. Pathol.* **28**, 313–323.

Tang, J.C., Ho, F.C. *et al.* (1998a). *Lab. Invest.* **78**, 1459–1466.

Tang, J.C., Ho, F.C. *et al.* (1998b). *Lab. Invest.* **78**, 205–212.

Vaage, J., Smith, G.H. *et al.* (1986). *Cancer Res.* **46**, 2096–2100.

Ward, J.M., Weisburger, J.H. *et al.* (1975). *Arch. Environ. Health* **30**, 22–25.

Ward, J.M., Anver, M.R. *et al.* (2000). In *Pathology of Genetically Engineered Mice* (eds J.M. Ward, J.F. Mahler, R.R. Maronpot and J.P. Sundberg), pp. 161–179. Iowa State University Press, Ames.

4

Infectious Agents and Diseases

Content

CHAPTER 23

Viral Infections

Cory Brayton
Center for Comparative Medicine, Baylor College of
Medicine, Houston, Texas, USA

Michael Mähler
Biomedical Diagnostics – BioDoc, Hannover, Germany

Werner Nicklas
Central Animal Laboratories, German Cancer
Research Center, Heidelberg, Germany

Introduction

In interpreting the microbiological status of laboratory animals, it must be understood that infection and disease are not synonymous. Infection refers to the invasion and multiplication of microorganisms in body tissues and may occur with or without apparent disease. Disease refers to interruption or deviation from normal structure and function of any tissue, organ, or system. Many of the infections with which we are concerned may not cause discernable disease in many strains of mice. However, they may cause inapparent or subclinical changes that can interfere with research. Such interference often remains undetected, and therefore modified results may be obtained and published.

The types of interference of an agent with experimental results may be diverse. There is no doubt that research complications due to overt infectious disease are significant and that animals with clinical signs of disease should not be used for scientific experiments. But also clinically inapparent infections may have severe effects on animal experiments. There are numerous examples of influences of microorganisms on host physiology and hence of the interference of inapparent infections with the results of animal experiments. Many microorganisms have the potential to induce activation or suppression of the immune system or both at the same time but on different parts of the immune system, regardless of the level of pathogenicity. All infections, apparent or inapparent, are likely to increase inter-individual variability and hence result in increased numbers of animals necessary to obtain reliable results. Microorganisms, in particular viruses, present in an animal may contaminate biological materials such as sera, cells, or tumours (Collins and Parker, 1972; Nicklas *et al.*, 1993). This may interfere with *in vitro* experiments conducted with such materials and may also lead to contamination of animals (Lipman *et al.*, 2000b). Mouse antibody production (MAP) testing or polymerase chain reaction (PCR) testing of biologics to be inoculated into mice is an important component of a disease prevention programme. Finally, latent infections may be activated by environmental factors, by experimental procedures, or by the combination and interaction between various microorganisms. For all these reasons, prevention of infection, not merely prevention of clinical disease, is essential.

Unfortunately, research complications due to infectious agents are usually considered artefacts and published only exceptionally. Information on influences of microorganisms on experiments is scattered in diverse scientific journals, and many articles are difficult to detect. To address this problem, several congresses were held on viral complications on research. The knowledge available was summarized in conference proceedings (Melby and Balk, 1983; Bhatt et al., 1986b; Hamm, 1986) and has later repeatedly been reviewed (Lussier, 1988; National Research Council, 1991; Baker, 1998; Nicklas et al., 1999).

This chapter covers only viral infections of laboratory mice. Viral infections of mice have been studied in detail, and comprehensive information on their pathogenic potential, their impact on research, and the influence of host factors such as age, genotype, and immune status on the response to infection is available. Bacterial agents may be similarly important, but with few exceptions (e.g. *Helicobacter* species) little is known about their potential to influence host physiology and experiments. Even less is known about most parasites in this regard. Among fungal agents, only *Pneumocystis carinii* can be expected to play a significant role in contemporary mouse colonies. The nomenclature and taxonomy of viruses described are based on recent nomenclature rules by the International Union of Microbiological Societies (2000) and the Universal Virus Database of the International Committee on the Taxonomy of Viruses (http://www.ictvdb.iacr.ac.uk). Retroviruses are not covered in this chapter because they are not included in routine health surveillance programmes and cannot be eradicated with presently available methods. This is because most of them are incorporated in the mouse genome as proviruses and thus are transmitted via germline.

The ability to accurately determine whether or not laboratory animals or animal populations have been infected with virus depends on the specificity and sensitivity of the detection methods used. Most viral infections in immunocompetent mice are acute or short-term, and lesions are often subtle or subclinical. The absence of clinical disease and pathological changes has therefore only limited diagnostic value. However, clinical signs, altered behaviour, or lesions may be the first indicator of an infection and often provide clues for further investigations.

Serology is the primary means of testing mouse colonies for exposure to viruses, largely because serological tests are sensitive and specific, are relatively inexpensive, and allow screening for a multitude of agents with one serum sample. They are also employed to monitor biological materials for viral contamination using the MAP test. Serological tests detect specific antibodies, usually immunoglobulin G (IgG), produced by the host against the virus and do not actually test for the presence of virus. An animal may have been infected, mounted an effective antibody response, and cleared the virus, but remains seropositive for weeks or months or forever, even though it is no longer infected or shedding the agent. Active infection can only be detected by using direct diagnostic methods such as virus isolation, electron microscopy, or PCR. Meanwhile, PCR assays have been established for the detection of almost every agent of interest. They are highly sensitive and depending on the demands, they can be designed to broadly detect all members of a genus or only one species. However, good timing and selection of the appropriate specimen is critical for establishing the diagnosis. In practice, combinations of diagnostic tests are often necessary including the use of sentinel animals or immunosuppression to get clear aetiological results or to avoid consequences from false-positive results.

Reports on the prevalence of viral infections in laboratory mice throughout the world have been published frequently. In general, the microbiological quality of laboratory mice has constantly improved during the last decades, and several agents (e.g. herpes- and polyomaviruses) have been essentially eliminated from contemporary colonies due to advances in diagnostic methodologies and modern husbandry and rederivation practices (Jacoby and Lindsey, 1998; Zenner and Regnault, 2000; Livingston and Riley, 2003). They may, however, reappear, since most have been retained or are still being used experimentally. Furthermore, the general trend towards better microbiological quality is challenged by the increasing reliance of biomedical research on genetically modified and immunodeficient mice, whose responses to infection and disease can be unpredictable. Increasing numbers of scientists are creating genetically modified mice, with minimal or no awareness of infectious disease issues. As a consequence, they are more frequently infected than 'standard' strains of mice coming from commercial breeders, and available information on their health status is often insufficient. Frequently, they are exchanged between laboratories, which amplifies the risk of introducing infections from a range of animal facilities. Breeding cessation strategies that have been reported to eliminate viruses from immunocompetent mouse colonies may prove to be costly and ineffective in genetically modified colonies of uncertain or incompetent immune status. It must also be expected that new agents will be detected, although only occasionally. Infections therefore remain a threat to biomedical research, and users of laboratory mice must be cognizant of infectious agents and the complications they can cause.

DNA viruses

Herpesviruses

Two members of the family Herpesviridae can infect mice (*Mus musculus*). Mouse cytomegalovirus 1 (MCMV-1) or murid herpesvirus 1 (MuHV-1) belongs to the subfamily Betaherpesvirinae, genus Muromegalovirus. Murid herpesvirus 3 (MuHV-3) or mouse thymic virus (MTV) has not yet been assigned to a genus within the family Herpesviridae. Both viruses are enveloped, double-stranded DNA viruses that are highly host-specific and relatively unstable to environmental conditions such as heat and acidic pH. Both agents are antigenically distinct and do not cross-react in serological tests, but their epidemiology is similar (Cross *et al.*, 1979).

Seropositivity to MCMV-1 was reported in less than 5% of specified pathogen-free (SPF) mouse colonies in the USA in 1996 (Jacoby and Lindsey, 1998), and some institutions reported to have mice 'on campus' that were positive for MTV. In a more recent study, a low rate (0.1%) of samples was found to be positive for MCMV-1 whereas no sample tested positive for MTV (Livingston and Riley, 2003). The data available suggest that the prevalence of both viruses in contemporary colonies and thus their importance for laboratory mice is negligible. However, both MCMV-1 and MTV are frequently found in wild mice, which may be coinfected with both viruses (National Research Council, 1991; Singleton *et al.*, 2000).

MCMV-1 or MuHV-1

Natural infection with MCMV-1 causes subclinical salivary gland infection in mice. The virus persists in the salivary glands (particularly in the submaxillary glands) and also in other organs (Osborn, 1982; Kercher and Mitchell, 2002; Lenzo *et al.*, 2002).

Most information concerning the pathogenesis of MCMV-1 infection is based on experimental infection studies. These results are very difficult to summarize because the outcome of experimental infection in laboratory mice depends on various factors such as mouse strain and age, virus strain and passage history, virus dose, and route of inoculation (Osborn, 1982). In general, newborn mice are most susceptible to clinical disease and to lethal infection. Virus replication is observed in newborn mice in many tissues (for details, see Osborn, 1982) and appears in the salivary glands towards the end of the first week of infection when virus concentrations in liver and spleen have already declined. Resistance develops rapidly after weaning between days 21 and 28 of age. Experimental infection of adult mice results in mortality only in susceptible strains and only if high doses are administered. Not even intravenous or intraperitoneal injections of adult mice usually produce signs of illness in resistant strains (Shanley *et al.*, 1993). Mice of the $H-2^b$ (e.g. C57BL/6) and $H-2^d$ (e.g. BALB/c) **haplotype** are more sensitive to experimental infection than are mice of the $H-2^k$ haplotype (e.g. C3H), which are approximately 10-fold more resistant to mortality than are those of the *b* or *d* haplotype (Osborn, 1986).

Subclinical or latent infections can be activated by immunosuppression (e.g. with cyclophosphamide or cortisone). Reactivation of MCMV-1 occurs also after implantation of latently infected salivary glands into $Prkdc^{scid}$ mice (Schmader *et al.*, 1995). Immunodeficient mice lacking functional T cells or natural killer (NK) cells, such as $Foxn1^{nu}$ and $Lyst^{bg}$ mice, are more susceptible than are immunocompetent animals. Experimental infection in $Prkdc^{scid}$ mice causes severe disease or is lethal, with necrosis in spleen, liver, and other organs, and multinucleate syncytia with inclusion bodies in the liver (Reynolds *et al.*, 1993). Similar to AIDS patients infected with human cytomegalovirus, athymic $Foxn1^{nu}$ mice experimentally infected with MCMV also develop adrenal necrosis (Shanley and Pesanti, 1986). The virus also replicates in the lungs leading to pneumonitis whereas in heterozygous ($Foxn1^{nu}/^{+}$) littermates replication and disease are not seen (Shanley *et al.*, 1997).

The most prominent histological finding of cytomegaloviruses is enlarged cells (cytomegaly) of salivary gland epithelium with eosinophilic nuclear and cytoplasmic inclusion bodies. The inclusion bodies contain viral material and occur in other organs such as liver, spleen, ovary, and pancreas (Osborn, 1982). Depending on inoculation route, dose, strain, and age of mice, experimental infections may result in inflammation or cytomegaly with inclusion bodies in a variety of tissues, pneumonitis, myocarditis, meningoencephalitis, or splenic necrosis in susceptible strains (National Research Council, 1991; Osborn, 1982; Percy and Barthold, 2001).

Virus is transmitted oronasally by direct contact and is excreted in saliva, tears, and urine for several months. Wild mice serve as a natural reservoir for infection. The virus is most frequently transmitted horizontally through mouse-to-mouse contact but does not easily spread between cages.

It is generally assumed that MCMV-1 has a very low prevalence in contemporary colonies of laboratory mice. The risk of introduction into facilities housing laboratory mice is very low if wild mice are strictly excluded. Monitoring is necessary if populations of laboratory mice may have been contaminated by contact with wild mice. As for other viruses, enzyme-linked immunosorbent assay (ELISA) and indirect immunofluorescence assay (IFA) are the most appropriate tests for detecting antibodies. As the virus persists, direct demonstration of MCMV-1 in infected mice is possible by PCR (Palmon et al., 2000) or by virus isolation using mouse embryo fibroblasts (3T3 cells).

Although MCMV-1 does not play a significant role as a natural pathogen of laboratory mice, it is frequently used as a model for human cytomegalovirus infection (Bolger et al., 1999). However, the virus is known to influence immune reactions in infected mice and may therefore have impact on immunological research (Osborn, 1986; National Research Council, 1991; Baker, 1998).

MTV or MuHV-3

Mouse thymic virus was detected during studies in which samples from mice were passaged in newborn mice. Unlike other herpesviruses, the virus can not yet be cultured in vitro and is propagated by intraperitoneal infection of newborn mice. The thymus is removed 7–10 days later, and thymus suspensions serve as virus material for further studies. The prevalence of MTV is believed to be low in laboratory mice, and for this reason and also due to the difficulties in virus production for serological assays, it is not included in many standard diagnostic or surveillance testing protocols. Limited data are available indicating that it is common in wild mice, and it is also found in laboratory mice (Osborn, 1982; Morse, 1987; National

Research Council, 1991). Further, MTV obviously represents a significant source of contamination of MCMV-1 (and vice versa) if virus is prepared from salivary glands since both viruses cause chronic or persistent salivary gland infections and can coinfect the same host.

All mouse strains are susceptible to infection, but natural or experimental infection of adult mice is subclinical. Gross lesions appear only in the thymus and only if experimental infection occurs at an age of less than about 5 days. Virus is present in the thymus but may also be found in the blood and in salivary glands of surviving animals. Salivary glands are the only site yielding positive virus isolations if animals are infected as adults. Mouse thymic virus also establishes a persistent infection in athymic Foxn1nu mice, but virus shedding is reduced compared to euthymic mice and virus recovery is possible only in a lower percentage of mice (Morse, 1988).

Pathological changes caused by MTV occur in the thymus, and reduced thymus mass due to necrosis in suckling mice is the most characteristic gross lesion (Percy and Barthold, 2001). Lymphoid necrosis also may occur in lymph nodes and spleen (Wood et al., 1981), with necrosis and recovery similar to that in the thymus. In mice infected during the first 3 days after birth, necrosis of thymus becomes evident within 3–5 days, and its size and weight are markedly reduced at day 12–14. Intranuclear inclusions may be present in thymocytes between days 10–14 post infection. The thymus and the affected peripheral tissues regenerate within 8 weeks after infection. Regardless of the age of mice at infection, a persistent infection is established in the salivary glands, and infected animals shed virus for life.

Several alterations of immune responses are associated with neonatal MTV infection. There is transient immunosuppression, attributable to lytic infection of T lymphocytes, but activity (e.g. response of spleen cells to T cell mitogens) returns to normal as the histological repair progresses (Wood et al., 1981). Selective depletion of CD4+ T cells by MTV results in autoimmune disease (Morse and Valinsky, 1989; Morse et al., 1999). Information about additional influences on the immune system is given by Osborn (1982), National Research Council (1991), and Baker (1998).

In experimentally infected newborn mice, oral and intraperitoneal infections similarly result in thymus necrosis, **seroconversion**, and virus shedding suggesting that the oral–nasal route is likely to be involved in natural transmission (Morse, 1989). The virus spreads to cage mates after long periods of contact. It is transmitted between mice kept in close contact, and transmissibility from cage to cage seems to be low. Mouse thymic virus is not transmitted to foetuses by the transplacental route, and intravenous infection of pregnant mice does not lead to congenital damage, impairment in size or development, or abortion (St-Pierre *et al.*, 1987).

Mouse thymic virus and MCMV-1 do not cross-react serologically (Cross *et al.*, 1979). Serological monitoring of mouse populations for antibodies to MTV is possible by IFA testing, which is commercially available; ELISA tests have also been established (Morse, 1990b). ELISA and complement fixation yield similar results (Lussier *et al.*, 1988). It must be noted that the immune response depends on the age at infection. Antibody responses are not detectable in mice infected as newborns whereas adult mice develop high titres that are detectable by serological testing. If neonatal infection is suspected, homogenates of salivary glands or other materials can be inoculated into pathogen-free newborn mice followed by gross and histological examination of thymus, lymph nodes, and spleens for lymphoid necrosis (Morse, 1987). Alternatives to the *in vivo* infectivity assay for detecting MTV in infected tissues include a competition ELISA (Prattis and Morse, 1990) and MAP testing, although this is slightly less sensitive than infectivity assays (Morse, 1990a).

Very little experience exists on eradication methods for MTV due to its low prevalence in contemporary mouse colonies. Methods that eliminate other herpesviruses likely will eliminate MTV. Procurement of animals of known negative MTV status is an appropriate strategy to prevent infection. Strict separation of laboratory mice from wild rodents is essential to avoid introduction into laboratory animal facilities.

Mousepox (ectromelia) virus

Mousepox (ectromelia) virus (ECTV) is a member of the genus Orthopoxvirus belonging to the family Poxviridae. It is antigenically and morphologically very similar to vaccinia virus and other orthopoxviruses. Poxviruses are the largest and most complex of all viruses with a diameter of 200 nm and a length of 250–300 nm. Mousepox (ectromelia) virus contains one molecule of double-stranded DNA with a total genome length of 185,000 nucleotides. It is the causative agent of mousepox, a generalized disease in mice. Experimental transmission to young rats (up to 30 days of age) is possible (Jandasek, 1968; Buller *et al.*, 1986).

The virus is resistant to desiccation, dry heat, and many disinfectants. It is not consistently inactivated in serum heated 30 min at 56°C (Lipman *et al.*, 2000b)

and persists for 26 weeks when maintained at 4°C in foetal bovine serum (Bhatt and Jacoby, 1987a). Effective disinfectants include vapour-phase formaldehyde, sodium hypochlorite, and iodophores (Small and New, 1981; National Research Council, 1991).

Historically, ECTV has been an extremely important natural pathogen of laboratory mice. The virus was widespread in mouse colonies worldwide and can still be found in several countries. Between 1950 and 1980 almost 40 individual ectromelia outbreaks were reported in the USA. The last major epizootic in the USA occurred in 1979–80 and has been described in great detail (e.g. Wagner and Daynes, 1981). Severe outbreaks were also described in various European countries (Deerberg *et al.*, 1973; Owen *et al.*, 1975; Osterhaus *et al.*, 1981). A more recent outbreak in the USA, which resulted in the eradication of almost 5000 mice in one institution, was described by Dick *et al.* (1996). The most recent well-documented case of mousepox was published by Lipman *et al.* (2000b). Few additional but unpublished cases of ectromelia have been observed thereafter. In a recent survey conducted in the USA, one population was reported to be seropositive for mousepox (Jacoby and Lindsey, 1998).

Natural infections manifest differently depending on many factors. Mousepox may occur as a rapidly spreading outbreak with acute disease and deaths, or may be inconspicuous with slow spreading and mild clinical signs. The mortality rate can be very low in populations in which the virus has been present for long periods. The infection usually takes one of three clinical courses: acute asymptomatic infection, acute lethal infection (systemic form), or subacute to chronic infection (cutaneous form; Fenner, 1981, 1982; Manning and Frisk, 1981; National Research Council, 1991; Dick *et al.*, 1996). The systemic or visceral form is characterized clinically by facial oedema, conjunctivitis, multisystemic necrosis, and usually high mortality. This form is less contagious than the cutaneous form because the animals die before there is virus shedding. The cutaneous form is characterized by typical dermal lesions and variable mortality. The outcome of infection depends on many factors including strain and dose of virus; route of viral entry; strain, age, and sex of mouse; husbandry methods; and duration of infection in the colony. While all mouse strains seem to be susceptible to infection with ECTV, clinical signs and mortality are strain-dependent (Fenner, 1982; Wallace and Buller, 1985, 1986; Brownstein *et al.*, 1989b). Acute lethal (systemic) infection occurs in highly susceptible inbred strains such as DBA/1, DBA/2, BALB/c, A, and C3H/HeJ. Immunodeficient mice may also be very susceptible (Allen *et al.*, 1981). Outbreaks among susceptible mice can be explosive, with variable morbidity and high mortality (>80%). Clinical disease may not be evident in resistant strains such as C57BL/6 and AKR, and the virus can be endemic in a population for long periods before being recognized. Furthermore, females seem to be more resistant to disease than males, at least in certain strains of mice (Wallace *et al.*, 1985; Brownstein *et al.*, 1989b).

The mechanisms determining resistance versus susceptibility are not fully understood but appear to reflect the action of multiple genes. The genetic loci considered to be important include *H-2D^b* (termed *Rmp-3*, resistance to mousepox, on chromosome 17; O'Neill *et al.*, 1983), the *C5* genes (*Rmp-2*, on chromosome 2), *Rmp-1*, localized to a region on chromosome 6 encoding the NK cell receptor NKR-P1 **alloantigens** (Brownstein and Gras, 1997), the *nitric oxide synthase 2* locus on chromosome 11 (Karupiah *et al.*, 1998), and the *signal transducer and activator of transcription 6* locus on chromosome 10 (Mahalingam *et al.*, 2001). Clearance of the virus by the immune system is absolutely dependent upon the effector functions of CD8+ T cells while NK cells, CD4+ T cells, and macrophages are necessary for the generation of an optimal response (Niemialtowski *et al.*, 1994; Delano and Brownstein, 1995; Karupiah *et al.*, 1996).

Mousepox (ectromelia) virus usually enters the host through the skin with local replication and extension to regional lymph nodes (Fenner, 1981, 1982; Wallace and Buller, 1986; National Research Council, 1991). It escapes into the blood (primary viraemia) and infects splenic and hepatic macrophages resulting in necrosis of these organs and a massive secondary viraemia. This sequence takes approximately 1 week. Many animals die at the end of this stage without premonitory signs of illness; others develop varying clinical signs including ruffled fur, hunched posture, swelling of the face or extremities, conjunctivitis, and skin lesions (papules, erosions, or encrustations mainly on ears, feet, and tail; Figure 23.1). Necrotic amputation of limbs and tails can sometimes be seen in mice that survive the acute phase, hence the original name of the disease '**ectromelia**' (meaning absent or short limbs; Figure 23.2).

Common gross lesions of acute mousepox include enlarged lymph nodes, Peyer's patches, spleen, and liver; multifocal to semiconfluent white foci of necrosis in the spleen and liver; and haemorrhage into the small intestinal lumen (Allen *et al.*, 1981; Fenner, 1982; Dick *et al.*, 1996; Percy and Barthold, 2001). In animals that survive, necrosis and scarring of the spleen

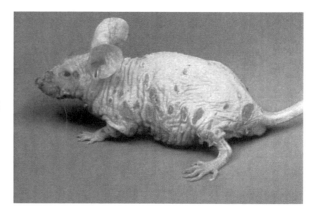

Figure 23.1 The rash of mousepox in a hairless (*hr*) mutant mouse (from Deerberg *et al.*, 1973; used with permission from Verlag M. & H. Schaper).

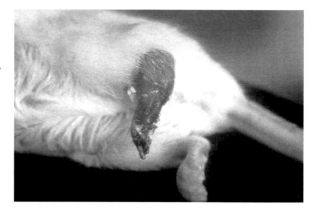

Figure 23.2 Dry gangrene of the left hind foot of a mouse infected with ECTV.

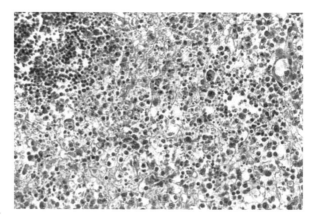

Figure 23.3 Section of the spleen of a mouse infected with ectromelia virus. There is marked parenchymal necrosis with extensive cellular debris and only few lymphoid cells left (H&E stain, magnification 200×; courtesy of Dr. A. D. Gruber).

can produce a mosaic pattern of white and red-brown areas that is a striking gross finding.

The most consistent histological lesions of acute mousepox are necroses of the spleen (Figure 23.3),

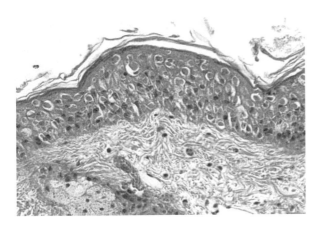

Figure 23.4 Section of the skin of a mouse infected with ECTV. Cutaneous hyperplasia with epithelial cell degeneration and numerous large intraepithelial cytoplasmic viral inclusion bodies (Cowdry type A) are seen (H&E stain, magnification 400 ×; courtesy of Dr. A. D. Gruber).

lymph nodes, Peyer's patches, thymus, and liver (Allen et al., 1981; Fenner, 1982; Dick et al., 1996; Lipman et al., 2000b; Percy and Barthold, 2001). Occasionally, necrosis may also be observed in other organs such as ovaries, uterus, vagina, intestine, and lungs. The primary skin lesion, which occurs about a week after exposure at the site of inoculation (frequently on the head), is a localized swelling that enlarges from inflammatory oedema. Necrosis of dermal epithelium provokes a surface scab and heals as a deep, hairless scar. Secondary skin lesions (rash) develop 2–3 days later as the result of viraemia (Figure 23.1). They are often multiple and widespread and can be associated with conjunctivitis. The skin lesions also can ulcerate and scab before scarring. Mucosal and dermal epithelial cells may have characteristic intracytoplasmic eosinophilic (Cowdry type A) inclusion bodies (Figure 23.4). Basophilic (Cowdry type B) inclusions may be found in the cytoplasm of all infected cells, especially in hepatocytes.

Natural transmission of ECTV mainly occurs by direct contact and **fomites** (Fenner, 1981; Wallace and Buller, 1986; National Research Council, 1991). The primary route of infection is through skin abrasions. Faecal–oral and aerosol routes may also be involved (Werner, 1982). In addition, the common practice of cannibalism by mice may contribute to the oral route of infection (Bhatt and Jacoby, 1987b). Intrauterine transmission is possible at least under experimental conditions (Schwanzer et al., 1975). Virus particles are shed from infected mice (mainly via scabs and/or faeces) for about 3–4 weeks, even though the virus can persist for months in the spleen of an occasional mouse

(Bhatt and Jacoby, 1987b; National Research Council, 1991). Cage-to-cage transmission of ECTV and transmission between rooms or units is usually low and largely depends on husbandry practices (e.g. mixing mice from different cages). Importantly, the virus may not be transmitted effectively to sentinel mice exposed to dirty bedding (Lipman et al., 2000b).

Various tests have been applied for the diagnosis of ectromelia. Previous epidemics were difficult to deal with because of limited published data and information on the biology of the virus and the lack of specific and sensitive assays (Wallace, 1981). In the 1950s, diagnosis relied on clinical signs, histopathology, and animal passages of tissues from moribund and dead animals. Culture of the virus on the chorioallantoic membrane of embryonated eggs was also applied. Serology is currently the primary means of testing mouse colonies for exposure to ECTV. The methods of choice are ELISA and IFA; they are more sensitive and specific than the previously used haemagglutination inhibition (HI) assay (Collins et al., 1981; Buller et al., 1983; ACLAD, 1991). Both tests detect antibodies to orthopoxviruses and do not distinguish between ECTV and vaccinia virus. Vaccinia virus is commonly used as antigen for serological testing to avoid the risk of infection for mice. Thus, false-positive serological reactions may be found after experimental administration of replication-competent vaccinia virus. It has been shown that even cage contact sentinels may develop antibodies, and vaccinia virus leading to seroconversion may even be transmitted by dirty bedding (Gaertner et al., 2003). Confirmation of positive serological results is important before action is taken because vaccinia virus is increasingly prevalent in animal facilities as a research tool (e.g. for vaccination or gene therapy). As observed in different outbreaks, serological testing is of little value in the initial stages of the disease. For example, in the outbreak described by Dick et al. (1996) depopulation was nearly completed before serological confirmation was possible. For this reason, negative serological results should be confirmed by direct detection methods (PCR, immunohistochemistry, virus isolation) or by histopathology, especially when clinical cases suggestive of mousepox are observed. Polymerase chain reaction assays to detect different genes of poxviruses in infected tissues have been described by Dick et al. (1996), Neubauer et al. (1997), and Lipman et al. (2000b).

The key to prevention and control of mousepox is early detection of infected mice and contaminated biological materials. All institutions that must introduce mice from other than commercial barrier facilities should have a health surveillance programme and test incoming mice. Perhaps even more important than living animals are samples from mice (tumours, sera, tissues). The virus replicates in lymphoma and hybridoma cell lines (Buller *et al.*, 1987), and such cells or material derived from them may therefore be a vehicle for inadvertent transfer between laboratories. The last two published outbreaks of ectromelia were both introduced into the facilities by mouse serum (Dick *et al.*, 1996; Lipman *et al.*, 2000b). Lipman *et al.* (2000b) found that the contaminated serum originated from a pooled lot of 431 that had been imported from China. Because mouse serum commonly is sold to the end user in small aliquots (few millilitres), it has to be expected that aliquots of the contaminated lot are still stored in numerous freezers. Both cases provide excellent examples of why MAP or PCR testing should be performed on all biological materials to be inoculated into mice.

Eradication of mousepox usually has been accomplished by elimination of the affected colonies, disinfection of rooms and equipment, and disposal of all infected tissues and sera. While culling of entire mouse colonies is the safest method for eradication of mousepox, it is not a satisfactory method due to the uniqueness of numerous lines of genetically modified animals housed in many facilities. Several studies indicate that mousepox is not highly contagious (Wallace and Buller, 1985; Bhatt and Jacoby 1987a,b) and that it may be self-limiting when adequate husbandry methods are applied. Therefore, strict quarantine procedures along with cessation of breeding (to permit resolution of infection) and frequent monitoring with removal of clinically sick and seropositive animals are a potential alternative. The period from the last births before the break until the first matings after the break should be at least 6 weeks (Bhatt and Jacoby, 1987b). Sequential testing of immunocompetent contact sentinels for seroconversion should be employed with this option.

In the past, immunization with live vaccinia virus was used to suppress clinical expression of mousepox. Vaccination may substantially reduce the mortality rate, but it does not prevent virus transmission or eradicate the agent from a population (Buller and Wallace, 1985; Bhatt and Jacoby, 1987c). After vaccination, typical pocks develop at the vaccination site, and infectious vaccinia virus is detectable in spleen, liver, lungs, and thymus (Jacoby *et al.*, 1983). Vaccination also causes seroconversion so that serological tests are not applicable for health surveillance in vaccinated populations. It is therefore more prudent to control mousepox by quarantine and serological surveillance than by relying on vaccination.

Mortality and clinical disease are the major factors by which ECTV interferes with research. Severe disruption of research can also occur when drastic measures are taken to control the infection. The loss of time, animals, and financial resources can be substantial.

Murine adenoviruses

Murine adenoviruses (MAdV) are non-enveloped, double-stranded DNA viruses of the family Adenoviridae, genus Mastadenovirus. Two distinct strains have been isolated from mice. The FL strain (MAdV-1) was first isolated in the USA as a contaminant of a Friend leukaemia (Hartley and Rowe, 1960); the K87 strain (MAdV-2) was first isolated in Japan from the faeces of a healthy mouse (Hashimoto *et al.*, 1966). Both strains are now considered to represent different species (Hamelin and Lussier, 1988; Jacques *et al.*, 1994a,b). In laboratory mice, seropositivity to adenoviruses was reported in 2% of SPF colonies and in 8% of non-SPF colonies in the USA (Jacoby and Lindsey, 1998). Antibodies were also detected at a low prevalence rate in French colonies (Zenner and Regnault, 2000), but the virus strain used as antigen is not mentioned. A similar range of positive samples was reported by Livingston and Riley (2003). Antibodies to MAdV were also found in wild mice (Smith *et al.*, 1993b) and in rats (Otten and Tennant, 1982; Smith *et al.*, 1986).

Both viruses are not known to cause clinical disease in naturally infected, immunocompetent mice. However, MAdV-1 can cause a fatal systemic disease in suckling mice after experimental inoculation (Hartley and Rowe, 1960; Heck *et al.*, 1972; Wigand, 1980). Disease is characterized by scruffiness, lethargy, stunted growth, and often death within 10 days. Experimental infection of adult mice with MAdV-1 is most often subclinical and persistent (Richter, 1986) but can cause fatal haemorrhagic encephalomyelitis with neurological symptoms, including tremors, seizures, ataxia, and paralysis, in susceptible C57BL/6 and DBA/2J mice (Guida *et al.*, 1995). BALB/c mice are relatively resistant to this condition. Athymic *Foxn1^{nu}* mice experimentally infected with MAdV-1 develop a lethal wasting disease (Winters and Brown, 1980). Similarly, *Prkdc^{scid}* mice succumb to experimental infection with MAdV-1 (Pirofski *et al.*, 1991).

Gross lesions in response to natural MAdV infections are not detectable. Occasional lesions observed after experimental infection with MAdV-1 include small surface haemorrhages in the brain and spinal cord of C57BL/6 and DBA/2J mice (Guida *et al.*, 1995), duodenal haemorrhage in *Foxn1^nu* mice (Winters and Brown, 1980), and pale yellow livers in *Prkdc^scid* mice (Pirofski *et al.*, 1991).

Histologically, experimental MAdV-1 infection of suckling mice is characterized by multifocal necrosis and large basophilic intranuclear inclusion bodies in liver, adrenal gland, heart, kidney, salivary glands, spleen, brain, pancreas, and brown fat (Heck *et al.*, 1972; Margolis *et al.*, 1974; National Research Council, 1991; Percy and Barthold, 2001). In experimentally induced haemorrhagic encephalomyelitis, multifocal petechial haemorrhages occur throughout the brain and spinal cord, predominantly in the white matter, and are attributed to infection and damage to the vascular epithelium of the central nervous system (CNS; Guida *et al.*, 1995). Histopathological manifestations in MAdV-1-infected *Prkdc^scid* mice are marked by microvesicular fatty degeneration of hepatocytes (Pirofski *et al.*, 1991). In contrast to MAdV-1, the tissue tropism of MAdV-2 is limited to the intestinal epithelium. Naturally or experimentally infected mice develop intranuclear inclusions in enterocytes, especially in the ileum and caecum (Takeuchi and Hashimoto, 1976; Otten and Tennant, 1982; National Research Council, 1991; Percy and Barthold, 2001).

Transmission of MAdV primarily occurs by ingestion. MAdV-1 is excreted in the urine and may be shed for up to 2 years (Van der Veen and Mes, 1973). Murine adenovirus-2 infects the intestinal tract and is shed in faeces for only a few weeks in immunocompetent mice (Hashimoto *et al.*, 1970); immunodeficient mice may shed the virus for longer periods (Umehara *et al.*, 1984).

Murine adenovirus infections are routinely diagnosed by serological tests. However, there is a one-sided cross reactivity of MAdV-1 with MAdV-2 (Wigand *et al.*, 1977). Serum from mice experimentally infected with MAdV-1 yielded positive reactions in serological tests with both viruses while serum from mice infected with MAdV-2 reacted only with the homologous antigen (Lussier *et al.*, 1987). Smith *et al.* (1986) reported that sera may react with MAdV-1 or MAdV-2 or both antigens. Occasional reports of mice with lesions suggestive of adenovirus infections and negative serology (with MAdV-1) indicate that the infection may not be detected if only one virus is used as antigen (Luethans and Wagner, 1983). It has therefore become standard practice to test sera for antibodies to both MAdV-1 and MAdV-2. The common methods are IFA and ELISA, and both are more sensitive than the previously used complement fixation test.

The low prevalence in colonies of laboratory mice indicate that MAdV can easily be eliminated (e.g. by hysterectomy derivation or embryo transfer) and that barrier maintenance has been very effective in preventing infection.

The low pathogenicity and the low prevalence in contemporary mouse populations are the main reasons why adenoviruses are considered to be of little importance. However, immunodeficient mice are increasingly used and candidates for natural infections and wasting disease (Richter, 1986), and the viruses might easily be spread by the exchange of genetically modified mice and therefore re-emerge. Only few influences on research attributable to MAdV have been published. For example, it has been shown that MAdV-1 significantly aggravates the clinical course of scrapie disease in mice (Ehresmann and Hogan, 1986). Natural infections with MAdV could also interfere with studies using adenovirus as a gene vector.

Polyomaviruses

Polyomaviridae are enveloped, double-stranded DNA viruses. Two different agents of this family exclusively infect mice (*Mus musculus*), and both belong to the genus Polyomavirus. Murine pneumotropic virus (MPtV) has formerly been known as 'newborn mouse pneumonitis virus' or 'K virus' (named after L. Kilham who first described the virus). The second is murine polyomavirus (MPyV). Both are related but antigenically distinct from each other (Bond *et al.*, 1978). They are **enzootic** in many populations of wild mice but are very uncommon in laboratory mice. Even older reports indicate that both have been eradicated from the vast majority of contemporary mouse colonies, and their importance is negligible (National Research Council, 1991). Seropositivity to these viruses was not reported in a survey conducted in the USA (Jacoby and Lindsey, 1998). In a retrospective study in French facilities, antibodies to MPyV were found in 1 of 69 colonies, and all samples tested for MPtV were negative (Zenner and Regnault, 2000). Comparable data were reported by Livingston and Riley (2003). Due to their low prevalence, both viruses are not included in the list of agents for which testing is recommended on a regular basis by FELASA (Nicklas *et al.*, 2002).

Although polyomavirus genes, especially those of SV40 are used widely in gene constructs for insertional mutagenesis, very few reports have been published on spontaneous or experimental disease due to MPyV or MPtV in the last 10–15 years. The reader is therefore referred to previous review articles for details (Eddy 1982; Parker and Richter, 1982; Richter, 1986; Shah and Christian, 1986; National Research Council, 1991; Orcutt, 1994; Porterfield and Richter, 1994).

MPtV

Natural infections with MPtV are subclinical. The prevalence of infection is usually low in an infected population. The virus may persist in infected animals for months and perhaps for life depending on the age at infection and is reactivated under conditions of immunosuppression. Virus replicates primarily in endothelial cells, but renal tubular epithelial cells are the major site of viral persistence (Greenlee *et al.*, 1991, 1994).

Clinical signs are observed only after infection of infant mice less than 6–8 days of age. Infected pups suddenly develop respiratory symptoms after an incubation period of approximately 1 week, and many die within a few hours of onset of symptoms with an interstitial pneumonia caused by productive infection of and damage to pulmonary endothelium. Endothelial cells in other organs are involved in virus replication also (Ikeda *et al.*, 1988; Greenlee *et al.*, 1994). In older suckling mice, MPtV produces a more protracted infection, and the virus or viral antigen can be detected for as long as 4 months. In adult animals, the virus produces a transient asymptomatic infection. Even in immunodeficient *Foxn1^{nu}* mice, experimental infection of adults is clinically asymptomatic although virus is detectable for a period of several months (Greenlee, 1986).

In vitro cultivation of MPtV is difficult. No susceptible permanent cell line is known to support growth. It can be cultured in primary mouse embryonic cells, but viral titres are not sufficient for use in serological assays (Greenlee and Dodd, 1987). For this reason, the **HI test** using homogenates of livers and lungs of infected newborn mice is still frequently used, but IFA and ELISA tests are also available (Groen *et al.*, 1989). Furthermore, a PCR test for demonstration of MPtV in biological samples has been published (Carty *et al.*, 2001).

MPyV

Murine polyomavirus was first detected as a contaminant of murine leukaemia virus (MuLV) when sarcomas developed in mice after experimental inoculation of contaminated samples. It has later been shown to be a frequent contaminant of transplantable tumours (Collins and Parker, 1972). Natural infection of mice is subclinical, and gross lesions including tumours are usually not found. Tumour formation occurs if mice are experimentally infected at a young age or if they are inoculated with high virus doses. Development of tumours may be preceded by multifocal necrosis and mortality during the viraemic stage (Percy and Barthold, 2001). Parotid, salivary gland, and mammary tumours are common, and sarcomas or carcinomas of kidney, subcutis, adrenal glands, bone, cartilage, teeth, blood vessels, and thyroid occur also. Virus strains vary with regard to the tumour types or lesions that they induce, and mouse strains vary in their susceptibility to different tumour types. Those of C57BL and C57Br/cd lineage are considered to be the most resistant strains; athymic *Foxn1^{nu}* mice are considered to be most susceptible; C3H mice are particularly susceptible to adrenal tumours and A mice tend to develop bone tumours. Immunosuppression or inoculation into immunodeficient strains (e.g. *Foxn1^{nu}*) also support the growth of tumours. On the other hand, experimental infection of adult immunocompetent mice does not result in tumour formation because the immune response suppresses tumour growth, and newborn immunocompetent mice develop runting only if inoculated with high virus doses (Atencio *et al.*, 1995).

After experimental intranasal infection, MPyV initially infects the respiratory tract followed by a systemic phase in which liver, spleen, kidney, and the colon become infected (Dubensky *et al.*, 1984). The virus is shed in faeces and in all body fluids, and transmission occurs rapidly by direct contact between animals, but also between cages in a room. Further, intrauterine transmission has been documented after experimental infection (McCance and Mims, 1977). Murine polyomavirus persists in all organs in *Prkdc^{scid}* mice while viral DNA is detectable in immunocompetent mice after experimental infection for only a limited period of about 4 weeks (Berke *et al.*, 1994). However, virus may persist and can be reactivated by prolonged immunosuppression (Rubino and Walker, 1988) or during pregnancy, at least in young mice (McCance and Mims, 1979). Biological materials of mouse origin are likely to be the most common source of contamination of laboratory mice emphasizing the importance of MAP or PCR screening of biological materials to be inoculated into mice.

The most frequently used tests for health surveillance of mouse colonies are ELISA and IFA (ACLAD, 1991); in addition, the HI test is still used. Latent infections can be detected by intracerebral inoculation of neonate mice or by MAP testing, but direct demonstration of virus in biological samples is also possible by PCR testing (Porterfield and Richter, 1994; Carty *et al.*, 2001).

Parvoviruses

Parvoviruses are non-enveloped small viruses (approximately 20 nm in diameter) with a single-stranded DNA genome of approximately 5000 nucleotides. Murine parvoviruses are members of the family Parvoviridae, genus Parvovirus. They are remarkably resistant to environmental conditions like heat, desiccation, acidic and basic pH-values. Two distinct serotypes infect laboratory mice: the mice minute virus (MMV) and the mouse parvovirus 1 (MPV). Nonstructural proteins (NS-1 and NS-2) are highly conserved among both viruses whereas the capsid proteins (VP-1, VP-2, VP-3) are more divergent and determine the serogroup (Ball-Goodrich and Johnson, 1994). Both viruses require mitotically active cells for replication. Severe infections are therefore not found in mature animals due to the lack of a sufficient number of susceptible cells in tissues. General aspects of rodent parvovirus infections and their potential effects on research results have been reviewed (Tattersall and Cotmore, 1986; National Research Council, 1991; Jacoby and Ball-Goodrich, 1995; Jacoby et al., 1996).

MPV

Already in the mid-1980s mouse colonies were identified that gave positive reactions for MMV by IFA but not by HI tests. It was subsequently shown that these colonies were infected with a novel parvovirus, initially referred to as 'mouse orphan parvovirus'. The first isolate of MPV was detected as a contaminant of cultivated T-cell clones interfering with in vitro immune responses (McKisic et al., 1993) and was named 'mouse parvovirus'. It does not replicate well in currently available cell cultures, and sufficient quantities of virus for serological tests are difficult to generate. Hitherto, only very few isolates of MPV have been cultured and characterized on a molecular basis (Ball-Goodrich and Johnson, 1994; Besselsen et al., 1996).

At present, MPV is among the most common viruses in colonies of laboratory mice. The prevalence of sera positive for parvoviruses was nearly 10% in a study from Livingston et al. (2002), with the majority of sera being positive for MPV. This is consistent with a recent survey conducted in the USA showing that almost 40% of non-SPF colonies were seropositive (Jacoby and Lindsey, 1998). Similar results were obtained for genetically modified mice in Japan (Yamamoto et al., 2001), in contrast to earlier studies indicating that the infection was rare in Japan (Ueno et al., 1998).

Clinical disease and gross or histological lesions have not been reported for mice naturally or experimentally infected with MPV. Infections are subclinical even in newborn and immunocompromised animals (Smith et al., 1993a; Jacoby et al., 1995). In contrast to many other viruses infecting mice, viral replication and excretion is not terminated by the onset of host immunity. Tissue necrosis has not been observed at any stage of infection in infected infant or adult mice (Smith et al., 1993a; Jacoby et al., 1995). Humoral immunity to MPV does not protect against MMV infections and vice versa (Hansen et al., 1999).

Serological surveys have indicated that MPV naturally infects only mice. Differences in mouse strain susceptibility to clinical MPV infection do not exist. However, seroconversion seems to be strain-dependent. After experimental infection, seroconversion occurred in all C3H/HeN mice, fewer BALB/c, DBA/2, and ICR mice, and seroconversion could not be detected in C57BL/6 mice (Besselsen et al., 2000). Diagnosis of MPV infection by PCR testing of small intestine and mesenteric lymph nodes also depended on the mouse strain. MPV DNA was detected in all mouse strains evaluated except DBA/2 even though seroconversion was detected in these mice.

After oral infection, the intestine is the primary site of viral entry and replication. The virus spreads to the mesenteric lymph nodes and other lymphoid tissues, where it persists for more than 2 months (Jacoby et al., 1995), and seems to be excreted via the intestinal and the urinary tract. After experimental inoculation of weanling mice, MPV is transmitted to cagemates by direct contact for 2–4 weeks (Smith et al., 1993a), and transmission by dirty bedding is also possible. These results implicate a role for urinary, faecal, and perhaps respiratory excretion of virus. Another study showed that naturally infected mice may not transmit the virus under similar experimental conditions (Shek et al., 1998).

Serology is a useful tool to identify MPV infections in immunocompetent hosts, but reaching a diagnosis based on serological assays may be difficult and requires a good knowledge of the available techniques. Neither the virion ELISA nor HI are practical screening tests for MPV because they require large quantities of purified MPV which is difficult to obtain. Diagnosis of MPV infections has long been made on the basis of an MMV HI-negative result coupled with an MMV IFA-positive result. A generic rodent parvovirus ELISA using a recombinant NS-1 protein as antigen has been developed (Riley et al., 1996), but MPV IFA and MPV HI assays are more sensitive techniques than the NS-1

ELISA and the MMV IFA (Besselsen *et al.*, 2000). Recently, ELISA tests have been described that use recombinant VP-2 and provide sensitive and serogroup-specific assays for the diagnosis of MPV infections in mice (Ball-Goodrich *et al.*, 2002; Livingston *et al.*, 2002). In immunodeficient mice that do not generate a humoral immune response, PCR assays can be used to detect MPV (Besselsen *et al.*, 1995; Redig and Besselsen, 2001) and other parvoviruses. MPV has been shown to persist for at least 9 weeks in the mesenteric lymph nodes (Jacoby *et al.*, 1995). This tissue is considered the best suited for PCR analysis, but spleen and small intestine can also be used with good success (Besselsen *et al.*, 2000). The virus persists sufficiently long in mesenteric lymph nodes so that PCR assays may also be used as a primary screening tool for laboratories that do not have access to specific MPV antigen-based serological assays. Polymerase chain reaction is further a good confirmatory method for serological assays and has also been described for the detection of parvoviruses in cell lines and tumours (Yagami *et al.*, 1995). In addition, the MAP test has been reported as a sensitive tool to detect MPV (Shek *et al.*, 1998).

Given the high environmental stability of the virus and the potential **fomite** transmission together with the long virus persistence in infected animals, spontaneous disappearance from a mouse population (e.g. by cessation of breeding) is very unlikely. Eradication of infection is possible by elimination of infected animals and subsequent replacement with uninfected mice, and the agent can be eliminated from breeding populations only by embryo transfer or by hysterectomy.

Although there are few published reports of confounding effects of MPV on research, it is lymphocytotropic and may perturb immune responses *in vitro* and *in vivo*. Infections with MPV have been shown to influence rejection of skin and tumour grafts (McKisic *et al.*, 1996, 1998).

MMV

Mice minute virus is the type species of the genus Parvovirus. The virus was formerly called 'minute virus of mice' (MVM) and was renamed recently (International Union of Microbiological Societies, 2000). It was originally isolated by Crawford (1966) from a stock of mouse adenovirus, and this prototype isolate was later designated MVMp. Its allotropic variant was detected as a contaminant of a transplantable mouse lymphoma (Bonnard *et al.*, 1976) and designated MVMi because it exhibits immunosuppressive properties *in vitro*. Both variants have distinct cell tropisms *in vivo* and *in vitro*.

The MMVp infects fibroblast cell lines and does not cause clinical disease (Kimsey *et al.*, 1986, Brownstein *et al.*, 1991). The MMVi grows lytically in T cells and inhibits various functions mediated by these cells *in vitro*. Both strains are apathogenic for adult mice, but the immunosuppressive variant is more pathogenic for neonatal mice than is MMVp.

Serological surveys show that the mouse is the primary natural host (Parker *et al.*, 1970; Smith *et al.*, 1993b; Singleton *et al.*, 2000), but the virus is also infective for rats, hamsters (Garant *et al.*, 1980; Ward and Tattersall, 1982), and *Mastomys* (Haag *et al.*, 2000) during foetal development or after parenteral inoculation.

Natural infections are usually asymptomatic in adults and infants, and the most common sign of infection is seroconversion. Kilham and Margolis (1970) observed mild growth retardation a few days after experimental infection of neonatal mice with MMVp. Studies of transplacental infection yielded no pathological findings in mice (Kilham and Margolis, 1971). The immunosuppressive variant but not the prototype strain is able to produce a runting syndrome after experimental infection of newborn mice (Kimsey *et al.*, 1986). Depending on the host genotype, experimental infections of foetal and neonatal mice with MMVi produce various clinical presentations and lesions. Infection in C57BL/6 mice is asymptomatic, but the virus causes lethal infections with intestinal haemorrhage in DBA/2 mice. Infection of strains such as BALB/c, CBA, C3H/He, and SJL is also lethal and mice have renal papillary haemorrhage (Brownstein *et al.*, 1991). The MMVi also infects haematopoietic stem cells and mediates an acute **myelosuppression** (Segovia *et al.*, 1991, 1995). Due to their dependency on mitotically active tissues, the foetus is at particular risk for damage by parvoviruses. Mice minute virus and other parvoviruses may have severe teratogenic effects and cause foetal and neonatal abnormalities by destroying rapidly dividing cell populations, often resulting in foetal death. Adult *Prkdc*^scid mice develop an acute leukopenia 1 month after experimental infection with MMVi and die within 3 months. The virus persists lifelong in the bone marrow of these mice (Segovia *et al.*, 1999).

Mice minute virus is shed in faeces and urine. Contaminated food and bedding are important factors in viral transmission because the virus is very resistant to environmental conditions. Direct contact is also important and the virus does not easily spread between cages.

Routine health surveillance is usually conducted by serological methods. Unlike MPV, MMV can easily be cultured in cell lines so that antigen production for HI and ELISA (using whole purified virions) is easy. Haemagglutination inhibition is a highly specific diagnostic test whereas IFA always exhibits some degree of cross reactivity with MPV and other closely related parvoviruses. Enzyme-linked immunosorbent assay is probably the most frequently used test, but depending on the purity of the antigen preparation, cross reactions with MPV may occur due to contamination with non-structural proteins that are common to both viruses. This problem can be avoided by the use of recombinant VP-2 antigen (Livingston et al., 2002). Viral detection is also possible by PCR in biological materials and in organs (intestines, kidney, spleen) from infected animals (Yagami et al., 1995; Chang et al., 1997; Redig and Besselsen, 2001). In contrast to MPV, PCR is not appropriate as a confirmatory method for serology because MMV has not been shown to persist in immunocompetent animals for sufficiently long periods.

The virus can be eliminated from infected breeding populations by caesarean derivation or by embryo transfer. In experimental colonies, elimination of infected animals and subsequent replacement with uninfected mice is practical if careful environmental sanitation is conducted by appropriate disinfection procedures. It is important that reintroduction is avoided by exclusion of wild mice and by strict separation from other infected populations and potentially contaminated materials in the same facility. Admission of biological materials must be restricted to samples that have been tested and found free from viral contamination.

Both allotropic variants of MMV have been used as models for molecular virology, and their small size and simple structure have facilitated examination of their molecular biology and expedited understanding of cell tropism, viral genetics, and structure. The significance for laboratory mouse populations was considered low or uncertain because natural infections are inapparent. However, various effects on mouse-based research have been published (Tattersall and Cotmore, 1986; Jacoby et al., 1996; Baker, 1998; Nicklas et al., 1999). Due to their predilection for replicating in mitotically active cells, they are frequently associated with tumour cells and have a marked **oncosuppressive** effect (Rommelaere and Cornelis, 1991). Special attention is also necessary for immunological research and other studies involving rapidly dividing cells (embryology, teratology). In addition, MMV is a common contaminant of transplantable tumours, murine leukaemias, and other cell lines (Collins and Parker, 1972; Nicklas et al., 1993; Garnick, 1996).

RNA viruses

Lactate dehydrogenase-elevating virus

Lactate dehydrogenase-elevating virus (LDV) is a single-stranded RNA virus of the genus Arterivirus belonging to the family Arteriviridae. Lactate dehydrogenase-elevating virus has repeatedly been detected in feral mice (*Mus musculus*), which are considered to be a virus reservoir (Rowson and Mahy, 1975; Li et al., 2000). Only mice and primary mouse cells are susceptible to infection with LDV. After infection, virus titres of 10^{10}–10^{11} particles per ml serum are found within 12–14 h after infection. The virus titre drops to 10^5 particles per ml within 2–3 weeks and remains constant at this level for life. Lactate dehydrogenase-elevating virus replicates in a subpopulation of macrophages in almost all tissues and persists in lymph nodes, spleen, liver, and testes tissues (Anderson et al., 1995a). The virus can be stored in undiluted mouse plasma at $-70°C$ without loss of infectivity, but it is not stable at room temperature and is very sensitive to environmental conditions.

Lactate dehydrogenase-elevating virus was first detected during a study of methods that could be used in the early diagnosis of tumours (Riley et al., 1960). It produces a persistent infection with continuous virus production and a lifelong viraemia despite LDV-specific immune reactions of the host (Van den Broek et al., 1997). Lactate dehydrogenase-elevating virus has been found in numerous biological materials that are serially passaged in mice such as transplantable tumours including human tumours (Nicklas et al., 1993; Ohnishi et al., 1995), monoclonal antibodies or ascitic fluids (Nicklas et al., 1988), or infectious agents (e.g. haemoprotozoans, K virus, *Clostridium piliforme*). These materials are contaminated after passage in an infected and viraemic animal. Contamination with LDV leads to the infection of each sequential host and to transmission of the virus by the next passage and remains associated with the specimen. It is therefore the most frequently detected contaminant in biological materials (Collins and Parker, 1972; Nicklas et al., 1993).

Infection with LDV is usually asymptomatic, and there are no gross lesions in immunocompetent as well as in immunodeficient mice. The only exception is polyomyelitis with flaccid paralysis of hind limbs developing in C58 and AKR mice when they are immunosuppressed either naturally with aging or experimentally (Anderson *et al.*, 1995b; Monteyne *et al.*, 1997). It has been shown that only mice harbouring cells in the CNS that express a specific endogenous MuLV are susceptible to poliomyelitis (Anderson *et al.*, 1995c).

The characteristic feature of LDV infection is the increased activity of lactate dehydrogenase (LDH) and other plasma enzymes (Brinton, 1986; National Research Council, 1991), which is due to the continuous destruction of permissive macrophages that are responsible for the clearance of LDH from the circulation. As a consequence, the activity of plasma LDH begins to rise by only 24 h after infection and peaks 3–4 days after infection at 5–10-fold normal levels, or even be up to 20-fold in SJL/J mice. The enzyme activity declines during the next 2 weeks but remains elevated throughout life.

Antigen–antibody complexes produced during infection circulate in the blood and are deposited in the glomeruli (Brinton, 1986; National Research Council, 1991). In contrast to other persistent virus infections (e.g. lymphocytic choriomeningitis virus LCMV), these complexes do not lead to immune complex disease and produce only a very mild glomerulopathy. The only gross finding associated with LDV infection is mild splenomegaly. Microscopically, necrosis of lymphoid tissues is visible during the first days of infection. In mouse strains that are susceptible to poliomyelitis, LDV induces lesions in the grey matter of the spinal cord and the brain stem (Brinton, 1986).

Lactate dehydrogenase-elevating virus is not easily transmitted between mice, even in animals housed in the same cage. Fighting and cannibalism increase transmission between cage mates most likely via blood and saliva. Infected females transmit the virus to their foetuses if they have been infected few days prior to birth and before IgG anti-LDV antibodies are produced, but developmental and immunological factors (e.g. gestational age, timing of maternal infection with LDV, placental barrier) are important in the regulation of transplacental LDV infection (Haven *et al.*, 1996; Zitterkopf *et al.*, 2002). Maternal immunity protects foetuses from intrauterine infection. Immunodeficient *Prkdc^{scid}* mice transmit virus to their offspring also during chronic infection (Broen *et al.*, 1992). An important means of transmission is provided by experimental procedures such as mouse-to-mouse passage of contaminated biological materials or the use of the same needle for sequential inoculation of multiple mice.

In principal, serological methods such as IFA may be used for detecting LDV infection (Hayashi *et al.*, 1992) but they are not of practical importance.

Circulating virus–antibody complexes interfere with serological tests (ACLAD, 1991), and sufficient quantities of virus for serological tests are difficult to generate because LDV replicates only in specific subpopulations of primary cultures of murine macrophages and monocytes for one cell cycle (Brinton, 1986). Therefore, diagnosis of LDV infection is primarily based on increased LDH activity in serum or plasma of mice. Lactate dehydrogenase-elevating virus activity in serum or plasma can be measured directly, or samples (e.g. plasma, cell or organ homogenates) are inoculated into pathogen-free mice and the increase in LDH activity within 3–4 days is measured. An 8–10-fold increase is indicative of LDV infection. Detection of infectivity of a plasma sample by the induction of increased LDH activity in the recipient animal is the most reliable means of identifying an infected animal. However, it is important to use clear nonhaemolysed samples because haemolysis will (falsely) elevate activities of multiple serum or plasma enzymes, including LDH. While this assay may be included in a commercial 'MAP test', it does not involve antibody detection. Persistent infection makes LDV an ideal candidate for PCR detection in plasma or in organ homogenates (van der Logt *et al.*, 1994; Chen and Plagemann, 1997). However, reports exist that PCR may produce false-negative results and should be used cautiously (Lipman *et al.*, 2000a). Similarly important as detecting LDV in animals is its detection in biological materials. This may be done by assay for increased LDH activity after inoculation of suspect material into pathogen-free mice (Collins and Parker, 1972; Nicklas *et al.*, 1993) or by PCR (Goto *et al.*, 1998; Bootz and Sieber, 2002).

Lactate dehydrogenase-elevating virus spreads slowly in a population because direct contact is necessary. Therefore LDV-negative breeding populations can easily be established by selecting animals with normal plasma LDH activity. Embryo transfer and hysterectomy derivation are also efficient. The presence of LDV in experimental populations is indicative of contaminated biological materials. In such cases, it is essential that the virus is also eliminated from these samples. This is easily achieved by maintenance of cells by *in vitro* culture instead of by animal-to-animal passages (Plagemann and Swim, 1966). Due to the extreme host specificity of the virus, contaminated tumour samples can also be sanitized by passages in nude rats or other animal species.

Lactate dehydrogenase-elevating virus is a potential confounder of any research using biological materials that are passaged in mice. Once present in an animal, the virus persists lifelong. The most obvious signs are increased levels of plasma LDH and several other enzymes. Lactate dehydrogenase-elevating virus may also exhibit numerous effects on the immune system (thymus involution, depression of cellular immunity, enhanced or diminished humoral responses, NK cell activation, development of autoimmunity, and suppression of development of diabetes in NOD mice; Cafruny and Hovinen, 1988; Nicklas et al., 1988; Takei et al., 1992; Markine-Goriaynoff et al., 2002; Gomez et al., 2003) and enhance or suppress tumour growth (Brinton, 1982; Baker, 1998; Nicklas et al., 1999).

LCMV

Lymphocytic choriomeningitis virus is an enveloped, segmented single-stranded RNA virus of the genus Arenavirus, family Arenaviridae. Its name refers to the condition that results from experimental intracerebral inoculation of the virus into adult mice and is not considered to be a feature of natural infections. Mice (*Mus musculus*) serve as the natural virus reservoir (Salazar-Bravo et al., 2002), but Syrian hamsters are also important hosts (Ackermann, 1977). Additional species such as rabbits, guinea pigs, squirrels, monkeys, and humans are susceptible to natural or experimental infection. Infection in hamsters is considered to be asymptomatic (National Research Council, 1991). Natural infection of callitrichid primates (marmosets and tamarins) leads to a progressive hepatic disease that is known as 'callitrichid hepatitis' (Montali et al., 1995; Asper et al., 2001; Lukashevich et al., 2003). Antibodies to LCMV have been found in wild mice in Europe (Ackermann et al., 1964), Africa (El Karamany and Imam, 1991), Asia (Morita et al., 1991, 1996), Australia (Smith et al., 1993b), and America (Childs et al., 1992). Thus, it is the only arenavirus with worldwide distribution. Infection with LCMV is rarely found in laboratory mice (Smith et al., 1984). Seropositivity to LCMV was reported in approximately 5% of non-SPF mouse colonies in the USA in 1996 (Jacoby and Lindsey, 1998) and in 4% of French colonies in 1996–97 (Zenner and Regnault, 2000). Recent studies confirm that only a small percentage of mice tested are positive for LCMV (Livingston and Riley, 2003). In addition to laboratory mice and other vertebrate hosts, the virus has frequently been found in transplantable tumours and tissue culture cell lines from mice and hamsters (Bhatt et al., 1986a; Nicklas et al., 1993).

Despite the low prevalence in laboratory mice, seropositivity to this **zoonotic** agent should raise serious concern for human health. Lymphocytic choriomeningitis virus is frequently transmitted to humans from

wild mice (Childs et al., 1991) and is also endemic to a varying degree in the human population (Childs et al., 1991; Marrie and Saron, 1998; Lledo et al., 2003) due to contact with wild mice. Lymphocytic choriomeningitis virus is further transmitted to humans by domestic Syrian hamsters (Bowen et al., 1975; Rousseau et al., 1997). In addition, infected laboratory mice (Dykewicz et al., 1992) and contaminated biological materials are important sources of infections for humans, and several outbreaks of LCM among laboratory personnel have been traced to transplantable tumours (Hinman et al., 1975; Biggar et al., 1977; Mahy et al., 1991).

In mice, clinical signs of LCMV infection vary with strain and age of mouse, strain and dose of virus, and route of inoculation (Lehmann-Grube, 1982; National Research Council, 1991). Two forms of natural LCMV infection are generally recognized: a persistent tolerant and an (acute) nontolerant form. The persistent form results from infection of mice that are immunotolerant. This is the case if mice are infected *in utero* or during the first days after birth. This form is characterized by lifelong viraemia and shedding. Mice may show growth retardation, especially during the first 3–4 weeks, but appear otherwise normal. Infectious virus is bound to specific antibodies and complement, and these complexes accumulate in the renal glomeruli, the choroid plexus, and sometimes also in synovial membranes and blood vessel walls. At 7–10 months of age, immune complex nephritis develops with ruffled fur, hunched posture, ascites, and occasional deaths. This immunopathological phenomenon is called 'late onset disease' or 'chronic immune complex disease'. The incidence of this type of disease varies between mouse strains. Gross lesions include enlarged spleen and lymph nodes due to lymphoid hyperplasia. Kidneys affected with glomerulonephritis may be enlarged with a granular surface texture or may be shrunken in later stages of the disease process. Microscopically, there is generalized lymphoid hyperplasia and immune complex deposition in glomeruli and vessel walls, resulting in glomerulonephritis and plasmacytic, lymphocytic perivascular cuffs in all visceral organs (Percy and Barthold, 2001).

The nontolerant acute form occurs when infection is acquired after the development of immunocompetence (in mice older than 1 week). These animals become viraemic but do not shed virus and may die within a few days or weeks. Natural infections of adults are usually asymptomatic. Surviving mice are seropositive and in most cases clear the virus to below detection levels of conventional methods. However, virus may persist at low levels in tissues (particularly spleen, lung, and kidney) of mice for at least 12 weeks after infection as determined by sensitive assays such as nested reverse transcriptase–polymerase chain reaction (RT-PCR) or immunohisto-chemisty (Ciurea et al., 1999). Such nonlethal infection leads to protection against otherwise lethal intracerebral challenge. Protection from lethal challenge is also achieved by maternally derived anti-LCMV antibodies through nursing or by the administration of anti-LDV monoclonal IgG2a antibodies (Baldridge and Buchmeier, 1992).

In experimentally infected animals, the route of inoculation (subcutaneous, intraperitoneal, intravenous, intracerebral) also influences the type and degree of disease (Lehmann-Grube, 1982; National Research Council, 1991). Intracerebral inoculation of adult immunocompetent mice typically results in tremors, convulsions, and death due to **meningo-encephalitis** and hepatitis. Neurological signs usually appear on day 6 postinoculation, and animals die within 1–3 days after the onset of symptoms or recover within several days. The classic histological picture is of dense perivascular accumulations of lymphocytes and plasma cells in meninges and choroid plexus. While infection following subcutaneous inoculation usually remains inapparent, reaction of mice to intraperitoneal or intravenous inoculation depends on the virus strain and on the mouse strain. Infection by these routes primarily causes multifocal hepatic necrosis and necrosis of lymphoid cells. Athymic *Foxn1^{nu}* mice and other immunodeficient mice do not develop disease but become persistently viraemic and shed virus.

As a general rule, all pathological alterations following LCMV infection are immune-mediated; and mice can be protected from LCMV-induced disease by immunosuppression (Gossmann et al., 1995). Lymphocytic choriomeningitis virus disease is a prototype for virus-induced T-lymphocyte-mediated immune injury and for immune complex disease. For detailed information on the pathogenesis of LCMV infection, the reader is referred to a recent review article by Oldstone (2002). Extensive information on the clinical and pathological features of LCMV infection in mice has been assembled by Lehmann-Grube (1982).

In nature, carrier mice with persistent infection serve as the principal source of virus. Intrauterine transmission is very efficient, and with few exceptions all pups born from carrier mice are infected. Furthermore, persistently infected mice and hamsters can shed large numbers of infectious virions primarily in urine, but also in saliva and milk. The virus can replicate in the gastric mucosa after intragastric infection (Rai et al., 1996, 1997). Gastric inoculation elicits antibody

responses of comparable magnitudes as intravenous inoculation and leads to active infection with LCMV indicating that oral infection is possible, e.g., by ingestion of contaminated food or cannibalism. A self-limiting infection frequently results from infection of adult mice. The virus does not spread rapidly after introduction in populations of adult mice, and the infectious chain usually ends. However, if the virus infects a pregnant dam or a newborn mouse, a lifelong infection results, and soon a whole breeding colony of mice may become infected if the mice live in close proximity (which is the case under laboratory conditions).

Lymphocytic choriomeningitis virus is most commonly diagnosed by serological methods. Methods of choice are IFA and ELISA, which have replaced the relatively insensitive complement fixation test. It is important that bleeding of mice is done carefully because of a potential risk due to viraemic animals. Historically, direct viral detection was performed by inoculating body fluids or tissue homogenates into the brain of LCMV-free mice or by subcutaneous injection into mice and subsequent serological testing (MAP test). More recently, PCR assays have been developed for the direct detection of viral RNA in clinical samples or animals (Park et al., 1997, Besselsen et al., 2003). Both MAP test and PCR can also be used to detect contamination of biological materials (Bootz and Sieber, 2002).

Vertical transmission of LCMV by transuterine infection is efficient so this virus cannot reliably be eliminated by caesarean rederivation. Caesarean derivation may be effective if dams acquired infection after the development of **immunocompetence** (nontolerant acute infection) and subsequently eliminated the virus, but such a strategy is difficult to justify in light of LCMV's zoonotic potential. In breeding colonies of great value, virus elimination might be possible soon after introduction into the colony by selecting nonviraemic breeders. This procedure is expensive and time consuming and requires special safety precautions.

Fortunately, infections of laboratory mice with LCMV are very uncommon. However, once LCMV has been detected in animals or in biological materials, immediate destruction of all contaminated animals and materials is advisable to avoid risk of human infection. *Foxn1^{nu}* and *Prkdc^{scid}* mice may pose a special risk because infections are silent and chronic (Mahy et al., 1991). Cages and equipment should be autoclaved, and animal rooms should be fumigated with disinfectants such as formaldehyde, vaporized paraformaldehyde, and hydrogen peroxide.

Appropriate precautions are necessary for experiments involving LCMV, or LCMV-infected animals or materials. Biological safety level (BSL) 2 will be considered to be sufficient in most cases. Biological safety level 3 practices may be considered when working with infected animals owing to the increased risk of virus transmission by bite wounds, scratching, or aerosol formation from the bedding. Animal Biosafety Level (ABSL) 3 practices and facilities are generally recommended for work with infected hamsters. Appropriate precautions have been defined for different BSLs or animal biology safety levels by CDC (1999).

Lymphocytic choriomeningitis virus is an important zoonotic agent. It has been transmitted to humans working with infected animals or with contaminated biological materials and can cause mild to serious or fatal disease in humans (Dykewicz *et al.*, 1992; Barton *et al.*, 1995; Barton and Hyndman, 2000). Congenital infection in humans may result in hydrocephalus, or foetal or neonatal death (Barton *et al.*, 2002). Lymphocytic choriomeningitis virus is also frequently utilized as a model organism to study virus–host interactions, immunological tolerance, virus-induced immune complex disease, and a number of immunological mechanisms *in vivo* and *in vitro* (Slifka, 2002; Zinkernagel, 2002). Accidental transmission may have a severe impact on various kinds of experiments (for details, see Lehmann-Grube, 1982; Bhatt *et al.*, 1986b; National Research Council, 1991; Baker, 1998; Nicklas *et al.*, 1999).

Mammalian orthoreovirus serotype 3

Mammalian orthoreoviruses (MRV) are nonenveloped, segmented double-stranded RNA viruses of the family Reoviridae, genus Orthoreovirus. They have a wide host range and are ubiquitous throughout the world. The designation *reo* stands for *respiratory enteric orphan* and reflects the original isolation of these viruses from human respiratory and intestinal tract without apparent disease. The term 'orphan' virus refers to a virus in search of a disease. Mammalian orthoreovirus can be grouped into three serotypes (1, 2, 3). Mammalian orthoreovirus-3 (synonyms: hepatoencephalomyelitis virus; ECHO 10 virus) infection remains prevalent in contemporary mouse colonies and has been reported in wild mice (Smith *et al.*, 1993b; Barthold, 1997a). Seropositivity to MRV-3 was found in less than 5% of SPF colonies and in approximately 20% of non-SPF mouse colonies in the USA in 1996 (Jacoby and Lindsey, 1998). A study in France reported antibodies to MRV-3 in 9% of mouse colonies examined (Zenner

and Regnault, 2000). More recently, a study in North America found a low rate (0.2%) of mouse sera to be positive for antibodies against this virus (Livingston and Riley, 2003). In addition, contamination of mouse origin tumours and cell lines by MRV-3 has been reported many times (National Research Council, 1991; Nicklas *et al.*, 1993; Barthold, 1997a). Experimentally, MRV-3 infection of infant mice has been used to model human hepatobiliary disease, pancreatitis, diabetes mellitus, and lymphoma (Kraft, 1982; National Research Council, 1991; Fenner *et al.*, 1993).

The literature on MRV-3 infections in mice is dominated by studies on experimentally infected animals. The virus can cause severe **pantropic infection** in infant mice (Kraft, 1982; Tyler and Fields, 1986; Barthold, 1997a). After parenteral inoculation, virus can be recovered from the liver, brain, heart, pancreas, spleen, lymph nodes, and blood vessels. Following oral inoculation, reoviruses gain entry by infecting specialized epithelial cells (M cells) that overlie Peyer's patches. The virus then becomes accessible to leukocytes and spreads to other organs by way of the lymphatic system and the bloodstream. Neural spread to the CNS has also been well documented (Morrison *et al.*, 1991). The mechanisms of viral pathogenesis and their interactions with the host cell are reviewed in detail by Tyler (2001) and Tyler *et al.* (2001).

Natural infection by MRV-3 in a mouse colony usually is subclinical although diarrhoea or steatorrhoea and oily hair effect in suckling mice may be noted (Kraft, 1982; Tyler and Fields, 1986; National Research Council, 1991; Barthold, 1997a; Percy and Barthold, 2001). The latter term has been used to describe the matted, unkempt appearance of the hair coat that results from **steatorrhoea** due to pancreatitis, maldigestion, and **biliary atresia**. In addition, runting (attributed to immune-mediated destruction of cells in the pituitary gland that produce growth hormone), transient alopecia, jaundice (due to excessive bilirubin in the blood, which is attributed to the liver pathology, especially biliary atresia), and neurological signs such as incoordination, tremors, or paralysis may develop. When present in natural infections, clinical signs and lesions are similar to but milder than in experimental neonatal infections. Early descriptions of naturally occurring disease may have been complicated by concurrent infections such as MHV or murine rotavirus A (MuRV-A)/epizootic diarrhoea of infant mice (EDIM) virus that contributed to the severity of the lesions especially in liver, pancreas, CNS, and intestine. The outcome of MRV-3 infection depends on age and immunological status of mouse, dose of virus, and route of inoculation. Adult immunocompetent mice typically show no clinical signs and have no discernible lesions even in experimental infections. Mucosal and maternally conferred immunity are considered to be important in protection from or resolution of disease (Cuff *et al.*, 1990; Barthold *et al.*, 1993b). Experimental infection of adult *Prkdc^scid* mice is lethal (George *et al.*, 1990). Depending on the route of inoculation, experimental infection of adult *Foxn1^nu* mice is subclinical or results in liver disease (Carthew, 1984; George *et al.*, 1990).

Histological findings reported to occur after experimental MRV-3 infection of neonatal mice include inflammation and necrosis in liver, pancreas, heart, adrenal, brain, and spinal cord; lymphoid depletion in thymus, spleen, and lymph nodes; and hepatic fibrosis with biliary atresia (Papadimitriou and Robertson, 1976; Tyler and Fields, 1986; Barthold *et al.*, 1993b; Barthold, 1997a; Percy and Barthold, 2001).

Transmission of reoviruses probably involves the aerosol as well as the faecal–oral route (National Research Council, 1991). Fomites may play an important role as passive vectors because reoviruses resist environmental conditions moderately well.

Serological screening with ELISA or IFA is in widespread use for detection of antibodies to MRV-3 in diagnostic and health surveillance programmes. Both ELISA and IFA detect cross-reacting antibodies to heterologous MRV serotypes that can infect mice (ACLAD, 1991). The HI test does not detect such cross-reacting antibodies but is prone to give false positive results due to nonspecific inhibitors of haemagglutination (Kraft and Meyer, 1986; Van Der Logt, 1986; ACLAD, 1991). Reverse transcriptase–polymerase chain reaction methods for the detection of MRV-3 RNA (Steele *et al.*, 1995) or MRV RNA (Leary *et al.*, 2002) are also available. Reports on contamination of mouse origin tumours and cell lines by MRV-3 and its interference with transplantable tumour studies (Bennette, 1960; Nelson and Tarnowski, 1960) emphasize the importance of screening of biological materials to be inoculated into mice by MAP test or PCR. Natural seroconversion to MRV-3 without clinical disease is also observed in laboratory rats, hamsters, and guinea pigs (National Research Council, 1991; Barthold, 1997a).

Caesarean derivation and barrier maintenance have proven effective in the control and prevention of MRV-3 infection (Kraft, 1982; National Research Council, 1991).

The virus may interfere with research involving transplantable tumours and cell lines of mouse origin. It has the potential to alter intestinal studies and multiple immune response functions in mice. In enzootically infected colonies, protection of neonates by maternal antibody could complicate or prevent experimental infections with reoviruses. It could further complicate experiments that require evaluation of liver, pancreas, CNS, heart, lymphoid organs, and other tissues affected by the virus.

Murine hepatitis virus

The term murine hepatitis virus (MHV; commonly referred to as 'mouse hepatitis virus') designates a large group of antigenically and genetically related, single-stranded RNA viruses belonging to the family Coronaviridae, genus Coronavirus. They are surrounded by an envelope with a corona of surface projections (spikes). Murine hepatitis virus is antigenically related to rat coronaviruses and other coronaviruses of pigs, cattle, and humans. Numerous different strains or isolates of MHV have been described. They can be distinguished by neutralization tests that detect strain-specific **spike (S) antigens**. The best studied strains are the prototype strains MHV-1, MHV-2, MHV-3, JHM (MHV-4), A59, and S, of which MHV-3 is regarded as the most virulent. Murine hepatitis virus, like other coronaviruses, mutates rapidly, and strains readily form recombinants, so that new (sub)strains are constantly evolving. Strains vary in their virulence, organotropism, and cell tropism (Homberger, 1997). Based on their primary organotropism, MHV strains can be grouped into two biotypes: respiratory (or polytropic) and enterotropic. However, intermediate forms (enterotropic strains with tropism to other organs) exist. Murine hepatitis virus is relatively resistant to repeated freezing and thawing, heating (56°C for 30 min), and acid pH but is sensitive to drying and disinfectants, especially those with detergent activity (National Research Council, 1991).

Mus musculus is the natural host of MHV. It can be found in wild and laboratory mice throughout the world and is one of the most common viral pathogens in contemporary mouse colonies. While polytropic strains have historically been considered more common, this situation is thought to have reversed. A survey conducted in the USA in 1996 reported antibodies to MHV in more than 10% of SPF mouse colonies and more than 70% of non-SPF colonies (Jacoby and Lindsey, 1998), though very recent monitoring results for research institutions across North America indicate that the prevalence of MHV has decreased during the past few years (Livingston and Riley, 2003). A retrospective study in France covering the period from 1988 to 1997 reported antibodies to MHV in 67% of mouse colonies examined (Zenner and Regnault, 2000). Suckling rats inoculated experimentally with MHV had transient virus replication in the nasal mucosa and seroconversion but no clinical disease (Taguchi *et al.*, 1979). Similarly, deer mice seroconverted but showed no clinical disease after experimental infection (Silverman *et al.*, 1982). Murine hepatitis virus is also a common contaminant of transplantable tumours (Collins and Parker, 1972; Nicklas *et al.*, 1993) and cell lines (Sabesin, 1972; Yoshikura and Taguchi, 1979).

The pathogenesis and outcome of MHV infections depend on interactions among numerous factors related to the virus (e.g. virulence and organotropism) and the host (e.g. age, genotype, immune status, and microbiological status; Kraft, 1982; Barthold, 1986; National Research Council, 1991; Compton *et al.*, 1993; Homberger, 1997; Percy and Barthold, 2001). Murine hepatitis virus strains appear to possess a **primary tropism** for the upper respiratory or enteric mucosa. Those strains with respiratory tropism initiate infection in the nasal mucosa and then may disseminate via blood and lymphatics to a variety of other organs because of their polytropic nature. Respiratory (polytropic) strains include MHV-1, MHV-2, MHV-3, A59, S, and JHM. Infection of mice with virulent polytropic MHV strains, infection of mice less than 2 weeks of age, infection of genetically susceptible strains of mice, or infection of immunocompromised mice favour virus dissemination. Virus then secondarily replicates in vascular endothelium and parenchymal tissues, causing disease of brain, liver, lymphoid organs, bone marrow, and other sites. Infection of the brain by viraemic dissemination occurs primarily in immunocompromised or neonatal mice. Additionally, infection of adult mouse brain can occur by extension of virus along olfactory neural pathways, even in the absence of dissemination to other organs. In contrast, enterotropic MHV strains (e.g. LIVIM, MHV-D, and MHV-Y) tend to selectively infect intestinal mucosal epithelium, with no or minimal dissemination to other organs such as mesenteric lymph nodes or liver.

All ages and strains are susceptible to active infection, but disease is largely age-related. Infection of neonatal mice results in severe necrotizing enterocolitis with high mortality within 48 h. Mortality and lesion severity diminish rapidly with advancing age at infection. Adult mice develop minimal lesions although replication of equal or higher titres of virus occurs compared with neonates. The age-dependent decrease in severity of enterotropic MHV disease is probably related to the higher mucosal epithelium turnover in older mice,

allowing more rapid replacement of damaged mucosa. Another factor that is of considerable importance to the outcome of MHV infections is host genotype. For example, BALB/c mice are highly susceptible to enterotropic MHV disease while SJL mice, at the other end of the spectrum, are highly resistant (Barthold *et al.*, 1993a). Unlike in polytropic MHV infection where resistance is correlated with reduced virus replication in target cells (Barthold and Smith, 1987), enterotropic MHV grows to comparable titres in SJL and BALB/c mice at all ages (Barthold *et al.*, 1993a). Therefore, the resistance of the SJL mouse to disease caused by enterotropic MHV seems to be mediated through an entirely different mechanism than resistance to polytropic MHV. Furthermore, mouse genotypes that are susceptible to disease caused by one MHV strain may be resistant to disease caused by another strain (Barthold, 1986). It is therefore not possible to strictly categorize mouse strains as susceptible or resistant. The genetic factors determining susceptibility versus resistance in MHV infections are as yet poorly understood. Both polytropic and enterotropic MHV infections are self-limiting in immunocompetent mice. Immune-mediated clearance of virus usually begins about a week after infection, and most mice eliminate the virus within 3–4 weeks (Barthold, 1986; Barthold and Smith, 1990; Barthold *et al.*, 1993a). Humoral and cellular immunity appear to participate in host defences to infection, and functional T cells are an absolute requirement (Williamson and Stohlman, 1990; Kyuwa *et al.*, 1996; Lin *et al.*, 1999; Haring and Perlman, 2001). Therefore, immunodeficient mice such as *Foxn1nu* and *Prkdc scid* mice cannot clear the virus (Barthold *et al.*, 1985; Compton *et al.*, 1993). Similarly, some genetically modified strains of mice may have deficits in antiviral responses or other alterations that allow the development of persistent MHV infection (Rehg *et al.*, 2001). Recovered immune mice are resistant to reinfection with the same MHV strain but remain susceptible to repeated infections with different strains of MHV (Barthold and Smith, 1989a,b; Homberger *et al.*, 1992). Similarly, maternal immunity protects suckling mice against homologous MHV strains but not necessarily against other strains (Homberger and Barthold, 1992; Homberger *et al.*, 1992). However, maternal immunity, even to homologous strains, depends on the presence of maternally acquired antibody in the lumen of the intestine (Homberger and Barthold, 1992). Therefore, the susceptibility of young mice to infection significantly increases at weaning.

Most MHV infections are subclinical and follow one of two epidemiological patterns in immunocompetent mice (National Research Council, 1991; Homberger, 1997). Enzootic (subclinical) infection, commonly seen in breeding colonies, occurs when a population has been in contact with the virus for a longer period (e.g. several weeks). Adults are immune (due to prior infection), sucklings are passively protected, and infection is perpetuated in weanlings. **Epizootic** (clinical) infection occurs when the virus is introduced into a naive population (housed in open cages). The infection rapidly spreads through the entire colony. Clinical signs depend upon the virus and mouse strains and are most evident in infant mice. Typically, they include diarrhoea, poor growth, lassitude, and death. In infections due to virulent enterotropic strains, mortality can reach 100% in infant mice. Some strains may also cause neurological signs such as flaccid paralysis of hind limbs, convulsions, and circling. Adult infections are again usually asymptomatic. As the infection becomes established in the colony, the epizootic pattern is replaced by the enzootic pattern. In immunodeficient (e.g. *Foxn1^{nu}* and *Prkdc^{scid}*) mice, infection with virulent polytropic MHV strains often is rapidly fatal while less virulent strains cause chronic wasting disease (Compton *et al.*, 1993). In contrast, adult immunodeficient mice can tolerate chronic infection by enterotropic MHV, with slow emaciation and diarrhoea, or minimal clinical disease (Barthold *et al.*, 1985; Barthold, 1986). Subclinical MHV infections can be activated by a variety of experimental procedures (e.g. thymectomy, whole body irradiation, treatment with chemotherapeutic agents, halothane anaesthesia) or by co-infections with other pathogens (e.g. *Eperythrozoon coccoides*, K virus; reviewed by Kraft, 1982; National Research Council, 1991).

In most natural infections, gross lesions are not present or are transient and not observed. Gross findings in neonates with clinical signs include dehydration, emaciation, and in contrast to EDIM, an empty stomach (Ishida *et al.*, 1978; Barthold *et al.*, 1982; Kraft, 1982). The intestine is distended and filled with watery to mucoid yellowish, sometimes gaseous contents. Haemorrhage or rupture of the intestine can occur. Depending on the virus strain, necrotic foci on the liver (Ishida *et al.*, 1978; Kraft, 1982; Percy and Barthold, 2001) and thymus involution (Barthold *et al.*, 1982; Godfraind *et al.*, 1995) may also be seen in susceptible mice. Liver involvement may be accompanied by jaundice and haemorrhagic peritoneal exudate. Splenomegaly may occur as a result of compensatory haematopoiesis (Fox *et al.*, 1977).

Histopathological changes in susceptible mice infected with polytropic MHV strains include acute necrosis with syncytia in liver, spleen, lymph nodes, gut-associated lymphoid tissue, and bone marrow (Kraft, 1982; Barthold, 1986; National Research Council, 1991; Percy and Barthold, 2001). Neonatally infected mice can have vascular-oriented necrotizing (meningo)encephalitis with demyelination in the brain stem and peri-ependymal areas. Lesions in peritoneum, bone marrow, thymus, and other tissues can be variably present. Mice can develop nasoencephalitis due to extension of infection from the nasal mucosa along olfactory pathways to the brain, with meningoencephalitis and demyelination, the latter of which is thought to be largely T cell-mediated (Haring and Perlman, 2001). This pattern of infection regularly occurs after intranasal inoculation of many MHV strains but is a relatively rare event after natural exposure. Syncytia arising from endothelium, parenchyma, or leukocytes is a hallmark of infection in many tissues including intestine, lung, liver, lymph nodes, spleen, thymus, brain, and bone marrow. Lesions are transient and seldom fully developed in adult immunocompetent mice, but they are manifest in immunocompromised mice. Highly unusual presentations can occur in mice with specific gene defects. For example, granulomatous peritonitis and pleuritis were found in interferon-γ-deficient mice infected with MHV (France *et al.*, 1999).

Histopathological changes caused by enterotropic strains of MHV are mainly confined to the intestinal tract and associated lymphoid tissues (Kraft, 1982; Barthold, 1986; National Research Council, 1991; Percy and Barthold, 2001). The most common sites are terminal ileum, caecum, and proximal colon. The severity of disease is primarily age-dependent, with neonatal mice being most severely affected. These mice show segmentally distributed areas of villus attenuation, enterocytic syncytia (balloon cells), and mucosal necrosis accompanied by leukocytic infiltration. Intracytoplasmic inclusions are present in enterocytes. Erosions, ulceration, and haemorrhage may be seen in more severe cases. Lesions can be fully developed within 24–48 h, but are usually more severe at 3–5 days after infection. Surviving mice may develop compensatory mucosal hyperplasia. Mesenteric lymph nodes usually contain lymphocytic syncytia, and mesenteric vessels may contain endothelial syncytia. Pathological changes in older mice are generally much more subtle and may only consist of transient syncytia. An occasional exception seems to occur in immunodeficient animals such as *Foxn1^{nu}* mice, which can develop chronic hyperplastic **typhlocolitis** of varying severity (Barthold *et al.*, 1985), but other agents such as *Helicobacter* species may have been involved. In general, enterotropic

MHV strains do not disseminate, but hepatitis and encephalitis can occur with some virus strains in certain mouse genotypes.

Murine hepatitis virus is highly contagious. It is shed in faeces and nasopharyngeal secretions and appears to be transmitted via direct contact, aerosol, and fomites (Kraft, 1982; National Research Council, 1991). Vertical (*in utero*) transmission has been demonstrated in experimental infections (Katami *et al.*, 1978) but does not seem to be of practical importance under natural conditions.

Diagnosis during the acute stage of infection can be made by histological demonstration of characteristic lesions with syncytia in target tissues, but clinical signs and lesions can be highly variable and may not be prominent. Suckling, genetically susceptible or immunocompromised mice are the best candidates for evaluation. Active infection can be confirmed by immunohistochemistry (Brownstein and Barthold, 1982) or by virus isolation. Virus recovery from infected tissues is difficult but can be accomplished using primary macrophage cultures or a number of established cell lines such as NCTC 1469 or DBT (ACLAD, 1991). These cells, however, may not be successful substrates for some enterotropic MHV strains. Virus in suspect tissue can also be confirmed by bioassays such as MAP testing or infant or *Foxn1^nu* mouse inoculation (De Souza and Smith, 1989; ACLAD, 1991). Amplification by passage in these mice increases the likelihood of detection of lesions and antigen, or virus recovery. Other direct diagnostic methods that have been successfully utilized to detect MHV in faeces or tissue of infected mice include monoclonal antibody solution hybridization assay (Casebolt and Stephensen, 1992) and a number of RT-PCR assays (Homberger *et al.*, 1991; Kunita *et al.*, 1992; Yamada *et al.*, 1993; Besselsen *et al.*, 2002). Because of the transient nature of MHV infection in immunocompetent mice, serology is the most appropriate diagnostic tool for routine monitoring. Enzyme-linked immunosorbent assay and IFA are well established and sensitive, and all known MHV strains cross-react in both tests (Smith, 1983; ACLAD, 1991). The magnitude of antibody response depends on MHV strain and mouse genotype (Nakanaga *et al.*, 1983; Barthold and Smith, 1987). DBA/2 mice are poor antibody responders whereas C57BL/6 mice produce a high antibody titre and are therefore good sentinels. Antibody titres remain high over a period of at least 6 months (Barthold and Smith, 1989b; Homberger *et al.*, 1992). Infected mice may not develop detectable antibodies for up to 14 days after initial exposure (Smith, 1983). In such cases, a direct diagnostic method as discussed above may be useful. Another drawback of serology is that mice weaned from immune dams can have maternal antibodies until they are 10 weeks of age (Homberger, 1992). This may impact serological monitoring because the possibility must be considered that low positive results are due to maternally-derived passive immunity. Because the virus can be transmitted by transplantable tumours and other biological materials from mice, including hybridomas (Holmes *et al.*, 1986) and embryonic stem cells (Okumura *et al.*, 1996; Kyuwa, 1997), these materials should also be routinely screened for MHV contamination. Mouse inoculation bioassay, MAP test, and RT-PCR can be used for this purpose.

The best means of MHV control is to prevent its entry into a facility. This can be accomplished by purchase of mice from virus-free sources and maintenance under effective barrier conditions monitored by a well-designed quality assurance programme. Control of wild mouse populations, proper husbandry and sanitation, and strict monitoring of biological materials that may harbour virus are also important measures to prevent infection. If infection occurs, the most effective elimination strategy is to cull the affected colony and obtain clean replacement stock. However, this is not always a feasible option when working with valuable mice (e.g. genetically modified lines, breeding stocks). Caesarean derivation or embryo transfer can be used to produce virus-free offspring, and foster-nursing also has been reported to be effective (Lipman *et al.*, 1987). Quarantine of an affected colony with no breeding and no introduction of new animals for approximately 2 months has been effective in immunocompetent mice (Weir *et al.*, 1987). The infection is likely to be terminated because MHV requires a constant supply of susceptible animals. This method works best when working with small numbers of mice. Large populations favour the development of new MHV strains that may result in repeated infections with slightly different strains (Adami *et al.*, 1995). It may be practical to select a few future breeders from the infected population and quarantine them for approximately 3 weeks (Compton *et al.*, 1993). This can be achieved in isolators, or in individually ventilated cages if proper handling is guaranteed. After this interval, breeding can resume. The 3-week interval should permit recovery from active infection, and the additional 3-week gestation period effectively extends the total quarantine to 6 weeks. It is advisable to select seropositive breeders because the possibility of active infection is lower in such animals. The breeding cessation strategy may not be successful if immunodeficient mice are used because they are susceptible to chronic infection and viral excretion (Barthold *et al.*, 1985). Genetically engineered mice of unclear, unknown or deficient immune status pose a special challenge because they may develop unusual manifestations of infection or may be unable to clear virus. Rederivation likely is the most cost effective strategy in such situations. Along with the measures described, proper sanitation and disinfection of caging and animal quarters as well as stringent personal sanitation are essential to eliminate infection. Careful testing with sentinel mice should be applied to evaluate the effectiveness of rederivation. If transplantable tumours are contaminated with MHV, virus elimination can be achieved by passage of tumours in athymic Whn^{rnu} rats (Rülicke *et al.*, 1991).

Murine hepatitis virus is one of the most important viral pathogens of laboratory mice and has been intensively studied from a number of research perspectives (e.g. as a model organism for studying coronavirus molecular biology or the pathogenesis of viral-induced demyelinating disease). Numerous reports document the effects of natural and experimental infections with MHV on host physiology and research, especially in the fields of immunology and tumour biology (reviewed by Barthold, 1986; National Research Council, 1991; Compton *et al.*, 1993; Homberger, 1997; Baker, 1998; Nicklas *et al.*, 1999).

Murine pneumonia virus

Murine pneumonia virus, commonly referred to as 'pneumonia virus of mice' (PVM), is an enveloped, single-stranded RNA virus of the family Paramyxoviridae, genus Pneumovirus. It is closely related to human respiratory syncytial virus (HRSV). The virus name is officially abbreviated as 'MPV' according to the International Union of Microbiological Societies (2000); however, the former designation 'PVM' will be used in this chapter to avoid confusion with the official abbreviation of mouse parvovirus 1 (MPV). 'Pneumonia virus of mice' infection is relatively common in colonies of mice and rats throughout the world. Seropositivity to PVM was reported in less than 5% of SPF mouse colonies and in approximately 20% of non-SPF mouse colonies in the USA (Jacoby and Lindsey, 1998). A serological survey in France demonstrated antibodies to PVM in 16% of mouse colonies examined (Zenner and Regnault, 2000). In a more recent study in North America, such antibodies were found in only 0.1% of mice monitored (Livingston and Riley, 2003). Antibodies to PVM have also been detected in

hamsters, gerbils, cotton rats, guinea pigs, and rabbits (Parker and Richter, 1982; Richter, 1986; National Research Council, 1991). Experimentally, PVM infection of mice is used as a model for HRSV infection (Domachowske *et al.*, 2000).

In immunocompetent mice, natural infection with PVM is transient and usually not associated with clinical disease or pathological findings (Parker and Richter, 1982; National Research Council, 1991; Brownstein, 1996b). However, natural disease and persistent infection may occur in immunodeficient mice (Carthew and Sparrow, 1980; Richter *et al.*, 1988; Weir *et al.*, 1988). In particular, athymic $Foxn1^{nu}$ mice seem to be susceptible to PVM infection, which can result in dyspnoea, cyanosis, emaciation, and death due to pneumonia (Richter *et al.*, 1988; Weir *et al.*, 1988). Similar clinical signs have been reported for experimentally infected, immunocompetent mice (Cook *et al.*, 1998).

Necropsy findings in naturally infected $Foxn1^{nu}$ mice include **cachexia** and diffuse pulmonary oedema or lobar consolidation (Weir *et al.*, 1988). Pulmonary consolidation (dark red or grey in colour) also has been found after experimental infection of immunocompetent mice (Brownstein, 1996b).

Histologically, natural infection of $Foxn1^{nu}$ mice with PVM presents as interstitial pneumonia (Richter *et al.*, 1988; Weir *et al.*, 1988). Experimental intranasal inoculation of immunocompetent mice can result in rhinitis, erosive bronchiolitis, and interstitial pneumonia with prominent early pulmonary eosinophilia and neutrophilia (Brownstein, 1996b; Domachowske *et al.*, 2000). Hydrocephalus may result from intracerebral inoculation of neonatal mice (Lagace-Simard *et al.*, 1980). Susceptibility to infection is influenced by age of mouse, dose of virus, and a variety of local and systemic stressors (Parker and Richter, 1982; National Research Council, 1991).

Pneumonia virus of mice is labile in the environment and rapidly inactivated at room temperature (Parker and Richter, 1982; National Research Council, 1991). The virus is tropic for the respiratory epithelium (Carthew and Sparrow, 1980; Cook *et al.*, 1998), and transmission is exclusively horizontal via the respiratory tract, mainly by direct contact and aerosol (Parker and Richter, 1982; National Research Council, 1991). Therefore, transmissibility in mouse colonies is low, and infections tend to be focal enzootics.

Serology (ELISA, IFA, or HI) is the primary means of testing mouse colonies for exposure to PVM. Immunohistochemistry has been applied to detect viral antigen in lung sections (Carthew and Sparrow, 1980; Weir *et al.*, 1988), however, proper sampling (see Chapter on Health Monitoring) is critical for establishing the diagnosis due to the focal nature of the infection. An RT-PCR assay to detect viral RNA in respiratory tract tissues has also been reported (Wagner *et al.*, 2003). However, the use of direct methods requires good timing because the virus is present for only up to about 10 days in immunocompetent mice (Brownstein, 1996b).

Embryo transfer or caesarean derivation followed by barrier maintenance can be used to rear mice that are free of PVM. Because active infection is present in the individual immunocompetent mouse for only a short period, strict isolation of a few (preferably seropositive) mice with the temporary cessation of breeding might also be successful in eliminating the virus (Richter, 1986; National Research Council, 1991).

Pneumonia virus of mice could interfere with studies involving the respiratory tract or immunological measurements in mice. In addition, PVM can have devastating effects on research using immunodeficient mice because they are particularly prone to develop fatal disease (Richter *et al.*, 1988; Weir *et al.*, 1988) or become more susceptible to the deleterious effects of other agents such as *Pneumocystis carinii* (Roths *et al.*, 1993).

MuRV-A/EDIM

Murine rotavirus-A/EDIM (commonly referred to as 'mouse rotavirus' or 'epizootic diarrhoea of infant mice virus') is a nonenveloped, segmented double-stranded RNA virus of the family Reoviridae, genus Rotavirus. It is antigenically classified as a group A rotavirus, similar to rotaviruses of many other species that cause neonatal and infantile gastroenteritis (Fenner *et al.*, 1993). Murine rotavirus-A/EDIM infection remains prevalent in contemporary mouse colonies and appears to occur worldwide. Seropositivity to MuRV-A/EDIM was reported in approximately 5% of SPF colonies and in almost 30% of non-SPF mouse colonies in the USA in 1996 (Jacoby and Lindsey, 1998). More recently, Livingston and Riley (2003) found a low rate (1%) of mouse sera to be positive for antibodies against MuRV-A/EDIM. Experimentally, MuRV-A/EDIM infection in mice is used as a model for human rotavirus infection, especially in investigations on the mechanisms of rotavirus immunity and in the development of vaccination strategies (Ward and McNeal, 1999).

Clinical symptoms following MuRV-A/EDIM infection range from inapparent or mild to severe, sometimes fatal, diarrhoea. 'Epizootic diarrhoea of infant mice' describes the clinical syndrome associated with natural or experimental infection by MuRV-A/EDIM during the first 2 weeks of life (Kraft, 1982; Sheridan and Vonderfecht, 1986; National Research Council, 1991; Barthold, 1997b; Percy and Barthold, 2001). Diarrhoea usually begins around 48 h after infection and persists for about 1 week. Affected suckling mice have soft, yellow faeces that wet and stain the perianal region. In severe instances, the mice may be stunted, have dry scaly skin, or are virtually covered with faecal material. Morbidity is very high but mortality is usually low.

Gross lesions in affected mice are confined to the intestinal tract. The caecum and colon may be distended with gas and watery to paste-like contents that are frequently bright yellow. The stomach of diarrheic mice is almost always filled with milk, and this feature has been reported to be a reliable means to differentiate diarrhoea caused by rotavirus from the diarrhoea caused by MHV infection.

Histopathological changes may be subtle even in animals with significant diarrhoea. They are confined to the small intestine and are most prominent at the apices of villi, where rotaviruses infect and replicate within epithelial cells. Hydropic change of villous epithelial cells is the hallmark finding of acute disease. The villi become shortened, and the cells that initially replace the damaged cells are less differentiated, typically cuboidal instead of columnar, and lack a full complement of enzymes for digestion and absorption, resulting in diarrhoea due to maldigestion and malabsorption. Undigested milk in the small intestine promotes bacterial growth and exerts an osmotic effect, exacerbating damage to the villi. Intestinal fluid and electrolyte secretion is further enhanced by activation of the enteric nervous system (Lundgren *et al.*, 2000) and through the effects of a viral enterotoxin called NSP4 (for nonstructural protein 4; Ball *et al.*, 1996). It is hypothesized that NSP4 is released from virus-infected cells and then triggers a signal transduction pathway that alters epithelial cell permeability and chloride secretion.

Susceptibility to EDIM depends on the age of the host and peaks between 4 and 14 days of age (Kraft, 1982; Sheridan and Vonderfecht, 1986; National Research Council, 1991; Barthold, 1997b; Percy and Barthold, 2001). Mice older than about 2 weeks can still be infected with MuRV-A/EDIM, but small numbers of enterocytes become infected, there is little replication of virus, and diarrhoea does not occur. The exact reason for this age-related resistance to disease is unknown. Pups suckling immune dams are protected against EDIM during their period of disease susceptibility

(Rosé *et al.*, 1998). In general, the infection is self-limiting and resolves within days. Successful viral clearance is promoted by an intact immune response (Feng *et al.*, 1997; McNeal *et al.*, 1997; Rosé *et al.*, 1998), and some immunodeficient mice (e.g. *Prkdc^{scid}* and *Rag2^{tm1Fwa}* mice) may shed virus for extended periods or become persistently infected (Riepenhoff-Talty *et al.*, 1987; Franco and Greenberg, 1995). Protection against MuRV-A/EDIM reinfection is primarily mediated by antibodies (Feng *et al.*, 1997; Rosé *et al.*, 1998).

Murine rotavirus-A/EDIM is highly contagious and transmitted by the faecal–oral route (Kraft, 1982; Sheridan and Vonderfecht, 1986; National Research Council, 1991). Dissemination of the virus occurs through direct contact or contaminated fomites and aerosols. MuRV-A/EDIM is stable at −70°C but otherwise tends to be susceptible to extreme environmental conditions, detergents, and disinfectants.

Enzyme-linked immunosorbent assay and IFA are in widespread use for detection of serum antibodies to MuRV-A/EDIM in diagnostic and health surveillance programmes; other assay systems such as those using latex agglutination are also utilized (Ferner *et al.*, 1987). Rotazyme II is a commercially available ELISA for detection of rotavirus antigen in faeces; however, great care must be used in interpreting the results because some feeds have been reported to cause false positive reactions (Jure *et al.*, 1988). Electron microscopy of faeces of diarrheic pups should reveal typical wheel-shaped rotavirus particles, 60–80 nm in diameter. Reverse transcriptase-polymerase chain reaction also can be used to detect rotavirus RNA in faecal samples (Wilde *et al.*, 1990). Good timing is critical for establishing the diagnosis from faeces because virus is shed for only a few days in immunocompetent mice.

Embryo transfer or caesarean derivation followed by barrier maintenance is recommended for rederivation of breeding stocks (Kraft, 1982; National Research Council, 1991). In immunocompetent mice in which infection is effectively cleared, a breeding suspension strategy combined with excellent sanitation, filter tops, and conscientious serological testing of offspring may also be effective.

Murine rotavirus-A/EDIM has the potential to interfere with any research utilizing suckling mice. It may have a significant impact on studies where the intestinal tract of neonatal or infant mice is the target organ. The infection also poses a problem for infectious disease and immune response studies, particularly those involving enteropathogens in infant mice (Newsome and Coney, 1985). In addition, runting could be interpreted erroneously as the effect of genetic manipulation or other experimental manipulation.

Sendai virus

Sendai virus (SeV) is an enveloped, single-stranded RNA virus of the family Paramyxoviridae, genus Respirovirus. It is antigenically related to human parainfluenza virus 1. The virus was named for Sendai, Japan, where it was first isolated from mice. Infections of mice and rats are relatively common and occur worldwide. In addition, there is evidence that hamsters, guinea pigs, and rabbits are susceptible to infection with SeV (Machii *et al.*, 1989; ACLAD, 1991; National Research Council, 1991; Percy and Palmer, 1997); however, some apparently seropositive guinea pigs may in fact be seropositive to other parainfluenza viruses instead of SeV. Seropositivity to SeV was reported to be absent from SPF mouse colonies and to be approximately 20% in non-SPF mouse colonies in the USA (Jacoby and Lindsey, 1998). A study in France reported antibodies to SeV in 17% of mouse colonies examined (Zenner and Regnault, 2000). A low rate of seropositive mice (0.2%) was found in a recent survey in North America (Livingston and Riley, 2003). Furthermore, SeV can contaminate biological materials (Collins and Parker, 1972).

Sendai virus is pneumotropic and the leading cause of viral respiratory disease in mice. The pneumotropism is partially a consequence of the action of respiratory serine proteases such as tryptase Clara, which activate viral infectivity by specific cleavage of the viral fusion glycoprotein (Tashiro *et al.*, 1999). In addition, the apical budding behaviour of SeV may hinder the spread of virus into subepithelial tissues and subsequently to distant organs via the blood.

Two epidemiologic patterns of SeV infection have been recognized, an enzootic (subclinical) and epizootic (clinically apparent) type (Parker and Richter, 1982; National Research Council, 1991; Brownstein, 1996a). Enzootic infections commonly occur in breeding or open colonies, where the constant supply of susceptible animals perpetuates the infection. In breeding colonies, mice are infected shortly after weaning as maternal antibody levels wane. Normally, the infection is subclinical, with virus persisting for approximately 2 weeks, accompanied by seroconversion that persists for a year or longer. Epizootic infections occur upon first introduction of the virus to a colony and either die out (self-cure) after 2–7 months or become enzootic depending on colony conditions. The epizootic form is generally acute, and morbidity is very high resulting in nearly all susceptible animals becoming infected within a short time. Clinical signs vary and include rough hair coat, hunched posture, chattering, respiratory distress, prolonged gestation, death of neonates and sucklings, and runting in young mice. Breeding colonies may return to normal productivity in 2 months and thereafter maintain the enzootic pattern of infection. Factors such as strain susceptibility, age, husbandry, transport, and copathogens are important in precipitating overt disease. DBA and 129/J strains of mice are very susceptible to SeV pneumonia whereas SJL/J and C57BL/6/J strains and several outbred stocks are relatively resistant. A/J, BALB/c, and SWR/J are among the strains that show intermediate susceptibility. There is no evidence for persistent infection in immunocompetent mice, but persistent or prolonged infection may occur in immunodeficient mice and can result in wasting and death due to progressive pneumonia (Ward *et al.*, 1976; Iwai *et al.*, 1979; Percy *et al.*, 1994). Clearance of a primary SeV infection is mediated by CD8$^+$ and CD4$^+$ T cell mechanisms (Kast *et al.*, 1986; Hou *et al.*, 1992).

Heavier than normal, consolidated, plum-coloured or grey lungs are a characteristic gross finding in severe SeV pneumonia (Parker and Richter, 1982; National Research Council, 1991; Brownstein, 1996a; Percy and Barthold, 2001). Lymphadenopathy and splenomegaly reflect the vigorous immune response to infection.

Histologically, three phases of disease can be recognized in susceptible immunocompetent mice: acute, reparative, and resolution phases (Brownstein, 1996a; Percy and Barthold, 2001). Lesions of the acute phase, which lasts 8–12 days, are primarily attributed to the cell-mediated immune response that destroys infected respiratory epithelial cells and include necrotizing rhinitis, tracheitis, bronch(iol)itis, and alveolitis. Epithelial syncytiae and cytoplasmic inclusion bodies in infected cells may be seen early in this phase. Alveoli contain sloughed necrotic epithelium, fibrin, neutrophils, and mononuclear cells. Atelectasis, bronchiectasis, and emphysema may occur as a result of damage and obstruction of airways. The reparative phase, which may overlap the acute phase but continues through about the third week post infection, is indicated by regeneration of airway lining epithelium. Adenomatous hyperplasia and squamous metaplasia (with multilayered flat epithelial cells instead of normal columnar cells) in the terminal bronchioles and alveoli are considered to be a hallmark of SeV pneumonia. Mixed inflammatory cell infiltrates in this phase tend to be primarily interstitial rather than alveolar as they are in the acute phase. The resolution phase may be complete by the fourth week post infection and lesions may be difficult to identify subsequently. Residual, persistent lesions that may occur include organizing

alveolitis and bronchiolitis fibrosa obliterans. Alveoli and bronchioles are replaced by collagen and fibroblasts, foamy macrophages, and lymphoid infiltrates, often with foci of emphysema, cholesterol crystals, and other debris, which represent attempts to organize and wall off residual necrotic debris and fibrin. Lesions are more severe and variable when additional pathogens such as *Mycoplasma pulmonis* are present (National Research Council, 1991). Otitis media has also been reported in natural infections with SeV although some of these studies have been complicated by the presence of other pathogens (Ward, 1974). Sendai virus has been detected in the inner ear after experimental intracerebral inoculation of neonatal mice (Shimokata *et al.*, 1977).

Sendai virus is extremely contagious. Infectious virus is shed during the first 2 weeks of infection and appears to be transmitted by direct contact, contaminated fomites, and respiratory aerosol (Parker and Reynolds, 1968; Parker and Richter, 1982; National Research Council, 1991).

Serology (ELISA, IFA, or HI) is the approach of choice for routine monitoring because serum antibodies to SeV are detectable soon after infection and persist at high levels for many months, although active infection lasts only 1–2 weeks in immunocompetent mice. The short period of active infection limits the utility of direct methods such as immunohistochemistry (Carthew and Sparrow, 1980) and RT-PCR (Hayase *et al.*, 1997; Wagner *et al.*, 2003). Although SeV is considered to be highly contagious, studies have shown that dirty bedding sentinel systems do not reliably detect the infection and that outbred stocks may not seroconvert consistently (Dillehay *et al.*, 1990; Artwohl *et al.*, 1994). Mouse antibody production test and RT-PCR can be used to detect SeV in contaminated biological materials.

Sendai virus infection in mouse colonies has proven to be one of the most difficult virus infections to control because the virus is highly infectious and easily disseminated. Depopulation of infected colonies probably is the most appropriate means to eliminate the virus in most situations. Embryo transfer followed by barrier maintenance has also been used successfully in eliminating the virus (National Research Council, 1991). A less effective alternative is to place the infected animals under strict quarantine, remove all young and pregnant mice, suspend all breeding, and prevent addition of other susceptible animals for approximately 2 months until the infection is extinguished and then breeding and other normal acitivities are resumed (Parker and Richter, 1982; National Research Council,

1991). Vaccines against the virus have been developed (Brownstein, 1986; National Research Council, 1991), but these probably do not represent a practical means to achieve or maintain the seronegative status of colonies that is in demand today.

Sendai virus has the potential to interfere with a wide variety of research involving mice. Reported effects include interference with early embryonic development and foetal growth; alterations of macrophage, NK cell, and T and B cell function; altered responses to transplantable tumours and respiratory carcinogens; altered isograft rejection; and delayed wound healing (reviewed by National Research Council, 1991; Baker, 1998; Nicklas et al., 1999). Pulmonary changes during SeV infection can compromise interpretation of experimentally induced lesions and may lead to opportunistic infections by other agents. They could also affect the response to anaesthetics. In addition, natural SeV infection would interfere with studies using SeV as a gene vector.

Theiler's murine encephalomyelitis virus

Theiler's murine encephalomyelitis virus (TMEV) or murine poliovirus is a member of the genus Cardiovirus in the family Picornaviridae. Members of this genus are nonenveloped viruses with single-stranded RNA. The virus is rapidly destroyed at temperatures above 50°C. It is considered to be a primary pathogen of the CNS of mice and can cause clinical disease resembling that due to poliomyelitis virus infections in humans. Antibodies to TMEV have been identified in mouse colonies and feral populations worldwide, and *Mus musculus* is considered to be the natural host of TMEV (Lipton et al., 2001). The most well-known and most frequently mentioned TMEV strain is GDVII, which is virulent for mice. Infant or young hamsters and laboratory rats are also susceptible to intracerebral infection. The original isolate is designated TO (Theiler's original) and represents a group of TMEV strains with low virulence for mice. Many additional virus strains have been isolated and studied, and they all fall in the broad grouping of TO and GDVII. A similar virus strain has also been isolated from rats, but in contrast to mouse isolates this virus is not pathogenic for rats and mice after intracerebral inoculation (Hemelt et al., 1974). Recently, another rat isolate has been characterized and shown to be most closely related to but quite distinct from other TMEV viruses (Ohsawa et al., 2003). Antibodies to TMEV

(strain GDVII) have been detected in guinea pigs and are considered to indicate infection with another closely related cardiovirus (Hansen et al., 1997).

Seropositivity to TMEV was reported in approximately 5% of SPF mouse colonies and approximately 35% of non-SPF mouse colonies in the USA (Jacoby and Lindsey, 1998). Zenner and Regnault (2000) reported a prevalence rate of 9% in French mouse colonies in a retrospective study, and it has been one of the most common virus infections in rodent colonies. In a recent study, antibodies were found in 0.2% of mice monitored (Livingston and Riley, 2003) indicating that TMEV, like most viruses, has meanwhile been eliminated from the majority of mouse colonies.

Theiler's murine encephalomyelitis virus is primarily an enteric pathogen, and virus strains are enterotropic. In natural infections, virus can be detected in intestinal mucosa and faecal matter, and in some cases it is also found in the mesenteric lymph nodes. However, histological lesions in the intestine are not discerned. Virus may be shed via intestinal contents for up to 22 weeks, sometimes intermittently (Brownstein et al., 1989a), and transmission under natural conditions is via the faecal–oral route by direct contact between mice as well as by indirect contact (e.g. dirty bedding). The host immune response limits virus spread, but it does not immediately terminate virus replication in the intestines. Virus is cleared from extraneural tissues, but it persists in the CNS for at least a year.

Clinical disease due to natural TMEV infection is rare, with a rate of only 1 in 1000–10,000 infected immunocompetent animals (Percy and Barthold, 2001). In immunodeficient mice, especially in weanlings, clinical signs may be more common and mortality may be higher (Rozengurt and Sanchez, 1993). This group of viruses usually causes asymptomatic infections of the intestinal tract. They may spread to the CNS as a rare event where they cause different neurological disease manifestations. The most typical clinical sign of TMEV infection is flaccid paralysis of hind legs. The animals appear otherwise healthy, and there is no mortality.

Experimental infection in mice provides models of poliomyelitis-like infection and virus-induced demyelinating disease including multiple sclerosis (McGavern et al., 2000). After experimental infection, TMEV causes a biphasic disease in susceptible strains of mice. The acute phase is characterized by early infection of neurons in the grey matter. Encephalomyelitis may develop during this phase and may be fatal, but most animals survive and enter the second phase of the disease at 1–3 months after the acute phase. This phase is characterized by viral persistence in the spinal cord white matter, mainly in macrophages, and leads to white matter demyelination. Persistence and demyelination occur only in genetically susceptible mouse strains while resistant strains clear the infection after early grey matter encephalomyelitis through a cytotoxic T lymphocyte response. For this reason, the nude mutation (*Foxn1^nu*) confers susceptibility on mice with an otherwise resistant background.

The severity and nature of disease depend on virus strain, route of inoculation, host genotype and age (Downs, 1982; Lipton and Rozhon, 1986; National Research Council, 1991; Percy and Barthold, 2001). In general, virus isolates with low virulence produce persistent CNS infection in mice whereas virulent strains are unable to cause persistent infection. Intracerebral inoculation results in the most severe infections, but the intranasal route is effective also. Experimental intracerebral infections with virulent FA and GDVII strains of TMEV are more likely to cause acute encephalomyelitis and death in weanling mice 4–5 days after inoculation ('Early Disease'). Death may be preceded by neurological manifestations of encephalitis such as hyperexcitability, convulsions, tremors, circling and rolling, and weakness. Animals may develop typical flaccid paralysis of hind limbs, and locomotion is possible only by use of the forelimbs. Interestingly, the tail is not paralysed. Experimental infections with low virulence virus strains (e.g. TO, DA, WW) are more likely to cause persistent infection with development of mild encephalomyelitis followed by a chronic demyelinating disease after a few months ('late disease'). These virus strains infect neurons in the grey matter of the brain and spinal cord during the acute phase of viral growth, followed by virus persistence in macrophages and glial cells in the spinal cord white matter. SJL, SWR, and DBA/2 strains are most susceptible to this chronic demyelinating disease. CBA and C3H/He are less susceptible strains, and strains A, C57BL/6, C57BL/10, and DBA/1 are relatively resistant (Lipton and dal Canto, 1979). Differences in humoral immune responses play a role in resistance to TMEV infection (Pena Rossi et al., 1991a), but genetic factors are also important. Several genetic loci implicated in susceptibility to virus persistence, demyelination, or clinical disease have been identified, including the H-2D region of the major histocompatibility complex (Brahic and Bureau, 1998). Furthermore, the age at infection influences the severity of clinical disease. In infant mice, intracerebral infection with low virulence virus strains (e.g. TO) is often lethal. Young mice develop paralysis after an incubation period of 1–4 weeks while adult mice often show no clinical signs of infection (Downs, 1982).

The only gross lesions are secondary to the posterior paralysis and may include urine scald or dermatitis due to incontinence of urine and trauma to paralysed limbs, or wasting or atrophy of the hind limbs in long term survivors.

Theiler's murine encephalomyelitis virus infects neurons and glial cells, and histological changes in the CNS include nonsuppurative meningitis, perivasculitis, and poliomyelitis with neuronolysis, neuronophagia, and microgliosis in the brainstem and ventral horns of the spinal cord (Percy and Barthold, 2001). Demyelination in immunocompetent mice is considered to be immune-mediated. Susceptible strains develop a specific delayed-type hypersensitivity response which is the basis for inflammation and demyelination. This reaction is mediated by cytotoxic T lymphocytes (Lindsley et al., 1991; Pena Rossi et al., 1991b) and by the activation of cytokines as a consequence of infection of macrophages and other cells of the CNS (Rubio and Capa, 1993; Sierra and Rubio, 1993; Palma et al., 2003). Protection from chronic demyelinating disease is possible by vaccination with live virus given previously by subcutaneous or intraperitoneal inoculation (Crane et al., 1993; Kurtz et al., 1995). Early immunosuppression at the time of infection, e.g. by treatment with cyclophosphamide or antithymocyte serum, inhibits or diminishes demyelination. Immunosuppression in mice chronically infected with TMEV leads to remyelination of oligodendrocytes (Rodriguez and Lindsley, 1992). Further details related to the pathogenesis of TMEV infections and the role of immune mechanisms have been reviewed by Yamada et al. (1991).

Experimental infection of *Foxn1^nu* mice results in acute encephalitis and demyelination. Demyelination associated with minimal inflammation and neurological signs including the typical hind limb paresis develop 2 weeks after inoculation, and most animals die within 4 weeks. In *Foxn1^nu* mice, demyelination is caused by a direct lytic effect of the virus on oligodendrocytes (Rosenthal et al., 1986). Demyelination and lethality are reduced after administration of neutralizing antibodies (Fujinami et al., 1989). Histopathological changes in *Prkdc^scid* mice are very similar to those in *Foxn1^nu* mice (Rozengurt and Sanchez, 1992).

Young mice born in infected populations usually acquire infection shortly after weaning and are almost all infected by 30 days of age. Intrauterine transmission to foetuses is possible during the early gestation period, but a placental barrier develops during gestation and later prevents intrauterine infection (Miyamae, 1990; Abzug *et al.*, 1991).

All TMEV isolates are closely related antigenically and form a single serogroup as determined by complement fixation and HI (Lipton and Rozhon, 1986). Hemelt *et al.* (1974) demonstrated cross reactions among four strains used in experimental infections, but differences were evident in homologous and heterologous titres. The viral strain most commonly used as antigen for serological testing is GDVII. This strain agglutinates human type 0 erythrocytes at 4°C, and HI has been the standard test for routine screening of mouse populations. Meanwhile, HI has been replaced by ELISA or IFA, both of which are more sensitive and specific. Virus isolation is possible from brains or spinal cords of mice with clinical disease or from the intestinal contents of asymptomatic mice. PCR techniques also are available to test for virus-specific nucleotide sequences in biological samples (Trottier *et al.*, 2002).

Mice that have been shown to be free from TMEV by serological testing can be selected for breeding populations. If the virus is introduced into a mouse population, depopulation of infected colonies may be the most appropriate means to eliminate TMEV. Embryo transfer or caesarean derivation are the methods of choice for eliminating virus from valuable breeding populations. Foster-nursing has been reported to be effective in generating virus-free offspring (Lipman *et al.*, 1987) although transplacental transmission has been demonstrated with experimental infection early in gestation.

Lesions of demyelination in CNS of mice with clinically inapparent chronic infection may interfere with investigations that require evaluation of the CNS (Krinke and Zurbriggen, 1997). Conceivably, such lesions also could affect neuromuscular responses or coordination, and affect neurological and behavioural evaluations.

References

Abzug, M.J., Rotbart, H.A., Magliato, S.A. and Levin, M.J. (1991). *J. Infect. Dis.* **163**, 1336–1341.

Ackermann, R. (1977). *Dtsch. Med. Wochenschr.* **102**, 1367–1370.

Ackermann, R., Bloedhorn, H., Kupper, B., Winkens, I. and Scheid, W. (1964). *Zentralbl. Bakteriol. Orig.* **194**, 407–430.

ACLAD (American Committee on Laboratory Animal Disease) (1991). *Lab. Anim. Sci.* **41**, 199–225.

Adami, C., Pooley, J., Glomb, J., Stecker, E., Fazal, F., Fleming, J.O. and Baker, S.C. (1995). *Virology* **209**, 337–346.

Allen, A.M., Clarke, G.L., Ganaway, J.R., Lock, A. and Werner, R.M. (1981). *Lab. Anim. Sci.* **31**, 599–608.

Anderson, G.W., Rowland, R.R., Palmer, G.A., Even, C. and Plagemann, P.G. (1995a). *J. Virol.* **69**, 5177–5185.

Anderson, G.W., Even, C., Rowland, R.R., Palmer, G.A., Harty, J.T. and Plagemann, P.G.W. (1995b). *J. Neurovirol.* **1**, 244–252.

Anderson, G.W., Palmer, G.A., Rowland, R.R., Even, C. and Plagemann, P.G. (1995c). *J. Virol.* **69**, 308–319.

Artwohl, J.E., Cera, L.M., Wright, M.F., Medina, L.V. and Kim, L.J. (1994). *Lab. Anim. Sci.* **44**, 73–75.

Asper, M., Hofmann, P., Osmann, C., Funk, J., Metzger, C., Bruns, M., Kaup, F.J., Schmitz, H. and Gunther, S. (2001). *Virology* **284**, 203–213.

Atencio, I.A., Belli, B., Hobbs, M., Cheng, S.F., Villareal, L.O. and Fan, H. (1995). *Virology* **212**, 356–366.

Baker, D.G. (1998). *Clin. Microbiol. Rev.* **11**, 231–266.

Baldridge, J.R. and Buchmeier, M.J. (1992). *J. Virol.* **66**, 4252–4257.

Ball, J.M., Tian, P., Zeng, C.Q.-Y., Morris, A.P. and Estes, M.K. (1996). *Science* **272**, 101–104.

Ball-Goodrich, L.J. and Johnson, E. (1994). *J. Virol.* **68**, 6476–6486.

Ball-Goodrich, L.J., Hansen, G., Dhawan, R., Paturzo, F.X. and Vivas-Gonzalez, B.E. (2002). *Comp. Med.* **52**, 160–166.

Barthold, S.W. (1986). In *Viral and Mycoplasmal Infections of Laboratory Rodents: Effects on Biomedical Research* (eds P.N. Bhatt, R.O. Jacoby, H.C. Morse and A.E. New), pp. 571–601. Academic Press, Orlando.

Barthold, S.W. (1997a). In *Monographs on Pathology of Laboratory Animals: Digestive System* (eds T.C. Jones, J.A Popp and U. Mohr), 2nd edn, pp. 196–200. Springer-Verlag, Berlin.

Barthold, S.W. (1997b). In *Monographs on Pathology of Laboratory Animals: Digestive System* (eds T.C. Jones, J.A Popp and U. Mohr), 2nd edn, pp. 384–388. Springer-Verlag, Berlin.

Barthold, S.W. and Smith, A.L. (1987). *Virus Res.* **7**, 225–239.

Barthold, S.W. and Smith, A.L. (1989a). *Arch. Virol.* **104**, 187–196.

Barthold, S.W. and Smith, A.L. (1989b). *Arch. Virol.* **107**, 171–177.

Barthold, S.W. and Smith, A.L. (1990). *Lab. Anim. Sci.* **40**, 133–137.

Barthold, S.W., Smith, A.L., Lord, P.F.S., Bhatt, P.N., Jacoby, R.O. and Main, A.J. (1982). *Lab. Anim. Sci.* **32**, 376–383.

Barthold, S.W., Smith, A.L. and Povar, M.L. (1985). *Lab. Anim. Sci.* **35**, 613–618.

Barthold, S.W., Beck, D.S. and Smith, A.L. (1993a). *Lab. Anim. Sci.* **43**, 276–284.

Barthold, S.W., Smith, A.L. and Bhatt, P.N. (1993b). *Lab. Anim. Sci.* **43**, 425–430.

Barton, L.L. and Hyndman, N.J. (2000). *Pediatrics* **105**, E35.

Barton, L.L., Peters, C.J. and Ksiazek, T.G. (1995). *Emerg. Infect. Dis.* **1**, 152–153.

Barton, L.L., Mets, M.B. and Beauchamp, C.L. (2002). *Am. J. Obstet. Gynecol.* **187**, 1715–1716.

Bennette, J.G. (1960). *Nature* **187**, 72–73.

Berke, Z., Dalianis, T., Feinstein, R., Sandstedt, K. and Evengard, B. (1994). *In Vivo* **8**, 339–342.

Besselsen, D.G., Besch-Williford, C.L., Pintel, D.J., Franklin, C.L., Hook, R.R. and Riley, L.K. (1995). *J. Clin. Microbiol.* **33**, 2859–2863.

Besselsen, D.G., Pintel, D.J., Purdy, G.A., Besch-Williford, C.L., Franklin, C.L., Hook, R.R. and Riley, L.K. (1996). *J. Gen. Virol.* **77**, 899–911.

Besselsen, D.G., Wagner, A.M. and Loganbill, J.K. (2000). *Comp. Med.* **50**, 498–502.

Besselsen, D.G., Wagner, A.M. and Loganbill, J.K. (2002). *Comp. Med.* **52**, 111–116.

Besselsen, D.G., Wagner, A.M. and Loganbill, J.K. (2003). *Comp. Med.* **53**, 65–69.

Bhatt, P.N. and Jacoby, R.O. (1987a). *Lab. Anim. Sci.* **37**, 11–15.

Bhatt, P.N. and Jacoby, R.O. (1987b). *Lab. Anim. Sci.* **37**, 23–27.

Bhatt, P.N. and Jacoby, R.O. (1987c). *Lab. Anim. Sci.* **37**, 610–614.

Bhatt, P.N., Jacoby, R.O., and Barthold, S.W. (1986a). *Lab. Anim. Sci.* **36**, 136–139.

Bhatt, P.N., Jacoby, R.O., Morse, H.C. and New, A.E. (eds) (1986b). *Viral and Mycoplasmal Infections of Laboratory Rodents: Effects on Biomedical Research*. Academic Press, Orlando.

Biggar, R.J., Schmidt, T.J. and Woodall, J.P. (1977). *J. Am. Vet. Med. Assoc.* **171**, 829–832.

Bolger, G., Lapeyre, N., Rheaume, M., Kibler, P., Bousquet, C., Garneau, M. and Cordingley, M. (1999). *Antiviral Res.* **44**, 155–165.

Bond, S.B., Howley, P.M. and Takemoto, K.K. (1978). *J. Virol.* **28**, 337–343.

Bonnard, G.D., Manders, E.K., Campbell, D.A., Herberman, R.B. and Collins, M.J. (1976). *J. Exp. Med.* **143**, 187–205.

Bootz, F. and Sieber, I. (2002). *Altex* **19**(Suppl. 1), 76–86.

Bowen, G.S., Salisher, C.H., Winkler, W.G., Kraus, A.L., Fowler, E.H., Garmann, R.H., Fraser, D.W. and Hinman, A.R. (1975). *Am. J. Epidemiol.* **102**, 233–240.

Brahic, M. and Bureau, J.F. (1998). *Bioessays* **20**, 627–633.

Brinton, M. (1982). In *The Mouse in Biomedical Research, Vol. II, Diseases* (eds H.L. Foster, J.D. Small and J.G. Fox), pp. 193–208. Academic Press, New York.

Brinton, M. (1986). In *Viral and Mycoplasmal Infections of Laboratory Rodents: Effects on Biomedical Research* (eds P.N. Bhatt, R.O. Jacoby, H.C. Morse and A.E. New), pp. 389–420. Academic Press, Orlando.

Broen, J.B., Bradley, D.S., Powell, K.M. and Cafruny, W.A. (1992). *Viral Immunol.* **5**, 133–140.

Brownstein, D.G. (1986). In *Viral and Mycoplasmal Infections of Laboratory Rodents: Effects on Biomedical Research* (eds P.N. Bhatt, R.O. Jacoby, H.C. Morse and A.E. New), pp. 37–61. Academic Press, Orlando.

Brownstein, D.G. (1996a). In *Monographs on Pathology of Laboratory Animals: Respiratory System* (eds T.C. Jones, D.L. Dungworth and U. Mohr), 2nd edn, pp. 308–316. Springer-Verlag, Berlin.

Brownstein, D.G. (1996b). In *Monographs on Pathology of Laboratory Animals: Respiratory System* (eds T.C. Jones, D.L. Dungworth and U. Mohr), 2nd edn, pp. 317–321. Springer-Verlag, Berlin.

Brownstein, D.G. and Barthold, S.W. (1982). *Lab. Anim. Sci.* **32**, 37–39.

Brownstein, D.G. and Gras, L. (1997). *Am. J. Pathol.* **150**, 1407–1420.

Brownstein, D., Bhatt, P., Ardito, R., Paturzo, F. and Johnson, E. (1989a). *Lab. Anim. Sci.* **39**, 299–301.

Brownstein, D., Bhatt, P.N. and Jacoby, R.O. (1989b). *Arch. Virol.* **107**, 35–41.

Brownstein, D.G., Smith, A.L., Jacoby, R.O., Johnson, E.A., Hansen, G. and Tattersall, P. (1991). *Lab. Invest.* **65**, 357–364.

Buller, R.M.L. and Wallace, G.D. (1985). *Lab. Anim. Sci.* **35**, 473–476.

Buller, R.M., Bhatt, P.N. and Wallace, G.D. (1983). *J. Clin. Microbiol.* **18**, 1220–1225.

Buller, R.M., Potter, L.M. and Wallace, G.D. (1986). *Curr. Top. Microbiol. Immunol.* **127**, 319–322.

Buller, R.M.L., Weinblatt, A.C., Hamburger, A.W. and Wallace, G.D. (1987). *Lab. Anim. Sci.* **37**, 28–32.

Cafruny, W.A. and Hovinen, D.E. (1988). *J. Gen. Virol.* **69**, 723–729.

Carthew, P. (1984). *J. Pathol.* **142**, 79–85.

Carthew, P. and Sparrow, S. (1980). *Br. J. Exp. Pathol.* **61**, 171–175.

Carty, A.J., Franklin, C.L., Riley, L.K. and Besch-Williford, C. (2001). *Comp. Med.* **51**, 145–149.

Casebolt, D.B. and Stephensen, C.B. (1992). *J. Clin. Microbiol.* **30**, 608–612.

CDC (1999). *Biosafety in Microbiological and Biomedical Laboratories (BMBL)*, 4th edn. U.S. Department of Health and Human Services, Centers for Disease Control and Prevention and National Institutes of Health, U.S. Government Printing Office, Washington. http://www.cdc.gov/od/ohs/biosfty/bmbl4/bmbl4toc.htm

Chang, A., Havas, S., Borellini, F., Ostrove, J.M. and Bird, R.E. (1997). *Biologicals* **25**, 415–419.

Chen, Z. and Plagemann, P.G.W. (1997). *J. Virol. Meth.* **65**, 227–236.

Childs, J.E., Glass, G.E., Ksiazek, T.G., Rossi, C.A., Oro, J.G. and Leduc, J.W. (1991). *Am. J. Trop. Med. Hyg.* **44**, 117–121.

Childs, J.E., Glass, G.E., Korch, G.W., Ksiazek, T.G. and Leduc, J.W. (1992). *Am. J. Trop. Med. Hyg.* **47**, 27–34.

Ciurea, A., Klenerman, P., Hunziker, L., Horvath, E., Odermatt, B., Ochsenbein, A.F., Hengartner, H. and Zinkernagel, R.M. (1999). *Proc. Natl. Acad. Sci. USA* **96**, 11964–11969.

Collins, M.J. and Parker, J.C. (1972). *J. Natl. Cancer Inst.* **49**, 1139–1143.

Collins, M.J., Peters, R.L. and Parker, J.C. (1981). *Lab. Anim. Sci.* **31**, 595–598.

Compton, S.R., Barthold, S.W. and Smith, A.L. (1993). *Lab. Anim. Sci.* **43**, 15–28.

Cook, P.M., Eglin, R.P. and Easton, A.J. (1998). *J. Gen. Virol.* **79**, 2411–2417.

Crane, M.A., Yauch, R., dal Canto, M.C. and Kim, B.S. (1993). *J. Neuroimmunol.* **45**, 67–73.

Crawford, L.V. (1966). *Virology* **26**, 602–612.

Cross, S.S., Parker, J.C., Rowe, W.P. and Robbins, M.L. (1979). *Infect. Immun.* **26**, 1186–1195.

Cuff, C.F., Lavi, E., Cebra, C.K., Cebra, J.J. and Rubin, D.H. (1990). *J. Virol.* **64**, 1256–1263.

Deerberg, F., Kaestner, W., Pittermann, W. and Schwanzer, V. (1973). *Dtsch. Tierärztl. Wochenschr.* **80**, 78–81.

Delano, M.L. and Brownstein, D.G. (1995). *J. Virol.* **69**, 5875–5877.

De Souza, M.S. and Smith, A.L. (1989). *J. Clin. Microbiol.* **27**, 185–187.

Dick, E.J., Kittell, C.L., Meyer, H., Farrar, P.L., Ropp, S.L., Esposito, J.J., Buller, R.M.L., Neubauer, H., Kang, Y.H. and McKee, A.E. (1996). *Lab. Anim. Sci.* **46**, 602–611.

Dillehay, D.L., Lehner, N.D. and Huerkamp, M.J. (1990). *Lab. Anim. Sci.* **40**, 367–370.

Domachowske, J.B., Bonville, C.A., Dyer, K.D., Easton, A.J. and Rosenberg, H.F. (2000). *Cell. Immunol.* **200**, 98–104.

Downs, W.G. (1982). In *The Mouse in Biomedical Research, Vol. II, Diseases* (eds H.L. Foster, J.D. Small and J.G. Fox), pp. 341–352. Academic Press, New York.

Dubensky, T.W., Murphy, F.A. and Villareal, L.P. (1984). *J. Virol.* **50**, 779–783.

Dykewicz, C.A., Dato, V.M., Fisher-Hoch, S.P., Howarth, M.V., Perez-Oronoz, G.I., Ostroff, S.M., Gary, H., Schonberger, L.B. and McCormick, J.B. (1992). *J. Am. Med. Assoc.* **267**, 1349–1353.

Eddy, B.E. (1982). In *The Mouse in Biomedical Research, Vol. II, Diseases* (eds H.L. Foster, J.D. Small and J.G. Fox), pp. 293–311. Academic Press, New York.

Ehresmann, D.W. and Hogan, R.N. (1986). *Intervirology* **25**, 103–110.

El Karamany, R.M. and Imam, I.Z. (1991). *J. Hyg. Epidemiol. Microbiol. Immunol.* **35**, 97–103.

Feng, N., Franco, M.A. and Greenberg, H.B. (1997). *Adv. Exp. Med. Biol.* **412**, 233–240.

Fenner, F. (1981). *Lab. Anim. Sci.* **31**, 553–559.

Fenner, F. (1982). In *The Mouse in Biomedical Research, Vol. II, Diseases* (eds H.L. Foster, J.D. Small and J.G. Fox), pp. 209–230. Academic Press, New York.

Fenner, F.J., Gibbs, E.P.J., Murphy, F.A., Rott, R., Studdert, D.W. and White, D.O. (1993). In *Veterinary Virology*, 2nd edn, pp. 537–552. Academic Press, San Diego.

Ferner, W.T., Miskuff, R.L., Yolken, R.H. and Vonderfecht, S.L. (1987). *J. Clin. Microbiol.* **25**, 1364–1369.

Fox, J.G., Murphy, J.C. and Igras, V.E. (1977). *Lab. Anim. Sci.* **27**, 173–179.

France, M.P., Smith, A.L., Stevenson, R. and Barthold, S.W. (1999). *Aust. Vet. J.* **77**, 600–604.

Franco, M.A. and Greenberg, H.B. (1995). *J. Virol.* **69**, 7800–7806.

Fujinami, R.S., Rosenthal, A., Lampert, P.W., Zurbriggen, A. and Yamada, M. (1989). *J. Virol.* **63**, 2081–2087.

Gaertner, D.J., Batchelder, M., Herbst, L.H. and Kaufmann, H.L. (2003). *Comp. Med.* **53**, 85–88.

Garant, P.R., Baer, P.N. and Kilham, L. (1980). *J. Dent. Res.* **59**, 80–86.

Garnick, R. L. (1996). *Dev. Biol. Stand.* **88**, 49–56.

George, A., Kost, S.I., Witzleben, C.L., Cebra, J.J. and Rubin, D.H. (1990). *J. Exp. Med.* **171**, 929–934.

Godfraind, C., Holmes, K.V. and Coutelier, J.-P. (1995). *J. Virol.* **69**, 6541–6547.

Gomez, K.A., Longhi, S.A., Marino, V.J., Mathieu, P.A., Loureiro, M.E., Coutelier, J.P., Roguin, L.P. and Retegui, L.A. (2003). *Scand. J. Immunol.* **57**, 144–150

Gossmann, J., Lohler, J., Utermöhlen, O. and Lehmann-Grube, F. (1995). *Lab. Invest.* **72**, 559–570.

Goto, K., Takahura, A., Yoshimura, M., Ohnishi, Y. and Itoh, T. (1998). *Lab. Anim. Sci.* **48**, 99–102.

Greenlee, J.E. (1986). *J. Gen. Virol.* **67**, 1109–1114.

Greenlee, J.E. and Dodd, W.K. (1987). *Arch. Virol.* **94**, 169–173.

Greenlee, J.E., Phelps, R.C. and Stroop, W.G. (1991). *Microb. Path.* **11**, 237–247.

Greenlee, J.E., Clawson, S.H., Phelps, R.C. and Stroop, W.G. (1994). *Comp. Pathol.* **111**, 259–268.

Groen, J., Broeders, H., Spijkers, I. and Osterhaus, A. (1989). *Lab. Anim. Sci.* **39**, 21–24.

Guida, J.D., Fejer, G., Pirofski, L.A., Brosnan, C.F. and Horwitz, M.S. (1995). *J. Virol.* **69**, 7674–7681.

Haag, A., Wayss, K., Rommelaere, J. and Cornelis, J.J. (2000). *Comp. Med.* **50**, 613–621.

Hamelin, C. and Lussier, G. (1988). *Experientia* **44**, 65–66.

Hamm, T.E. (ed.) (1986). *Complications of Viral and Mycoplasmal Infections in Rodents to Toxicology Research and Testing.* Hemisphere Publ. Co., Washington, DC.

Hansen, A.K., Thomsen, P. and Jensen, H.J. (1997). *Lab. Anim.* **31**, 212–218.

Hansen, G.M., Paturzo, F.X. and Smith A.L. (1999). *Lab. Anim. Sci.* **49**, 380–384.

Haring, J. and Perlman, S. (2001). *Curr. Opin. Microbiol.* **4**, 462–466.

Hartley, J.W. and Rowe, W.P. (1960). *Virology* **11**, 645–647.

Hashimoto, K., Sugiyama, T. and Sasaki, S. (1966). *Jpn. J. Microbiol.* **10**, 115–125.

Hashimoto, K., Sugiyama, T., Yoshikawa, M. and Sasaki, S. (1970). *Jpn. J. Microbiol.* **14**, 381–395.

Haven, T.R., Rowland, R.R.R., Plagemann, P.G.W., Wong, G.H.W., Bradley, S.E. and Cafruny, W.A. (1996). *Virus Res.* **41**, 153–161.

Hayase, Y., Tobita K., Kii, M., Hakamada, Y. and Arai, T. (1997). *Exp. Anim.* **46**, 307–310.

Hayashi, T., Mori, I., Noguchi, Y., Itoh, T. and Saitoh, M. (1992). *J. Comp. Pathol.* **107**, 179–183.

Heck, F.C., Sheldon, W.G. and Gleiser, C.A. (1972). *Am. J. Vet. Res.* **33**, 841–846.

Hemelt, I.E., Huxsoll, D.L. and Warner, A.R. (1974). *Lab. Anim. Sci.* **24**, 523–529.

Hinman, A.R., Fraser, D.W., Douglas, R.G., Bowen, G.S., Kraus, A.L., Winkler, W.G. and Rhodes, W.W. (1975). *Am. J. Epidemiol.* **101**, 103–110.

Holmes, K.V., Boyle, J.F. and Frana, M.F. (1986). In *Viral and Mycoplasmal Infections of Laboratory Rodents: Effects on Biomedical Research* (eds P.N. Bhatt, R.O. Jacoby, H.C. Morse and A.E. New), pp. 603–624. Academic Press, Orlando.

Homberger, F.R. (1992). *Arch. Virol.* **122**, 133–141.

Homberger, F.R. (1997). *Lab. Anim.* **31**, 97–115.

Homberger, F.R. and Barthold, S.W. (1992). *Arch. Virol.* **126**, 35–43.

Homberger, F.R., Smith, A.L. and Barthold, S.W. (1991). *J. Clin. Microbiol.* **29**, 2789–2793.

Homberger, F.R., Barthold, S.W. and Smith, A.L. (1992). *Lab. Anim. Sci.* **42**, 347–351.

Hou, S., Doherty, P.C., Zijlstra, M., Jaenisch, R. and Katz, J.M. (1992). *J. Immunol.* **149**, 1319–1325.

Ikeda, K, Dörries, K., and ter Meulen, V. (1988). *Acta Neuropathol.* **77**, 175–181.

International Union of Microbiological Societies (2000). In *Virus Taxonomy. Seventh Report of the International Committee on Taxonomy of Viruses* (eds M.H.V. van Regenmortel, C.M. Fauquet, D.H.L. Bishop, E.B. Carstens, M.K. Estes, S.M. Lemon, J. Maniloff, M.A. Mayo, D.J. McGeoch, C.R. Pringle and R.B. Wickner), pp. 311–323. Academic Press, San Diego.

Ishida, T., Taguchi, F., Lee, Y.-S., Yamada, A., Tamura, T. and Fujiwara, K. (1978). *Lab. Anim. Sci.* **28**, 269–276.

Iwai, H., Goto, Y. and Ueda, K. (1979). *Jpn. J. Exp. Med.* **49**, 123–130.

Jacoby, R.O. and Ball-Goodrich, L.J. (1995). *Sem. Virol.* **6**, 329–337.

Jacoby, R.O. and Lindsey, J.R. (1998). *ILAR J.* **39**, 266–271.

Jacoby, R.O., Bhatt, P.N., Johnson, E.A. and Paturzo, F.X. (1983). *Lab. Anim. Sci.* **33**, 435–441.

Jacoby, R.O., Johnson, E.A., Ball-Goodrich, L.J., Smith, A.L. and McKisic, M.D. (1995). *J. Virol.* **69**, 3915–3919.

Jacoby, R.O., Ball-Goodrich, L.J., Besselsen, D.J., McKisic, M.D., Riley, L.K. and Smith A.L. (1996). *Lab. Anim. Sci.* **46**, 370–380.

Jacques, C., Cousineau, L., D'Amours, B., Lussier, G. and Hamelin, C. (1994a). *J. Gen. Virol.* **75**, 1311–1316.

Jacques, C., D'Amours, B. and Hamelin, C. (1994b). *FEMS Microbiol. Lett.* **115**, 7–11.

Jandasek, L. (1968). *Zentralbl. Bakt. I. Abt. Orig.* **207**, 172–177.

Jure, M.N., Morse, S.S. and Stark, D.M. (1988). *Lab. Anim. Sci.* **38**, 273–278.

Karupiah, G., Buller, R.M.L., van Rooijen, N., Duarte, C.J. and Chen, J. (1996). *J. Virol.* **70**, 8301–8309.

Karupiah, G., Chen, J.-H., Nathan, C.F., Mahalingam, S. and MacMicking, J.D. (1998). *J. Virol.* **72**, 7703–7706.

Kast, W.M., Bronkhorst, A.M., de Waal, L.P. and Melief, C.J. (1986). *J. Exp. Med.* **164**, 723–738.

Katami, K., Taguchi, F., Nakayama, M., Goto, N. and Fujiwara, K. (1978). *Jpn. J. Exp. Med.* **48**, 4981–490.

Kercher, L. and Mitchell, B.M. (2002). *J. Virol.* **76**, 9165–9175.

Kilham, L. and G. Margolis (1970). *Proc. Soc. Exp. Biol. Med.* **133**, 1447–1452.

Kilham, L. and G. Margolis (1971). *Teratology* **4**, 43–62.

Kimsey, P.B., Engers, H D., Hirt, B. and Jongeneel, C. B. (1986). *J. Virol.* **59**, 8–13.

Kraft, L. (1982). In *The Mouse in Biomedical Research, Vol. II, Diseases* (eds H.L. Foster, J.D. Small and J.G. Fox), pp. 159–191. Academic Press, New York.

Kraft, V. and Meyer, B. (1986). *Lab. Anim. Sci.* **36**, 271–276.

Krinke, G.J. and Zurbriggen, A. (1997). *Exp. Toxicol. Pathol.* **49**, 501–503.

Kunita, S., Terada, E., Goto, K. and Kagiyama, N. (1992). *Lab. Anim. Sci.* **42**, 593–598.

Kurtz, C.I., Sun, X.M. and Fujinami, R.S. (1995). *Microb. Pathog.* **18**, 11–27.

Kyuwa, S. (1997). *Exp. Anim.* **46**, 311–313.

Kyuwa, S., Machii, K., and Shibata, S. (1996). *Exp. Anim.* **45**, 81–83.

Lagace-Simard, J., Descoteaux, J.P. and Lussier, G. (1980). *Am. J. Pathol.* **101**, 31–40.

Leary, T.P., Erker, J.C., Chalmers, M.L., Wetzel, J.D., Desai, S.M., Mushahwar, I.K. and Dermody, T.S. (2002). *J. Virol. Methods* **104**, 161–165.

Lehmann-Grube, F. (1982). In *The Mouse in Biomedical Research, Vol. II, Diseases* (eds H.L. Foster, J.D. Small and J.G. Fox), pp. 231–266. Academic Press, New York.

Lenzo, J.C., Fairweather, D., Cull, V., Shellam, G.R. and James Lawson, C.M. (2002). *J. Mol. Cell. Cardiol.* **34**, 629–640.

Li, K., Schuler, T., Chen, Z., Glass, G.E., Childs, J.E. and Plagemann, P.G. (2000). *Virus Res.* **67**, 153–162.

Lin, M.T., Hinton, D.R., Marten, N.W., Bergmann, C.C. and Stohlman, S.A. (1999). *J. Immunol.* **162**, 7358–7368.

Lindsley, M.D., Thiemann, R. and Rodriguez, M. (1991). *J. Virol.* **65**, 6612–6620.

Lipman, N.S., Newcomer, C.E. and Fox, J.G. (1987). *Lab. Anim. Sci.* **37**, 195–199.

Lipman, N.S., Henderson, K. and Shek, W. (2000a). *Comp. Med.* **50**, 255–256.

Lipman, N.S., Perkins, S., Nguyen, H., Pfeffer, M. and Meyer, H. (2000b). *Comp. Med.* **50**, 426–435.

Lipton, H.L. and dal Canto, M.C. (1979). *Infect. Immun.* **26**, 369–374.

Lipton, H.L. and Rozhon, E.J. (1986). In *Viral and Mycoplasmal Infections of Laboratory Rodents: Effects on Biomedical Research* (eds P.N. Bhatt, R.O. Jacoby, H.C. Morse and A.E. New), pp. 253–275. Academic Press, Orlando.

Lipton, H.L., Kim, B.S., Yahikozawa, H. and Nadler, C.F. (2001). *Virus Res.* **76**, 79–87.

Livingston, R.B. and Riley, L.K. (2003). *Lab Anim. NY* **32**(5), 44–51.

Livingston, R.S., Besselsen, D.G., Steffen, E.K., Besch-Williford, C.L., Franklin, C.L. and Riley, L.K. (2002). *Clin. Diagn. Lab. Immunol.* **9**, 1025–1031.

Lledo, L., Gegundez, M.I., Saz, J.V., Bahamontes, N. and Beltran, M. (2003). *J. Med. Virol.* **70**, 273–275.

Luethans, T.N. and Wagner, J.E. (1983). *Lab. Anim. Sci.* **33**, 270–272.

Lukashevich, I.S., Tikhonov, I., Rodas, J. D., Zapata, J.C., Yang, Y., Djavani, M. and Salvato, M.S. (2003). *J. Virol.* **77**, 1727–1737.

Lundgren, O., Peregrin, A.T., Persson, K., Kordasti, S., Uhnoo, I. and Svensson, L. (2000). *Science* **287**, 491–495.

Lussier, G. (1988). *Vet. Res. Contrib.* **12**, 199–217.

Lussier, G., Smith, A.L., Guenette, D. and Descoteaux, J.P. (1987). *Lab. Anim. Sci.* **37**, 55–57.

Lussier, G., Guenette, D., Shek, W.S. and Desoteaux, H.P. (1988). *Lab. Anim. Sci.* **38**, 577–579.

Machii, K., Otsuka, Y., Iwai, H. and Ueda, K. (1989). *Lab. Anim. Sci.* **39**, 334–337.

Mahalingam, S., Karupiah, G., Takeda, K., Akira, S., Matthaei, K.I. and Foster, P.S. (2001). *Proc. Natl. Acad. Sci. USA* **98**, 6812–6817.

Mahy, B.W., Dykewicz, C., Fisher-Hoch, S., Ostroff, S., Tipple, M. and Sanchez, A. (1991). *Dev. Biol. Stand.* **75**, 183–189.

Manning, P.J. and Frisk, C.S. (1981). *Lab. Anim. Sci.* **31**, 574–577.

Margolis, G., Kilham, L. and Hoenig, E.M. (1974). *Am. J. Pathol.* **75**, 363–372.

Markine-Goriaynoff, D., Hulhoven, X., Cambiaso, C.L., Monteyne, P., Briet, T., Gonzalez, M.D., Coulie, P. and Coutelier, J.P. (2002). *J. Gen. Virol.* **83**, 2709–2716.

Marrie, T.J. and Saron, M.F. (1998). *Am. J. Trop. Med. Hyg.* **58**, 47–49.

McCance, D.J. and Mims, C.A. (1977). *Infect. Immun.* **18**, 196–202.

McCance, D.J. and Mims, C.A. (1979). *Infect. Immun.* **25**, 998–1002.

McGavern, D.B., Murray, P.D., Rivera-Quinones, C., Schmelzer, J.D., Low, P. and Rodriguez, M. (2000). *Brain* **123**, 519–531.

McKisic, M.D., Lancki, D.W., Otto, G., Padrid, P., Snook, S., Cronin, D.C. II, Lohmar, P.D., Wong, T. and Fitch, F.W. (1993). *J. Immunol.* **150**, 419–428.

McKisic, M.D., Paturzo, F.X. and Smith, A.L. (1996). *Transplantation* **61**, 292–299.

McKisic, M.D., Macy, J.D., Delano, M.L., Jacoby, R.O., Paturzo, F.X. and Smith, A.L. (1998). *Transplantation* **65**, 1436–1446.

McNeal, M.M., Rae, M.N. and Ward, R.L. (1997). *J. Virol.* **71**, 8735–8742.

Melby, E.C. and Balk, M.W. (1983). *The Importance of Laboratory Animal Genetics, Health, and the Environment in Biomedical Research.* Academic Press, Orlando.

Miyamae, T. (1990). *Microbiol. Immunol.* **34**, 841–848.

Montali R.J., Connolly B.M., Armstrong D.L., Scanga C.A. and Holmes K.V. (1995). *Am. J. Pathol.* **147**, 1441–1449.

Monteyne, P., Meite, M. and Coutelier, J.P. (1997). *J. Neurovirol.* **3**, 380–384.

Morita, C., Matsuura, Y., Kawashima, E., Takahashi, S., Kawaguchi, J., Iida, S., Yamanaka, T. and Jitsukawa, W. (1991). *J. Vet. Med. Sci.* **53**, 219–222.

Morita, C., Tsuchiya, K., Ueno, H., Muramatsu, Y., Kojimahara, A., Suzuki, H., Miyashita, N., Moriwaki, K., Jin, M.L., Wu, X.L. and Wang, F.S. (1996). *Microbiol. Immunol.* **40**, 313–315.

Morrison, L.A., Sidman, R.L. and Fields, B.N. (1991). *Proc. Natl. Acad. Sci. USA* **88**, 3852–3856.

Morse, S.S. (1987). *Lab. Anim. Sci.* **37**, 717–725.

Morse, S.S. (1988). *Virology* **163**, 255–258.

Morse, S.S. (1989). *Lab. Anim. Sci.* **39**, 571–574.

Morse, S.S. (1990a). *J. Virol. Meth.* **28**, 15–24.

Morse, S.S. (1990b). *Lab. Anim.* **24**, 313–320.

Morse, S.S. and Valinsky, J.E. (1989). *J. Exp. Med.* **169**, 591–596.

Morse, S.S., Sakaguchi, N. and Sakaguchi, S. (1999). *J. Immunol.* **162**, 5309–5316.

Nakanaga, K., Ishida, T. and Fujiwara, K. (1983). *Lab. Anim.* **17**, 90–94.

National Research Council, Committee on Infectious Diseases of Mice and Rats. (1991). *Infectious Diseases of Mice and Rats.* National Academy Press, Washington, DC.

Nelson, J.B. and Tarnowski G.S. (1960). *Nature* **188**, 1086–1089.

Neubauer, H., Pfeffer, M. and Meyer, H. (1997). *Lab. Anim.* **31**, 201–205.

Newsome, P.M. and Coney, K.A. (1985). *Infect. Immun.* **47**, 573–574.

Nicklas, W., Giese, M., Zawatzky, R., Kirchner, H. and Eaton, P. (1988). *Lab. Anim. Sci.* **38**, 152–154.

Nicklas, W., Kraft, V. and Meyer, B. (1993). *Lab. Anim. Sci.* **43**, 296–300.

Nicklas, W., Homberger, F.R., Illgen-Wilcke, B., Jacobi, K., Kraft, V., Kunstyr, I., Mähler, M., Meyer, H. and Pohlmeyer-Esch, G. (1999). *Lab. Anim.* **33**(Suppl. 1), S1:39–S1:87.

Nicklas, W., Baneux, P., Boot, R., Decelle, T., Deeny, A.A., Fumanelli, M. and Illgen-Wilcke, B. (2002). *Lab. Anim.* **36**, 20–42.

Niemialtowski, M.G., Spohr de Faundez, I., Gierynska, M., Malicka, E. Toka, F.N., Schollenberger, A. and Popis, A. (1994). *Acta Virol.* **38**, 299–307.

Ohnishi, Y., Yoshimura, M. and Ueyama, Y. (1995). *J. Natl. Cancer Inst.* **87**, 539–539.

Ohsawa, K., Watanabe, Y., Miyata, H. and Sato, H. (2003). *Comp. Med.* **53**, 191–196.

Okumura, A., Machii, K., Azuma, S., Toyoda, Y. and Kyuwa, S. (1996). *J. Virol.* **70**, 4146–4149.

Oldstone, M.B.A. (2002). *Curr. Top. Microbiol. Immunol.* **263**, 83–117.

O'Neill, H.C., Blanden, R.V. and O'Neill, T.J. (1983). *Immunogenetics* **18**, 255–265.

Orcutt, R. (1994). In *Manual of Microbiologic Monitoring of Laboratory Animals* (eds. K. Waggie, N. Kagiyama, A.M. Allen and T. Nomura), pp. 79–86. National Institutes of Health, NIH publication No. 94-2498.

Osborn, J.E. (1982). In *The Mouse in Biomedical Research, Vol. II, Diseases* (eds H.L. Foster, J.D. Small and J.G. Fox), pp. 267–292. Academic Press, New York.

Osborn, J.E. (1986). In *Viral and Mycoplasmal Infections of Laboratory Rodents: Effects on Biomedical Research* (eds P.N. Bhatt, R.O. Jacoby, H.C. Morse and A.E. New), pp. 421–450. Academic Press, Orlando.

Osterhaus, A.D.M.E., Teppema, J.S., Wirahadiredja, R.M.S. and van Steenis, G. (1981). *Lab. Anim. Sci.* **31**, 704–706.

Otten, J.A. and Tennant, R.W. (1982). In *The Mouse in Biomedical Research, Vol. II, Diseases* (eds H.L. Foster, J.D. Small and J.G. Fox), pp. 335–340. Academic Press, New York.

Owen, D., Hill, A. and Argent, S. (1975). *Nature* **254**, 598–599.

Palma, J.P., Kwon, D., Clipstone, N.A. and Kim, B.S. (2003). *J. Virol.* **77**, 6322–6331.

Palmon, A., Tel-Or, S., Rager-Zisman, B. and Burstein, Y. (2000). *J. Virol. Meth.* **86**, 107–114.

Papadimitriou, J.M. and Robertson, T.A. (1976). *Am. J. Pathol.* **85**, 595–608.

Park, J.Y., Peters, C.J., Rollin, P.E., Ksiazek, T.G., Gray, B., Waites, K.B. and Stephensen, C.B. (1997). *J. Med. Virol.* **51**, 107–114.

Parker, J.C. and Reynolds, R.A. (1968). *Am. J. Epidemiol.* **88**, 112–125.

Parker, J.C. and Richter, C.B. (1982). In *The Mouse in Biomedical Research, Vol. II, Diseases* (eds H.L. Foster, J.D. Small and J.G. Fox), pp. 109–158. Academic Press, New York.

Parker, J.C., Collins, M.J., Cross, S.S. and Rowe, W.P. (1970). *J. Natl. Cancer Inst.* **45**, 305–310.

Pena Rossi, C., Cash, E., Aubert, C. and Coutinho, A. (1991a). *J. Virol.* **65**, 3895–3899.

Pena Rossi, C., McAllister, A., Fiette, L. and Brahic, M. (1991b). *Cell. Immunol.* **138**, 341–348.

Percy, D.H. and Barthold, S.W. (2001). In *Pathology of Laboratory Rodents and Rabbits*, 2nd edn, pp. 3–106. Iowa State Press, Ames.

Percy, D.H. and Palmer, D.J. (1997). *Lab. Anim. Sci.* **47**, 132–137.

Percy, D.H., Auger, D.C. and Croy, B.A. (1994). *Vet. Pathol.* **31**, 67–73.

Pirofski, L., Horwitz, M.S., Scharff, M.D. and Factor, S.M. (1991). *Proc. Natl. Acad. Sci. USA* **88**, 4358–4362.

Plagemann, P.G.W. and Swim, H.E. (1966). *Proc. Soc. Exp. Biol. Med.* **121**, 1142–1146.

Porterfield, P.D. and Richter, C.B. (1994). In *Manual of Microbiologic Monitoring of Laboratory Animals* (eds K. Waggie, N. Kagiyama, A.M. Allen and T. Nomura), pp. 25–29. National Institutes of Health, NIH publication No. 94-2498.

Prattis, S.M. and Morse, S.S. (1990). *Lab. Anim. Sci.* **40**, 33–36.

Rai, S.K., Cheung, D.S., Wu, M.S., Warner, T.F. and Salvato, M.S. (1996). *J. Virol.* **70**, 7213–7218.

Rai, S.K., Micales, B.K., Wu, M.S., Cheung, D.S., Pugh, T.D., Lyons, G.E. and Salvato, M.S. (1997). *Am. J. Pathol.* **151**, 633–639.

Redig, A.J. and Besselsen, D.G. (2001). *Comp. Med.* **51**, 326–331.

Rehg, J.E., Blackman, M.A. and Toth, L.A. (2001). *Comp. Med.* **51**, 369–374.

Reynolds, R.P., Rahija, R.J., Schenkman, D.I. and Richter, C.B. (1993). *Lab. Anim. Sci.* **43**, 291–295.

Richter, C.B. (1986). In *Viral and Mycoplasmal Infections of Laboratory Rodents: Effects on Biomedical Research* (eds P.N. Bhatt, R.O. Jacoby, H.C. Morse and A.E. New), pp. 137–192. Academic Press, Orlando.

Richter, C.B., Thigpen, J.E., Richter, C.S. and Mackenzie, J.M. (1988). *Lab. Anim. Sci.* **38**, 255–261.

Riepenhoff-Talty, M., Dharakul, T., Kowalski, E., Michalak, S. and Ogra, P.L. (1987). *J. Virol.* **61**, 3345–3348.

Riley, V., Lilly, F., Huerto, E. and Bardell, D. (1960). *Science* **132**, 545–547.

Riley, L.K., Knowles, R., Purdy, G., Salome, N., Pintel, D., Hook, R.R., Franklin, C.L. and Besch-Williford, C.L. (1996). *J. Clin. Microbiol.* **34**, 440–444.

Rodriguez, M. and Lindsley, M.D. (1992). *Neurology* **42**, 348–375.

Rommelaere, J. and Cornelis, J.J. (1991). *J. Virol. Methods* **33**, 233–251.

Rosé, J., Franco, M. and Greenberg, H. (1998). *Adv. Virus Res.* **51**, 203–235.

Rosenthal, A., Fujinami, R.S. and Lampert, P.W. (1986). *Lab. Invest.* **54**, 515–522.

Roths, J.B., Smith, A.L. and Sidman, C.L. (1993). *J. Exp. Med.* **177**, 1193–1198.

Rousseau, M.C., Saron, M.F., Brouqui, P. and Bourgeade, A. (1997). *Eur. J. Epidemiol.* **13**, 817–823.

Rowson, K.E.K. and Mahy, B.W.J. (1975). *Lactic Dehydrogenase Virus*. Virology Monographs 13, Springer Verlag, Wien.

Rozengurt, N. and Sanchez, S. (1992). *J. Comp. Pathol.* **107**, 389–398.

Rozengurt, N. and Sanchez, S. (1993). *Lab. Anim.* **27**, 229–234.

Rubino, M.J. and Walker, D. (1988). *Virus Res.* **9**, 1–10.

Rubio, N. and Capa, L. (1993). *Cell. Immunol.* **149**, 237–247.

Rülicke, T., Hassam, S., Autenried, P. and Briner, J. (1991). *J. Exp. Anim. Sci.* **34**, 127–131.

Sabesin S.M. (1972). *Am. J. Gastroenterol.* **58**, 259–274.

Salazar-Bravo, J., Ruedas, L.A. and Yates, T.L. (2002). *Curr. Top. Microbiol. Immunol.* **262**, 25–63.

Schmader, K., Henry, S.C., Rahija, R.J., Yo, Y., Daley, G.G. and Hamilton, J.D. (1995). *J. Infect. Dis.* **172**, 531–534.

Schwanzer, V., Deerberg, F., Frost, J., Liess, B., Schwanzerova, I. and Pittermann, W. (1975). *Z. Versuchstierk.* **17**, 110–120.

Segovia, J.C., Real, A., Bueren, J.A. and Almendral, J.M. (1991). *Blood* **77**, 980–988.

Segovia, J.C., Real, A., Bueren, J.A. and Almendral, J.M. (1995). *J. Virol.* **69**, 3229–3232.

Segovia, J.C., Gallego, J.M., Bueren, J.A. and Almendral, J.M. (1999). *J. Virol.* **73**, 1774–1784.

Shah, K.V. and Christian, C. (1986). In *Viral and Mycoplasmal Infections of Laboratory Rodents: Effects on Biomedical Research* (eds P.N. Bhatt, R.O. Jacoby, H.C. Morse and A.E. New), pp. 505–527. Academic Press, Orlando.

Shanley, J.D. and Pesanti, E.L. (1986). *Arch. Virol.* **88**, 27–35.

Shanley, J.D., Biczak, L. and Forman, S.J. (1993). *J. Infect. Dis.* **167**, 264–269.

Shanley, J.D., Thrall, R.S. and Forman, S.J. (1997). *J. Infect. Dis.* **175**, 309–315.

Shek, W.R., Paturzo, F.X., Johnson, E.A., Hansen, G.M. and Smith, A.L. (1998). *Lab. Anim. Sci.* **48**, 294–297.

Sheridan, J.F. and Vonderfecht, S. (1986). In *Viral and Mycoplasmal Infections of Laboratory Rodents: Effects on Biomedical Research* (eds P.N. Bhatt, R.O. Jacoby, H.C. Morse and A.E. New), pp. 217–243. Academic Press, Orlando.

Shimokata, K., Nishiyama, Y., Ito, Y., Kimura, Y. and Nagata, I. (1977). *Infect. Immun.* **16**, 706–708.

Sierra, A. and Rubio, N. (1993). *Immunology* **78**, 399–404.

Silverman, J., Paturzo F. and Smith A.L. (1982). *Lab. Anim. Sci.* **32**, 273–274.

Singleton, G.R., Smith, A.L. and Krebs, C.J. (2000). *Epidemiol. Infect.* **125**, 719–727.

Slifka, M.K. (2002). *Curr. Top. Microbiol. Immunol.* **263**, 67–81.

Small, J.D. and New, A.E. (1981). *Lab. Anim. Sci.* **31**, 616–629.

Smith, A.L. (1983). *Lab. Anim. Sci.* **33**, 157–160.

Smith, A.L., Paturzo, F.X., Gardner, E.P., Morgenstern, S., Cameron, G. and Wadley, H. (1984). *Can. J. Comp. Med.* **48**, 335–337.

Smith, A.L., Winograd, D.F. and Burrage, T.G. (1986). *Arch. Virol.* **91**, 233–246.

Smith, A.L., Jacoby, R.O., Johnson, E.A., Paturzo, F. and Bhatt, P.N. (1993a). *Lab. Anim. Sci.* **43**, 175–182.

Smith, A.L., Singleton, G.R., Hansen, G.M. and Shellam, G. (1993b). *J. Wildl. Dis.* **29**, 219–229.

Steele, M.I., Marshall, C.M., Lloyd, R.E. and Randolph, V.E. (1995). *Hepatology* **21**, 697–702.

St-Pierre, Y., Potworowski, E.F. and Lussier, G. (1987). *J. Gen. Virol.* **68**, 1173–1176.

Taguchi, F., Goto, Y. and Fujiwara, K. (1979). *Arch. Virol.* **59**, 275–279.

Takei, I., Asaba, Y., Kasatani, T., Maruyama, T., Watanabe, K., Yanagawa, T., Saruta, T. and Ishii, T. (1992). *J. Autoimmun.* **5**, 665–673.

Takeuchi, A. and Hashimoto, K. (1976). *Infect. Immun.* **13**, 569–580.

Tashiro, M., McQueen, N.L. and Seto, J.T. (1999). *Front. Biosci.* **4**, D642–D645.

Tattersall, P. and Cotmore, S.F. (1986) In *Viral and Mycoplasma Infections of Laboratory Rodents: Effects on Biomedical Research* (eds. P.N. Bhatt, R.O. Jacoby, H.C. Morse and A.E. New), pp. 305–348. Academic Press, Orlando.

Trottier, M., Schlitt, B.P. and Lipton, H.L. (2002). *J. Virol. Methods* **103**, 89–99.

Tyler, K.L. (2001). In *Fields Virology* (eds D.M. Knipe, P.M. Howley, D.E. Griffin, R.A. Lamb, M.A. Martin, B. Roizman and S.E. Straus), 4th edn, pp. 1729–1745. Lippincott Williams & Wilkins, Philadelphia.

Tyler, K.L. and Fields, B.N. (1986). In *Viral and Mycoplasmal Infections of Laboratory Rodents: Effects on Biomedical Research* (eds P.N. Bhatt, R.O. Jacoby, H.C. Morse and A.E. New), pp. 277–303. Academic Press, Orlando.

Tyler, K.L., Clarke, P., DeBiasi, R.L., Kominsky, D. and Poggioli, G.J. (2001). *Trends Microbiol.* **9**, 560–564.

Ueno, Y., Iwama, M., Ohsima, T., Sugiyma, F., Takakura, A., Itoh, T. and Yagami, K. (1998). *Exp. Anim.* **47**, 207–210.

Umehara, K., Hirakawa, M. and Hashimoto, K. (1984). *Microbiol. Immunol.* **28**, 679–690.

Van den Broek, M.F., Spörri, R., Even, C., Plagemann, P.G.W., Hänseler, E., Hengartner, H. and Zinkernagel, R.M. (1997). *J. Immunol.* **159**, 1585–1588.

Van der Logt, J.T.M. (1986). *Sci. Tech. Anim. Lab.* **11**, 197–203.

Van der Logt, J.T., Kissing, J. and Melchers, W.J. (1994). *J. Clin. Microbiol.* **32**, 2003–2006.

Van der Veen, J. and Mes, A. (1973). *Arch. Ges. Virusforsch.* **42**, 235–241.

Wagner, J.E. and Daynes, R.A. (1981). *Lab. Anim. Sci.* **31**, 565–569.

Wagner, A.M., Loganbill, J.K. and Besselsen, D.G. (2003). *Comp. Med.* **53**, 173–177.

Wallace, G.D. (1981). *Science* **211**, 438.

Wallace, G.D. and Buller, R.M.L. (1985). *Lab. Anim. Sci.* **35**, 41–46.

Wallace, G.D. and Buller, R.M.L. (1986). In *Viral and Mycoplasmal Infections of Laboratory Rodents: Effects on Biomedical Research* (eds P.N. Bhatt, R.O. Jacoby, H.C. Morse and A.E. New), pp. 539–556. Academic Press, Orlando.

Wallace, G.D., Buller, R.M.L. and Morse, H.C. (1985). *J. Virol.* 890–891.

Ward, J.M. (1974). *Lab. Anim. Sci.* **24**, 938–942.

Ward, M.L. and McNeal, M.M. (1999). *Handbook of Animal Models of Infection* (eds O. Zak and M.A. Sande), pp. 1049–1060. Academic Press, London.

Ward, D.C. and Tattersall, P.J. (1982). In *The Mouse in Biomedical Research, Vol. II, Diseases* (eds H.L. Foster, J.D. Small and J.G. Fox), pp. 314–334. Academic Press, New York.

Ward, J.M., Houchens, D.P., Collins, M.J., Young, D.M. and Reagan, R.L. (1976). *Vet. Pathol.* **13**, 36–46.

Weir, E.C., Bhatt, P.N., Barthold, S.W., Cameron, G.A. and Simack, P.A. (1987). *Lab. Anim. Sci.* **37**, 455–458.

Weir, E.C., Brownstein, D.G., Smith, A.L. and Johnson, E.A. (1988). *Lab. Anim. Sci.* **38**, 133–137.

Werner, G.T. (1982). *Zentralbl. Vet. Med. B* **29**, 401–404.

Wigand, R. (1980). *Arch. Virol.* **64**, 349–357.

Wigand, R., Gelderblom, H. and Ozel, M. (1977). *Arch. Virol.* **54**, 131–142.

Wilde, J., Eiden, J. and Yolken, R. (1990). *J. Clin. Microbiol.* **28**, 1300–1307.

Williamson, J.S.P. and Stohlman, S.A. (1990). *J. Virol.* **64**, 4589–4592.

Winters, A.L. and Brown, H.K. (1980). *Proc. Soc. Exp. Biol. Med.* **164**, 280–286.

Wood, B.A., Dutz, W. and Cross, S.S. (1981). *J. Gen. Virol.* **57**, 139–147.

Yagami, K. Goto, Y., Ishida, J., Ueno, Y., Kajiwara, N. and Sugiyama, F. (1995). *Lab. Anim. Sci.* **45**, 326–328.

Yamada, M., Zurbriggen, A. and Fujinami, R.S. (1991). *Adv. Virus Res.* **39**, 291–320.

Yamada, Y.K., Yabe, M., Yamada, A. and Taguchi, F. (1993). *Lab. Anim. Sci.* **43**, 285–290.

Yamamoto, H., Sato, H., Yagami, K., Arikawa, J., Furuya, M., Kurosawa, T., Mannen, K., Matsubayashi, K., Nishimune, Y., Shibahara, T., Ueda, T. and Itoh, T. (2001). *Exp. Anim.* **50**, 397–407.

Yoshikura, H. and Taguchi F. (1979). *Intervirology* **11**, 69–73.

Zenner, L. and Regnault, J.P. (2000). *Lab. Anim.* **34**, 76–83.

Zinkernagel, R.M. (2002). *Curr. Top. Microbiol. Immunol.* **263**, 1–5.

Zitterkopf. N.L., Haven, T.R., Huela, M., Bradley, D.S. and Cafruny, W.A. (2002). *Placenta* **23**, 438–446.

PART

5

Animal Husbandry and Production

Contents

CHAPTER 24

Housing and Maintenance

Hans J Hedrich
Institut für Versuchstierkunde, Medizinische
Hochschule, Hannover, Germany

Horst Mossmann
Max-Planck-Institut für Immunbiologie,
Freiburg i. Br., Germany

Werner Nicklas
Zentrales Tierlabor, Deutsches Krebsforschungszentrum
Heidelberg, Germany

Introduction

Proper housing taking into account the physical and social environment of the animals, well organised colony management and correctly followed animal care regulations are indispensable prerequisites for animal experiments of high quality and reproducibility. Appropriate conditions for breeding, maintenance and experimentation will be described with respect to the different hygienic levels. In animal experimentation the established (national) guidelines for the care and use of laboratory animals in line with the local animal welfare regulations have to be followed. For details, see the specific textbooks (e.g. ILAR, 1996; van Zutphen et al., 2001), but also Chapters 25 and 29 in this handbook.

Husbandry

The husbandry shall provide a standardised macro- and micro-environment as basis of reliable and reproducible research results but should also take into account the welfare of the animals. Several recommendations have been published which serve as guidelines for proper (e.g. GV-SOLAS, 1988; ILAR, 1996; Jennings et al., 1998; van Zutphen et al., 2001), but not in any case comfortable (Sherwin, 2002) housing of mice. These guidelines refer to requirements on ventilation, temperature, humidity, lighting, noise levels, health status (see Chapter 27 on Health Monitoring; Nicklas et al., 2002), feeding, water supply, animal enclosures, handling and experimentation including

The Laboratory Mouse
Copyright 2004 Elsevier
ISBN 0-1233-6425-6

TABLE 24.1: Environmental requirements of mice

Temperature	20–24°C
Relative humidity	50 ± 10%
Ventilation (air change/h)	8–20 (in IVCs 30–80)
Photoperiod (light/dark)	12/12, 14/10
Light intensity	60–400 lx
Noise	~50 to ≤85 dB
Water intake[a]	5–8 ml/day
Food intake[a]	4–8 g/day

See also Clough (1992), Fox et al. (1984), GV-SOLAS (1988).

[a]According to Harkness and Wagner (1989): 15 ml/100 g BW/day, respectively 15 g/100 g BW/day.

anaesthesia, analgesia and euthanasia (see Chapter 34). The environmental requirements of mice are summarised in Table 24.1, the space requirements according to ILAR (1996) and the European standards (ETS 123, Appendix A presently under revision) are listed in Table 24.2.

In general, the personnel in charge of handling the animals should be educated and well trained in the care of laboratory animals (e.g. according to FELASA (1995) guidelines). The standard procedures have to be strictly defined such as (i) hygienic procedures for personnel, (ii) daily monitoring of animals for adequate environmental conditions and general health, (iii) food (standard diet) and water control, (iv) regular changes of adequate cages and bedding, (v) cleaning and sterilisation of cages, water bottles, racks and other equipment, and (vi) sanitation programmes.

TABLE 24.2: Space requirements for mice

	Body weight (g) EU[a]	Body weight (g) ILAR[b]	Minimum floor area (cm²) EU[a]	Minimum floor area (cm²) ILAR[b]	Minimum cage height (cm) EU[a]	Minimum cage height (cm) ILAR[b]	Floor area per animal (cm²) EU[a]	Floor area per animal (cm²) ILAR[b]
		<10				12.7		39.24
In		<15				12.7		51.6
stock		<20	330		12		60	
and	<25	<25	330		12	12.7	70	77.4
during	<30		330		12		80	
procedure		>25				12.7		96.75
	>30		330		12		100	
Breeding			330[c]	[d]	12	12.7		

[a]European Convention for the Protection of Vertebrate Animals used for Experimental and other Scientific Purposes, ETS No.: 123, Appendix A (Draft, 2003); in the current version a minimum floor area of 200 cm² is required (http://conventions.coe.int/treaty/en/Treaties/Html/123.htm#APPENDIX-A).

[b]ILAR (1996). Guide for the Care and Use of Laboratory Animals. National Academy Press, Washington, DC.

[c]For a monogamous pair or a trio; for each additional female plus litter 180 cm² should be added.

[d]Mice are typically bred in male to female ratios of 1:1 or 1:2. If bred permanently monogamous (1:1), the male, female and any litter of pups may be kept together until the pups are weaned. In case of a trio (1 male and 2 females) kept in a micro-isolator cage, it is recommended that one of the two females be removed to a separate cage once observed to be pregnant. This facilitates compliance with housing standards and furthermore permits unobscured pedigree documentation. It is, however, permissible to keep the trio and the pups of one or two litters together providing that when the eldest litter turns 14 days of age there are no more than 12 pups in total in the cage (for reference see Reeb-Whitaker et al., 2001). Higher female ratios (1:3, 1:4) are not recommended unless pregnant females are removed once observed to be pregnant. Weaning of pups is recommended by ILAR at 18–21 days of age especially when intensive mating is done where the male is kept with the female(s) permitting breeding on the postpartum oestrus. Weaning of mouse pups latest should be done by 28 days of age or if a new litter is borne by the still nursing female.

Preventive care for immunocompromised breeders

Treating immunocompromised breeders with immunocompetent cells (from syngeneic donors such as an immunocompetent genetic background or F1 hybrids of the strain to be reconstituted and the immunocompetent strain) can assist in the propagation of highly immunocompromised strains (Wang *et al.*, 1997; Kawachi *et al.*, 2000). For the reconstitution of nude mice, *Foxn1^nu^*, thymus homogenates can be injected intraperitoneally to reconstitute their T-cell defect. As a preventive measure for homozygous *Prkdc^scid^*, *Rag1^tm1^*, *Rag2^tm1^*, etc. mice an injection of syngeneic or F1 spleen cells i.p. ($1–2 \times 10^7$), or bone marrow cells ($2–5 \times 10^6$) i.v. into juvenile mice has been shown to be very efficient (Mossmann, unpublished). This treatment improves the constitution of the individual mice and thus predestines those (however only) as breeders.

Housing conditions

Earlier descriptions of housing systems for small rodents have not lost their principal validity (see e.g. ILAR Committee, 1976; Spiegel, 1976; Otis and Foster, 1983; Heine, 1998) although many refinements have been introduced meanwhile. According to the animal house premises, the scientific demands, international standards and the risk of contamination within the building, different hygienic areas should be established. In principle, these could be:

Germfree or **axenic**, designating a status in which no microorganisms are present except those integrated in the genome.

Gnotobiotic or **gnotoxenic** where the animals are colonised with a fully defined flora which may induce some resistance to ubiquitous microorganisms. This could be on the basis of a pre-stimulation of the innate immune response and/or on an interference between bacterial strains including yeast in the resident flora of the intestinal tract.

Specified pathogen free (SPF) describes by definition the respective animals as being free from the pathogens specified and for which the colony is regularly monitored (Nicklas *et al.*, 2002). Suggestions for the exclusion list as well as methods of monitoring can be found in Chapter 25 on Gnotobiology and Breeding Techniques and in Chapter 27 on Health Monitoring in this handbook.

Quarantine originally used to cover the potential latency period of infections by opportunistic or pathogenic agents, describes in this context a status in which, in particular, newly introduced strains from other institutions have to be imported, maintained and put through repeated and thorough hygienic examination (screening for viral, bacterial and endo- ectoparasitic infections) before the animals may be transferred into the respective animal quarters or the new strain undergoes a rederivation process.

Infectious denominates the status of animals either being infected naturally or within experimental research. They are a risk for the other animals in the facility and/or for humans. Animals with undefined microbiological status are usually treated like infectious animals.

Animal facilities

In the last years, individually ventilated cage (IVC) systems came into increased use. This new set up not only affects facility construction, but also animal care management and health monitoring. While about $20\,m^2$ proved to be optimal for rooms with the conventional cage system in scientific institutions $50–70\,m^2$ per room are recommended for an effective use of IVCs providing a maximum rate of cages respectively mouse holding capacity per square metre of floor space. In IVCs the ventilation of the single cage is several fold higher in comparison to the conventional open caging, allowing some extension of the cage-changing interval, which may at least in part compensate for the higher labour intensity. If correctly designed, the use of IVCs considerably decreases the allergen and odour level in the animal rooms.

SPF-units

By definition, animals are free of specified pathogens. However, no declaration on residual microorganisms is given, implying the probability of extensive differences from one SPF-unit to another (Heine, 1980; O'Rourke *et al.*, 1988; Boot *et al.*, 1996). Therefore, it must be considered that when transferring animals from one SPF-unit into another, additional microorganisms may be introduced which can have an effect on the microbiological equilibrium. This is particularly to be expected if immunocompromised animals are being maintained (Ohsugi *et al.*, 1996). SPF-units are either used as breeding stations, with restricted access for researchers, or for long-term experiments. These are protected by strict hygienic barrier systems with sterile supply of air, food, cages, bedding and any other material, while all personnel including researchers have to pass a water or air shower as entry lock (Otis and Foster, 1983; ILAR Committee, 1989; Heine, 1998). A conventional open caging system or IVC systems may be used within the SPF-unit. After microbial decontamination with e.g. formaldehyde or hydrogen peroxide (Krause *et al.*, 2001), gnotobiotic animals can be introduced via a chemical entry lock. If standardised diets are introduced by steam-sterilisation the diet has to be enriched ('fortified') with heat sensitive vitamins to ensure that sufficient amounts remain after the heat treatment. As an alternative, gamma-irradiated standardised diet (25 kGy) packed in waterproof evacuated

bags can be transferred in into the SPF-unit after sufficient disinfection of external surfaces. The drinking water should be sterilised either by heat, ultrafiltration, and/or fluorescent (UV) light and then acidified (using hydrochloric acid or acidic acid) to a pH of 3.0–2.5, or chlorinated at a pH of 5 using stabilized hydrochloride to 6–8 ppm free chlorine (Leblanc, pers. comm.) in order to minimise the microbial growth. Bedding material should be dust-free (<1% dust) and must be steam-sterilised prefilled in cages or separately in bags (Table 24.3).

The personnel entering the barrier bring about the highest risk in terms of microbial contamination. Therefore, only well-trained and highly motivated staff (FELASA, 1995), having had no contact to external rodents for several days and being free of infections should be allowed to enter the SPF-unit.

The *microbiological status* of the SPF-unit should be monitored on a regular basis (see Chapter 27 on Health Monitoring). All sick animals have to be removed from the unit (unless the disease status is part of the animal's genetic constitution or the experiment) and submitted to health monitoring (microbiological examination, necropsy). In addition sentinel (germfree or gnotobiotic)

TABLE 24.3: Suggestions for autoclaving temperatures and times				
Material	Pre-vacuum (Cycles × min)	Sterilisation		Vacuum for drying
		°C	min	
cage (empty), hoods,	1 × 5	120[a]	10	10 min
bottles (empty), lids in stacks		134	5	
bedding in bags[b]	3 × 10	120	30	30 min
		134	20	
bedding in cages	3 × 10	120	15	15 min
		134	10	
food[c]	1 × 5	120	15	5–10 min
		134	5	5–10 min
water[d]	1 × 5	120	20	about 10 h pressure kept at
		134	10	3.0 bar
supply cylinder:		120	30	
cages with bedding,	3 × 5			20 min
bottles (empty), lids, food		134	15	
IVCs fitted with cages	3 × 5	120	10	

The proposed procedures have to be adapted to the respective autoclave because of enormous differences between products.

[a]120°C for polycarbonate (macrolon[R]) material. 134°C may be used for polysulfonate material.

[b]Bedding in perforated plastic bags (15 kg); one layer on shelf.

[c]In perforated paper bags (15 kg); one layer on shelf; 15 min warming before pre-vacuum.

[d]Water in bottles (not tightly fixed cover) with measurement of reference temperature in one vessel.

animals should be checked at fixed intervals. Regular disinfection of floors, walls and racks should be a routine operation procedure in order to minimize the possibility of infections. If properly managed such units may be maintained free of unwanted microorganisms for many years. In case of an infection within the unit it is unlikely that this can be restricted to a few cages, if conventional caging is used. This may be prevented only by the use of IVCs when handled properly.

IVC systems

The individual ventilation of the cages with HEPA-filtered air allows long-term bio-containment on the cage level when the handling of the cage's interior follows aseptic rules. Individually ventilated cage systems are used to breed and maintain animals in order to reduce the risk of miscellaneous microbial contamination and to improve environmental conditions, which may be of special interest when maintaining immunocompromised animals. Individually ventilated cage systems are particularly useful when a barrier-sustained (SPF) unit is not available and in experimental units, where easy access to the animals by the scientists is indispensable. In addition, IVCs with positive pressure are also ideal for a preliminary or timed containment of animals derived from different sources to maintain their respective hygienic status, if the area is barrier protected. For quarantine purposes without an additional barrier system and for infectious experiments these racks should be used with negative pressure. A problem of IVCs is the health monitoring requiring *sensu strictu* testing of each individual cage.

Ventilation of the cages can be performed by blowers in the animal room. In this case the supply air is drawn from the room's air; the exhaust air duct should be loosely interconnected to the room's exhaust. For facilities to be newly constructed, separate channels for air supply/exhaust for the room and the IVCs, respectively, might be installed. This avoids blower units within the room and allows decreasing room ventilation and bio-filter capacity and thus may reduce running costs. In addition, the stocking rate can be increased because it is no more limited to the ventilation capacity of the room.

For most IVC-rack types the intra-cage pressure can be adjusted to be either positive or negative. This makes the units versatile in their use under different hygienic conditions. Ventilation with *positive* pressure in the cages counteracts the leakiness of the system. *Negative* cage-pressure may prevent the escape of microorganisms from the IVCs

and further reduces allergen escape. The different aspects of ventilated cage systems are described in an overview by Lipman (1999) and special topics are presented by Baumans *et al.* (2002), Chaguri *et al.* (2001), Clough *et al.* (1994), Gordon *et al.* (2001), Hasegawa *et al.* (1997), Höglund and Renström (2001), Perkins and Lipman (1996), Reeb-Whitaker *et al.* (2001), Renström *et al.* (2001), Tu *et al.* (1997) and Novak and Sharpless (2003). Different products are commercially available, in which the air is delivered either directly into the cage or distributed by a wide filter mesh in the hood to reduce the intracage air velocity. In many systems the exhaust air passes a filter in the cover to retain dust from the exhaust pipes. A critical event in IVCs is the failure of air supply either by an electrical defect, blower break down or disconnection of tubes. It is therefore recommended to install an air flow controller in the supply air duct (positive pressure) or exhaust duct (negative pressure) respectively, which is connected to the alarm system of the animal house (Huerkamp *et al.*, 2003).

Handling of IVCs

The procedure of sterile handling of IVCs is labour intensive. This can be compensated – at least in part – by extending the cage change interval. Due to the higher intra-cage ventilation the ammonia concentration will be lower and the bedding stays drier for a longer period. In addition, increasing the change interval will reduce the stress for the animals (Duke *et al.*, 2001) and may increase breeding efficacy (Reeb-Whitaker *et al.*, 2001).

The proper management of IVCs is critical and quite often underestimated. There are three different hygienic levels that have to be considered: (i) The sterility level of the autoclaved material: *cage* with bedding, lid and hood (sterilised as a not tightly closed unit, or separately in a container or foil), *diet* (autoclaved or gamma-irradiated at 25 kGy) and *water* bottles, transferred sterile into the laminar flow changing station (e.g. within a filter hood covered cage); (ii) the hygienic level of the animal room (i.e. the direct environment of the animal enclosure termed 'room-clean'); and (iii) the hygienic status of the animals. In a correctly manipulated setting, these three levels have to be strictly discriminated. To maintain the health status of the animals within the individual cages it is advisable to run the regular caretaking routines by two animal technicians (a 'clean' and a 'room-clean' person; Table 24.4). In certain IVC systems with water bottles outside the hood, there is a high risk of contaminating the mice when bottle-change is not performed in connection with regular cage changing in the laminar flow cage changing station.

> **TABLE 24.4: Handling of IVC racks (two persons)**
>
> 1. Laminar flow cage changing station or laminar flow hood is running (30 min in advance).
> 2. A fast acting sterilisation compound (recommended: Clidox) is freshly diluted for gloves, bench worktop, (in separate beakers) for the forceps to handle the mice (alternately used) and a pincers to handle the lid.
> 3. The animal technicians handling the cages and animals should at least wear gloves; however it is recommended to also wear a surgical gown, cap and mask.
>
>   *Room-clean person*: The cage to be changed is placed into the changing station, the cage tag is removed and hood and lid partially displaced backward. An autoclaved cage with bedding, lid and hood is placed into the bench and opened partially without contacting the inner side of the cage.
>
>   *Clean person*: Sterilised diet is transferred into the clean cage from the outside sterilised, irradiated food-bag. Sterile water bottle is inserted. Mice are transferred in the sterile cage with the disinfected forceps and the lid is closed with the pincers.
>
> 4. *Room-clean person*: Hood is replaced, cage tag remounted and both cages removed.
> 5. Worktop and gloves are disinfected after each step with the sterilising solution.
>
> *Note*: In case of infected animals, the contaminated cages have to be autoclaved with the filter tops in place.

Static micro-isolators

'Static micro-isolators' are non-ventilated, partially perforated boxes with a tight filter medium, covering the perforation and a cover, which has to be sealed (Kraft, 1958). They provide a microenvironment, which is protected from adventitious contamination from outside. Micro-isolators are still in use (e.g. Han-Gnotocage) for the transport of germfree, SPF-founder and quarantine (Otto and Tolwani, 2002). They can also be used for short-term housing of germfree or gnotobiotic foster mothers (in a laminar flow cabinet). The disadvantage of these micro-isolators, however, is their impeded intra-cage ventilation which results in an increase in humidity, carbon dioxide and ammonia concentrations. With increasing animal density static micro-isolators become intolerable. However, static filter top cages prevent the release of allergens and are therefore well suited for transportation of mice within an institution.

Isolators (positive pressure)

They consist of a closed construction with a germ-tight air inlet filter, a liquid air outlet trap or a germ-tight outlet filter, long-arm gloves and a chemically sterilisable lock for the supply via an autoclave cylinder (see e.g. Trexler, 1983; Heine, 1998). Freshly mixed buffered peracetic acid (Kesla Pharma Wolfen, D-06803 Greppin, Germany), diluted to 2.0% is commonly used to sterilise the interior of the isolator and the entry lock. For details on how to generally operate isolators the reader is referred to Chapter 25 on Gnotobiology and Breeding Techniques, or Heine (1998). As a consequence of sterilisation for 30 min at 134°C the diet may become less palatable (Porter and Festing, 1970) and masticable than a surrogate gamma-irradiated diet (50 kGy) supplied in evacuated bags.

Isolators (negative pressure)

The equipment of most commercially available isolators allows the alternative use with negative pressure. In this version isolators protect the environment (other mouse colonies or man) from infections that are present in the isolator by a HEPA-filter in the exhaust valve. Their use is obligatory for experiments with high-risk pathogens. It is furthermore recommended for new strains from other sources being infected with agents imposing a high infectious risk to the existing colonies. All materials of isolators or quarantine have to be autoclaved when leaving the infectious area. In this context, it should be mentioned that repeated autoclaving of cages with soiled bedding will destroy polycarbonate (macrolon), while polysulfone is much more resistant.

Quarantine

As a consequence of the genetic manipulation, the exchange of breeding stocks between institutions has rapidly increased. Because of the presumably different hygienic constitutions, the single stocks should be preserved in their own microbiological status until rederivation can be performed. This can be achieved by the use of IVCs in positive pressure within a separate barrier-unit in negative pressure. Quarantine precautions should also be established when cellular material with unknown microbiological status has to be inoculated into experimental animals. If infected this biological material could be of risk for the animal facility (Yoshimura *et al.*, 1997) contaminating the whole unit with a viral infection, such as **ectromelia** (Lipman *et al.*, 2000), or mouse minute virus (Tietjen, 1992). Animals that have been gamma-irradiated should also be submitted to a quarantine-type area and not returned to their previous quarters (due to the same reasons). However, specific containers (sealed to avoid contamination of animals, with sufficient air volume to warrant oxygen supply for a certain period of time) have been developed for irradition which may allow reintroduction of these mice after whole-body irradiation into a barrier unit.

Therapeutic treatment

The administration of therapeutics can influence the outcome of animal experiments in different ways and should not become a routine procedure, nor is it a means to substitute for improved hygienic standards. Antibiotic treatment guided by microbial resistance testing may be a necessity in certain natural and induced mutants (e.g. *Ncf1*, due to their defective granulocyte bactericidal activity; Jackson *et al.*, 1995) unless maintained under germfree or strict gnotobiotic conditions. The treatment of parasitic invasions is in particular dependent on the accompanying hygienic procedures e.g. use of gloves, chemical and/or physical disinfection of the animal rooms, cages, lids, bottles etc. For drug dosages see Hawk and Leary (1995). It should be mentioned, however, that any treatment might interfere with the experimental outcome or might be associated with toxic effects (Scopets *et al.*, 1996, Toth *et al.*, 2000), as e.g. some knockout mice (targeted disruption of the multi-drug resistance gene) and CD-1 mice have been shown to be very sensitive to ivermectin, with resulting mortality.

Permanent antibiotic treatment of infections may induce bacterial resistance to antibiotics especially when used on a large scale. This will not only affect the animal colony (Hansen, 1995) by a shift in the gut flora (Morris, 1995) and other derangements in physiological functions (el Ayadi and Errami, 1999), it may lead to a contamination of the environment with multi-resistant bacteria that could be a hazard to humans. All bedding material of treated animals should therefore be sufficiently steam-sterilised prior to disposal.

Presumptions for genetic manipulation

The possibility of genetically engineering mice has brought about an exponential increase of new strains/stocks specifically constructed for the different aspects of research. In general, donor mice of oocytes or early embryonic stages are bought from commercial breeders. If these mice are maintained separately and manipulation of the oocytes/embryos (laminar flow, medium, washing, handling) is designed to interrupt a potential transmission of microorganisms, the risk of transmission of an infection should be low. Decisive for the hygienic status of the genetically manipulated progeny is, by far, the status of the foster mothers and their handling being of special importance when creating immunodeficient strains.

Identification systems

Regulations contained in the ILAR (1996) *Guide for the Care and Use of Laboratory Animals* stress the importance of proper animal identification in sound research and humane animal care. The identification of single mice has become an indispensable tool even more so as a consequence of genetic manipulation, where genotyping results have to comply with the respective individual. Normally, correct labelling of the cages gives the first hint to strain, sex, age and individual numbering. The ILAR (1996) *Guide* as well as a number of other sources, list many acceptable identification methods for most common laboratory animal species. Several methods are being used:

Metal ear tags are an easy method of permanently marking laboratory animals. The procedure is simple using an ear punch and a specific applicator and does not require anaesthesia. The tags are available with a three digit numbering system up to 999. This implies a consecutive numbering of all mice, irrespective of the strain or a numbering of up to 999 within a strain. Ear tags with different colours per strain within a unit may be of help. Unfortunately, mice sometimes lose their tags, requiring remarking.

Tattooing of footpads, tail etc. is an acceptable alternative, which may be especially helpful in marking new borne mice.

Ear punching: Exterior markings of the pinna can be done on mice after two weeks of age without anaesthesia. Ear punch identification may be obliterated after several weeks because of wound healing or by fighting between individuals. A system of holes and notches permits a numbering from 1 to 9 respectively 10 to 90 on both ears (see Figure 24.1), allowing for an individual numbering system of up to 100 individual mice.

Toe-clipping: One method of identification that has been used for rodents is toe-clipping. This method involves removal of the first phalanges of one toe of up to all four limbs, corresponding to a predetermined numbering code (ALAT Manual, 1998). The different digits removed code the identifier. Because toe-clipping is a potentially painful procedure that can also alter the gait or weight-bearing ability of a rodent's rear limbs, the ILAR (1996) *Guide for the Care and Use of Laboratory Animals* limits its use only to justified instances (toe-clipping protocol must be discussed by the IACUC, North America), and requires anaesthesia for animals older than 1 week. In other countries this method is not permitted at all (e.g. German Animal Welfare Act §6).

Transponder: The subcutaneous implantation of a transponder is without doubt the most reliable, but most expensive method for identification. This system is practically not limited in the number of individuals that can be differentiated and is rather feasible, when animals have to be identified very often, e.g. if monitoring of body weight or other parameters like body temperature have to be recorded together with the identification number (Kort *et al.*, 1998).

Refinement

Animal welfare regulations (see also Chapter 29 on Legal Regulations for the Protection of Animals Used for Scientific Experiments) require that the environment of the animals maintained should meet their physiological and behavioural needs and that the housing conditions must facilitate the performance of natural behavioural patterns and allow for adequate social contacts (Jennings *et al.*, 1998). Current debate on enrichment of the animals' environment and space requirements has not come to a general conclusion. There is no doubt that cages with pregnant females should be equipped with nesting material like nestlets, paper towels etc. The published opinion is rather contradictory with respect to the effects imposed on research data obtained under 'standard' (sometimes called 'impoverished') vs. structurally and/or spatially enriched housing systems. While several groups report on distinct effects on behaviour, such as increased locomotion, exploration, learning ability, reduced anxiety etc. (Widman and Rosellini, 1989; Prior and Sachser, 1995; Boehm *et al.*, 1996; Sherwin and Nicol, 1997; Powell *et al.*, 1999; Sachser, 2001), other studies indicate

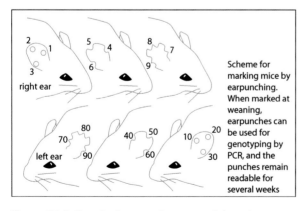

Figure 24.1 Standard pattern for ear notch/punch numbering in mice.
Source: From http://iacuc.cwru.edu/training/mousecolonymngmnt.html

that an enrichment of the housing cage leads to an increase in aggressive behaviour (McGregor and Ayling, 1990; Haemisch and Gärtner, 1994; Haemisch *et al.*, 1994) respectively to an increase in the coefficient of variation of several parameters depending on the strains analysed (Mering *et al.*, 2001; Tsai *et al.*, 2002, 2003). The latter result may suggest that due to a higher variation of different parameters from one experiment to another and between different environments the estimations of the appropriate number of animals to be used have to be increased. Marashi *et al.* (2003) report on an increase of aggressive behaviour in male mice and elevated levels of stress hormones in mice housed in structurally (plastic inset and a wooden scaffolding) and a structurally as well as spatially enriched environment (richly structured by a complete, passable enriched cage, extra plains, plastic stairs, wooden footpaths, hemp ropes, and a climbing tree). Interestingly, it is concluded that stress is a most important prerequisite to achieve good welfare. This controversy with a thus far agreed notion that animals should be housed with the goal of maximising species-specific behaviour and minimising stress-induced behaviours needs further analysis. For a detailed bibliography on mouse enrichment and refinement see the UCCAA database providing access to regularly updated information that can be addressed to improve or refine research with mice (http://www.vetmed. ucdavis.edu/Animal_Alternatives/miceR3.htm).

Genetic monitoring

Mutations as well as differential fixation of alleles at early generations of inbreeding may alter the genetic constitution and thus the phenotype of an inbred strain. Many of the phenotypic differences detected between substrains have been shown to be due to these factors. Inadvertent outcrossing will alter a strain seriously, questioning its further use for research, since results are no longer comparable and repeatable. It is thus of utmost importance to separate strains that are not immediately distinguishable by their phenotypic appearance. In most facilities due to shortage in shelf space and separate animal rooms, many strains have to

TABLE 24.5: Principles of proper colony management

1. During regular handling only one cage at a time should be managed. This will prevent accidental exchange of animals from different cages.

2. Animals that have escaped or dropped to the floor must never be returned to the suspected cage. Animals caught outside the cage should be isolated, if identification is possible, otherwise killed.

3. Cages and hoods should be in sufficient condition that no animal can escape or enter another cage, a problem more often encountered in mouse than in rat breeding units.

4. For ease of identification and in order to prevent an inadvertent mix-up, cage tags should have a strain-specific colour code and a strain-specific number (code).

5. Cage tags should always be filled out properly, including the strain name, strain number, identification numbering of the animals in the cage, parentage, date of birth, generation, and in case of experimental use the name and licence number of the scientist.

6. If a cage tag is lost, one should not redefine the cage except in the case of definite proof of identity through marked animals within the cage.

7. If at weaning the number of animals is larger than that recorded at birth the whole litter should be discarded or submitted to genetic monitoring.

8. Any change in phenotype and/or increase in productivity should immediately be reported to the colony supervisor. The latter change should always be considered suspect for a possible genetic contamination.

9. Regular training programmes on basic Mendelian genetics, systems of mating and the reproductive physiology of the animals maintained should make animal technicians and caretakers conscious of the consequences any mistake will impose on the colonies. Further training should stress the importance of a search for deviants as potentially new models for biomedical research.

be housed in one room, making regular screenings for strain discriminating markers as well as the differentiating locus (in case of congenic strains) indispensable.

As repeated handling of animals during regular caretaking cannot be avoided, there is always the risk of mistakes. Proper colony management is the first step towards the provision of authentic laboratory animals. Either inadvertently an animal might be placed into a wrong cage, or a false entry is put on the label. Assigning this type of work to well trained and highly motivated animal technicians should be a matter of course. This can be summarised in nine principles of a proper colony management (see Table 24.5). The colony set-up and structuring with nucleus colonies in a single (Festing, 1979) or a (modified) parallel line system (Hedrich, 1990), pedigreed expansion colonies and multiplication colonies should be self evident, but strictly monitored. There are several publications dealing with the set-up of colonies for maintenance and large-scale production (Green, 1966; Lane-Petter and Pearson, 1971; Hansen *et al.*, 1973; Festing, 1979). In general, permanent monogamous mating is to be given preference, as this provides a constant colony output by minimal disturbance of the litters during the early post natal period and by utilising the chance that females are fertilised at the post-partum oestrus.

The measures required for genotyping a strain have to be adjusted to the specific needs and may depend on the scientific purpose, the physical maintenance conditions and the laboratory equipment. Nevertheless, there are specific demands (although unfortunately not stringent rules) on how to authenticate a strain or to verify its integrity.

For any authentication it is necessary to determine a genetic profile that is to be compared with published data (as far as available), and which allows to distinguish between (all) strains/stocks maintained in one unit. In general this profile is composed of monogenetic polymorphic markers, which may be further differentiated by the method of detection into immunological, biochemical, cytogenetical, morphological and DNA markers. Microsatellite markers (Simple Tandem Repeats, STRs) have almost fully replaced the classical genetic markers in routine applications. A large number of primer pairs is available on the world wide web (e.g. through Research Genetics Inc., Huntsville, AL, USA, http://www.resgen.com). However, as with the classical markers it is indispensable to set up a genetic profile representing a random sample of the genome, which should be evenly spaced on the chromosomes, and which will allow to discriminate between all strains maintained per separate housing unit. There are numerous publications and textbooks with protocols for polymerase chain reaction (PCR) amplification and electrophoretic separation of the **amplicons**. Moreover, commercial suppliers of primers (e.g. Research Genetics) and of genetically modified animals (e.g. The Jackson Laboratory's Induced Mutant Resource genotyping protocols, http://www.jax.org/imr /geno_index.html) do provide PCR protocols. In any laboratory setting, it might be necessary to adjust temperature conditions as well as Mg^{++} concentrations for each microsatellite marker. For routine screening separation in agarose gel and visualisation by ethidium bromide will suffice. If separation of the amplicons is insufficient one should run a polyacrylamide gel electrophoresis. Instead of radioactive labelling with ^{32}P using a kinase reaction and since the half-life of the isotope is relatively short, a silver staining procedure is recommended. Information on restriction fragment length polymorphisms (RFLPs) as determined by a Southern blot (Sambrook *et al.*, 1989) using a specific probe is partly provided through mouse genome informatics (MGI) maintained by The Jackson Laboratory.

Despite the ease of analysis and the abundant number of markers, classical methods are still relevant and may need to be verified, and sometimes may even allow for a faster and less expensive phenotyping. In this context immunological markers are of prime importance. This group is composed of cell surface markers, such as the major histocompatibility antigens (*H2*), lymphocyte differentiation antigens, red blood cell antigens, minor histocompatibility antigens, allotypes (immunoglobulin heavy chain variants) which can be determined using specific antibodies by Trypan blue dye exclusion test, flow cytometry, immunodiffusion and enzyme-linked immunosorbent assay (ELISA). Further methods that can be applied easily and determine a specific phenotype may also be used, as e.g. the lysosomal trafficking regulator (*Lyst^bg*, beige, expressing a pigmentation and platelet storage pool defect). The phenotype of homozygous beige mice can be demonstrated by a prolonged bleeding time (20 min in homozygous *Lyst^bg* vs. 6 min in unaffected wild-type or heterozygous controls), or a histochemical staining (checking for abnormal giant lysosomal granules detectable in all tissues with granule-containing cells; see Novak *et al.*, 1985).

The set-up of a profile is time consuming and expensive, but strongly recommended as an initial check. In case of a segregating background genetic profiling is pointless, as the typing results will only assist in determining the degree of heterogeneity. However, these results may provide hints on modifying genes, if the stock is incipient inbred and nearly homozygous. In case several strains with an identical coat colour phenotype, but different genetic modifications, are maintained in one room, all (induced) genetic modifications need to be tested for by PCR or other appropriate methods.

Still one needs easy measures to distinguish between strains that are co-maintained and for which it is important to clearly identify an outcrossing event. A critical subset of the markers, at least the differentiating (genetically modified) markers for a given strain panel will provide reasonable information on the genetic quality of a strain and may partially serve to authenticate each strain within the unit. Unfortunately, with each strain added to such a unit the number of markers in the critical subset increases. These critical subsets need to be verified at regular intervals (every 3–6 months). The intervals and the number of animals to be tested is increment to the number of strains co-maintained and to the size of each colony.

Irrespective of these methods one of the most powerful information on an inbred strain is the demonstration of its **isohistogeneity**, only demonstrated by skin grafting. The technique is easy to perform, but time consuming due to an observation period of 100 days (for a description of the techniques see Hedrich, 1990). In certain immunodeficient mutants (e.g. *Foxn1^{nu}, Prkdc^{scid}, Rag1^{tm}, Rag2^{tm}*) a direct demonstration of isohistogeneity is impossible, as these animals are incapable of mounting an **allorecognition** response. Transferring grafts from these immunodeficient animals to their syngeneic background strains will circumvent this.

Genetic problems in engineered strains

Due to the enormous number of genetically modified strains their phenotypic discrimination is fairly impossible. In addition to the risk of mixing identities in the course of cage changing, false genotyping or false identification of the animals may be a source of mistakes. Therefore, regular screening of homozygous or still segregating strains/stocks is indispensable. In the process of backcrossing a transgene or targeted mutation onto a defined background it is advisable to select and analyse markers sufficiently differentiating between the background strain and the donor stock of the genetic modification. A genetic monitoring using markers similar to a marker-assisted selection protocol (speed congenics production; see: Markel *et al.*, 1997, Wakeland *et al.*, 1997, Visscher, 1999) is recommended.

Computer assisted management of animal facilities[1]

The production and breeding of gene-manipulated mouse strains in most institutions has exponentially increased in recent years. In addition, the need to document animal breeding and experimentation for governmental and scientific purposes is also increasing. A computer database can be very helpful to facilitate the efficient management of this information. With such a tool, it is possible to maintain an overview of many aspects of breeding management such as: mouse strains, breeding status and availability, services provided, registered users and ethically reviewed projects (e.g. by the Institutional Animal Care and Use Committee, IACUC). The database should be multi-user compatible, and platform-independent within the computer network of the institution and should be designed in such a manner, that it helps the facility administrator, researchers, animal technicians and the animal welfare officer to maintain an overview of the work, as well as to support communication with each other and to reduce the workload.

Typical functions of such a database could include: *Mouse strains*: Data of all (sub)strains and sublines maintained using proper nomenclature as well as internal laboratory denominations and a coding system, strain owner, responsible investigator, genotype, phenotype, breeding information, breeding/holding unit etc. can be stored. An *ordering system* would allow users to submit jobs to the breeding facilities, such as ordering of animals for experimentation or shipment, ordering of tissues, cryopreservation, immunisation, embryo transfer, isolator breeding etc. The animal technicians in the breeding facilities could process these jobs and the status of each job should be viewable at any time. *Animal welfare*: Projects approved by governmental authorities for work with experimental animals could be stored, as well as the type of experimental procedures permitted, the number of animals allowed, registered

---

[1]Author of this paragraph: Peter Nielsen, Max-Planck-Institut für Immunbiologie, Freiburg i.Br., Germany.

Database	Vendor	Address
Colony and Facility	Locus Technology Inc.	http://www.locustechnology.com/pages/company.html
MateBase and TierBase	Abase	nielsen@immunbio.mpg.de
MacMice and Mendel's Lab	Mendel's Software	https://web.princeton.edu/sites/lsilver/links/mendel.htm
MICE	IFR72 Centre de Biochimie, Université de Nice	pognonec@unice.fr
PyRAT	Scionics Computer Innovation GmbH and MPI-CBG	pyrat@scionics.de or helppi@mpi-cbg.de
Scion and Granite	Topaz Inc	http://www.topazti.com/

persons etc. *User list*: All users working with animals could be stored, along with strain access information. *Breeding analysis*: It should be possible to register the litters and offspring for all breeding for each strain. This information permits a clear overview on all breeding including animal and cage inventory, sex, phenotype, genotype, genealogy etc. For the purpose of archiving or further analysis, the data should be exportable to other programmes like MS Word® or MS Excel®. *Inventory*: It could be useful to know, at any time, the number of animals or cages present in the breeding facility. This can be useful for allocating breeding and holding space, as well as for cost calculations. *Invoices*: If desired, it should be possible to create invoices for animals, cage or special animal care costs, for jobs (specific diagnostics, sample collection etc.), as well as for other services rendered. *Online information*: Data sheets or comments could also be integrated, which describe the use of the database and aspects of animal work, some of which may be specific for the institute. *Language preference*: The employees in many research institutes may not have a common mother language. For this reason, it could be useful if the database can accommodate a preferred language for each user.

Such a database should help to maintain an overview of experimentation with animals. It should stimulate an efficient communication between the different usergroups and save work! *Researchers* should be able to see which strains and ethically reviewed projects they are responsible for, including the current number of animals used and the status of all jobs they have submitted. They may also be allowed to run a detailed analysis of genotype frequencies, pedigrees, reproductive data etc. *Animal technicians* could check which animals have been ordered for a specific date and where (holding unit/laboratory) these should be delivered/ shipped. *Animal welfare officers* should be able to verify how many animals have been ordered for which purpose, the duration of the treatment etc. They could use this information to easily create a report providing a summary of the animals used in any experiment. This report may be submitted directly to the government authorities, if required. *Breeding administrators* can identify which animals belong to whom and how many mice have been used. They can thus insure that sufficient numbers per sex and per strain will be available within the institution. If desired, invoices for the services provided could also be created.

For additional information, several databases, which have been developed for colony management, animal breeding and experimentation are listed above.

References

ALAT Manual (1998). American Association for Laboratory Animal Science, p. 57.

Baumans, V., Schlingman, F., Vonck, M. and van Lith, H.A. (2002). *Contemp. Top.* **41**(1), 13–19.

Boehm, G.W., Sherman, G.F., Hoplight, I.I., BJ, Hyde, L.A., Waters, N.S., Bradway, D.M., Galaburda, A.M. and Denenberg, V.H. (1996). *Brain Res.* **726**, 11–22.

Boot, R., van Herck, H. and van der Logt, J. (1996). *Lab. Anim.* **30**, 42–45.

Chaguri, L.C.A.G., Souza, N.L., Teixeira, M.A., Mori, C.M.C., Carissimi, A.S. and Merusse, J.L.B. (2001). *Contemp. Top.* **40**(5), 25–30.

Clough, G. (1992). *ASLAS Newslett.*, Summer, 5–10.

Clough, G., Wallace, J., Gamble, M.R., Merryweather, E.R. and Bailey, E. (1994). *Lab. Anim.* **29**, 139–151.

Duke, J.L., Zammit, T.G. and Lawson, D.M. (2001). *Contemp. Top.* **40**(4), 17–20.

el Ayadi, A. and Errami, M. (1999). *Therapie* **54**, 595–599.

FELASA (1995). *Lab. Anim.* **29**, 121–131.

Festing, M.F.W. (1979). *Inbred Strains in Biomedical Research*. Macmillan Press, London.

Fox, J.G., Cohen, B.J. and Loew, F.M. (1984). *Laboratory Animal Medicine*. Academic Press, Orlando.

Green, E.L. (1966). *Biology of the Laboratory Mouse*. McGraw-Hill, New York.

Gordon, S., Fisher, S.W. and Raymond, R.H. (2001). *J. Allergy Clin. Immunol.* **108**, 288–294.

GV-SOLAS Publikation Nr. 1 (1988). *Planung, Struktur von Versuchstierbereichen tierexperimentell tätiger Institutionen*. http://www.gv-solas.de/publ/pub.html

Haemisch, A. and Gärtner, K. (1994). *J. Exp. Anim. Sci.* **36**, 101–116.

Haemisch, A., Voss, T. and Gärtner, K., 1994. *Physiol. Behav.* **56**, 1041–1048.

Hansen, A.K. (1995). *Lab. Anim.* **29**, 37–44.

Hansen, C.T., Judge, F.J. and Whitney, R.A. (1973). *Catalogue of NIH Rodents*, DHEW Publication No. 74–606. US Department of Health, Education and Welfare, Washington, DC.

Harkness, J.E. and Wagner, J.E. (1989). *The Biology and Medicine of Rabbits and Rodents*, 3rd edn. Lea and Febiger, Philadelphia.

Hasegawa, M., Kurabayashi, Y., Ishii, T., Yoshida, K., Uebayashi, N., Sato, N. and Kurosawa, T. (1997). *Exp. Anim.* **46**, 251–257.

Hawk, C.T. and Leary, S.L. (1995) *Formulary for Laboratory Animals*. Iowa State University Press, Ames, IA.

Hedrich, H.J. (1990). *Genetic Monitoring of Inbred Strains of Rats*. Gustav Fischer Verlag, Stuttgart.

Heine, W.O.P. (1980). *Z. Versuchstierk.* **22**, 262–266.

Heine, W.O.P. (1998). *Umweltmanagement in der Labortierhaltung. Technisch-hygienische Grundlagen, Methoden und Praxis*. Pabst Science Publishers, Lengerich.

Höglund, A.U. and Renström, A. (2001). *Lab. Anim.* **35**, 51–57.

Huerkamp, M.J., Thomson, W.D. and Lehner, N.D.M. (2003). *Contem. Top.* **42**(3), 44–45.

ILAR (1996). *Guide for the Care and Use of Laboratory Animals*. National Academy Press, Washington.

ILAR Committee on Long-Term Holding of Laboratory Rodents. (1976). *ILAR News XIX* **4**, L1–L25.

ILAR Committee on Immunologically Compromised Rodents. (1989). *Immunodeficient Rodents. A Guide to their Immunobiology, Husbandry, and Use*. National Academic Press, Washington, DC.

Jackson, S.H., Gallin, J.I. and Holland, S.M. (1995). *J. Exp. Med.* **182**, 751–8.

Jennings, M., Batchelor, G.R., Brain, P.F., Dick, A., Elliot, H., Francis, R.J., Hubrecht, R.C., Hurst, J.L., Morton, D.B., Peters, A.G., Raymond, G.D., Sales, G.D., Sherwin, C.M. and West, C. (1998). *Lab. Anim.* **32**, 233–259.

Kawachi, S., Morise, Z., Jennings, S.R., Conner, E., Cockrell, A., Laroux, F.S., Chervenak, R.P., Wolcott, M., van der Heyde, H., Gray, L., Feng, L., Granger, D.N., Specian, R.A. and Grisham, M.B. (2000). *Inflamm. Bowel Dis.* **6**, 171–180.

Kort, W.J., Hekking-Weijma, J.M., TenKate, M.T. Sorm, V. and van Strik, R. (1998). *Lab. Anim.* **32**, 260–269.

Kraft, L.M. (1958). *Yale J. Biol. Med.* **31**, 121–127.

Krause, J., McDonnell, G. and Riedesel, H. (2001). *Contemp. Top.* **40**(6), 18–21.

Lane-Petter, W. and Pearson, A. E. G. (1971). *The Laboratory Animal – Principles and Practice*. Academic Press, London.

Lipman, N.S. (1999). *Contemp. Top.* **38**(5), 9–17.

Lipman, N.S., Perkins, S., Nguyen, H., Pfeffer, M. and Meyer, H. (2000). *Comp. Med.* **50**, 426–435.

McGregor, P.K. and Ayling, S.J. (1990). *Appl. Anim. Behav. Sci.* **26**, 277–281.

Marashi, V., Barnekow, A., Ossendorf, E. and Sachser, N. (2003). *Horm. Behav.* **43**, 281–292.

Markel, P., Shu, P., Ebeling, C., Carlson, G.A., Nagle, D.L., Smuko, J.S. and Moore, K.J. (1997). *Nat. Genet.* **17**, 280–284.

Mering, S., Kaliste-Korhonen, E. and Nevalainen, T. (2001). *Lab. Anim.* **35**, 80–90.

Morris, T. H. (1995). *Lab. Anim.* **29**, 16–36.

Nicklas, W., Baneux, P., Boot, R., Decelle, T., Deeny, A.A., Fumanelli, M. and Illgen-Wilcke, B. (2002). *Lab. Anim.* **36**, 20–42.

Novak, E.K., McGarry, M.P. and Swank, R.T. (1985). *Blood* **66**, 1196–1201.

Novak, G.R. and Sharpless, L.C. (2003). *Lab. Anim. Eur.* **32**, 41–47.

Ohsugi, T., Kiuchi, Y., Shimoda, K., Oguri, S. and Maejima, K. (1996). *Lab. Anim.* **30**, 46–50.

O'Rourke, J., Lee, A. and McNeill, J. (1988). *Lab. Anim.* **22**, 297–303.

Otis, A.P. and Foster, H.L. (1983). In *The Mouse in Biomedical Research*, Vol. 3 (eds H.L. Foster, J.D. Small and J.G. Fox), pp. 18–35. Academic Press, New York.

Otto, G. and Tolwani, R.J. (2002). *Contemp. Top.* **41**(1), 20–27.

Perkins, S.E. and Lipman, N.S. (1996). *Contemp. Top.* **35**(2), 61–65.

Porter, G. and Festing, M. (1970). *Lab. Anim.* **4**, 203–213.

Powell, S.B., Newman, H.A., Pendergast, J.F. and Lewis, M.H (1999). *Physiol. Behav.* **66**, 355–363.

Prior, H. and Sachser, N. (1995). *J. Exp. Anim. Sci.* **37**, 57–68.

Reeb-Whitaker, C.K., Paigen, B., Beamer, W.G., Bronson, R.T., Churchill, G.A., Schweitzer, I.B. and Myers, D.D. (2001). *Lab. Anim.* **35**, 58–73.

Renström, A., Björing, G. and Höglund, A. U. (2001). *Lab. Anim.* **35**, 42–50.

Sachser, N. (2001). In *Coping with Challenge: Welfare in Animals Including Humans* (ed. D.M. Broom), pp. 31–48. Dahlem Workshop Report 87, Dahlem University Press, Berlin.

Sambrook, J., Fritsch, E.F. and Maniatis, T. (1989). *Molecular Cloning: A Laboratory Manual*, 2nd edn. Cold Spring Harbor Laboratory Press, New York.

Scopets, B., Wilson, R.P., Griffith, J.W. and Lang, C.M. (1996). *Lab. Anim. Sci.* **46**, 111–112.

Sherwin C.M. (2002). In *Comfortable Quarters for Laboratory Animals*. (ed. V. Reinhardt and A. Reinhardt) pp. 6–17. Animal Welfare Institute, Washington, DC.

Sherwin, C.M. and Nicol, C.J. (1997). *Anim. Behav.* **53**, 67–74.

Spiegel, A. (1976). *Versuchstiere: Eine Einführung in die Grundlagen ihrer Zucht und Haltung.* Gustav Fischer Verlag, Stuttgart.

Tietjen, R.M. (1992). *Lab. Anim. Sci.* **42**, 422.

Toth, L.A., Oberbeck, C., Straign, C.M., Frazier, S. and Rehg, J.E. (2000). *Contemp. Top.* **39**(2), 18–21.

Trexler, P.C. (1983). In *The Mouse in Biomedical Research* (eds H.L. Foster, J.D. Small and J.G. Fox), pp. 1–16. Academic Press, New York.

Tsai, P.P., Pachowsky, U., Stelzer, H.D. and Hackbarth, H. (2002). *Lab. Anim.* **36**, 411–419.

Tsai, P.P., Oppermann, D., Stelzer, H.D., Mähler, M. and Hackbarth, H. (2003). *Lab. Anim.* **37**, 44–53.

Tu, H., Diberadinis, L.J. and Lipman, N.S. (1997). *Contemp. Top.* **36**(1), 69–73.

van Zutphen, L.F.M., Baumans, V. and Beynen, A. (2001). *Principles of Laboratory Animal Science.* Elsevier, Amsterdam.

Visscher, P.M. (1999). *Genet. Res.* **74**, 81–85.

Wakeland, E., Morel, L., Achey, K., Yui, M. and Longmate, J. (1997). *Immunol. Today* **18**, 472–477.

Wang, B., Simpson, S.J., Hollander, G.A. and Terhorst, C. (1997). *Immunol. Rev.* **157**, 53–60.

Widman, D.R. and Rosellini, R.A. (1989). *Physiol. Behav.* **47**, 57–62.

Yoshimura, M., Endo, S., Ishihara, K., Itoh, T., Takakura, A., Ueyama, Y. and Ohnishi, Y. (1997) *Exp. Anim.* **46**, 161–164.

CHAPTER 25

Gnotobiology and Breeding Techniques

Patrick Hardy

Charles River Laboratories, Lyon, France

Introduction and historical background, evolution of applications

Gnotobiology

Interaction between environmental microbism and animal health, either beneficial or deleterious, could be evidenced very early in laboratory animals. Pasteur and, a few years later, Nencki and Metchnikoff investigated the survival of several species in the absence or the presence of various bacteria, as cited by Luckey in 1963 or Pleasants in 1973. The first germfree animals were produced after overcoming the preparation of sterile diet, the design of the first isolator systems and the use of germicide products (Luckey, 1963). Breeding of germfree rats was reported from 1946 by Gustafsson, Trexler and Reynolds, using different types of isolators

and sterilization systems. The use of aseptic hysterectomy or hysterotomy, the reproduction and maintenance of gnotobiotic rat and mouse colonies became a common practice in the late fifties. The first research applications addressed the effects of microbial flora on animal physiology, nutrition and metabolism (Trexler, 1967, 1968; Pleasants, 1968; Gordon and Pesti, 1971). Later developments included extended experimental aspects in the fields of leukaemia, carcinogenesis, disease resistance, immunological response, cell line differentiation etc. (Coates, 1968; Salomon, 1968; Wostmann, 1968; Mirand and Back, 1969; Pleasants, 1973). Later on, when the breeding and experimental use of 'specific pathogen free' (SPF) rodents became the rule, the techniques developed in gnotobiology were used for rederivation of contaminated lines or colonies to gnotoxenic conditions before maintaining them under SPF conditions, i.e. under less demanding bio-exclusion conditions, not aiming anymore to preserve an axenic or oligoxenic definition but to prevent the contamination by a well defined list of species-specific pathogenic agents ('bio-exclusion list'). These animals were named 'disease-free', 'heteroxenic' or 'caesarean-obtained and barrier sustained' (COBS) (Schaedler et al., 1965; Raibaud et al., 1966;

The Laboratory Mouse
Copyright 2004 Elsevier
ISBN 0-1233-6425-6

Perrot, 1976; Balk, 1986). Their use in biomedical research grew quickly and outpaced germ-free or gnotoxenic animals. Last but not least, concurrently with the development of murine virology and serology as well as molecular biology techniques, the SPF definitions were progressively extended to the absence of zoonotic, pathogenic and research interfering agents (White, 1991; ILAR, 1998; FELASA, 2002; Hansen *et al.*, 2000). The extensive use of immuno-compromised research models, beginning with the *nude* mouse before a later extension to various spontaneous or target mutations, paved the way to SPF definitions extended to opportunistic organisms applied to immunodeficient or fragilised human disease mouse models (Perrot, 1976; Franklin and Riley, 1993; Moore, 1997; Shomer *et al.*, 1998). Considering the evolutions of research applications in mice, the next step could be an extension of the exclusion list to some murine retroviruses, transmissible spongiform encephalopathy agents or any other micro-organism known or suspected to interfere with research generally or under special conditions (ILAR, 1998). The greatest colony contamination risk is linked to the introduction of strains from the outside (directly, by introducing live individuals and using murine biological products or indirectly via humans and fomites). However, as the mouse is certainly the best defined laboratory species, potential contamination from other species introduced in the same resources should also be addressed in the biosecurity and health monitoring programmes.

Health standards: definitions and categories

Gnotobiology

The original meaning of gnotobiology designates the 'study of gnotobiotic animals in order to elucidate associated biological phenomena' (Luckey, 1963).

Holoxenic/'Conventional'

This category includes mice with an uncontrolled, undefined or unknown associated flora, generally observed in wild rodents freely exposed to their natural environment, in opposition to heteroxenic mice, originally obtained by aseptic hysterectomy before controlled flora transfer (gnotoxenic step) and subsequent maintenance in a barrier room where their original implantation flora is gradually enriched by microbial species originating from this new barrier environment, frequently from human origin (Table 25.1; Dabard *et al.*, 1977).

Even when a colony is under close veterinary supervision to guarantee the absence of clinical signs of disease or lesions, one can never exclude a carrier state with potentially pathogenic, zoonotic or interfering agents.

TABLE 25.1: Sterile and non-sterile organs

Normally sterile organs	Normally septic organs and normal flora
Respiratory tract (under vocal cords)	Skin: $10–10^5$/cm^2, (S. epidermidis, S. aureus, fungi)
Sinuses at middle ear	Naso-pharynx: 10^9/ml (anaerobic/aerobic: 1000/1, S. pneumoniae,
Pleura and peritoneum	H. influenzae, S. aureus, N. meningitidis, etc.)
Liver and biliar vesicle	Oesophagus/stomach: $10^2–10^3$/ml (H. pylori)
Urinary tract (except terminal part of urethra)	Small intestine: $10^2–10^3$/ml (Lactobacilli)
Bones, joints	Bowel: $10^{10}–10^{12}$/ml (anaerobic: Eubacterium,
Cephalo-rachidian fluid	Clostridium, Basteroïdes, Veilonella, etc. >> aerobic: E. coli, etc.)
Muscles, blood	Vagina: 10^8/ml (anaerobic and lactobacilli, faecal flora)
	Terminal urethra: 10^2/cm^2 (S. epidermidis, lactobacilli, E. coli).

Source: Hart and Shears, 1996.

In consequence, the use of non controlled animal models is *not* consistent with good scientific practices, quality assurance standards and ethical principles in animal experimentations (Russel and Burch, 1959) which are now integrated in regulations governing the use of animals for scientific purposes (European Directive, 1986).

Gnotoxenic, axenic, polyxenic, oligoxenic

'Gnotoxenic' or 'gnotobiotic' refers to animals living in the absence of detectable micro-organism (axenic or germfree) or associated with a well defined flora in the digestive tract (harbouring normally the highest number of associated micro-organisms) but also on the skin and in various cavities and organs (Table 25.2; Trexler, 1987; Hart and Shears, 1996).

One can easily understand the complexity to create and manage such a gnotoxenic colony. Technically, it requires the creation of an axenic (or germ-free) colony, as described later. When the research purpose requires the use of this category, it will be kept under strict bio-exclusion conditions in order to maintain the animals in a sterile environment. For other applications, an artificially selected bacterial flora is implanted to create a

TABLE 25.2: Gnotobiology: comparative definitions, bio-exclusions systems and obtention techniques

	Gnotobiotic		Agnotobiotic		
	Axenic germ-free	Gnotoxenic microbiologically defined	Heteroxenic (specific and opportunistic pathogen free, SOPF)	Heteroxenic (SPF)	Holoxenic conventional non defined
Definition	No detectable micro-organism, 'sterility'	Implantation and maintenance of a strictly defined flora, positive definition	SOPF: SPF exclusion list plus opportunistic agents	SPF: SPF exclusion list includes zoonotic, pathogenic and interfering agents	No control, definiton or knowledge of microbism; contamination possible
Flora	Strict control and maintenance of original gnotoxenic microbism		Control of aerobic opportunistic flora	Progressive enrichment of gnotoxenic flora with environmental and human microbism	Uncontrolled and unknown flora, microbism as found in the wild
Type of bio-exclusion system	Isolator only	Isolator only	Isolator or filter-top and laminar hood cage system in barrier room	Barrier room, open cages or micro-isolation cages with less strict procedures	None
Original obtention technique	Aseptic hysterectomy	Aseptic hysterectomy Embryo-transfer	Aseptic hysterectomy Embryo-transfer	Aseptic hysterectomy Embryo-transfer	None

polyxenic or oligoxenic colony (depending on the number of associated microbial species). All operations and housing conditions also require the use of isolators, with a very strict management of bio-exclusion conditions to avoid any source of microbial contamination: sterile water, feed, various supplies and transfers, high-efficiency air filtration (Sacquet, 1968; Heine, 1978). A classical technical limitation is the difficulty to control and to detect all categories of micro-organisms. One example is the anaerobic flora: testing for the absence of these very oxygen-sensitive agents requires specific and strictly anaerobic culture conditions both for isolation and for identification, when culture is possible. With recent molecular biology techniques (such as 16S rRNA based PCR, Temporal Temperature Gradient Gel Electrophoresis profiles), it is possible to assess the diversity and the composition of the dominant faecal microflora (Dore *et al.*, 1998; Dewhirst *et al.*, 1999; Zoetendal *et al.*, 1998), since viruses are included in the definition (originally gnotobiology was exclusively addressing bacterial and parasitic agents), it's likely that most of those affecting rodents are still to be discovered and some are almost impossible to exclude, e.g. vertically or epigenetically transmitted micro-organisms (Richoux *et al.*, 1989; Kajiwara *et al.*, 1996; Hesse *et al.*, 1999).

TABLE 25.3: Examples of implantation flora
Original Schaedler flora
Bacteroïdes distasonis (Schaedler 19X Bacteroïdes)
Lactobacillus acidophilus (Schaedler L1 Lactobacillus)
Lactobacillus salivaris (Schaedler L3 Lactobacillus)
Schaedler fusiform-shaped Bacterium (# 356)
CRAS (Charles River Altered Schaedler) flora
Schaedler flora plus:
3 strains of CRL fusiform – spiral shaped Bacterium
(# 492 500 and 502)
CRL Mouse Spirochete (# 457)
New SOPF flora (Charles River – Iffa Credo flora, 1999;
Adapted from *CRAS flora*)
CRAS flora plus:
Escherichia coli (non haemolytic, # ICO-IM 1803)
Streptococcus faecalis (Enterococcus spp., # ATCC
10541)
Original SOPF flora (Iffa Credo, 1970s)
Lactobacillus acidophilus
Lactobacillus plantarum
Escherichia coli
Streptococcus faecalis

Agnotoxenic, heteroxenic, SPF standards

More commonly, and depending on the breeders' rederivation procedures, mice are kept under axenic conditions only for a limited period of time, before being associated with a defined flora as illustrated in Table 25.3 (Schaedler *et al.*, 1965; Perrot, 1976).

For practical and economical reasons, a vast majority of rodents used in bio-medical research are not strictly 'gnotoxenic' i.e. with a 'positive' and exhaustive definition of their resident flora including any infectious, opportunistic or commensal agent: they are rather defined as 'agnotoxenic' but 'pathogen free', COBS or SPF. After transfer of an implantation flora, they are transferred in barrier-rooms (Figure 25.1) where they will progressively be exposed to environmental and human-born micro-organisms (Table 25.1), hence the 'heteroxenic' definition (Dabard *et al.*, 1977). As shown in Table 25.4, these SPF animals are defined according to an 'exclusion list' i.e. a 'negative' definition, detailing all organisms being excluded from a pre-determined SPF definition including parasites (uni- or pluri-cellular), bacteria, viruses, more recently identified agents (Van Zwieten *et al.*, 1980; Ward *et al.*, 1994a,b; Hansen

et al., 2000; FELASA, 2002) or even, for more specific applications, ecotropic retroviruses and non conventional transmissible agents.

Each breeder or user has to establish, validate and guarantee a more or less restrictive SPF definition matching specific expectations and research activities, with a clear reference to the methods used to control the exclusion and to monitor for its absence (FELASA, 2002). However, a SPF exclusion list should at least include the primary species pathogens, the zoonotic and the major interfering agents (variable list according to the type of activities). Commercial breeders generally adopt more comprehensive health definitions in order to meet the global expectations of the research community. The wording '*specific* pathogen free' refers originally to the screening and exclusion of *species*-specific pathogens. Considering the progressive extension of the definition to include an increasing number of purely interfering agents (Baker, 1988; Jacoby *et al.*, 1995, 1996; McKisic *et al.*, 1995; Ermel and Richard 1997; Ueno *et al.*, 1997) it is also common to refer to '*Specified* Pathogen Free' colonies in order to stress on the need to associate this wording to an accurately defined exclusion list. At this stage, when defining the health standard and monitoring techniques, it is critical to stress on the difference

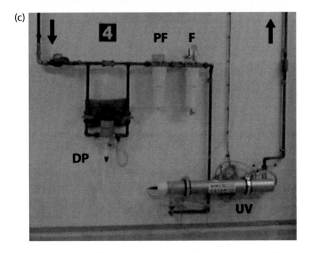

Figure 25.1 Full-barrier room. (a) Barrier-room, general view; (b) barrier: chemical lock (CL) and autoclave (A); (c) barrier: water pre-filtration (PF), 0.2 μm filtration (F) and UV treatment (UV) and disinfection pump (DP).

(NRC, 1991; GV-SOLAS, 1992, 1999; FELASA, 2002). Some commercial breeders, mainly in the USA, registered SPF 'trade marks' such as VAF (for 'Virus Antibody Free', focusing on the absence of viruses which are mostly screened by serological methods) or MPF (for 'Murine Pathogen Free'). Ideally, each breeder should include details about:

- the rederivation methods to create the breeding colonies,
- the management policy of security and barrier production colonies,
- the screening list with frequencies of testing and sampling methods,
- the exclusion list with the recycle policy for each agent: 'immediate recycle' for pathogenic and major interfering agents, 'planned recycle' for minor interfering agents (i.e. with a limited number of experimental projects or research activities) or no action for the purely commensal micro-organisms (see Table 25.4).

Agnotoxenic, heteroxenic, isolator-reared and extended SPF definitions

With genetic or experimental generation of immuno-compromised or fragilised mice, additional bio-exclusion techniques (Figures 25.2 and 25.3; Tables 25.5 and 25.6) are necessary not only to exclude the specific pathogens but also an extended list of opportunistic agents (mainly human-born and environmental bacteria) potentially able to cause infection either locally (subcutaneous or prepucial abscesses, conjunctivitis,) or systemically (septicemia and 'wasting syndrome, respiratory infections, reduced life span or breeding life). In other cases the goal will be to avoid experimental interferences with more sensitive or fragilised strains (Moore, 1997; Perrot, 1978) or to investigate genetic–microbial interactions in genetically modified mice (Mähler et al., 2002). In order to stress on an extension of the exclusion list and, by contrast, with more classical SPF definitions, some use the 'isolator reared' wording in opposition to 'barrier-reared' (see Tables 25.2 and 25.4). Obviously, it is much more relevant to rely on a definition relying on the exclusion list rather than on the bio-exclusion system. Example are the SOPF (Perrot et al., 1987) or 'Restricted Flora' definitions. When monitoring these 'opportunistic-free' colonies, for economical and practical reasons, and to avoid fastidious routine screenings of all human-born and environmental bacteria, the selection and use of a

between the *screening* list and the *exclusion* list. The first one can be purely informative (i.e. to include the monitoring of the resident flora, thus the global efficiency of the bio-exclusion system). The second will generate, in the case of a positive result, a decision to invalidate experimental results and to recycle the colony with the associated ethical, practical and economical consequences

TABLE 25.4: Examples of heteroxenic definitions (SPF and SOPF)

A. Example of SPF list

Bacteria and fungi	Parasites	Viruses
Mycoplasma pulmonis	All ecto-parasites	Reovirus 3
Bordetella bronchiseptica	All endo-parasites	Mouse Hepatitis Virus (MHV)
Pasteurella multocida		Sendaï Virus
Pasteurella pneumotropica		Pneumonia Virus of Mouse
Corynebacterium mirium		GDVII – Theiler Virus
Streptococcus pneumoniae		Hantaan Virus
Salmonella sp.		Lymphocitic
Helicobacter hepaticus/bilis		Choriomeningitis Virus (LCMV)
Citrobacter freundii		Ectromelia Virus
Streptococci beta hemolitica		Minute Virus of Mouse
Clostridium piliforme (Tyzzer's disease)		Mouse Parvovirus
Streptobacillus moniliformis		Mouse Adenovirus 1 and 2
Helicobacter hepaticus		Mouse Rotavirus
Dermatophytes		

Pathology: absence of lesion and clinical signs

B. Example of SOPF
'SPF' definition plus:

Staphylococcus aureus[a]

Proteus sp[a]

Pseudomonas aeruginosa[a]

Klebsiella pneumoniae/oxytoca[a]

Pneumocystis carinii

Corynebacterium bovis

Helicobacter sp.

All protozoans

Any micro-organism associated to lesions as a causative agent in immunodeficient lines

Experimentally-based extension of the SPF virus list (Mouse Cytomegalovirus)

[a]These frequent opportunistic agents are also used as indicators for *other* opportunistic contaminations which may occur following of a breach in aseptic handling of cages and/or animals.

C. Example of pre-defined 'policies' after contamination of SPF colonies

Breeding colonies	Experimental colonies
Immediate stop of shipments and recycle	Invalidation of all studies and immediate recyle
Shipment upon documented acceptation and, planned recyle	Limited/selective study invalidation and planned recycle
Reporting only, no recycle	Reporting only, no recycle

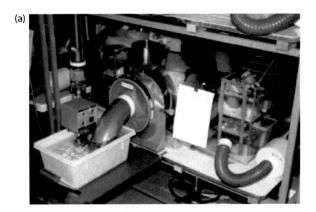

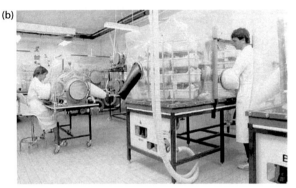

Figure 25.2 Isolator technology. (a) Small isolator equipped for aseptic hysterectomy (see figure 25.4); (b) standard flexible-film isolators; (c) macro-isolator (half suit type with quick-transfer door).

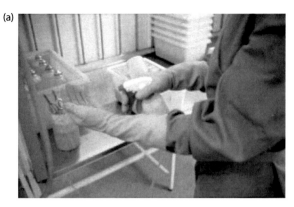

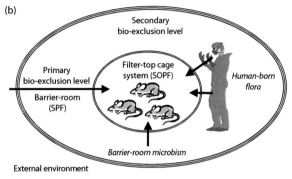

Figure 25.3 Filter-top cage system. (a) Aseptic cage opening and handling in laminar-flow working cabinet, in a barrier-room environment; (b) Filter-top cage system: a two-level bio-exclusion. The first level (barrier-room) is addressing external contaminants to guarantee an 'SPF' status. The second is an additional barrier, also excluding opportunistic agents (mainly human-born) to reach an 'SOPF' definition.

potential effects on immuno-deficient mice. In addition to these indicators, other agents may be more directly involved in lesions in immunodeficient colonies or cause very specific research interferences. Of course, any opportunistic agent associated to lesions as a causative agent in immunocompromised strains or in synergy with other organism should be considered for introduction in a SOPF exclusion list (Brennan *et al.*, 1969; Saito *et al.*, 1978; Moore, 1997).

Flora, microbism and healthy carriers

The flora associated with metazoan organisms is normally very complex, highly variable and incredibly rich: a normal body and particularly the intestinal tract contains up to 10 times more bacteria than endogenous cells (Hart and Shears, 1996). The commensal flora is of course environment and species dependant. When addressing its definition or control, defining a health standard (exclusion and screening lists) and a monitoring

set of 'indicator' agents can be a very convenient alternative. Each indicator should ideally be representative for a specific contamination route or the most common causes of 'barrier breach' (personnel, supplies, water etc.). Table 25.4B illustrates an example of SOPF standard in extension of a SPF exclusion list and including a set of 'marker' opportunistic bacteria selected because of their frequency in the environment or on humans, their common detection in barrier-maintained mouse colonies, their easy detection and identification by routine bacteriological techniques (without having to sacrifice the animals) and their

TABLE 25.5: Barrier bio-exclusion processes

Autoclave cycles

Theoretical conditions:

    15 min at 121°C

    10 min at 126°C

    3 min at 134°C

In practice, for bio-exclusion:

    20 min at 121°C

    10 min at 134°C

Application in rodent units (plateau temperatures and durations)

    Feed and bedding: 30 min at 107°C

    Stainless steel equipment: 10 min at 134°C

    Polycarbonate cages: 10 min at 118°C

    Waterbottles (filled): 10 min at 115°C (plateau)

Gamma radiation doses (applied to feed and bedding)

Vacuum packaging

Exposition time/dose adapted to type and density of irradiated product

Type of radiation (gamma rays)

Doses for:

    Germfree rodents: 40–50 kGy (5 Mrad)

    SOPF or SPF rodents: 25 kGy (2.5 Mrad)

    Heat-sensitive feed formula: 10 kGy (1 Mrad)

Other types of barrier processes

Chemical disinfection (peracetic acid, peroxides, formalin)

    isolator spray port (isolator with acetic acid, barrier)

    chemical lock or dunk tank

Air: pre-filtration and HEPA filtration (different levels)

Water: autoclaved in bottles, chlorination, acidification, pre-filtration, filtration at 0.1 or 0.2 μm, ultra-violet treatment and combinations)

Dirty and clean gowning locks, water or air shower etc.

TABLE 25.6: Example of bio-exclusion principles in a filter-top cage system

- Dedicated full-barrier unit
- Exclusive introduction and housing of SOPF rodents
- Introduction after rederivation and successful health monitoring
- Exclusive use of filter-top cages (no open cages)
- All operations in laminar flow workstations
- Aseptic handling after opening filter-top cages
- Skilled and trained staff
- Autoclaved caging equipment (cage with beddind, lid/feeder, bottle)
- 2.5 Mrad irradiated feed
- Sterilized drinking water (0.1 μm end-filter with daily disinfection)

programme it is critical to be aware of the various categories of micro-organisms colonizing metazoan individuals.

Commensal or resident flora

Normal, stable flora found associated with the host organism on the long-term. The commensal flora can include the following categories.

Synergistic flora

Flora having a demonstrated benefit to the host (e.g. digestion etc.) or a critical importance to the research applications. As an example, the quality and composition of the intestinal flora is a key regulating factor of host immune response, both locally (activation, modulation, regulation of the intestinal immune system) and systematically (modulation and regulation of different sub-classes of IgG) (Moreau and Corthier, 1988; Moreau and Gaboriau-Routhiau, 2001; Moreau, 2001; Wannemuehler et al.,1982).

Barrier flora

Gastro-intestinal, skin and other bacteria which are ecologically so adapted to the host organism that they may prevent colonization by other less adapted micro-organisms (including pathogens and opportunists) and thus contribute significantly to the stability of a healthy status. For this reason, a breeder will pay a particular attention to the colonization flora when creating isolator security colonies.

Opportunistic flora

Components of the commensal flora, not belonging to the SPF exclusion list but able to create either pathology or experimental interferences under special conditions (Perrot, 1978; Moore, 1997): immunodeficiency, stress, poor environmental conditions or fragilisation (disease models, ageing etc.). It is the responsibility of the breeder and the investigator to decide about the relevance to add them to the exclusion or the screening list (ILAR, 1998). Agents interfering with limited research application or playing a synergistic role when associated with other viruses or other bacteria should be addressed in a similar way. Both concepts of 'interfering' and 'opportunistic' agent are relative concepts, varying with the research environment and sometimes with variable documented evidence in peer-reviewed literature (Brennan et al., 1969; Royston et al., 1983; White, 1991; Shomer et al., 1998).

Transit flora

In addition to the stable commensal flora, some environmental or human-born micro-organisms can be detected over a short period of time before disappearing or reaching a non detectable level. The revised version of the FELASA recommendations duly considered this situation in accepting to stop reporting these agents when non detected after 18 months and six successive quarterly screenings (FELASA, 2002).

Pathogenic agents

All viruses, bacteria, fungi, protozoans, metazoan endo and ectoparasites known as major or minor pathogens should be listed as such, in the SPF exclusion list. It is also the practice to add to this 'primary list' additional micro-organisms causing well documented experimental interferences in the principal research fields (ILAR, 1998).

Zoonotic agents

To preserve personnel health, the zoonotic risk should be managed in a very specific way. The main concern, relying on bio-containment measures, is to prevent any possibility of accidental transmission of these animal-born human pathogens. Their absence should always be guaranteed by the strict implementation of bio-security and health monitoring programmes. Any suspicion should be reported and addressed immediately.

Healthy carriers

It is critical to bear in mind that pathogenic, parasitic or zoonotic agents can also be found in 'healthy carriers' i.e. individuals not displaying any clinical sign or lesion. Even if transmission to humans and clinical cases are rare, especially in barrier-bred laboratory mice, an example of zoonotic agent in mouse is the LCMV, an Arenavirus causing severe flu-like symptoms in man. A relevant health monitoring programme based on serology is the only way to guarantee its absence.

Rederivation techniques

When a mouse strain is confirmed as 'contaminated' i.e. not conforming to the exclusion list, several techniques are usable to eliminate non acceptable agents or, to use a common wording, to 'rederive' it. The selection of a technique will depend on the one hand on the number and the nature of the agent(s) to eliminate and, on the other hand, on the expected use of the animals. To be accurate, the wording 'rederivation' is generally used for the most efficient and sophisticated techniques such as 'aseptic hysterectomy' or 'embryo transfer'. Other techniques are also briefly described because they may help saving a breeding project when no other alternative is acceptable for time, budget or technical reasons. Of course, they do not present the same level of reliability and efficiency as aseptic hysterectomy and embryo transfer which rely on a very simple (but not absolute) principle: the sterility of the reproductive tract during the pregnancy until the onset of delivery and opening of the cervix. What is true for most but not all parasitic and infectious agents, particularly in mice, depends on the risk of 'false' vertical transmission, due to a contamination during the process (Hill and Stalley, 1991) or 'real' vertical transmission (transfer before or after embryo nidation and placentation). These limitations will be discussed later.

Quarantine, isolation and eradication time

Although being poorly reliable, having unpredictable results, and mainly reported in large species, some purely opportunistic bacteria originating from another species (mainly humans) and having a poor ecologic tropism for the mouse may spontaneously 'disappear' from the flora, i.e. becoming at least undetectable by the usual techniques over long period of times. The colony has to be kept under strict bio-containment, to avoid not only the spreading of the agent and the contamination of neighbour naïve individuals until getting a clear evidence of negativity after repeated screenings over at least 6–12 months (FELASA, 2002).

Quarantine, isolation and seroconversion

This 'burnout' procedure is valid only in non-breeding colonies. The most common example is with the frequently isolated MHV strains. Being highly infective (the rate of infection approaches 100%), they spread rapidly and disappear when the entire colony has recovered from the infection and seroconverted, if it is small, mature and immunocompetent, with no introduction of naïve mice and no breeding activity (Homberger and Thomann, 1994; Homberger, 1997). Depending on the exposure of mice to the virus and various factors, the whole process takes from a few weeks (minimum 6) to several months. The only reliable indicator of success is the absence of infection when introducing naïve mice into the colony. One

success factor is the poor resistance of MHV in the environment (a few hours). In addition to the stringent isolation of the colony, it is critical to get rid of any potential chronic carrier and excretors such as immuno-compromised or naïve individuals. All introduction of external animals should be prohibited during the process as well as any breeding animals. Naïve and immunologically immature pups keep spreading the virus.

Quarantine, isolation, selection and time

This variant may be used for a variety of agents (one at a time) that are not infecting immediately all individuals in a colony. It requires a sufficient number of mice in the contaminated colony, a highly sensitive screening method on live mice and it is based on individual separation in micro-isolation cages and procedures preventing inter-cage cross contamination. Each cage is assessed and followed by an immediate stamping-out in case of positivity. With some chance, if enough animals are found negative after repeated screenings, it is possible to re-start a negative colony. One important draw-back is the reduction of outbred breeding colonies, potentially causing a prohibitive 'genetic bottleneck' effect (critical loss of genetic polymorphism). A variant for Helicobacter spp. eradication has been described, using rapid neonatal transfer to Helicobacter negative foster dams on the first day of life (Truett et al., 2000).

Medical treatment

Most internal and external parasites can be eliminated using a well designed and conducted global anti-parasitic treatment, targeting all stages of parasitic development, e.g. ivermectin for Syphacia muris or obvelata (Flynn et al., 1989; Leblanc et al., 1993; Zenner, 1998). This may also require parallel drastic hygienic measures to eliminate all eggs from the environment, knowing that parasite eggs are highly resistant to usual disinfectants. Again, the colony should be carefully isolated during this treatment until repeated screening demonstrates the success of the process (Huerkamp, 1993).

Aseptic hysterectomy (see Figures 25.2 and 25.4)

Technical details are available in several publications (Heine, 1978). A classical method can be summarised as follows. Future foster mothers originating from a gnotoxenic or SPF colony are time-mated in order to

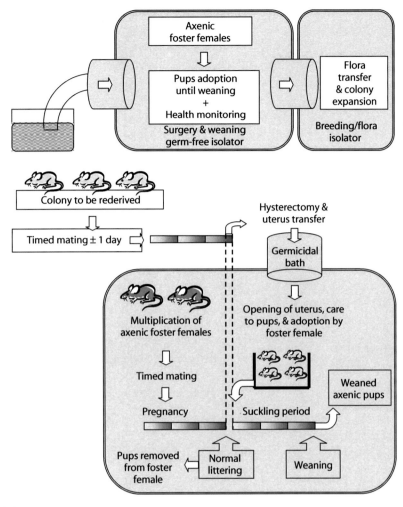

Figure 25.4 Aseptic hysterectomy.

become pregnant in synchronisation with contaminated donor females. Matings are scheduled to carry out the aseptic hysterectomy on donor females a few hours before their normal littering time and a few hours after the foster females. Hysterectomy is carried out after euthanasia of the donor females in order to introduce aseptically the whole gravid uterus into a sterile isolator through a germicidal bath. Then, after incision, the pups are removed from the uterus, warmed-up and substituted to the foster mother's litter. Isolator health screening is conducted at least once by weaning age before validating the success of the process. A double check is highly recommended a few weeks later. Depending on the breeding objective and genetic issues (outbred versus inbred) the total number of rederived females may vary significantly (from a few to over one hundred pairs). Despite the technical difficulties when using germ-free mice, one major advantage is the easy and quick detection of any contaminant micro-organism due to a breach in the aseptic process.

Embryo transfer

Technical details are available in specialized handbooks (*Manipulating the Mouse Embryo*, 1994), publications (Suzuki *et al.*, 1996; Morrell, 1999; Okamoto and Matsumoto, 1999; Chin and Wang, 2001) and in Chapter 26 on Assisted Reproduction and Cryopreservation. In a few words, contaminated donor females are mated in synchronisation with SPF or SOPF recipient females mated to vasectomised males to become pseudo-pregnant. Technical options are the use of super-ovulation or *in vitro* fertilization. Donor females are anaesthetised in order to collect 2-cell-embryos which are washed and morphologically selected before implantation into the oviduct of recipient females. After transfer into an isolator and weaning, the recipient females are used for health screening to assess the success of the rederivation, with specific concern for any undesirable agents identified beforehand. One advantage of embryo transfer when compared with hysterectomy is to avoid post-implantational vertical transmission (Reetz *et al.*, 1988).

Aseptic hysterectomy or embryo transfer associated with other procedures

To eliminate a vertically transmitted agent such as LCMV or to avoid intra-uterine contamination with agents like Mycoplasma pulmonis (Hill and Stalley, 1991), depending on the nature of the contaminant, one can combine the use of classical rederivation techniques with separation of individuals and selection by serological screening, embryo washing, antibiotic treatment etc. To be reliable, several screenings should be conducted before validating the success. A specific issue and difficulty is the vertical transmission described for agents like endogenous ecotropic retrovirus such as murine leukaemia virus (MuLV) (Richoux *et al.*, 1989; Hesse *et al.*, 1999).

Genetic issues when rederiving a colony

The number of pairs and origins used to obtain the pups composing the next generation, after rederivation, is a key issue with outbred stocks. It is critical to pay attention to this number to avoid a genetic bottleneck (Rapp and Burow, 1979). The rederivation series should at least use the progeny from 80 pairs, each pair contributing equally to the next generation.

Bio-security, bio-safety, health monitoring

The maintenance of a defined health standard depends on the proper implementation of a 'bio-security' policy,

i.e. all resources used in a consistent and integrated programme aiming at a 'zero contamination' level in the animal colonies, relying on facility design, finishes, housing and handling equipments, various procedures, personnel education and training. Mice benefiting from an adequate health standard are maintained in 'bio-exclusion' conditions in order to exclude any biological contaminant or source of contamination from their microbial environment. The strictest conditions are required to maintain a germfree environment, which can only be achieved in an isolator, with transfer procedures and supplies consistent with the expected sterility level (Figure 25.2).

Bio-exclusion equipment and procedures used for flora associated colonies (in isolator, filter-top cage system or barrier rooms; Table 25.7) are less demanding both for and depending on the barrier level. A comprehensive health monitoring programme should be designed and duly implemented (see Chapter 27 on Health Monitoring). Other mice, microbiologically undefined or contaminated, should be kept under 'bio-containment' conditions in order to avoid the dissemination of their undesirable or hazardous flora components to cleaner animals or to humans. When the 'bio-containment' aims to protect operators and/or environment from a biological hazard, one speaks about 'bio-safety' policy and procedures. Various specific regulations and guidelines are addressing such bio-containment situations both for infectious agents and genetically modified organisms (classification of pathogens, level and type of bio-containment, bio-safety cabinets, waste management). Practically, it is necessary to rely both on bio-containment (reception, quarantine and health screening steps) and on bio-exclusion (rederivation and subsequent housing) techniques.

TABLE 25.7: Comparison of two bio-exclusion systems usable for SOPF status		
Characteristics	**Isolator**	**Filter-top cage system (static)**
Capacity limitation	yes	no
Handling procedures	difficult	easier
Fixed and variable costs	higher	lower
Capital expenditure	higher	lower
Adaptability/versatility	lower	higher
Maintenance of SOPF status	yes	yes
Maintenance of germ-free status	yes	no
Bio-containment of contaminants	yes	limited
Staff education for basic procedures	quick, easy	longer
Health monitoring (sampling)	easy	complex

Colony termination and recycling policy

Some examples of situations when breeding colony termination and recycling has to be considered:

- in case of major contamination (i.e. with an agent of the exclusion list established upon colony creation),
- less urgently, in case of opportunistic contamination(s) causing a drift of the colony microbism, associated with occasional lesions,
- after a genetic contamination,
- before repopulation with breeders from the genetic reference colony, in accordance to the genetic management system and breeding schemes.
- more commonly, for management reasons, when the different stocks, and strains have to be re-organized and re-allocated to different barrier-rooms.

Breeding techniques

A prerequisite is to have a good understanding of reproduction biology as well as genetics and strain/ stock definitions. Housing, maintenance and nutrition are critical factors in the management of a breeding facility as well as access to health and genetic monitoring resources. Assisted reproduction techniques and cryo-preservation are also a must. They are addressed in another chapter.

Genetic standards and breeding systems

Reference or foundation colony

The goal of such colonies is to serve as a genetic reference in order to guarantee a long-term genetic stability (i.e. to minimise the 'genetic drift') over generations and inter-colony homogeneity when several breeding colonies are extant. (Figure 25.5). In other words, for 'outbred' stocks, the objective is to ensure that the population

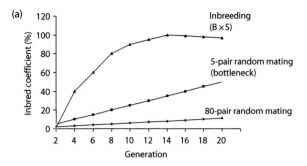

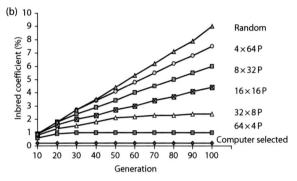

Figure 25.5 Evolution of inbreeding coefficient with different mating systems (White and Lee, 1998). (a) Brother × sister; 5-pair and 80-pair random matings; (b) various mating systems in a 256-pair colony, divided in 4–64 mating groups (P = pairs).

remains constant in all characteristics for as many generations as possible (Festing *et al.*, 1972), to maintain the highest level of heterozygocity and to preserve the population's original allelic forms and frequencies as stable as possible over generations (Rapp, 1972; Green, 1981; Hartl and Clark, 1997; White and Lee, 1998; Hartl, 2000). For 'inbred' strains (Figure 25.5), the primary purpose is to preserve isogenicity and a maximum inbreeding coefficient (Festing, 1979; The Staff of Jackson Laboratory, 1997).

This last goal is achieved by avoiding the three main causes of divergence: genetic contamination, mutation and residual heterozygocity. By definition, both for inbred strains or outbred stocks, the Foundation Colony is unique. When it cannot produce enough animals for experimental use, one or several Production Colonies are created. In consequence, a genetic management programme should address:

- the genetic management and breeding system both in the Foundation Colony and in the Production Colonies
- migrations between Foundation and Production Colonies (Figures 25.6 and 25.7).

Of course, inbred and outbred management systems are totally different. Factors which may rapidly affect

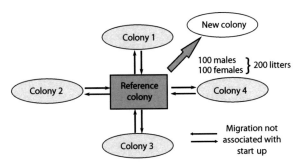

Figure 25.6 New outbred breeding colony start-up using the foundation colony (White and Lee, 1998).

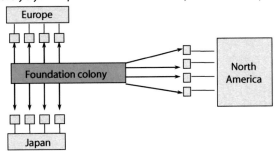

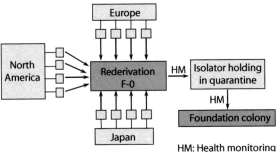

Figure 25.7 Outbred breeding system, forward and backward migrations from and to foundation colony (White and Lee, 1998).

the genetic quality of a population are (Rapp, 1972; Festing, 1979):

- the introduction of external breeders, accidentally or purposefully (inbred and outbred),
- the population size, population and founder effects (outbred),
- the type of mating scheme (inbred and outbred),
- mutations (inbred and outbred),
- conscious or unconscious selection (outbred).

Outbred stocks

The long-term goals of an outbred foundation colony and a genetic management system are:

- to maintain a maximum genetic variability/polymorphism,
- to guarantee production colonies similar to the foundation colony (i.e. the geographic uniformity),
- to avoid the introduction of new allelic forms,
- to maintain stable allelic frequencies from generation to generation,
- to minimize the increase of inbreeding from generation to generation.

General recommendations for the long term maintenance of genetic polymorphism are listed in Table 25.8 (Rapp, 1972), as well as an example of practical application.

Before adopting a breeding system, one should evaluate its performance. Simplified formulas are usable for such simulation, as illustrated in Tables 25.9 and 25.10.

As a minimum goal, the inbreeding coefficient of an outbred stock should increase by far less than 1% per generation. A monogamous system is highly recommended in foundation colonies, with one pedigreed and identified pair per cage, in order to allow genealogical records and selection of future breeding pairs. The ideal result is obtained by using a computer-based mating system, selecting the less related males and females to create new pairs. For practical reasons, other simpler systems are generally used (Falconer, 1960, 1972; Poiley, 1960; Rapp, 1972). Figure 25.8 is illustrates two of them.

A long generation interval is another critical factor helping to minimize the actual 'drift' on the long term. This can be achieved by keeping the breeding pairs as long as their performances are acceptable. A common practice is to use pups from the fifth litter to create the breeding pairs of next generation. Purposeful or unconscious selection should be carefully avoided at this step (each individual should have the same chance to contribute to the next generation), but a very limited number of individuals displaying abnormal phenotypic characteristics can be excluded. On the other hand, conscious or unconscious selection criteria may result in a tendency towards inbreeding and should be carefully avoided (large litter size, fecundity, ability to bring a litter to term, behaviour). When creating a new outbred colony one should guarantee the highest level of genetic polymorphism, using a high number of pairs from several independent origins, with the least possible common ancestorship. The Foundation Colony is used to

TABLE 25.8: Genetic management of an outbred stock foundation colony (guidelines and example of practical implementation)

Long term maintenance of genetic variability/polymorphism (based on Falconer & Rapp publications and on Festing *et al.* in ICLA Bulletin, 1972)

- closed colony, no introduction of new breeders (i.e. foreign allelic forms)
- population size (number of breeding pairs, male and females) as large as possible
- equal chance for each individual of participating to the next generation (random mating)
- no selection criteria (i.e. loss of allelic forms)
- high interval between generations (to slow down genetic drift)
- mating scheme using sub-groups to minimise risk of inbreeding

Prevention of mutations is not under control.

Example of application in rodent colonies: genetic management system of a foundation colony

- closed colony, no introduction of new breeders
- minimum 200 permanent breeding pairs
- overlapping generations, constant mean age of colony
- fixed reproduction life: 30 weeks
- mating scheme : 4-unit-Robertson's mating scheme
- future breeders from fourth litter, maximum 3 F.B. per pair
- minimum 84% of pairs contribute to the next generation
- no selection to increase productivity!

TABLE 25.9: Outbred foundation colony: simulation of genetic drift/inbreeding increase (simplified formulas)

A. Increase of homozygocity over generation in an outbred population
(Wright formula in Rapp, 1972)

$$F_x = S\,[0.5^{n\,+\,n'\,-\,1}\cdot(1 + F_a)]$$

F_x: inbreeding coefficient of the individual x

S: symbol for sum if repeated inbreeding towards the same or several commonancestors occur

n: number of generations between the individual x and the common ancestor fromthe father's side y, without adding x and y

n': number of generations between the individual x and the common ancestor fromthe mother's side y, without adding x and y

F_a: inbreeding coefficient of a common inbred ancestor

B. Increase of inbreeding per generation (Festing et al., 1972)

Mating scheme	Inbreeding increase per generation (%)
Random mating	I.I. = $1/8\,N_{ma}$ + $1/8\,N_{fe}$
Maximum avoidance mating	I.I. = $1/16\,N_{ma}$ + $1/16\,N_{fe}$

I.I. = Increase of Inbreeding per generation

N_{ma} = Number of males

N_{fe} = Number of females

TABLE 25.9: (*Continued*)

Example of simplified simulation, showing the influence of the less represented sex (males) with a maximum avoidance mating system:

2 males, 20 females = 3.53%

12 males, 24 females = 0.78%

40 males, 200 females = 0.19%

120 males, 120 females = 0.10%

TABLE 25.10: Outbred foundation colony: simulation of genetic drift/inbreeding increase – long-term influence of the 'bottleneck'/founder effects due to a temporary reduction of the colony size (simplified formulas)

Mean population size, $n_e = g/\{1/n_1 + 1/n_2 + \ldots + 1/n_t\}$

n_e = mean population size calculated over n generations

g = number of generations

$n_1, n_2, \ldots, n_t$ = actual population size from generations 1 to t

Examples of simplified simulation, showing the influence of a genetic bottleneck of 40 individuals (20 pairs) at N2, before colony re-expansion

$n_1 = 240 - n_2 = 40 - n_3 = 80 - n_4 = 160 - n_5 = 240 \rightarrow n_e = 96$

$n_1 = 240 - n_2 = 40 - n_3 = 40 - n_4 = 80 - n_5 = 160 \rightarrow n_e = 68$

$n_1 = 20 - n_2 = 100 - n_3 = 1000 - n_4 = 1000 \rightarrow n_e = 64$

Source: Rapp and Burow, 1979.

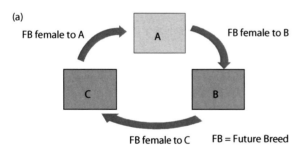

(a)

FB female to A   A   FB female to B

C   B

FB female to C   FB = Future Breed

Production colonies: A + B + C used for experimental groups

(b) Male origin   Female origin   New pair

U 1   ×   U 2   →   U 1

U 3   ×   U 4   →   U 2

U 2   ×   U 1   →   U 3

U 4   ×   U 3   →   U 4

Figure 25.8 Examples of outbred mating systems. (a) Three-group outbred mating system, single sex rotation (White and Lee, 1998); (b) Robertson's four-group outbred mating system.

create the different Production Colonies and to guarantee a long-term inter-colony homogeneity. This goal can be achieved either by restarting Production Colonies on a regular basis with pairs from the Foundation Colony and/or by the use of a 'migration system' as illustrated in Figure 25.7.

Inbred strains

Genetic contamination should be prevented by strict maintenance of breeding records and cage identification. Strains having the same coat colours should not be kept in the same physical enclosure. Any escaped mouse should be removed from breeding and euthanised. Any abnormal observation (phenotypic deviants) should be immediately reported as a potential sign of genetic contamination or mutation and investigated (increase of litter size as a sign of heterosis, coat colour, behaviour). Animal technicians should be trained to the surveillance of breeding colonies and record keeping. In opposition to outbred populations, the size of an inbred foundation colony is not critical. In theory, only one pedigreed pair from a selected reference origin could be used. Practically, for security reasons and to speed up the colony expansion process, a higher number of cages and pairs is generally used. All are pedigreed and individually identified, hence the name 'Foundation and Pedigree Identified Pairs'. To avoid a decrease of the inbreeding coefficient, this Foundation Colony will be managed with strict application of continuous inbreeding. Individuals from the Foundation Colony will be used to propagate lineal descendants for a defined and limited number of generations (e.g. 10) of brother–sister matings from the common ancestor in the Foundation Stock, plus a strictly limited number of generations (2–3) in Production Colonies. (Festing, 1979, 2002; The Staff of Jackson Laboratory, 1997). When several production rooms exist for the same strain, pairs from the Foundation Colony are transferred to create a Pedigreed Expansion Stock (using brother × sister matings) for each Production Colony.

As shown in Figure 25.9, the Foundation Colony is self-supporting and is used to replace all breeders in any existing Pedigree Expansion Colony. At the next step, in Production Stocks 1, brother × sister matings or trios can be used. Only one level of random-mating is acceptable: in Production Stock 2. At this level, trios or even small harem groups are generally allowed in order to maximise the productivity and generate homogenous groups of animals for research. For strict compliance to a genetic standard, it is necessary to keep introducing new pedigreed pairs on a regular basis (every 10–20 generation) from the Foundation Colony into Production Stocks. A bank of cryo-preserved embryos should be kept as a security and back-up to the Foundation Colony. Such a bank can also be used

to better control the genetic drift and prevent cumulative change by thawing back embryos on a regular basis (e.g. every 5 generation) in order to renew the Foundation Colony. Sufficient embryo number should be frozen to act as a reservoir on the long term.

Microbiological security colony

In addition to the availability of a 'genetic security' (i.e. the Foundation Colony), a 'health security' should also be created in order to restart breeding groups after a microbiological contamination. Considering the limited size of the mouse Foundation Colony, it can be easily kept in a large isolator, hence being also usable as a health security colony (Figure 25.2b and c).

Production colonies

Outbred stocks

The most common practice is to use harem matings (number of females per male adapted to the cage dimensions and to the stock reproductive capabilities). Around the 14th day of pregnancy, when visually detectable, pregnant females are removed from the mating cages to be kept in female groups. At birth, smaller groups are generally created. Except for specific requests, such as time-mated females or dated births, there is no urgent need to sex the pups at this stage yet but rather to guarantee a growth homogeneity. For this reason, it is recommended to give exactly the same number of suckling pups (born on the same day) to each female (10 or 11 per female). At weaning age (18–21 days), pups are sexed, males and females are separated and stock groups are created for growth in cages identified with the birth date. Weekly, the caging density of these groups is adapted to the caging standards. From these cages, experimental groups are either composed by weight or age bracket or both.

Inbred strains (Figure 25.9)

As mentioned already, a 'Pedigree Expansion Stock' (PES) is created and maintained in each production room, with pairs directly originating from the Foundation Colony. These pairs are used to create a 'Production Stock 1' (brother × sister matings or trios), then a 'Production Stock 2'. This last level is the only one at which harem mating (1 male with 3–4 females) is allowed. In Production Stocks 1 and 2, the mating

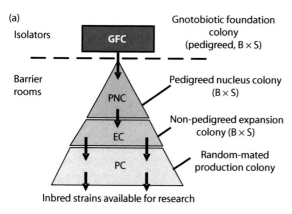

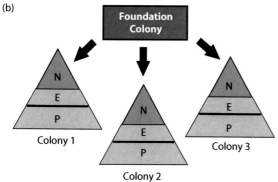

Figure 25.9 Global organisation of an inbred breeding (White and Lee, 1998). (a) System with one inbred production colony; (b) system with three inbred production colonies.

system (pairs, trios or harems) is also adapted to strain characteristics and zootechnical issues, in order to optimise the productivity. Some sensitive strains may be bred in pairs even in Production Stock 2.

F1 'hybrids'

F1 'hybrids' are produced by mating two inbred strains together (in opposition to 'real' hybrids which are the product of an interspecies mating). This type of breeding, associated with inbred colonies, doesn't present any specific requirement except that by definition the colony is not self-perpetuating and requires the availability of the two parental strains in the same room. In consequence, the risk of mis-mating and genetic contamination should be carefully addressed, especially when strains with similar coat colours are kept in the same breeding room (example: B6SJLF1 and their parental strain C57BL/6). As a rule when strains or stocks with similar coat colours cannot be separated, identification and strict procedures should be adapted in consequence, as well as genetic monitoring and surveillance.

Oligogenic and polygenic traits

Polygenic traits (corresponding to 'quantitative trait loci' i.e. set of genes that together control a phenotypic trait not completely determined by one gene acting alone) can be subject to selection, starting from a group of individuals expected to contain as many allelic forms as possible (Okamoto and Aoki, 1963; Sabourdy, 1967). For this reason, it's recommended to create a breeding colony with breeders from various and independent origins. After creating this new highly 'allele polymorphic' colony, each new generation will be created with males and females displaying the expected phenotype (arterial hypertension, body size or weight, growth, fatness, food intake, alcohol preference etc.) until phenotype stabilisation. The selection can be mono-directional or bi-directional. Such a selection also causes a severe increase of the inbreeding coefficient. To maintain an acceptable level of breeding performances, a minimum productivity level has also to be included in the selection process and to be maintained on the long-term, even after obtaining the final phenotypic expression. At this stage an inbred breeding system is generally used, with a continuous phenotypic selection of the future breeders (ILAR, 1976).

Mutations and transgenes (Table 25.1)

For practical or experimental reasons, such as the size of the colony, the influence of modifier genes, the phenotype control, most mutations (spontaneous, induced chemically or by homologous recombination) are maintained on an inbred background. One critical issue, too often neglected, is to control the genetic background and to guarantee the isogenicity between individuals carrying the gene of interest and their control 'wild type' line (Linder, 2001). This can be achieved (i) by selecting the optimal genetic background, then creating a strain

(a)   'Random' or standard backcrossing
Simplified formulas (Falconer, 1972)

Generation	Inbr. Coef.
N0	0
N1	0.5
N2	0.75
N3	0.875
N4	0.938
N5	0.969
N6	0.984
N7	0.992
N8	0.996
N9	0.998
N10	0.999

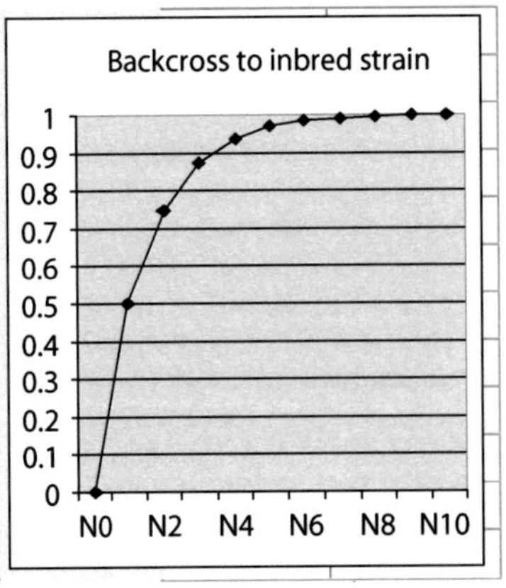

(b)   Marker-assisted / speed congenics and equivalent number of backross generations (BC)

Inbreeding coefficient	Equivalent number of standard BC	Number of marker-assisted BC
0.984	(N6)	~ 3
0.999	(N10)	~ 5

Figure 25.10 Evolution of the inbreeding coefficient and congenicity in backcrossing protocols, using an inbred strain. (a) Random or traditional backcrossing protocol: simplified formulas (Festing, 1979); (b) marker-assisted speed congenic protocols (N3: inbreeding coefficient ~ 98.5%; N5: inbreeding coefficient ~ 99.9%).

that is co-isogenic to its control strain (i.e. generating transgenics or mutants with a defined genetic background such as C57BL/6 or FVB, (ii) by choosing a 129 substrain matching the ES cell line used for homologous recombination or using a C57BL/6 or BALB/cJ ES cell line for co-isogenic background or (iii) by generating congenic strains, using traditional backcross or marker-assisted speed congenic protocols as shown in Figure 25.10 (Silver, 1995; Markel *et al.*, 1997; Wakeland *et al.*, 1997; Linder, 2001). For homozygous matings with an inbred genetic background, before establishing a close colony, a minimum residual heterozygocity should be reached (i.e. an inbreeding coefficient of at least 99.9 %). For outbred backgrounds and heterozygous or hemizygous production, it may be more convenient to use a mating scheme with regular infusions from the reference congenic colony (Table 25.11B). Depending on the type of mutation (recessive or dominant, autosomal or not) and the phenotype (lethality or sterility), one has to choose the most productive and adapted mating scheme. A critical step is the selection of the genetic background to be used with any gene of interest, in order to avoid any major phenotypic or experimental interference (i.e. early blindness in behaviour studies, spontaneous tumour incidence with oncogene expression . . .).

The identification of some genotypes may require the use of a genotyping method: Southern blot, polymerase chain reaction (PCR), protein analysis (Table 25.11D). The maintenance of an outbred background requires an outbred management system or any other scheme guaranteeing a similar result (see example in Table 25.11B for with a hemizygous transgene over-expression). After random gene integration, one should make sure that the individuals used to start a colony are deriving from a common founder selected beforehand to carry a unique, identical and stable integration site, with the same number of copies.

Mating systems and breeding techniques

Reproduction data and mating systems

Breeding data

Some breeding and development data are detailed in Table 25.12.

Mating systems

Different systems can be used (Table 25.13) depending on several decision parameters related to the type of colony or the breeding step (foundation colony, expansion, full production). One should first consider the principal goal pursued by the mating system. In foundation colonies, priority is given to pedigree follow-up and individual identification, long-term genetic management ('minimum drift') as described above, with little concern about productivity. In production colonies, priority is given to productivity to benefit from availability and large groups of homogenous animals (i.e. within a given weight and age bracket).

Timed pregnant females or timed births, use of 'plugged' females

After mating, a vaginal plug is formed by the secretions of the male vesicular and coagulating glands, filling the vagina in a variable way (often from cervix to vulva) and persisting from 8 to 24 h. It is a convenient and easily visible indicator that mating has occurred. With most strains 80–90% of 'plug positive' females are pregnant. Ovulation and mating usually occur over night. The first day of gestation is considered to be the day after observing the vaginal plug. For large production programmes, oestrus synchronisation can be used prior to mating (see section on Practical Applications of Pheromone Effects).

Superovulation

See Chapter 26 on Assisted Reproduction and Cryopreservation.

Pseudopregnancy

Mating a female with a sterile or vasectomised male will cause a pseudopregnancy hormonally similar to pregnancy and lasting a few days, before the next oestrus cycle about 11 days later.

Post-partum oestrus

A first post-partum oestrus cycle with an ovulation about 14–28 h after parturition (Runner and Ladman, 1950).

Practical applications of pheromone effects

Crowding females causes oestrus suppression, followed by oestrus synchronisation when removing this inhibition,

TABLE 25.11: Mating schemes for mutations and transgenes

A. Production of CBy. Cg- *Foxn1^{nu}*/*Foxn1^{nu}* (homozygous)

Autosomal and recessive mutation: *Foxn1^{nu}*

Male	Female	Progeny
nu/nu	*nu/+*	*nu/nu* (50%)
		nu/+ (50%)

After 10 generations a backcross is carried out: nu/nu males are mated with +/+ females originating from the BALB/cByJ reference colony.

B. Production of C57BL/6J-*Tg*/+ (hemizygous)

Example for a single insertion locus

Male	Female	Progeny
Tg/+	*+/+*	*Tg/+* (50%)
		+/+ (50%)

Infusion with C57BL/6J females is carried at each generation.

C. Production of BKS.Cg-m *Lepr^{db}*/*Lep^{ob}* (homozygous)

Autosomal and recessive mutation: *Lepr^{db}*

Example of double heterozygote repulsion mating with *m* mutation (*misty* coat colour marker)

Male	Female	Progeny
db +/+ m	*db +/+ m*	*db +/db +* (25%)
		db +/+ m (50%)
		+ m/+ m (25%)

db +/db + (fat and black)   *db +/+ m* (lean and black)

+ m/+ m (lean and dark grey)

D. Production of BKS.V-*Lep^{ob}*/*Lep^{ob}* (homozygous)

Autosomal and recessive mutation: *Lep^{ob}*

Use of PCR genotyping to genotype (lean) mice

Male	Female	Progeny
ob/+	*ob/+*	*ob/ob* (25%)
		ob/+ (50%)
		+/+ (25%)

ob/ob (fat and black)

ob/+ (lean and black) ⟨══ PCR ══⟩ *+/+* (lean and black)

nu, nude, current nomenclature, *Foxn1^{nu}*

db, diabetic, current nomenclature, *Lepr^{db}*

ob, obese, current nomenclature, *Lep^{ob}*

before placing one to three females with one male. As the presence of a male in a female group is not only overriding the negative female stimuli under crowded conditions but is regularising and accelerating the oestrus, another technique is to use a double-compartment cage, separating the male from 2–3 mature females (aged minimum 8 weeks) by a removable grid-type partition for 2–3 days. Then, the partition is removed, matings take place and plugs are checked. One should be aware that crowding can also cause increased pre- and post-implantational mortality or male fighting (van der Lee and Boot, 1956; Ross, 1961; Bruce, 1970).

TABLE 25.12: Some mouse breeding and development data
Ovulation: polyoestrus species, spontaneous, during oestrus, usually around midnight
Oestrus cycle: 4–5 days
Oestrus duration: ~ 12 h
Gestation period: 18–22 days (average: 19–20)
Birth: most births occur at night or late afternoon
Litter size: varies greatly depending on the strain/stock, the age, the order of the litter, the environmental conditions:
4–9 (with extremes from 2 to 12)
Post-natal development:
Post-partum oestrus : 12–24 h
External ear opening : 3 days
Coat : well developped at 2 weeks
Vagina opening: at 3–5 weeks
First oestrus: 5–6 weeks
Post-partum oestrus: generally about 12–24 h after parturition, causing simultaneous gestation and lactation in permanent breeding groups (i.e. when the male is permanently present in the maternity cage).

Record keeping (cages cards and pedigree charts)

Depending on the type of breeding system and colony, a relevant record keeping system has to be designed and duly used. It relies mainly on the use of cage and individual identification, cages card records (Figure 25.11) and pedigree charts, at least in Foundation Colonies. These records should make it possible to trace the ancestors, the progeny and the collateral relatives of any individual as well as the number of pups at birth and at weaning, the parturition dates in order to monitor breeding performances. They are also critical indicators of breeding environmental conditions and adequate nutrition.

Birth, sexing and suckling-mouse groups, weaning

At birth, pups are examined and may be sexed at this stage. In outbred stocks, to benefit from more homogenous growth, a common practice is to balance the size of litters at birth, giving back 10–11 pups to each female. Pups are sexed at the latest upon weaning, usually by 19–21 days, or a few days earlier when required.

Post-weaning and growth, future breeders

After weaning, male and female growth groups are created, according to experimental and cage dimension specifications. An important indicator calculated at weaning age from data recorded on the cage card (Figure 25.11) is the productivity index or the 'average number of pups weaned per female and per week', integrating the litter size at weaning (i.e. litter size at parturition minus pre-weaning mortality) and the parturition interval. This index varies according to the strain, the breeding system and numerous environmental influences. Under defined standard conditions (caging, density, feed), the breeder should update the growth curve as well as other selected biological data.

Breeders' retirement

The breeding life is generally longer in foundation colonies in order to increase the generation time and to minimise the long-term genetic drift. In production colonies, breeder retirement takes place when the litter size starts decreasing significantly. With outbred stocks, it is usually after 7–8 litters in foundation and PES colonies or 4–6 litters in production colonies. Early

TABLE 25.13: Different types of matings systems

Pros	Cons	Applications
Monogamous permanent pairs (1 male + 1 female per cage, male not removed)		
Genealogical records	Large number of males	Mutations, transgene
Female zootechnical records	Space consuming	Project start
Post-partum oestrus	Labour consuming	Foundation colonies
	Male present at littering (can increase pup losses[a])	Inbred PES
		Backcross/progeny test
		Breeding to homozygocity
		Sensitive strains
Permanent trios (1 male + 2 females per cage, male not removed)		
Genealogical records	Competition between litters	Same than monogamous
Productivity per m^2	Male present at littering	Intermediate production level
Post-partum oestrus		Inbred production stock 1 and 2
Olfactive stimulation		
Permanent harem mating (females remain with males for littering and suckling, not recommended)		
Productivity per m^2	Fights (strain-dependant)	Large scale production
Post-partum oestrus	No genealogical records	Outbred production colony
Olfactive stimulation	Cage size	(1 male + up to 13 females)
Synchronisation of littering	Male breeding capacity	
	Competition between litters	
Boxing-out harem mating, 1 male and 4–13 females per cage (females removed when pregnant)		
Productivity per m^2	No post-partum oestrus use	Large scale production
Olfactive stimulation		Timed births (pregnant females removed to individual cages)
Synchronisation of littering	Fights (strain-dependant)	
Lower pup mortality	Cage size	
Genealogical records (if 1 pregnant female per cage)	Male breeding capacity	Outbred production colony (1 male + up to 13 females)
Less competition between litters (homogenous birth dates)		Inbred production stock 2 (1 male + up to 4 females)

[a]Example: B6.129P2-*Apoe*[tm1Unc] productivity index ~ 5.0 without male and ~ 3.5 with male.

Figure 25.11 Examples of cage cards used for mutant (left) and inbred (right) strains (in French; *Naissances*: births; *Sevrages*: weanings; *Sou.*: strain; *Acc.*: mating)

retirement is also carried out in case of abnormal observations (excessive weight, skin lesion, poor health conditions). All these individuals should be used for diagnostic purposes. In inbred strains, females are usually mated at 7–8 weeks of age and retired after about 26 weeks of breeding in production stocks, or 28–32 weeks in PES.

Genetic monitoring

This topic is addressed in another chapter. Genetic monitoring is mainly used to detect genetic contamination in inbred strains, using routine monitoring not only of DNA markers such as micro-satellites/short tandem repeats or SNP's (single nucleotide polymorphism) but also genetically determined and relevant phenotypic traits or markers. These markers should be spread throughout the genome and be as discriminant as possible between the different lines bred on site. If short marker scans aim to detect genetic contamination, extended scans (80 to 120 markers) are used in accelerated backcrossing or in gross strain comparison. It is critical to bear in mind that minor genetic drifts and sub-strain differences are generally not detectable by genetic monitoring: the quality and rigor of the genetic management system is the only guarantee of the purity of a genetic standard. With transgenes or mutations, it is usually necessary to rely on routine genotyping to identify the gene of interest, its zygosity or to evaluate the number of transgene copies and their integration site(s) using standard or real-time PCR, Southern blot, fluorescent *in situ* hybridisation, restriction enzymes. Identification and reporting of phenotypic deviants is critical, especially in Foundation and Expansion Colonies.

Husbandry practices, caging and environmental conditions

Nutrition, husbandry practices, caging and environmental conditions are of course critical both for ethical and breeding performance reasons and may have to be adapted to the line sensitivity and its specific requirements (Jennings *et al.*, 1998).

Acknowledgments

Acknowledgments and thanks to my colleagues Georges Canard, Jean-Paul Champier, Marc Mercier, Michel Rizoud and François Veillet for their kind contributions and help as well as to Jacques Bonnod, Pierre Forissier and Alain Perrot who are now enjoying retirement.

References

Baker, D.G. (1988). *Clin. Microbiol. Rev.* **11**(2), 231–266.
Balk, M.W. (1986). In *Complications of Viral and Mycoplasmal Infections in Rodents to Toxicology Research*

and Testing (ed T.E. Hamm Jr), pp. 161–174. Hemisphere Publishing Corporation, Washington, DC.

Brennan, P.C., Fritz, T.E. and Flynn, R.J. (1969). *J. Bacteriol.* **97**, 337–349.

Bruce, H.M. (1970). *Br. Med. Bull.* **26**, 10–13.

Chin, E.Y. and Wang, C.K. (2001). *Genesis* **30**(2), 77–81.

Coates, M.E. (1968). *Proc. Nutr. Soc.* **27**(2),143–148.

Dabard, J. Dechambre, R.P., Ducluzeau, R., Gosse, C., Guillon, J.C., Perrot, A., Raibaud, P., Sabourdy, M., Sacquet, E. and Tancrède, C. (1977). *Sci. Techn. Anim. Lab.* **2**(1), 7–8.

Dabard, J., Dechambre, R.P., Ducluzeau, R., Gosse, C., Guillon, J.C., Perrot, A., Raibaud, P., Sabourdy, M., Sacquet, E. and Tancrède, C. (1977). *Sci. Techn. Anim. Lab.* **2**(1), 9–11.

Dewhirst, F.E. *et al.* (1999). *Appl. Environ. Microbiol.* **65**(8), 3287–3292.

Dore, J., Schir, A., Hannequart-Gramet, G., Corthier, G. and Pochart, P. (1998). *Syst. Appl. Microbiol.* **21**(1), 65–71.

Ermel, R.W. and Richard, D.R. (1997). *Anim. Technol.* **48**(1), 17–26.

European Directive. (1986). 86/609/EC, article 7.

Falconer, D.S. (1960). In *Introduction to Quantitative Genetics* (ed D.S. Falconer). Oliver and Boyd, Edinburgh.

Falconer, D.S. (1972). In *The U.F.A.W. Handbook* (ed U.F.A.W.), pp. 5–25. Churchill Livingstone, Edinburgh.

FELASA: Nicklas, W., Baneux, P., Boot, R., Decelle, T., Deeny, A.A., Fumanelli, M. and Illgen-Wilcke, B. (2002). *Lab. Anim.* **36**, 20–42.

Festing, M.F.W. (1979). In *Inbred Strains in Biomedical Research* (ed M.F.W. Festing), pp. 1–132. MacMillan Press, London and Basingstoke.

Festing, M., Kondo, K., Loosli, R., Poiley, M. and Spiegel, A. (1972). *ICLA Bull.* **30**, 4–17.

Festing, M.F.W. (2002). In *Mouse Genome Informatics*, version 2.8. http://www.informatics.jax.org/external/festing/search_form.cgi

Flynn, M.B., Brown, P.A., Eckstein, J.M. and Strong, D. (1989). *Lab. Anim. Sci.* **39**(5), 461–463.

Franklin, C.L. and Riley, L.K. (1993). http://www.criver.com/techdocs/93win_rp/

Franklin, C.L., Riley, L.K., Hoof, R.R. and Besch-Williford, C. (1994). In *Charles River Laboratories Technical Bulletin*, On-Line Literature (http://www.criver.com/techdocs).

Gordon, H.A. and Pesti, L. (1971). *L. Bact. Rev.* **35**, 390.

Green, E.L. (1981). In *Breeding Systems in the Mouse in Biomedical Research*, Vol. 1 (eds H.L. Foster, J.D. Small and J.G. Fox), pp. 91–104, Academic Press, San Diego, CA.

GV-SOLAS (1992). In *Diagnostic Microbiology for Laboratory Animals* (ed Ivo Kunstyr). Gustav Fischer Verlag, Stuttgart.

GV-SOLAS (1999). *Lab. Anim.* **33**(Suppl. 1), 39–87.

Hansen, A.K., Velschow, S., Clausen, O., Amtoft-Neubauer, H., Kristensen, K. and Jørgensen, P.H. (2000). *Scand. J. Lab. Anim. Sci.* **2**(27), 65–84.

Hart, T. and Shears, P. (1996). In *Color Atlas of Medical Microbiology*, Introduction/tables. Mosby-Wolfe, Times Mirror International Publishers.

Hartl, D.L. (2000). In *A primer of Population Genetics*, 3rd edn. Sinauer Associates, Sunderland, MA.

Hartl, D.L. (2001). In *Genetic Management of Outbred Laboratory Rodent Populations*. Charles River On-Line Literature. http://www.criver.com/techdocs

Hartl, D.L. and Clark, A.C. (1997). In *Principles of Population Genetics*, 3rd edn. Sinauer Associates, Sunderland, MA.

Heine, W. (1978). *Afr. Vet. Assoc.* **49**(3), 175–177.

Hesse, I., Luz, A., Kohleisen, B., Erfle, V. and Schmidt, J. (1999). *Lab. Anim. Sci.* **49**(5), 488–495.

Hill, A.C. and Stalley, G.P. (1991). *Lab. Anim. Sci.* **41**(6), 563–566.

Homberger, F.R. (1997). In *Enterotropic Mouse Hepatitis Virus*. Habilitationsschrift, University of Zurich.

Homberger, F.R. and Thomann, P.E. (1994). *Lab. Anim.* **28**, 113–120.

Huerkamp, M.J. (1993). *Lab Anim. Sci.* **43**(1), 86–90.

ILAR (1976). *ILAR News* **19**(3).

ILAR (1998). *ILAR Journal* **32**(4).

Jacoby, R.O., Gaertner, D.J., Johnson, E.A., Paturzo, F.X. and Smith, A.L. (1995). *Lab. Anim. Sci.* **45**(3), 249–253.

Jacoby, R.O, Ball-Goodrich, L.J., Besselsen, D.G., McKisic, M.D., Riley, L.K. and Smith, A.L. (1996). *Lab. Anim. Sci.* **46**(4), 370–380.

Jennings, M., Batchelor, G.R., Brain, P.F., Dick, A., Elliott, H., Francis, R.J., Hubrecht, R.C., Hurst, J.L., Morton, D.B., Peters, A.G., Raymond, R., Sales, G.D., Sherwin, C.M. and West, C. (1998). *Lab. Anim.* **32**, 233–259.

Kajiwara, N., Ueno, Y., Takahashi, A., Sugiyama, F., Sugiyama, Y. and Yagami, K. (1996). *Exp. Anim.* **45**(3), 239–244.

Leblanc, S.A., Faith, R.E. and Montgomery, C.A. (1993). *Lab. Anim. Sci.* **43**(5), 526–528.

Linder, C. (2001). *Lab. Anim.* **30**(5), 34–39.

Luckey, T.D. (1963). In *Germ-Free Life and Gnotobiology*. Academic Press, New York.

Mahler, M., Janke, C., Wagner, S. and Hedrich, H.J. (2002). *Scand. J. Gastroenterol.* **37**(3), 267–278.

Manipulating the Mouse Embryo, A Laboratory Manual (1994). 2nd edn (eds B. Hogan, R. Beddington, F. Costantini and E. Lacy). Cold Spring Harbor Laboratory Press, New York.

Markel *et al.* (1997). *Nat. Genet.* **17**, 280–284.

McKisic, M.D., Paturzo, F.X., Gaertner D.J., Jacoby R.O. and Smith, A.L. (1995). *J. Immunol.* **155**(8), 3979–3986.

Mirand, E.A. and Back, N. (1969). In *Germ-Free Biology, Experimental and Clinical Aspects*. Plenum Press, New York.

Moore, D.M. (1997). In *Charles River Laboratories Reference Paper*, **10**(4), On-Line Literature. www.criver.com/techdocs

Moreau, M.C. and Corthier, G. (1988). *Infect. Immun.* **56**(10), 2766–2768.

Moreau, M.C. and Gaboriau-Routhiau, V. (2001). *Microb. Ecol. Health Dis.* **13**(2), 65–86.

Moreau, M.C. (2001). *J. Pediatric. Puéric.* **14**(3), 135–139.

Morrell, J.M. (1999). *Lab. Anim.* **33**(3), 201.

NRC, National Research Council. (1991). *Companion Guide to Infectious Diseases of Mice and Rats.* National Academy Press, Washington, DC.

Okamoto, K. and Aoki, K. (1963). *Jpn. Circ. J.* **27**, 282–293.

Okamoto, M. and Matsumoto, T. (1999). *Exp. Anim.* **48**(1), 59–62.

Perrot, A. (1976). *Lab. Anim.* **10**, 143–156.

Perrot, A. (1978). *Sci. Techn. Anim. Lab.* **3**(1), 49–56.

Perrot, A., Canard, G. and Champier, J.P. (1987). *Microecology and Therapy,* **17**(29), 112–116.

Pleasants, J.R. (1968). In *The Germ-Free Animal in Research* (ed. M.E. Coates), pp. 113, Academic Press, London and New York.

Pleasants, J.R. (1973). *Endeavour* **32**(117), 112–116.

Poiley, S.M. (1960). *Proc. Anim. Care Panel* **10**, 159–166.

Raibaud, P., Dickinson, A.B., Sacquet, E., Charlier, H. and Mosquot, G. (1966). *Ann. Inst. Pasteur* **111**, 193–210.

Rapp, K.G. (1972). *Z. Versuchstierk. Bd.* **14**(S), 133–142.

Rapp, K.G. and Burow, K. (1979). In *Clinical and Experimental Gnotobiotics*, Suppl. 7 (eds Fliedner *et al.*). Gustav Fischer Verlag, Stuttgart, New York.

Reetz, I.C., Wullenweber-Schmidt, M., Kraft, V. and Hedrich, H.J. (1988). *Lab. Anim. Sci.* **38**(9), 696–701.

Richoux, V., Panthier, J.J., Salmon, A.M. and Condamine, H. (1989). *J. Exp. Zool.* **252**(1), 96–100.

Ross, M. (1961). *J. Anim. Techn. Assoc.* **XII**(1).

Royston, D., Minty, B.D., Needham and J.R., Jones, J.G. (1983). *Lab. Anim.* **17**(3), 227–229.

Russel, W.M.S. and Burch, R.L. (1959). In *The Principles of Humane Experimental Technique*, pp. 283. Methuen, London.

Sabourdy, M., Sacquet, E. and Tancrede, C. (1977). *Sci. Techn. Anim. Lab.* 2(1), 7–8

Sabourdy, M. (1967). In *L'Animal de Laboratoire dans la Recherche Biologique et Medicale* (ed. M. Sabourdy), pp. 38–57, Presses Universitaires de France, Paris.

Sacquet, E. (1968). In *The Germ-Free Animal in Research* (ed. M.E. Coates), pp. 1–21. Academic Press, London and New York.

Saito, M., Nakagawa, M., Kinoshita, K. and Imaizumi, K. (1978). *Jpn. J. Vet. Sci.* **40**, 283–290.

Salomon, J.C. (1968). In *The Germ-Free Animal in Research* (ed. M.E. Coates), pp. 113. Academic Press, London and New York.

Schaedler, R.W., Dubos, R. and Costello, R. (1965). *J. Exper. Med.* **122**, 77.

Shomer, N.H., Dangler, C.A., Marini, R.P. and Fox, J.G. (1998). *Lab. Anim. Sci.* **48**(5), 455–459.

Silver, L.M. (1995). In *Mouse Genetics: Concepts and Applications*, pp. 42–49. Oxford University Press, New York.

Suzuki, H., Yorozu, K., Watanabe, T., Nakura, M. and Adachi, J. (1996). *Exp. Anim.* **45**(1), 33–38.

The Staff of The Jackson Laboratory. (1997). In *Handbook on Genetically Standardized JAX® Mice*, 5th edn (eds R.R. Fox and B.A. Witham), pp. 107–129. The Jackson Laboratory, Bar Harbor, MA.

Trexler, P.C. (1967). *Vet. Rec.* **81**(19), 474–478.

Trexler, P.C. (1968). *Z. Versuchstierk.* **10**(1), 121–130.

Trexler, P.C. (1987). In *The UFAW Handbook on the Care and Management of Laboratory Animals*, 6th edn (ed. T. Poole), pp. 85–98. Longman Scientific & Technical, Harlow.

Truett, G.E., Walker, J.A. and Baker, D.G. (2000). *Comp. Med.* **50**(4), 444–451.

Ueno, Y., Sugiyama, F., Sugiyama, Y., Phsawa, K., Sato, H. and Yagami, K. (1997). *J. Vet. Med. Sci.* **59**(4), 265–269.

Van Zwieten, M.J., Solleveld, H.A., Lindsey, J.R., de Groot, F.G. and Hollander, C.F. (1980). *Lab. Anim. Sci.* **30**, 215–221.

Wakeland *et al.* (1997). *Immunol. Today* **18**, 472–477.

Wannemuehler, M.J. *et al.* (1982). *J. Immunol.* **129**(3), 959–965.

Ward, J.M., Anver, M.R., Haines, D.C. and Benveniste, R.E. (1994a). *Am. J. Pathol.* **145**, 959–968.

Ward, J.M., Fox, J.G., Anver, M.R., Haines, D.C., *et al.* (1994b). *J. Nat. Cancer Inst.* **86**, 1222–1227.

White, W.J. (1991). In *A Review of Pasteurella pneumotropica.* Charles River On-Line Literature. www.criver.com/techdocs

White, W.J. and Lee, C.S. (1998). In *Biological Reference Data on CD(SD) IGS Rats* (eds T. Matsuzawa and H. Inoue), pp. 8–14. CD(IGS) Study Group, Yokohama.

Wostmann, B.S. (1968). In *The Germ-Free Animal in Research* (ed. M.E. Coates), pp. 113. Academic Press, London and New York.

Zenner, L. (1998). *Lab. Anim.* **32**, 337–342.

Zoetendal, E.G., Akkermans, A.D.L. and De Vos, W.M. (1998). *Appl. Environ. Microbiol.* **64**(10), 3854–3859.

Assisted Reproduction and Cryopreservation

Martina Maria Dorsch

Zentrales Tierlaboratorium, Medizinische Hochschule, Hannover, Germany

Embryotransfer

General remarks

Efficient transfer of preimplantation embryos into pseudopregnant surrogate dams is an important step for the rederivation of cryopreserved or contaminated mouse stocks, for the generation of genetically modified mice, as well as for *in vitro* fertilisation (IVF) programmes.

Preferably providing a natural copulation stimulus by mating with sterile, but sexually active males should induce pseudopregnancy. Vasectomised males are used for this purpose.

In principle, embryo transfer to pregnant females is possible, but the transferred embryos have to compete with the recipient's own embryos.

Mouse embryos from one-cell to the morula stages are transferred into the oviduct of 0.5–2.5 pseudopregnant recipients, blastocysts have to be transferred into the uterus of day 2.5 or latest at day 3.5 of pseudopregnancy.

To complete development, embryos transferred to the oviduct must be protected by the zona pellucida (Bronson and McLaren, 1970), whereas this is not a prerequisite if the embryos are transferred into the uterus.

The number of embryos transferred to a foster mother depends on the donor strain and on the kind of manipulations of the embryos prior to transfer. Embryos of some inbred strains are less viable (e.g. BALB/c; Mobraaten, 1986), whereas most of the transferred embryos of hybrid or outbred stocks complete their development (50–70%). Freshly collected and unmanipulated embryos will develop with more success than frozen–thawed or microinjected embryos. It needs some experience to transfer the ideal number of embryos to give a litter-size of approximately 5–6 pups. In case a litter is too small (one or two pups only) or too big (more then 12 pups) the mother sometimes may take no care of them.

Outbred mice, such as Swiss Webster or ICR and F1 hybrid females (e.g. C57BL/6 × DBA or BALB/c × DBA) make good foster mothers. They should be aged between 10 and 12 weeks. It is also advantageous to use experienced mothers. Immediately after ejaculation the secretions of the male accessory sexual organs will coagulate and close the vagina temporarily. This 'plug'

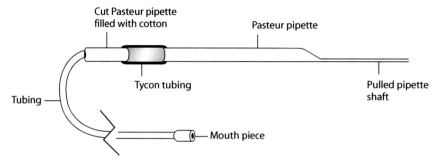

Figure 26.1 Drawn out Pasteur pipette with a mouth-pipetting device for embryo handling (according to Rafferty, 1970).

is used as a sign that copulation has occurred. To produce enough plugs we mate between 20 and 30 females to 20–30 sterile males.

Pipettes for embryo handling and embryo transfer

Pipettes for embryo handling are pulled from Pasteur pipettes as described by Rafferty (1970). The pipette is held over the blue flame of a Bunsen burner. As soon as the glass begins to soften, it is removed from the flame and pulled out immediately. The narrow part should be 7.0 cm in length. The tip should have an inner diameter between 150 and 200 μm. To polish the tip, hold it into the blue flame of the Bunsen burner for less than a second. The pipettes are attached to a silicone tube connected to a mouthpiece (Figure 26.1).

Pipettes for embryo transfer are torn out from the narrow end of a Pasteur pipette to an inner diameter of 120–150 μm. The resulting capillary is shortened to 8–10 cm and inserted into a Pasteur pipette, 2–3 cm should project. The capillary is fixed in position with sealing wax (Figure 26.2).

Loading the pipettes with embryos

To abolish capillary forces, the micropipettes are briefly tipped into the medium, then a small air bubble is drawn in. Now you can easily suck the embryos from the medium. At least one more air bubble and a small amount of medium will follow. When the embryos are introduced into the reproductive tract of a foster mother, the second air bubble allows control of a successful transfer. Figure 26.3 illustrates how the embryos must be sucked into the handling- and transfer capillaries.

Preparation of recipient females

Females are mated to sterile (vasectomised) males in the afternoon at about 4:00 p.m. With a light/dark cycle of 12 h each (light: 6:00 a.m. to 6:00 p.m.; dark: 6:00 p.m.

(a) (b)

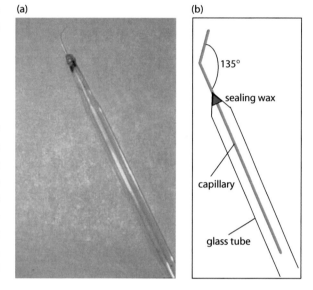

Figure 26.2 Embryo transfer capillary. (a) Photograph; (b) schematic view. **(See also Colour Plate 7.)**

to 6:00 a.m.), copulation with females in oestrus will take place approximately in the middle of the dark period. With an oestrus cycle of 4–5 days, 20–25% of the females should mate. However, in practice fewer females will do so. Whitten (Scharman and Wolff, 1980) found that the oestrus cycle of the females is strongly influenced by the odour of the males. With this in mind, the copulation rate can be raised up to 45% when the females are exposed to male odours at least one night prior to mating ('Whitten-effect'; see Ch. 18 about "Priming effects of mouse odours"). In the morning after mating, the females are checked for the presence of a vaginal plug (day 0.5 of pseudopregnancy, see section on General Remarks). Unplugged females should not be mated again until their normal oestrus cycle has been re-established (in approximately 10 days). This is necessary because the vaginal plug sometimes is lost and the female was judged as 'false negative'. The cervical stimulation during copulation induces an increased secretion of hormone prolactin, which is responsible for the maintenance of (pseudo-) pregnancy.

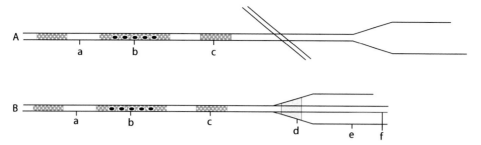

Figure 26.3 Tips of capillaries loaded with embryos. (A) Narrow part of a drawn out Pasteur pipette for embryo handling; (B) tip of a transfer capillary; (a) air bubble; (b) medium containing embryos; (c) medium; (d) sealing wax; (e) glass tube; (f) capillary.

Anaesthesia

Usually anaesthesia by injection is used for embryo transfer and vasectomy in mice. The regimen described below uses surgical anaesthesia that lasts for about 30 min, giving enough time for both kinds of operations.

A freshly prepared mixture of 0.5 ml Ketamine HCl 10% (WDT, Garbsen, Germany) and 0.1 ml xylazine HCl 2% (Rompun®, Bayer, Leverkusen, Germany) diluted with sterile saline solution (Merck, Darmstadt, Germany) to a total volume of 5.0 ml is used.

The mice are anaesthetised by intraperitoneal (i.p.) injection of 0.1 ml of the mixture per 10 g body weight (bwt) equal to a dose of 100 mg ketamine and 4 mg/kg xylazine. Dexpanthenol eye ointment (Bepanthen®, Hoffmann-La Roche, Leverkusen, Germany) must be administered for corneal protection. In order to prevent hypothermia, cages with anaesthetised mice should be placed on a warming plate (25–30°C) until recovery.

Transfer into the oviduct

To improve the pregnancy success rate, we perform unilateral embryo transfers to double the number of surrogate dams. The methods described below can be used for both sides of the body.

Anaesthetise the female as described. After loosening the toe pinch reflex, the lower back is shaved and disinfected (Braunol® 2000, Braun Melsungen AG, Melsungen, Germany). Place the mouse under a stereomicroscope (6 × 20 magnification) with the tail lying away from you. A transverse incision (<1 cm) is made across the lumbar area with fine dissection scissors. The skin is loosened from the tissue with a pair of scissors (either to the left or right side) and you will see the ovary or the fat pad through the body wall. The body wall is lifted with fine forceps and a small transverse incision (0.5 cm) is made. The fat surrounding the ovary and oviduct is grasped with fine forceps and pulled out. The

ovary and oviduct are placed on the back by fixing it on the fat pad with a small 'Bulldog'-clamp (Aesculap, Tuttlingen, Germany). Pick up the mouse and place it with its head lying to the left. Select the 16 × 20 magnification on the microscope. With two fine watchmaker's forceps carefully tear open the bursa over the infundibular opening of the oviduct. Try to avoid any blood vessels. Now the transfer micropipette has to be loaded with the embryos as described in the section on Loading Pipettes with Embryos. The pipette with embryos is gently inserted into the infundibular opening and held in place with the tips of watchmaker's forceps (Figure 26.4a). The embryos are blown into the oviduct with a minimum of fluid. Carefully withdraw the pipette and squeeze the infundibular opening with the forceps for a few seconds. This prevents back flow of the embryos. Remove the 'Bulldog'-clamp and reposition the ovary and oviduct back into the abdominal cavity. There is no need to suture the body wall on condition that the incision is less than 0.5 cm. The skin is closed with Michel clips (7.5 × 1.75 mm, Medicon EG, Tuttlingen, Germany).

Transfer into the uterus

The procedure is almost the same as for the oviduct transfer. Here, not only ovary and oviduct are excised, but also the upper part of the corresponding uterine horn. A 26-G needle is used to make a hole into the uterus. Blood vessels should carefully be avoided. After removing the needle, the transfer pipette is inserted into the hole and the embryos (blastocysts) are blown into the uterus lumen towards the tail (Figure 26.4b). Continue as described above.

Vasectomy of mice

Outbred mice, such as Swiss Webster or ICR and F1 hybrid males (e.g. C57BL/6 × DBA or BALB/c × DBA) are ideal partners for the production of surrogate dams.

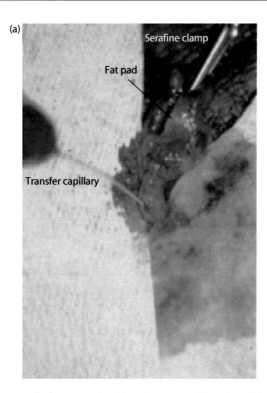
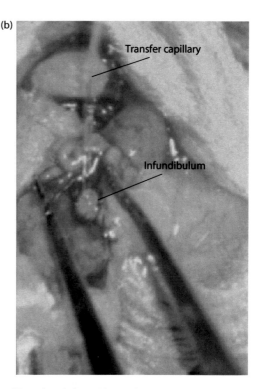

Figure 26.4 Embryo transfer (a) to the uterus; (b) to the oviduct. **(See also Colour Plate 8.)**

One efficient and fast method for vasectomy of mice is the cauterisation of the vas deferens. Beside the surgical instruments you need a Bunsen burner and a pair of 'old' watchmaker's forceps.

The male mice should be between 8 and 10 weeks old. For surgery, injection anaesthesia as described in the section on Anaesthesia is used. After loosening the toe pinch reflex, the mouse is placed on its back and the ventral abdomen is shaved and disinfected (Baronial® 2000, Braun Melsungen AG, Melsungen, Germany). The shaving of the ventral abdomen is not absolutely necessary but useful. A middle incision of the skin (1 cm), 0.5 cm cranial to the preputial orifice is made with fine dissection scissors. The skin is loosened from the tissue with a pair of scissors (to the left and right side). The body wall is lifted with fine forceps and a small transverse incision (0.5 cm) is made. The fat pad surrounding the right testes is grasped with fine forceps and drawn out until the vas deferens is visible. If possible, avoid exteriorisation of the testes to reduce postoperative pain. Use a pair of fine forceps to separate the vas from the mesenteric membrane. The tips of the 'old' watchmaker forceps are made red-hot in the flame of the Bunsen burner and a small piece of the vas deferens (0.5 cm) is removed by cauterisation (Figure 26.5). Replace the tissue into the abdominal cavity. Repeat the procedure with the left side. The body wall is sutured with an absorbable suture material (Vicryl®-rapid, 4-0, Ethicon, Norderstedt, Germany), the skin is sutured with non-absorbable material (Mersilene, 3-0, Ethicon, Norderstedt, Germany). During surgery, the testes are normally separated from the scrotum. By gently pressing the ventral abdomen with your thumbs, the testis will slide back into the correct position. One week later the males can be mated. If possible, the vasectomised males should be mated twice, to ensure their sterility before using them for the production of pseudopregnant females.

The males can be used for about 1 year. For the permanent production of pseudopregnant females, 30 sterile, sexually active males should be available. To ensure this, we exchange 50% of the males every 6 months.

Collection of preimplantation embryos

Media

Most media used for embryo handling and culture are modifications of a Krebs Ringer solution.

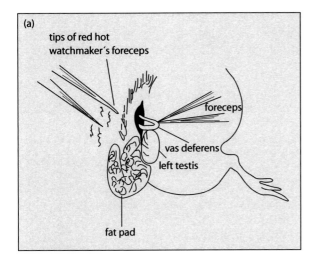

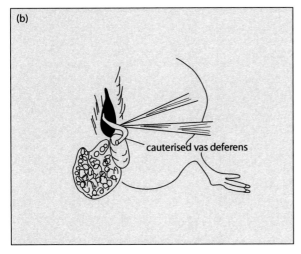

Figure 26.5 Vasectomy through cauterisation.
(a) Immobilisation of the vas deferens; (b) cauterisation.

For embryo collection and embryo handling we prefer to prepare a phosphate buffered medium (PB1, Table 26.1) ourselves according to the recipe recommended by Whittingham, 1974 from stock solutions. For *in vitro* culture and IVF we use M16 (Table 26.2; Hogan *et al.*, 1994). Table 26.3 summarises the concentrations for the stock solutions. Medium M16 should be preincubated at least 2 h prior to use (preferably overnight) at 37°C, 95% humidity and 5% CO_2. As the volume of the embryo-culture-medium is relatively small it is covered with paraffin oil to avoid drying out.

Superovulation

To increase the number of embryos that can be collected from one donor female and to reduce the number of animals needed for the production of embryos, exogenous gonadotrophins are applied. The reproductive cycle of the treated females will then be synchronised. This allows

TABLE 26.1: Phosphate-buffered medium (PB1)	
Substance	**Endconcentration (mM)**
H_2O	
NaCl	100.93 mM
KCl	2.73 mM
Na-Pyruvat	0.3 mM
KH_2PO_4	1.42 mM
$MgCl_2 \times 6H_2O$	0.48 mM
$Na_2HPO_4 \times 12H_2O$	8.05 mM
$CaCl_2 \times 2H_2O$	0.94 mM
Phenolred	0.01 mg/ml
Penicillin G	1 mg/ml
Streptomycin	0.05 mg/ml
Glucose	1 mg/ml
BSA	3 mg/ml

Substances have do be mixed in the given order from stock solutions. PB1 can be stored for 3 weeks at 4°C.
Source: Whittingham *et al.*, 1974.

TABLE 26.2: Embryo culture medium M16	
Substance	**Endconcentration**
H_2O	
NaCl	94.66 mM
Na-pyruvate	0.33 mM
Na.lactate	23.28 mM
KCl	4.7 mM
KH_2PO_4	1.19 mM
$MgSO_4 \times 7H_2O$	1.19 mM
$CaCl_2 \times 2H_2O$	1.71 mM
$NaHCO_3$	25.00 mM
Penicillin G, potassium salt	100 units/ml
Streptomycin sulfate	50 mg/ml
Phenolred in $NaHCO_3$	0.01 mg/ml
Glucose	1 mg/ml
BSA	

Substances have do be mixed in the given order from stock solutions. Filter medium through a millipore filter (0.45 μm; Schleicher & Schüll). Osmolarity should be between 288 and 292 mosmoles. M16 should not be stored longer than 1 week at 4°C.
Source: Hogan *et al.*, 1994.

TABLE 26.3: Stock solution for M16 and PB1	
Component	**Concentration**
Na-pyruvate	suspend 85 mg in 0.154 mM NaCl, store at −20°C as 1 ml aliquots; for use melt aliquot and dilute to 50 ml with 0.154 mM NaCl
Penicillin G (Na-salt)	suspend 10.000/U in 20 ml 0.154 mM NaCl, store at −20°C in 1 ml aliquots
Streptomycin (sulfate)	suspend 50 mg in 10 ml 0.154 mM NaCl, store at −20°C in 1 ml aliquots
Phenol red	dissolve 13 mg in 10 ml 0.154 mM NaCl
Na-lactate	dissolve 863 mg in 50 ml H_2O
NaCl	0.154 mM (4.5 g/500 ml aqua dest.)
KCl	0.154 mM (574 mg/50 ml aqua dest.)
KH_2PO_4	0.154 mM (1.048 g/50 ml aqua dest.)
$MgCl_2 \times 6H_2O$	0.154 mM (1.565 g/50 ml aqua dest.)
$Na_2HPO_4 \times 12H_2O$	0.154 mM (5.515 g/100 ml aqua dest.)
$CaCl_2 \times 7H_2O$	0.110 mM (8.096 mg/50 ml aqua dest.)
$MgSO_4 \times 7H_2O$	0.154 mM (1.898 g/50 ml aqua dest.)
$NaHCO_3$	0.154 mM (647 mg/50 ml aqua dest.)

If not mentioned otherwise these stock solutions can be stored at 4°C for 4–5 weeks in the amount given in the table. Use 'aqua injectabile' (Braun, Melsungen, Germany) for the preparation of solutions.

timed mating and collection of defined developmental stages of embryos. Whittingham (1971) suggested a scheme for the superovulation of mice that works for most of the common mouse in- and outbred strains. However some inbred (e.g. CBA) strains and some genetically modified strains do not respond to exogenous hormones, whereas others (e.g. NMRI) will produce a multiple of the normal amount (Helfer and Gregg, 2002). Besides genetic factors, some others influence the success of superovulation. These are age of the female, bodyweight and state of health. Despite the fact that the animals are maintained in a controlled environment under standardised conditions, the season will also influence the success of superovulation. The quality of the embryos may decrease with the number produced after superovulation. NMRI females sometimes produce more than 40 oocytes. A high proportion however, will not develop beyond the first 2 or 3 days. Ideally the virgin females should be between 6 and 8 weeks of age, but older females may also succeed. With a light/dark cycle of 12 h each (light from 6:00 a.m. to 6:00 p.m.; dark from 6.00 p.m. to 6.00 a.m.) the females receive an i.p. injection of 5–10 units of pregnant mare's serum gonadotrophin (PMSG: Intergonan®, Intervet GmbH, Tönisvorst, Germany) at 4:00 p.m. and 5 units of human chorionic gonadotrophin (hCG: Primogonyl-1000, Schering – Deutschland GmbH, Berlin, Germany) 48 h later. They are mated to males immediately after the

second injection. Next morning (day 0.5 of pregnancy) the females are checked for the presence of a vaginal plug. Table 26.4 shows the time course of the early embryonic development. The pregnant females are sacrificed at the appropriate developmental stage of the embryos.

As hormones are natural products, brand-new samples may vary in hormone content. Therefore PMSG in particular should be tested before routine use to find the optimal dose for superovulation. For this purpose we superovulate three groups (n = 5) of good responding females (such as Swiss Webster or ICR and F1 hybrids) with different doses (5–7–10 units). The hormones are then diluted in 0.9% NaCl to the optimal concentration, aliquots can be stored at −20°C for at least 1 year.

Collection of oocytes

On day 0.5 of pregnancy, fertilised oocytes can be collected. They can be used for the generation of transgenic animals through pronucleus injection of DNA. Where unfertilised oocytes are needed for IVF, the females are superovulated as described above, except for mating to males.

Kill the females by CO_2 inhalation or by cervical dislocation. The animal is laid on its back and the ventral abdomen is disinfected. A small lateral middle incision of the skin is made and can then be pulled towards

TABLE 26.4: Early development (data of a cross of C57BL/6 female and a CBA male)			
Time(h)	Stage	Appearance	Location
0	fertilisation	ova surrounded by cumulus cells	ampulla
10	pronuclei become visible	male and female pronuclei are clearly visible	ampulla
		second polar body is visible	
		fewer cumulus ceolls surround the ova	
20	state of first mitotic cleavage	cumulus cells have disappeared fusion of pronuclei,	1–2 loop of oviduct
24	two-cell stage	two blastomeres of equal size, surrounded by the zona pellucida	2 loop of oviduct
48	four-cell stage	two blastomeres of equal size, surrounded by the zona pellucida	lower half of the oviduct
52	eight-cell stage (stage of compaction)	The contours of the blastomeres are no more distinctly visible	lower half of the oviduct
72	morula	The blastomeres are not equal in size	pars intramuralis tubae (connection from oviduct to uterus)
96	blastocyst	the zona pellucida has disappeared	uterus
120	beginning of implantation	distinct differences in the degree of development in the same litter	blastocyst adheres to the uterine epithelium

Source: According to Theiler, 1989.

the head and tail respectively using your fingers. The body wall is opened by a lateral incision beneath the xiphoid cartilage from right to left. The body wall is then laid flat towards the head and tail respectively. Push the gut to one side and find the reproductive tract (the two horns of the uterus, oviduct and ovary). Now grasp one uterine horn about 1 cm proximal to the oviduct with fine forceps. Remove the fat pad and mesometrium with the shank of fine scissors. Carefully make a first cut between ovary and oviduct. With a second cut 0.5 cm proximal to the oviduct, the oviduct with a small part of the uterine horn is removed. Both oviducts of one female are transferred to a Petri dish (30 mm, Greiner, Solingen, Germany) containing 2.0 ml PB1 medium at room temperature. By using a stereomicroscope one can easily find the swollen ampulla, the upper part of the oviduct (Figure 26.6). The swelling of the ampulla at this stage of development is due to the presence of the cumulus cells that surround the oocytes. With two fine watchmaker's forceps tear open the ampulla. The oocyte–cumulus-complexes will pour out

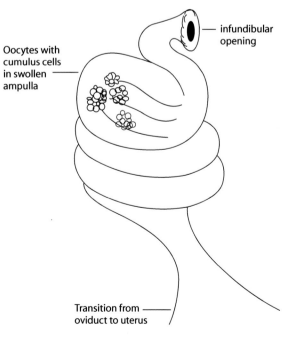

Oocytes with cumulus cells in swollen ampulla

infundibular opening

Transition from oviduct to uterus

Figure 26.6 Oviduct with the ampulla containing cumulus–oocyte complexes.

into the medium. Remove the oviducts from the Petri dish. The cumulus cells are removed by adding 20 μl of hyaluronidase (10,000 IU/ml in PB1). Under the microscope you can watch the disaggregation of the cumulus cell mass within 2–3 min. The embryos must then be washed three times in fresh PB1 medium. They can then be transferred to M16 medium for *in vitro* culture until further micromanipulation takes place.

Collection of two-cell embryos to morulae

Embryos from the two-cell stage to morulae can be used for cryopreservation or for the rederivation of microbiologically contaminated animals. These developmental stages are flushed out from the oviduct. At that time of development the embryos have migrated from the ampulla to approximately the second loop of the oviduct, which is said to exhibit peristaltic contractions to transport the embryos along the oviduct (Rugh, 1968).

The females are killed at the appropriate time of embryo development. The abdominal cavity is opened and the oviducts are excised as describe above. A blunt end gauge 33 needle, attached to a syringe filled with PB1medium, is inserted into the infundibular opening. Hold the needle in position with a pair of watchmaker's forceps. The embryos are flushed out of one oviduct with 3–4 drops of PB1 medium into an inner well of a four-well Petri dish (30 mm, Greiner, Solingen, Germany). The morphologically intact embryos are washed three times with fresh PB1 medium and than transferred into the freezing medium for cryopreservation (described below) or they will be directly transferred to foster mothers (described above) for hygienic sanitation. Reetz *et al.* (1988) showed that all germs could be removed through washing as long as the zona pellucida is intact.

Collection of blastocysts

As mentioned in the sectiion on Superovulation the quality of embryos may decrease as a result of overproduction due to applied hormones. In our experience, this reduced quality becomes evident at the blastocyst level. In the case of a huge overproduction of embryos and reduced quality, the female will absorb most of them and render the procedure of superovulation invalid. To avoid this, we do not superovulate females for the collection of this developmental stage.

Blastocysts are needed for targeted mutation through injection of embryonic stem cells into the blastocoel. Ideally the blastocysts are flushed out from the two

uterine horns. The females are killed on day 2.5–3.5 (depending on the strain). And prepared as described above. The uterus (both horns) is removed by a first cut across the cervix and a second cut below the junction to the oviducts and transferred to a 30 mm Petri dish containing 2 ml PB1 medium. Each horn is flushed with five drops of PB1 medium towards the cervix. The blastocysts are washed 2–3 times in fresh PB1 medium and transferred to M16 until further use.

Cryopreservation and revitalisation of preimplantation embryos and gametes

Reasons for freezing embryos and gametes

The reasons for freezing gametes and embryos are multifarious. One reason is the preservation of the genetic pool of different strains. An embryo- or gamete bank serves as an important back up of vitally maintained strains in case of diseases, reduced reproduction, genetic contamination or incidental environmental events. By storage in liquid nitrogen (LN_2) at $-196°C$ the genetic drift (mutations) is abolished as different groups (Whittingham *et al.*, 1977; Mobraaten and Bailey, 1987) have shown. In addition, the exponential increase in the number of genetically modified strains exceeds the capacities of most institutes. Cryopreservation of strains that are not used reduce actual shelf costs.

Today there are numerous protocols for the cryopreservation of gametes and embryos. Most success is achieved by cryopreservation of preimplantation embryos and spermatozoa.

Cryopreservation and revitalisation of preimplantation mouse embryos

All preimplantation developmental stages can be cryopreserved and revitalised with greater or lesser success.

The results for cryopreservation of fertilised and non-fertilised oocytes however, are not yet satisfactory (Wood *et al.*, 1992). Results from Bouquet *et al.* (1993) also suggest that the complete cycle of cryopreservation might alter the oocyte and, more particularly, induce DNA damage. Preservation of two-cell embryos gives the best results after revitalisation. Ultrarapid freezing or vitrification as described by Rall and Fahy (1985) is a technique, which does not require a cryostat. However, the revitalisation rates are not as satisfactory as for the modified two-step method described below. Genetic differences may affect the survivability of embryos after freezing (Schmidt *et al.*, 1985; Mobraaten, 1986).

The choice of the type of cryocontainers determines the type of the storage system. We use plastic straws (Minitüb®, Minitüb, Tiefenbach, Germany) as cryocontainers. Minitübs can be permanent labelled with strain, batch etc. using a labelling-apparatus (Grauel, Berlin, Germany). Loaded Minitübs® are collected in magazines (Figure 26.7) that fit into a storage system as shown in Figure 26.8. The freezing medium is PB1 supplemented with 1,2-propandiol (PROH, Sigma) to a final concentration of 1.5 M. 1,2-propandiol is a small molecule that penetrates the cell membrane very rapidly. Its toxicity is rather low compared to other cryoprotectants (Shaw *et al.*, 1991, Nowshari *et al.*, 1995, Mukaida *et al.*, 1998).

The Minitüb® is sealed with a metal bulb on one side and filled with 150 µl freezing medium (Figure 26.7). The embryos (20/Minitüb®) are transferred with a drawn out Pasteur pipette into the cryocontainer, which is sealed immediately by a glass bulb. The filled straws are placed into a programmable automatic cooling bath with ethanol as coolant (Model F8, Haake, Karlsruhe, Germany, Figure 26.9). After an equilibration time of 10 min at 0°C the freezing programme is started. The straws are cooled to −6°C at 1°C/min. Mechanical seeding is induced with a pre-cooled (−196°C) metal die after 5 min. At 0.4°C/min the straws are then cooled to −32°C. After 10 min the straws are transferred to LN₂ for storage.

Precise record keeping should include the following items: physical conditions of the freeze run, type of freezing procedure, temperature recording, results of viability tests from control batches, description of the strain (strain history, phenotype, reproductive performance, pedigree information, genotype, generation), identification and storage location, number of the embryo batch, number and developmental stage of embryos.

For thawing, the cryocontainer is taken from the storage tank (−196°C) to room temperature. The metal bulb is cut off. Within 20–30 s the ice will melt.

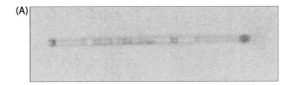

Figure 26.7 Plastic straws (Minitübs®) used as cryocontainers for embryo freezing. (A) Photograph; (B) schematic view; (a) glass bulb, (b) air space, (c) freezing medium containing the embryos, (d) seeding point, (e) metal bulb; (C) magazines for 10 straws each. **(See also Colour Plate 9.)**

Figure 26.8 View into the LN₂ storage tank with a storage system adapted for Minitübs®.

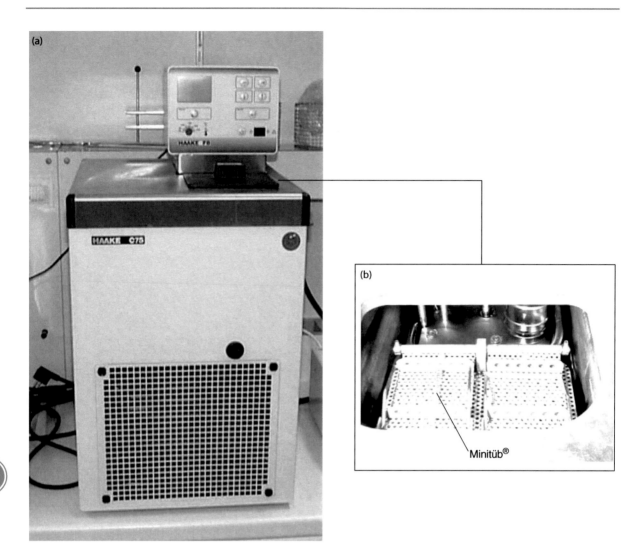

Figure 26.9 Programmable automatic cooling bath. (a) Chamber containing the coolant (ethanol) with programmer on top; (b) view into the chamber showing a horizontally holding device for the Minitübs®.

Now attach the straw to a 2 ml syringe and cut off the glass bulb. The content (including the embryos) is flushed into a 30 mm-Petri dish and an equal volume of fresh PB1 is added immediately to dilute the cryoprotectant. After 10 min add another volume PB1. Repeat this step to a final volume of 3 ml. The embryos must be washed three times with fresh buffer. After examination using a stereomicroscope, morphologically intact embryos are used for further *in vitro* culture in M16 medium or transferred to foster mothers.

Cryopreservation of epididymal sperm

The cryopreservation of spermatozoa offers a means of fast storage of the genetic pool of different mouse strains and is used routinely. Cryopreservation of sperm became important for mutagenesis programmes as well as for the storage of genetically modified strains that are not actually used. However, one must keep in mind, that IVF and embryo-transfer to foster mothers have to be performed after thawing. For that, appropriate donor females for the oocytes must be available. The resulting animals are heterozygous for the modified gene or mutation. Backcross programmes are needed to produce animals with the desired genetic status. This additional time may be considerably long.

Quite a number of protocols are published. A combination of the protocols published by Tada *et al.* (1990) and Nakagata and Takeshima (1993) gave the best results in our hands and is described below.

As cryopreservation medium (CPM) raffinose (18%) and skim milk (3%) is suspended in H_2O at

60°C. After centrifugation at 10×10^3 g for 15 min the supernatant is filtered using a sterile, pyrogen free disposable filter holder (0.45 μm; Schleicher & Schüll). The filtrate is used as CPM.

Adult males that were not mated 3 days prior to sperm preparation are used to maximise the sperm quantity and quality. The last mating may not be more than 2 weeks previously. For the preparation the male is sacrificed by inhalation of CO_2 to prevent ejaculation. The abdominal cavity is opened and the epididymis (caput, cauda and vas) is cut free. Transfer both epididymides of one male to a 2 ml plastic reaction tube (Eppendorf, Hamburg, Germany) containing 0.5 ml CPM. The tissue is cut into pieces with fine scissors and mixed gently. After incubation at room temperature for 15 min, the sperm will swim up and cell fragments will sink to the bottom. One drop of the supernatant, containing the sperm is checked under a stereomicroscope for morphological integrity, motility and sperm number using a haemocytometer. Songsasen and Leibo (1997) suggest using only those preparations in which at least 60% of the spermatozoa are motile with vigorously progressive movements and with a concentration of $2–4 \times 10^7$ total sperm/ml.

Dilute the supernatant, containing the sperm to a concentration of approximately 1×10^7 sperm/ml with CPM.

For freezing, melt holes (Ø 0.5 cm, depth 0.3 cm) in a block of dry ice (−70°C) using a steel die (Figure 26.10). Pipette 50 μl of the sperm suspension (diluted supernatant) into one hole and incubate 5 min. The sperm suspension will form 'pellets' during this time. Transfer the pellets into LN_2 for 5 min and transfer them to pre-cooled 1 ml cryotubes (Greiner, Solingen, Germany). The 'sperm-pellets' of one male should be divided among two or three vials. The cryotubes are stored in LN_2 for storage.

IVF

Freshly collected oocytes can be fertilised by a method described by Hogan *et al.* (1994). It can be used for unfrozen and freeze-thawed sperm. Superovulation and collection of oocytes are described above in the sections on Superovulation and Collection of Oocytes respectively.

One day prior to IVF, plastic culture dishes are prepared (Ø 3 cm, with four inner wells, for embryo

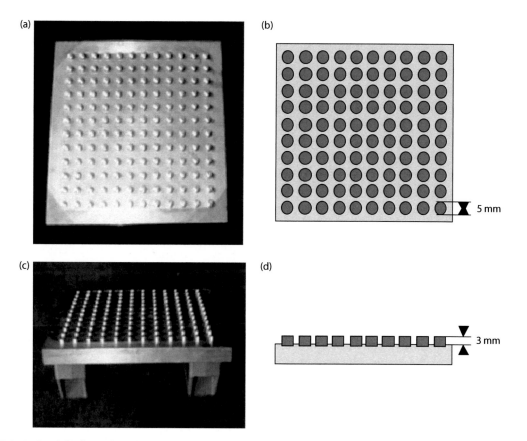

Figure 26.10 Steel die for melting holes into a block of dry ice for the freezing of mouse sperm. (a) and (b) Top view (photograph left, schematic right); (c) and (d) side view (photograph left, schematic right).

culture and Ø 5 cm with one inner well for the IVF, Greiner, Solingen, Germany) as described in the section on Media. Fill the wells with M16 culture medium (150 µl for Ø 3 cm and 500 µl for Ø 5 cm) and cover with paraffin oil. Incubate overnight at 37°C, 5% CO_2 for equillibration.

Approximately 12 h after hCG injection of the females, the males are killed and sperm is prepared from the epididymis as described. Transfer the epididymides of males into a 2 ml plastic reaction tube containing 500 µl PB1 and continue as described for cryopreservation. An aliquot of at least 2×10^5 spermatozoa is transferred to the prepared one-well dish and incubated for 1.5 h at 37°C, 5% CO_2 for capacitation.

Meanwhile, 12.5 h after injection of hCG, female mice are killed and oocytes are prepared (see section on Collection of Oocytes). For IVF the cumulus cells should not be removed because cumulus-intact eggs are more efficiently fertilised *in vitro* than cumulus-free eggs as was shown by Bleil (1993) and from our own experience. Incubate oocytes in the pre-prepared four-well dishes for 1 h at 37°C and 5% CO_2 then add 100 µl of the freshly collected sperm suspension. Four hours later, the cumulus cell free fertilised oocytes are transferred to fresh M16 medium. Fertilised oocytes can be recognised by the visible second polarbody that is generated during the second maturation division of the egg when the sperm penetrates the vitelline membrane.

For IVF with frozen sperm, melt one sperm pellet in the pre-prepared one-well culture dishes and incubate for 1 h at 37°C and 5% CO_2, then add the cumulus-oocyte complexes from one oviduct.

Ovary Cryopreservation and transplantation

Freezing and thawing of ovaries

The method described is adapted from Sztein *et al.* (1998) and Harp *et al.* (1994). Ideally donor females should be between 3 and 7 weeks old. They are killed by cervical dislocation and the ovaries are aseptically removed. The ovaries are transferred to a 30 mm Perti dish containing 2 ml PB1 at room temperature. Fat pad and bursa are removed using a stereomicroscope. The ovaries are than transferred to 2 ml cryotubes containing 200 µl PB1 supplemented with cryoprotectant (PROH) at a final concentration of 1.5 M and 10% foetal bovine serum. After 10 min at room temperature the cryotubes are placed on ice for 45 min. They are transferred to a controlled-rate freezing chamber

(model F8, Haake, Karlsruhe, Germany, Figure 26.9), pre-cooled to −6°C and ice crystallisation is induced on the surface of the medium, using a Pasteur pipette containing frozen medium, 5 min later. Cooling continues to −80°C at 0.5°C/min. After 10 min at −80°C the cryotubes are transferred to LN_2 (−196°C) for storage.

For thawing, the cryotubes are removed from the storage tank to room temperature. As soon as the ice is melted, the cryoprotectant has to be removed immediately using a Pasteur pipette. It is replaced by 200 µl fresh PB1. After 10 min the ovaries can be used for transplantation.

Transplantation of ovaries

The technique described here is adapted from Jones and Krohn (1960).

Recipient females are anaesthetised as described. The surgical field is shaved and disinfected (Braunol® 2000, Braun Melsungen AG, Melsungen, Germany) and a single transverse incision of the skin dorsally, across the lumbar area, gives access to the ovaries on both sides. The right uterine horn is closed with a single ligature using absorbable suture material (Vicryl®-rapid, 4-0, Ethicon, Norderstedt, Germany) and the ovary is removed with a single cut between the oviduct and the uterine horn. The left ovary is carefully removed through a small incision made into the bursa with microsurgical scissors and replaced by the donor ovary. Sztein *et al.* (1998) recommend cuting the ovary into two halves, to ensure that the donor ovary will fit into the bursa of the recipient. In addition one donor will give grafts for four recipients. The peritoneum is closed by continuous stitches with absorbable material and the skin is closed with Michel clips (7.5 × 1.75 mm, Medicon EG, Tuttlingen, Germany). In order to prevent hypothermia, the cage with the anaesthetised animals is placed on a warming plate (25–30°C) after surgery until recovery from anaesthesia.

Setting up a frozen storage

General remarks

Table 26.5 summarises what can be stored in LN_2 and revitalised with greater or lesser success. The table also shows additional steps that are needed after thawing of the respective gametes or developmental stages to start

TABLE 26.5: Cryopreservation of gametes and assosiated methods				
Cryopreserved storage of	Methods needed prior to crypreservation	After thawing	Additional demands	Possible problems associated with cryopreservation
ovaries	• ovarectomy	• orthotopic transplantation tof ovarectomised recipients • backcross programmes • genetic monitoring	• need for histo compatible or immune-deficient recipients	possibility of transmission of viruses
preimplantation embryos	• superovulation • collection of preimplantation embryos • embryoculture	• embryotransfer • embryoculture • genetic monitoring	• need for pseudopregnant dams • need for vasectomised males	
sperm	• collection of spermatozoa	• IVF • superovulation • oocyte recovery • embryoculture • embryotransfer • backcross programmes • genetic monitoring	• need for egg-donors • need for pseudopregnant dams • need for vasectomised males	possibility of transmission of viruses

a new breeding nucleus. These 'additional' steps may not be neglected and are very time consuming.

Another important question is the number of embryos, spermatozoa or ovaries that need to be frozen to ensure a high probability that a new breeding colony can be built up after thawing.

Number of preimplantation embryos that should be frozen for back up

Between 25–30% of the embryos such as Swiss Webster or ICR and F1 hybrids, frozen by the two-step method previously described will result in live offspring. Depending on these results at least ten batches, containing 10 embryos each, have to be frozen to produce one breeding pair with a probability of 99.9% after thawing. This number may be much lower for some inbred stains, mutants, genetically modified strains, congenic or isogenic strains. We recommend freezing 'control' batches of the specific strain and revitalising these embryos for calculation of the appropriate number of embryos needed for a frozen back up. This revitalisation should include the whole programme of thawing and transfer as embryos that appear morphologically intact after thawing may

also fail to finish their development in the reproductive tract of a foster mother.

Depending on the type and genetic requirements of the strain that should be frozen, different selection principles for the embryo donors have to be followed. Table 26.6 summarises these principles.

Number of spermatozoa that should be frozen for a back up

The fertilising ability of frozen–thawed sperm strongly depends on the strain used as sperm donor as well as oocyte donor and can vary considerably. The developmental ability of *in vitro* fertilised embryos can also be reduced. For freezing of preimplantation embryos, we recommend freezing 'control' batches of the specific strain. With the method described here, at least 10 sperm-pellets per male are frozen. With ten pellets, the eggs of five females can be fertilised. Nevertheless the sperm of 5–10 males with proven fertility should be frozen for a genetic back up.

A still unsolved problem associated with the freezing of spermatozoa and the necessity of IVF is the possibility of vertical transmission of viruses with the

TABLE 26.6: Selecting principles for embryo donors		
Type of strain	Genetic requirements	Origin of embryos of one batch
outbred	random sample of the strain	n embryos from n donors
inbred	B × S pairs	all embryos from one B × S pair
congenic	maintenance of differentiating gene	pool of n donors
mutant and genetically modified	maintenance of the mutation/ modification	pool of n donors

B × S pair: brother x sister pair.

sperm suspension. Therefore direct descendants of mice produced from eggs fertilised *in vitro* by sperm from males with unknown microbiological status should be reared in quarantine to exclude contamination of healthy breeding colonies.

Number of ovaries that should be frozen for a back up

Since the ovary is a complex structure of different cell types and containing follicles at various stages of folliculogenesis, a large proportion of follicles may be damaged during the freezing and thawing of ovarian tissue. Sztein *et al.* (1998) compared the outcome of transplanting fresh versus frozen mouse ovaries. Two out of four females that received frozen–thawed grafts became pregnant compared to three out of four females that received unfrozen ovaries. Experience with frozen ovary-repositories is still too low to give recommendations about the number that should be frozen.

Two more problems associated with the transfer of ovaries are still unsolved. The first is the possibility of vertical transmission of viruses along with the ovary. Therefore the same precautions as for the freezing and revitalisation of sperm should be taken.

The second problem is the need for immune-incompetent recipients to prevent graft rejection in case no recipients of the appropriate genetic background are available.

References

Bleil, J.D. (1993). *Methods Enzymol.* **225**, 253–263.

Bouquet, M., Selva, J. and Auroux, M. (1993). *Biol. Reprod.* **49**, 764–799.

Bronson, R.A. and McLaren, A. (1970). *J. Reprod. Fertil.* **22**, 129–37.

Harp, R., Leibach, J., Black, J., Keldahl, C. and Karow, A. (1994). *Cryobiology* **31**, 336–343.

Hefler. L.A. and Gregg, A.R. (2002). *Fertil. Steril.* **77**, 147–151.

Hogan,B., Beddington, R., Costantini, F. and Lacy, E. (1994). In *Manipulating the Mouse Embryo. A Laboratory Manual*, 2nd edition. Cold Spring Harbor Laboratory Press, New York.

Jones, E.C. and Krohn, P.L. (1960). *J. Endocrinol.* **20**, 135–146.

Mobraaten, L.E. (1986). *J. In Vitro Fertil. Embryo Transfer* **3**, 28–32.

Mobraaten, L.E. and Bailey, D.W. (1987). *Cryobiology* **24**, 586.

Mukaida, T., Wada, S., Takahashi, K., Pedro, P.B., An, T.Z. and Kasai, M. (1998). *Hum. Reprod.* **13**, 2874–2879.

Nakagata, N. and Takeshima, T. (1993). *Exp. Anim.* **42**, 317–320.

Nowshari, M.A., Nayudu ,P.L. and Hodges, J.K. (1995). *Hum. Reprod.* **10**, 3237–3242.

Rafferty, Jr. K.A. (ed.) (1970). In *Methods in Experimental Embryology of the Mouse*. Johns Hopkins Press, Baltimore.

Rall, W.F. and Fahy, G.M. (1985). *Nature* **313**, 573–575.

Reetz, I.C., Wullenweber-Schmidt, M., Kraft, V. and Hedrich, H.J. (1988). *Lab. Anim. Sci.* **38**, 696–701.

Rugh, R. (ed.) (1968). In *The Mouse: Its Reproduction and Development*, p. 430. Burgess, Minneapolis.

Scharmann, W. and Wolff, D. (1980). *Lab. Anim. Sci.* **30**, 206–208.

Schmidt, P.M., Hansen, C.T. and Wildt, D.E. (1985). *Biol. Reprod.* **32**, 507–514.

Schroeder, A.C., Champlin AK., Mobraaten, L.E. and Eppig, JJ (1990). *J. Reprod. Fertil.* **89**, 43–50.

Shaw, J.M., Kola, I., MacFarlane, D.R. and Trounson, A.O. (1991). *J. Reprod. Fertil.* **91**, 9–18.

Songsasen, N. and Leibo, S.P. (1997). *Cryobiology* **35**, 240–254.

Sztein, J., Sweet, H., Farley, J. and Mobraaten, L. (1998). *Biol. Reprod.* **58**, 1071–1074.

Tada, N., Sato, M., Yamanoi ,J., Mizorogi, T., Kasai, K. and Ogawa, S. (1990). *J. Reprod . Fertil.* **89**, 511–516.

Theiler, K. (ed.) (1989). In *The House Mouse: Atlas of Embryonic Development*. Springer Verlag, Heidelberg.

Whittingham, D.G. (1971). *Nature* **233**, 125–126.

Whittingham, D.G. (1974). *Genetics* **78**, 395–402.

Whittingham, D.G., Lyon, M.F. and Glenister, P.H. (1977). *Genet. Res.* **29**, 171–181.

Wood, M.J., Whittingham, D.G. and Lee, S.H. (1992). *Biol. Reprod.* **46**, 1187–1195.

Health Monitoring

Michael Mähler
Biomedical Diagnostics – BioDoc, Hannover, Germany

Werner Nicklas
Central Animal Laboratories, German Cancer
Research Centre, Heidelberg, Germany

Introduction

In animal research validity and reproducibility of the data are critically influenced by the microbiological status of the experimental animals. Only animals of good microbiological quality will give any kind of guarantee of an experiment undisturbed by health hazards. The use of such animals reduces the number of animals needed and makes an important contribution to animal welfare.

Health monitoring aims to produce animals that meet preset requirements of microbiological quality and to aid in the maintenance of this quality during experiments. The microbiological quality of laboratory animals is a direct result of colony management practices. Health monitoring by itself does not influence the microbiological quality of animals but provides an after-the-fact assessment of the adequacy of those practices and is therefore an important management tool. It is a prerequisite for microbiological standardization of laboratory animals. Health monitoring is therefore an integrated part of any quality assurance system, e.g. good laboratory practice (GLP), the accreditation programme of the Association for Assessment and Accreditation of

Laboratory Animal Care International (AAALAC), or the International Standards Organization (ISO).

Significance of infectious agents

Several groups of microorganisms (viruses, bacteria, fungi, and parasites) are responsible for infections in mice. They may affect animals (or humans) in various ways (Table 27.1). Some are pathogenic and may induce clinical signs with variable morbidity or mortality (see Chapter 23 on Viral Infections). Concomitant morphological changes in organs can confound diagnostic pathology, particularly in toxicological studies or in the phenotypic evaluation of genetically altered mice. In mice, infectious pathology is most commonly found in the respiratory tract and the digestive system. Most infectious diseases are multifactorial, i.e., an infectious agent alone or in insufficient quantities is usually not able to elicit the disease, and support by other factors is

TABLE 27.1: Possible consequences of infections in laboratory mice

- Outbreak of clinical disease, eventually with lethal outcome
- Hazardous for personnel if the agent is zoonotic
- Morphological changes in organs
- Changed behaviour
- Lower growth rate
- Reduction of lifespan
- Reduction of breeding efficacy
- Increase in interindividual variation
- Impact on physiological parameters (immunology, haematology, enzymology, clinical chemistry)
- Modulation of oncogenesis (induction of tumours, reduction of the incidence of tumours, enhancement or suppression of chemical or viral carcinogenesis, altered growth rate of transplantable tumours)
- Contamination of samples and tissue specimens (transplantable tumours, cells, sera, monoclonal antibodies)

necessary. Silent infections are often activated by experimental procedures (immunosuppression, toxic substances, tumours), environmental influences (physical, social, nutritional stress) or emergence of a second or more infectious agent(s) (interaction of microorganisms). In addition, genetic factors are important determinants of the susceptibility of mice to disease. For example, infection with ectromelia virus results in high mortality in the CBA, C3H, DBA/2, and BALB/c strains, whereas other strains such as C57BL/6 and C57BL/10 are almost resistant to clinical disease. Similarly, various microorganisms (e.g. *Corynebacterium bovis, Staphylococcus aureus, Pneumocystis carinii*) that usually do not cause clinical signs in immunocompetent animals can be highly pathogenic for immunodeficient animals. The outcome of infection is also dependent on specific properties of the infectious agent. There are different strains of many viruses with different organotropism (e.g. enterotropic and polytropic strains of murine hepatitis virus (MHV)). This influences the disease rate and the mortality as well as the type and severity of pathological changes.

It has to be emphasized that most infections in laboratory mice are subclinical. The absence of clinical manifestations has therefore only limited diagnostic value. However, modifications of research results due to natural infections often occur in the absence of clinical disease. Such modifications may be devastating for experiments

because they often remain undetected. The types of interference of an agent with experimental results may be diverse. There are numerous examples of influences of microorganisms on the physiology of the laboratory animal and hence of the interference of inapparent infections with results of animal experiments. Infected mice may show altered behaviour, suppressed body weight, or reduced life expectancy, which may, for example, influence the tumour rate. Many microorganisms have the potential to induce functional suppression or activation of the immune system or both at the same time but on different parts of the immune system. Sometimes only T cells, or B cells, or specific subpopulations, are influenced. Therefore, most infections are detrimental for immunological research and infectious disease studies. Microorganisms, in particular viruses, present in animals can also lead to contamination of biological materials such as transplantable tumours, cells and sera. This may interfere with experiments conducted with such materials. For all these reasons, monitoring the health status of laboratory animals is of crucial importance.

As an example of interference with research, a detailed list of the potential influences of MHV, a frequently occurring mouse pathogen, is given in Table 27.2. More information about the considerable effects on research due to infectious agents can be found in various review articles (Bhatt *et al.*, 1986b; Hamm, 1986; Lussier, 1988; National Research Council, 1991; Baker, 1998; Nicklas *et al.*, 1999; Hansen, 2002).

Definitions of microbiological status

Laboratory animals can be arbitrarily classified according to a number of different microbiological qualities. Unfortunately, a universal reporting terminology for clear and consistent definition of pathogen status in mouse populations does not exist. Three categories (gnotobiotic, specified pathogen-free (SPF), and conventional) representing the extremes of microbiological status, have clear definitions that are generally accepted and understood. Additional terms such as 'barrier-reared', 'virus-antibody-free (VAF)', 'clean conventional', 'pathogen-free' or 'murine pathogen-free (MPF)' and 'health-monitored' are used; however, they rather describe concepts than the microbiological quality of animals.

TABLE 27.2: Examples of interference with research: MHV

Immunology

- Virus replication in macrophages, T cells, and B cells; dysfunctions of macrophages, T cells, and B cells
- Activation of natural killer cells and induction of interferon
- Immunosuppression and immunostimulation depending on the time of infection
- Reduced *in vitro* cytokine production by spleen cells
- Permanent decrease of skin graft rejection and T cell-dependent antibody responses after recovery from infection

Microbiology

- Reduced susceptibility to viral infections (Sendai virus, murine pneumonia virus)
- Enhanced resistance to *Salmonella typhimurium* infection
- Confusion about the origin of Tettnang virus isolates

Physiology

- Alteration of liver enzyme levels and protein synthesis
- Changes in peripheral blood cell counts
- Increased monocyte procoagulant activity
- Decrease of the incidence of diabetes in non-obese diabetic mice

Oncology

- Abnormal tumour passage intervals and tumour invasion pattern
- Rejection of human xenografts in *Foxn1^{nu}* mice
- Contamination of transplantable tumours

Gnotobiotic animals harbour a microflora and fauna that is entirely known (*gnotos* = known; *biota* = flora and fauna). They may be germfree (axenic) or associated with one or more nonpathogenic microorganism(s). Endogenous retroviruses, e.g. murine leukemia viruses that occur in all mice, are also present in axenic mice. Gnotobiotic animals are derived by hysterectomy or embryo transfer, reared, and maintained in an isolator by germfree techniques.

A frequently used term to describe the microbiological quality is *specified pathogen-free*. This term is often misused and requires explicit definition every time it is used. It means that the absence of individually listed microorganisms has been demonstrated by regular monitoring of a sufficient number of animals of appropriate ages and by accepted methods. Specified pathogen-free animals originate from gnotobiotic animals and subsequently lose their gnotobiotic status by contact with environmental and human microorganisms. Such animals are bred and housed under barrier conditions that prevent the introduction of unwanted microorganisms. Specified pathogen-free animals are morphologically and physiologically normal, well-suited for modelling the situation of a human population.

Conventional animals have an undefined or nominally defined microbiological status that includes common pathogens. They are generally housed without special precautions to prevent entry of infectious agents. However, this does not mean that they are necessarily infected with a number of pathogens. Furthermore, it is considered less prudent to monitor for the presence of unwanted agents in conventional animals than in their rederived, barrier-maintained counterparts.

Sources of infections

Laboratory mice can become infected with unwanted microorganisms by various routes and materials. Important sources of infections are: other laboratory animals, biological materials, personnel, vermin, and equipment. The risk of inadvertently introducing microorganisms into experimental units is generally higher than for breeding units, as animals and various experimental materials need to be introduced into experimental facilities. In addition, a number of personnel may have access to animals due to the requirements of the experiments.

Apart from constructive measures, an appropriate management system is essential for the prevention of infections as well as for their detection and control. It is important for the management of an animal facility to understand how microorganisms might be introduced or spread under the specific conditions given. Management of all animal facilities in an institution is best centralized. This means that all information dealing with the purchase of animals, use of experimental materials and equipment as well as the performance of animal experiments, flows through one office, thereby reducing the opportunity for failures of communication. A centralized management can best establish comprehensive monitoring programmes to evaluate important risk factors, such as animals and/or biological materials, before they are introduced into a facility. Contamination of animals can generally happen in two ways: (1) the introduction of microorganisms coming from outside and, (2) transmission of microorganisms within a colony. Both can be influenced by the management and the housing system.

Animals

The greatest risk of contamination arises from infected animals of the same or closely related species (e.g. rats). The importance of animals as sources of infections becomes obvious from a survey of biomedical research institutions funded in 1996 by the National Institutes of Health (Jacoby and Lindsey, 1998). In this survey, MHV, parvoviruses, and ecto- and endoparasites were reported in 10% to more than 30% of the 'SPF' mouse colonies. Among non-SPF mice, MHV was present in more than 70% of the reporting institutions, pinworms in about 70%, parvoviruses and ectoparasites in about 40%, and Theiler's murine encephalomyelitis virus in more than 30%.

Most animal facilities are multipurpose and must therefore house a variety of species coming from various sources. The risk of introducing pathogens via animals from external sources is relatively low when animals are available from very few sources of well-known microbiological status and when these animals have been protected from contamination during shipment. In many cases, direct transfer of such animals (without quarantine) into an experimental unit is necessary; however, spot checks should be performed from time to time to redefine the status upon arrival. While it may be acceptable to introduce such animals into experimental units, they should never be transferred into a breeding unit, especially if many different strains are co-maintained. In the latter case, new animals should only be introduced via embryo transfer or hysterectomy. Most breeders implement health monitoring programmes and supply their test results, indicating that many commercial breeding colonies are free of pathogens. This, however, may not always be possible; for example, in the case of mice of transgenic strains that cannot be obtained from commercial sources. Transgenic mice are usually produced and bred in experimental facilities, where less attention is paid to preventive hygienic measures as compared with breeding units, and available information on their health status is often insufficient. Frequently, they are exchanged between institutions, with a high risk of introducing pathogens from a range of animal facilities. As a consequence, they are much more frequently infected than mice coming from commercial breeders. In addition, only a few animals are usually sent and at short notice so that a proper evaluation of their health status upon arrival is not always possible. It must also be emphasized that a specific risk of transmitting microorganisms may arise from immunodeficient animals. Many virus infections (e.g. MHV, Sendai virus, murine pneumonia virus) are limited in immunocompetent mice and the virus may be eliminated completely. Immunodeficient mice, however, may shed infectious virus for longer periods of time or may be infected persistently. As a general rule, all animals coming from sources of unknown microbiological status should be regarded as infected unless their status has been defined. This is especially important when animals are introduced from other experimental colonies. These animals must be separately housed from others in quarantine areas. If possible, quarantine facilities should be physically isolated from the rest of the facility. If an isolated area is not available, this can best be achieved in isolators or, if proper handling is guaranteed, in individually ventilated cages (IVCs) or in filter top (microisolator) cages. Here, the animals should be maintained until health monitoring has been performed to define their status. If this health monitoring shows the absence of unwanted microorganisms, they may be transferred to the experimental facility. In cases of hazardous infection, rederivation, further separation, or other forms of risk management of animals must be considered.

Biological materials

Biological materials represent a high risk of contamination if they originate or have been propagated in animals (Petri, 1966; Collins and Parker, 1972; Smith et al., 1983; Bhatt et al., 1986a; Nicklas et al., 1988, 1993a; Nicklas, 1993; Dick et al., 1996; Lipman et al., 2000). In particular, tumours, viruses, or parasites that are serially passaged in animals often pick up infectious agents, and therefore a high percentage of these are contaminated. The infectious agents may survive for years or decades when contaminated samples are stored frozen or freeze-dried.

In principle, biological materials can contain the same, notably intracellular microorganisms, that are present in live animals. Most contaminations are viral, but also bacteria (e.g. *Mycoplasma pulmonis*, *Pasteurella pneumotropica*) and protozoans (e.g. *Encephalitozoon cuniculi*) have been found to be contaminants. The problem of viral contamination in biological materials becomes obvious from studies by Nicklas et al. (1993a). They found 70% of mouse tumours propagated *in vivo* positive for murine viruses. The most frequent contaminant was lactate dehydrogenase-elevating virus (LDV) followed by mammalian orthoreovirus-3, lymphocytic choriomeningitis virus, mice minute virus (MMV), and MHV. Many organisms disappear under *in vitro* conditions, so that the contamination rate of tumour cells after *in vitro* passages is lower. Pathogenic microorganisms can also be transmitted by other contaminated biological materials such as monoclonal antibodies (Nicklas *et al.*, 1988) and viruses (Smith *et al.*, 1983). Two recent outbreaks of ectromelia in the USA both resulted from use of contaminated serum samples (Dick *et al.*, 1996; Lipman *et al.*, 2000). In colonies of genetically engineered mice, embryonic stem (ES) cells, sperm, and embryos should be considered as potential sources of infection. ES cells especially are at increased risk of infection because they require growth factors that are usually supplemented by co-culture with primary mouse cells. Recent reports (Okumura *et al.*, 1996; Kyuwa, 1997) show that ES cells are susceptible to persistent infection with MHV and may produce virus although they appear to be functionally intact.

It is recommended that biological materials be considered as contaminated and that animal experiments be performed under conditions of strict containment (isolation), unless the biological materials have been tested and found free of contamination. Monitoring is usually done by the mouse antibody production (MAP) test. This test is based on the serum antibody response to microorganisms which is stimulated in pathogen- and antibody-free animals if the material injected is contaminated. Meanwhile, polymerase chain reaction (PCR) assays have been established to replace the MAP test as the preferred method for detecting viral contaminants in biological materials (Compton and Riley, 2001; Bootz *et al.*, 2003).

Personnel

Humans are unlikely to be an appropriate host where mouse pathogens can reside and replicate. However, the importance of humans as mechanical vectors should not be underestimated, and several microorganisms of human origin (e.g. *Staphylococcus aureus*) can cause infectious disease or research complications in mice, particularly in immunodeficient mice. Transmission from humans to animals (or vice versa) certainly cannot be avoided in barrier-maintained colonies, even by wearing gloves, surgical masks, and taking other precautions. It may only be avoided by establishing strict barriers as provided by isolator maintenance. Immunodeficient mice, at least mice used for breeding or long-term experiments, which are known to have an increased sensitivity to infection with microorganisms of human origin should therefore be housed in isolators, microisolators, or IVCs, respectively.

Little published information is available on the role of humans as mechanical vectors. There is no doubt that microorganisms can be transmitted by handling (La Regina *et al.*, 1992). Microorganisms can even be transported from pets to laboratory animals by human vectors (Tietjen, 1992). Such examples emphasize the need for proper hygienic measures and the importance of education and positive motivation of staff. It is an important task of the management of an animal facility to ensure that personnel coming into contact with animals have no contact with animals of a lower microbiological quality.

Vermin

Vermin is another potential source of infection. Flying insects do not present a serious problem because they can easily be removed from the incoming air by air filters or by insect-electrocuting devices. Crawling insects like cockroaches are more difficult to control. The most serious problem arises from wild rodents, which are frequently carriers of infections. Wild as well as escaped rodents are attracted by animal diets, bedding and waste. Usually, the design of a modern animal facility in combination with proper hygienic measures are able to control vermin and to reduce their importance to a minimum.

Materials and equipment

Husbandry supplies, research equipment, and other materials that have been in contact with infected animals may be contaminated and may act as potential vectors. However, many of them (cages, food, water, bedding, etc.) can easily be decontaminated by disinfection, sterilization, or other hygienic procedures so that the risk of contamination is relatively low. The effect of these measures should be monitored regularly. For example, autoclaving can be monitored by recording time, temperature, and pressure or by examining whether test organisms were killed by the treatment or not.

The interested reader is directed to other publications (Bhatt *et al.*, 1986b; National Research Council, 1991; Nicklas, 1993; Rehg and Toth, 1998; White *et al.*, 1998; Hansen, 2002) for more detailed information about possible routes of infection and bioexclusion practices.

Health monitoring programme

Health monitoring of laboratory animals can be defined as the science of evaluating representative sample groups from given units against a specific listing of infectious agents to define the microbiological status of the source colony. Health monitoring procedures in laboratory animal populations differ from procedures used in human medicine. Especially in populations of laboratory rodents, a single animal usually has only limited value. Health monitoring of laboratory rodents aims at detecting health problems or defining the pathogen status in a population rather than in an individual. Therefore, systematic laboratory investigations (health monitoring programmes) are necessary to determine the colony status and, most importantly, to prevent influences on experiments. Disease monitoring differs from routine health monitoring in that abnormalities are the subject of testing. This testing is not scheduled, and tests are directed towards identifying those pathogens most likely to cause the lesion or change in established patterns of reaction to experimental protocols.

While the need for health monitoring programmes is generally accepted, there is a great diversity of opinions about their design. Each institution selects its own list of pathogens, test procedures, sampling strategy, frequency of sampling, and reporting terminology, and the terms used vary greatly in precision and meaning (National Research Council, 1991; Lindsey, 1998; Jacoby and Homberger, 1999). An individual programme is usually tailored to the conditions it is to serve. The type of programme will differ between institutions or between different units of the same facility in its dependence on (a) research objectives, (b) physical conditions and the layout of the animal facility, (c) husbandry methods, (d) sources of animals, (e) staff quantity and qualification, (f) diagnostic laboratory support, and (g) finances. Professional guidance is necessary to shape the monitoring to meet specific institutional needs. The Federation of the European Laboratory Animal Science Associations (FELASA) recommends that each facility appoints a person with sufficient understanding of the principles of health monitoring, who is responsible for devising and maintaining the health monitoring programme (Nicklas et al., 2002). It should also be noted that health monitoring is not confined to laboratory reporting. In any animal facility, there should be a culture of communication between animal care staff, facility managers, veterinarians, and researchers so that observed abnormalities in breeding animals and experimental data can rapidly be evaluated and acted upon.

Detailed recommendations for the health monitoring of rodent and rabbit colonies in breeding and experimental units have been published by FELASA (Nicklas et al., 2002). This paper sets common standards for which agents to test for, which methods and samples to use, how many samples to test, how frequently this should be done, and how test results should be reported. The FELASA recommendations are periodically reviewed and amendments are published as necessary (www.felasa.org). Furthermore, FELASA has issued a set of guidelines for the accreditation of laboratory animal diagnostic laboratories (Homberger et al., 1999).

Selection of agents

A difficult problem which often leads to discussions between breeders and users of laboratory animals is the selection of relevant microorganisms for which animals should be monitored. Lists of infectious agents to be monitored in routine programmes have been published by various organizations (Kunstyr, 1988; National Research Council, 1991; Waggie et al., 1994; Nicklas et al., 2002) and can be used for guidance. Regular monitoring for all the agents mentioned is neither realistic nor necessary. The most important microorganisms are those that are indigenous and pose a threat to research or to the health of animals and humans and, in addition, those which can be eliminated. Therefore, oncogenic retroviruses are excluded as they integrate into the host genome, and thus cannot be eradicated by the presently available methods. Other microorganisms may be less important as they are unlikely to occur in good quality mice due to repeated rederivation procedures (e.g. murine polyoma virus, *Eperythrozoon coccoides*, *Leptospira* species (spp.), *Yersinia pseudotuberculosis*). Most cestodes are unlikely to be found because they require an intermediate host. On the other hand, too strict adherence to existing lists may also bear the risk that monitoring does not include important (recently detected, or not of general relevance) agents. Monitoring for additional agents is advised when they are associated with disease or when there is evidence that they affect physiological parameters or breeding performance. Especially in immunocompromised mice or in infectious experiments, monitoring for a comprehensive list of microorganisms is reasonable. Various microorganisms (e.g. *Corynebacterium bovis*, *Helicobacter hepaticus*, *Klebsiella pneumoniae*, *Pseudomonas aeruginosa*, *Staphylococcus aureus*, *Pneumocystis carinii*) that usually do not cause disease in immunocompetent animals may cause serious problems in immunodeficient animals. It is therefore necessary that immunodeficient mice are not only monitored for strong or weak pathogens, but also for opportunistic pathogens or commensals. In other cases, microorganisms of low clinical importance may cause disease or have a severe impact on research if animals are concomitantly infected with several agents (e.g. Sendai virus and *Pasteurella pneumotropica* (Jakab, 1981)).

Each institution should define a list of those microorganisms which are not acceptable in the facility or only in parts of it. This list is easiest to establish for viruses. A large amount of information is available on their pathogenic potential and on their ability to compromise the object of research. It is generally accepted that mouse colonies should be free from viruses, even if they are usually not pathogenic. For some viruses (e.g. mouse thymic virus, murine polyoma virus), the only question is whether or not monitoring is necessary because they have been eradicated from the vast majority of mouse colonies many years or even decades ago. Very few new mouse viruses have been detected during the last few years, e.g. mouse parvovirus 1 (MPV; Ball-Goodrich and Johnson, 1994), and it has to be expected that new viruses will be isolated, although only occasionally.

Less is known about the ability of most parasites to influence research results. They are regarded as an indicator of a low hygiene level and are therefore eradicated from mouse colonies. Some apathogenic flagellates such as trichomonads are occasionally detectable, even in mice from commercial breeders. At present, no evidence exists that they have any impact on the physiology of their host. They might, however, be an indicator of a leak in the barrier system or of direct or indirect contact with wild rodents.

The most complex problems exist for bacteria. In contrast to viruses, their importance for laboratory mice is usually estimated on the basis of their ability to induce pathological changes or clinical disease since very little is known about most bacterial species with regard to their potential to cause other effects on their hosts and on experiments. Insufficient information exists on the taxonomy and proper identification for various rodent-specific bacterial species such as *Pasteurella pneumotropica* or other members of the Pasteurellaceae (e.g. *Actinobacillus muris, Haemophilus influenzaemurium*). Lack of detailed information on the characteristics of these organisms, together with the presently unclear taxonomic situation, often leads to misidentification, and the lack of knowledge about species-specificity impedes their elimination. Therefore FELASA decided to recommend that rodents should be monitored for all Pasteurellaceae (Nicklas *et al.*, 2002). There is, however, evidence that some growth factor-dependent Pasteurellaceae found in rodents are closely related to *Haemophilus parainfluenzae* and

might therefore be transmitted by humans (Nicklas *et al.*, 1993b). It is unclear if these bacteria can be successfully eradicated from barrier units because exposure of barrier-produced animals to humans represents a permanent risk for reinfection. The same is true for several members of the Enterobacteriaceae (*Escherichia coli, Klebsiella* spp., *Proteus* spp.), *Pseudomonas aeruginosa*, and *Staphylococcus aureus* for which humans are the reservoir. Another problem arises from the fact that many bacteria are presently being reclassified resulting in changes of their names. For example, the mouse-specific organism known as '*Citrobacter freundii* 4280' has been reclassified as *Citrobacter rodentium* (Schauer *et al.*, 1995). Whole genera have been renamed, and additional bacterial species have been detected, e.g. *Helicobacter (H.) hepaticus* (Fox *et al.*, 1994), *H. bilis* (Fox *et al.*, 1995), and *H. typhlonicus* (Franklin *et al.*, 1999). Some of these fastidious organisms are not detected or not properly identified by each monitoring laboratory.

The above examples show that the whole spectrum of microorganisms as a concept is not a permanent list for all times; it rather represents a moving boundary in which old pathogens are eradicated and new pathogens are added. In practice, such lists of agents do not differ much between different facilities or commercial breeders, at least for viruses and parasites. The most important viruses, bacteria and parasites for which mice should be monitored according to FELASA (Nicklas *et al.*, 2002) are listed in Appendix (FELASA-approved health monitoring report).

Sampling

Proper sampling is necessary to detect an infection in a given population as early as possible. In general, the animals are the most crucial point in a monitoring programme. Animals coming from outside have to be checked to assess or exclude the risk of introducing unwanted microorganisms, and animals already within the unit need to be monitored to define their status and to obtain information on the presence or absence of infectious agents in the colony. In order to know the actual health status of a colony, it would be necessary to examine all the animals within the colony. Because this procedure is usually not possible, only a fraction of the animals are selected for examination and the results are used to describe the entire colony. Therefore, if one animal is infected with a certain microorganism, the entire colony is considered infected with that particular organism, and if the infection is not found in any of the animals sampled, the entire colony is considered free of that organism. For this practice, it is essential to define the microbiological entity (unit), i.e., a definition of the group of animals for which a sample is predictive, which is a very complicated matter. One isolator, one IVC (if handled properly), or a simple barrier-protected animal room used for breeding may be defined as one microbiological entity each, as the idea of the system is to prevent entry of unwanted agents and also because there is close contact between the animals within the system. In experimental facilities, however, it is often difficult to define the microbiological entity due to less stringent barrier measures. As a consequence, some microorganisms may easily spread from one room to another. Depending on the actual measures taken, and on the professional judgement by the person responsible for health monitoring, the total facility may be considered as multiple units or a single unit.

It is obvious that a sufficient number of animals has to be monitored to obtain relevant information on a given population. Guidelines for determining sample sizes have been provided by using various formulas. In essence, these formulas show that as the prevalence of infection decreases, the sample size required to detect infection with a high level of confidence increases (Dubin, 1991). The prevalence that a certain infection reaches depends on many factors, e.g. contact among the animals, susceptibility of the animals, and characteristics of the agent itself. Based on a recommendation by the Institute for Laboratory Animal Resources Committee on Long-Term Holding of Laboratory Rodents (ILAR, 1976), it has become common practice to monitor at least nine randomly sampled animals per (microbiological) unit, which results in a 95% chance of detecting an infection if at least 30% of a population (with 100 or more animals) is infected. The ILAR formula to determine the sample size for an estimated prevalence rate is given in Table 27.3. This formula is applicable only in populations of at least 100 animals, if the infection is randomly distributed in the unit, and if the animals are randomly sampled. However, because the distribution of an infection may be dependent on sex and age, attention should also be given to sampling animals of both sexes and of different ages. Younger animals often have a greater parasite burden, whereas older animals ($\geq$3 months) are more

---

TABLE 27.3: Calculation of the number of animals to be monitored

Assumptions
1. Both sexes are infected at the same rate
2. Population size > 100 animals
3. Random sampling
4. Random distribution of infection

Formula

$$\frac{\log (1 - \text{desired confidence level})}{\log (1 - \text{assumed prevalence rate})} = \text{no. of animals to be monitored}$$

Number of animals required to detect an infection

Assumed prevalence rate (%)	Sample sizes at different confidence levels		
	95%	99%	99.9%
10	29	44	66
20	14	21	31
30	9	13	20
40	6	10	14
50	5	7	10

Example: Nine animals should be monitored to be 95% confident of finding at least one positive animal if the suspected prevalence rate of an infection is $\geq$30%.

suitable for detecting viral infections. Similarly, there may be strain differences in susceptibility to infection and serological response to agents. Therefore, if more than one strain is present, all strains should be screened and each strain should be monitored at least once a year, where possible. Alternatively, the use of sentinel animals (see below) may be considered. It is also important to consider that the sample size needed to detect a specific agent in a colony is influenced by the sensitivity of the diagnostic method applied. For example, PCR might require a smaller sample size to detect or exclude infections with *Helicobacter* spp. than (less sensitive) bacterial culture (Mähler *et al.*, 1998). Independent from animals which are scheduled for monitoring, all clinically sick or dead animals should be examined; they are a valuable source of information about the hygienic status of the colony.

Sampling strategies and calculation of sample sizes have been discussed in more detail by Selwyn and Shek (1994) and Clifford (2001).

Sentinels

Random sampling for health monitoring is not a serious problem in breeding colonies, but it may be impossible in some experimental units and colonies of genetically modified or immunodeficient mice. It may also be inappropriate to carry out health monitoring in such colonies; for example, serological testing of immunodeficient mice may yield false-negative results because these mice do not always produce sufficient amounts of antibodies. Health monitoring may then be carried out on sentinel mice, which act as surveillance substitutes. However, the use of sentinels may not be covered by the ILAR formula (Table 27.3) for the sampling of animal colonies.

If sentinels are not bred within the colony that is being monitored, they must be obtained from a breeding colony of known microbiological status, i.e., they must be free from all agents to be monitored and free from antibodies to these agents. The sentinel animals must be housed for a sufficiently long time in the population that is to be monitored to develop detectable antibody titres (for serology) or parasitic stages. It is common to house sentinels within a population for at least 4 to 6 weeks prior to testing. Longer periods are even better. Strain may also be a consideration in some circumstances due to the well-known differences in susceptibility to various microorganisms that exist even among immunocompetent strains of mice. Sentinels should be kept in a manner that maximizes their exposure to potential infections. Provided that the animals in the general population are

in open cages, exposure of sentinels to possible infectious agents might be enhanced by putting them into open cages throughout the unit in locations where possible exposure to infectious agents is known or thought to be maximal (usually on the bottom shelves). The transmission of infectious agents may be further enhanced by exposing the sentinel animals to soiled bedding, water and feed taken from the cages of the experimental animals, and by exposing sentinel animals directly to experimental animals by placing them in the same cage. It should be noted, however, that transfer of soiled bedding, the customary method for sentinel exposure, may be inefficient in transmitting certain agents (e.g. Sendai virus (Artwohl *et al.*, 1994), *Pasteurella pneumotropica* (Scharmann and Heller, 2001)).

During the last decade, additional housing systems such as filter cabinets, microisolator cages, and IVCs have emerged. They offer the advantage of separating small populations from each other and are frequently used for housing immunocompromised or infected mice. If handled properly, they very efficiently prevent transmission of infectious agents and must therefore be considered as self-contained microbiological entities. Health monitoring under such housing conditions as well as monitoring isolator-housed animals can only be conducted by the use of sentinel animals. In many cases, less than the recommended number of animals is available, due to limited spaces which is acceptable if sentinels are properly housed. In the case of isolators, a realistic number of sentinel animals is housed in one or several cages (depending on the isolator size) on soiled bedding taken from as many cages as possible. In most cases, only 3 to 5 animals per isolator will be available for monitoring. If animals are housed in microisolators or IVCs, sentinels must be housed in filter-topped cages like other animals. When cages are changed in changing cabinets, soiled bedding from several cages is transferred into a separate cage which is used to house sentinels. Weekly changes of donor cages may then give a representative insight into the microbiological status of the whole population. Other examples of methods that may be considered for monitoring are the use of contact sentinels and the testing of exhaust filters or cage surfaces using PCR. Agents are very likely to be transmitted to cage contact sentinels which, therefore, are very useful if small populations are to be monitored (e.g. few breeding pairs of genetically modified mice during quarantine).

Frequency of monitoring

Health monitoring must be performed on a regular basis to detect unwanted microorganisms in good time. The frequency of monitoring mainly depends on the specific purpose of the population in question, the importance of a pathogen or other contaminant to the use of the population, the risk of introducing agents, and economic considerations. It is stated in the FELASA recommendations (Nicklas *et al.*, 2002) that monitoring for the most relevant infections be conducted at least quarterly. Most commercial breeders of laboratory mice monitor more frequently (every 4 to 6 weeks). In most multipurpose experimental units where animals are regularly bought and introduced, more frequent monitoring is also preferable as this results in earlier detection of an infection. As a general rule, monitoring small numbers of animals from each unit at high frequency is more reliable in picking up infections than monitoring larger numbers taken at less frequent intervals (e.g. three to five animals every 4 to 6 weeks instead of 10 animals every 3 months) (Kunstyr, 1992). Under practical conditions, not every animal may be monitored for all microorganisms. Depending on the factors already mentioned, the frequency of testing may be different for different agents. Monitoring for frequently occurring organisms or for agents that have a serious impact on research should be performed more frequently (e.g. monthly), whereas testing for unusual organisms like mouse cytomegalovirus 1 can be done less frequently (e.g. annually). If germfree or gnotobiotic mice are housed in isolators, monitoring for bacteria (environmental organisms) should be conducted more often than monitoring for viruses and parasites due to the higher risk of the former being introduced in the event of a barrier breakdown.

Test methods

Health monitoring of laboratory mice is accomplished using a variety of methodologies. Traditional methods used to detect infectious agents or disease processes include pathology, parasitological examinations, bacterial cultures and serological tests. Molecular methods have become increasingly important over the last few years. The technique used to identify a specific microorganism depends on a number of factors, including the type of organism, the fastidiousness of the organism, the immunological status of the host, the ecology of the organism, and the particular tests that have been developed to detect and identify the organism.

Testing of mice usually starts with routine necropsy and blood sampling for serology, followed by microscopic examination for parasites and sampling of organs for bacteriology, pathology, and in rare cases, virological examinations. All major organs should be inspected macroscopically to decide whether further investigation by histopathology is needed, though histopathological examination is seldom the method of choice for routine monitoring and is more efficient as a tool for disease diagnosis and validation of the impact of a certain microorganism.

Microscopic methods such as stereomicroscopy are commonly used for monitoring for ectoparasites. Endoparasites (enteric helminths, protozoans) may be diagnosed by direct microscopy after flotation of intestinal contents, on smears from intestinal contents, and on adhesive tapes used for sampling around the anus.

Culture is the diagnostic method of choice for most bacteria and fungi because they can be readily grown on artificial media. Samples for bacterial culture are typically taken from the respiratory tract (nasopharynx, trachea, lungs), the intestinal tract (caecal contents or faeces), and the urogenital tract (vagina or prepuce, kidney). In the case of pathological changes, additional organs (liver, spleen, mammary gland, lymph nodes, conjunctiva, skin, etc.) should be monitored. Culture of bacterial agents is generally accomplished by inoculating non-selective and selective media with the specimen and incubating the media in appropriate conditions to allow growth. Morphologically distinct colonies are then evaluated by Gram stain and subjected to a panel of biochemical tests to determine the specific genus and species of each colony type. Commercial identification kits for human and veterinary pathogenic bacteria are sometimes not suitable to correctly identify bacteria from mice, e.g. Pasteurellaceae and *Citrobacter rodentium*. Some bacterial species are difficult to grow *in vitro* because they require special media, environmental conditions, or the presence of mammalian cells. For example, *Helicobacter* spp. require a microaerophilic atmosphere and *Clostridium piliforme* requires cultivation in certain cell cultures. For these agents, alternative diagnostic approaches such as PCR or serological assays are often used.

Serology is the most widely used methodology for monitoring infections in mice and relies on the detection of specific antibodies, usually immunoglobulin G (IgG), produced during an infection. It is applied for all viruses except LDV, which is monitored by testing for lactate dehydrogenase activity or PCR. Also, a few bacteria such as *Clostridium piliforme* and *Mycoplasma* spp. are screened by serological assays. Suitable methods include the enzyme-linked immunosorbent assay (ELISA), the indirect immunofluorescence assay (IFA), and the haemagglutination inhibition (HI) test. Serological methods must be selected properly as they may differ in sensitivity and specificity, and unexpected serological results should always be confirmed by a second method, by a second laboratory, or by monitoring additional animals. In general, ELISA and IFA are more sensitive than HI and so should be used as primary tests. The specificity of the tests is primarily determined by the antigen chosen and the methods used for antigen preparation. Immunofluorescence assay, for example, measures cross-reacting antibodies to various parvoviruses, whereas HI is specific for the virus (MMV, MPV). While extremely useful, serological methods have certain limitations. Serology represents historical evidence of prior exposure to a microorganism and is not necessarily indicative of the continuing presence of the organism. Serology is also inadequate for detecting exposure to microorganisms in immunodeficient mice and in infant mice with immature immune systems. Serological methods for detection of antibodies in the IgG class will not detect exposure of an animal to a pathogen early in the course of infection. It takes some time (generally a number of days, sometimes much longer) after the initial contact with an antigen, before the amount of antibodies in the blood serum will exceed the detection limit of a test. Finally, some pathogenic microorganisms do not reliably induce an antibody response, particularly if cellular immunity is required to clear the organism (e.g. *Pneumocystis carinii*). In the case of acute infection or where antibodies are not developed, a direct assay such as PCR may be the best option.

Molecular methods have become an integral part of most health monitoring programmes, supplementing traditional techniques, and are available for most pathogenic organisms of interest. They are aimed at detecting specific nucleic acid (DNA or RNA) sequences from infectious agents. The most common molecular methods used to detect infectious agents utilize PCR methodology. Polymerase chain reaction assays represent an attractive choice for the detection of microorganisms which cannot be reliably detected by traditional methods and for monitoring biological materials. Major advantages of PCR are its high sensitivity, which allows detection of minute levels of infectious agents, and its high specificity, which allows differentiation of closely related organisms. Another advantage of PCR is that it is rapid. Disadvantages of PCR-based testing are the expense and the potential for false-positive or false-negative results. Diagnosis of infectious agents by PCR requires careful attention to the selection of animals and tissues for evaluation since the organism must be present at the time of testing and in the specimen evaluated. Improper selection of animals or specimens can result in false-negative results.

False-negative reactions may also be the result of inadequate purification of the nucleic acids from the specimen resulting in samples containing substances that inhibit polymerase.

Monitoring for microorganisms is usually done by commercial laboratories, and is thus determined by their capabilities (some of the larger research institutes and commercial breeders have dedicated diagnostic laboratories). It is important that all investigations be performed in laboratories with sufficient expertise in microbiology or pathology of laboratory mice. Serological and molecular tests also require technical competence to ensure sufficient standardization of tests (including controls) and accurate interpretation of results. Accreditation increases the trustworthiness of a laboratory. Federation of the European Laboratory Animal Science Associations advocates accreditation of laboratory animal diagnostic laboratories according to ISO Guide 25, in which special emphasis is placed on competency of the staff, validation of in-house test methods, and participation in interlaboratory testing (Homberger *et al.*, 1999).

For more information about the diagnostic detection and identification of microorganisms in laboratory mice, the reader is referred to special publications (Kunstyr, 1992; Owen, 1992; Weisbroth *et al.*, 1998; Compton and Riley, 2001; Feldman, 2001; Nicklas *et al.*, 2002) and textbooks (e.g. Flynn, 1973; Hansen, 1999).

Health report

A detailed health report should be required prior to the introduction of animals from outside sources in order to better assess the hygienic risk from them. These reports are usually made available by the vendor for all purchasers of the animals, e.g. by publishing them on the Internet. Similarly, health monitoring data should be made available to the researchers in experimental facilities. The data are part of the experimental work and should therefore be evaluated for their influence on the results of experiments, and included in scientific reports and publications as part of the animal specification.

A health report must contain sufficient data to provide reliable information on the quality of a population. Table 27.4 provides a checklist of the basic information that should be included in a health report. Usually, each animal facility or breeder has its own style of report sheets which are sometimes difficult to read and to interpret. In order to easily compare monitoring reports from different breeders and users, FELASA has developed report forms for all common species of laboratory rodents and rabbits (Nicklas *et al.*, 2002). The form

TABLE 27.4: Information that should be included in a health report

- Date of issue
- Exact location (designation) of the unit
- Housing condition (non-barrier, barrier, IVC, isolator)
- Species and strains present within the unit
- Names of agents for which monitoring is recommended
- Frequency of monitoring
- Date of latest monitoring
- Results of latest monitoring and during the last 18 months (expressed as number of positive animals/number of animals examined)
- Name(s) of laboratory/ies involved in monitoring
- Methods used (clinical signs, gross pathology, microscopy, culture, serology, PCR)
- Treatment, vaccination, etc.

Appendix : Health report form recommended by FELASA (Nicklas *et al.*, 2002)

Health Monitoring in Accordance with FELASA Recommendations

Date of issue:

Location: Housing: (Barrier/Non-Barrier/IVC/Isolator)

Species: Mouse   Strain:

Species and strains present within the unit:

	Test frequency	Lastest test date	Latest results	Testing laboratory	Test method	Historical results (≤18 months)
Viruses						
Murine hepatitis virus	3 months					
Murine rotavirus (EDIM)	3 months					
Parvoviruses						
Mice minute virus	3 months					
Mouse parvovirus 1	3 months					
Murine pneumonia virus	3 months					
Sendai virus	3 months					
Theiler's murine encephalomyelitis virus	3 months					
Ectromelia virus	Annually					
Lymphocytic choriomeningitis virus	Annually					
Murine adenovirus type 1 (FL)	Annually					
Murine adenovirus type 2 (K87)	Annually					
Mouse cytomegalovirus 1	Annually					
3	Annually					
Additional organisms tested:						
Bacteria, mycoplasma and fungi						
Citrobacter rodentium	3 months					
Clostridium piliforme (Tyzzer's disease)	3 months					

Appendix : (*Continued*)

Corynebacterium kutscheri	3 months
Mycoplasma spp.	3 months
Pasteurellaceae	3 months
Salmonella spp.	3 months
Streptococci β-haemolytic (not group D)	3 months
Streptococcus pneumoniae	3 months
Helicobacter spp.	Annually
Streptobacillus moniliformis	Annually
Additional organisms tested:	
Parasites	
Ectoparasites:	3 months
Species designation	
Endoparasites:	3 months
Species designation	
Pathological lesions observed	3 months

Data are expressed as number positive/number tested

Positive findings in other species in the same unit:

Abbreviations used in this report:

ELISA = enzyme linked immunosorbent assay, MICR = microscopy, IFA = immunofluorescence assay, CULT = culture, PATH = gross pathology, PCR = polymerase chain reaction, HIST = histopathology, NT = not tested

recommended for reporting health monitoring data on mice is shown in Appendix.

References

Artwohl, J.E., Cera, L.M., Wright, M.F., Medina, L.V. and Kim, L.J. (1994). *Lab. Anim. Sci.* **44**, 73–75.

Baker, D.G. (1998). *Clin. Microbiol. Rev.* **11**, 231–266.

Ball-Goodrich, L.J. and Johnson, E. (1994). *J. Virol.* **68**, 6467–6468.

Bhatt, P.N., Jacoby R.O. and Barthold, S.W. (1986a). *Lab. Anim. Sci.* **36**, 136–139.

Bhatt, P.N., Jacoby R.O., Morse, H.C. and New, A. (eds) (1986b). *Viral and Mycoplasma Infections of Laboratory Rodents: Effects on Biomedical Research*. Academic Press, New York.

Bootz, F., Sieber, I., Popovic, D., Tischhauser, M. and Homberger, F.R. (2003). *Lab. Anim.* **37**, 341–351.

Clifford, C.B. (2001). *Lab. Anim. (NY)* **30**(10), 26–31.

Collins, M.J. and Parker, J.C. (1972). *J. Natl Cancer Inst.* **49**, 1139–1143.

Compton, S.R. and Riley, L.K. (2001). *Comp. Med.* **51**, 113–119.

Dick, E.J., Kittell, C.L., Meyer, H., Farrar, P.L., Ropp, S.L., Esposito, J.J., Buller, R.M.L., Neubauer, H., Kang, Y.H. and McKee, A.E. (1996). *Lab. Anim. Sci.* **46**, 602–611.

Dubin, S. (1991). *Lab. Anim. (NY)* **20**(3), 29–33.

Feldman, S.H. (2001). *Lab. Anim. (NY)* **30**(10), 34–42.

Flynn, R.J. (1973). *Parasites of Laboratory Animals*. Iowa State University Press, Ames.

Fox, J.G., Dewhirst, F.E., Tully, J.G., Paster, B.J., Yan, L., Taylor, N.S., Collins, M.J., Gorelick, P.L. and Ward, J.M. (1994). *J. Clin. Microbiol.* **32**, 1238–1245.

Fox, J.G., Yan, L.L., Dewhirst, F.E., Paster, B.J., Shames, B., Murphy, J.C., Hayward, A., Belcher, J.C. and Mendes, E.N. (1995). *J. Clin. Microbiol.* **33**, 445–454.

Franklin, C.L., Riley, L.K., Livingston, R.S., Beckwith, C.S., Hook, R.R., Besch-Williford, C.L., Hunziker, R., and Gorelick, P.L. (1999). *Lab. Anim. Sci.* **49**, 496–505.

Hamm, T.E. (ed.) (1986). *Complications of Viral and Mycoplasmal Infections in Rodents to Toxicology Research and Testing*. Hemisphere Publ. Co., Washington, DC.

Hansen, A.K. (1999). *Handbook of Laboratory Animal Bacteriology*. CRC Press, Boca Raton.

Hansen, A.K. (2002). In Essential Principles and practices *"Handbook of Laboratory Animal Science, Second Edition."* Vol. 1, (eds J. Hau and G.L. Van Hoosier), pp. 233–279. CRC Press, Boca Raton.

Homberger, F.R., Boot, R., Feinstein, R., Hansen, A.K. and van der Logt, J. (1999). *Lab. Anim.* **33**(Suppl. 1), S1:19-S1:38.

ILAR Committee on Long-Term Holding of Laboratory Rodents (1976). *ILAR News* **19**(4), L1-L25.

Jacoby, R.O. and Homberger, F.R. (1999). *Lab. Anim. Sci.* **49**, 230.

Jacoby, R.O. and Lindsey, J.R. (1998). *ILAR J.* **39**, 266–271.

Jakab, G.J. (1981). *Lab. Anim. Sci.* **31**, 170–177.

Kunstyr, I. (ed.) (1988). *List of Pathogens for Specification in SPF Laboratory Animals.* Society for Laboratory Animal Science Publication No. 2, Biberach.

Kunstyr, I. (ed.) (1992). *Diagnostic Microbiology for Laboratory Animals*, GV-SOLAS Vol. 11. Gustav Fischer Verlag, Stuttgart.

Kyuwa, S. (1997). *Exp. Anim.* **46**, 311–313.

La Regina, M., Woods, L., Klender, P., Gaertner D.J. and Paturzo, F.X. (1992). *Lab. Anim. Sci.* **42**, 344–346.

Lindsey, J.R. (1998). *Lab. Anim. Sci.* **48**, 557–558.

Lipman, N.S., Perkins, S., Nguyen, H., Pfeffer, M. and Meyer, H. (2000). *Comp. Med.* **50**, 426–435.

Lussier, G. (1988). *Vet. Res. Contrib.* **12**, 199–217.

Mähler, M., Bedigian, H.G., Burgett, B.L., Bates R.J., Hogan, M.E., and Sundberg, J.P. (1998). *Lab. Anim. Sci.* **48**, 85–91.

National Research Council, Committee on Infectious Diseases of Mice and Rats (1991). *Infectious Diseases of Mice and Rats.* National Academy Press, Washington, DC.

Nicklas, W. (1993). *Scand. J. Lab. Anim. Sci.* **20**, 53–60.

Nicklas, W., Giese, M., Zawatzky, R., Kirchner, H. and Eaton, P. (1988). *Lab. Anim. Sci.* **38**, 152–154.

Nicklas, W., Kraft, V. and Meyer, B. (1993a). *Lab. Anim. Sci.* **43**, 296–300.

Nicklas, W., Staut, M. and Benner, A. (1993b). *Zentralbl. Bakt.* **279**, 114–124.

Nicklas, W., Homberger, F.R., Illgen-Wilcke, B., Jacobi, K., Kraft, V., Kunstyr, I., Mähler, M., Meyer, H. and Pohlmeyer-Esch, G. (1999). *Lab. Anim.* **33**(Suppl. 1), S1:39-S1:87.

Nicklas, W., Baneux, P., Boot, R., Decelle, T., Deeny, A.A., Fumanelli, M. and Illgen-Wilcke, B. (2002). *Lab. Anim.* **36**, 20–42.

Okumura, A., Machii, K., Azuma, S., Toyoda, Y. and Kyuwa, S. (1996). *J. Virol.* **70**, 4146–4149.

Owen, D.G. (1992). *Parasites of Laboratory Animals. Laboratory Animal Handbooks No. 12.* Laboratory Animals Ltd, London.

Petri, M. (1966). *Acta Pathol. Microbiol. Scand.* **66**, 13–30.

Rehg, J.E. and Toth, L.A. (1998). *Lab. Anim. Sci.* **48**, 438–447.

Scharmann, W. and Heller, A. (2001). *Lab. Anim.* **35**, 163–166.

Schauer, D.B., Zabel, B.A., Pedraza, I.F., O'Hara, C.M., Steigerwalt, A.G. and Brenner, D.J. (1995). *J. Clin. Microbiol.* **33**, 2064–2068.

Selwyn, M.R. and Shek W.R. (1994). *Contemp. Top. Lab. Anim. Sci.* 33, 55–60.

Smith, A.L., Casals, J. and Main, A.J. (1983). *Am. J. Trop. Med. Hyg.* **32**, 1172–1176.

Tietjen, R.M. (1992). *Lab. Anim. Sci.* **42**, 422 (abstract).

Waggie, K., Kagiyama, N., Allen, A.M. and Nomura, T. (eds) (1994). *Manual of Microbiologic Monitoring of Laboratory Animals*, 2nd edn. U.S. Department of Human Health and Human Services, NIH Publication No. 94–2498. Bethesda, MD.

Weisbroth, S.H., Peters, R., Riley, L.K. and Shek, W. (1998). *ILAR J.* **39**, 272–290.

White, W.J., Anderson, L.C., Geistfeld, J. and Martin, D.G. (1998). *ILAR J.* **39**, 291–311.

Nutrition of Laboratory Mice

Merel Ritskes-Hoitinga

Laboratory Animal Science and Comparative
Medicine, Biomedical Laboratory, University of
Southern Denmark, Odense, Denmark

Introduction

Laboratory mice and rats have always been a large percentage of the total number of animals used for biomedical research purposes. This percentage usually varies around 80–90% of the total number of animals used. Due to this, these species have been well-characterised in many ways. The use of the laboratory mouse (*Mus musculus*) has increased even more dramatically over the last decades, due to the possibility of studying gene function *in vivo* by the use of genetic modification techniques which have resulted in many newly established mouse strains.

The nutrition of the laboratory mouse (and rat) has also been well-studied and -defined in comparison with other species. This chapter will concentrate on important aspects when feeding laboratory mice. One needs to be aware of how nutrition and feeding as an environmental factor can interact with experimental results and animal welfare when nutrition is not the main focus of study. Moreover, when the mouse is used as an animal model for human nutritional conditions, specific experimental conditions need to be taken care of. This is important in order to obtain reliable experimental results and optimal welfare of the animals simultaneously.

Nutritional requirements

Energy

Under *ad libitum* feeding conditions animals usually eat an amount of food that is determined by the animals' energy requirements (Beynen and Coates, 2001). In Table 28.1 the energy need is given according to the stage of life the mouse (or other animal species) is in. During lactation an animal has a much higher energy need, than during the maintenance phase where the animal has stopped growing and is not pregnant or lactating. Metabolic kilos ($kg^{0.75}$) are used in order to be able to compare species of different sizes. One cannot compare species on the basis of kilograms, as depending on

TABLE 28.1: Food intake under *ad libitum* conditions is based on the energy requirement, which is related to the stage of life	
Stage of life	**Energy need (MJ/kg$^{0.75}$)**
Growth	1.20
Maintenance	0.45
Pregnancy	0.60
Lactation	1.30
Source: Beynen and Coates, 2001.	

the size of the animal, a different metabolic rate per kg body weight is present. By using metabolic kilos one compensates for these differences, so that a reliable comparison between species can be made.

That part of the energy in the diet capable of transformation by the body is called metabolisable energy (Keenan *et al.*, 2000). Usually the metabolisable energy content (MEC) of diets can be obtained from the diet manufacturers. In case it is not provided, it can also be estimated. For that purpose one uses the levels of energy producing substances in the diets, which are fat, carbohydrates and protein. Fats have an MEC of about 37 kJ/g, and protein and carbohydrates about 17 kJ/g. As, for example, different types of carbohydrates have different MECs, metabolic studies are required in case one needs to know the exact MEC for specific studies. The contribution made by fibre to dietary energy content is usually negligible (Beynen and Coates, 2001).

The need for energy not only depends on the stage of life of the mice, it also depends on other factors such as the environmental temperature and the amount of activity. As laboratory conditions are usually standardised to a certain room temperature and cage size, which prevents an unusually high activity, these general formulae for energy requirements according to the stage of life, can be used reliably in the current laboratory setting. A mouse showing stereotyped behaviour like circling, will have a higher energy need than the cage mate that does not exert this behaviour. This will increase variation in results. In order to prevent stereotyped behaviour, environmental conditions must meet the animals essential needs (Council of Europe Resolution, 1997).

Food intake according to energy need will only hold when the rest of the diet lives up to the minimal needs. In case there is a shortage of an essential nutrient, the animal may become (sub)clinically affected, which may lead to a reduced food intake. A reduced food intake may also occur in the case of a test substance with a bad taste being added to the diet. This can interfere with the reliability of the experimental results. On the other hand, where diets have a very good taste, mice are expected to ingest more than their energy need, leading to obesity.

Nutrient requirements

The National Research Council (NRC, 1995) has given scientific documentation on nutritional requirements for mice and other species. There is a lot of data published on these requirements, and the NRC establishes committees to review these data as new scientific data becomes available. On the basis of this review process, these committees then publish guidelines on the estimated nutrient requirements (NRC, 1978, 1995). In Table 28.2 the estimated nutrient requirements for the mouse are shown (NRC, 1995). The nutrient amounts per kg diet as well as the nutrient amounts per 100 kJ of diet are shown. As mice eat according to energy need under *ad libitum* conditions, it is advantageous to present the diets in amounts per kJ. This makes it easier to judge whether experimental diet compositions live up to all the essential nutrient needs under ad libitum conditions. The guidelines of the NRC (1995) recommend nutrient allowances, that are greater than the minimum requirements, as they are often based on the criterion for obtaining maximum growth (Beynen and Coates, 2001). This is not necessarily the best situation for obtaining optimal (longterm) health. However, as these recommendations are the best scientifically documented requirements of essential nutrients available at the moment, it is advisable to use these, until new scientific proof becomes available.

The recommendations do not take into account that there can be differences in minimum requirements between different strains (Beynen and Coates 2001). Interactions between the nutritional requirements and genetic background of mouse strains can occur. Dystrophic cardiac calcification is a post mortem finding in various strains, which coincides with calcifications in the tongue, lungs and diaphragms (Van den Broek *et al.*, 1997). The inbred strains DBA/2Ola and C3H/Ola are susceptible for the development of soft tissue calcifications, whereas C57BL/6Ola and BALB/cOla are resistant towards the disorder (Table 28.3; Van den Broek *et al.*, 1997). The region on chromosome 7 containing the gene *Hrc* (coding for the histidine-rich calcium binding protein in the sarcoplasmatic reticulum) is likely to be associated with soft tissue calcifications in DBA/2 mice (Van den Broek *et al.*, 1998). Nutritional measurements can be taken to prevent excessive calcifications. The diet of susceptible

TABLE 28.2: Estimated nutrient requirements of mice (NRC, 1995)		
Nutrient	**Amount, per kg diet**	**Amount, per 100 kJ**
Metabolisable Energy	16,500 kJ	
Lipid	50.0 g	0.30 g
Linoleic acid	6.8 g	0.04 g
Protein (growth)	180.0 g	1.09 g
Amino Acids		
Arginine	3.0 g	18.18 mg
Histidine	2.0 g	12.12 mg
Isoleucine	4.0 g	24.24 mg
Leucine	7.0 g	42.42 mg
Valine	5.0 g	30.30 mg
Threonine	4.0 g	24.24 mg
Lysine	4.0 g	24.24 mg
Methionine	5.0 g	30.30 mg
Phenylalanine	7.6 g	46.06 mg
Tryptophan	1.0 g	6.06 mg
Minerals		
Calcium	5.0 g	0.03 g
Chloride	0.5 g	3.03 mg
Magnesium	0.5 g	3.03 mg
Phosphorus	3.0 g	18.18 mg
Potassium	2.0 g	12.12 mg
Sodium	0.5 g	3.03 mg
Copper	6.0 mg	36.36 µg
Iron	35.0 mg	0.21 mg
Manganese	10.0 mg	60.61 µg
Zinc	10.0 mg	60.61 µg
Iodine	150.0 µg	0.91 µg
Molybdenum	150.0 µg	0.91 µg
Selenium	150.0 µg	0.91 µg
Vitamins		
A (retinol)	0.72 mg	4.36 µg
D (cholecalciferol)	0.03 mg	0.15 µg
E (RRR-α-tocopherol)	22.0 mg	0.13 mg
K (phylloquinone)	1.0 mg	6.06 µg
Biotin (d-biotin)	0.2 mg	1.21 µg
Choline (bitartrate)	2.0 g	12.12 mg
Folic acid	0.5 mg	3.03 µg
Niacin (nicotinic acid)	15.0 mg	90.91 µg
Panthothenate	16.0 mg	96.97 µg
Riboflavin	7.0 mg	42.42 µg
Thiamin	5.0 mg	30.30 µg
B6 (pyridoxine-HCl)	8.0 mg	48.48 µg
B12	10.0 µg	0.06 µg

TABLE 28.3: Cardiac calcification in four mouse strains				
Strain/ parameter	**DBA/2**	**BALB/c**	**C3H**	**C57BL/6**
Heart histology				
Incidence[a]	7/7	0/7	7/7	0/7
Score[b]	0.3–2.6	0.0–0.0	0.1–2.8	0.0–0.0
Mineral content (µmol/g dry wt)				
Calcium[c]	43.6 ± 16.7	17.2 ± 1.8	42.9 ± 33.6	17.1 ± 2.2

[a]The number of mice that were scored positive in histological sections for the presence of cardiac calcification out of a group of seven mice.
[b]Range of average scores per mouse.
[c]Mean ± standard deviation.
Source: Van den Broek *et al.*, 1997.

mice, especially at a young age, should contain adequate amounts of magnesium and fluoride (Van den Broek *et al.*, 2000), whereas excessive phosphorus and vitamin D intake should be avoided (Van den Broek, 1998).

Genetic modification may also alter nutrient requirements. In order to make sure that one lives up to the requirements of a particular genetically modified strains, it may be necessary to add special nutrients to the diets.

The recommendations do not necessarily hold for germ-free mice either. Vitamins K and B12 for examples, are synthesized by the gut flora of conventional mice and will be sufficiently ingested as a result of coprophagy. Grit floors do not prevent **coprophagy**, as the mice can still eat the faeces directly from the anus. For germ-free and SPF (Specified Pathogen Free) mice it is advised to include higher vitamin B and K levels in the diets, as the microflora of germ-free or SPF animals may not contain all vitamin-synthesising organisms (Beynen and Coates, 2001).

Toxic levels

In order to use scientifically sound diets, they must live up to the minimum requirements of essential nutrients, unless marginal levels of specific essential nutrients are

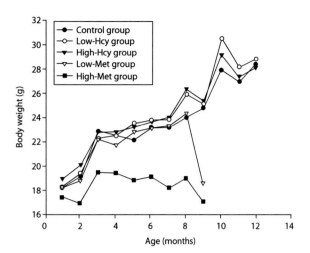

Figure 28.1 Body weight development in apoE deficient mice fed five different experimental diets (Zhou et al., 2001). A 'high' (4.4%, High-Met group) and 'low' (2.2%, Low-Met group) dietary Methionine level fed to apoE deficient mice caused weight loss and death before the scheduled end of the study. Reproduced with permission from Zhou, J., Moeller, J., Danielsen, C.C., et al. (2001). *Arterioscler. Thromb. Vasc. Biol.* **21**, 1470–1476.

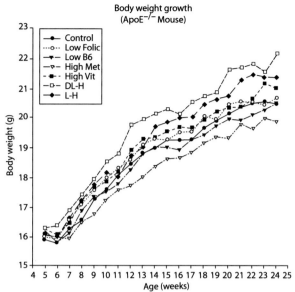

Figure 28.2 The body weight development af ApoE deficient mice on seven different purified diets. The 'high' methionine level of 1.4% did not appear to cause adverse effects during the course of the study (Zhou, 2001).

the topic of research. Even in this last situation, it is not advisable to leave out essential nutrients from the diet completely, as this will lead to sickness and premature death of the mice.

The same situation occurs when investigating the effects of nutrients in higher dietary concentrations. The NRC document (1995) gives the literature documentation for each essential nutrient concerning the minimum level needed, as well as the toxic levels. It is not wise to exceed toxic levels, when wanting to study the influence of the effects of higher nutrient levels, as this may lead to early death, compromised welfare and unreliable results.

Hyperhomocysteinaemia in **ApoE deficient** mice was induced – among other experimental factors – by adding extra methionine to the diets (Zhou *et al.*, 2001). Hyperhomocysteinemia is considered an independent risk factor for atherothrombosis. As is shown in Figure 28.1, both the 'low' and 'high' methionine groups did not live until the scheduled end of the study (Zhou *et al.*, 2001). The total dietary methionine levels fed to the mice in these groups were 2.2% and 4.4% respectively. These methionine levels were obviously toxic for mice. The minimum recommended dietary methionine level for mice during growth is only 0.3% (NRC, 1995). As a dietary methionine level of 2.3% caused a clearly reduced body weight in Wistar rats (Fau *et al.*, 1988) higher dietary concentrations must be chosen with care, in order to avoid toxic effects. In

order to reach desired dietary nutrient levels precisely, it may be necessary to use purified diets. By using a purified diet, a high dietary methionine level of 1.4% was used, which allowed the mice to survive until the end of the study without obvious clinical or toxicity problems (Figure 28.2; Zhou, 2001). By carefully screening the literature before starting animal experiments, the most optimal dietary levels can be calculated and chosen in accordance with the purpose of the study.

Contaminants

Contaminants can be defined as undesirable substances (usually foreign) which, when present at a sufficiently high concentration in the food, may affect the animal and therefore the outcome of the experiments (BARQA, 1992). Possible contaminants include industrial chemicals, pesticides (e.g. DDT), plant toxins, mycotoxins (e.g. aflatoxin), heavy metals, nitrates, nitrites, bacteria and bacterial toxins (BARQA, 1992). Various guidelines on maximum permitted levels of dietary contaminants have been published (Environmental Protection Agency, 1979; BARQA, 1992; GV-Solas, 2002). Toxicologists all over the world usually refer to the guidelines issued by the Environmental Protection

Agency (1979). Table 28.4 compares various guidelines on the maximum permitted levels of contaminants and illustrates how variable these guidelines can be. The specific experimental parameters that are critical, may vary from user to user and depend upon the objectives of the study in which the diet is to be used. The user must determine what is critical to the particular study, as it is not feasible to identify and analyse for every possible contaminant. On the basis of this information a researcher can establish which contaminants at what maximum levels he or she wants to allow in the diet and/or which guidelines he or she wants to adhere to.

Those contaminants that will influence the health of the mice and interfere with the specific experimental purpose in particular will be the focus of attention.

Most commercial manufacturers provide diets with batch analysis certificates. This means that a number of selected nutrients and contaminants of each specific batch are analysed. This provides the buyer with the opportunity to judge whether this specific batch is suitable for the purpose of his or her particular study or to order a new batch. After the completion of the study, the batch analysis certificate is also valuable for the interpretation of the results.

TABLE 28.4: Comparison of three different guidelines on maximum allowed dietary levels of contaminants for mice (and rats)

Contaminant	BARQA[a]	GV Solas[b]		EPA[c]	
Fluorine mg/kg	40	150			
Nitrate mg/kg	100				
Nitrite mg/kg	5.0	15.0			
Nitrosamines mg/kg		NDEA[d]	0.01	0.01	
		NDMA[e]	0.01		
Lead mg/kg	3.0	1.5[f]		1.5	
Arsenic mg/kg	1.0	1.0		1	
Cadmium mg/kg	0.5	0.4		0.15	
Mercury mg/kg	0.1	0.1		0.1	
Selenium mg/kg	0.5			minimum	0.1
				maximum	0.5
Aflatoxins μg/kg	5.0	B1	10	5[g]	
		B2	5		
		G1	5		
		G2	5		
PCB μg/kg	50	50		50	
DDT (total) μg/kg	100	50[h]		100	
Dieldrin μg/kg	20	10[i]		20	
Endrin μg/kg		10			
Lindane μg/kg	100	100		20	
Heptachlor μg/kg		10[j]		20	
HCB μg/kg		10			
α,β,δ-HCH μg/kg		20			
α, γ-chlordan μg/kg		20			
α, β-endosulfan and -sulfate μg/kg		100			
Malathion μg/kg	500	1000		2500	
Fenitrothion μg/kg		1000			
Pirimiphos (-methyl) μg/kg		1000			
Chlorpyriphos (-methyl) μg/kg		1000			
Other phosphoric acid esters μg/kg		500			

Contaminant	BARQA[a]	GV Solas[b]		EPA[c]
TABLE 28.4: (Continued)				
Estrogenic activity µg/kg				1
TVO per g[k]	20,000	fibre <7%	100,000	
		fibre >7%	500,000	
Mesophilic Spores per g	20,000			
Salmonellae per g	none	none		
E. Coli per g	none	10		
Fungal units per g	200	fiber <7%	1000	
		fiber >7%	5000	
Fusarium toxins mg/kg		deoxynivalenol	0.50	
		ochratoxin	0.10	
		zearalenone	0.10	
A/B activity per g	none			none

[a]British Association of Research Quality Assurance (1992).
[b]German Association for Laboratory Animal Science (2002).
[c]Environmental Protection Agency (1979).
[d]Nitrosodiethylamine.
[e]Nitrosodimethylamine.
[f]With a dietary protein level of over 20% or with a crude fibre level of over 12%, values of up to 2.5 mg/kg feed are possible.
[g]Aflatoxin B1, B2, G1, G2.
[h]DDT + DDE + DDD.
[i]Dieldrin + Aldrin.
[j]heptachlor and heptachlor epoxy.
[k]total viable organisms.

Types of diets

Natural-ingredient diets

The two mostly used types of mouse diets in the laboratory are natural-ingredient and purified diets. A third category of diets, the chemically defined diets have major disadvantages and are therefore not widely used and will not be discussed in this chapter (further reading: Beynen and Coates, 2001). The pelleted diets that are often the standard diets in most laboratory animal facilities, are usually produced from natural ingredients. Manufacturers can produce diets according to variable or fixed formulas. In a variable formula, the final product levels are kept as constant as possible, i.e. aimed at keeping the level of nutrients in the endproduct as constant as possible. This means that the amount of individual ingredients is adjusted for variation in nutrient levels of the raw material (BARQA, 1992). In a fixed formula, the recipe does not change for a particular type of diet, i.e. the same proportions of raw material ingredients are used each time a batch is produced (BARQA, 1992). As natural ingredients can differ, the commercially available pelleted natural-ingredient diets are subject to variation.

Natural-ingredient diets can also be divided in open- and closed-formula diets (Thigpen *et al.*, 1999). In open-formula diets, all dietary ingredients and their concentrations are reported and should not vary from batch to batch. In closed-formula diets, the dietary ingredients used are reported, however, the concentration of each dietary ingredient is not stated by the manufacturing company. The concentration of dietary ingredients may vary from batch to batch or with availability of ingredients (Thigpen *et al.*, 1999).

Manufacturers provide information on the diets in their catalogues and on their websites. The amount of information provided and the way the catalogue values for nutrient concentrations have been established is not standardised. One has to contact the individual firms to find out the details. Between different brands of diets large variations can occur, even though these diets are all so-called 'standard' diets for mice and rats (Ritskes-Hoitinga *et al.*, 1991; Beynen and Ritskes-Hoitinga, 1992). On comparing the catalogue values with the actual batch analyses, large deviations can also occur, the so-called between-batch variation (Ritskes-Hoitinga *et al.*, 1991). Due to these variations, the standard deviations of results can become quite considerable, increasing the need for using larger numbers of animals in order to find statistically significant findings. Within the striving for the 3 Rs (Reduction, Refinement and Replacement; Russell and Burch, 1959), one of the goals is to limit the number of animals used. From the point of view of standardization it is advisable to use purified diets instead of natural-ingredient diets. In case natural-ingredient diets are used, it is advisable to buy them with a batch-analysis certificate, so that one at least has information on the exact levels of selected nutrients and contaminants of the specific batch used.

Purified diets

Purified or semi-purified diets (also named synthetic or semi-synthetic diets) are defined as being formulated with a combination of natural ingredients, pure chemicals and ingredients of varying degrees of refinement (Beynen and Coates, 2001). This results in diets having a much more standardised composition than natural-ingredient diets, consequently leading to (more) reproducible results. This will therefore lead to a more responsible use of laboratory animals, as the number of animals needed for reaching statistically significant results can be reduced and experimental results between studies and laboratories can be compared more directly.

In order to obtain a global standard for a purified mouse diet, it is advisable to follow the American Institute of Nutrition guidelines from 1993 (the so-called AIN-93 diet; Reeves *et al.*, 1993). For toxicity studies it is generally accepted that the AIN dietary composition is used as the 'golden standard' for reasons of standardisation and comparison. The AIN93 guidelines fulfill nearly all nutrient requirements for mice (NRC, 1995), only the dietary vitamin B12 concentration in the AIN93 diet is 50% lower than the NRC recommendations. Therefore, it is advisable to add twice as much vitamin B12 to the AIN diet (final concentration should be 50 mcg/kg diet). It is also advisable to add more selenium, so that the final concentration becomes 0.15 mg/kg diet (NRC, 1995). The AIN93 diet contains 0.10 mg/kg selenium (Reeves *et al.*, 1993). Table 28.5 gives a typical example of a diet designed according to AIN 93 guidelines, with the above-mentioned modifications (Zhou, 2001; Zhou *et al.*, 2003).

It may be impossible to pellet purified diets, as this depends on which types of carbohydrates are chosen

TABLE 28.5: Composition of a purified mouse diet, according to AIN-93 guidelines	
Ingredient	Amount (g)
Casein	14.0
Corn flour	71.5
Solkafloc (cellulose)	5.0
Corn oil	4.0
Choline bitartrate	0.2
L-cystine	0.2
Vitamin and mineral mix	5.0
Methionine	0.1
Total	100.0

Source: Reeves *et al.*, 1993; Zhou, 2001; Zhou *et al.*, 2003.

and how much fat is included in the diet. The example of the diet given in Table 28.5 is of such a composition. When pelleting is not possible, the use of special feeding devices becomes necessary (Ritskes-Hoitinga and Chwalibog, 2003).

By using purified diets it becomes possible to aim precisely at obtaining certain (low) concentrations of particular nutrients. Often this cannot be reached when using natural-ingredient diets, due to the relatively high concentrations naturally present in the raw materials. Table 28.6 illustrates the analysed values versus the targeted levels in purified diets (Zhou *et al.*, 2001).

Storage conditions

In order to ensure that the nutrient contents of the natural ingredient diets remain within the specifications until the recommended expiry date, diets must be stored in a cool and dry place and free from pests. Keeping the diets at −20°C instead of 5°C, will prolong shelf life. In order to avoid cross-contamination, the storage area must be dedicated to non-medicated diets. Exact storage conditions must be provided by the manufacturer (BARQA, 1992). The purified AIN93 diet is best stored at −20°C. During a 3 months' storage period at −20°C, the peroxide value (measure of oxydation) did not increase in contrast to storage at +5°C (unpublished observations). When storing at −20°C, addition of antioxidants is not considered necessary. As these may interfere with the purpose of studies, e.g. atherosclerosis induction, it is advisable to store diets at −20°C and omit the addition of an antioxidant whenever possible. Where highly unsaturated oils like fish oil are to be used, the addition of this fish oil needs careful consideration as they are easily oxidised,

TABLE 28.6: Critical nutrients in five purified diets in concentrations as aimed for (in brackets) and according to actual analysis

Type of diet	Vitamin B6 (mg/kg)	Vitamin B12 (μg/kg)	Folate (mg/kg)	Methionine (g/kg)
Control	8.9 (8)	6.4 (10)	0.53 (0.5)	4.3 (5)
Low Folate	7.2 (8)	10 (10)	0.02[a]	4.5 (5)
Low vitamin B6	3.7 (4)	5.3 (10)	0.86 (0.5)	4.1 (5)
Low Methionine	6.8 (8)	6.4 (10)	0.8 (0.5)	13.7 (15)
High vitamins	18 (24)	31 (30)	2.3 (1.5)	4.4 (5)

[a] As low as possible.
Source: Zhou, 2001.

thereby reducing vitamin E levels. When adding fish oil to diets, the best procedure is to keep the fish oil stored under liquid nitrogen at −80°C, and then mix the fish oil freshly through the (purified) diet each day, just before feeding (Ritskes-Hoitinga et al., 1998).

Pellet hardness

In case of feeding dietary pellets, the hardness of pellets should not be over a certain limit, as this was found to reduce the growth of preweaned mice (Koopman et al., 1989a). A too high pellet hardness will make it difficult for the young preweaned mice to obtain enough food, which will reduce growth. Part of this effect is suggested to be mediated through the effect on the mothers (Koopman et al., 1989b). Hardness of pellets is measured as the amount of pressure, in kp weight, that is required to crush a pellet. In one type of diet the measured pellet hardness could vary between 4 and 50 kp/cm^3 (Koopman et al., 1989a). A value higher than 20 is considered problematic.

Autoclaving/irradiation

Diets that are to be used behind barriers are usually autoclaved or gamma irradiated. These treatments can diminish the levels of certain amino acids (lysin, methionine and cystin) and certain vitamins (e.g. A and B1) for example. The longer the time used for heating, the more the vitamin levels are reduced. Minerals, spore elements, crude fibre, crude protein, crude fat and crude ashes do not appear to be measurably influenced by autoclaving. Fatty acid patterns do not appear to be influenced by autoclaving either. Manufacturers provide special diets with increased levels of vitamins in order to guarantee that nutrient levels still live up to the requirements after

the autoclaving or sterilising process. For irradiation it is advisable to use 25 kGy as a minimal dose for diets to be used in SPF units, and 50 kGy for germ free animals (O'Doherty, 2001). Lower doses can be used, when an analysis of the microbial quality prior to irradiation indicates that this will be sufficient (O'Doherty, 2001).

Ad libitum feeding versus food restriction

Ad libitum feeding means that the diet is available at all times. Restricted feeding refers to restricting the amount of food while still insuring nutritional adequacy (Hart et al., 1995). This implies that only the amount of energy has been restricted. It is still common practice to feed laboratory mice (and rats) ad libitum, even though this is undesirable from the point of view of animal welfare as well as the validity of the experimental results. Keenan et al. (1999) have stated that ad libitum feeding is currently our worst standardised experimental factor in laboratory units. In some toxicology laboratories it has become common practice to restrict feed intake of rodents to 75–80% of ad libitum intake, as this gives better standardization and longer survival. This implies, however, that animals are housed individually as there are no appropriate housing and feeding systems available for the restricted feeding of group housed rodents. As rodents are social species, the resolution of the Council of Europe (1997) requires that social species need to be group-housed whenever possible. For that reason it is necessary that appropriate feeding systems

Tumour type and survival	Males AL	Males FR	Females AL	Females FR
TABLE 28.7: Some results of control groups of B6C3F1 mice on *ad libitum* (AL) or food restriction (FR) schedules				
Liver tumours %	54	22**	45	8**
Lung tumours %	28	17*	2	12
Malignant lymphoma %	9	8	16	4*
Hemangio(sarco)ma %	5	2	8	2
Survival %	80	88	65	94**

*(p < 0.05),** (p < 0.01) significantly different as compared to AL fed controls.
Source: Hart *et al.*, 1995.

and/or methods be developed that both guarantee restricted feeding as well as social housing conditions. Until then it should be considered – whenever possible – to feed the animals individually, and then provide social housing as soon as the feeding period is over.

When comparing *ad libitum* feeding versus food restriction, it becomes clear that *ad libitum* feeding has a negative impact on the health of rodents. *Ad libitum* feeding will lead to more obesity, a shorter survival time, increased degenerative kidney and heart diseases, a shorter latency time and higher incidence of cancer as compared with restricted feeding (Hart *et al.*, 1995). Table 28.7 shows a comparison between *ad libitum* and food restricted B6C3F1 mice that served as control groups in longterm toxicity studies (Hart *et al.*, 1995). Food restriction gave a reduced body weight as compared with *ad libitum* feeding, and also a decreased incidence of liver tumours in both sexes, decreased lung tumour incidence in males and decreased malignant lymphomas in females. Food restriction led to an improved survival time, especially in females.

During the 1980s and 1990s the variation in results of longterm bioassays in rodents that were on *ad libitum* feeding schemes increased (Keenan *et al.*, 1999). Therefore the feeding of a standardised restricted amount of food to each individual is considered a necessity in order to obtain standardised results without unnecessarily large variations. This will also increase the reliability of the interpretation of the toxicity of test substances, as the amount of the test substance ingested through the diet becomes standardised for each individual. Even though food restriction will increase the amount of labour and will pose practical problems, it will give more reliable results and should therefore become our feeding method of choice.

Besides a better standardisation, another advantage of food restriction is that animals become 'more robust', i.e. they are better in coping with stress-factors (Keenan *et al.*, 1999). That way they can be exposed to test substances in higher concentrations and/or longer periods without compromising the animal's health, physiology or metabolic profile (Hart *et al.*, 1995). As food restriction to 75% of *ad libitum* intake improves longevity (as seen above), animals can be exposed to test compounds for a longer period, thereby improving the sensitivity of bioassays to detect compound-specific chronic toxicity and carcinogenicity (Hart *et al.*, 1995).

The level of nutrients in the natural-ingredient diets is usually far above the levels needed for fulfilling the nutrient requirements (NRC, 1995). Manufacturers do this in order to guarantee that no deficiencies will occur, not even after longer storage periods. Moreover, the NRC requirements support maximum growth. Therefore limiting food intake to 75% of *ad lib* intake will still make sure that the diet lives up to the minimum levels of essential nutrients.

Pair feeding

Where a cancerogenic test substance is mixed through the diet and has a bad taste, this will reduce food intake, particularly when fed *ad libitum*. The control group will ingest a higher amount of food than the test group, thereby developing cancer at a higher frequency and at an earlier age. The outcome could thus be false-negative, i.e. that one does not judge the cancerogenic test substance to be cancerogenic. In order to avoid these problems, pair-feeding is necessary. How much food the test animals ingest is measured, and then the same amount of food is given to the control group/animals the next day. This assumes that food intake of each individual is measured (Beynen and Coates, 2001; Ritskes-Hoitinga and Chwalibog, 2003).

Normal feeding behaviour versus food restriction

Rodents are nocturnally active animals and ingest most of their food during the dark period. Food restriction (60% of *ad libitum* intake) as compared with *ad libitum* feeding will affect physiologic variables, circadian rhythms, activity and feeding behaviour (Hart *et al.*, 1995). Table 28.8 illustrates the effects of *ad libitum* feeding and food restriction during the dark or the light period, on various physiological and behavioural variables in 28 month old male B6C3F1 mice (Duffy *et al.*, 1990; Hart *et al.*, 1995). Total food and caloric consumption was decreased in food restricted groups, as expected. Water consumption remained nearly the

TABLE 28.8: Effects of diet and feeding time on physiologic and behavioural variables in old male B6C3F1 mice								
Measurement	AL group Mean ± S.E.	LF restricted group Mean ± S.E.	% of AL	DF restricted group Mean ± S.E.	% of AL	Significance levels		
						AL vs. LF	AL vs. DF	LF vs. DF
Total food consumption (g)	5.21 ± 0.10	3.35 ± 0.02	64.3	3.41 ± 0.08	65.5	A***	B***	–****
Caloric consumption (kcal/g)	22.66 ± 0.43	14.57 ± 0.09	64.3	14.83 ± 0.35	65.5	A***	B***	–****
Total water consumption (g)	3.64 ± 0.38	5.15 ± 1.07	141.5	3.72 ± 0.07	102.2	A*	–****	C*
Number of feeding episodes	16.51 ± 0.69	2.80 ± 0.23	17.0	4.32 ± 0.37	25.6	A***	B***	C*
Number of drinking episodes	11.13 ± 1.28	9.03 ± 0.44	81.1	8.99 ± 0.67	80.8	–***	–***	–***
Average body temperature (°C)	36.78 ± 0.08	35.54 ± 0.15	96.6	35.11 ± 0.18	95.5	A***	B***	–***
Max – min body temperature (range; °C)	37.98 – 35.81 (2.18)	38.15 – 32.24 (5.91)	271.1	37.52 – 32.93 (4.59)	210.6	A***	B***	–***
Average activity (pulse/hr)	10.54 ± 2.30	18.26 ± 1.71	173.2	26.50 ± 5.57	251.4	A*	B**	–***
Average O_2 consumption (g/LBM) (mL g^{-1} hr^{-1})	3.34 ± 0.16	3.44 ± 0.16	103.0	3.19 ± 0.06	95.5	–***	–***	–***
Max – min respiratory quotient (range)	0.95 – 0.86 (0.09)	0.99 – 0.80 (0.19)	211.0	1.01 – 0.77 (0.24)	267.0	A***	B***	–***

Results of Student's t-test analysis:

A = AL × LF restricted comparison (significant effect), B = AL × DF restricted comparison (significant effect),

C = LF restricted × DF restricted comparison (significant effect; adapted from Duffy *et al.*, 1990).

AL = *ad libitum*, LF = restricted group fed during light period, DF = restricted group fed during dark period,

LBM = lean body mass, Age = 28 months.

*$p < 0.05$.

**$p < 0.01$.

***$p < 0.001$.

****$p > 0.05$.

Source: Duffy *et al.*, 1990; Hart *et al.*, 1995.

same in the restricted group fed during the dark period as compared with the *ad libitum* fed B6C3F1 mice, so that total water consumption increased in relation to food intake (Table 28.8). However, when fed restrictedly during the light period, water intake increased to a level of about 141% of the *ad libitum* fed mice. This may be related to the fact that the mice do not have food available during their normal active period in which they ingest food, thereby possibly using drinking water as a substitute. The duration of food consumption was compressed by restriction, so that most food was consumed during the first few hours immediately after feeding commenced. Restricted mice ate fewer meals (feeding episodes) but spent more time feeding per meal and consumed more food per meal than the *ad libitum* fed mice. The number of drinking episodes was also decreased by food restriction. Average body temperature decreased, whereas the range in body temperature increased by dietary restriction. Spontaneous activity was increased by food restriction. Average oxygen consumption increased in the restricted group fed during the light period and decreased in the restricted group fed during the dark period. This may be explained by the fact that in the first group, the mice were actively searching for food during the entire dark period, which is their normal feeding time. The daily variations in the respiratory coefficient (RQ) were increased by dietary restriction, indicating rapid substrate-dependant shifts in metabolic pathways from carbohydrate metabolism (immediately after feeding) to fatty acid metabolism (several hours before feeding). This may be related to an enhanced metabolic efficiency (Hart *et al.*, 1995).

Because mice are nocturnally active animals, it is advisable that when animals are fed restrictedly, food is provided during the normal feeding time, i.e. the dark period in order to fulfil the animals basic needs. As the relationship between circadian rhythms and nutrition is beyond the scope of this chapter, further reading can be found in Ritskes-Hoitinga and Jilge (2001).

Individual housing versus group housing

After the 1970s, in mice studies performed in the National Toxicology Program, body weights showed the tendency to increase (Haseman, 1992; Haseman *et al.*, 1994; Hart *et al.*, 1995). This may be related to the fact that the protocol changed from group housing to individual housing. Individually housed animals had higher body weights and higher tumour incidence than group-housed mice (Haseman *et al.*, 1994). The variance in body weight in group-housed animals (two, four or eight per cage) was smaller than in individually housed mice (Claassen, 1994). According to the National Toxicology Program, female mice are now housed again in groups, but male mice are still housed individually to avoid fighting problems among group-housed males (Haseman *et al.*, 1994; Hart *et al.*, 1995).

Different types of studies based on the route of administration (e.g. inhalation, feed, corn oil gavage etc.) are registered in the National Toxicology Program database (Hart *et al.*, 1995). Each of these type of studies treats control animals differently, resulting in different results, for example in average tumour incidence in mice (Table 28.9; Hart *et al.*, 1995). In inhalation studies a clear relationship is seen between body weight biomarkers and tumour incidence. Mice are almost always singly housed in inhalation chambers, so that

TABLE 28.9: Average body weight at 12 months on test (BW12) and tumour incidences for different study types for B6C3F1 mice								
Type of test	i		c		g		f	
	M	F	M	F	M	F	M	F
BW12 (g)	40.1	34.6	44.0	36.2	43.1	38.0	40.9	38.1
Liver tumour incidence (%)	39.6	18.6	41.4	12.2	38.8	10.6	33.2	15.5
Lymphoma incidence (%)	8.7	21.1	16.7	39.1	15.6	43.6	10.0	37.2
Pituitary tumour incidence (%)	0.5	23.2	1.1	33.3	2.6	23.6	0.3	20.0

i = inhalation studies (n = 8); c = corn oil gavage studies (n = 21); g = water gavage studies (n = 5); f = feed studies (n = 17); M = Males; F = Females.

Tumor incidences were total incidences after 24 months on test.

Source: Hart *et al.*, 1995.

TABLE 28.10: Average survival and lymphoma incidence in single and group-housed male B6C3F1 mice

Type of housing	Single-housed	Group-housed
Survival %	65 ± 9	84 ± 10
Lymphoma incidence %	15 ± 5	7 ± 7

Single-housed combines results of 9 studies, group-housed 12 studies.
Source: Hart *et al.*, 1995.

any dietary modulation of body weight and spontaneous tumourigenesis is not confounded by social interactions among the animals. Some feed and corn oil gavage studies also use single housing in the case of males, in order to prevent fighting. When compared with similar studies that used group housing, group housing increased the percentage of animals surviving and decreased the incidence of lymphomas (Table 28.10; Hart *et al.*, 1995). The clear relation between body weight and survival and (liver) tumour incidences is found in singly housed animals only, in group housing this relationship is confounded (Hart *et al.*, 1995).

Isocaloric exchange

When designing experiments in which the dietary fat content is intended to be altered in the test groups, it is important to consider the basic facts in order to reach properly controlled and standardised diet compositions. The MEC of fat (37 MJ/kg) is about twice as high as those in carbohydrates and protein (17 MJ/kg). So when the dietary fat content is increased, the dietary energy concentration will increase. In case of *ad libitum* food intake, the animals ingest according to energy need, so the food intake in grams will decrease. When an exchange with carbohydrates for example is done on the basis of weight, test and control groups will ingest different amounts of all nutrients (Table 28.11; Beynen and Coates, 2001). In order to reach a full standardisation, the exchange of fats and carbohydrates needs to be done on the basis of calories (isocaloric exchange; see Table 28.9). In that way, the intake in grams of all nutrients will be similar in control and test groups, only the intake in fats and carbohydrates will be different as this is inherent to the design of the diets.

When intending to examine the influence of certain dietary fatty acids, it is advisable to keep the total fat content similar, and vary the types of fats used, in order to obtain variable concentrations of fatty acids only. An example of how to design two different dietary fat levels with varying fatty acid concentrations within each fat level is illustrated by Ritskes-Hoitinga *et al.* (1996a).

Fasting

Fasting is often used in pharmacokinetics studies, as an empty stomach is considered a prerequisite for bioavailability and drug absorption studies. Fasting causes severe changes in the physiological and biochemical processes of the animal, which become more severe with longer duration of food withdrawal (Claassen, 1994). Also the often used fasting periods of 16–24 h causes important changes, which may significantly affect the responsiveness towards experimental stimuli (Claassen, 1994). During fasting there is a decrease in body and liver weight, usually the loss in liver weight is more pronounced than the loss in body weight (Claassen, 1994). After overnight fasting the glutathione content per gram liver diminished by 49–57% in mice, as reported by various investigators (Claassen, 1994). Strubelt *et al.* (1981) reported a 33% decrease of the specific cytochrome p-450 content in the mouse after 24 h of food deprivation. Strains and individuals may also react to fasting in different ways. Some strains of mice (C3H/HeJ) will increase their water intake, others (DBA/2J, BALB/cJ, A/J, C57BL/6J) will decrease water intake and some (CBA/J, SWR/J) have an unchanged water intake after food deprivation (Kutscher, 1974; Claassen, 1994). Claassen (1994) therefore advises using fasting only when the feeding condition functions as an experimental factor in the study. Other solutions may be to provide the animals with sucrose, 10% glucose or maltose during the period of fasting, as has been suggested for rats (Levine and Saltzman, 2000; Gyger, 2001) and hamsters (Elste, 2002).

Mice models in nutrition research

Animal models can contribute to the understanding of (parts of) human processes. Russell and Burch (1959)

TABLE 28.11: Isocaloric exchange				
	Diet 1	Diet 2	Diet 3	Diet 4
	Low-fat	High-fat	High-fat, adjusted	High-fat, adjusted
Diet ingredient				
Protein (g)	20	20	20	20
Carbohydrate (g)	60	40	15	15
Fat (g)	10	30	30	30
Fibre (g)	4	4	4	4
Mineral mix (g)	4	4	4	4
Vitamin mix (g)	1	1	1	1
Test compound (g)	1	1	1	1
'Inert' compound (g)	—	—	—	25
TOTAL (g)	100	100	75	100
Energy value (kcal/g)	4.10	5.10	5.47	4.10
Expected intake				
Energy (kcal/day)	82	82	82	82
Food (g/day)	20	16	15	20
Protein (g/day)	4	3.2	4	4
Carbohydrate (g/day)	12	6.4	3	3
Fat (g/day)	2	4.8	6	6
Fibre (g)	0.8	0.64	0.8	0.8
Mineral mix (g/day)	0.8	0.64	0.8	0.8
Vitamin mix (g/day)	0.2	0.16	0.2	0.2
Test compound (g/day)	0.2	0.16	0.2	0.2
'Inert' compound (g/day)	—	—	—	5

Source: Beynen and Coates, 2001.

distinguished in their famous book on *The Principles of Humane Experimental Technique* between two types of animal models: the High fidelity and the Discrimination animal models. In the High Fidelity model all characteristics resemble those in humans, whereas in a discrimination model only one characteristic is reproduced. Researchers need to be aware as to which type of model they are choosing and using, in order to make reliable comparisons with the human situation (Ritskes-Hoitinga, 1998). A part of what is studied is independent of the interference of scientists, but another part is influenced by the choices scientists make (Jacobsen, 2001). The choices that are made regarding the model, the design and environmental conditions, are of major importance for the outcome of the study (Ritskes-Hoitinga et al., 1996b). Results of animal studies should never be extrapolated directly to the human situation, but need discussion and critical evaluation. Literature reviews of other animal studies, epidemiological studies and clinical trials must be part of the entire evaluation.

The influence of dietary linoleic acid on mammary cancer development

In a study by De Wille *et al.* (1993) the influence of dietary linoleic acid concentration on mammary tumour development in transgenic mice (MMTV/v-Ha-*ras*) was studied. Three levels of dietary linoleic acid were given, 0%, 1.2% and 6.7%. There was a significant reduction of mammary tumour development on the 0 dietary level of linoleic acid, as compared with the other two dietary groups (Figure 28.3). As linoleic acid is an essential fatty acid necessary for the development

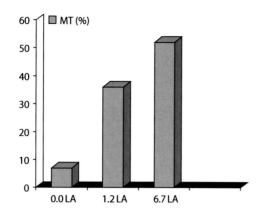

Figure 28.3 Dietary linoleic acid level (LA) and Mammary tumour frequency (MT; De Wille, 1993).

of cell membranes (NRC, 1995), this cannot be considered a reliable control group. Where the diet contains no linoleic acid at all, general health is expected to be compromised. In the study by De Wille *et al.* (1993) it was reported that there were initially 25 animals in the 0 level linoleic acid group, however, in the results section only data from 15 animals were presented. It was not described what happened to the missing 10 animals. It is essential to mention all details in scientific reports and publications, in order to fully understand the course of events and to be able to reproduce studies. The control group in the study by De Wille *et al.* (1993) should have contained the minimum necessary level of linoleic acid in the diet (0.7%; NRC, 1995), in order to obtain control animals in good health and reliable results. By leaving out dietary linoleic acid completely, one would not expect that cell membranes or tumours could develop.

In a literature review by Ritskes-Hoitinga *et al.* (1996b) the results of studies examining the influence of dietary linoleic acid in different rodent models were compared. The results were dependent on the type of animal model used. In Table 28.12 the results from the mouse models are given. When using dimethyl-benz(a)anthracene (DMBA) to induce mammary tumours, higher dietary linoleic acid concentrations were associated with increased mammary tumour incidences (Fischer *et al.*, 1992; Craig-Schmidt *et al.*, 1993). In spontaneous (Brown, 1981) and the BALB/c-MMTV mouse models (Ritskes-Hoitinga *et al.*, 1996a), no clear association was detected between dietary linoleic acid concentrations and mammary tumour development. As mentioned above, De Wille *et al.*'s (1993) study showed a higher mammary tumour incidence in the higher dietary linoleic acid groups. This may have been an artefactual finding, as the control group lacked linoleic acid completely. Depending

on the model, the latency period differed (Ritskes-Hoitinga *et al.*, 1996b). It may be that the amount of linoleic acid needed for tumour development is dependent on the latency period: by using DMBA, a quick tumour development is induced, which may be associated with a high linoleic acid 'requirement' for tumour growth. This illustrates, that the choice of animal models and set of experimental conditions, as well as the interpretation of results, needs careful consideration.

Atherosclerosis-inducing diets

Mice are historically resistant to the development of atherosclerosis (Moazed, 1998). On a normal chow diet (about 4.5% fat and 0.02% cholesterol) most of the plasma cholesterol is in the form of high density lipid (HDL), the anti-atherogenic fraction of cholesterol (Moazed, 1998). High fat diets will result in the development of atherosclerotic lesions over time in susceptible strains such as the C57BL/6. After 7 weeks on the high fat diet, these mice develop fatty streaks and progress to more complicated lesions by 14 weeks. Other strains, such as the BALB/c and C3H strains never develop atherosclerotic lesions (Moazed, 1998). Special diets used to induce atherosclerosis are a 'western-type' diet, that contains about 21% fat and 0.15% cholesterol, and an 'atherogenic' diet, which contains 15% fat, 1.25% cholesterol and 0.5% cholic acid (Moazed, 1998). This last diet is also referred to as the 'Paigen's diet'. Historically, this diet was used to induce gallstones. It is known to be hepatotoxic and to induce a proinflammatory state (Moazed, 1998). When using the Paigen's diet, atherosclerotic plaques can be induced, however, hepatotoxicity and gallstones are induced simultaneously. Hepatotoxicity may interfere with the development of atherosclerotic lesions, as was seen in rabbits (Ritskes-Hoitinga *et al.*, 1998). Therefore it is considered necessary to at least evaluate the condition of the liver and publish this, when atherosclerosis studies are performed. As gallstones are painful in humans, it may be expected to be the same for mice as well.

The use of transgenic mouse models instead of wildtype mice models may be a good alternative solution, as it then becomes possible to leave out the cholate from the diet. On a normal chow diet, apoE-deficient mice develop plasma cholesterol levels that are at least 10 times as high as in wild type mice and most of the cholesterol is in the form of the highly atherogenic VLDL form (Moazed, 1998). Apo-E deficient mice develop atherosclerotic lesions on a normal chow and on a Western type diet, however, lesions develop

Reference	Strain	Type of model	Dietary fat %	Dietary linoleic acid level %	Mammary tumour incidence %
Fischer 92	Sencar mouse	DMBA[a] induction	15	0.8	23
			15	4.5	43
			15	8.4	50
Craig-Schmidt 93	Balb/c mouse	DMBA induction	20	1.4	36
			20	1.5	45
			20	11.6	77
Brown 81	C3H mouse	Spontaneous	5	0.5	13
			5	0.9	3
			5	1.1	2
			5	2.9	3
			5	3.0	7
			17	1.6	8
			17	3.2	15
			17	3.6	12
			17	9.7	8
			17	10.1	5
De Wille 93	MMTV/v-Ha-*ras* mouse	MMTVirus[b]	0	0.0	7
			2	1.2	36
			11	6.7	52
Ritskes-Hoitinga 96a	Balb/c-MTV	MMTVirus	7	0.9	40
			7	1.3	30
			7	2.6	44
			7	4.3	32
			16	0.9	52
			16	1.3	30
			16	2.6	52
			16	4.3	38

TABLE 28.12: Mammary tumour incidence in different mouse models

[a]Dimethylbenz(a)anthracene.

[b]mouse mammary tumour virus.

Source: Ritskes-Hoitinga *et al.*, 1996b.

more rapidly and at an earlier age on the Western type diet. The atherosclerotic lesions in apoE-deficient mice have strikingly similar pathological characteristics and anatomical distributions as compared with humans (Moazed, 1998). The low density lipid (LDL)-receptor-deficient mouse does not develop atherosclerosis on a normal chow diet, but this can be induced by feeding a high fat diet. Lesion characteristics are the same as in the apoE-deficient mouse, but lesion formation is better controllable by dietary changes (Moazed, 1998). Plasma cholesterol levels are lower than in the apoE-deficient mice and thereby more human-like.

In case an atherosclerosis-inducing test diet is used next to a control diet, care must be taken that a proper isocaloric exchange is made between the control and test diet. By making isocalorically exchanged diets, only the fat and carbohydrate intake (in grams) will be different between the two groups. The intake of all other nutrients will be similar, thereby allowing a more reliable interpretation of results due to optimal standardisation.

Welfare considerations and enrichment

The use of genetically modified mice has increased dramatically. When using these models, the possible health and welfare problems of the specific models must be evaluated in order to prevent unnecessary suffering. Mouse models of muscle dystrophy died at a very young age, as these mice were physically unable to reach for the food in the feeding trough (unpublished observations). By giving the food pellets on the bottom of the cages and a longer drinking nipple, the animals could survive until a much older age without any obvious clinical problems.

After surgery the mice are sometimes unable to stretch their bodies and/or heads towards the feeding trough and drinking nipple. In this case a small feeding device with feeding pellets mixed with water must be provided on the bottom of the cage, until the animals are completely recovered.

Currently, the enrichment of housing conditions receives a lot of attention. Much focus is given to environmental factors, e.g. space, cage furniture and nesting material. As animals use a large proportion of their active time searching for food and foraging, modifying the way we feed our laboratory animals has a large potential for enrichment. This has not yet been investigated to a large extent, and it is the challenge of the coming years to investigate and use this potential. For rabbits it was found that when feeding a restricted amount of food at a time point approaching the 'normal' feeding hours, stereotyped behaviour was registered as significantly decreased as compared with *ad libitum* or feeding restrictedly at another time point (Krohn *et al.*, 1999). This may imply, that when feeding a restricted amount of food at the 'proper time point', better welfare, better health and more standardised results can be obtained.

In case single housing is a necessity for the experimental design, welfare considerations become necessary as mice are social species. According to the Resolution of the Council of Europe (1997) social species must be housed socially, unless there are good reasons to do otherwise. In certain cases mice may be housed individually at the time of feeding, and housed socially outside the feeding period. Another possibility is to divide the cage by bars, so that the mice can be in close contact without being able to reach each others food. Human contact may also compensate to a certain degree for a shortage in social contact with conspecifics. During procedures, e.g. weighing the animals, individually housed animals may be temporarily housed in a group. It goes without saying that this needs effective individual identification. By providing social contact during procedures, this may be regarded by the animals as a positive reward, thereby making the handling and procedures more easy to execute.

References

BARQA: British Association for Research Quality Assurance (1992). *Guidelines for the Manufacture and Supply of GLP Animal Diets.* http://www.barga.com BARQA, 3 Wherry Lane, Ipswich IP4 1LG, UK.

Beynen, A.C. and Ritskes-Hoitinga, J. (1992). *Scand. J. Lab. Anim. Sci.* **19**, 93–94.

Beynen, A.C. and Coates, M.E. (2001). *In Principles of Laboratory Animal Science* (eds L.F.M. van Zutphen, V. Baumans and A.C. Beynen), pp. 111–127. Elsevier, Amsterdam.

Brown, R.R. (1981). *Cancer Res.* **41**, 3741–3742.

Claassen, V. (1994). *Neglected Factors in Pharmacology and Neuroscience Research.* Elsevier, Amsterdam.

Council of Europe (1997, 30 May). *Resolution on the Accomodation and Care of Laboratory Animals.*

Craig-Schmidt, M., White, M.T., Teer, P., *et al.* (1993). *Nutr. Cancer* **20**, 99–106.

De Wille, J.W., Waddell, K., Steinmeyer, C. and Farmer, S.T. (1993). *Cancer Lett.* **69**, 59–66.

Duffy, P.H., Feuers, R.J. and Hart, R.W. (1990). *Chronobiol. Int.* **7**, 113–124.

Elste, V. (2002). FELASA Symposium, Aachen, Germany, Poster 35, 17–20 June.

Environmental Protection Agency (1979). GLP Standards for Health Effects in Federal Register Vol. 44 No. 91, 27353–27354.

Fau, D., Peret, J. and Hadjiisky P. (1988). *J. Nutr.* **118**, 128–133.

Fischer, S.M., Conti, C.J., Locniskar, M., *et al.* (1992). Cancer Res. **52**, 662–666.

GV-Solas: Gesellschaft für Versuchstierkunde – German Society for Laboratory Animal Science (2002). *Guidelines* for the Quality-Assured Production of Laboratory Animal Diets. EN A-06-2002. http://www.gv-solas.de/auss/aus.html

Gyger, M. (2001, spring). *SGV Newsletter* no. 24: http://www.sgv.org/Newsletter/news-24.htm

Hart, R.W., Neumann, D.A. and Robertson, R.T. (1995). *Dietary Restriction: Implications for the Design and Interpretation of Toxicity and Carcinogenicity Studies.* ILSI Press, Washington, DC.

Haseman, J.K. (1992). *Drug Inf. J.* **26**, 191–200.

Haseman, J.K., Bourbina, J. and Eustis, S.L. (1994). *Fundam. Appl. Toxicol.* **23**, 44–52.

Jacobsen, B. (2001). *Hvad er god forskning?* (in Danish). Hans Reitzels Forlag, Copenhagen.

Keenan, K.P., Ballam, G.C., Soper, K.A., Laroque, P., Coleman, J.B. and Dixit, R. (1999). *Toxicol. Sci.* **52** (2 Suppl.), 24–34.

Keenan, K.P., Ballam, G.C., Haught, D.G. and Laroque, P. (2000). *The Laboratory Rat,* pp. 57–75. Academic Press Ltd, London.

Koopman, J.P., Scholten, P.M., Roeleveld, P.C., Velthuizen, Y.W.M. and Beynen, A.C. (1989a). *Zeitschrift für Versuchstierkunde* **32**, 71–75.

Koopman, J.P., Scholten, P.M. and Beynen, A.C. (1989b). *Z. Versuchstierkunde* **32**, 257–260.

Krohn, T.C., Ritskes-Hoitinga, J. and Svendsen, P. (1999). *Lab. Anim.* **33**, 101–107.

Kutscher, C.L. (1974). *Physiol. Behav.* **13**, 63–70.

Levine, S. and Saltzman, A. (2000). *Lab. Anim.* **34**, 301–306.

Moazed, T.C. (1998). Continuing Education Seminar, The American Society of Laboratory Animal Practitioners, October, Cincinnati, USA.

National Research Council (1978). *Nutrient Requirements of Laboratory Animals,* 3rd rev. edn. National Academy Press, Washington, DC.

National Research Council (1995). *Nutrient Requirements of Laboratory Animals,* 4th rev. edn, pp. 80–102. National Academy Press, Washington, DC.

O'Doherty (2001). ScandLAS Symposium, Aarhus, Denmark, 13–15 May.

Reeves, Ph.G., Nielsen, F.H. and Fahey, G.C. Jr. (1993). *J. Nutr.* **123**, 1939–1951.

Ritskes-Hoitinga, J. (1998). Professorial seminar. *Animals and Models: Who Cares?* ISBN 87-986909-3-0. University of Southern Denmark, Odense.

Ritskes-Hoitinga, J. and Chwalibog, A. (2002, in press). In *Nutrient Requirements, Experimental Design and Feeding Schedules in Animal Experimentation* (eds G. van Hoosier and J. Hau), CRC Press, Boca Raton.

Ritskes-Hoitinga, J. and Jilge, B. (2001). FELASA – quick reference paper on laboratory animal feeding and nutrition. www.felasa.org

Ritskes-Hoitinga, J., Mathot, J.N.J.J., Danse, L.H.J.C. and Beynen, A.C. (1991). *Lab. Anim.* **25**, 126–132.

Ritskes-Hoitinga, J., Meijers, M., Timmer, W.G., Wiersma, A., Meijer, G.W. and Weststrate, J.A. (1996a). *Nutr. Cancer* **25**, 161–172.

Ritskes-Hoitinga, J, Meijers, M., Meijer, G.W. and Weststrate, J.A. (1996b). *Scand. J. Lab. Anim. Sci.* **23**, 463–469.

Ritskes-Hoitinga, J., Verschuren, P.M., Meijer, G.W., Wiersma, A., van de Kooij, A., Timmer, W.G., Blonk, C.G. and Weststrate, J.A. (1998). *Food Chem. Toxicol.* **36**, 663–672.

Russell, W.M.S. and Burch, R.L. (1959). *The Principles of Humane Experimental Technique.* Methuen & Co. Ltd, London.

Strubelt, O., Dost-Kempf, E., Siegers, C.P., Younes, M., Völpel, M., Preuss, V. and Dreckmann, J.G. (1981). *Toxicol. Appl. Pharmacol.* **60**, 66–77.

Thigpen, J.E., Setchell, K.D.R., Alhmark, K.B., Locklear, J., Spahr, T., Caviness, G.F., Goelz, M.F., Haseman, J.K., Newbold, R.R., and Forsythe, D.B. (1999). *Lab. Anim. Sci.* **49**, 530.

Van den Broek, F.A.R. (1998). Dystrophic carciac calcification in laboratory mice. Thesis, Utrecht University, The Netherlands.

Van den Broek, F.A.R., Beems, R.B., van Tintelen G., *et al.* (1997). *Lab. Anim.* **31**, 74–80.

Van den Broek, F.A.R., Bakker, R, den Bieman, M., *et al.* (1998). *Biochem. Biophys. Res. Commun.* **253**, 204–208.

Van den Broek, F.A.R., Ritskes-Hoitinga J. and Beynen A.C. (2000). *Biol. Trace Elem. Res.* **78**, 191–203.

Zhou, J. (2001). Hyperhomocysteinemia and atherosclerosis in ApoE-deficient mice. Thesis, Faculty of Health Sciences, Aarhus University, Denmark.

Zhou, J., Moeller, J., Danielsen, C.C., Bentzon, J., Ravn, H.B., Austin, R.C. and Falk, E. (2001). *Arterioscler. Thromb. Vasc. Biol.* **21**, 1470–1476.

Zhou, J., Moller, J., Ritskes-Hoitinga, M., Larsen, M.L., Austin, R.C. and Falk, E. (2003). *Atherosclerosis* **168**, 255–262.

PART 6

Procedures

Contents

CHAPTER 29

Legal Regulations for the Protection of Animals Used for Scientific Experiments

W. A. de Leeuw

Food and Consumer Safety Authority, Directorate of Inspection, Department of Veterinary Public Health, Animal Diseases, Animal Welfare and Feed, The Hague, The Netherlands

K. Gärtner

Zentrales Tierlaboratorium, Medizinische Hochschule Hannover, Germany

Introduction: Historical developments, public and scientific claims

The conflict between the wish to protect animals and, on the other hand, to use them for various experimental purposes is governed by different national laws and international agreements. This applies particularly when animals are held captive or subjected to possible pain, distress and harm.

When animals are used for scientific experiments, a number of stipulations are imposed besides those complying with the legislation for protection of animals. These include regulations corresponding to drug safety, consumer protection, medical and other applied methods of research and basic research. The following contribution describes the legal regulations concerning the controversy between the interests of biomedical research and the protection of animals.

The national laws in force today differ from one to another due to the historical and regional discrepancies in the human–animal relations.

The Laboratory Mouse
Copyright 2004 Elsevier
ISBN 0-1233-6425-6

Experiments with animals date back to at least the fifth century BC. The experiments conducted in those days were mainly for the purpose of describing biological systems. In the second century AD, animal experiments were performed that were to form the basis of medicine for the following ages. As Christianity progressed, the growth of experimental science stagnated until the fifteenth century. It was then that scientists again came to recognise that experimental research was more important than theoretical debate. Following freedom of religion in the nineteenth century, the constitutional laws laid down the freedom of science, research and teaching in the European countries. It is embodied expressis verbis in their today's constitutional laws e.g. Italy, Austria, Spain, Germany.

Since the eighteenth century animal experiments have been considered necessary for progress in medical, biological and basic life sciences. However, inflicting pain, harm or suffering on animals used for experimental purposes, in particular, vertebrates, left by no means unaffected the people performing those experiments. Other people who are not directly involved also suffer a similar feeling when hearing about the experiments performed. The feelings of protection and care towards the animals are violated, feelings which have been nurtured over the centuries as a result of living closely with domestic animals and which became the basis of using them. At the latest, around the beginning of the nineteenth century, the scientists concerned gave their reasons for conducting animal experiments and acknowledged their actions. Five prerequisites were laid down:

1. Animal experiments may only be performed in the case of real scientific issues.
2. Pain relief and anaesthesia must be administered whenever possible.
3. Pain, suffering and harm and the number of animals used should be reduced to an indispensable amount.
4. Animal experiments may only be conducted by experienced people.
5. Preference is given for the use of animals of less developed species than animals of higher developed species.

In parallel with the self-liability undertaken by the scientists, at the beginning of the nineteenth century organisations were established with the aim of protecting animals from laboratory research. Over the years, the debate gradually reached the political arena and thereby, also the discussion concerning the regulation of the use of animals for experimental purposes. In the

beginning, laws were laid down to confirm the five codices to be heeded by the researchers (London, 1822; Dresden, 1838; Württemberg, 1839; Berlin, 1886; Bretscheider, 1962). Major disputes ensued concerning adherence to these codices. A number of measures for control were introduced into the animal protection laws. This was first an issue in England as early as 1876, or in Germany repeatedly since 1933.

The main purpose of the laws introduced to control animal experiments was laid down by the UK Royal Commission as early as 1875 and was enshrined in the British Cruelty to Animals Act (1986). It was the result of a long debate between scientists, politicians and animal protectionists. This debate was also held in other western countries and continued in the twentieth century. Over the last three decades in particular, specific legislation has been issued and revised on international, federal and national levels around the world.

Essential parts of most legal documents are based on the 3R-principles of Refinement, Reduction and Replacement, as described by Russell and Burch (1959) in their book 'The Principles of Humane Experimental Technique'. The strength of the book lies in the fascinating manner the authors link the humane approach to experiments on animals, animal welfare and optimal science. In the 1980s, its real value was recognised by policy makers as well as by people involved directly or indirectly in the performance of animal experiments. The above mentioned five basic prerequisites were extended as follows:

- Performance of an experiment is not permissible if the result can be reasonably and practically obtained without the use of animals.
- Maintenance and care of the animals before and during the experiments must satisfy the animal's needs and welfare.
- Special attention must be paid to the process of reviewing protocols, providing training courses for people involved, in the performance of experiments on animals and in encouraging research concerning the development and validation of alternatives.

The following sections aim to clarify the legal aspects for the realisation of these goals. However, it is impossible to condense in a brief chapter all information on the regulations and laws in force today. Attempts have been made to mention the most stringent regulations laid down and to describe some of the national specific features of these laws and the status of the international vote in Europe. For those undertaking animal experiments, particularly in a foreign laboratory, information given has to be supported by acquainting the national and regional regulations in detail. Violation of regulations could be liable to a severe penalty (e.g. a German fine of up to 5000 Euro and imprisonment of up to 3 years; Tierschutzgesetz, 1998).

The legal frameworks

Legislation and regulations are laid down in all countries to protect the animals on the one hand and scientific research and teaching on the other. Their relations are difficult to describe overall. Therefore, both will be described separately.

Legal rules for the protection of animals

Although the respective national regulations for the protection of animals that is valid may differ from one country to another they have a common basis laid down in two European documents, which serve as a guideline for the national laws. These documents are the Council of Europe Convention for the Protection of Vertebrate Animals used for Experimental and other Scientific Purposes (ETS123, 1986) and the Council Directive (86/609/EEC) on the Approximation of Laws, Regulations and Administrative Provisions of the Member States regarding the Protection of Vertebrate Animals used for Experimental and other Scientific Purposes (1986). Their provisions must be seen as minimum requirements. States are free to regulate specific issues more strictly if desired.

The Convention of the Council of Europe is a treaty between Member States. When the national Parliament has approved this Convention, the Member State is Party to the Convention and legally bound under international law to implement the Convention. Regularly, meetings are held to discuss the progress of the implementation of the Convention ETS123 and the need for revision or extension of any of its provisions on the basis of new facts or developments. In addition to this Convention, there are four other Conventions concerning animal welfare. These cover the International Transport of Animals (ETS65),

Farm Animals (ETS87), Slaughter (ETS102) and Pet Animals (ETS125).

In 1986, the previously mentioned Directive 86/609/EEC was adopted by the Council of Ministers of the European Community. This document was based upon ETS123, but its text is more concise and its requirements more stringent and aims to avoid affecting the establishment and functioning of the Common Market, in particular, by distortions of competition or barriers to trade. Member States are compelled to implement the provisions of the EU Directive through their national legislation.

In the United States of America (USA), the federal Animal Welfare Act (AWA, 1996) is the central legal document. Research establishments covered under the AWA must be registered with and inspected by the Animal and Plant Health Inspection (APHIS) of the United States Department of Agriculture (USDA). All institutions receiving grants from the US Public Health Service (PHS) must adhere to the Health Research Extension Act (PHS policy). This policy also covers birds, rats and mice. One of the elements of the PHS policy is that the standards as laid down in the *Guide for the Care and Use of Laboratory Animals* (National Research Council, NRC, 1996) must be followed. Under the AWA as well as the PHS policy, institutions that are covered by it are required to publish a yearly report. Some institutions are covered by both policies, whereas others are not covered at all. These are, for example, colleges or biotechnological companies that only use mice and rats and do not receive any PHS funding. Some states have additional laws.

Both Canada and Australia have no national legislation controlling the use of animals for experimental and other scientific purposes. In both countries, however, (some) states have their own laws and regulations. In Australia, the *Australian Code of Practice for the Care and Use of Animals for Scientific Purposes* (National Health and Medical Research Council, NHMRC, 1997) is a guiding document. Canadian researchers using vertebrates and cephalopods must comply with the requirements of the Canadian Council on Animal Care (CCAC, 1984). The CCAC is a peer review organisation that comprises 20 member organisations.

In New Zealand, the AWA (1999) regulates the use of animals for research, testing or teaching. No research, testing or teaching may be carried out on any live animal unless the person or organisation involved holds an approved code of ethical conduct. A national *Code of Recommendations and Minimum Standards for the Care and Use of Animals for Scientific Purposes* is issued by the Animal Welfare Advisory Committee (1995).

In Japan, there are two laws regulating animal care and use. These include the Law for the Protection and Management of Animals (1973) and the Standards Relating to the Care and Management of Experimental Animals (1980). The Japanese Association for Laboratory Animal Science (JALAS) has published guidelines on animal experimentation. These are used by various institutes. Overall, the Japanese laws and regulations are basically ethical codes that rely on the good sense of researchers and do not contain strict punitive measures.

Worldwide, the responsibility for the protection of laboratory animals is linked to different parts of the government. In some countries (e.g. USA, New Zealand, Finland, Germany, Norway and Portugal), the Minister of Agriculture is responsible for the protection of laboratory animals. In other countries, the Minister responsible can be the Home Office (UK), the Minister of Justice (Denmark), the Minister of Science and Education (France), the Minister of Public Health (the Netherlands) or the Prime Minister's Office (Japan).

Legal rules for the protection of scientific research

Freedom of opinion is a core element of national constitutional laws in democratic countries. The constitutional laws in European countries safeguard freedom of opinion. This applies not only to freedom of the press or art but to the same extent and expressis verbis to freedom of science, research and teaching.

The Charter of European Constitutional Rights (2000) stated in Article 13 'Art and research are free. Academic freedom is acknowledged'. This freedom applies to all EU countries, even if not stated in their respective constitutional laws. However, it involves diverse restrictions towards other constitutional standards, as integrity of a human individual or his/her property, etc. But, in relation to German laws, freedom of research has to remain unrestricted concerning the choice of a scientific problem, the chosen scientific method and the manner of publishing the results. (Deutsches Hochschulrahmengesetz, 1994; Maunz-Dühring-Herzog-Scolz, 1999).

Restrictions on the use of animals for experiments, such as featured in the national regulations for the

protection of animals or in the EU Directive 86/609/EEC, must comply with the right of freedom for science and research. The laws and international regulations are concerned in maintaining this balance. This sometimes leads to misunderstandings with regard to the formulation of the regulations and, thus, complications in their application. This has given rise to controversies and in Germany, on two occasions, the leading Constitutional Courts have had to enforce the demand for freedom for scientific research and teaching.

Freedom of science and research in a legal sense is granted to the researcher, but, is no personal privilege for him to do everything. This freedom is based on the obligations pertaining to scientific search processes (Singer, 1995; DFG, 1996; Gärtner, 1995). The obligations are laid down in the internal regulations of scientific responsibility – these are rationality as the sole justification; Lucidity and acknowledgement of sources, methods, protocols, goals, hypotheses and other elements pertaining science; Outspokenness as opposed to critical assessment etc. – and they are laid down in external regulations with regard to political–social responsibility (Mohr, 1987; Nida-Rümelin, 1996). In relation to the external regulations, statements have been made by scientific organisations, sponsors and academies in connection with experiments using animals. The above-mentioned 3R-concept has gained unrestricted support. Animal-ethically based conditions laid down by governmental authorities should be taken into consideration whenever possible (e.g. DFG, 1991; Swiss Academy of Medical Sciences, 1995; European Science Foundation, 2001).

This understanding of science covers the legal balance between the protection of animals and science.

The assessment of scientific necessity and the consideration of the concept of animal-ethics determine the official authorisation to perform experiments on animals. The ethical concept of utilitarism has been appointed for this bioethical consideration in national regulations (Germany, Austria, Switzerland) and the EU Directive 86/609/EEC. It states: if an animal is expected to experience severe pain …, the experiment … is only permitted if it is of sufficient importance in meeting the essential needs of man or animal.

However, the ethical concept of utilitarianism can apply only to projects of applied scientific fields. The utilitarian concept has limitations if used for projects of basic research, if no simple reference can be established as to medical or other uses to the advantage of man or animal. A few national laws for the protection of animals (Germany, Austria) and the European regulations solve this problem in the following way. They allege that the solution of scientific problems by means of basic research is of great importance for man and animal. They state 'the experiment is only permitted if it is of sufficient importance in meeting the essential needs of man or animal, including the solution of scientific problems'. In the case that this allegation not be heeded, science was taking action successfully against this restriction of freedom of science.

Aim and scope of legislation of animal experimentation

The overall aims of legislation of animal experimentation are: (1) scientific procedures likely to cause pain, suffering, distress or lasting harm to animals must be carried out with the minimum cost in animal suffering and in animal numbers, (2) they have to be well justified by its scientific quality and by potential benefits of men or animals, (3) they are not permitted, if the result can be reasonably and practically obtained without the use of animals, (4) alleviation of pain and anaesthesia has to be applied whenever possible, (5) preference is given for using animals of less developed species instead of animals of more highly developed species.

The regulatory systems, described in the following sections in more detail, aim at enforcing these principles by the following topics and are based on the pathocentric concept of animal protection:

- to define legitimate purposes for which laboratory animals may be used;
- to ensure the competence of all people involved in animal experimentation;
- to exert control over permitted levels of pain or other forms of distress;
- to provide for inspection of facilities and procedures;
- to ensure humane standards of animal husbandry and care;
- to ensure that animal experiments are only performed if there are no sound alternatives.

Special attention is paid to providing training courses for people involved in the performance of experiments on animals and encouraging research into the development and validation of alternatives.

The concept of pathocentric animal protection and killing of laboratory animals

The pathocentric concept says: it is forbidden to subject the animals to pain, suffering and harm. Exceptions to this rule when performing animal experiments require detailed scientific reasons and ethical consideration between the extent of pain and suffering inflicted and the gaining of scientific findings.

However, the pathocentric concept only offers a very limited degree of protection for the life of an animal. From the point of view of pathocentric protection, the loss of life expectancy does not subject an animal to any harm if the animal is killed in a painless manner. All that is required is a suitable reason for killing an animal. In some countries (Austria, Germany, Switzerland), killing of laboratory animals without any prior treatment, but for the use for scientific purposes, is not considered an animal experiment as defined by law, whereas in some other countries such as France, Sweden and the Netherlands, it is considered as an experiment.

The Council of Europe Convention ETS123 and the EU Directive 86/609/EEC apply only to any experimental or other scientific use of an animal which may cause it pain, suffering, distress or lasting harm, but excluding the least painful methods accepted in modern practice of killing an animal.

Protection of animals differ among the species

The pathocentric concept for the protection of animals entails differing protection warranties among the species, in particular, between the invertebrates and vertebrates or the foetal and developed forms of vertebrates. Invertebrates and undeveloped forms of vertebrates are mostly considered to experience no or less pain and distress. On the other hand the pathocentric concept applies also to gentechnological procedures performed on the genome or on the germ cells which themselves are insensitive to pain, if pain, suffering or harm is inflicted on the genotypical-changed, developed animals or animals carrying foetuses or germ cells.

The range of animals protected by legislation differs between the countries. Some countries offer protection to all species including invertebrates (Germany, Austria, Switzerland). However, invertebrates are only given a low degree of protection. A high degree of protection is given to vertebrates, however including cephalopods and decapods (Germany), and higher still to warm-blooded species. In some countries such as the UK and the Netherlands, the use of great apes is banned. In some countries legislation covers only vertebrate animals. The provisions of the USA AWA, aiming at the protection of animals used for scientific and experimental purposes, does not cover all the vertebrate species. The US House of Representatives approved the Farm Bill, which includes an amendment to exclude birds, rats and mice from the definition of 'animal' in the AWA. The PHS policy applies also to these species. On the other hand, in some countries, such as New Zealand and the UK the law also covers non-vertebrate animals, such as the Octopus vulgaris, as an experimental animal.

The EU Directive 86/609/EEC states that protection be given to each living vertebrate used for experimental purposes no matter what species. In this context, 'animal' means any live non-human vertebrate, including free-living and/or reproducing larval forms, but excluding other foetal or embryonic forms.

The legitimate purposes of animal experimentation

In a few countries, the purposes of animal experimentation have been categorised into positive and negative lists. Animal experimentation is accepted on the grounds that the animals serve to recognise, treat or prevent illnesses, harm to body or suffering, also to detect physiological functions both in humans and in animals. Experimentation is also permitted to detect endangering aspects to the environment and for testing

suspected components in substances and products and the efficacy against animal pests. Finally, animal experimentation in basic research is accepted. However, a few national laws (Germany, Austria etc.) do not authorise animal experimentation for the production and testing of weapons, munition and the corresponding devices, also for the development of tobacco products, detergents and cosmetics.

The use of animals for educational purposes is not covered by all national laws. In most countries, animal experiments are only accepted when no alternative methods are possible in education, advanced training and further training in universities and in colleges for medicine and science.

The EU Directive 86/609/EEC is restricted in its application to experiments undertaken for the development and safety testing of drugs and other products, together with protection of the natural environment. However, in 1987, the European Parliament passed a resolution stating that its provisions should also apply, through national legislation, to all animal experiments, including those undertaken for fundamental or basic research and for educational purposes.

In a few countries (Germany, Austria etc.), gentechnological procedures performed on the genome or on the germ cells of vertebrates are regulated by the animal protection laws as well as by other rules, which are laid down specifically for gen-technological procedures. Animal protection rules are concerned, whenever the procedures may cause pain, suffering or harm for the genotypical-changed animals or for animals carrying such manipulated foetuses or germ cells.

Evaluation of the scientific and bio ethical quality of animal experiments

The following guiding principle is broadly accepted nowadays: 'An animal experiment may only be performed when it has been decided that the experiment is justified weighing the scientific and educational value or social significance against the potential effects on the health, welfare and integrity of the animals used'. This cost–benefit approach can be found explicitly or implicitly in most of the specific legislation on animal experiments. The process of review and authorisation of protocols varies substantially between the individual countries.

Canada was the first country where, in 1968, institutional Animal Care Committees (ACC) were introduced. The CCAC programme requires the establishment of these ACCs and they are regularly being assessed by CCAC assessment panels.

In the USA, it is required by the AWA as well as by the PHS policy that each institution establishes an Institutional Animal Care and Use Committee (IACUC). This IACUC reviews the protocols, site visits the facility and is responsible for the evaluation of the total animal care programme of the institution. It has the power to reject protocols or to stop ongoing research if it believes that USDA standards are not being met.

Both the Canadian ACCs and the American IACUCs are faced with an extensive task. Among other things, they review new protocols and ongoing research, ensure that animal facilities are adequate, that the personnel are experienced and well trained and that housing conditions meet the specific requirements of the species concerned.

In New Zealand, no project may proceed without the approval of an Animal Ethics Committee (AEC) established under a Code of ethical conduct. Only the Minister has the right to approve research or testing in the national interest without AEC review. At regular intervals, the committees are subjected to independent review.

In most of the European countries there is some form of ethical review, although this is not required by law in all of these countries. There are differences between countries with regard to the composition of the AECs, the expertise on board, the localisation (central, regional or local), their method of function, their legal status, the criteria used in the evaluation, the scope of the tasks of the AECs, the level of assessment (project level versus procedure level) and the competence of the AECs (advisory versus mandatory). Nowadays, funding agencies also require an ethical review before proposals can be financially supported. The authorisation procedure, whether this is done by applying for a project licence and discussing it with a government inspector, or explaining a research proposal to an Institutional Ethics Committee, focuses the mind of the researcher on the need to keep animal use to the minimum, and also encourages a consideration of alternatives and ways to minimise pain or distress, both during and after the experiment.

There are no specific provisions concerning the ethical review of experiments on animals in the EU Directive 86/609/EEC and the CoE Convention ETS123. However it is to be foreseen that the issue of ethical review of proposed animal experiments will be taken on board when EU Directive 86/609/EEC is revised. Scientific research, be it fundamental or

applied, is an international domain where exchange of information as well as co-operation between laboratories in different countries are key factors. Furthermore, be it for publications or for funding, respect of ethical standards for the use of animals is now required. In this context, a common reflection at international level, on the ethical evaluation processes of experimental projects and moreover, the definition of a minimum harmonised basis seems essential.

Competence of persons involved in animal experimentation

In the last 50 years, the quality of the methodical procedures in all biomedical disciplines has improved enormously. Scientific acceptance nowadays includes methods that demand accurate handling by technical, supervisory or academic experts. The assessment of results by means of appropriate statistical calculations is also of great importance. Similar improvements are also indicated in most procedures entailing animal experimentation. Biomedical disciplines have established particular forms of training and further education for the qualification of specialists in the specific field of animal experimentation and laboratory animal husbandry, care and breeding. This first took place within the various disciplines and then on a national level. Today, there are interdisciplinary and international forms of training. Specialist examinations and diplomas at a technical, supervisory or academic level conclude such education courses. This development in biomedical science corresponds extensively with the demands laid down for the protection of animals used for experiments in that only experts are permitted to carry out the procedures and that the number of animals used and the pain inflicted must be reduced to an indispensable minimum. Such education has developed optimally in Europe, particularly in the Netherlands, the Scandinavian countries, Switzerland, Germany and Austria.

In most of the national and international legislation concerning experiments on animals there are provisions relating to the competence of people involved in these experiments. Article 7 of the EU Directive 86/609/EEC states that animal experiments should be performed solely by a person considered as competent or under the direct responsibility of such a person. This provision is emphasised in Article 14, which states that people undertaking experiments, participating in procedures or caring for experimental animals (including supervision) should have the appropriate education and training. It is essential that people involved in the design and conduct of experiments should have received an education in a scientific discipline relevant to the experimental work. They also need to be capable of handling and taking care of laboratory animals. It has become more and more recognised that competence is based upon a combination of skills, experience and attitude.

The Federation of European Laboratory Animal Science Associations has formulated standards for four categories of people. These are: Cat. A: persons taking care of animals (FELASA, 1995); Cat. B: persons carrying out animal experiments (FELASA, 2000); Cat. C: persons responsible for designing and conducting animal experiments (FELASA, 1995); Cat. D: laboratory animal specialists (FELASA, 1999). In the last two decades, in many countries, much effort has been put into setting up training courses for the different categories of people involved in the performance of experiments on animals. Although at this moment there are substantial differences between countries concerning the training courses for the same categories of people and criteria used for competence, it is to be expected that, at least within the EU, this will be gradually harmonised in the future.

these guidelines are based on legal provisions, but are issued as separate documents that lack legal force as such. Attached to the CoE Convention ETS123 and the EU Directive 86/609/EEC in the early 1980s appendices were formulated specifying guidelines on housing and care which reflect the experience and knowledge as it was two decades ago. Appendix A of Convention ETS123 is being revised and it is to be expected that it will also replace the Annex A of the EU Directive 86/609/EEC. Seven expert groups have set up proposals for the different species. The revised Appendix A will cover more species and emphasis will be placed on social and environmental improvement with the aim of maintaining laboratory animals in a physical and mental state where they suffer least and the procedures give the best results. The next paragraph gives some general starting points that are used for the formulation of new guidelines for the husbandry and care of laboratory mice.

Laboratory mice should be allocated sufficient space of adequate complexity to express a wide range of normal behaviour. Mice are social animals and should only be kept isolated for welfare, veterinary or experimental reasons. The cages must be equipped to ensure that the basic requirements of the mice are met (food, water, temperature, mobility, resting area etc.) and that the mice are kept under conditions which resemble as close as possible their natural environment. In order for the mice to demonstrate certain behavioural traits, social or material triggering signals must be installed within the cage. These are described as enrichments. Bedding and nesting material must be made available.

Humane standards of animal husbandry and care

The housing and care of the animals is of great importance mainly for the health and welfare of the animals and consequently, also for the quality of the research performed. Specialised personnel and others are often required to ensure the animals receive maximal care and minimal disturbance. Scientific associations are constantly publishing detailed recommendations for the maintenance and care of each species of animal used for experiments and consideration for the conduct of the procedures.

In most countries there are reference documents or Codes that contain guidelines on the housing and care of animals used for experimental purposes. In most cases,

Control over permitted levels of pain or other forms of distress and the number of animals used

Refinement – relief and prevention of pain

A general significant point included in the regulations more or less explicitly requires that experiments be carried out under general or local anaesthesia, unless anaesthesia is judged to be more traumatic than the experiment itself or is incompatible with the aims of the experiment. If anaesthesia is not possible, then pain, distress or suffering must be limited and analgesics or other appropriate methods should be used. In the last decade, more emphasis has been placed on the use of humane end points in experiments. This is an important way of refining experiments on animals.

Euthanasia

Another significant point, that is generally accepted, states that at the end of the experiment a veterinarian or other competent person must decide whether or not the animals be kept alive or humanely killed. No animal is to be kept alive if it is likely to remain in permanent pain or distress or if its wellbeing is jeopardised in any way. It is not permissible to use animals more than once in experiments entailing severe pain, distress or suffering.

Reduction in experimentation and the number of animals used to an indispensable minimum

The most frequent aim of animal experimentation is the evaluation of experimental conditions on quantitative biological characteristics. In which direction and by what amount do they differ from the normal range? The large biological variability of these quantitative characteristics requires repeated investigations on different animals in order to gain accurate scientific statements. The more often the experiments are repeated, the more certain is the scientific statement. However, it must be ensured that only the indispensable smallest number of animals is used. Only a few repeated investigations are sometimes sufficient for a pilot study in clinical research. Reliable scientific evidence requires frequent repeated investigations, the number of which has to be biostatistically calculated. Determining points are: (1) the extent of variability of the observed characteristics, (2) the wanted degree of probability for the scientific statement based on the experiment (significance of the investigation), and (3) assessment of the degree of difference between a characteristic measured in control and experimental groups, which has scientific relevance.

Reducing the variability in the observed biological characteristics reduces the number of repeated investigations in different animals. This can be accomplished by biological standardisation (genotype, age, sex, health state etc.) of the animals as well as their maintenance (caging and care) conditions and the performance of the experiment. Special consideration should be given to the potential impact of enrichment programmes on the outcome of scientific studies to avoid the generation of invalid data and consequent animal wastage.

The scientific community gives the degree of probability which this community expects for accepting a scientific statement. Scientific statements are only acceptable if they are characterised by an uncertainty smaller than 5% (significance, $p<0.05$). If smaller degrees ($p<0.01$; $p<0.001$) are wanted many more animals are needed for such experimentation. Therefore, such demands require special bioethical considerations.

Scientific relevance is given for most of the questions arising even in the case of minimal differences between controls and experimental groups. Should this not be the case, the determination of larger differentials contributes considerably towards the reduction of necessary repetition of investigations and number of animals needed.

Permission for animal experimentation

Permission for animal experimentation is granted on the basis of a written detailed application, the formulation of which is already rigidly laid down in some countries (Germany, UK, Sweden, The Netherlands). As a minimum, information is required, concerning:

- a detailed and precisely described scientific issue,
- the intended individual methodical procedures,
- the prospective species of animal needed and number of animals,
- conduct of experiments by specialised personnel,
- the anaesthesia and analgesic procedures,
- the location of the intended experimentation.

Permission is usually on a restricted time basis. Once permission has been granted, the conduct of the animal experimentation is subject to further local monitoring and control.

Inspection of facilities and procedures

Linked to legislation are always the concepts of monitoring and enforcement. In this context, two levels can be distinguished, namely the institutional monitoring and the external, regional or national monitoring and enforcement. In several European countries at an institutional level, one or more persons are appointed to monitor the welfare of the animals and, in some cases, also the implementation of regulations. Examples are the Named Veterinary Surgeon and named Day-to-Day Care Person (UK), the Animal Welfare Officer (the Netherlands), and the authorised person responsible for protection of animals (Germany). These people have also an important advisory task.

Furthermore, in most of the European countries many organisations conducting animal experiments are subjected to regular inspection of their locations by external supervisory authorities. In this way, the conduct of animal experiments can be kept under control.

In the USA, Canada, Australia and New Zealand, the institutional committees are responsible for the evaluation and implementation of the total animal care programme. These committees visit the premises on a more or less regular basis. Both the named people and the committees play a very significant intermediate role, namely, being advocate of the animals and also the advisor and counterpart of the scientists, animal technicians and caretakers, and the director of the institutes. In the United States, the institutes that are covered by the AWA are inspected by the APHIS of the USDA.

Assurance of public accountability

Statistics have been published in England for over 100 years stating the number of animals required for experiments, the degree of stress involved for each animal, and the scientific fields of work for which the animals are used. These data are an important indication for control of the animals used for experimentation in biomedical research. Statistics have also been published for a number of years in several other European countries with the same objectives as mentioned above.

EU Member States must collect information on the number of animals used for specific purposes. As far as possible this must be made available to the public in the form of published statistics. Specific EU tables have been designed and distributed for the collection on such data. From 1999, data on animals used for scientific purposes has been collected in all EU Member States. Every 2 years, the European Commission sends a report of animals used for scientific purposes in Europe to the European Parliament. In many countries, national statistics are published and sent to the National Parliament. Often these reports include more information than is required for the EU tables. Statistical data can be considered as a valuable tool. It provides insight into the relation between the species, the number of animals used, and the type of research for which the animals are used. They provide up to date information on the trends. Thus, it provides an account of the animals used for experiments and is a basis for discussions on the use of animals for research and the effects of alternative methods.

In the USA, institutions covered by the AWA must report yearly to the USDA about the number of animals (not including mice, rats or birds) used and the pain category of each animal. In this way, around 5% of the animals used in the USA is disclosed. In Canada, the statistics also include rats and mice.

Alternatives

Animal experimentation is only permissible if there are no sound alternative methods not requiring the use of live animals. When applying for authorisation for animal experimentation, it must be made clear by the applicant that the scientific issue can only be dealt with by means of using animals for the experiments. In the last 30 years, molecular biology and cell-biology have created possibilities for dealing successfully with certain biomedical issues using *in vitro* methods and not animals. False expectations arose among the public in that it was hoped that all biomedical issues could be solved with methods not requiring the use of animals. The organisations named below, dedicated to the control and development of such alternatives, are concerned with other suitable methods available which do not entail the use of animals.

The EU Directive 86/609/EEC includes a provision stating that the Commission of the European Union and its Member States should encourage research into the development and validation of alternatives. Consequently, the EU has set up the European Centre for the Validation of Alternative Methods (ECVAM). Several European countries have set up national centres that have a role in stimulating the development, validation, application and implementation of alternative methods. In the USA, the Johns Hopkins Centre for Alternatives to Animal Testing has a leading role in this field. In addition to this, many countries have private funds and industries that support substantially the development and implementation of alternative methods. Nowadays, there are many operational networks in which universities, governmental bodies, industries and private funds co-operate. In 2002, the European Consensus Platform for Alternatives (ECOPA) was founded. It unites a number of consensus platforms that are active in different European countries. In these platforms representatives of academia, industry, animal welfare organisations and governments participate.

References

Animal Welfare Act (1996). In *United States Code*, Title 7, Sections 2131–2156.

Animal Welfare Act (1999). In *Public Act 1999*, number 142, New Zealand.

Animal Welfare Advisory Committee (1995). *Code of Recommendations and Minimum Standards for the Care and Use of Animals for Scientific Purposes.* Code of Animal Welfare No. 17, c/o Ministry of Agriculture and Forestry, PO Box 2526, Wellington, New Zealand.

Bretscheider, H. (1962). *Der Streit um die Vivisektion im 19. Jahrhundert*, Fischer Verlag, Stuttgart.

British Cruelty to Animals Act (1876/1986). 1st edn 1876, valid of to-day: Animals (Scientific Procedures) Act 1986. In *Guidance on the Operation of the Animals (Scientific Procedures) Act 1986*. HMSO, London.

Canadian Council on Animal Care (1984). *Guide to the Care and Use of Experimental Animals*, Vols 1 and 2. Ottawa.

Charter of European Constitutional Rights (2000). Nice 8.12. 2000. cit. A. Weber, Die Charta der Europäischen Union, München, Verlag Sellier, 2002.

Council Directive (86/609/EEC) of 24 November 1986 on the approximation of laws, regulations and administrative provisions of the Member States regarding the protection of animals used for experimental and other scientific purposes. *Off. J. Eur. Commun.* L 358, Volume 29, 18 December 1986.

Deutsches Hochschulrahmengesetz (1994). Deutschland §3 (2), Bundesgesetzblatt 1087.

DFG (1991). Novellierung des Tierschutzgesetzes 1986, Informationen für den Forscher. VCH Verlagsgesellschaft Weinheim, ISBN 3-527-27387-5.

DFG (1996). *Deutsche Forschungsgermeinschaft, Forschungsfreiheit*. VCH Verlagsgesellschaft Weinheim, ISBN 3-527-27211-9.

European Convention for the Protection of Vertebrate Animals Used for Experimental and Other Scientific Purposes ETS123. (1986). Council of Europe, Strasbourg.

European Science Foundation (2001). *Use of Animals in Research: ESF Statement on the Use of Animals in Research.* European Science Foundation, Strasbourg.

Federation of European Laboratory Animal Science Associations (1995). *Lab. Anim.* **29**, 121–131.

Federation of European Laboratory Animal Science Associations (1999). *Lab. Anim.* **33**, 1–15.

Federation of European Laboratory Animal Science Associations (2000). *Lab. Anim.* **34**, 229–235.

Gärtner, K. (1996). *Biologie in unserer Zeit* **26**, 81–85.

Law for the Protection and Management of Animals (1973/1982). *Exp. Anim.* **31**, 221–224.

Maunz-Dühring-Herzog-Scholz (1999). Kommentar zum Grundgesetz der Bundesrepublik Deutschland, RN 101 zu Art. 5 Abs. 3; Beck.Verlag, München.

Mohr, H. (1987). Natur und Moral, Ethik in der Biologie, Wissenschaftliche Buchgesellschaft Darmstadt. Seite 42.

National Health and Medical Research Council (1997). *Australian Code of Practice for the Care and Use of Animals for Scientific Purposes.* Publication number 1960, Australian Council, Australian Government Publishing Service, Canberra, Australia.

National Research Council (1996). *Guide to the Care and Use of Laboratory Animals.* National Academic Press, Washington, DC.

Nida-Rümelin, J. (1996). In *Angewandte Ethik* (ed. J. Nida-Rümelin). Alfred Kröner Verlag, Stuttgart.

Russell, W.M.S. and Burch, R.L. (1992/1959). *The Principles of Humane Experimental Technique* (originally published in 1959). Universities Federation of Animal Welfare, Wheathampstead, England.

Swiss Academy of Medical Sciences (1995). *Ethical Principles and Guidelines for Scientific Experiments on Animals.* Swiss Academy of Medical Sciences, Petersplatz 13, CH-4051, Basel.

Singer, W. (1995). *Neugier als Verpflichtung, Warum der Mensch unentwegt forschen muß.* Frankfurter Allgemeine Zeitung 301, N1, Frankfurt/Main-Germany.

Standards Relating to the Care and Management of Experimental Animals (1980/1982). *Exp. Anim.* **31**, 228–231.

Tierschutzgesetz (1998). Fünfter Abschnitt, Paragraph 7–9, BGBl. I S. 1105, 1818, Bundesministerium für Verbraucherschutz, Ernährung und Landwirtschaft, Berlin.

Updated information with regard to Legislation worldwide can be found by http://www.frame.org.uk/Links/ Worldlaws.htm

Necropsy Methods

Rosemarie Seymour, Tsutomu Ichiki,[1]
Igor Mikaelian, Dawnalyn Boggess,
Kathleen A. Silva and John P. Sundberg
The Jackson Laboratory, Bar Harbor, ME

Introduction

The methodology used to systematically evaluate bodies to determine the cause of death is commonly called the autopsy procedure. Use of the word autopsy should be limited to performing a post mortem examination on humans. In contrast, necropsy is the correct term used in veterinary medicine for post mortem examination of non-human species. A necropsy is a highly technical procedure when done correctly and can yield a great deal of information. Proper handling of the animal antemortem and postmortem are necessary to optimize results. Detailed analysis of clinically ill or genetically engineered mice (GEMs) is necessary to obtain the correct diagnosis or characterization of the clinical features (phenotype). Simply looking at the organ of interest, or selections of organs (e.g. lungs, kidney, and/or liver) is not adequate and has resulted in many errors in the literature.

[1]Current address: Dainippon Ink and Chemicals, Inc. Chiba, Japan.

Biological characterization of a new mutation

Individual mice of an inbred strain are essentially the same; therefore normal gross anatomy and histology should be the same for all mice within a strain. If maintained properly in controlled specific pathogen free (SPF) facilities, only a few mutant and littermate control mice from a new line need to be examined. Because pathogen status, diet, water, room conditions, and many other factors may change over time, using appropriate controls for each study is imperative. Preliminary studies can be done using small groups to identify important lesions. Information obtained in this manner can be used to plan more in-depth studies.

The numbers and ages of mice needed for any given study should be carefully considered. Some groups select ages based on convenience. A common age is 6–8 weeks of age because mice are fully developed (adult) and have been weaned. Understanding the biology of the laboratory mouse provides a more logical approach. Major changes in the life of a mouse, as in all animals, provide reasons to choose specific ages for following the development and progression of lesions. Major life changes include birth, weaning (3–4 weeks), sexual maturity (6–8 weeks), sexual quiescence (6–8 months), and geriatric stages (1–2 years +; Sundberg *et al.*, 1998). Ranges for sexual quiescence are given to allow for differences between inbred strains (Green and Witham, 1991). It is important to understand these strain specific differences and determine specific ages for the strain of interest at the onset of the study. The cost of maintaining mice through geriatric age is prohibitive for screening studies, so this is rarely done. Many mice with genetic mutations often do not live to geriatric age so it may also be impossible to obtain mutant mice for such a group.

Rodents, including mice, are altricial, meaning their young are born in a relatively undeveloped state. For example, hair fibers in mice do not emerge from the skin until 5 days of age because hair follicles have not completed development. Eyelids open around 12 days of age. Glomeruli in the kidneys do not fully develop for several weeks after birth. Each organ has a specific developmental process postpartum. As such, specific modifications in protocols must be made to conduct detailed studies on each organ. The skin is used as an example in this text. Readers should refer to books and review articles to find details on other organs. During the first 3 weeks of a mouse's life the skin and hair undergo dramatic changes. The epidermis of a normal newborn mouse is relatively thick and becomes thinner by 2 weeks of age (Ward *et al.*, 2000; Smith *et al.*, 2002). Hair follicles develop completely during the first week after birth and the first hair fibers begin to emerge at around 5 days of age (Paus *et al.*, 1999; Muller-Rover *et al.*, 2001). The hair cycle is synchronous and short during the first 3 weeks after birth, making it easy to study details of the hair cycle in mice. To evaluate these cutaneous changes systematically, skin is collected from mice at 2–3-day intervals (birth, 3 days, 6 days, etc. to 21 days). Interval-specific postpartum collection ages can be defined for each organ using this approach.

The numbers of mice in this type of study can rapidly become quite large. However, since most mutations occur, or are induced, on inbred mouse strains maintained in controlled environments, free of specific pathogens, it is possible to do a complete study on as few as two mutants and two controls of each sex at each of the time points mentioned above (N = 8 per age group). Appropriate housing, as described above, helps to ensure the accuracy of the study. Mice collected for the major life event points should have complete sets of organs harvested for study (Table 30.1). Injecting the mice with bromodeoxyuridine or tritiated thymidine prior to euthanasia makes it possible to use serial sections of tissues taken from individual mice for kinetic analysis. Mice with some skin mutations exhibit a positive **Köbner's reaction** following injury, so the quality of multiple biopsies over time from the same mouse may be less than optimal (Nanney *et al.*, 1996; Muller-Rover *et al.*, 2001). This is why individual mice are necropsied at specific time points rather than taking successive biopsies from the same animal.

In order to ensure optimal standardization, the same technician should collect tissues from the same anatomic sites from each mouse. Whenever possible, the same technician should be used for a particular study, thus keeping the technique and any inherent errors consistent.

Clinical evaluation

Live mice should be carefully examined for behavior and external abnormalities. Most **homozygous** recessive mutations (*m/m*) are available with **heterozygous** (+/*m*) or wild type (+/+) age and sex matched controls on the same genetic background (where '*m*' is the mutant gene being studied, and '+' is the normal or wild type gene). Controls should be examined side by side with mutants as a basis for comparison. Familiarity with the normal phenotype of the inbred strain being used is essential in assessing phenotypic variations of the mutant. Many infectious diseases in mice can present as behavioral abnormalities, such as circling or torticollis associated with middle ear infections (otitis media). When present in both mutant and control mice, it is prudent to do infectious disease surveillance studies on the colony. Some mutations have clinical phenotypes that resemble infectious diseases such as cutaneous scaling (as in ringworm) or focal alopecia (hair loss potentially due to ectoparasite infestation). Disease issues are beyond the scope of this chapter and are covered in detail elsewhere (Percy and Barthold, 1993; Mohr *et al.*, 1996).

Mice should be allowed to walk around in their box so that their patterns of behavior can be observed.

TABLE 30.1: Sample corresponding table	
Cassette	Organs or tissues
1–3	Small intestinal rolls
4	Colon roll with anus
5	Longitudinal section of stomach
6	Longitudinal section of cecum (or cecum and stomach)
7	Cross section of medial lobes of liver with gall bladder
	Cross section of left lateral lobe of liver (largest piece)
	Cross section of spleen and attached pancreas
8	Longitudinal section of left kidney and adrenal
	Cross section of right kidney and adrenal
9	Reproductive organs and urinary bladder
10	Longitudinal section of heart, showing all chambers
11	Longitudinal sections of lungs (center of lobes on both sides)
12	Coronal sections of brain
13	Dorsal skin, tail skin, and ear
14	Ventral skin, muzzle, and eyelid
15	Longitudinal section of hind leg, showing long bones (foot removed)
16	Longitudinal section of front leg, showing long bones (foot removed)
17	Longitudinal section of hind foot, both halves, showing bones and nails
18	Longitudinal section of front foot, both halves, showing bones and nails
19	Foot pad from hind foot
20	Coronal sections of the skull
21	1–2 cross sections of the tail, 1 longitudinal section of the tail
22	1 cross section, 1 longitudinal section of the cervical and thoracic spine
23	1 cross section (at the hip joint) and 1 longitudinal section of the lumbar spine
24	3 sections of the lower jaw: 1 cross section at the caudal end including the thyroid and parathyroid
	1 longitudinal section showing the molars
	1 longitudinal section including the tongue and front teeth
	Other tissues such as salivary glands, thymus
25	Tumor or other abnormality

Normal mice will constantly explore their environment and be alert and active. They should respond to external stimuli without abnormal reactions. Some inbred strains like DBA/2J respond to loud noises by developing sudden and sometimes prolonged seizures, often ending in death (Hall, 1947). The mouse should have a uniform hair coat that lays flat. Vibrissae, the long hairs around the eyes and muzzle, should be straight and prominent. Ears should be erect and light pink in color for albino mice. Pale white ears in albino mice are suggestive of anemia. A blood sample can be taken to determine packed cell volume or hematocrit if needed. Eyes should be bilaterally symmetrical and clear. Incisors (front teeth) are commonly overgrown in many strains (malocclusion of incisors), but this may be part of the phenotype if it is consistently observed in association with a particular mouse mutation.

Body openings should be checked for any gross abnormalities, and any secretions or excretions produced should appear normal. For example, mouse feces are about the size and shape of a rice grain, of firm consistency, and dark brown in color. Perianal matting of feces or light yellow colored feces might indicate the phenotype of inflammatory bowel disease or other intestinal disease (Sundberg et al., 1994). Several commonly used inbred strains of mice may have the appearance of extra testicles (three or four) due to cysts of the bulbourethral glands (Kiupel et al., 2000). The extremities should be visually inspected for any obvious deformity or swelling. The nails should be short and slightly curved. Many abnormalities can be identified at the clinical and gross evaluation of the mouse. Thorough evaluation of the first mutants in a study will provide guidelines on what to look for more specifically as the study progresses.

Animal identification information should be recorded appropriately. This includes age (dates of birth and necropsy), sex, strain, genotype if known, pedigree numbers, animal identification numbers including ear tag or punch (Figure 30.1), or toenail amputation, source (room), reason for submission, and name of submitting technician or scientist. If the mouse is part of an ongoing study, a special code number for that study should be assigned (Sundberg and Sundberg 1990, 2000).

Clinical pathology

Routine collection of biological fluids is done as part of physical examinations for humans and domestic animals

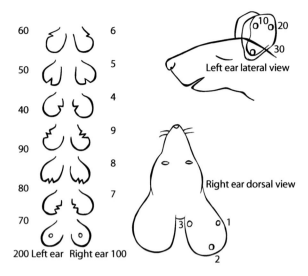

Figure 30.1 Mouse identification by ear notching.

in sickness or health. The methods are identical for mice but microassays have had to be developed (Car and Eng, 2001). Specimen collection and analysis can be done prospectively, as a routine procedure throughout all studies, or retrospectively, once a series of abnormalities are identified, to monitor or define the pathogenesis of the disease. The former approach requires a broad screen while the latter can focus on parameters specific to the organ of interest, potentially reducing costs. Since mutant mice usually provide a readily renewable population to study, the latter approach is commonly used in most research laboratories.

Blood collection

Blood is collected by **retro-orbital** bleeding, tail tip amputation, cardiac puncture, or decapitation. The reasons to use each method vary with age, purpose of the study, volume needed, and methods approved at each institution. These methods are covered in detail in Chapter 33 on Collection of Body Fluids.

Blood handling

For serum collection, blood can be held at room temperature for an hour and then centrifuged. The serum should be decanted and stored in plastic tubes (Eppendorf, Brinkmann Instruments, Inc., Westbury, NY; Nunc, Nalge Nunc International, Fisher Scientific, Pittsburg, PA) and frozen at $-80°C$ until used. Blood may sit for several hours before centrifuging, if stored at $4°C$.

Plasma is obtained by collecting blood in tubes containing ethylenediaminetetraacetic acid (EDTA) or heparin to prevent clotting. The blood is then centrifuged and frozen for future use.

Whole blood should be handled according to the diagnostic laboratory's instructions.

Feces

Feces are collected at the time of necropsy and can be frozen in plastic tubes for a variety of assays. Most mice defecate upon handling so a few fresh samples can be obtained from a defined individual. This is a simple resource for *Helicobacter* spp. surveillance using polymerase chain reaction (PCR) methods (Mahler *et al.*, 1998) or for fecal IgA quantification (Bristol *et al.*, 2000).

Urine

Urine is usually expelled when a mouse is handled (for details see Chapter 33 on Collection of Body Fluids). Urine can be collected in a clean plastic tube or tests done directly with a variety of chemically impregnated strips. Chemstrip® (Boehringer Mannheim Diagnostics, Indianapolis, IN) and Ames Multistix® (Miles Inc. Diagnostic Division, Tarrytown, NY) are two urine analysis reagent strips that test for numerous urine components including glucose, ketones, protein content, and pH. For more specific tests that require larger volumes of urine or urine collected over defined intervals, metabolic cages are commercially available (Columbus Instruments, Columbus, OH). Urine specific gravity is measured using a hand held refractometer. Some companies offer refractometers especially made for urine testing, such as the Fisherbrand UriSystem Refractometer (Fisher, Pittsburgh, PA).

Gross pathological examination

Abnormalities are phenotypic deviations from known traits. Any abnormalities should be recorded in as much detail as possible. Anatomic and pathologic terms should be used if known, but careful descriptions using lay terms can often be translated into anatomic nomenclature by a medically trained collaborator.

Simple and specific descriptive terms commonly used by pathologists are summarized in Tables 30.2 and 30.3. Combined with detailed anatomic location, these descriptions provide a great deal of information on lesions observed. Gross anatomy of the mouse is

TABLE 30.2: Basic components of a description

Organ or tissue name

Specific site (i.e. duodenum versus small intestine)

   Medial or lateral

   Dorsal or ventral

   Cranial or caudal

   Site specified by anatomic proximity (i.e. lumbar versus thoracic spinal cord)

   Pattern and/or number

   Focal – circumscribed process

   Patchy – alterations that are multiple and poorly delineated

   Multifocal – indication of the specific number of foci contributes to the visual picture

   Diffuse – total involvement of the structure

Specific alteration and/or morphologic diagnosis (i.e. hemorrhages, abscesses, edema, pneumonia)

   Color

   Shape

Size and/or severity

   Enlargement or decrement – if uniform size change

   Degree – mild, moderate, severe

Etiology

   Gross examination – identify parasites if present

   Impression smears – identify infectious agents by appearance

TABLE 30.3: Basic information needed for all necropsy worksheets

Signalment (species/mice, breed/strain/pedigree, color, sex, age (birth date), weight, animal identification number)

Clinical History

Laboratory Data (clinical chemistry, special tests)

Time of Death (if submitted dead)

Mode of Death (method of euthanasia)

---

detailed in several books (Feldman and Seely, 1988; Popesko *et al.*, 1992).

Fixatives

Numerous chemical solutions are available to preserve tissues for histologic, immunohistochemical, and ultrastructural studies (Table 30.4). Each has advantages and disadvantages. The choice depends upon the goals of the study, tissue processing available in the histology laboratory your group uses, experience, and preference of the investigators, especially of the pathologist who will interpret the results. Fixatives should be prepared in advance and be available in adequate volume in appropriately sized containers before the necropsy is started. There are two general rules for histology. (1) Fixatives have various penetrating abilities, and should be individually tested before use. Penetration is usually between 1–2 mm on any cut surface, so specimens should generally be trimmed and kept small. (2) Approximately 20 times the volume of fixative should be used to the volume of tissues to obtain optimal preservation. Excessive amounts of blood and feces will limit the usefulness of the fixative. At the end of the necropsy, the fixative can be drained and replaced with fresh fixative to minimize or eliminate this problem. If the solution is clear it is probably adequate. However, if the solution has a red or brown discoloration it should probably be replaced by fresh chemicals.

Examples of commonly used fixatives for mouse histopathology are described below (see Appendix).

Euthanasia

Laboratory mice are usually provided live for necropsy. Mice have a very high metabolic rate and decompose rapidly after death making histology useless in most other circumstances. A variety of euthanasia methods are approved by the American Veterinary Medical Association (1993). Care should be taken to ensure that humane treatment is provided. Commonly used methods include the following.

Carbon dioxide asphyxiation

This is a rapid and humane form of euthanasia for mice over the age of 7 days. It utilizes a container designed to allow gas to enter rapidly and remove room air. These can be easily manufactured out of plexiglass sheets and tubing, or a large clear glass jar may be used. Adequate ventilation should be available for the technicians performing the necropsies. Gas is provided from a compressed gas cylinder attached to a wall or cabinet. The container is lined by a disposable plastic bag and prefilled with carbon dioxide (CO_2) by opening the valve on

TABLE 30.4: Formulations of commonly used fixatives

Tellyesniczky/Fekete

70% ETOH	100 ml
Glacial Acetic Acid	5 ml
37–40% Formalin	10 ml

Bouin's Solution

Sat. Ag. Picric Acid	85 ml
Glacial Acetic Acid	5 ml
37–40% Formalin	10 ml

10% Neutral Buffered Formalin

37–40% Formalin	100 ml
Distilled Water	900 ml
Sodium Phosphate-Monobasic	4 g
Sodium Phosphate-Dibasic	6.5 g

B-5 Fixative

Mercuric Chloride	6 g
Sodium Acetate-Anhydrous	1.25 g
Distilled Water (hot)	90 ml
Just before use add:	
37–40% Formalin	10 ml

Carnoy's Fixative

Absolute ETOH	60 ml
Chloroform	30 ml
Glacial Acetic Acid	10 ml

4% Paraformaldehyde fixative

16% Paraformaldehyde	10 ml
PBS pH 7.2	30 ml

Glutaraldehyde Fixative

25% Glutaraldehyde	10 ml
0.2 M Phosphate Buffer pH 7.2	50 ml
Distilled Water	40 ml

Karnovsky's Fixative

25% Glutaraldehyde	8 ml
16% Paraformaldehyde	12.5 ml
0.2 M Phosphate Buffer pH 7.4	50 ml
Distilled Water	29.5 ml

JB4 Fixative

25% Glutaraldehyde	16 ml
16% Paraformaldehyde	12.5 ml
1 M Cacodylate Buffer	20 ml
Distilled H$_2$O	151.5 ml

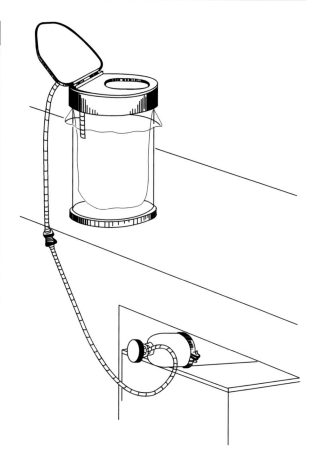

Figure 30.2 Plexiglass container attached to a CO_2 cylinder for euthanizing mice.

the attached cylinder to fill the container (Figure 30.2). The mouse is placed on the bottom of the container and the unit is refilled with carbon dioxide gas. The mouse will die within 1–2 minutes. Only one mouse should be euthanized at a time. Multiple mice placed in the container at the same time may result in those at the bottom not being euthanized. When the bag is removed and disposed of with the mice still in it, the underlying mice may revive. This is clearly an inhumane situation that must be prevented.

Neonatal mice placed in these containers will appear to be killed during the same time interval as for adults. However, neonatal mice are not euthanized by the gas unless it is done as follows. Place the pups in a plastic bag and without squeezing them, close down the bag around the pups and the inlet gas tube of the CO_2. Allow the bag to refill with CO_2 gas while maintaining a loose grip on the neck of the bag against the tube. Then close down the bag against the pups and the tube and refill again. Repeat this a third time. This assures the atmosphere in the bag is virtually 100% CO_2. The rate of gas entry into the bag should be slow enough so the gas temperature is not so cool that it chills the pups and makes them quiescent, while they are still alive. The pups soon become motionless and lose their color. Pups should routinely be decapitated with a pair of scissors immediately upon removal from a CO_2 filled container to ensure euthanasia.

Dry ice is the solid form of carbon dioxide. It undergoes sublimation to carbon dioxide gas. Dry ice should not be used to euthanize mice. Adequate amounts of gas cannot be generated rapidly enough to ensure humane euthanasia. Furthermore, the mice can suffer from thermal injury if they come into direct contact with the material.

Barbiturate overdose

This is an effective and humane method that is described in detail under perfusion methods below.

Decapitation of adults

This method should be avoided unless the experiment has very special requirements and the procedure has been approved by the institution's animal care and use committee. For description of procedure, see Chapter 33 on Collection of Body Fluids.

Cervical dislocation

This method involves separating the vertebrae in the cervical area with a firm pinch to the neck and pull of the tail. It is a quick, efficient method often used for routine diagnostic work. However, this method is not often recommended for research mice, since it results in damage to tissues in the cervical area, as well as releasing large amounts of blood into the body cavity which can make observation and collection of some organs more difficult.

Perfusion methods

Perfusion combines euthanasia with fixation, providing the quickest way to get organs into fixative, resulting in the freshest tissues for study, with minimal autolytic changes. Special protocols must be followed to ensure this is done in a humane manner.

Materials needed include two 10 cc syringes with 23 G needles, one 1 cc syringe with a 23 G needle, stock solution of pentobarbital sodium (50 mg/ml; Nembutal®, Abbott Laboratories, North Chicago, IL), phosphate

buffered saline (PBS), fixative of choice, and 0.85% saline.

First, prepare the working solution of Nembutal® by diluting 1.6 ml of the stock solution with 8.4 ml PBS. Then fill one 10 cc syringe with 0.85% saline, and the other with your chosen fixative. Label the syringes to avoid confusion during the procedure.

The mouse must be anesthetized using an intraperitoneal injection of Nembutal® working solution (0.1 ml/10 g body weight) using the 1 cc syringe. After the mouse is completely anesthetized, dip the mouse into a mixture of water and disinfectant and pin it to a dissection board, ventral side up (see section on Necropsy Procedure). Incise the ventral skin, undermine it using blunt dissection, and reflect it away from the incision site from the thorax to the mandible to reveal the jugular veins located under both salivary glands. Following instructions for the necropsy procedure, immediately open the thoracic cavity. If a blood sample is needed, a heart puncture must be done at this time (see section on Blood Collection). After blood is collected, carefully cut the jugular veins. Blood will start to flow from the vessels. Insert the needle of the syringe containing the saline into the left ventricle of the heart. With gentle but constant pressure, perfuse the saline into the heart while observing the area where the blood vessels were severed. The saline should flush the blood out of the vascular system. It is important that the pressure exerted on the syringe is enough to push the blood through and out of the vascular system, but not so excessive as to cause damage to any of the organs. After injecting 4–8 cc of saline, depending on the size of the mouse, repeat the same procedure with the fixative. If fixation is successful, the body will stiffen from the tip of the tail to the nose and all organs will blanch. Tissue collection may then proceed.

Necropsy procedure

Once the mouse has been euthanized it can be superficially disinfected by submersion in a dilute solution of a germicidal detergent such as Calgon Vestal Process NPD One Step Germicidal Detergent (ConvaTec, St. Louis, MO), or a solution of 95% ethanol. When performing necropsy of a mouse with an abnormality of the skin or hair, it is important to collect samples of the hair before the mouse is dipped in the disinfectant. Hairs can be plucked and used to inoculate fungal culture media (Dermatophyte Test Medium, MG Scientific,

Pleasant Prairie, WI; Sabouraud's medium, MG Scientific). To examine hair fibers for structural abnormalities or for ectoparasites the hairs can be plucked manually using the thumb and forefinger. Forceps may damage the hair fiber. Gently plucking hairs will remove those hair fibers from follicles in the telogen (exogen) stage of the hair cycle. Anagen follicles are actively growing, deep, and firmly attached to the dermis. Plucking hairs from follicles in anagen may result in damage or induction of abnormalities in fibers already weakened by structural defects. Plucking fibers from telogen follicles allows for examination of the whole hair shaft from root (club) to tip. Hair fibers should be stored in a clean cryopreservation tube (Nalge Nunc International, Denmark). When studying mutant mice, standardized collection techniques for skin and hair are as important as for other organs. Hair samples should be collected from the same area on every mouse in a study such as from the left lateral skin surface from shoulder to pelvic region. Avoid areas where full thickness skin will be collected for histologic examination as plucking hairs will artifactually distort the inner root sheath and other structures in the hair follicle. If the vibrissae (very long straight hair fibers around the eyes and muzzle) appear abnormal, samples of these should be plucked as well, from the same side as the hair is plucked. The vibrissae should be stored in a separate Nunc tube. Other specialized hair fiber and follicle types can be collected and examined if necessary.

If a mouse has a skin abnormality, remove hair prior to disinfection. This is usually accomplished by shaving the mouse with electric hair clippers such as the Oster Finisher Trimmer (Cat. # 76059-030, Oster Professional Products, McMinnville, TN). These clippers are easy to handle and have a small blade that is ideal for mice or other small mammals. If complete hair removal is desired, there are commercially available depilatory products (Nair, Carter-Wallace Inc, New York, NY; Neet, Reckitt & Coleman Inc, Wayne, NJ) that can be applied to the mouse after shaving. These products should be left on for 2–3 min, then rinsed off under warm running water, which will wash away the hair as well.

At this point, the mouse may be disinfected. The disinfectant washes off loose hairs and mats hair down on an unshaven mouse for ease of examination. Allow the disinfectant to drain from the mouse, and then place the mouse on one to two layers of absorbent paper towel on a cork dissection board. The cork board should be approximately $14\,cm \times 21.5\,cm$ in size and $1.0\,cm$ thick.

Skin should be collected for histology at this point. With the mouse ventral surface down on the board, gently grasp a fold of dorsal skin from the caudal region

and make a small incision with the scissors. Carefully cut out a rectangular piece of skin along the dorsal midline from the thoracolumbar junction to the interscapular region. Place the skin on a piece of unlined index type card or aluminum foil. Orient the skin sample cranial-caudally and trim lengthwise across this axis to optimize the orientation of the hair follicles. Fix by immersion in the appropriate fixative (see section on Clinical Pathology).

After collecting dorsal skin, collect a sample of ventral skin in a similar fashion. Fresh samples of both dorsal and ventral skin may be frozen in OCT compound (Tissue-Tek®, Sakura Finetek U.S.A., Inc., Torrance, CA) for immunofluorescence or *in situ* hybridization studies. The skin from the head should also be collected, including the pinnae (ears), eyelids, and muzzle. Each of these sites includes specialized glands and/or hair follicles. There are mucocutaneous junctions present and the epidermis varies slightly from truncal skin in these sites as well. The skin can be carefully peeled and trimmed from the skull as a unit and mounted flat on a piece of index card or foil for fixation. Tail skin may be collected by severing the tail from the body and incising it lengthwise, using the tip of a pair of scissors. The loose skin is grasped from the base of the tail and stripped away from the bone and tendons. Tail skin is mounted flat on a piece of foil as with other skin samples. Tail and head skins are often collected toward the end of the necropsy.

After skin has been collected, place the mouse ventral side up and pin each limb firmly to the board. The rear feet may be pinned between the gastrocnemius tendon and the bone, while the front feet may be pierced, through the skin, between the metacarpal bones, in order to do the least damage to the tissues being collected. During the necropsy, the board may be rotated easily to adjust the position of the mouse, providing various angles of access for organ collection.

With a #12 scalpel blade and #3 handle, make a ventral midline longitudinal incision through the skin, from the external genitalia to the ramus of the mandible, then cut from the genitalia laterally toward the rear feet, along the medial surface of the rear legs. Avoid using scissors as they are quickly blunted. On female mice, this incision passes between the fourth and fifth nipples of each side. Grip the skin on either side of the incision and pull gently outward, or use light strokes of the scalpel, to separate the skin from the abdominal muscles. Reflect the skin far enough so that it does not interfere with the rest of the necropsy.

With the skin reflected back, collect peripheral lymph nodes located on either side of the salivary gland (cervical lymph nodes), under each of the front legs (axillary lymph nodes), and on the medial side of each rear leg (inguinal lymph nodes). Unless they are enlarged due to disease, the peripheral lymph nodes may be difficult to locate, especially the axillary lymph nodes. Since the cervical lymph nodes are collected attached to the salivary gland, the cluster of glands and lymph nodes can be removed *in toto* and fixed. The inguinal lymph nodes are located in the fat pad on the medial surface of the rear leg. They are slightly darker in color than the fat and are usually no larger than 0.1–0.2 cm. The easiest way to collect them is to remove the fat pad with the embedded lymph node and trim away as much of the fat as possible before placing the lymph node in the fixative.

Using a new set of sterile instruments, grip the abdominal muscles at the inguinal region with forceps, lift firmly, make a small incision to allow air into the abdomen, and the viscera will not be injured or contaminated. Cut through the abdominal muscles on each side, extending from the inguinal midline to the lateral thorax exposing the viscera. The coelom can be cultured with a sterile swab before proceeding further.

Another set of sterile instruments can be used at this point if it is important to culture visceral organs. The longer the animal is dead before tissues are collected and fixed the more autolysis will occur which will make interpretation of histologic sections more difficult. As a general rule, no more than 15 min should pass between euthanasia and collection of tissues. This should allow adequate time for photography, microbiological culture collection, and other procedures. Throughout the necropsy, visceral organs should be evaluated to determine if they are in their proper anatomic orientation. Some mutations such as *situs inversus* (*iv*) can cause the orientation to be reversed (Hummel and Chapman, 1956, 1959; Layton, 1976).

All tissues and organs should be carefully checked for abnormalities and all observations should be noted. Gross photographs of any external or internal abnormalities are important to obtain in the study of a new mutant.

The intestines should be collected first because they undergo rapid autolysis. Separate the stomach from the esophagus at the diaphragm. Gently retract the stomach while cutting the mesenteric attachment with scissors or scalpel. Before separating the duodenum, open it near the stomach and gently express the gall bladder (found in the liver). Bile should flow out of the Ampula of Vater indicating that the common bile duct is patent. Continue to retract the intestine and cut the mesentery until the point is reached where the colon enters the pelvic girdle. Cut the pubic symphysis with a pair of heavy scissors.

Before the colon and attached cecum can be dealt with, one must first remove the reproductive organs, including the preputial gland (males)/clitoral gland (females) and urinary bladder.

If the bladder/urinary system is of interest, inflate the urinary bladder with formalin (0.5 cc) via a syringe using a 27 G needle (Figure 30.3). The urethra should be bound tightly with string to prevent outflow of fixative from the bladder. This method is effective for evaluation of the bladder wall, especially in early neoplasia or epithelial hyperplasia.

In the male mouse, testes may be gently grasped by the inguinal fat pad, cut away from the other viscera and placed on a piece of cardstock. Orient the testicles so that the epididymis and testis are in the same plane and may be trimmed simultaneously for histological presentation (Figure 30.4). The preputial gland is a paired organ located subcutaneously between the penis and anus. It may be collected by grasping one edge of

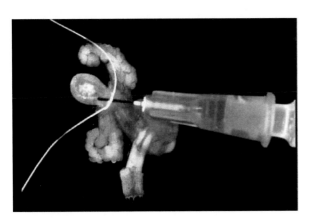

Figure 30.3 Urinary bladder fixation technique. **(See also Color Plate 10.)**

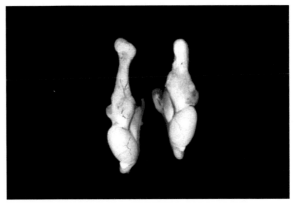

Figure 30.4 Orientation of the testes and epididymides. **(See also Color Plate 11.)**

the gland with forceps, cutting it away from the abdominal wall, and placing it directly into fixative.

The seminal vesicles, coagulating gland, prostate gland, urinary bladder, and penis may be removed as a unit by gently grasping the apex of the seminal vesicles with forceps and lifting it away from the colon. Insert the tip of heavy duty scissors between the colon and the pelvis and cut the bone, on both sides, then lift the reproductive tract further and cut away the connective tissue between it and the colon. Remove the reproductive tract and arrange it on a card, then place it in the fixative (Figure 30.5).

The female reproductive tract is removed in a similar fashion. Clitoral glands are normally not easily visible, but are located subcutaneously, cranial to the vulva. The best way to collect these glands is to cut a small square of 0.5–0.8 cm of abdominal muscle and overlying skin immediately anterior to the clitoris. This section will include the clitoral glands. Smooth this piece of tissue gently onto a piece of white cardboard and, using parallel pencil marks, indicate the area where the glands are expected to be. After placing the clitoral glands in fixative, grasp the fat pad of one ovary and cut it free from the mesentery, laying that ovary and uterine horn over on the other side of the colon. Remove the entire female reproductive tract as was done for the male organs, arrange it on a card to maintain orientation, then fix by immersion (Figure 30.6).

Retract the colon and sever the skin around the anus. The entire gastrointestinal tract can now be separated from the mouse, placed on a towel moistened with physiological saline, and examined. Mesenteric lymph nodes can be identified as small brown nodules at the root of the mesentery and along the intestine. These should be collected and fixed. At the duodenojejunal flexure, the pancreas is firmly adherent to both the small and large intestine. Care must be taken to separate the two parts of the intestine from the pancreas, while preserving the integrity of all three. The spleen and pancreas are connected. They are often removed as a unit unless it is necessary to examine one or the other separately, as in the case of a mouse model for diabetes where special fixative (Bouin's solution) and stains (aldehyde fuchsin) are used to examine the beta cells of the pancreatic islets. Grasp the pancreas gently with the forceps and pull upward, cutting away any mesenteric tissue that adheres to the spleen or pancreas to free the structures, then place them directly in the fixative.

For proper fixation, the intestine must be inflated with fixative before it is rolled or otherwise prepared for histologic preparation. This procedure is simple and rapid but runs the risk of personal injury if fixative splashes. Safety goggles and a fume hood are recommended since fixatives are noxious and toxic. Use a 10 cc syringe with a 17–22 G needle. Fill the syringe with fixative and introduce the needle into either end of the intestine. Gently depress the plunger on the syringe and the intestine will slowly inflate (Figure 30.7). Make several injections along the length of the intestine if it

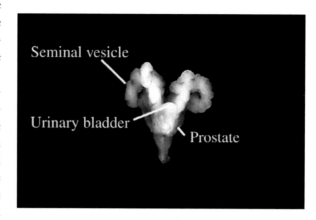

Figure 30.5 Male reproductive tract with urinary bladder. **(See also Color Plate 12.)**

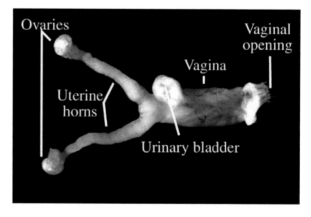

Figure 30.6 Female reproductive tract with urinary bladder. **(See also Color Plate 13.)**

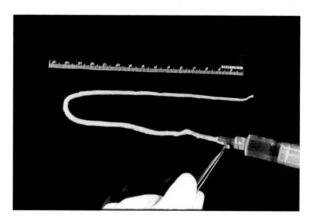

Figure 30.7 Intestines are removed and inflated by injection with fixative. **(See also Color Plate 14.)**

does not fill completely due to the presence of feces that form obstructions (Feldman and Seely, 1988). To make a second injection, simply pierce the wall of the intestine and clamp it gently with your fingers around the tip of the needle to prevent back flow of the fixative and proceed as before (Sundberg *et al.*, 1994). An intestine that has been over inflated is much more difficult to roll. Some laboratories prefer to open the entire intestine and remove digested food material and feces prior to rolling (Moolenbeek and Ruitenberg, 1981). This approach yields good mucosal fixation but may also cause damage to the mucosa.

Intestines can be rolled lengthwise for histologic presentation (Moolenbeek and Ruitenberg, 1981; Sundberg *et al.*, 1994). However, each roll must be able to fit comfortably in a cassette. For an average sized adult mouse this requires the small intestine to be cut into three equal pieces (Figure 30.8).

Another approach to presenting intestines is similar to the way they are routinely collected in larger animals. Representative segments are cut and fixed by immersion. More precision comes by laying out the entire gastrointestinal tract and cutting segments out at specified distances from anatomic structures such as the anus, cecum, or pylorus.

To create so-called 'Swiss Rolls' of intestine, roll the inflated intestine in concentric, centrifugal circles on a piece of unlined index card (Percy and Barthold, 1993). If the fixative drains, the intestine may flatten, making it more difficult to roll. We use large index cards cut into strips approximately the width of cassettes to mount tissues. Orientation of the segments is important and must be agreed upon with the pathologist. It is commonly accepted to keep the end of the intestine proximal to the stomach toward the center of the roll. Once the intestines have been rolled onto the paper, let them firm up to prevent the roll from unwinding, then fix by immersion. Intestinal rolls take practice to master. The important objectives are (1) do not over-inflate the intestine, (2) maintain proper orientation, (3) make the rolls smaller than the cassettes, and (4) minimize loss of fixative while rolling the segments.

The stomach and cecum are collected separately, inflated with fixative and placed *in toto* in fixative. Modifications of this approach can be made to evaluate specific structures (Sundberg *et al.*, 1994; Mahler, 2000). Histologic grading systems are available for the intestinal rolls and cecum (Sundberg *et al.*, 1994; Mahler *et al.*, 1999), that cannot be done easily when only segmental cross sections of bowel are available for examination, since lesions are often segmental and not uniformly distributed.

Kidneys may be removed by grasping the surrounding fat and pulling upward while cutting around the organs. The adrenal gland is a small white structure that lies within the perirenal fat pad just anterior to the kidneys. It should be left within the fat that clings to the kidney and the two should be presented to histology as a unit. It may be important for the pathologist to be able to distinguish between the right and left kidneys. Since this may only become important after the fact, the right kidney should be cut transversely while the left is cut lengthwise (left/long) prior to fixation.

The liver is the last organ to be removed from the abdominal cavity. The liver can be manipulated using the diaphragm to avoid damage. The organ is cut from its vascular attachments with scissors and placed on a towel moistened with physiologic saline. Separate the right and left medial lobes of the liver as a unit along with the gall bladder. This may be accomplished by folding these lobes back onto the work surface and trimming the connections to the remaining liver lobes. The left lateral lobe is the largest and should be separated from all others in a similar fashion. The remaining smaller lobes may be placed in the fixative together. When separating

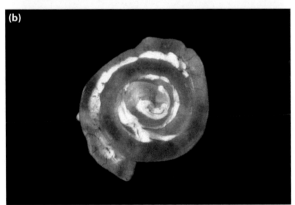

Figure 30.8 'Swiss roll' of intestines. It is important to make the rolls smaller than the cassettes (a), and minimize loss of fixative while rolling the segments (b). **(See also Color Plate 15.)**

the liver lobes, take care to handle them gently, by the edges, as they are easily damaged. Acid–alcohol–formalin fixatives penetrate quickly and deeply into this tissue. Other fixatives do not penetrate deeply even on cut surfaces so the liver lobes are best sectioned with a scalpel blade and separated to optimize fixation.

To enter the thorax, grasp the xiphoid process firmly with the forceps and lift upward. If the chest cavity was not previously opened for blood collection or by cutting the diaphragm, this will create a negative pressure within the thorax, and cautiously cutting through the ribs and diaphragm on one side allows air to enter the thoracic cavity. When negative pressure in the chest cavity is lost, the lungs will shrink away from the ribs and diaphragm. If a thoracic microbiological culture is to be taken, carefully extend the cut through the diaphragm and rib cage to expose the lungs, then use sterile scissors and forceps to take a tissue sample before proceeding. The thoracic cut may now be extended through the rest of the ribs at the costochondral junction to just short of the internal thoracic vein and artery on both sides. Often the mediastinum between the area of the heart and thymus continues pulling on the excised part of the rib cage, so it must be carefully trimmed away to prevent the rib cage from falling back over the thoracic cavity.

To remove the heart and lungs, turn the cork board around so that the head is facing you. Cut the mandibular symphysis, push the mandible laterally, retract the tongue and gently cut underlying soft tissue of the hyoid bones that extend dorsal to the larynx. Being careful not to puncture the trachea or the esophagus, continue cutting on either side of the neck until the clavicles are encountered. Using the tips of the scissors, sever the bones and continue to dissect carefully through the ribs, avoiding the trachea, esophagus, and lung. Repeat this on the other side. Once all the surrounding tissue has been removed, gently retract the tongue, trachea, and esophagus, with the lungs, heart, and thymus as a unit (sometimes referred to as the 'pluck'). The lungs should be inflated with fixative with a syringe to ensure proper histologic preparation (Figure 30.9). Slip the tip of a needle into the trachea via the glottis, which is normally the most apparent opening at the base of the tongue. Clamp down around the needle with a pair of forceps and slowly depress the plunger of the syringe. The lungs should begin to expand and blanch. If the fixative flows out in a puddle between the lungs, try again. Inflate the lungs very slowly, stopping when they are about the size that they would normally be on inhalation. Over-inflating the lungs can damage the alveoli.

The next organ to be collected is the brain. To access the brain, cut the vertebral column and spinal cord at the base of the skull. Slowly retract the skin from the skull if it has not already been removed. If the eyes are of interest, you may wish to remove them from the skull. To remove the eye, insert a pair of curved forceps behind the globe, grasp the optic nerve and pull outward until the eye has been freed from the orbit. Eyes may be fixed and embedded separately. However, histologic presentation of the eye within the skull is often sufficient for viewing many abnormalities.

Cut any remaining vertebrae away from the skull. The brain should be visible through the foramen magnum. This is the spinal medulla. Slip one blade of your scissors in between the neural tissue and the bone and make two small longitudinal cuts in the occipital bone, one on each side of the spinal medulla. Hold the skull between thumb and forefinger and, with forceps, grasp the edge of the occipital bone and pull upward to neatly break it off (Figure 30.10). Then gently insert the scissors tip between the brain and the skull to make a cut in the skull along the sagittal suture. Continue breaking away

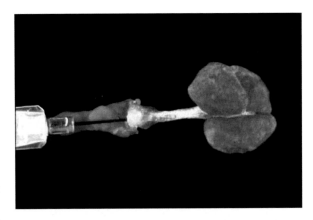

Figure 30.9 Lung inflation with fixative. **(See also Color Plate 16.)**

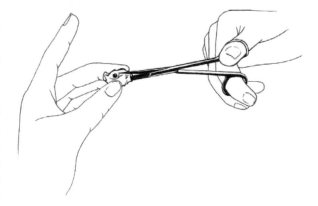

Figure 30.10 Initial cuts in the occipital bone to access the brain.

the interparietal and the parietal bones in the same manner, being careful not to harm the delicate brain below. The frontal bone comes to a slight point at the intersection of the sagittal and coronal sutures. This area should be broken off as well, to allow the brain to be removed cleanly from the cranial vault. There may be a thin reddish membrane around the brain, particularly in the area of the cerebrum. This is the meninges, and must be removed carefully with forceps or it will cut into the brain as it is removed from the skull. Carefully separate the olfactory bulbs from the cerebrum to avoid damaging the turbinates (the olfactory bulbs will remain in the skull when the brain is removed). Turn the skull upside down over the jar of fixative. Gently work the forceps between brain and bone and pull away the connective tissues, freeing the brain from the cranial vault. The brain will fall into the fixative. Place the skull into the fixative as well.

The spinal column may now be collected. Grasp the proximal end of the spine and lift it away from the skin, cutting away the fascia that holds the two together. Cut through the pelvis to sever the hind limbs from the distal vertebrae. Cut the tail away if this was not already done. Cut the ribs away from the spine, as closely as possible, without damaging the vertebrae. Place the spinal column in fixative, making sure to keep it straight so it is oriented correctly for trimming.

Trim the front and rear limbs from the skin and place them in fixative. If skin was not collected earlier in the necropsy, it may still be important to save it. Skin can be removed from a defined location for consistency. We usually remove a rectangular area over the thorax. The skin is flattened on a card or foil, cranial/caudal orientation is marked, and the tissue is fixed by immersion.

In summary, when evaluating a new mutation, it is important to collect total tissues in a methodical, standardized fashion to avoid diagnostic discrepancies. From these study sets, a more focused tissue collection protocol may be developed, concentrating on those tissues known to exhibit abnormalities in a particular mutant. Standard criteria for tissue collection, agreed upon by the technicians, researchers, and the pathologist who will be reading the slides, are essential.

Trimming tissues for histology

After the tissues have been adequately fixed (see section on Fixatives), they must be trimmed for histologic processing. Proper tissue trimming will optimize interpretation by the pathologist. Presentation is critical when trying to identify any variation from normal. For most fixatives, tissues can be trimmed and processed after approximately 12 h of fixation. Before bones are trimmed they must be decalcified. Overnight soaking in a dilute hydrochloric acid based decalcifying solution such as Cal-EX- (Fisher, Pittsburgh, PA) may be adequate. Bouin's and Fekete's solutions contain dilute acetic acid that also aids in decalcification when tissues are stored in them for days to weeks.

After decalcification, bones may be trimmed, following an initial rinse with water. However, the decalcified tissues must be continually rinsed in running water for at least 3 or 4 h before histologic processing. Failure to thoroughly rinse decalcified tissues may result in inadequate staining of the tissue sections. Over-use of the decalcifiers may also result in suboptimal staining of tissues. Optimal times for decalcification and washing should be customized for every laboratory, and is often specific to particular organs or studies.

To trim tissues use, a firm clean cutting surface such a cork board, teflon cutting board, or paraffin filled petri dish. Each tissue, as it is trimmed, should be placed into a labeled and numbered histology cassette (OmniSette® Tissue Cassettes, Fisher Scientific, Pittsburgh, PA). Cassettes can be labeled with a #2 pencil, solvent resistant marker (HistoPrep® Pen, Fisher Scientific), or mechanical labeling machine (Carousel Cassette MicroWriter®, Thermo Shandon, Pittsburgh, PA) to label the front and/or side of each cassette. Alternatively, identification information may be written on a small card and placed in the cassette. Indicate the mouse's accession number and, if there will be more than one cassette of tissue per mouse, number each cassette for that animal in sequential order. Sequential numbering will aid with identification and help determine whether all tissues sent to histology are returned. A corresponding table (such as that shown in Table 30.1) can be used to standardize the combinations of tissues and numbering of cassettes.

Cassettes containing tissues that were fixed and not decalcified should be placed into a container of 70% ethanol, while those containing decalcified tissues not previously rinsed should be placed into water. After thorough rinsing, as described above, decalcified tissues may be placed into ethanol with other tissues and delivered to the histology laboratory for processing and embedding. Any remaining tissues not submitted for embedding may be stored in 70% ethanol for future use, if necessary. Tissues stored in fixatives for long periods may be altered (refer to section on Fixatives and

Appendix). When handling fixed tissues, it is still important to be gentle. Soft parenchymal organs, such as the liver, brain, lungs, kidneys, etc., remain delicate after fixation and should be manipulated using a pair of wide wooden forceps or a similar tool. Trimming is best done with a sharp, single edge razor blade to produce a clean cut with minimal tissue damage. Residual chemicals on the tissues can dull a razor blade rapidly, so it is important to change blades frequently while trimming.

All tissues should be trimmed to a thickness of approximately 1–2 mm to permit adequate penetration of solvents and paraffin. If it is necessary to present a particular facet of a trimmed tissue to the pathologist, this is indicated by marking the side of the tissue one does *not* want presented. A blue colored pencil works well (Venus col-erase, #1276 blue, Eberhard Faber, Inc., Lewisburg, TN). The blue pigment will not wash off in alcohol and will clearly indicate the desired orientation to the histologist embedding the tissue. If improperly included in the section, the blue pigment will contaminate the field, making it difficult for a pathologist to find clean fields for photography. Small, related tissues of similar densities may be placed together in the cassettes (i.e. kidneys with spleen and/or liver, reproductive tract tissues together, etc.; refer to Table 30.1). The heart, lungs, brain, bones, and any possible tumors found should each go into an individual cassette.

The following is a description of trimming methods for each organ. After trimming, remaining tissues can be placed with 70% ethanol in heavy-duty, heat sealed stand-alone plastic bags for archival purposes (Kapak/Scotchpak, Kapak Corporation, Minneapolis, MN).

Large and small intestine

Intestines collected in rolls as described above need no further trimming at this point. Each segment should be carefully removed from its card backing, if rolled, and placed individually in a histocassette. If unrolled segments were collected, representative areas are cut in cross section. These may be anatomic areas from each major section of the intestine or areas where there were grossly evident abnormalities (e.g. intestinal polyps from the multiple intestinal neoplasia mutant mouse, *Apc*^{Min}).

Stomach and cecum

These organs should be cut in half longitudinally. The stomach should be cut in a manner that presents both the esophageal and duodenal openings. The cecum

should be cut to show both the ileocecal junction and the ampulla of the colon. Submit the half of each that best shows the desired features. If space allows, the cecum and stomach may be submitted in the same cassette. If lesions of the cecum or stomach are of interest, it may be important to submit both halves of these organs. If this is the case, each organ should be placed in its own cassette.

Liver (with gallbladder)

Cross sections of only the left lateral lobe of the liver and the medial lobes with the gall bladder are sufficient for histology unless pathologic changes are obvious on the accessory lobes. Lay the left lateral lobe flat on the trimming surface and cut a crosswise section near the center of the lobe. Trimming of the medial lobes must be a bit more precise, as they must be cross-sectioned to include a portion of the gall bladder in the section (Figure 30.11). This is usually accomplished by cutting across the medial lobes just below the juncture between the lobes, then just above that juncture, where the falciform and teres ligaments hold the two lobes together. The first cut should reveal a portion of the gall bladder. Mark the opposing side with a blue pencil and place both this section and the left lateral lobe section in the cassette.

Kidneys

The left kidney should be cut longitudinally down the center and should include a segment of the adrenal gland, which was left attached to the kidney at the time of fixation. The right kidney will be identified by a small transverse incision, if handled properly at the

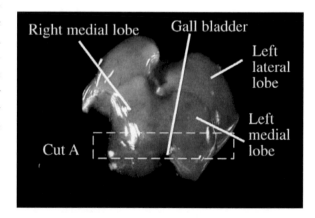

Figure 30.11 Example of trimming sites for a mouse liver. (a) left lateral lobe, (b) left medial lobe, (c) right medial lobe, and (d) gallbladder. **(See also Color Plate 17.)**

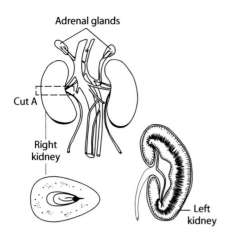

Figure 30.12 Anatomic location of both kidneys and adrenal glands. Right kidney is trimmed transversely while left is cut longitudinally so that histologic sections can be identified. Lower panels illustrate features of cut sections.

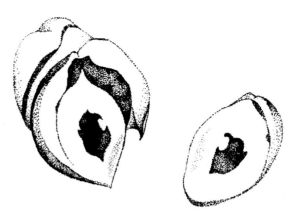

Figure 30.13 The heart is trimmed lengthwise to include all four chambers.

time of fixation. It should be presented in a lateral cross section, cut through the central area near the renal pelvis (Figure 30.12). Use the blue pencil to mark the side of this section furthest from the pelvis, as it is important to present the area nearest the center of the kidney to the pathologist.

Spleen and pancreas

Unless the pancreas is collected separately for focused study, the spleen and pancreas are fixed as a unit. Trim them together in cross section at any point along the length of the spleen.

Lungs

The lungs are collected with the heart and thymus as a unit at the time of necropsy, but are submitted separately for histology. Using a pair of wooden forceps, gently push apart the heart and lungs so that the lungs lay out flat. Cut a longitudinal section from the center of the lobes of the lung on each side. The individual lobes will separate, but should all be collected and placed in the same cassette. There is generally no need to separately identify each lobe of the lung, unless gross lesions are evident.

Heart and thymus

Carefully remove any remaining lung or tracheal tissue from around the heart, being certain not to separate the thymus from the heart. Place the heart on its base and begin the cut at the heart's apex. Angle the cut so that it bisects each of the four chambers of the heart. In some instances, the great vein of the heart (*vena cordis magna*) may be visible on the epicardium of the fixed heart. Making a cut along the line of this vein will often bisect the chambers properly (Figure 30.13). If space allows, both halves of the heart may be submitted. If only one half is submitted, ensure that it contains a portion of the thymus.

Salivary glands

Trim the base of the salivary glands to present a clean cut edge. Make the second cut approximately 4 mm from this first cut. Mark the face of the second cut with a blue pencil and place the cross section in the cassette.

Trachea and thyroid/parathyroid

Cut a cross section of the trachea at the point where the thyroid and parathyroid glands are attached. This is located in an area 1–2 mm below the epiglottis.

Lymph nodes

There are many lymph nodes located throughout the body, but it may not be necessary to submit all of these for processing unless involved in the disease process. Representative nodes from several areas may be chosen (i.e. mesenteric, axial, inguinal, and cervical). If not enlarged, lymph nodes may be submitted whole, after the surrounding fat has been removed. Severely enlarged lymph nodes must be cut in cross section.

Urinary bladder

The urinary bladder is often fixed as a unit with the reproductive organs. A lengthwise cut made down the center of the urinary bladder should also include the uterine body, vagina, and cervix in the female, and the prostate, bulbourethral glands, penis, and prepuce in the male. After this first cut is made, trim the opposite side of one half of the tissue to the appropriate width for a cassette (~4–5 mm), cutting off the uterine horn or seminal vesicle and any excess adipose tissue. Mark this side with a blue pencil and place the section in a cassette. If the urinary bladder is collected separately (see section on Necropsy Procedure), it should be trimmed in cross section or longitudinally under water to prevent spraying of fixative and excess deformation of tissues.

Reproductive organs

Female

Cut one ovary away from either uterine horn, trim away any excess fat and place the entire ovary and its associated uterine tube in a cassette. Place a cross section of one of the uterine horns into the same cassette. Trimming of the uterine body, vagina, and cervix is discussed above with the urinary bladder (Figure 30.14).

Male

Separate the testes from the card on which they were fixed. The testes and epididymides should be slightly flattened and lie in the same plane on the surface that was attached to the card. Mark the rounded side of the testis with a blue pencil and place into a cassette. Place a cross section from one side of the seminal vesicles into the same cassette. The penis, prepuce, and accessory organs are discussed above with the urinary bladder (Figure 30.15).

In both male and female, all reproductive organs, as well as the urinary bladder, may be submitted in the same cassette.

Clitoral/preputial glands

The clitoral glands of the female mouse are normally small, and embedded in a segment of inguinal fascia and fat. This segment should be cut in cross section and embedded on edge. It may be submitted in the same cassette as the reproductive organs. The preputial gland of the male mouse is larger and is collected individually. It should be cut in cross section as well and submitted with the reproductive organs.

Brain

The brain is submitted to the histology laboratory cut in three cross sections rostral to caudal (Figure 30.16). The first cut should be made through the cerebrum, ~1–2 mm from the most rostral surface. The second cut through the cerebrum, 2–4 mm from the first, will create the first section. This first section should present a view of the central portion of both cerebral hemispheres, so mark the face of the *first* cut with a blue pencil. The third cut should be made 2–4 mm from the second cut, just to the cerebral side of the confluence of sinuses. This will create the second cross section, of which the face created by the *second* cut should be marked with a blue pencil. A fourth cut should be made at the transverse sinus to separate the remaining

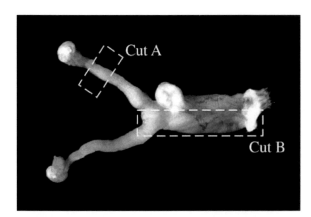

Figure 30.14 Female reproductive tract. Dotted lines indicate where to cut tissue for histologic processing. **(See also Color Plate 18.)**

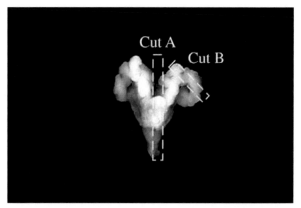

Figure 30.15 Male reproductive tract. Dotted lines indicate sampling sites for histology. **(See also Color Plate 19.)**

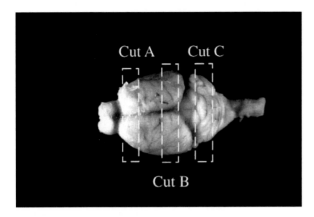

Figure 30.16 Trimming sites marked for sectioning a mouse brain. **(See also Color Plate 20.)**

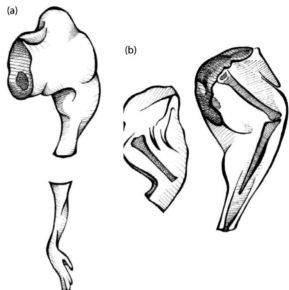

Figure 30.17 (a) Preliminary preparation of leg includes removal of skin and amputation of distal segment as indicated. (b) Decalcified limb cut lengthwise to expose joints.

portions of the cerebral hemispheres from the cerebellum, after which a fifth cut is made which approximately bisects the cerebellum laterally to create the third section. On this section, the cut face closest to the cerebrum should be marked with a blue pencil. Alternatively, serial fine coronal sections can be obtained using one of a variety of Brain Matrices (Kent Scientific Corporation, Torrington, CT). The brain should be submitted in an individual cassette.

Tongue

The tongue should be cut longitudinally down the midline. It is usually only necessary to submit half of the tongue to histology.

Legs (long bones)

One each of the fore and hind legs should be cut longitudinally to show the long bones and major joints of each leg (Figure 30.17). Excess fat should be trimmed away; on the hind legs it may be necessary to trim away some of the bulk of the muscle in order to fit the section properly into the cassette. Feet should also be separated from the leg, and submitted separately. The longitudinal cut should be made using the major joint of each leg as a reference to bisect the long bones. Choose the half of each leg that best shows the desired view of the bones and place it in its own cassette.

Feet

One each of the front and back feet should be cut longitudinally to show the skin, foot pads, and bones of the feet. The front foot should be cut so two toes are present on each half. The back foot should be cut

directly through the middle toe on that foot. Each foot is placed into a separate cassette. Both halves of each foot may be sent to the histology laboratory.

Spinal column

The spinal column should be trimmed to present both the cross-section and longitudinal section of the thoracic and lumbar regions. First, cut the spine laterally (cross section) between the thirteenth thoracic and first lumbar vertebrae (just below the thirteenth rib). Next, cut a lateral section 4–5 mm in width, from the distal end of each section. Bisect the remaining long segments longitudinally, placing the best half of each into a separate cassette with its related lateral section.

Skull

The skull should be cut into three cross sections, similar to the brain. The sections should show the eyes, nasal passages, ear canals, and pituitary gland (Figure 30.18). The first cut should be made through the posterior edge of the pituitary, identified as a whitish mass located in the area between the occipital bone and basisphenoid bone on the inner surface of the skull. This is the crucial cut, and should present a view both of the pituitary and the middle ear. The second cut may be made 4–5 mm anterior to the first cut to create the first section. The third cut should be made through

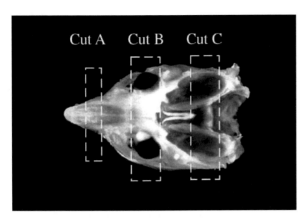

Figure 30.18 Three sections are cut in the decalcified skull. This exposes: (A) the nasal cavity, (B) eyes and associated glands, and (C) inner, middle, and external ear as well as the pituitary gland. **(See also Color Plate 21.)**

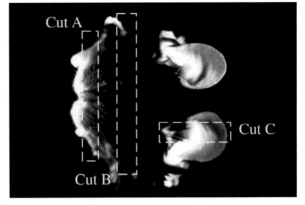

Figure 30.19 Skin of the head trimmed to study (A) muzzle and vibrissae, (B) eyelids, cilia, Meibomian gland, and conjunctiva, and (C) pinna of the ear. **(See also Color Plate 22.)**

the posterior edge of the visible portion of the eyes, with the following cut made just anterior to the eyes. The third section should present a view of the sinuses, and may be cut from the approximate center of the remaining portion of the snout. Each section should be marked with a blue pencil on its anterior surface. All sections of the skull may be submitted in the same cassette.

Skin

Trim portions of the dorsal and ventral skin longitudinally in the direction of the hair growth, into pieces approximately 0.3×1.0 cm. Cut 2–3 sections of both dorsal and ventral skin in this manner and mark one long edge with a blue pencil to indicate that the pieces should be embedded on the opposite edge. Depending on the focus of the study, you may also want to cut a piece of skin approximately 0.7×0.7 cm square to be submitted horizontally, haired side down, to view the hair follicles in horizontal section. Mark the underside of the section with a blue pencil. Tail skin should be trimmed in the same orientation as the longer pieces of dorsal and ventral skin. Cut the section from an area that was not handled as the tail skin was removed from the bone, and mark one long edge with a blue pencil. Eyelids may be presented by making a cut bisecting both lids then making a second cut just posterior or anterior to the corners of the eyelids so that the upper and lower eyelids remain attached to one another. Mark the face of the second cut with a blue pencil. A section of the muzzle skin may be obtained by making a cut approximately 3–4 mm in from the front edge of the muzzle. Mark the outer, uncut, edge with a blue pencil. Trim a section out of one ear by first cutting one of the ears in half lengthwise and then cutting one of

the halves completely off from the scalp. Lay this half flat on the cutting surface and make a second cut parallel to the first to obtain a section similar in size and shape to that of the dorsal and ventral skin sections. Put a blue mark on one of the long sides of the section of ear (Figure 30.19). The various skin sections may be combined in cassettes. However, the same sections should be combined every time (refer to Table 30.1) to aid with identification. We combine the long sections of dorsal skin with the sections of ear and tail skin in one cassette; long sections of ventral skin with the sections of eyelid and muzzle in a second cassette, and each of the square sections of dorsal and ventral skin in individual cassettes. Each cassette is also labeled with a 'D' or 'V' to aid in identification.

Hematoxylin and Eosin (H&E) stain is usually requested on all tissues sent to the histology laboratory. Special stains may be requested on specific tissues if this is necessary to verify any suspected pathologic changes not sufficiently disclosed by the H&E stain. Protocols for various stains, what they stain, and what colors they stain are subjects of various histology and pathology text books (Luna, 1960; Smith and Bruton, 1977).

Skeletal staining of whole mice

The alizarin staining technique is an old technique (Schultze, 1897) that has been extensively used to investigate skeletal abnormalities in various strains of mutant (Kaufman *et al.*, 1995; Sweet *et al.*, 1996) and genetically engineered (Kawaguchi *et al.*, 2001) mice. This technique has the advantage of being performed on whole bodies, which provides excellent three-dimensional visualization of skeletal lesions. The bones are stained red while the other tissues do not stain and are translucent to pale blue. Continual improvement in X-ray technology has reduced the usage of alizarin staining. However, this technique still provides incomparable three-dimensional representations of the skeleton and is still extensively used, especially for animals with a minimally mineralized skeleton, such as fetuses and young mice. Also, it has the potential to be used to assess the number and location of skeletal metastases. The techniques presented here are modified from P.B. Selby (1987).

The mouse should be skinned, except for the skin below the digits to avoid damaging them. The mouse should be eviscerated. Any subcutaneous, retro-orbital, mesenteric, and mediastinal fat, as well as the trachea, esophagus, salivary glands, tongue, and eyes should be removed. The mice should not be fixed but may be frozen.

Solution A	1% weight/volume (w/v) aqueous KOH.
Solution B	2% (w/v) aqueous KOH.
Solution C	1.9% (w/v) aqueous KOH containing 0.040 g alizarin red S per liter
Solution D	1.6% (w/v) aqueous KOH containing 0.033 g alizarin red S per liter
Solution E	clearing solution made of 400 ml white glycerin, 200 mL benzyl alcohol, and 400 ml 70% ethanol.

Four working solutions are used. Deionized water is used in all solutions.

Solutions A, B, and E can be stored indefinitely. Solutions C and D should be discarded when they precipitate, which may start a few weeks after preparation.

The mice can be stained in 3 days (works best when mice are 6–9 weeks old), or in 11–14 days (when mice are more than 12 weeks old). The bones of mice stained with the 3-day procedure may be brittle, especially with mice older than 12 weeks.

The staining procedures are presented below. Water rinses are with tap water. Room temperature should be 20–25°C.

Time in solution C should always be 2 days. Times in solutions A, B, and E can be varied to accommodate working schedule.

Three day procedure

Day 0 1. Skin and eviscerate specimen.
2. Cover specimen with solution A. Use 0.5% aqueous KOH solution if mice are 3–5 weeks old, and 2% aqueous KOH solution for mice 10–17 week-old.

Day 1 1. Pour out solution; rinse jar and carcass with water.
2. Remove any loose fat and muscle.
3. Cover specimen with solution D.

Day 3 1. Pour out solution; rinse jar and carcass with water.
2. Clean off tail and immerse specimen in solution E.
3. After at least 3 h, heat at 45°C for 1 h in a water bath.
4. Pour out solution; drain jar and specimen, do not rinse.
5. Cover specimen with white glycerin.

Eleven to fourteen day procedure:

Day 0   1. Skin and eviscerate specimen.
2. Cover specimen with solution A.

Day 1–2   1. Pour out solution. Rinse jar and carcass.
2. Cover specimen with solution A.

Day 4–7   1. Pour out solution. Rinse jar and carcass.
2. Cover specimen with solution B.

Day 6–8   1. Pour out solution. Rinse jar and carcass.
2. Remove loose fat and muscles and clean tail.
3. Cover specimen with solution C for 2 days.

Day 8–10 1. Pour out solution. Rinse jar and carcass.
2. Cover specimen with solution E.

Day 11–14 1. Pour out solution; drain jar and specimen, do not rinse.
2. Cover specimen with white glycerin.

Conclusions

A research quality necropsy and interpretation of gross and microscopic changes requires a great deal of skill on the part of both the technician and the pathologist. The two need to work together to coordinate their efforts and to optimize protocols to achieve consistent, high quality results. This chapter provides an overview on how to achieve these results but practice is required to develop the skills.

Acknowledgments

This work was supported by grants from the National Institutes of Health (CA34196, AR 43801, and RR173), The National Alopecia Areata Foundation, PXE International, the Council for Nail Disorders, and Transgenic, Inc. The authors thank I. K. Sundberg for producing the line drawings used in this chapter.

Appendix: Examples of commonly used fixatives for mouse histopathology

Fekete's acid alcohol formalin (Tellyesniczky's/Fekete's solution)

This is a commonly used fixative in mouse laboratories (Sundberg et al., 1997). It provides rapid and surprisingly deep tissue fixation. Specimens are transferred to 70% ethanol following overnight fixation after which they are trimmed and processed. This fixative yields high quality specimens in histologic sections. Long term storage can be a problem since ethanol is flammable and evaporates easily, which can render specimens useless. An artifact is that erythrocytes (red blood cells) are leached so that they appear only as pink ghosts within vessels.

Bouin's solution

This fixative uses picric acid (stains everything permanently yellow), acetic acid, and formalin (Bristol et al.,

2000; Car and Eng, 2001). Delicate detail is not often well preserved but this fixative is preferred by some pathologists and researchers. Penetration in tissue is moderate. Bouin's fixed specimens must be washed in running tap water for 2–4 h after initial overnight fixation and stored in 70% ethanol or they become very brittle. Bouin's solution can be used for bone decalcification since the acids will demineralize specimens that are left in the fixative for several days or weeks. If used for decalcification, the Bouin's solution must be changed weekly to optimize demineralization.

Neutral buffered 10% formalin

This is the most commonly used fixative in most pathology laboratories and is the fixative of choice for participants in the Human Mouse Models Cancer Consortium that represents M.D. and D.V.M. pathologists (Cardiff et al., 2000). Specimens can be fixed overnight and left indefinitely in the fixative. Specimens can be processed at any time as long as they remain wet, sometimes many years after initial collection. This fixative is particularly useful for retrospective evaluation of lipids or other substances that are soluble in ethanol and would be lost during tissue processing. For example, the presence of fat in adipocytes can be demonstrated if you take wet tissue fixed in neutral buffered 10% formalin, trim the tissue, cut frozen sections, and stain the sections with oil-red O or other lipid histochemical stains. Most of the other fixatives are alcohol based and remove lipid.

Long term storage in neutral buffered 10% formalin causes continued cross-linking of amino groups resulting in changes in the tertiary structure of proteins. As a result, many antigenic epitopes are lost or changed, making immunohistochemistry problematic. Transfer of tissues into 70% ethanol after overnight (12 h) fixation can reduce this effect.

Neutral buffered 10% formalin is noxious and should be used in a fume hood. Buffering is needed to reduce acid hematin formation, an artifact of fixation.

B5 fixative

This fixative is used for immunohistochemistry and often yields the most accurate and reproducible results of any fixative. It is difficult to use because it has to be prepared immediately before use, is based on mercury salts (difficult to dispose of), and fine precipitates can form (a particular problem when using Gomori's methenamine silver or similar stains). Tissues fixed in

B5 may be difficult to cut, and sections require treatment with Lugol's Iodine to remove pigments.

Carnoy's fixative

This treatment fixes tissues rapidly. It preserves glycogen and enhances the staining of mast cell granules. Nissle granules are also well preserved. Because it is an alcohol/acid based fixative, it lyses red blood cells and acid soluble granules.

4% paraformaldehyde

This is used as a fixative for electron microscopy and *in situ* hybridization and has become popular as a general histologic fixative in laboratories that employ these techniques. The paraformaldehyde should be prepared in a buffered solution at pH 7 and refrigerated until use. It will keep for several weeks this way.

Zinc based fixatives

A variety of proprietary fixatives based on variations of the classic fixatives have been developed and marketed not so much as fixatives for routine, high quality histopathology but rather to optimize antigenic epitopes for immunohistochemistry (Tome *et al.*, 1990). Such fixatives are the basis for large-scale tissue arrays that provide large numbers of small tissues on individual slides. However, a variety of proprietary zinc based fixatives (such as IHC Zinc, BD Biosciences PharMingen, San Diego, CA) yield good immunohistochemical results while providing adequate fixation for histopathology.

Glutaraldehyde

This fixative is commonly used for ultrastructural studies. Tissue penetration is minimal, approximately 1 mm on any cut surface, so specimens have to be finely minced with a sharp razor blade to achieve adequate fixation. Several different buffers can be used. The most common are phosphate and cacodylate based buffers. Phosphate buffers are safe and yield good results when used fresh. Fine electron dense precipitates may form that will render a specimen useless if the buffer used is old. Cacodylate buffer is arsenic based, which is toxic and can be difficult to dispose of properly.

Karnovsky's fixative

This fixative is used for plastic embedding for transmission electron microscopy. Karnovsky's is a general term for any fixative combining glutaraldehyde and paraformaldehyde in a phosphate buffer. Glutaraldehyde has minimal penetration ability. Paraformaldehyde penetrates deeper, but fixation is unstable. Karnovsky's fixative combines the positive points of both these chemicals.

JB4 fixative

This fixative is used for plastic embedding for 1 μm sections used in light microscopy. It combines glutaraldehyde and paraformaldehyde in a cacodylate buffer.

O.C.T. compound

O.C.T. Compound (Tissue-Tek®, Sakura Finetek U.S.A., Inc., Torrance, CA) is an embedding medium for frozen sections. It is a thick, clear fluid used in conjunction with plastic base molds such as CMS Tissue Path-Disposable Plastic Base Molds (Curtis Matheson Scientific, Inc. Houston, TX) to bind fresh tissues for freezing and sectioning. Tissues are floated in the mold and placed on dry ice. The O.C.T. solidifies and turns white as if freezes, after which the block has to be stored frozen at $-80°C$ until sectioning.

References

American Veterinary Medical Association (1993). *J. Am. Vet. Med. Assoc.* **202**, 229–249.

Bristol, I.J., Farmer, M.A., *et al.* (2000). *Inflamm. Bowel Dis.* **6**, 290–302.

Car, B. and Eng, V. (2001). *Vet. Pathol.* **38**, 20–30.

Cardiff, R.D., Anver, M.R., *et al.* (2000). *Oncogene* **19**(8), 968–988.

Feldman, D. and Seely, J. (1988). *Necropsy Guide: Rodents and the Rabbit.* CRC Press, Inc., Boca Raton.

Green, M.C. and Witham, B.A. (eds) (1991). *Handbook on Genetically Standardized JAX Mice.* The Jackson Laboratory, Bar Harbor, ME.

Hall, C. (1947). *J. Hered.* **38**(3).

Hummel, K. and Chapman, D. (1956). *Mouse News Lett.* **14**, 21.

Hummel, K. and Chapman, D. (1959). *J. Hered.* **50**, 9–13.

Kaufman, M.H., Chang, H.H., *et al.* (1995). *J. Anat.* **186**(Pt 3), 607–17.

Kawaguchi, J., Azuma, Y., *et al.* (2001). *J. Bone Miner. Res.* **76**(7), 1265–1271.

Kiupel, M., Brown, K.S., *et al.* (2000). *J. Exp. Anim. Sci.* **40**, 178–188.

Layton, W. (1976). *J. Hered.* **67**(6), 336–338.

Luna, L. (1960). *Manual of Histologic Staining Methods of the Armed Forces Institute of Pathology*. McGraw-Hill, Inc., New York.

Mahler, J.F. (2000). In *Pathology of Genetically Engineered Mice* (eds J.M. Ward, J.F. Mahler, R.R. Maronpot and J.P. Sundberg), pp. 137–144. Iowa State University Press, Ames.

Mahler, M., Bedigian, H.G., *et al.* (1998). *Lab. Anim. Sci.* **48**(1), 85–91.

Mahler, M., Sundberg, J.P., *et al.* (1999). *Genomics* **55**(2), 147–156.

Mohr, U., Dungworth, D.L., *et al.* (1996). *Pathobiology of the Aging Mouse*. ILSI Press, Washington, DC.

Moolenbeek, C. and Ruitenberg, E. (1981). *Lab. Anim.* **15**(1), 57–59.

Muller-Rover, S., Handjiski, B., *et al.* (2001). *J. Invest. Dermatol.* **117**, 3–15.

Nanney, L.B., Sundberg, J.P., *et al.* (1996). *J. Invest. Dermatol.* **106**, 1169–1174.

Paus, R., Muller-Rover, S., *et al.* (1999). *J. Invest. Dermatol.* **113**, 523–532.

Percy, D. and Barthold, S. (1993). *Pathology of Laboratory Rabbits and Rodents*. Iowa State University Press, Ames.

Popesko, P., Rajitov, V., *et al.* (1992). *A Color Atlas of Anatomy of Small Laboratory Animals*. Wolfe Publishing, Ltd, London.

Schultze, O. (1897). *Verhandlungen der Anatomischen Gesellschaft. Anat. Anz.* **13**, 3–5.

Selby, P.B. (1987). *Stain Technol.* **62**, 143–146.

Smith, A. and Bruton, J. (1977). *Color Atlas of Histological Staining Techniques*. Year Book Medical Publishers, Inc., Chicago.

Smith, R.S., John, S.W.M., *et al.* (2002). *Systematic Evaluation of the Mouse Eye. Anatomy, Pathology, and Biomethods*. CRC Press, Boca Raton.

Spraycar, M. (ed.) (1995). *Stedman's Medical Dictionary*. Williams & Wilkins, Baltimore.

Sundberg, B.A. and Sundberg, J.P. (1990). *Lab. Anim.* **19**, 55–58.

Sundberg, B.A. and Sundberg, J.P. (2000). In *Systematic Approach to Evaluation of Mouse Mutations* (eds J.P. Sundberg and D. Boggess), pp. 47–55. CRC Press, Boca Raton.

Sundberg, J.P., Elson, C.O., *et al.* (1994a). *Gastroenterology* **107**, 1726–1735.

Sundberg, J.P., Elson, C.O., *et al.* (1994b). *Gastroenterology* **107**, 1726–1735.

Sundberg, J.P., Boggess, D., *et al.* (1997). *Am. J. Pathol.* **151**(1), 293–310.

Sundberg, J., Montagutelli, X., *et al.* (1998). *Cutaneous Appendages* (ed. M. Chuong), pp. 421–435. Molecular Biology Intelligence Unit I, Landes Company, Austin, TX.

Sweet, H.O., Bronson, R.T., *et al.* (1996). *J. Hered.* **87**(2), 87–95.

Tome, Y., Hirohashi, S., *et al.* (1990). *Histopathology* **16**(5), 469–474.

Ward, J., Mahler, J., *et al.* (2000). *Pathology of Genetically Engineered Mice*. Iowa State University Press, Ames.

Handling and Restraint

Thomas Buerge and Tilla Weiss
Novartis Pharma AG, Basel, Switzerland

Introduction

The use of the mouse in biomedical research can be traced back to the 1600s and since then this species has contributed to a vast number of scientific findings and to progress in basic biological and pharmaceutical research (Grieder and Strandberg, 2003). Nowadays, an enormous number of different inbred and outbred mouse strains, including genetically modified mouse lines are available and used in research laboratories worldwide. Within the scope of experimental work with rodents, mice are not necessarily regarded as a species with a strong drive to cooperate. Despite the long time of selective breeding in captivity, the natural behavioural pattern of the wild mouse – although less prevalent – still persists. Normally laboratory mice have to be 'involved' by means of professional handling/restraint in order to perform all procedures necessary during husbandry and/or experimentation. Nevertheless, mice (with strain differences) are usually not very aggressive and can be handled or restrained without major problems. Correct handling should not only be imperative during experimental work, but should already start at breeding sites and be continued as part of daily husbandry procedures in order to familiarize the animals with people and manipulations.

Although little scientific information on handling and restraint of mice is available, many general text books touch upon the technical approach of this topic. (Cunliffe-Beamer, 1983; Anderson and Edney, 1991; Biological Council, 1992; Harkness and Wagner, 1995; Hrapkiewicz *et al.*, 1998; Wolfensohn and Lloyd, 1998; Baumans, 1999; Suckow *et al.*, 2001; Pekow and Baumans, 2003). This chapter compiles comprehensive information about handling and restraint, including the personal experience of the authors with this species.

Occupational health and risks

Injuries

Work with laboratory mice does not usually bear the risk of severe injuries. Still, minor injuries through mouse bites, mainly into fingers, may occur, especially if the staff is not very experienced and/or adequate protective measures are not properly applied when mice are handled and restrained. Such events can be traced back to the fact that mice are extremely fast in their

The Laboratory Mouse
Copyright 2004 Elsevier
ISBN 0-1233-6425-6

movements and usually tend to escape or defend themselves if they are given the opportunity to do so. In addition to proper handling and restraint, the wearing of single layered synthetic hypo-allergenic gloves, or preferentially a double layer of both, cotton and synthetic gloves have considerable potential to reduce the number of mouse bites that perforate gloves and intact skin at the same time. Initial reluctance by staff to the use of double gloves is overcome eventually because it can increase comfort and well being of people handling and restraining animals.

Human infection and disease

Most purpose-bred laboratory mice from defined sources are specified pathogen free and their hygiene status should undergo repeated microbiological testing during housing and experimentation (*Baker, 1998; FELASA Working Group on Health Monitoring of Rodent and Rabbit Colonies*, 2002). Despite those precautions, the laboratory mouse and its excretions still harbour the potential to transmit opportunistic agents and cause human disease. Infection of skin scratches and bite wounds with mouse or human-borne opportunistic microbes demands attention through a strict occupational medical treatment program (National Research Council, 1997). Immediate cleansing and disinfection of the wound represents the first step in order to prevent infection. Special attention should be given to mice infected with human-pathogenic or zoonotic agents or genetically modified mice that harbour receptors for human pathogens. Both cases require work in higher biosafety level containments, additional screening methods and special guidelines for the handling and restraint of such animals. Tumour cells that are implanted into mice should be microbiologically screened for human and mouse pathogens and excluded if found positive before injection.

Allergies

A more severe problem in people working with mice has been observed for more than 25 years: the development of a human allergy to mice. This phenomenon, also called 'Laboratory Animal Allergy' (LAA), is a form of occupational allergic disease and includes a great number of laboratory animal species to which people may develop allergic reactions. After the phase of sensitization, resulting from complex processes within the immune system, allergy occurs and usually is represented by nasal symptoms (e.g. sneezing, watery

discharge etc.) eye reactions or skin rashes. Asthma and, rarely, bite related anaphylaxis, a life threatening allergic reaction may occur. The level of exposure to the laboratory animal allergen is crucial to the nature and intensity of the symptoms (Bush, 2001). In mice, the major allergen is MUP (Mus m 1) the major urinary protein which is a prealbumin and may be found in urine as well as in hair follicles and dander. As the level of production of this protein within the liver is testosterone dependent, it is predominant in adult male mice. The second mouse allergen Mus m 2 is a glycoprotein, found in hair and dander and the third one is albumin, a serum protein. Mouse allergens can be distributed and found throughout an animal facility and even spread into separate buildings adjacent to the facility. Although the wide distribution of particles may also cause problems of sensitization and allergy to people not directly working with mice, the highest exposure to the allergens has been reported in people dealing with cage cleaning and feeding of the animals (Wood, 2001). In order to reduce exposure to mouse allergens and prevent LAA, the following personal protection measures like (a) reducing skin contact with animal products such as urine, dander and serum by using long-necked, non-allergic gloves, laboratory coats and adequate respiratory protective equipment, (b) avoidance of wearing street clothes while working with animals as well as (c) leaving work clothes at the workplace, should be taken. Furthermore, processes and procedures in animal husbandry and handling can be adapted, for example, directing airflow away from workers, performing manipulations within ventilated hoods where possible, installing ventilated animal cage racks or filter-top cages, using absorbent pads for bedding etc. (Harrison, 2001). It has been shown that the combined use of ventilated micro-isolators, ventilated cage change stations, ventilated benches for procedures and robotics for automatic cage emptying and cleaning, together with the use of a centralized vacuum cleaning system, resulted in considerably lower exposure levels to allergens (Thulin *et al.*, 2002).

Definitions

Handling

Handling within this context is defined as dealing with a mouse by hands, in a direct or indirect way – with or without touching the animal. Handling should always be done in a species-specific, calm and firm way in order not to harm the animal and to provide as much safety as possible to the experimenter. In order to reduce the stressful component of any handling procedure to a minimum for both parties, the personnel involved should be dedicated to animals, motivated and well trained. Aims of training are attainment of sovereign handling skills as well as habituation of animals to people and manipulations with as little disturbance of their physical and psychological well being as possible. In the best case, animals can even be motivated to cooperate with their trainers, a fact which does not only facilitate work and enhance safety for people but also helps to reduce stress-induced changes in physiological parameters in animals under experimental conditions.

Restraint

Restraint is described as immobilization of an animal by keeping it or parts of it, in a comfortable but safe hold by hand or by means of a physical device. Physical restraint is performed on conscious animals undergoing manipulations, which do not require sedation or anaesthesia, but necessitate exact positioning of the animal as well as prevention of unexpected movements during the manipulation. Restraining measures therefore are indispensable for the performance of experimental work as they not only facilitate avoidance of injuries in animals but also provide an adequate level of safety to the participating members of staff. In instances where unacceptable stress or pain may occur to the animal, physical restraining measures may be facilitated by sedation or by general anaesthesia and analgesia of an animal, respectively.

Handling of mice

Despite the general non-aggressiveness of the laboratory mouse, only marginal success can be expected regarding cooperation even after weeks of training. Therefore, reinforced individual cooperative training has not become common practice when handling mice. Handling is generally restricted to individual or group transfer from cage to cage during cage change or to transfer of animals from and to the experimental environment. As with other species, hectic and jerky movements should be avoided. Time should be given to animals to investigate the handler's hand and become adapted to the smell of the gloves.

Transfer of groups of mice

By hand

Small groups of mice, often sitting together in a corner of the cage, can be surrounded from two sides with the palms of both hands cupped. Without exerting any pressure, the hands are then slid towards each other beneath the mice and the whole group is lifted up and transferred to, for example, another cage, where they are gently put back on to the bottom (Figures 31.1 and 31.2). This method is very effective when animals are not trained and/or the transfer must be time-efficient.

By means of a device

Another way of transferring groups of mice or individuals, is by using a glass or synthetic bowl. The vessel is brought close to the mice with its open end directed

Figure 31.1 Grasping a group of C57BL/6 mice.

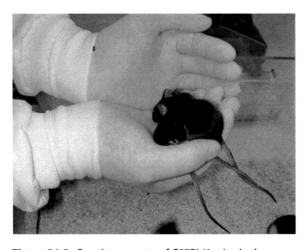

Figure 31.2 Carrying a group of C57BL/6 mice in the cupped hand.

towards the cage wall. Mice can then be encouraged to climb into the beaker (Figures 31.3 and 31.4). A reel may also serve as means of transfer, as mice like to crawl into the dark tube or climb onto the device. They can be placed back by allowing them to climb freely from the device into the cage (Figure 31.5).

Transfer of single mice

By hand

For a short transfer of less than 2–3 s, mice are gripped by the base of the tail, lifted up and carried to the new destination. This does not apply to very heavy, obese or pregnant mice, which have to be supported by the other hand. The tail remains held by one hand in order to prevent the animal from escaping (Figure 31.6). In case of transferring mice over a longer distance, they should be placed on the hand and must not be carried by the tail. Otherwise the overlying skin of the tail may become detached from the body due to the force

exerted on it. Again mice are put back into the cage gently. After weighing for example, they can be released directly from the scales pan into the cage.

By means of a device

In case of special hygienic precaution requirements (e.g. specific pathogen free (SPF) or immuno-compromised animals) where exposure of the animals to potential pathogens and opportunistic microbes should be kept to a minimum, mice can also be transferred by means of a pair of forceps (25–30 cm long, with rubber protected tips). The loose skin at the rear of the neck (neck fold) is grasped with the forceps. In order not to harm the animal, it is approached from behind with the forceps and carefully lifted (Figure 31.7). The animal is released gently by opening the forceps after putting it on to the bottom of the new cage. This method mimics the behaviour of a mouse pup carried by its mother by

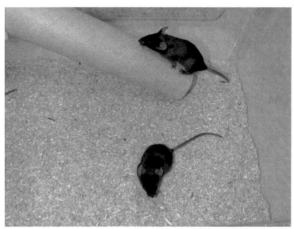

Figure 31.5 C57BL/6 mouse climbing onto a reel for further transfer.

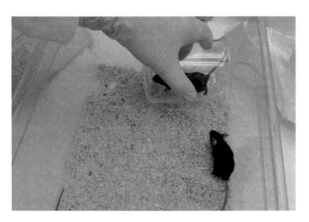

Figure 31.3 Climbing of a C57BL/6 mouse into a glass beaker.

Figure 31.4 Transferring a group of C57BL/6 mice within a glass beaker.

Figure 31.6 Carrying a C57BL/6 mouse within the hand while fixing the tail.

Figure 31.7 Gripping a C57BL/6 mouse by rubber-tipped forceps.

gripping of the pup's neck fold with its mouth. This relaxation can still be seen in adult animals when being handled (Figure 31.8a and b).

Transfer of litters and mother

In case of transferring a mother with her litter, the mother is removed first, in order not to provoke defensive reactions by her when the nest is taken out of the cage. The female is transferred according to the procedure described above for single mice and placed into the new cage. The litter, i.e. nesting material and pups together, is grasped with both hands forming a cup and sliding beneath the nest. The whole nest with its contents is then lifted up, carried to the new cage and gently placed back, preferably not touching the pups by unprotected hands. By transferring litters this way, the female usually immediately approaches the nest and accepts her pups without any problems (Figures 31.9 and 31.10).

Restraining of mice

Limited cooperation of the mouse, its unpredictable behaviour and continuous readiness to bite when being restrained demands careful action and proper restraining of each individual animal. This includes secure immobilizing that minimizes movements of the animal but still allows it to breath normally. Such action avoids casualties even in very sensitive strains and reduces animal accidents that may be caused by mouse bites and unexpected reflexes of the bitten person. Gentle release into the researcher's hands before return of the animal into the cage can contribute to adaption of the animals to restraining procedures.

(a)

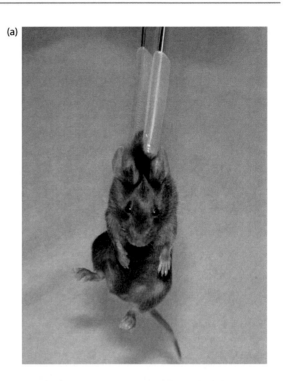

(b)

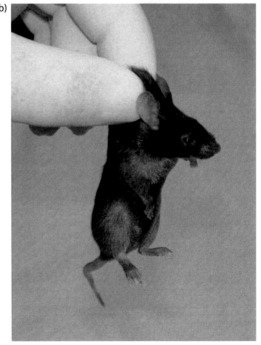

Figure 31.8 (a) Carrying a C57BL/6 mouse by rubber-tipped forceps in a relaxed position; (b) C57BL/6 mouse carried by hand at neck skin fold; mimics carriage by mother. (*Note*: Natural relaxation.)

Restraining by hand

The tail of the mouse is gripped at its base and the mouse is lifted onto the grid cage top. By gently pulling the tail backward, the animal tends to move forward and to hold on to the grid with its forelegs. At this

Figure 31.9 Grasping a nest with litter from the bottom of a cage. (Note: C57BL/6 mother has been transferred first.)

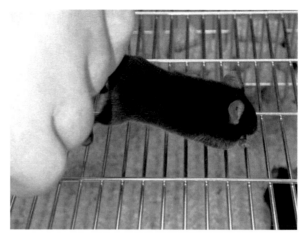

Figure 31.11 Fixation of a C57BL/6 mouse by the base of its tail.

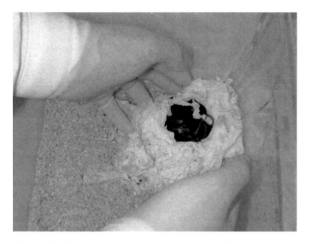

Figure 31.10 Placing a complete nest with litter and nesting material back into a cage.

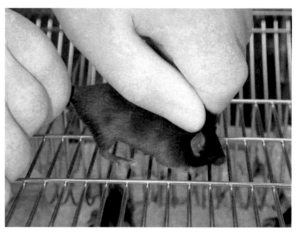

Figure 31.12 Grasping a skin fold at the rear of the neck with thumb and forefinger.

moment, the other hand approaches the rear of the neck and a skin fold, quite close to the ears, is grasped with the thumb and the forefinger, while the loose skin extending over the back is gripped with the other fingers. It is important to grip the loose skin in the rear of the neck properly, in order to prevent the animals from turning its head and biting into the handler's fingers. At the same time, care must be taken not to impair the animal's breathing and venous blood backflow from the head to the chest. By turning the hand upwards, the mouse is positioned with its ventral side uppermost. The tail is then gripped between the third finger and the ball of the thumb. The head and body of the animal are brought into a straight and comfortable position with its back being supported by the palm of the hand. In this position the mouse is held safely for any further manipulations (Figures 31.11–31.15).

Figure 31.13 Fixing loose skin along the back and tail of the C57BL/6 mouse.

Figure 31.14 Final fixation for further manipulations.

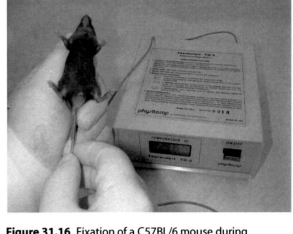

Figure 31.16 Fixation of a C57BL/6 mouse during temperature recording (note: adult animal).

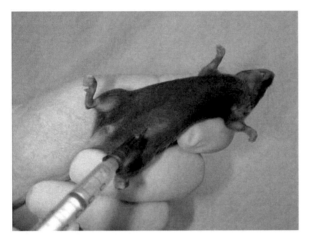

Figure 31.15 Fixation of a C57BL/6 mouse and ip injection into the left, caudal abdomen.

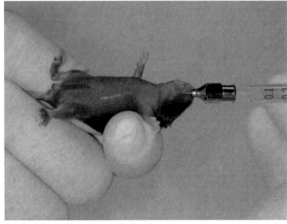

Figure 31.17 Restraining a mouse pup during oral dosing.

Mouse pups can be restrained in two ways: (A) Without any prior handling, the thumb and the first two fingers are placed around the shoulder and thorax region and the animal is picked up from the cage. It is then held in this way and can be positioned for physical examination or rectal temperature recording for example (Figure 31.16). (B) A skin fold in the dorsal neck/shoulder region is first grasped between the thumb and the index finger. Special care has to be taken not to restrict their breathing due to the small size of pups. After positioning the pup in the same way as described for the adult, oral administration of drugs by means of a ball-ended metal tube, for example, can be performed (Figure 31.17).

Two further indications for which manual restraint is performed frequently are tail marking and sexing. *Tail marking* is best performed in the following way: After lifting the mouse by the base of its tail and putting it onto the grid cage top, the tail is gently pulled backwards and different marks can be applied by means of a waterproof text marker. For *sexing*, the mouse is put onto the grid cage top and the tail is carefully pulled backward in the same way as described for tail marking. When the animal reaches an extended position due to its drive to move forward, its back is gently depressed with the third and fourth finger while the tail base and rear legs are lifted up in order to expose the genitalia. The sex can then be determined by checking the ano-genital distance, which is longer in male animals (Figures 31.18 and 31.19).

Restraining by means of a device

New mouse restraining devices are continuously being developed. They are self-made for special purposes or can be supplied from commercial sources. Materials

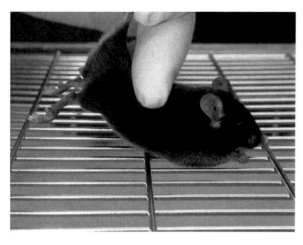

Figure 31.18 Animal positioning for sexing.

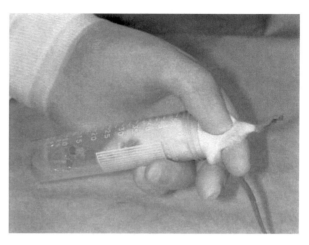

Figure 31.20 Tube restraining of a BALB/c mouse for blood sampling at the vena saphena.

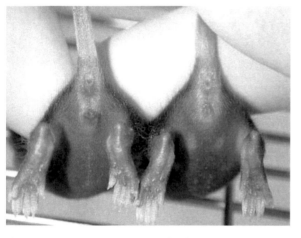

Figure 31.19 Exposing genitalia for determining gender of the animals by checking the ano-genital distance; left: male, right: female.

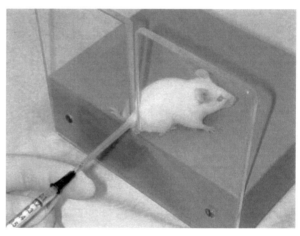

Figure 31.21 Using of a restraining wall for e.g. blood sampling on the lateral tail vein of a BALB/c mouse.

used include soft leather or plastic, hard plexy-glass or macrolon, metal among others. Unlimited design possibilities are restricted by demands on hygienic properties, harm- and stresslessness for animals and man, and optimally, the restraining device allows the experimenter to have both hands free for the execution of procedures on the animal. Long-term restraining devices should allow the animal to fulfil its basic physiological needs. Commercial catalogues are available from various suppliers or devices can be searched for on the internet (AALAS 2002, *Laboratory Animals Buyers' Guide*, 2002).

Some examples of commonly used restraining devices are shown on Figures 31.20–31.22. One example (Figure 31.20) shows a commercially available tube usually used for cell culture technique. The animal is

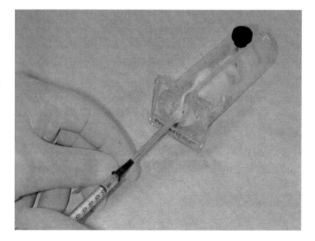

Figure 31.22 Restraining tube for manipulations on the tail of a BALB/c mouse.

gripped at the base of the tail and lifted up. After introducing the mouse into the tube, the device is turned into a horizontal position and the animal is gently pushed forward into the tube or the tail slightly drawn backward to motivate the animal to escape ahead into the tube. The tip and bottom of the tube are equipped with self-made holes, which the mouse tends to reach into. When the mouse is completely in the tube, a rear leg can be exposed e.g. for blood sampling at the vena saphena. Different restraining devices for blood sampling from or injections into the lateral tail vein are widely distributed and many of them are commercially available. Another example (Figure 31.21) of a 'restraining wall' has been used and modified. In order to restrain the mouse, the animal is grasped at the base of the tail and lifted up. It is then positioned in front of the wall with its tail being placed in the slit and the mouse is lowered to the underlying platform. When the mouse has reached the bottom, the tail can be gently pulled backwards and blood can be taken from the tail vein with the animal not being squeezed into a narrow tube but allowed to move freely. Figure 31.22 shows a more sophisticated version of a restraining device (Provet AG, Lyssach, CH) for exposing the tail of the animal. The mouse is lifted up by the base of the tail and then placed in front of the open end of the tubular device. It is introduced backwards with its tail being gently pulled along the open longitudinal slit. A head button is slid into the tube up to the animals head in order to prevent the mouse from moving forward. The animal is now ready for further manipulations.

Effect of handling and restraint on well being of mice

Little scientific background information on stress related to handling or short term restraint in mice is currently available. Stress is considered to be influenced by the combination of restraint and procedure and be dependent on the duration and frequency in which the animals are exposed to manipulations. The outcome of continuous restraint stress can be manifold and range from temporary weight loss to restraint induced pathology (Paré and Glavin, 1986). More recent restraint studies have shown that stress response can be more subtle. It has been shown that mice that were

restrained for 12–24 h in restraint cages and tubes showed reduction of lymphocyte cell numbers in lymphoid organs and suppression of *in vivo* antibody production (Fukui *et al.*, 1997), elevation of endogenous glucocorticoid and suppression of migration of granulocytes and macrophages to an inflammatory focus (Mizobe *et al.*, 1997), delay of cutaneous wound healing (Padgett *et al.*, 1998) and also impairment of bacterial clearance during wound healing (Rojas *et al.*, 2002). These findings suggest that handling and restraint should be carried out in a firm, confident and gentle manner and permanent care should be taken, not to crush or squeeze the animals (Rodent Refinement Working Party, 1998). There is still some controversy about the question as to whether frequent handling and restraint will reduce or increase stress in the mouse. Although Li *et al.* (1997) have shown that repeated restraint caused significant impairment of anti-tumour T cell responses, further studies are required to clarify the effect of repeated handling and restraint in the mouse. Different temperament, adaptability and stress sensitivity of strains must be taken into account before any final conclusion regarding stress response to handling and restraint can be made.

Summary and recommendation

Despite its limited friendliness and cooperative behaviour but for the many other benefits as e.g. its high reproduction rate, small size and vitality, the laboratory mouse has been most prevalent in the *in vivo* research laboratory. Unlike the rat, the mouse shows generally a less positive response to good handling. The risk of deep bite injuries, however is low. Nevertheless, the animal should be approached, handled and restrained with care and deep respect. All measures shall be taken to ensure competent and least stressful manipulation. This can be achieved by professional training of the experimenter and animal care staff. Proper handling and restraint contribute to refinement of animal research and validity of research data.

Although physical restraining alone can serve to achieve safe and efficient manipulations in the animal as, for example, subcutaneous, intra-peritoneal and intra-muscular injections or gavage applications, procedure-related stress and pain of an animal shall be evaluated carefully.

Safe and efficient anaesthetic agents providing fast onset of anaesthesia together with a short recovery phase may be used for chemical restraint of mice in situations, where physical restraint of conscious animals may not be appropriate for certain procedures from the animal welfare point of view. Such instances may not only be surgical events but also injection of transponders, tattooing of tails, ear punching and injections of compounds (for chemical restraint see Chapter 34 in this book).

Acknowledgements

We wish to thank our veterinary technicians M. Aeberhard and L. Fozard for sharing their experience with us and taking the pictures as presented in this chapter.

References

AALAS (2002). In *Reference Directory 2001/2002*, pp. 182–208. American Association for Laboratory Animal Science. http://www.aalas.org

Anderson, R.S. and Edney, A.T.B. (1991). *Practical Animal Handling*. Pergamon Press, Oxford, UK.

Baker, D.G. (1998). *Clin. Microbiol. Rev.* **11**(2), 231–266.

Biological Council. (1992). *Guidelines on the Handling and Training of Laboratory Animals*. Universities Federation of Animal Welfare, Potters Bar, Herts.

Baumans, V. (1999). In *The UFAW Handbook on the Care and Management of Laboratory Animals, Vol. I* (ed. T. Poole), pp. 282–312. Blackwell Science Ltd., Oxford.

Bush, R.K. (2001). *ILAR J.* **42**, 4–11.

Cunliff-Beamer, T.L. (1983). In *The Mouse in Biomedical Research, Vol. III* (eds H.L. Foster, J.D. Small and J.G. Fox), pp. 401–437. Academic Press, New York.

FELASA Working Group on Health Monitoring of Rodent and Rabbit Colonies. (2002). *Lab. Anim.* **36**, 20–42.

Fukui, Y., Sudo, N., Yu, X.-N., Nukina, H., Sogawa, H. and Kubo, C. (1997). *J. Neuroimmunol.* **79**, 211–217.

Grieder, F.B. and Strandberg, J.D. (2003). In *Handbook of Laboratory Animal Science, Vol. I* (eds J. Hau and G.L van Hoosier, Jr.), pp. 1–11. CRC Press, Boca Raton, FL USA.

Harkness, J.E. and Wagner, J.E. (1995). *The Biology and Medicine of Rabbits and Rodents*. pp. 58–68. Williams & Wilkins, Media, PA USA.

Harrison, D.J. (2001). *ILAR J.* **42**, 17–36.

Hrapkiewicz, K., Medina, L. and Holmes, D.D. (1998). *Clinical Medicine of Small Mammals and Primates*, pp. 3–30. Iowa State University Press.

Laboratory Animals Buyers' Guide 2002. Laboratory Animals Ltd. http://www.lal.org.uk

Li, T., Harada, M., Tamada, K., Abe, K. and Nomoto, K. (1997). *Anticanc Res.* **17**, 4259–4268.

Mizobe, K., Kishihara, K., EI-Naggar, R.E., Matkour, G.A., Kubo, C. and Nomoto, K. (1997). *J. Neuroimmunol.* **73**, 81–89.

National Research Council. (1997). *Occupational Health and Safety in the Care and Use of Research Animals*. National Academy Press, Washington, DC.

Padgett, D.A., Marucha, P.T. and Sheridan, J.F. (1998). *Brain Behav. Immun.* **12**, 64–73.

Paré, W.P. and Glavin, G.B. (1986). *Neurosci. Biobehav. Rev.* **10**, 339–370.

Pekow, C.A. and Baumans, V. (2003). In *Handbook of Laboratory Animal Science, Vol. I* (eds J. Hau and G.L. Van Hoosier, Jr.), pp. 351–390. CRC Press, Boca Raton, FL USA..

Rodent Refinement Working Party. (1998). *Lab. Anim.* **32**, 233–259.

Rojas, I.-G., Padgett, D.A., Sheridan, J.F. and Marucha, P.T. (2002). *Brain Behav. Immun.* **16**, 74–84.

Suckow, M.A., Danneman, P. and Brayton, C. (2001). *The Laboratory Mouse*. CRC Press.

Thulin, H., Björkdahl, M., Karlsson, A. and Renström, A. (2002). *Ann. Occup. Hyg.* **46**, 61–68.

Wolfensohn S. and Lloyd M. (1998). *Handbook of Laboratory Animal Management and Welfare*. Blackwell Science Ltd.

Wood, R.A. (2001). *ILAR J.* **42**, 12–16.

CHAPTER 32

Routes of Administration

Shinya Shimizu

National Institute of Animal Health, Tsukuba, Japan

General

Mice are the most widely used animals for a range of experiments including medical, chemical, pharmacological, toxicological, biological, and genetic. The administration of test substances, such as chemical elements, compounds, drugs, antibodies, cells or other agents, to mice is one of the major methods for evaluating their biological activity.

The route of administration is largely dependent on the property of the test substance and the objective of the experiment. All administration should be performed with knowledge of the chemical and physical characteristics of the substance. All routes have both demerit and merit, such as the absorption, bioavailability and metabolism of the substance. Consideration should be paid to the pH, viscosity, concentration, sterility, pyrogenicity, toxicity as well as the existence of hazardous substances. A knowledge of available methods and techniques of administration as well as knowledge of the deposition and fate of the administered substance will help the scientist/investigator to select the most appropriate route for her/his purpose. This route must be selected before the start of any experiment (Nebendahl, 2000).

Proper restraint is the most important technique when mice were treated as this decreases stress and increases successful treatment. Personnel using experimental animals should be well trained in handling and restraint, should obtain authentication for responsible use of experimental animals and attain a scientifically high standard (ETS 123, 1986; Nebendahl, 2000). Further experience will lead to repeatable and reliable results (see Chapter 31 on Handling and Restraint).

During administration mice should be protected from pain, suffering, distress or lasting harm or at least pain and distress shall be kept to a minimum (ETS 123, 1986). Some injections (such as footpad injection) are strongly discouraged and if required must be justified on a case by case basis (CCAC, 2002).

Principles of administration

Handling and restraint

Good handling and restraint is the most important technique for correct administration. Proper restraining leads to successful administration and varies with the routes of administration. Disposable gloves must be worn as manual restraint is frequently used for injections.

The Laboratory Mouse
Copyright 2004 Elsevier
ISBN 0-1233-6425-6

There are two styles of manual restraint, one uses both hands and the other is single handed. (Chapter 31 on Handling and Restraint is helpful; Donovan and Brown, 1991; Suckow *et al.*, 2000).

Double handed manual restraint

The mouse is lifted by the base of the tail and placed on the cage lid or other solid surface with one hand and then its tail is pulled gently back (Figure 32.1a). It is quickly and firmly picked up by the scruff of the neck behind the ears with the thumb and index finger of the other hand (Figure 32.1b). The tail is transferred from the first hand to between the palm and little or ring finger of the other hand, then fixed (Figure 32.1c). The mouse is restrained (Figure 32.1d).

Single handed restraint

The tail is picked up using thumb and fore finger of the chosen hand (Figure 32.2a), then the mouse is placed on the cage lid or other solid surface (Figure 32.2b). The tail is immediately grasped by the palm and middle finger, ring finger and/or little finger, and the thumb and forefinger released (Figure 32.2c). The fold of skin from the scruff of neck down the back is immediately gripped using the thumb and forefinger (Figure 32.2d and e). The mouse is then restrained (Figure 32.2f).

To prevent kicking by the hind legs, the tail is fixed using the palm and forefinger and then the left hind leg is held firmly between the ring and little finger (where the mouse is restrained by the left hand) (Figure 32.3).

(a)

(b)
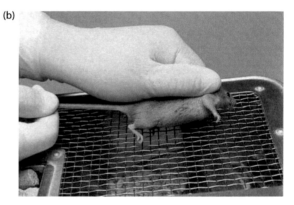

(c)
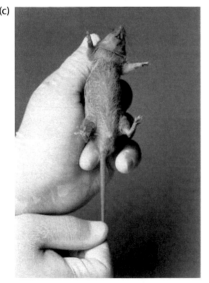

(d)

Figure 32.1 Manual restraint of a mouse using both hands. (a) The mouse is placed on the cage lid with the preferred hand. The tail is pulled gently back by the hand. (b) The mouse is quickly and firmly picked up by the scruff of the neck behind the ears with thumb and index finger of other hand. (c) The tail is transferred from the preferred hand to between palm and little or ring finger of the other hand, then held firmly. (d) The mouse is restrained.

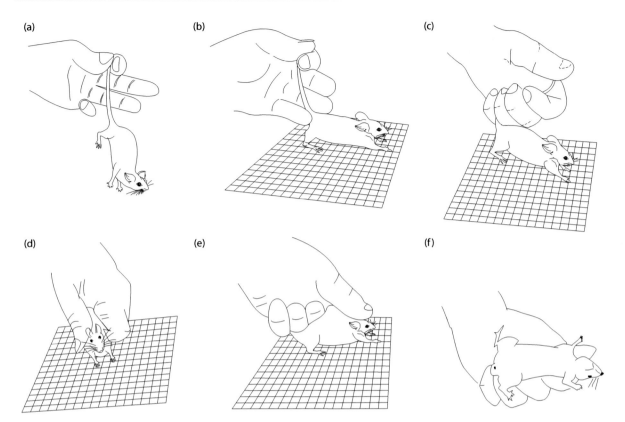

Figure 32.2 Single-handed restraint of the mouse. (a) The tail is picked up using thumb and forefinger of the preferred hand. (b) The mouse is placed on the cage lid or other solid surface pulling gently back by the hand. (c) The tail is immediately grasped by the palm and middle finger, ring finger and/or little finger and then, the tail held between thumb and forefinger is released. (d) and (e) The fold of skin from the scruff of the neck down the back is immediately gripped using the thumb and forefinger. (f) The mouse is restrained.

Site of administration

Among several possibilities for the administration of substances to mice, the most common routes are subcutaneous, intraperitoneal or intravenous injection. The intramuscular administration is not recommended, as the muscle of the mouse is too small. Some sites, such as footpad injection of Freund's complete adjuvant, intrasplenic injection and intra lymph node injection are unacceptable nowadays (CCAC, 2002), and should be restricted to cases where it is absolutely necessary.

Preparation of the site

The area for administration is clipped (Figure 32.4) or cleaned with warm water if necessary before cleaning the skin with alcohol- or disinfectant-moistened cotton. Where aseptic skin is necessary; the fur must be clipped followed by a three-stage surgical preparation: surgical soap, alcoholic rinse and surgical preparation solution. The skin is dried immediately before administration

(CCAC, 2002). In some cases a local anesthetic may be applied first to prevent pain.

Preparation, solubility and safety of solutions

Test substances, solutions and equipment should be prepared aseptically and free from pyrogens, especially for parenteral injections. Solutions can be sterilised by filtration (0.22 μm). Living organisms or cells must be free from contaminants when administered. The toxicity of the substance, the volume and the way of administration should be considered to prevent tissue damage and to give precise dosage.

The following solvents or vehicles have been found suitable in most instances and do not greatly affect drug action because of their own inherent properties: water, water with 0.85% sodium chloride, water with up to 50% polyethylene glycol, water with not over 10% Tween 80, water with up to 0.25% methylcellulose or carboxymethylcellulose, corn oil; vegetable oil;

peanut oil (oral and intramuscular route only). A low percentage of the lower alcohols, glycols, and acetone can also be used, provided the volume administered is kept small (Woodard, 1965). Phosphate buffered saline (PBS) or various culture media are also suitable vehicles (Nebendahl, 2000). Lipid-soluble substances can only be dissolved in oil but this delays absorption. Oil soluble drugs have been successfully given intravenously in 15% oil–water emulsions using lecithin as an emulsifier (Woodard, 1965).

Figure 32.3 Manual restraint of a mouse to prevent kicks by hind leg. The tail is held using the palm and forefinger and then the left hind leg is fixed between the ring and little finger (when the mouse is restrained by the left hand).

Unless experience has indicated otherwise, solutions or suspensions should be prepared as near to the time of use as possible because some substances will deteriorate in solution within a few hours (Woodard, 1965). When administering drugs, the solvent should ideally be the same as the one in which the drug is normally formulated (Nebendahl, 2000).

Although distilled water can be used under certain conditions, saline is preferable because water *ad injectionem* injected subcutaneously causes pain and intravenous injection produces hemolysis. Oil and viscous fluids cannot be injected intravenously (Nebendahl, 2000). If suspended material is to be used for intravenous injection, the particles should be removed by filtration to prevent embolism (Woodard, 1965).

The temperature of fluids must be raised at least to room temperature or better still up to body temperature before use, because the injection of cold fluids is painful (Baumans *et al.*, 1993).

Concentration of substances

The concentration can vary over a fairly wide range without greatly influencing the end result of the experiment. Lower concentrations are clearly desirable (Waynforth and Flecknell, 1992). Factors limiting the use of aqueous solutions for parenteral administration are probably related to their osmotic pressure. Low concentrations can be corrected by the addition of sodium chloride but ought not to be so high as to materially exceed the osmotic pressure of 0.15 M sodium chloride (Woodard, 1965). Highly concentrated solutions can be administered intravenously provided the rate of injection is kept slow and precautions are taken to avoid getting the solution outside the vein.

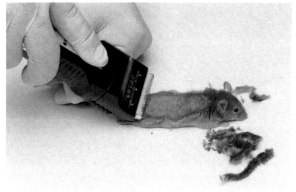

Figure 32.4 Clipping of hair on the back. Hair on the back is clipped by a cordless electric clipper.

pH of the injected solution

For most routes of administration, providing the solutions are not highly buffered, a pH range of 4.5–8.0 is satisfactory. For oral administration a pH as low as 3 can be tolerated, but alkaline solutions are very poorly tolerated. A rather wide range of pH is indicated for intravenous administration, because of the buffering effect of blood and dilution by blood flow, following use of the intramuscular and then subcutaneous routes. When low or high pH solutions are intravenously injected, the rate of injection is kept slow and again precautions taken to avoid getting solution outside the vein (Woodard, 1965).

Volume and frequency of administration

The injection volume is limited by any toxicity of the substance and by the size of the mouse. It should be kept as small as possible. Excess volumes of solution can startle the animal. The frequency of administration should be limited to a minimum, to avoid unnecessary stress. If solutions are administrated intravenously, hemodynamic changes and pulmonary oedema may occur while very rapid injections can produce cardiovascular failure and be lethal (Nebendahl, 2000). Maximum volumes are shown in Table 32.1. (Flecknell 1987; Reeves et al., 1991; Wolfensohn and Lloyd, 1994). For immunization, the maximum is still lower, because of the mixing with adjuvant. Maximum volumes for injection of antigen with or without adjuvant per route are indicated in the section on Immunization of mice in this chapter.

The rate of absorption and distribution of administrated substances

The blood flow to the site of administration, the nature of the substance and its concentration influence the rate of absorption (Wolfensohn and Lloyd, 1994; Nebendahl, 2000). The time-course of the effect of the substance is an important factor in determining the dosage and is influenced by the rate of absorption (Waynforth and Flecknell, 1992). Normally, injected substances must be absorbed from the site of administration into the blood. Therefore, the rate of absorption will be determined by the size of the absorbing surface, the blood flow and the solubility of the substance in the tissue fluid. The rate of absorption is also influenced by lipid solubility, physicochemical properties, degree of ionization and molecular size of the substance (Nebendahl, 2000). Compounds, which are highly soluble in the body fluids, will be absorbed quickly. Substances that are ionized and are not lipid soluble can only be absorbed if a specific carrier exists. In general, the rate of absorption is arranged in the order iv > ip > im > sc > po (Wolfensohn and Lloyd, 1994).

Needles and syringes

Usually, 26–27-G, 1/2- to 5/8-in. (12.5–15.6-mm) needles are satisfactory for injection. The smallest gauge should be selected as a fine needle prevents leakage of fluids and will help to minimize discomfort to the animal (Nebendahl, 2000). 1–2 ml syringes size is enough for most injections. When a small volume (less than 1.0 ml) is administered, an insulin syringe plus needle is convenient (Figures 32.5 and 32.6; 27–30-G, 5/16- to 1/2-in. (8.0–12.5-mm)) these syringes can be obtained from the companies (Terumo, Tokyo, Japan; Becton Dickinson, Franklin Lakes, USA). Intradermal needles are practical for intracerebral injections (Figure 32.7; Top, Tokyo, Japan). Plastic syringes cannot be used with solvents such as acetone.

The withdrawal of hazardous substances from bottles requires great care. An alcohol-moistened cotton pledget can be kept at the point where the needle enters the stopper in order to minimize the inadvertent formation of aerosols (Silverman, 1987). Because of the risk of embolism, air bubbles in fluid and syringe and

TABLE 32.1: Guidelines for maximal administration volumes (in milliliters) and needle size

Oral	Subcutaneous	Intraperitoneal	Intravenous	Intradermal	Intramuscular	Intracerebral	Intranasal
0.2	2–3 (scruff) 0.2 (inguinal)	2–3	0.2	0.05	0.05	<0.03	<0.02
<22 G	<25 G	<23 G	<25 G	<26 G	25–27 G	<27 G	

Source: Flecknell, 1987; Reeves et al., 1991; Wolfensohn and Lloyd, 1994.

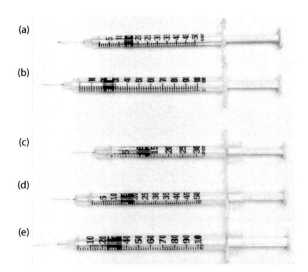

Figure 32.5 Insulin syringes. (a) 29 G × 1/2 in., 0.5 ml, Terumo; (b) 27 G × 1/2 in., 1.0 ml, Terumo; (c) 29 G × 1/2 in., 0.3 ml, Becton Dickinson; (d) 29 G × 1/2 in., 0.5 ml, Becton Dickinson; (e) 29 G × 1/2 in., 1.0 ml, Becton Dickinson.

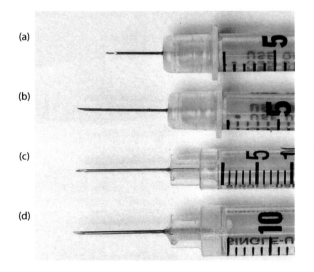

Figure 32.6 Needles for insulin syringes. (a) 30 G × 3/8 in., 0.3 ml, Becton Dickinson; (b) 29 G × 1/2 in., 0.5 ml, Becton Dickinson; (c) 29 G × 1/2 in., 0.5 ml, Terumo; (d) 27 G × 1/2 in., 1.0 ml, Terumo.

needle must be purged. Air bubbles can be purged by gently tapping the side of the syringe and slowly expelling the air into absorbent tissue to prevent any dispersion of the contents until fluid appears at the tip of the needle. The needle size will vary with the viscosity of the substance being used, the greater the viscosity, and the bigger the needle (Reeves *et al.*, 1991). If blood or body fluid flow back into the needle, it must be removed and a fresh attempt be made.

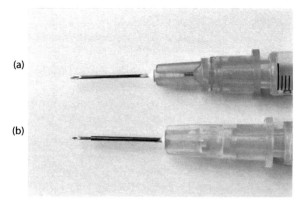

Figure 32.7 Intradermal needle. (a) 26 G × 1/2 in. needle, Terumo, Japan; (b) 1/2 in. intradermal needle, Top, Japan (tip is 27 G, base is 22 G).

Enteral administration

Enteral administration has the advantages that it is possible to give quite large amounts of non-sterile substances or solution and that a pH as low as 3 can be administrated by this route. On the other hand, alkaline solutions are very poorly tolerated by mouth (Woodard, 1965). When using the oral route it should be understood that substances can be destroyed by the gastric juices and that the food content of the stomach influences both rate and order of the gastric emptying. The rate of absorption is markedly influenced by its time of residence in the stomach and is also is directly related to the rate at which substances are passed from the stomach into the intestine (Levine, 1970). Enzymes of the host and microflora of the digestive tract can also metabolize the substance. On the other hand some insoluble substances become solubilized as the result of enzymatic activity during their passage through the stomach and intestine making absorption possible (Nebendahl, 2000). The two major methods for enteral administration are mixing the substance with food or water or direct administration using gavage. Rectal administration is also possible (Woodard, 1965).

Oral administration (per os, p.o.)

The simplest method for administration is to give the substance with food or drinking water. However, this is not practicable with those that are unpalatable, insoluble or chemically unstable in drinking water or when they irritate the mucosa of the gastrointestinal tract (Nebendahl, 2000). The daily food and water intake of mouse should be known before the experiment, to calculate the quantity of substance to be mixed (see Part 5 on Animal Husbandry and Production).

Because food and water wastage happens all the time, it is difficult to determine the precise amount of food and water intake and therefore the precise intake of the substance. The only way this can be done is by keeping mice in metabolic cages and recording the wastage.

Intragastric administration

Direct administration by oral gavage is preferred to mixing substances with food or drinking water because the intake of the substances is precisely measured. A ball tip needle is used to prevent damaging the oesophagus and from passing through the glottal opening into the trachea (Figure 32.8). A 22 G ball tip needle is suitable for administration to adult mice and can be obtained from Popper and Sons, Inc. (New Hyde Park, USA). The conscious mouse is manually restrained firmly by gripping a fold of skin from the scruff of neck down the back (Figure 32.9a), immobilization of the head is essential for this procedure (Figure 32.9b). When the neck is extended the position is vertical. A straight line is formed between the mouth and the cardiac sphincter through the oesophageal orifice (Figure 32.9b). The needle is passed gently through the mouth and pharynx into the oesophagus (Figure 32.9c).

The mouse usually swallows as the feeding needle approaches the pharynx, these swallowing movements can help so that the probe slips through the oesophageal opening. The substance is then administered slowly. If any obstruction is felt, if the mouse coughs, chokes or begins to struggle vigorously after the gavage begins, or if fluid is seen coming out through the nose, these may indicate that the needle has entered the lungs. Any of these signs would necessitate immediate withdrawal of the needle, and the mouse must be observed very carefully. If there is any sign that fluid has got into the lungs, the mouse should be euthanized. As soon as administration is finished, the needle must be withdrawn (Cunliffe-Beamer and

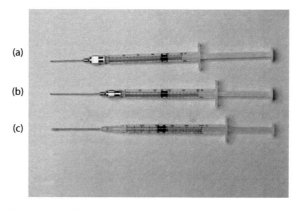

Figure 32.8 Syringes with a gavage needle. (a) 1.0 ml syringe with a 22 G × 1.0 in. feeding needle; (b) 1.0 ml syringe with a 20 G × 1½ in. feeding needle; (c) 1.0 ml syringe with a 20 G × 1½ in. disposable feeding needle.

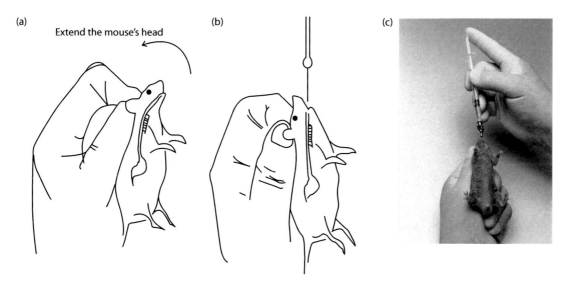

Figure 32.9 Procedure for intragastric administration using a ball tip needle. (a) First extend the neck; (b) A straight line is formed between the mouth and stomach; (c) Intragastric injection using 1.0 ml syringe with 22 G × 1.0 in. feeding needle is made.

Les, 1987; Suckow *et al.*, 2000). A volume of >2 ml is recommended.

Parenteral administration

Administration of substances other than via the alimentary canal to the body includes injection, infusion, topical application and inhalation, and implantation of an osmotic pump or a controlled-release drug delivery pellet. Small amounts of solution are injected, and large volumes are infused. In both cases the skin must be penetrated by a needle.

Subcutaneous, intraperitoneal and intravenous administration are the most common and major routes to inject substance solution or suspension in to the mouse. The rate of absorption is dependent on the route of administration. The substance will immediately disperse following intravenous injection; therefore the most rapid absorption is achieved by this route. The large surface area of the abdominal cavity and its abundant blood supply also facilitate rapid absorption, absorption from this route is usually one-half to one-quarter as rapid as that from the intravenous route (Woodard, 1965).

Subcutaneous administration (s.c.)

Subcutaneous administrations are easy. As they are rarely painful (Wolfensohn and Lloyd, 1994) a conscious mouse can usually be used. The rate of absorption is lower than from intraperitoneal or intramuscular injections (Simmons and Brick, 1970).

Subcutaneous administrations are made into the loose skin over the interscapular (Figure 32.10a) or inguinal area (Figure 32.10b). Subcutaneous administrations over the interscapular area are made as follows. The mouse is manually restrained and then placed on a clean towel or solid surface. The needle is inserted under the skin of the interscapular area tented by the thumb and forefinger and the substance then injected. A volume <3 ml is recommended. Subcutaneous administration over the inguinal area is made as follows. The mouse is restrained manually and the head tilted downwards. Holding the hind leg firmly helps this procedure (Figure 32.3). The needle is inserted into the lower left or right quadrant of abdomen avoiding the abdominal midline and the substance injected. A volume of <0.2 ml/site is recommended. To minimize leakage, the needle should be advanced several millimeters through the subcutaneous tissue (Cunliffe-Beamer and Les, 1987; Suckow *et al.*, 2000).

Intraperitoneal administration (i.p.)

This is the most common route being technically simple and easy. It allows quite long periods of absorption from the repository site. The rate of absorption by this route is usually one-half to one-fourth as rapid as from the intravenous one (Woodard, 1965). Limitations are the sensitivity of the tissue to irritating substances, less tolerance to solutions of non-physiological pH. These should be isotonic and quite large volume can be administered by this route.

(a)

(b)

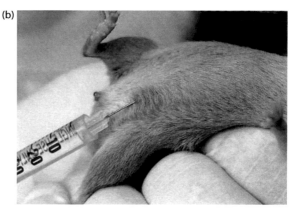

Figure 32.10 Subcutaneous injection. (a) Subcutaneous injection at the base of a fold of loose skin (area at the neck) using an Insulin syringe: 27 G × 1/2 in., 1.0 ml; (b) subcutaneous injection at the lower left quadrant using an Insulin syringe: 27 G × 1/2 in., 1.0 ml.

Figure 32.11 Intraperitoneal injection to lower left quadrant using an Insulin syringe: 27 G × 1/2 in., 1.0 ml.

The conscious mouse is manually restrained (Simmons and Brick, 1970) and is held in a supine position with its posterior end slightly elevated or the head can be tilted lower than the body (Figure 32.11). The needle and syringe should be kept almost parallel to its vertebral column in order to avoid accidental penetration of the viscera (Eldridge *et al.*, 1982). The needle is pushed in at an approximately 10° angle between the needle and the abdominal surface in the lower quadrant of the abdomen (Simmons and Brick, 1970). To avoid leakage from the puncture point, the needle is run through subcutaneous tissue in a cranial direction for 2–3 mm and then inserted through the abdominal wall (Cunliffe-Beamer and Les, 1987). The recommended volume is >3.0 ml.

Intravenous administration (i.v.)

Intravenous injection has advantages over other routes. Solutions at a high concentration, high or low pH or irritating can be administered intravenously provided that the rate of injection is kept slow and precautions are taken to avoid getting the solution outside the vein. Compounds that are poorly absorbed by the digestive tract may be given intravenously but intravenous administrations require technical expertise and skill. The syringe plus needle or the catheter must first be filled with the solution to remove air bubbles. Administrations are usually made into the lateral tail veins not into the dorsal tail vein (Figure 32.12a), as it is not straight.

The lateral veins are readily visualized, but have quite small diameters. If anaesthesia is not used, a restraining device is usually necessary (see Chapter 31 on Handling and Restraint; Reeves *et al.*, 1991; Suckow *et al.*, 2000; Weiss *et al.*, 2000).

The mouse is either placed in the restrainer or anesthetized and the tail is then warmed with a lamp or warm towel, or immersed in warm water (40–45°C) in order to dilate the vessels (Flecknell, 1987). The tail is swabbed with 70% alcohol on a gauze sponge or swab. Insert the needle parallel to the tail vein penetrating 2–4 mm into the lumen while keeping the bevel of the needle face upwards (Figure 32.12b). The solution is then injected slowly and no resistance should be felt if the solution is properly administered (Figure 32.12c). The injected solution temporarily replaces the blood but then should be washed away by the blood stream. If this does not happen the position of the needle is certainly not in the vein but in the surrounding tissue so it must be moved in the surrounding tissue in such a way that it then enters the vein or a new try must be made. When the intravenous administration is finished or the cannula is pulled out, the injection site must be pressed firmly with a swab or fingers to prevent backflow of the administered solution and/or blood (Nebendahl, 2000; Suckow *et al.*, 2000). If the same vein must be used several times the first administration should be made as distal as possible in relation to the heart and subsequent administrations should be placed progressively more proximally. Because venipuncture and the administration of substances can damage and/or block the vein, the distal part of vein may no longer be used (Nebendahl, 2000). The recommended volume is <0.2 ml.

The ophthalmic plexus route is also used for intravenous administration (Pinkerton and Webber, 1964). The technique resembles the blood collection by retroorbital sinus puncture (see Chapter 33 on Collection of Body Fluids). The mouse is anesthetized, and then manually restrained on a solid surface being held gently but firmly by the nape of the neck. By pressing down with the thumb and forefinger in the occipital area and pulling back the skin, the point of the needle can be directed toward the back of the orbit at a 20–40° angle. The needle is inserted medially through the conjunctiva on the inner side of the ocular cavity. If entry is blocked by bone, the needle is withdrawn slightly (Figure 32.13a). Fluid is injected slowly loosening the skin slightly (Figure 32.13b). Also this route is useful for rescuing mice showing anaphylaxis by administration of an isotonic solution.

Other routes for intravenous administration via the external jugular vein (Kassel and Levitan, 1953), the dorsal metatarsal vein (Nobunaga *et al.*, 1966) and the sublingual vein (Waynforth and Parkin, 1969) have been reported.

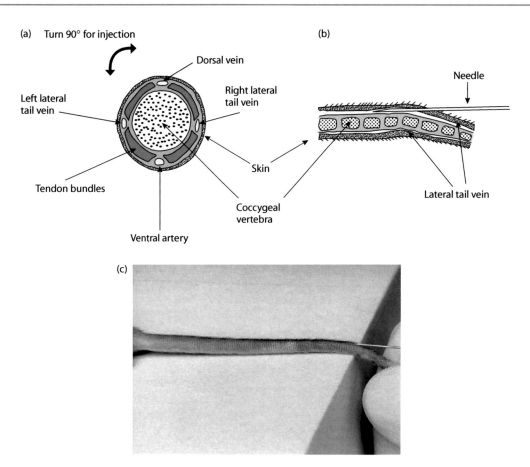

Figure 32.12 (a) Transverse section view of the mouse tail; (b) sagittal view of the mouse tail (the tail is turned 90°); (c) intravenous injection into the lateral tail vein of an anesthetized mouse using an Insulin syringe: 27 G × 1/2 in., 1.0 ml.

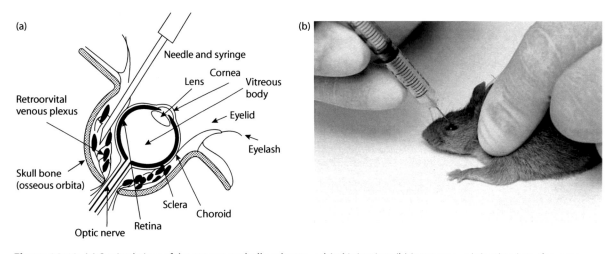

Figure 32.13 (a) Sagittal view of the mouse eyeball and retro-orbital injection; (b) intravenous injection into the retro-orbital sinus of an anesthetized mouse using an Insulin syringe: 27 G × 1/2 in., 1.0 ml.

Intramuscular administration (i.m.)

This should usually be avoided, as mouse muscles are small. If necessary, it may be given into the thigh muscle with injection volumes <0.05 ml. The tip of needle should be directed away from the femur and sciatic nerve (Figure 32.14). The mouse is anesthetized or is manually restrained by another person. The needle tip is inserted through the skin and into the muscle. Aspirate briefly with the syringe before injection. If blood or body fluid reverses, stop the procedure. The

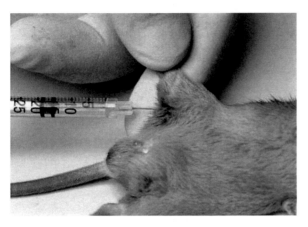

Figure 32.14 Intramuscular injection into the leg muscle.

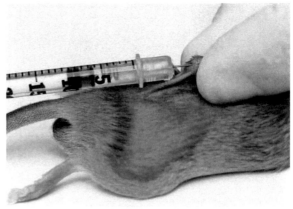

Figure 32.15 Intradermal injection into the back skin.

needle must be moved or a fresh attempt must be made. Good technique and restraint are necessary and intramuscular administration should only be performed by well-trained personnel (Woodard, 1965; Cunliffe-Beamer and Les, 1987; Donovan and Brown, 1991; Nebendahl, 2000).

Intradermal administration

This route is not recommended in general and should be restricted to cases of absolute necessity (Saloga *et al.*, 1993; CCAC, 2002). It is very difficult in the mouse due to the very thin skin. Using a fine needle (29 G or smaller) is recommended. The mouse is anesthetized, the fur clipped or hair removed from an area on the back, ventral abdomen, or hind footpad, which is wiped with 70% ethanol on a gauze sponge or swab. The skin is held tautly with thumb and index finger and the needle inserted, bevel up and at a shallow angle, just under the superficial layer of epidermis. The volume should be <0.05 ml per site. Resistance should be felt both as the needle is advanced and as the compound is injected. A hard bleb will be seen upon successful intradermal injection of even a small quantity of fluid (Figure 32.15; Suckow *et al.*, 2000). If multiple sites are injected, adequate separation is necessary to prevent coalescing of lesions.

Intracerebral administration

This is made as follows (Prier, 1966; Liu *et al.*, 1970). The mouse is anesthetized and then restrained manually on a solid surface. The site of injection is approximately half way between the eye and ear and just off the midline (Figure 32.16a). The recommended maximum volume per suckling mouse is 0.01 ml and that

for weanling or older mice is up to 0.03 ml. The needle directly pierces the cranium (Figure 32.16b). An intradermal needle (Figure 32.7) is convenient in order to prevent the needle from extending too deeply into the brain.

Intrathoracic administration

Intrathoracic injection is restricted to special experiments. It can be made in mice with a slightly bent or curved needle, which should be inserted between the ribs at approximately the midpoint of the rib cage. Caution must be taken to insert it at an angle, thus preventing injection directly into lung tissue. The speed of absorption is similar to the intraperitoneal route (Simmons and Brick, 1970).

Intranasal administration (i.n.)

These are usually performed with the mouse lightly anesthetized. The mouse is manually restrained and the tail anchored between the small finger and the palm (Simmons and Brick, 1970). The mouse is held in a supine position with the head elevated. The end of the micropipette is placed at or in the external nares, and then the solution is poured in slowly (Figure 32.17; Prier, 1966; Shen *et al.*, 2000, 2001). The volume should be <0.02 ml, excess volume or rapid injection will induce suffocation and death.

Topical application

It is not often realized that the skin is the largest organ of the body and survival depends on its patency perhaps more than for most other organs. An animal or man can survive with only about one-seventh of its liver or

(a)

Ear line   Eye line

(b)

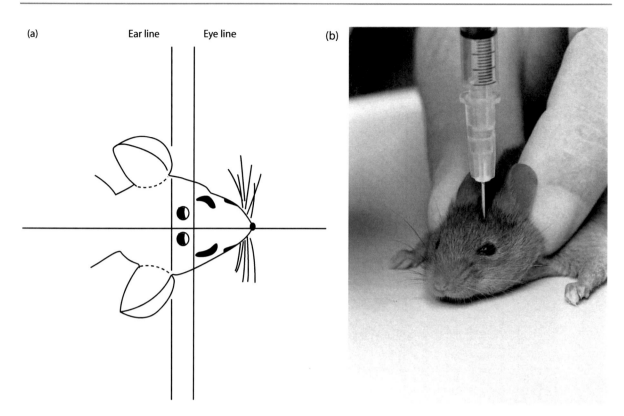

Figure 32.16 Intracerebral injection. (a) Injection site of head for intracerebral injection; (b) intracerebral injection into an anesthetized mouse using an intradermal needle.

Figure 32.17 Intranasal injection into an anesthetized mouse using a pipette (Gilson P-20).

one-fourth of its kidney functioning. In contrast, the destruction of more than 50% of the skin usually results in death (Woodard, 1965). The skin is also a convenient site for the administration of drugs. Numerous factors, such as the physicochemical properties of the substance, the attributes of the vehicle and the permeability of the skin, can affect the degree of percutaneous absorption. (Wester and Maibach, 1986; Franklin *et al.*, 1989). The ability of a substance to be absorbed through the skin and enter the systemic circulation is determined by its

ability to partition into both lipid and water phases (Nebendahl, 2000).

The usual site is the skin covering the back or the abdomen. After clipping the hair for topical administration (Figure 32.4), the hairless area should be cleaned from any fat and grease and other debris. The substance should be dissolved in a volatile solvent or mixed in a suitable cream before application and then applied with a dropper or smeared onto the skin with a swab (Nebendahl, 2000). Some precautions are usually necessary to prevent the animal from licking or scratching the application sites (Woodard, 1965).

Inhalation

This route is used for experiments on asthma, air pollution or respiration (Haddad el-B *et al.*, 2002; Hopfenspirger and Agrawal, 2002). The inhalation route incidentally is the nearest akin to an intravenous injection because of the relatively large area presented for absorption by a membrane that is separated from the blood by only one or two cell layers. Consequently, absorption of gases and aerosols that reach the alveoli is virtually complete. The greatest problems surrounding the use of the inhalation route are the generation of a suitable aerosol of the test substance, if it is not sufficiently volatile, a constant and suitable air level of the material under study and the determination of the dosage given. Both small and large particle sizes are not adequate, it is generally believed that particle size of 0.5–2.0 μm in diameter are optimum (Woodard, 1965). Equipment is available to purchase from Omuron, Kyoto, Japan; Buxco Electronics, Inc., Sharon, USA among others.

Other routes

Other routes of administration have been reported such as intra-arterial administration using the femoral artery (Simmons and Brick, 1970) or the carotid artery (Sugano and Nomura, 1963), intrathymic injection (Donovan and Brown, 1991), or intraspinal injection (Habel and Li, 1951).

Dosing and treatment of new born mice provides special problems because of their size or that their mother is apt to reject or cannibalize neonates that have been handled. Subcutaneous injections can be made over the neck and shoulders using a less than 30 G × 5/16 needle. Up to 0.1 ml (depending on the age of the infant mice) may be administered orally using a piece of plastic tubing inserted over a needle (Ujiie and Kobari, 1970; Cunliffe-Beamer and Les, 1987). The direct injection in to the stomach of infant mice can be made through the abdominal wall (Dean et al., 1972). Intravenous injection of infant mice has also been reported (Anderson et al., 1959; Barnes et al., 1963; Cunliffe-Beamer and Les 1987).

Implantable pump, controlled-release drug delivery pellet and cannulas

The delivery of substances at a slow, steady rate over a period of days, weeks or months without the need for external connection or frequent animal handling can be supplied by using an osmotic pump or controlled-release drug delivery pellets.

Osmotic pumps, ALZET pumps (Figure 32.18), can be used for systemic administration when implanted subcutaneously or intraperitoneally or can be attached to a catheter for intravenous, intracerebral or intra-arterial infusion. The pumps have been used to target delivery to a wide variety of sites including the spinal cord, spleen, liver, organ or tissue transplants and wound healing sites. ALZET pumps are supplied by DURECT Corporation (Cupertino, USA).

Controlled-release drug delivery pellets effectively and continuously release the active product in the animal. The pellets are intended for, although not limited to, simple subcutaneous implantation in laboratory animals. The pellets are available from Innovative Research of America (Sarasota, USA) and Southern BioSystems, Inc. (Birmingham, USA).

Implantable cannulas permit continuous access to the venous or arterial system for either intravenous substance administration or blood withdrawal. Using strict aseptic techniques, the cannula is inserted into a vein or artery (the femoral vessels, jugular vein, and

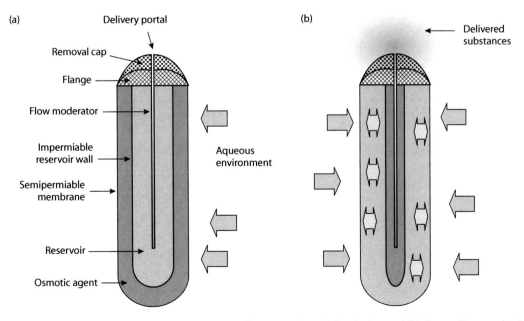

Figure 32.18 Cross section of an Alzet®-osmotic pump, demonstrating (a) the design and (b) the working method.

TABLE 32.2: Maximum volume (in milliliters) of administration of antigen with or without adjuvant per route					
	Subcutaneous	Intramuscular	Intraperitoneal	Intravenous	Intradermal
with adjuvant	0.1	not recommended	0.2	no	not recommended
without adjuvant	0.5	0.05	1.0	0.2	0.05

ID route should be restricted to the cases where it is absolutely necessary (CCAC, 2002).
Source: van Zutphen *et al.*, 1993; Iwarsson *et al.*, 1994.

carotid artery are common sites) and secured in place. The other end of the cannula is attached to a small port that is secured in a subcutaneous location, most often over the shoulders (Suckow *et al.*, 2000). See Desjardins (1986) for more information on implantable cannulas.

Immunization

Mice are not used for the production of polyclonal antibodies because the small amounts produced. On the other hand, mice are a good source of antibody producing lymphoid cells or the production of hybridomas (Kohler and Milstein, 1975). In general, immunization consists of two stages; primary and booster. The primary antigen is usually injected with adjuvant. Boosters are injected once or more with/without adjuvant depending on the immunogen. The footpad, intrasplenic (Nilson and Larsson, 1992) or intra-lymph (Goudie *et al.*, 1966) node injection is not recommended in general. If required the scientist/investigators should provide scientific justification to ethical committees for such protocols (such as the need to use extremely unique and irreplaceable antigens, or extremely small quantities of antigen). The injection of immunogens at the base of the tail or in the popliteal area substitute for footpad injections with much less distress to the animal because immunogens injected into the footpad are processed by the popliteal node. (Leenars *et al.*, 1999; CCAC, 2002). The intraperitoneal route for injection of Freund's complete adjuvant (FCA) is permitted in small rodents only. FCA should be administered only once, and be limited to minimal volumes of up to 0.1 ml. In the mouse, up to 0.1 ml with adjuvant may be administered in the neck region subcutaneously. The injection of oil-based or viscous gel adjuvant should not be given by the intramuscular route (CCAC, 1991). The intravenous route should not be used for oil-based adjuvants, viscous gel adjuvants or large particle antigens due to the risk of pulmonary embolism (Herbert, 1978). Though FCA is the strongest adjuvant, use of other adjuvants can be recommended. Mice must be closely monitored immediately following injection for any anaphylactic reactions, both after the primary and any booster injections (CCAC, 2002). The recommended route and volumes are shown in Table 32.2. Mice showing anaphylaxis will recover by the administration of the isotonic solution by Opthalmic plexus route (see this chapter on intravenous administration).

Acknowledgment

We are grateful to Mr. Andoh Y. for his skilful photography.

References

Anderson, N.F., Delorme, E.J., Woodruff, M.F.A. and Simpton, D.C. (1959). *Nature*, 1952–1953.

Barnes, D.W.H., Ford, C.E. and Harris, J.E. (1963). *Transplantation* 1, 574.

Baumans, V., ten Berg, R.G.H., Bertens, A.P.M.G., Hackbarth, H.J. and Timmermann, A. (1993). In *Principles of Laboratory Animal Science*, (eds L.F.M. van Zutphen, V. Baumans and A.C. Beynen), p. 389. Elsevier, Amsterdam.

CCAC (The Canadian Council on Animal Care) (1991). *CCAC Guidelines on Acceptable Immunological Procedures*. CCAC, Ottawa. http://www.ccac.ca/english/gui_pol/policies/IMMUNO.HTM

CCAC (The Canadian Council on Animal Care) (2002). *Guidelines on Antibody Production*, pp. 1–40. CCAC, Ottawa.

Cunliffe-Beamer, T.L. and Les, E.P. (1987). In *The UFAW Handbook on the Care and Management of Laboratory Animals*, 6th edn (ed. T. B. Poole), pp. 275–308. Longman Scientific and Technical, Essex.

Dean, A.G., Ching, Y.C., Williams, R.G. and Harden, L.B. (1972). *J. Infect. Dis.* 125, 407–411.

Desjardins, C. (1986). In *Research Surgery and Care of Small Laboratory Animals Part A. Patient Care, Vascular Access, and Telemetry* (eds W.I. Gay and J.E. Heavner), p. 143–194. Academic Press, Orlando.

Donovan, J. and Brown, P. (1991). In *Current Protocol in Immunology* (eds Coligan *et al.*) pp. 1.4.1–1.4.4. John Wiley & Sons, New York.

Eldridge, S.F., McDonald, K.E., Renne, R.A. and Lewis, T.R. (1982). *Lab. Anim.* **11**, 50–54.

ETS 123 (1986). *European Convention For The Protection of Vertebrate Animals used for Experimental and Other Scientific Purposes.* Council of Europe, Strasbourg.

Flecknell, P.A. (1987). In *Laboratory Animals: An Introduction for New Experimenters* (ed. Tuffery), pp. 225–260. John Wiley & Sons, Chichester, England.

Franklin, C.A., Somers, D.A. and Chu, I. (1989). *J. Am. Col. Toxicol.* **8**, 815–827.

Goudie, R.B., Home, C.H. and Wilkinson, P.C. (1966). *Lancet* **7475**, 1224–1226.

Habel, K. and Li, C. P. (1951). *Proc. Soc. Exp. Biol. Med.* **76**, 357–361.

Haddad, el-B., Underwood, S.L., Dabrowski, D., Birrell, M.A., McCluskie, K., Battram, C.H., Pecoraro, M., Foster, M.L. and Belvisi, M.G. (2002). *J. Immunol.* **168**, 3004–3016.

Herbert, W.J. (1978). In *Handbook of Experimental Immunology*, 3rd edn (ed. D.M. Weir), pp. A3, 1–3, 15. Blackwell Scientific Publications, Oxford.

Hopfenspirger, M.T. and Agrawal, D.K. (2002). *J. Immunol.* **168**, 2516–2522.

Iwarsson, K., Lindberg, L. and Waller, T. (1994). In *Handbook of Laboratory Animal Science*, Vol. 1 (eds P. Svensen and J. Hau), pp. 229–272. CRC Press, Boca Raton.

Kassel, R. and Levitan, S. (1953). *Science* **118**, 563–564.

Köhler, G. and Milstein, C. (1975). *Nature* **256**, 495–497.

Leenars, M.P.P.A., Hendriksen, C.F.M., De Leeuw, W.A., Carat, F., Delahaut, P., Fischer, R., Halder, M., Hanly, W.C., Hartinger, J., Hau, J., Lindblad, E.B., Nicklas, W., Outschoorn, I.M. and Stewart-Tull, E.S. (1999). *Alternatives Lab. Anim.* **27**, 79–102, http://altweb.jhsph.edu/publications/ECVAM/ecvam35.htm

Levine, R.R. (1970). *Am. J. Dig. Dis.* **15**, 171–188.

Liu, C., Voth, D.W., Rodina, P., Shauf, L.R. and Gonzalez, G. (1970). *J. Infect. Dis.* **122**, 53–63.

Nebendahl, K. (2000). In *The Laboratory Rat* (ed. G. Krinke), pp. 463–483. Academic Press, London.

Nilson, B.O. and Larsson, A. (1992). *Res. Immunol.* **143**, 553–557.

Nobunaga, T., Nakamura, K. and Imamichi, T. (1966). *Lab. Anim. Care* **16**, 40–49.

Prier, J. E. (1966). In *Basic Medical Virology* (ed. J. E. Prier), pp. 38–77. The Williams & Wilkins Company, Baltimore.

Pinkerton, W. and Webber, M. (1964). *Proc. Soc. Exp. Biol. Med.* **116**, 959–961.

Reeves, J.P., Reeves, P.A. and Chin, L.T. (1991). In *Current Protocol in Immunology* (eds. Coligan *et al.*), pp. 1.6.1–1.6.9. John Wiley & Sons, New York.

Saloga, J., Renz, H., Lack, G., Bradley, K.L., Greenstein, J.L., Larsen, G. and Gelfand, E.W. (1993). *J. Clin. Invest.* **91**, 133–140.

Shen, X., Lagergard, T., Yang, Y., Lindblad, M., Fredriksson, M. and Holmgren, J. (2000). *Infect. Immun.* **68**, 5749–5755.

Shen, X., Lagergard, T., Yang, Y., Lindblad, M., Fredriksson, M. and Holmgren, J. (2001). *Infect. Immun.* **69**, 297–306.

Simmons, M.L. and Brick, J.O. (1970). In *The Laboratory Mouse* (ed. A. Hollaender), pp. 127–129. Prentice-Hall Inc., Englewood Cliffs.

Silverman, J. (1987). In *Laboratory Hamsters* (eds G.L. Van Hoosier and C.W. McPherson), pp. 72–75. Academic Press, Orland.

Suckow, M.A., Danneman, P. and Brayton, C. (2000). In *The Laboratory Mouse* (ed. Suckow), pp. 120–125. CRC Press, Boca Raton.

Sugano, S. and Nomura, S. (1963). *Bull. Exp. Anim. (Jap. Eng. Abstract)* **12**, 1–5.

Ujiie, A. and Kobari, K. (1970). *J. Infect. Dis.* **121**, s50–s55.

van Zutphen, L.F.M., Baumans, V. and Beynen, A.C. (1993). In *Principles of Laboratory Animal Science* (eds L.F.M. van Zutphen, V. Baumans and A.C. Beynen), p. 389. Amsterdam, Elsevier.

Waynforth, H. B. and Parkin, R. (1969). *Lab. Anim.* **3**, 35–37.

Waynforth, H. B. and Flecknell, P. A. (1992). In *Experimental and Surgical Technique in the Rat* (eds H. B. Waynforth and P. A. Flecknell), pp. 1–67. Academic Press, London.

Weiss, J., Taylor, G. R., Zimmermann, F. and Nebendahl, K. (2000). In *The Laboratory Rat* (ed. G. Krinke), pp. 485–510. Academic Press, London.

Wester, R.C. and Maibach, H. I. (1986). In *Progress in Drug Metabolism*, Vol. 9 (eds J.W. Bridges and L.F. Chasseaud), pp. 95–109. Taylor and Francis, London.

Wolfensohn, S. and Lloyd, M. (1994). In *Handbook of Laboratory Animals Management and Welfare* (eds S. Wolfensohn and M. Lloyd), pp. 143–173. Oxford University Press, Oxford.

Woodard, G. (1965). In *Methods of Animal Experimentation*, Vol. 1 (ed. W.J. Gay), pp. 343–359. Academic Press, New York.

Additional information

Becton Dickinson, Franklin Lakes, USA: http://www.bd.com

Buxco Electronics, Inc., Sharon, USA: http://www.buxco.com/index.html

Canadian Council on Animal Care, Canada: http://www.ccac.ca/index.htm

DURECT Corporation, Cupertino, USA: http://www.alzet.com/index.html

Innovative Research of America, Sarasota, USA: http://www.innovrsrch.com

Omuron, Kyoto, Japan: http://www.healthcare.omron.co.jp/global/

Popper and Sons, Inc., New Hyde Park, USA: http://www.popperandsons.com

Southern BioSystems, Inc., Birmingham, USA: http://www.southernbiosystems.com/index.html

Terumo, Tokyo, Japan: http:// www.terumo.co.jp/English/index.html

Top, Tokyo, Japan: http://www.top-tokyo.co.jp/m_english.html

CHAPTER 33

Collection of Body Fluids

Katsuhiro Fukuta

Laboratory of Animal Morphology and Function,
Graduate School of Bioagricultural Sciences,
Nagoya University, Japan

Blood

Blood is composed of formed elements or blood cells consisting of erythrocytes, platelets and leukocytes, and fluid or plasma consisting of serum and coagulant. The alteration of blood cell components and plasma proteins such as albumin, globulin and fibrinogen reflects the healthy state of the animals. Therefore, blood collection and its analysis is indispensable for the maintenance of laboratory mice. Moreover, blood transports nutrients, metabolism residues, hormones, antibodies and enzymes. Antibodies indicate the history of infection of animals while polymorphism of serum enzymes is utilized to identify inbred strains. In a breeding facility, blood collection and analysis are executed periodically to check up the microorganism contamination of the facility and accidental mating of an inbred strain with others.

Blood collection is executed for (1) obtaining all the blood from sacrificed animals or (2) a proportion of the blood from living animals. Blood is taken from mice for a wide variety of scientific purposes and the anatomical sites of blood collection are varied. The hematological values, such as hematocrit and cell count may vary between different sampling sites (Sakaki *et al.*, 1961).

Therefore, investigators should select an adequate blood collecting technique to fit the purpose of their examination.

Total blood collection from heart under thoracotomy

Using this technique, a large amount of venous blood can be collected with certainty. A mouse is first anesthetized, sacrificed and its thorax is opened. Anesthesia of mice for blood collection is usually achieved by ether inhalation. Ether is considered to have no effect on hematological findings (Grice, 1964). A glass anesthetizing jar with ether soaked cotton and a partition of wire mesh is used. After putting the mouse into the jar, it soon exhibits rapid heart beating which comes slowly down. By tilting the jar in order to change the position of the mouse the heart rate can be seen clearly. Once the rate has become slow, deep anesthesia has been induced. If a mouse is left in the jar more than 1.5 min, it may die. The anesthetized mouse is picked up and operated on for blood collection according to the following procedure. Should the animal wake up during the procedure, additional anesthesia can be achieved by covering the snout of the mouse with a small cup containing ether soaked cotton.

The Laboratory Mouse
Copyright 2004 Elsevier
ISBN 0-1233-6425-6

(1) The anesthetized mouse should be laid on its back on an operating board, and restrained by adhesive tape or pins at the carpal and tarsal regions of its four limbs.

(2) The xiphoid cartilage is located at approximately the middle of the head and body length. The skin around the xiphoid cartilage is cut off using scissors and the muscular wall of the thorax and abdomen is exposed.

(3) Incise the abdominal wall just caudally to the xiphoid cartilage, then cut the diaphragm and thoracic wall at both sides of sternum with scissors (Figure 33.1). Paired blood vessels, the internal thoracic arteries, run on both sides of the sternum on the internal surface of the thoracic wall. Case should be taken to avoid cutting the arteries. In deeply anesthetized mice, bleeding from the cut thorax wall is minimal, though this will increase in insufficiently anesthetized mice and disturb the following procedure.

(4) The cut thoracic wall is gripped by artery forceps and pulled upwards utilizing weight of the forceps. Through the cut wall, the rapidly beating heart is exposed. Although respiration stops quickly following the opening of the thorax, heart continues to beat for several minutes. While the heart is still beating, blood can be collected successfully.

(5) The investigator prepares a disposable 1 ml plastic syringe with a needle gauge 21–23. By inserting the needle into the right ventricle of the heart and withdrawing the plunger, blood will enter into the syringe (Figure 33.1).

(6) Withdrawal of the plunger should be continued slowly, because rapid withdrawal leaves the ventricle empty and the tip of the needle suck on the wall of the ventricle.

By this procedure, more than 1.3 ml of venous blood can be obtained from an adult mouse. Although the total blood volume is roughly estimated as being 7% of the body weight, complete collection of the blood cannot be obtained. Quality is preferable to quantity. The collected blood should be transferred immediately into a test tube by removing the needle and depressing the plunger slowly so that the blood enters the test tube along its wall.

Total blood collection by bleeding from the femoral artery

This technique is easy to use for collection of arterial blood from mice. The mice must be anesthetized

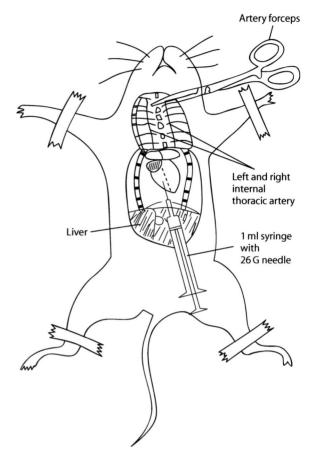

Figure 33.1 Venous blood collection from the right ventricles of the open thorax mouse.

deeply by ether inhalation as described above and then sacrificed.

(1) An anesthetized mouse is restrained on its back on the operating board. The skin around the inguinal region on either side of the mid line is cut off with scissors. The abdominal muscular wall and muscles of the medial side of the femur are exposed. The femoral artery and vein can be seen at the base of the thigh.

(2) Before bleeding, the operating board should be tilted slightly so as to raise the head of the mouse in order to collect blood effectively. Place a test tube on the exposed femoral artery, then incise the artery just above the test tube (Figure 33.2).

(3) Blood from the femoral artery then flows into test tube to give approximately 1 ml per adult mouse.

While collecting blood, bleeding sometimes stops before the animal dies. The investigator can then collect blood from the other side of the hind limb. If a considerable volume has been obtained already, the

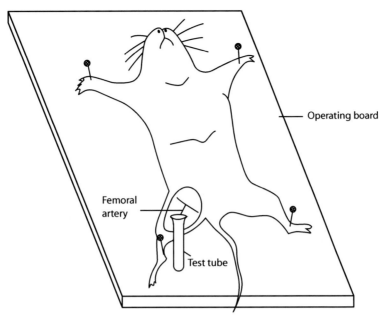

Figure 33.2 Arterial blood collection from the femoral artery. An operating board is tilted to collect blood effectively.

mouse should be sacrificed quickly. Further collection may induce hemolysis.

Collecting arterial blood from the axillary artery is done using a similar procedure at the axillary region (Young and Chambers, 1973).

Bleeding from the abdominal aorta

The abdominal aorta is also used for arterial blood collection (Lushbough and Moline, 1961). The deeply anesthetized mouse should be restrained on its back on the board as before. Incision of the abdominal wall along the midline is started just below the xiphoid cartilage, continued to near the pubic bone and then along the flank, so that the viscera are exposed. The intestine should be moved to left side of the mouse, and the connective tissues removed along the midline of the dorsal wall of the body. The caudal vena cava and the abdominal aorta can then be identified (Cook, 1965). The abdominal aorta is incised at the level near the divergence of the right and left common iliac arteries (Figure 33.3). The blood is collected by a Pasteur pipette and transferred into a test tube.

The plastic disposable 1 ml syringe with a 26-G needle can also inserted into the abdominal aorta and aspirated to collect blood. In this case the best place to make the puncture is at the divergence of right and left common iliac arteries (Figure 33.3). As the abdominal aorta of mice is very narrow at this level, the technique requires

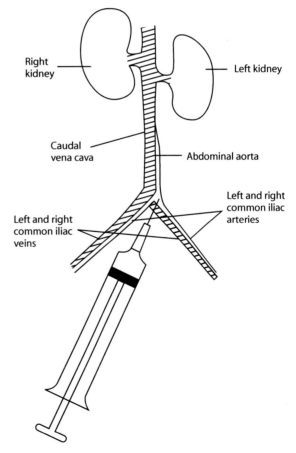

Figure 33.3 Abdominal aorta and caudal vena cava for blood collection. An incision is made in the abdominal aorta near to the divergence of the aorta into the right and left common iliac arteries, at which point a needle can be inserted for sampling.

skill and experience. Incision of the abdominal aorta is easier than puncturing the artery with the needle.

Decapitation

For certain experiments, blood should be collected by decapitation. In this procedure, a decapitation device or guillotine is used (Figure 33.4). Mice must be lightly anesthetized or stunned with a blow to the back of the head before they are placed on the guillotine, and their heads chopped off. The severed neck must be placed over a test tube for blood collection as quickly as possible. This is a fairly messy procedure, arterial blood is obtained, but may be contaminated with saliva and respiratory secretions.

To avoid contamination, the common carotid artery or jugular vein is surgically dissected, then either arterial or venous blood can be collected from the exposed vessels (Ambrus *et al.*, 1951; Kassel and Levitan, 1953).

Repeated blood collection from retroorbital venous plexus

In some experiments or during routine inspections, small amounts of blood may be collected repeatedly from the mice. For this purpose, the most popular technique is bleeding from the retroorbital venous plexus (Gray, 1979). Blood collection from the plexus can be repeated with intervals of 3–4 days. However, this technique requires training and experience, as untrained personnel cannot collect blood smoothly and may cause damage such as injured nerves. In order to collect blood, thin glass tubes, micro-hematocrit capillary tubes (length 75 mm, internal diameter 1–1.2 mm, wall thickness 0.2 ± 0.02 mm), are used exclusively. The bleeding and collecting procedure for this technique is as follows.

(1) A mouse should be anesthetized lightly by ether inhalation. It is then laid on the operating board on its ventral surface. The investigator holds the mouse by the left hand and depresses it softly on to the board while holding a glass capillary tube by the thumb and forefinger of right hand.

(2) The head is depressed and the skin pulled by the thumb and forefinger of the left hand so as to extrude the eyeball as shown in Figure 33.5. Before bleeding, the eye should be wiped gently with sterile gauze to remove lacrimal fluids.

(3) The glass capillary tube is inserted along the lower edge of the eyeball from the medial or nasal side towards the lateral side, and the tube pushed slightly deeper (3–4 mm) together with rotating by the thumb and forefinger of the right hand. As a result of this action, the blood vessel of the retroorbital venous plexus is incised and bleeding occurs. The capillary tube is withdrawn slightly, then blood enters the capillary tube by capillarity. When the

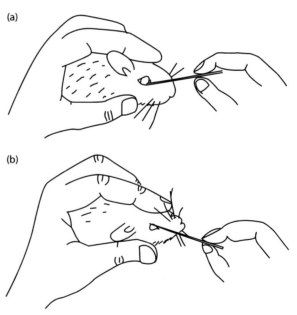

(a)

(b)

Figure 33.5 Collection of venous blood from the (a) right and (b) left retroorbital venous plexuses. A micro-hematocrit capillary tube is inserted along the lower edge of the eye ball from the nasal or medial side toward the side for a depth of 3–4 mm.

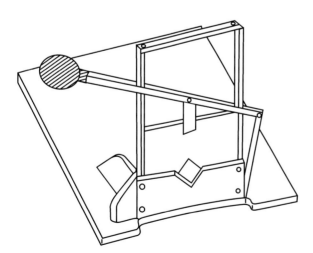

Figure 33.4 A guillotine. The neck of an unconscious mouse is placed on the diamond-shaped slit of the device and its head is chopped off.

opposite end of the tube is lowered, entrance of blood into the tube is promoted.

(4) After blood collection, the investigator should depress the eyeball using sterile gauze until the bleeding stops. The mouse can then be returned to the cage.

In this procedure, two or three micro-hematocrit capillary tubes can be filled with blood, i.e. a total of about 200 μl of blood per head obtained. Bleeding from the retroorbital venous plexus can be done on either side of both eyes alternately with an interval of 3 days, depending on the volume of sampling.

Bleeding from the surface veins of the hind limb

For repeated blood collection from mice, bleeding from a surface vein of a hind limb is also used. The dorsal metatarsal vein or medial saphenous vein may be incised. In this procedure the mouse does not have to be anesthetized there is an assistant available to hold the mouse or a restraining device is available (Figure 33.6). It is an easy procedure, but only very small volumes of blood are obtained. As the surface veins of the mouse hind limb are very fine, the mouse or at least the hind limb should be warmed before bleeding to dilate the blood vessels. Either the mouse is laid on a wire-mesh cage lid placed on warm water for 5–10 min or the limb itself is soaked in warm water at about 40°C.

To collect blood from the dorsal tarsal vein, either foot is extended using the left hand. The surface is cleaned with alcohol soaked cotton, wiped with dried sterilized gauze, and the vein pierced by a needle of 21 G or incised by a razor blade. Blood accumulates as a droplet on the dorsal side of the metatarsal region. By touching the droplet with a capillary tube very quickly it can be collected by capillarity before the droplet spreads out and becomes difficult to collect.

When bleeding from the medial saphenous vein, the procedure is almost the same as that from the dorsal metatarsal vein. The mouse or the hind limb should be warmed to dilate the blood vessel. The hind foot of the restrained mouse is extended, the skin cleaned with alcohol soaked cotton, the hair shaved and wiped with dry gauze, then the medial saphenous vein is pierced by a 21-G needle or incised by a razor blade. The dripping blood or the droplet is again touched with a capillary tube to induce collection by capillarity (Figure 33.7).

Blood collection from the surface vein of the hind limb gives only small quantities. The medial saphenous vein can yield about 80 μl, but less is obtained from the

Figure 33.6 A mouse in a restraining device. The tail and a hind limb are free for operation.

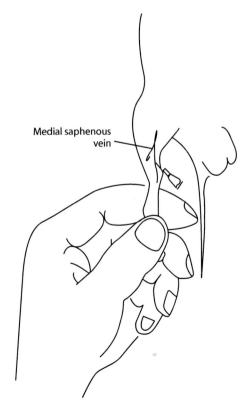

Medial saphenous vein

Figure 33.7 Puncture of the medial saphenous vein. About 80 μl of blood can be obtained.

tarsal vein. If the vein is not dilated, blow flow is slow and little obtained from bleeding. Therefore, warming up the mouse or its hind limb is indispensable in these procedures.

Bleeding from the tail vein

Tail veins are frequently used for collection of venous blood and intravenous injections. In the tail there are three veins and one artery, that is, paired lateral caudal veins, unpaired dorsal caudal vein and ventral caudal artery (Figure 33.8). As the tail veins are thicker than the dorsal metatarsal and medial saphenous veins, a larger quantity of venous blood can be obtained. The lateral caudal veins are preferentially used for both blood collection and intravenous injection. However, the tail should be also warmed up to dilate the blood vessel before bleeding. The mouse is warmed up as before or its tail is soaked directly in warm water at 40°C. The mouse need not be anesthetized, if the assistant holds the mouse or a restraining device is used. The investigator should grip near the tip of the tail of the restrained mouse by his left hand, and locate the veins running both sides of the tail. Clean from one third to the middle of the tail from its base with alcohol soaked cotton then wipe dry with sterile gauze. The vessel is then pierced with a needle (18–21 G) or incised by a razor blade. A blood droplet will accumulate on the tail which can then be collected using a glass capillary tube as before, the blood entering the tube by capillarity. Two or three capillary tubes are filled, giving a total of 200–300 μl of venous blood. After bleeding the incision site should be depressed with sterile gauze until the bleeding stops.

Larger volumes can be obtained in this procedure, i.e. between 0.5–1.0 ml (Grice, 1964) and 1.0–1.5 ml (Lewis et al., 1976). However, removal of this volume of blood will cause severe hypovolemic shock or death of the animals.

Bleeding from the tail artery

Collection of small amounts of arterial blood is executed by bleeding from a tail artery or ventral caudal artery. The mouse should be warmed up as before, the tail of the restrained mouse cleaned by alcohol soaked cotton and wiped by dried sterile gauze. Locate the artery on the ventral side of the tail, then pierce the artery with a 21 G needle. Incision by a razor blade is not adequate for this artery, because the bleeding continues longer from the incision site. A blood droplet can be aspirated by a capillary tube.

Cardiac puncture from the closed thorax

Cardiac puncture is used for direct blood collection from the heart through the thoracic wall of mice. The mouse must be anesthetized deeply by ether inhalation or intraperitoneal administration of sodium pentobarbital. This technique is repeatedly executed to obtain considerable amounts of blood from living mice. It is said that there is a slight difference between the components of blood obtained by open or closed thorax punctures (Cubitt and Barrett, 1978).

In the cardiac puncture, the approach to the heart can be done from three alternative directions: (1) directly insert a needle into the heart at a perpendicular angle to sternum; (2) insertion from the thoracic inlet; and (3) insertion below the xiphoid cartilage towards the head (Figure 33.9a, b and c). In any procedure, the investigator

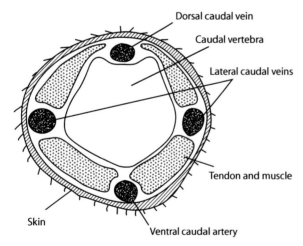

Figure 33.8 A schematic drawing of a transverse section of a tail showing the location of the dorsal caudal and lateral caudal veins and a ventral caudal artery. Paired lateral caudal veins are used for venous blood collection and intravenous injections.

Dorsal caudal vein

Caudal vertebra

Lateral caudal veins

Tendon and muscle

Skin

Ventral caudal artery

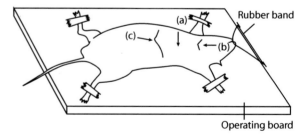

Rubber band

Operating board

Figure 33.9 Three approaches for cardiac puncture of the closed thorax. (a) The perpendicular approach against the sternum at the point of 9–10 mm caudally from the thoraccic inlet; (b) the approach from the thorax inlet; and (c) the cranial approach from just behind the xiphoid cartilage.

has to be thoroughly trained to get a perfect technique before the experiment.

In the perpendicular approach to the thoracic wall, a deeply anesthetized mouse is restrained on the operating board on its back. The xiphoid cartilage is approximately in the middle of the head and body length. The length from the anterior thoracic inlet to the xiphoid cartilage is about 20 mm. The point at 9–10 mm caudally from the thoracic inlet on midline is the site of the heart base. A 1 ml disposable plastic syringe with a needle of 26 G 1/2 in. (11 mm in length) is used. The needle is pushed in vertically to the thoracic wall just beside the sternum (Figure 33.9a). While withdrawing the plunger of the syringe, depress the needle. When blood appears in the syringe, keep the position of the needle steady and continue withdrawing the plunger slowly. In this procedure, 300–400 μl of blood can be obtained. The blood should be transferred into a test tube from the syringe after removing the needle and pouring slowly along the wall of the test tube to avoid hemolysis.

The approach from the thoracic inlet is more commonly used for cardiac puncture (Frankenberg, 1979; Ohwada, 1986). The mouse is deeply anesthetized and restrained on an operating board on its back. The neck must be fully extended by using a rubber band to pull on its upper incisors (Figure 33.9b). A 1 ml plastic syringe with a 26 G 1/2 in. needle is held ready. The anterior thoracic region is cleaned with alcohol soaked cotton, the needle is pushed in from the thoracic inlet and then depressed toward the base of the tail, parallel to the sternum or slightly downward. Continue withdrawing the plunger to aspirate while depressing the needle in the caudal direction, until blood appears in the syringe. When this occurs, depressing of the syringe is stopped but aspiration continued until the expected amount of blood is obtained. Skillful experienced investigators can grip an anesthetized animal in the left hand, push in the needle and collect blood from the heart, without restraining the mouse on a board.

The third is the cranial approach from just behind the xiphoid cartilage (Simmons and Brick, 1970). A deeply anesthetized mouse is restrained on the board and a 1 ml plastic syringe with a 21 G 1/2 in. needle held ready. The needle is inserted just caudally of the xiphoid cartilage slightly to the left of the midline and then the syringe is depressed cranially at an angle of 20° against the sternum. Depress the syringe and aspirate it by withdrawing the plunger. When blood appears in the syringe, stop depressing the syringe and continue aspirating until the expected blood volume is obtained.

After collection of the blood, put the mouse back into the cage and keep it warm until it awakes. This procedure can be done repeatedly in living animals. However, heart puncture induces considerable damage. Intervals of at least 7 days between collections must be allowed even if the small sampling volume does not affect the health of the animal.

Urine

Urine is examined for alterations in its components, the total volume, and for the appearance of abnormal materials to check the state of health and to identify certain diseases. Consecutive unrinalysis of an animal is adopted to chart the process of recovery from disease and the effect of an experiment.

Either the total urine produced over a certain length of time or fresh urine may be required. The first is executed using a metabolic cage (Figure 33.10). A mouse is placed in the metabolic cage and given food and water. The urine and feces are collected

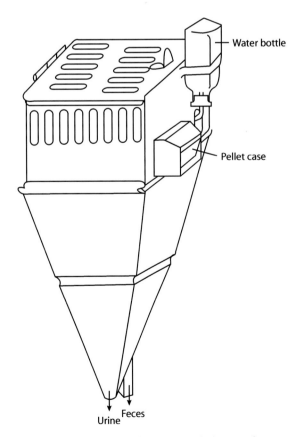

Figure 33.10 Drawings of a steel metabolic cage for mice.

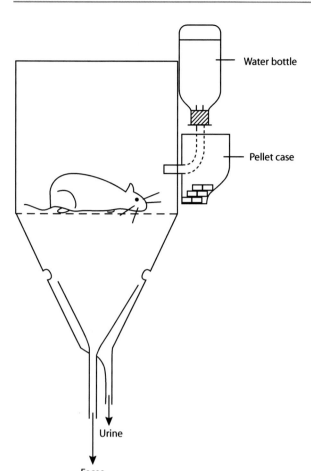

Figure 33.11 Illustration showing a mouse in a metabolic cage. Urine and feces are collected separately.

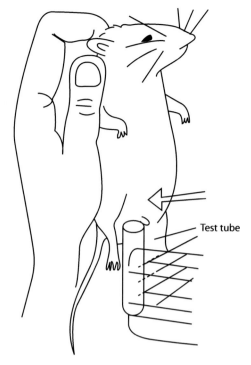

Figure 33.12 Collection of fresh urine. Grip the mouse firmly and place the external urethra over a test tube, then depress gently the lower abdomen with the forefinger of the right hand to induce excretion of urine.

separately (Figure 33.11). In this method, the urine volume produced/time is determined. However, the urine is not fresh and its components may change. Moreover, it may be contaminated with feces or residues of food. To avoid contamination of urine with feces, anal cups or plastic bags are used (Ryer and Walker, 1971; Roerig et al., 1980).

Fresh urine should be collected directly from mice. They excrete urine easily, when picked up from their cage, especially when gripped by hand. Therefore, the preparations for collection of urine must be made before taking the mouse out of the cage. A cage lid of wire-mesh is placed on a clean steel plate or a spread aluminum foil. A mouse is picked up gently from the cage by its tail, and placed on the wire-mesh lid. As the mouse will pull away by instinct when its tail is pulled back, the investigator places the palm of his free (usually left) hand on to the mouse's back and holds it. When the mouse is gripped, it excretes a little urine. Move the mouse near to a test tube stand, place the external urethra over a test tube and depress the lower abdomen softly where the urinary bladder is supposed

to be located, by the forefinger of the right hand (Figure 33.12). Then the mouse excretes urine into the test tube. Usually the amount of urine produced in this procedure is between one drop and 0.4 ml at any one time. If the quantity of urine is not sufficient, the investigator can try again some time. Excreted urine on the wire-mesh lid can be collected from the steel plate or aluminum foil. In this technique the bladder is not completely emptied. However, depressing the lower abdomen by forefinger is limited, because excessive depression induces change in the urine components by entering blood. The collected urine should be quickly stored in a freezer.

In the mouse, insertion of a catheter is not generally used for collecting urine. In case of laparotomy, the abdominal cavity is opened under anesthesia and the urine in the bladder can then be aspirated directly after puncture giving a total of 0.1–0.15 ml.

Milk

Milk is a complete food and indispensable for newly born pups. It is important for pups not only as a nutrient supply, but also as it contains immunoglobulins. Therefore, the milk is examined for its components and volume for nutritional and immunological studies. For the establishment of a first generation of clean mice, such as specific pathogen free (SPF) and germ free animals, the milk must be collected when a clean foster mother is not available.

To collect milk, there are two procedures; milking directly from the lactating mother or collecting milk from the stomach of suckling pups. In the case of large litter size, the pups can be sacrificed over several days after birth and any change in the components of the milk can be investigated. The lactating mother can be milked at certain days after parturition under anesthesia. On the other hand, the total volume of the milk secretion can be estimated indirectly by the sum of the increments in the pup's body weight during suckling. In this chapter milking from lactating mice is described.

Milking from lactating mice

In mice the quantity of secreted milk from the mother reaches a maximum at 8–10 days after parturition. Therefore, lactating mice of 8–12 days after parturition are used for milking. The ICR strain of mouse is frequently used, but other strains may be utilized.

To collect milk, a milking device must be prepared. Some milking machines with single or multiple teat cups have been designed for this purpose. (McBurney *et al.*, 1964; Feller and Boretos, 1967; Nagasawa, 1979). The device with a single teat cup is made of two disposable syringe needles, a test tube, vinyl tubing, a silicon plug and an aspirator (or vacuum system). These parts are connected as shown in Figure 33.13. A disposable syringe needle is adequate as a teat cup for the milking device. The aspirator should be connected with the water supply. The procedure for milking is as follows.

(1) Lactating mice 8 days after parturition should be prepared.
(2) Six to eight hours before milking the lactating mother must be separated from her pups.
(3) The mouse is anesthetized by intraperitoneal injection of sodium pentobarbital (0.5% sodium pentobarbital solution is administered at a dosage of 0.1 ml/10 g body weight).

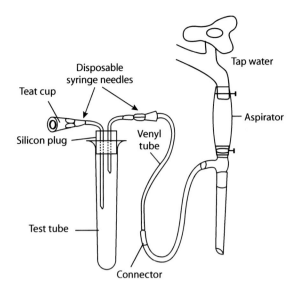

Figure 33.13 A milking device with single teat cup.

(4) After anesthesia, the mammary glands should be massaged gently with steam-warmed gauze. An injection of 0.1–0.18 IU oxytocin/kg body weight is given subcutaneously.
(5) Two to three minutes after oxytocin administration, milking should be started. Aspirating begins by turning on the tap. The investigator holds the mouse by the left hand and the milking device by the right hand.
(6) In the mouse there are three pairs of pelvic mammary glands and two pairs of inguinal ones. The teat cup of the milking device is placed over the teat and aspirated with weak vacuum. Milk enters the test tube of the device (Figure 33.14).
(7) Try aspirating for 2–3 min from a teat and then change to the next teat. Sometimes a teat itself may fill the hole of needle or the needle may get blocked. Therefore, check the state of the entering milk, then adjust the angle of the teat cup, control the vacuum pressure and remove the blockage.
(8) After milking the mouse should be returned gently into its cage of pups present when she was anesthetized. Soon the pups will suck on the teats of the anesthetized mother. This suckling stimulus is considered to have a good effect on the mother so she awakes early from the anesthesia.

Skillful investigators can collect 1.0–1.5 ml of milk/mouse.

Oxytocin begins to exert its effect on milk secretion 10 min after injection, but milking should be started 2 or 3 min after injection, i.e. much earlier.

Subsequent milking can be done at 3 day intervals, so that one lactating mother can be milked twice, 8 and

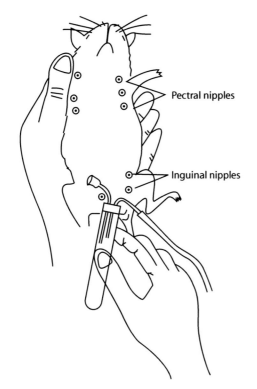

Figure 33.14 Milking from an anesthetized lactating mother. A teat cup is put on an inguinal nipple and the milk aspirated.

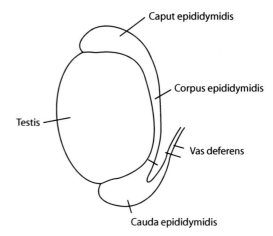

Figure 33.15 An epididymidis with testis and vas deferens. The cauda epididymidis is cut off at the broken line and placed in a drop of Whittingham's medium.

12 days after parturition. This milking is known not to affect the pup's growth. However, the investigator should handle lactating mice gently so as not to give unnecessary stress, as nursing mothers are delicate.

Semen

Semen is secreted from male reproductive organs, such as the testes, seminal vesicles, prostate and coagulating glands. It is composed of sperm and fluid, and is introduced into the female reproductive tracts to produce progeny. The purpose of the collection of semen is utilization for artificial insemination or *in vitro* fertilization, inspection of sperm and preservation of sperm as a genetic resource. Investigators can obtain ejaculated semen or that from the dissected epididymis. Ejaculated semen is gained from the uterine horns of the female following copulation and by electroejaculation. In mice, semen collection from the mated females is not normally executed. **Electroejaculation** can be done in mice (Scott and Dziuk, 1959). However, the electroejaculated semen of mice is coagulated by

secretions from the coagulating glands, and it is not adequate for the utilization of sperm. Semen obtained from the epididymis or vas deferens is not coagulated, and is generally used for fertilization (Hogan *et al.*, 1994). In this chapter the collecting procedure of semen from the epididymis is described.

Semen collection from the epididymis

Adult males of 2–6 months old are used. For the dilution of semen and incubation to get capacitation and storage, media such as Whittingham's, MKRB and others are used. The procedure of semen collection is as follows.

(1) Before the semen collection, male mice must be separated from females for 3 days. However, any male that has not mated for over 1 week is not suitable for semen collection.

(2) The male mouse is sacrificed by cervical dislocation. A small cut is made in the abdominal skin which is then peeled back aseptically to expose the muscular abdominal wall. The muscular wall is then incised in order to expose the viscera. The fatty tissues of inguinal regions are then drawn back so that a testis with an epididymis from the scrotum of both sides can be pulled out.

(3) Remove the epididymis along with the testes and place them on filter paper (Figure 33.15). Remove as much fat as possible and then separate the cauda epididymidis with vas deferens from the testes and the remaining part of the epididymis.

(4) Place the cauda epididymidis into a drop of Whittingham's medium plus 30 mg/ml bovine serum albumin (BSA).

(5) Cut open the cauda epididymidis with sharp fine scissors and gently squeeze out the sperm using watchmaker's forceps.

The sperm in the medium must be incubated in 5% CO_2 and 95% air at 37°C, for 1.5 h to **capacitate** them. After this, the density of the sperm is adjusted to $3–4 \times 10^6/\mu l$ for artificial insemination and for *in vitro* fertilization. For freeze preservation, the semen is diluted in an appropriate medium such as RPI and then frozen in liquid nitrogen. The frozen semen can be stored for a long period in liquid nitrogen below −195°C.

Saliva

Saliva is excreted from the salivary glands, such as the mandibular, parotid and sublingual glands. Small amounts of saliva can be obtained by soaking it up with a piece of filter paper. To collect more, the salivary glands should be stimulated by administration of pilocarpine hydrochloride, at a dosage of 3–5 mg/kg body weight. Excreted saliva is collected by a Pasteur pipette or a micro-hematocrit tube. As pilocarpine hydrochloride stimulates the lacrimal glands and nasal mucous glands at the same time, the stimulated saliva is contaminated to some extent with lacrimal fluid and secretions from nasal glands. The stimulated saliva is also soaked up by filter paper and analyzed.

Lacrimal fluid

The lacrimal fluid of mice is produced in very small amounts in intact animals. To collect lacrimal fluid, mice are stimulated by pilocarpine hydrochloride to induce excretion. The solution of 0.05% pilocarpine hydrochloride in saline is injected intraperitoneally at a dosage of 0.2 ml/animal. Several minutes after the injection, lacrimal fluid appears. This is collected by capillarity using a glass hematocrit capillary tube. Where small amounts are analyzed, the fluid is soaked up by a small piece of filter paper. The contamination of saliva and secretion from the nasal glands must be carefully avoided.

Peritoneal fluids

Peritoneal exudate fluid is very small in quantity and contains some neutrophilic leukocytes, macrophages and lymphocytes. Therefore collection of the peritoneal exudate is usually done to obtain the leukocytes, especially free macrophages. Resident peritoneal exuded cells in intact mice and the increased numbers of stimulated cells are collected and used for experimental purposes. Here the general procedure for the collection of the stimulated peritoneal exudate is described. Stimulants, such as 5% glycogen, 2.4% fluid thyoglycolate medium and 10% proteose peptone, are used. The procedure of collection of stimulated peritoneal exudate cells is as follows.

(1) Two milliliter of thyoglycolate medium (or 0.5 ml of 5% glycogen) is administered intraperitoneally.

(2) Three to four days after administration, the mouse is sacrificed by cervical dislocation, and placed on its back on the operating board. The skin of the abdomen is incised and flipped open in order to expose the abdominal muscular wall.

(3) 10 ml of Eagle's MEM are injected intraperitoneally by a 10 ml plastic syringe with a 21 G needle. The abdomen is then massaged gently using fingers.

(4) Removal and injection of the Eagle's MEM is repeated three times with the same syringe. Finally the medium is recovered. About 9 out of 10 ml of the medium is normally recovered.

(5) The recovered medium is centrifuged slowly and the cell pellet resuspended in fresh Eagle's medium supplemented with 10% fetal calf serum depending on subsequent treatments.

After stimulation, polymorphonuclear leukocytes, macrophages and lymphocytes appear in the peritoneal exudate in turn. Resident peritoneal exudate cells are also collected by same procedure without stimulation.

Macrophages are separated utilizing their adhesive nature. Other immune competent cells are collected from the spleen as described in another chapter.

References

Ambrus, J.L., Ambrus, C.M., Harrisson, J.W.E., Leonard, C. A., Moser, C.E. and Cravitz, H. (1951). *Am. J. Pharm.* **123**, 100–104.

Cook, M.J. (1965). In *The Anatomy of the Laboratory Mouse*. Academic Press, London and New York.

Cubitt, J.G.K. and Barrett, C.P. (1978). *Lab. Anim. Sci.* **28**, 347.

Cunliffe-Beamer, T.L. and Les, E.P. (1987). In *The UFAW Handbook on the Care and Management of Laboratory Animals*, 6th edn, pp. 275–308. Longman Scientific & Technical, Essex, UK.

Feller, W.F. and Boretos, J. (1967). *J. Nat. Cancer Inst.* **38**, 11–17.

Frankenberg, L. (1979). *Lab. Anim.* **5**, 311–312.

Gray, D.G. (1979). *J. Clin. Path.* **2**, 633–634.

Grice, H.C. (1964). *Lab. Anim. Care.* **14**, 483–493.

Hogan, B., Beddington, R., Constantini, F. and Lacy, E. (1994). In *Manipulating the Mouse Embryo: A laboratory Manual*, 2nd edn, pp. 146–147. Cold Spring Harbor Laboratory Press, New York.

Iwaki, T., Yamashita, H. and Hayakawa, T.A. (2001). In *Color Atlas of Sectional Anatomy of the Mouse*. Adothree, Tokyo.

Kassel, R. and Levitan, S. (1953). *Science* **118**, 563–564.

Lewis, V.J., Thacker, W.L., Mitchell, S.H. and Baer, G.M. (1976). *Lab. Anim. Sci.* **26**, 211–213.

Lushbough, C.H. and Moline, S.W. (1961). *Proc. Ani. Care Panel* **11**, 305–308.

McBurney, J.J., Meier, H. and Hoag, W.G. (1964). *J. Lab. Clin. Med.* **64**, 485–487.

Nagasawa, H. (1979). *Lab. Anim. Sci.* **29**, 633–635.

Ohwada, K. (1986). *Exp. Anim.* **35**, 353–355.

Roerig, D.L., Hasegawa, A.T. and Wang, R.I.H. (1980). *Lab. Anim. Sci.* **30**, 549–551.

Ryer, F.H. and Walker, D.W. (1971). *Lab. Anim. Sci.* **21**, 942–943.

Sakaki, K., Tanaka, K. and Hirasawa, K. (1961). *Bull. Exp. Anim.* **10**, 14–19.

Scott, J.V. and Dziuk, P.J. (1959). *Anat. Rec.* **133**, 655–664.

Simmons, M.L. and Brick, J.O. (1970). In *The Laboratory Mouse, Selection and Management*. Prentice-Hall Inc., Englewoood.

Young, L. and Chambers, T.R. (1973). *Lab. Anim. Sci.* **23**, 428–430.

Anesthesia, Analgesia and Euthanasia

Klaus Otto
Zentrales Tierlaboratorium, Medizinische Hochschule
Hannover, Germany

Anesthesia and analgesia in mice is particularly challenging because of a number of problems associated with the small body size. These include:

- Mice are very susceptible to hypothermia since the high surface area relative to body mass results in rapid heat loss. Much of the mortality during anesthesia may be due to hypothermia rather than to other physiologic effects of anesthetic agents.
- Considerable strain-, sex-, age-dependent variability in the effective dose of some injectable anesthetics may result in inadequate depth of anesthesia or lethal overdose.
- Lack of reliable respiratory and cardiovascular monitors.
- Rapid loss of even small amounts of blood as low as 0.5 ml may cause cardiovascular failure (Flecknell, 1993).
- Lack of objective measurements of pain in mice.

Anesthesia

Preanesthetic considerations

Health conditions

Any animal subjected to an experiment must be in excellent health conditions. Animals should be housed for a 1–2-week acclimatization period which allows daily health control. It is undesirable to withhold food or water before anesthetizing mice (Flecknell, 1993).

Hypothermia

Intra- and postoperative hypothermia in mice may be potentially lethal and needs to be a major concern before anesthesia. Any means sufficient to prevent a pronounced decrease in body temperature (e.g. heating blanket, infrared heat lamp etc.) rather than treatment of hypothermia should be considered. In addition, body temperature in the mouse is closely tied to locomotion activity (Gardner *et al.*, 1995). Therefore, the duration of anesthesia and of the recovery period should be kept as short as possible in order to minimize the decrease in body temperature postoperatively.

Anticholinergic premedication

Mice should be treated with atropine (0.04 mg/kg subcutaneously (s.c.)) about 30 min prior to induction of anesthesia (Green, 1979). Although atropine does not eliminate the copious secretions associated with ether or ketamine anesthesia the reduced secretion will help to maintain a patent airway (Tarin and Sturdee, 1972).

Anesthetic regimen

When selecting an anesthetic regimen the type and length of the procedure, advantages and disadvantages of the various anesthetics, and especially the aim of the study need to be considered in order to avoid interaction between the anesthetic and the experimental protocol. The main choices to be made are between administration of injectable anesthetics or inhalant anesthetics. For major surgery and other long-term procedures a combination of an injectable anesthesic with an inhalant anesthetic or inhalation anesthesia plus an opioid analgesic may be considered as well.

While in mice most injectable anesthetics are easy to administer by the intraperitoneal (i.p.) route this technique is more likely to produce an unpredictable depth of anesthesia and prolonged recovery periods. This is because drug absorption into the systemic circulation is slow after i.p compared with intravenous (i.v.) administration making it impossible to titrate the anesthetic to effect and overdosing may result. Furthermore, i.p. administration of relatively large total quantities of a drug are required to produce anesthesia, which in turn results in prolonged recovery (Flecknell, 1993).

On the other hand, for modern inhalation anesthetics such as halothane, isoflurane or sevoflurane, more sophisticated and expensive equipment (e.g. calibrated vaporizer, breathing systems etc.) are required. The major advantages of this technique include rapid recovery and immediate adjustment of the anesthetic concentration to the animal needs.

Injectable anesthetics

Injectable anesthetics (Table 34.1) can be administered i.p., s.c., intramuscularly (i.m.) or i.v. Acceptable volumes for i.p. and s.c. injection range from 0.1 to 0.5 ml, for i.v. from 0.05 to 0.1 ml and should not exceed 0.05 ml for i.m. injection for an adult mouse (Green, 1979; Flecknell, 1993).

Propofol

Propofol (2,6-diisopropylphenol) is a hypnotic agent that is used as induction agent and as a maintenance anesthetic delivered by continuous i.v. infusion or intermittent i.v. bolus (Branson and Gross, 1994). In mice, a single bolus i.v. injection of 20–30 mg/kg produces about 2–3 min of surgical anesthesia and loss of righting reflex for approximately 3–20.5 min (mean: 7.1 min; Glen, 1980; Flecknell, 1993; Koizumi *et al.*, 2002). Advantages of propofol include rapid onset and short duration of anesthesia, lack of accumulation after repetitive injections, no tissue damage following inadvertent perivascular infiltration and rapid, tranquil recovery after the cessation of i.v. administration (Wixson and Smiler, 1997). Rapid bolus injection in rodents may be followed by 5–10 s of **apnea** and hypotension while continuous infusion causes only minimal changes in heart rate and respiratory rate (Koizumi *et al.*, 2002).

Propofol is useful for very short-term procedures (e.g. tattooing, blood sampling). The disadvantages of propofol include poor analgesia and the requirement for i.v. access.

Alphaxolone/alphadolone

Slow i.v. administration of 10–20 mg/kg of the steroid combination alphaxalone/alphadolone produces deep anesthesia with analgesia and good muscle relaxation within 10 s and maintenance of the maximum effect for approximately 5–8 min (Green, 1979; Flecknell, 1993). The mean sleep time is 10 min and the time to full recovery ranges from 20 to 30 min. Prolonged anesthesia may be achieved either by further increments (6 mg/kg i.v.) at 15 min intervals or by continuous infusion of 0.25–0.75 mg/kg/min via an indwelling needle in the tail or jugular vein (Green, 1979; Flecknell, 1993). When used in rodents, a 1:10 dilution of the commercial preparation is recommended. The wide therapeutic index and the lack of cumulative effects or tolerance with repetitive i.v. doses make this combination an excellent i.v. anesthetic in mice

TABLE 34.1: Anesthetics and tranquilizers used in mice

Drug	Dosage (mg/kg)	Comments	Reference
α-Chloralose	114 i.p.	5% solution, only in combination with analgesics and/or other anesthetic agents	White and Field (1987)
Alphaxalone/ alphadolone (*Saffan, Althesin*)	10–20 i.v.	unpredictable anesthetic effect following i.p., volume too large for i.m.	Green *et al.* (1978), Flecknell (1993)
Chloral hydrate	60–90 i.p.	light surgical anesthesia, considerable strain differences	Green (1979)
	370–400 i.p.		Flecknell (1993)
Fentanyl/ fluanisone (*Hypnorm*)	0.4 ml/kg i.m.	muscle rigidity, pronounced respiratory depression, 1 : 10 dilution	Flecknell (1993)
Hypnorm/ midazolam	10 ml/kg i.p.	2 parts water for injection + 1 part *Hypnorm* + 1 part midazolam (5 mg/ml)	Flecknell (1993)
Fentanyl/ droperidol (*Innovar Vet*)	0.5 i.m.	irritant, tissue necrosis, self-trauma following i.m.	Flecknell (1993)
Ketamine	80–100 i.m.	sedation, muscle rigidity	Green *et al.* (1981a)
	100 i.p.		White and Field (1987)
	100–200 i.m.		Flecknell (1993)
Ketamine/ acepromazine	44/0.75 i.p.	Sedation only	Gardner *et al.* (1995)
	100/2.5 i.p.	marked respiratory depression	Flecknell (1993)
	100/5 i.p		Flecknell (1993)
Ketamine/ midazolam	100/5 i.p.	only light anesthesia	Flecknell (1993)
Ketamine/ medetomidine	75/1 i.p.	female mice	Flecknell (1993), Cruz *et al.* (1997)
	40/1 i.p.	male mice	Cruz *et al.* (1998)
	100–150/0.25 i.p.		Kilic *et al.* (2001)
Ketamine/xylazine	100/5 i.m.	excellent relaxation, sedation, analgesia	Erhardt *et al.* (1984)
	100/10 i.p.		Flecknell (1993)
	120/16 i.p.		Zeller *et al.* (1998)
Methohexital (*Brevital, Brevimytal*)	10 i.v.	short-term anesthesia	Flecknell (1993)
	44 i.p.		Dörr and Weber-Frisch (1999)
Metomidate/ fentanyl	60/0.06 s.c.		Green *et al.* (1981b), Flecknell (1993)
Pentobarbital (*Nembutal*)	45 i.p.	1 : 10 dilution, narrow safety margin, marked strain differences in response	Flecknell (1993)
	50 i.p.	severe respiratory depression	Erhardt *et al.* (1984)
	60 i.p.		Zeller *et al.* (1998), Koizumi *et al.* (2002)

TABLE 34.1: (Continued)			
Drug	Dosage (mg/kg)	Comments	Reference
Propofol (Rapinovet, Diprivan)	26 i.v.	short-term anesthesia, i.v. injection required	Flecknell (1993)
	30 i.v.		Koizumi et al. (2002)
Thiopental (Penthotal, Trapanal)	30 i.v.	short-term anesthesia, i.v. injection required, dose dependent hypothermia and respiratory depression	Flecknell (1993)
Tiletamine/ zolazepam (Telazol)	40 i.p.		Flecknell (1993)
	80–100 i.p.		Silverman et al. (1983)
Telazol/xylazine	7.5/45 i.m.	long-term anesthesia	Gardner et al. (1995)
	45/7.5 i.m.		Gardner et al. (1995)
Tribromoethanol (Avertin)	125–300 i.p.	1.2% solution, possible peritonitis, serositis	Flecknell (1993)
	240 i.p.		Zeller et al. (1998)

Intraperitoneally (i.p.), intramuscularly (i.m.), intravenously (i.v.), subcutaneously (s.c.).

(Green, 1979). However, the unpredictable anesthetic effect following i.p. administration and the large volume required for i.m. injection limit its usefulness in this species (Green, 1979).

Methohexital

Intravenous injection of 6–10 mg/kg of the ultrashort-acting barbiturate methohexital produces anesthesia within 10 s for about 3–5 min (Green, 1979; Flecknell, 1993). Time to full recovery is about 50 min. Repetitive dosing, however, may cause accumulation with subsequent prolongation of the recovery period. Methohexital produces moderate cardiovascular and respiratory depression. Intraperitoneal injection of 44 mg/kg methohexital (6.46 mg/ml) to female C3H/Neu mice induced anesthesia within 3.3 min (mean; Dörr and Weber-Frisch, 1999). Complete immobilization lasted for 1.5 min (mean) and recovery was completed between 10–15 min after injection. The study also revealed that methohexital at a dose <40 mg/kg did not result in chemical restraint while doses >50 mg/kg caused considerable lethality. Although no gross pathomorphological changes were found after repeated i.p. injections of methohexital in this study administration of methohexital by the i.p. route has not been recommended because of unpredictable effects (Flecknell, 1993).

Thiopental

Thiopental is irritating perivascularly and thus must be administered i.v. Slow injection of 30–40 mg/kg i.v. produces light surgical anesthesia within 10–15 s and with a duration of about 10–12 min (Flecknell, 1993; Wixson and Smiler, 1997). Thiopental produces moderate respiratory and cardiovascular depression. Incremental doses should not be used for prolonging anesthesia as accumulation is very marked with this agent (Green, 1979).

Ketamine

When used as a sole agent (100–200 mg/kg i.m. or i.p.), ketamine may produce surgical anesthesia for about 30 min, which, however, may be accompanied by insufficient analgesia, muscle rigidity and significant mortality (Green et al., 1981a; Flecknell, 1993; Wixson and Smiler, 1997). Therefore, administration of ketamine with an accompanying tranquilizer (e.g. α_2-adrenoceptor-agonist, phenothiazine) has been recommended (Flecknell, 1993). When using ketamine, a reduction of endotoxin-induced production of proinflammatory cytokines, including tumor necrosis factor-alpha (TNF) in monocytes and macrophages (Jones et al., 1999; Sakai et al., 2000) need to be considered. Ketamine alone (Peuler et al., 1975) or in combination with xylazine

(Zeller *et al.*, 1998) may cause exophthalmus due to increased intracerebral and intraocular pressure.

Ketamine/xylazine

In order to overcome the muscle rigidity and incomplete analgesia produced by ketamine alone, ketamine plus the α_2-adrenoceptor agonist xylazine (100 mg/kg + 10 mg/kg i.p.) has become the most widely used ketamine combination (Flecknell, 1993; Wixson and Smiler, 1997). Xylazine has sedative and analgesic properties. Both drugs may be given as a mixture in a single injection. A combination of ketamine with a lower dose of xylazine (100 mg/kg + 5 mg/kg i.m.) produces calm, rapid (2–3 min) induction, 80-min surgical anesthesia and a total anesthesia time of 110 min (Erhardt *et al.*, 1984). Surgical anesthesia was associated with excellent muscle relaxation, sedation and analgesia throughout the 80-min period of anesthesia. Zeller *et al.* (1998) recommended a ketamine/xylazine combination (120 mg/kg + 16 mg/kg i.p.) for embryo transfer in female CD1-, OF-1- and NMRI-mice. The combination provided adequate surgical anesthesia. In addition, all animals survived and no disadvantages were detected regarding the result of the embryo transfer.

Adverse effects of this combination on cardiovascular function include bradycardia, increase in preload and left ventricular fractional shortening along with a significant reduction in cardiac output and hypotension (Erhardt *et al.*, 1984; Chaves *et al.*, 2001; Hart *et al.*, 2001; Roth *et al.*, 2002). In addition, production of acute temporary cataracts in some animals were reported (Calderone *et al.*, 1986). Ketamine/xylazine may cause injury to lymphocytes and hepatic Kupffer and endothelial cells within 3 h of administration accompanied by an increase in activity of hepatic aspartate transaminase (AST) (Thompson *et al.*, 2002).

Ketamine/medetomidine

Ketamine may be combined with the newer α_2-agonist medetomidine (75 mg/kg + 1.0 mg/kg i.p.) (Flecknell, 1993). Medetomidine closely resembles xylazine in its effects, but possesses much higher affinity for the α_2-adrenoceptor (Virtanen, 1989). It causes sedation, excellent muscle relaxation and analgesia but cardiovascular and respiratory depression, hypothermia, hyperglycemia, diuresis and increased salivation may be also present (Flecknell, 1993; Cruz *et al.*, 1998). Cruz *et al.* (1998) used ketamine/medetomidine combinations for short-term (5–10 min) anesthesia in Swiss Webster mice of both sexes. Female mice needed a higher dose of ketamine (75 mg/kg + 1 mg/kg i.p.) than male mice (40 mg/kg + 1 mg/kg i.p.). When ketamine/medetomidine (100–150 mg/kg + 0.25 mg/kg i.p.) were given to NMRI mice, loss of righting reflex occurred within 5 min of drug administration and lasted for approximately 75–80 min (mean) (Kilic *et al.*, 2001).

Because of the depression of cardiovascular and respiratory function and the development of long-lasting hypothermia induced by these combinations, reversal of the xylazine or medetomidine effects by the α_2-adrenoceptor antagonist atipamezole is strongly recommended (Flecknell, 1993; Cruz *et al.*, 1998). Atipamezole (1 mg/kg i.m., i.p., s.c., i.v.) and, if needed, followed by increments of 0.25 mg/kg, also results in a much faster reappearance of righting reflex and a shorter total recovery time. However, the total dose of atipamezole required for satisfactory reversal was considerably higher (up to 2.5 mg/kg) in female mice than in male mice (Cruz *et al.*, 1998).

Ketamine/acepromazine

Ketamine combined with the phenothiazine tranquilizer acepromazine (44 mg/kg + 0.75 mg/kg i.p.) may fail to produce loss of righting reflex in some mice (Gardner *et al.*, 1995) while high doses of this combination (100 mg/kg + 2.5 mg/kg i.p.) can produce surgical anesthesia in some strains of mice but this may be accompanied by marked respiratory depression (Flecknell, 1993).

Ketamine/midazolam

Administration of ketamine plus the water-soluble benzodiazepine midazolam produces effects similar to that of ketamine/acepromazine. Since neither midazolam nor acepromazine possesses detectable analgesic action in mice, low doses produce only light anesthesia while high doses (100 mg/kg + 5 mg/kg i.p.) may cause deeper levels of anesthesia with pronounced respiratory depression (Flecknell, 1993).

Telazol™/xylazine

Telazol is a combination of the dissociative anesthetic tiletamine with the benzodiazepine zolazepam (Wixson and Smiler, 1997). Because Telazol alone is an adequate anesthetic in rats but not in mice, it has been combined with xylazine for anesthesia in mice (Gardner *et al.*, 1995). Following i.m. injection of Telazol/xylazine (7.5 mg/kg + 45 mg/kg i.m.) to male Hsd: ICR mice, onset of anesthesia occurred between 0.6 and 1.3 min of injection and duration of surgical anesthesia (absence of toe pinch reflex) ranged from approximately 46 to 164 min (mean: 99 min), suggesting that this combination could be useful for long-term anesthesia (Gardner *et al.*, 1995). However, total recovery time may last up to 14 h accompanied by respiratory depression and hypothermia. When 45 mg/kg i.m. of Telazol plus 7.5 mg/kg i.m. of xylazine were administered, mean duration of surgical anesthesia was decreased to 36.1 min and some mice did not lose toe pinch reflex at all.

Fentanyl/droperidol

For neuroleptanalgesia in mice a combination of the potent opioid analgesic fentanyl and the butyrophenone tranquilizer droperidol (*Innovar Vet*™) can be used. The usefulness of *Innovar Vet* in mice, however, is compromised by its irritant nature causing tissue necrosis and self-trauma to the digits following i.m. injection (Lewis and Jennings, 1972).

Fentanyl/fluanison

The combination of fentanyl and the butyrophenone fluanison (*Hypnorm*™) produces immobilization and profound analgesia accompanied by muscle rigidity and pronounced respiratory depression (Flecknell, 1993). Furthermore, nervous twitching, paddling, extreme hyperacusia and hyperaesthesia has been reported for *Hypnorm* in mice (Green, 1975). The undesirable effects can be overcome if the dose of the neuroleptanalgesic is reduced and a benzodiazepine is incorporated in the combination (Flecknell and Mitchell, 1984). The commercially available *Hypnorm* solution should be diluted 1:10 in normal saline prior to administration to mice (Wixson and Smiler, 1997).

For surgical anesthesia in mice 10 ml/kg i.p. of a mixture of 2 parts water for injection + 1 part *Hypnorm* + 1 part midazolam (5 mg/ml) were recommended (Flecknell, 1993). Anesthesia may last 45–60 min with a total sleep time between 2 and 4 h. Although the Hypnodil/benzodiazepine combination may cause respiratory depression, there is no cyanosis and mortality is very low provided due attention is paid to the body temperature and hydration (Green, 1979).

Fentanyl/metomidate

The combination of the hypnotic metomidate plus fentanyl (60 mg/kg + 0.06 mg/kg s.c.) produces stable, surgical anesthesia for 60–70 min (Green *et al.*, 1981b).

Etorphine/methotrimeprazine

This neuroleptanalgesic combination of the opioid etorphine and the phenothiazine tranquilizer methotrimaprazine (*Immobilon*™) can be used alone (1:10 dilution, 0.1–0.2 ml/mice s.c.) (Flecknell, 1987) or in combination with midazolam to produce deeper levels of anesthesia (0.1 ml/10 g i.p. of the mixture composed of 0.3 ml *Immobilon* + 1 ml midazolam + 8.3 ml sterile water for injection; Whelan and Flecknell, 1994). This combination produces prolonged anesthesia. The mean duration for absence of the righting reflex and pedal withdrawal reflex was 155 and 170 min, respectively. When compared with *Hypnorm* plus midazolam, *Immobilon* plus midazolam produces a twofold increase in the duration of surgical anesthesia accompanied by a severe respiratory depression and followed by a prolonged recovery of up to 6 h. Therefore, supplemental oxygen and, if necessary, positive pressure ventilation, should be provided.

Carfentanyl/etomidate

The use of this combination of the highly potent opioidanalgesic carfentanyl and the hypnotic etomidate (0.003 mg/kg + 15 mg/kg i.m.) may cause strong excitation during induction and prolonged recovery from anesthesia (Erhardt *et al.*, 1984). The recovery period may be accompanied by **tonic-clonic spasms** and trembling. Because of these adverse effects, the combination is not recommended for use in mice.

Chloral hydrate

Chloral hydrate (370–400 mg/kg i.p.) provides light surgical anesthesia for 45–60 min (Flecknell, 1993). There are, however, considerably strain differences in the depth of anesthesia.

Tribromoethanol

Tribromoethanol (125–250 mg/kg i.p., 1.2% v/v solution) can produce excellent surgical anesthesia in mice that is characterized by rapid induction, a 16-min duration of surgical anesthesia, good skeletal muscle relaxation, moderate degree of respiratory depression and full recovery within 40–90 min (Green, 1979; Papaioannou and Fox, 1993; Wixson and Smiler, 1997). However, while the use of tribromoethanol anesthesia in mice has been recommended by a number of investigators (Papaioannou and Fox, 1993; Wixson and Smiler, 1997; Weiss and Zimmermann, 1999) other advise against its use (Flecknell, 1993; Zeller et al., 1998). Exposure of tribromoethanol to light or improper storage at room temperature may cause decomposition of the anesthetic into its toxic by-products dibromoacetaldehyde and hydrobromic acid, which are potent gastrointestinal (GI) irritants, leading to fibrinous peritonitis, ileus and fatalities (Papaioannou and Fox, 1993). Therefore, only freshly prepared (on the day of administration), properly stored (4°C, dark conditions) solutions must be used (Papaioannou and Fox, 1993; Wixson and Smiler, 1997). However, i.p. injection of even freshly prepared solution of tribromoethanol to CDI-, OF-1- and NMRI-mice at either 240 mg/kg i.p., dissolved in tertiary amyl alcohol and distilled water to a solution of 1.2%, or 450 mg/kg i.p., diluted to a concentration of 2.5% in 0.9% NaCl, produced profound histopathological changes 24 h after injection (Zeller et al., 1998). Histopathological changes included focal to diffuse necrosis of subperitoneal muscle fibers associated with peritonitis and **serositis** in the spleen, the liver, the intestines and the stomach. In mice treated with the higher concentration an increased severity of necrotic and inflammatory changes was noticed. Because the lesions were not related to the solvent or strain-specific differences, the authors concluded that the histopathological changes were induced by tribromoethanol.

Pentobarbital

The barbiturate pentobarbital has been widely used in mice because of its nonirritant nature and modest costs. Intraperitoneal injection of 45 mg/kg of a 1 : 10 dilution of the commercially available solution may result in surgical anesthesia for 15–60 min and total sleep time of 2–4 h (Flecknell, 1993). However, the appropriate dose depends on the strain and within strain differences are also present for age, sex, dose level, litter size, diurnal periodicity, diet, fasting prior to anesthesia and type of bedding material (Green, 1979; Lovell, 1986a–c). Pentobarbital at 50–60 mg/kg i.p. may result in good sedation but inadequate surgical, moderate to severe circulatory and respiratory depression during anesthesia in some strains (e.g. BALB/c, ICR; Erhardt et al., 1984; Gardner et al., 1995) while it can produce deep anesthesia for up to 169 min in other strains (e.g. outbred ICR mice; Koizumi et al., 2002). For surgical anesthesia, pentobarbital at 50–90 mg/kg i.p. may be required (Green, 1979). Anesthesia may last for 20–40 min with a total

sleep time of approximately 120 min and time to full recovery between 6 and 24 h. Thus, supplemental oxygen and a recovery incubator are mandatory for deeply anesthetized mice.

Inhalation anesthesia

For induction of inhalation anesthesia the *bell jar* has been traditionally used. The liquid (diethylether, methoxyflurane) is volatilized by placing it on cotton balls or gauze squares in the bottom of the jar. The anesthetic-impregnated cotton should be covered with a woven mesh grid to prevent local irritation of the feet of the animal by direct contact with the liquid anesthetic. The animal is placed within the jar and visually observed for cessation of movement and recumbency, thereby signifying the onset of anesthesia. There are, however, major disadvantages associated with this delivery system. The anesthetic concentration within the jar cannot be controlled and lethal concentrations can rapidly accumulate when volatile agents with high vapor pressures (e.g. halothane, isoflurane, sevoflurane) are used. For this reason, the bell jar should be reserved for inhalant agents such as methoxyflurane, which reaches a maximum concentration of approximately 3% after full vaporization at room temperature in contrast to levels of approximately 30% inhalant gas which can be reached upon volatilization of halothane or isoflurane with this method (Wixson and Smiler, 1997). Furthermore, scavenging of waste anesthetic gas is difficult. Hence, high vapor pressure agents (e.g. halothane) should be delivered at *controlled concentrations* (Table 34.2) into a transparent induction chamber using an anesthesia machine with a calibrated vaporizer (Flecknell, 1993; Wixson and Smiler, 1997). The induction chamber should also have both an inlet for delivery of fresh gas and an outlet for effective removal of waste anesthetic gases.

Following induction of anesthesia, the mouse is removed from the induction chamber and very brief surgical procedures (<30 s) are possible. For longer periods of anesthesia, reduced concentrations of the inhalant anesthetic (Table 34.2) should be administered via a face mask/nose cone connected to the anesthesia machine. Face or head masks can be purchased or can be easily made from the proximal end of a 20-, 50- or 60-ml syringe (Wixson and Smiler, 1997). Alternatively, orotracheal intubation of the mouse and controlled ventilation at a rate of 60–100 breaths/min and a tidal volume of 0.15 ml/10 g body weight (bwt) can be performed (Flecknell, 1993).

TABLE 34.2: Inhalant anesthetics used in mice

Anesthetic	Induction[a] (Vol%)	Maintenance[a] (Vol%)	Reference
Methoxyflurane	3.5	0.4–1.0	Flecknell (1993)
(Metofane,Penthrane)	3.0	1.5–2.0	Sedgwick et al. (1992)
Halothane	3–4	1–2	Flecknell (1993)
(Fluothane)	2.4	1.0–1.5	Sedgwick et al. (1992)
Isoflurane	3.5–4.5	1.5–3	Flecknell (1993)
(Forane, Aerrane)	2.5	1.0–1.5	Sedgwick et al. (1992)
Enflurane	3–5	0.5–2	Flecknell (1993)
(Ethrane)	2.0	1.5–2.0	Sedgwick et al. (1992)
Diethylether	10–20	4–5	Flecknell (1993)
	5–10		Sedgwick et al. (1992)

[a] to effect

Ether

Ether (diethyl ether) is usually administered to mice by simple 'open-drop' methods using an ether-impregnated cotton ball in a bell jar for induction followed by inhalation via a simple face cone if prolonged anesthesia is required (Green, 1979). The simple induction method can be used to provide 5–10 min of anesthesia suitable for minor procedures. The advantages of ether anesthesia such as low costs, the wide margin of safety for the inexperienced investigator due to the slow induction and a lack of effect on hematological parameters (e.g. hematocrit, white and red blood cell count; Wixson and Smiler, 1997) are opposed by a number of disadvantages that need thorough consideration before ether is used. Induction of ether anesthesia is unpleasant, is irritant to the respiratory tract and may provoke excessive mucous secretions, pulmonary edema and airway obstruction (Tarin and Sturdee, 1972). Furthermore, ether is flammable and explosive. Although Green (1979) concluded that ether retains a useful place in mouse anesthesia when measured volumes (2–4 ml) are used for induction of anesthesia, other investigators considered ether to be highly unsuitable for this species (Tarin and Sturdee, 1972; Flecknell, 1993).

Methoxyflurane

Methoxyflurane combined with O_2 or air is the volatile agent of choice for inhalation anesthesia in mice (Green, 1979) and many investigators agreed that methoxyflurane is the safest agent to use whether by simple open-drop methods (bell jar) or with sophisticated apparatus

(Tarin and Sturdee, 1972; Green, 1979). Induction of anesthesia (time to loss of righting reflex) is smooth and rapid (1–3 min) if mice are introduced to an induction chamber in which methoxyflurane on gauze (0.5–1.0 ml in 1 L chamber) has been allowed approximately 10 min at room temperature to evaporate. Methoxyflurane reaches a maximum concentration of approximately 3% after full votilization at room temperature (Green, 1979; Gardner et al., 1995; Wixson and Smiler, 1997). Thereafter, the mouse is removed and short procedures (10–20 min) such as retroorbital bleeding or surgery about the head and neck without the encumbrance of masks or tubes can be performed. For further maintenance of anesthesia a methoxyflurane-soaked nose cone or a calibrated vaporizer with a nonrebreathing circuit at 100–500 ml/min fresh gas flow and 0.5–2% methoxyflurane can be used.

The advantages of methoxyflurane include the ease to maintain even prolonged surgical anesthesia, no need for an assistant to act as anesthetist, smooth recovery with some degree of analgesia and a postoperative survival rate of 100%. Furthermore, methoxyflurane produces less salivation than ether and less cardiovascular and respiratory depression than halothane (Wixson and Smiler, 1997). However, animals must be monitored closely for maintenance of body temperature and depth of anesthesia, and exposure must be quickly reduced in animals exhibiting slow and erratic respiration (Green, 1979; Gardner et al., 1995).

Recovery may be very prolonged (up to 24 h) depending on the duration of administration (Green, 1979). About 40% of methoxyflurane is subjected to metabolism resulting in inorganic fluoride ion release, which can cause renal damage (Murray and Fleming, 1972). Therefore, scavenging of anesthetic waste gases is important for personnel safety.

Halothane

Halothane is a nonflammable, nonexplosive, nonirritating volatile anesthetic but needs a calibrated vaporizer for delivery of controlled concentrations. It provides much more rapid induction and recovery than methoxyflurane, necessitating careful anesthetic monitoring to prevent overdose and requires the use of a calibrated vaporizer (Wixson and Smiler, 1997). Halothane provides excellent surgical conditions.

In newborn, spontaneously breathing mice, surgical anesthesia can be safely induced by inhalation of 3% halothane in 1 L/min fresh gas flow composed of an equal $N_2O : O_2$ mixture and maintained at 1–1.5% halothane (Dazert et al., 2000). During recovery, supplemental oxygen (2 L/min) should be applied. When halothane (0.25–0.75%) was compared with ketamine/xylazine (80 mg/kg + 10 mg/kg i.p.) in CF-1 mice, halothane anesthesia was more convenient and reliable with respect to rate of induction, reversal and control of anesthetic depth and produced much less cardiac depression (heart rate, left ventricular fractional shortening, cardiac output) than the injectable regimen (Chaves et al., 2001).

However, halothane causes a dose-dependent depression of the cardiovascular and respiratory system (Flecknell, 1993). Furthermore, halothane, like isoflurane, may inhibit immune function (interferon stimulation of natural killer (NK) cell activity) in mice (Markovic et al., 1993) and female CBI mice subjected to multiple exposures to halothane anesthesia before mating, may produce increased amounts of specific antibody secreting B cells accompanied by microscopic fatty changes in the liver. Although halothane does not alter reproductive performance, offspring survival may be lowered (Puig et al., 1999).

Isoflurane

Isoflurane is the anesthetic agent of choice particularly for procedures requiring low risk and reliable rapid recovery. Isoflurane produces stable hemodynamic conditions and has been recommended for anesthesia in Swiss, CD-1, and C57BL/6 strains (Zuurbier et al., 2002). Its effects on cerebral metabolism and cerebral blood flow offer some degree of protection to ischemic and hypoxic brains (Maekawa et al., 1986). Unlike methoxyflurane and many injectable anesthetics, only minimal amounts of isoflurane are subjected to hepatic metabolism, biotransformation and excretion (Wixson and Smiler, 1997) and do not affect hepatic enzyme activities (Thompson et al., 2002). After induction of anesthesia with 2% isoflurane in oxygen (700 ml/min) delivered to an induction chamber, long-term anesthesia (6 h) was maintained at 1.7% isoflurane in O_2 delivered via a face mask to spontaneously breathing mice (Wiersema et al., 1997). Although recovery was followed by a 12-h period of lethargy, all survivors (89%) returned to normal activities.

Potential disadvantages of isoflurane include transient postoperative immunosuppression (Markovic et al., 1993), increased frequencies of cleft palate, skeletal variations and fetal growth retardation in pregnant mice exposed to light anesthetic doses of isoflurane and an increased maternal blood loss secondary to depressed uterine muscle contractility (Mazze et al., 1985). Therefore, isoflurane should be avoided in surgical procedures directly preceding immunologic research studies in mice (Markovic et al., 1993).

Sevoflurane

Sevoflurane offers a significantly greater precision and control of maintenance of anesthesia and potentially a much more rapid induction and recovery from anesthesia than all the other inhalants (Wixson and Smiler, 1997). Its blood solubility is one-half to one-third of that shown by isoflurane, approaching that of nitrous oxide (Eger, 1994). Furthermore, sevoflurane is less pungent than isoflurane and rapid induction is not accompanied by struggling.

Sevoflurane resembles isoflurane in that it depresses ventilation and blood pressure in a dose-dependent manner, but maintains heart rate. This stability of heart rate provided by sevoflurane is desirable because it neither increases myocardial oxygen consumption nor decreases the time available for myocardial perfusion. Hypotension occurs due to a decrease in total peripheral resistance. At clinically relevant concentrations, sevoflurane preserves cardiac output, but excessive levels depress cardiac contractility and can produce cardiovascular collapse. It preserves splanchnic, including renal, blood flow. Sevoflurane can also decrease cerebral vascular resistance and cerebral metabolic rate.

Sevoflurane is degraded by soda lime or baralyme in a temperature-dependent manner (Eger, 1994). Although this breakdown may not significantly affect the course of anesthesia, the breakdown product *compound A* (an olefin) is lethal in rodents at high concentrations (400 ppm).

Nitrous oxide

The use of nitrous oxide as the sole anesthetic agent in rodents is discouraged because it is not a complete anesthetic (Flecknell, 1987). It is also a health hazard to humans.

Carbon dioxide

Carbon dioxide (CO_2) combined with oxygen can be used to produce very short-term anesthesia (e.g. retroorbital bleeding, cardiac puncture) in adult mice. Carbon dioxide is readily available, inexpensive, safe for personnel and provides rapid and smooth recovery. When mice were exposed to a mixture of 80% CO_2 plus 20% O_2 for about 120 s, anesthesia was induced within 10 s and surgical tolerance lasted for 19.5 s (mean) (Köhler et al., 1999). There has been, however, no consensus on the inspired CO_2 concentration and the duration of exposure to carbon dioxide in mice. Low (50%) inspiratory CO_2 concentrations may produce long induction times up to 16 min and may also lead to severe adverse effects including nasal bleeding, excessive salivation, seizures and death associated with moderate distress and discomfort for the animals (Wixson and Smiler, 1997). In contrast, high (100%) inspiratory CO_2 concentrations produce rapid anesthesia with fewer adverse effects but may not be reliable or appropriate for anesthesia where recovery is planned. Therefore, 70% CO_2 in oxygen may be the optimal concentration, based on practicality and humane acceptability (Wixson and Smiler, 1997). Furthermore, it is not known, how long CO_2/O_2 anesthesia can be safely maintained but exposure to CO_2 for no longer than 2 min may be appropriate (Green, 1979).

Anesthesia in neonates and pups

Anesthesia in neonatal mice and pups is critical not just because of the small size but also because only a paucity of information exists on drug metabolism, excretion and biotransformation in this species. The majority of reports deal almost exclusively with anesthetic regimens in neonatal rats.

Hypothermia has been judged to be an appropriate form of analgesia/anesthesia in neonatal mice (Green, 1979; Wixson and Smiler, 1997). Newborn mice are **poikilothermic** and adult thermoregulatory capabilities do not develop until the third week of life (Wixson and Smiler, 1997). Because of their small body mass, rapid

core cooling can be achieved by surface cooling. Infant rodents can tolerate extended periods of 1°C body temperature without known negative effects. Neonatal mice may be cooled in an ice-water slush at 1–2°C for 20–30 min, and retained with elastic bands on a piece of sponge soaked in ice cold water for the duration of the operation (Green, 1979). This technique is commonly used for thymectomy of 1–2 day old pups. On completion of the operation, the pups should be dried on paper tissue, warmed to 37°C in an incubator, and returned to their original cage as a group to join a single unoperated littermate. They should be rubbed in the original bedding to ensure that they acquire the dam's scent, thus minimizing the risk of cannibalism when she is eventually returned to them some 15 min later. Aggressive rewarming techniques such as heating pads or lamps, however, should not been used in order to avoid tissue damage (Wixson and Smiler, 1997). Hypothermia produced consistent and safe anesthesia for minor surgical procedures with rapid induction and recovery times, minimal mortality and no evidence of maternal rejection. However, hypothermia may be associated with an increased risk of ventricular fibrillation, tissue hypoxia and metabolic acidosis after rewarming need to be considered. Other techniques applicable to neonatal mice include pentobarbital (5 mg/kg i.p.) or a combination of an inhalant anesthetic (e.g. methoxyflurane) with hypothermia (Wixson and Smiler, 1997).

Monitoring anesthesia

Because *hypothermia* may be the most important cause of anesthetic death in mice, continuous monitoring of body temperature is mandatory and it is very important to prevent losses of body heat rather than to treat hypothermia.

For monitoring *respiratory function*, recording of the respiratory rate and pattern, color of the muzzle and footpads for evidence of cyanosis and pulse oximetry can be used (Wixson and Smiler, 1997; Kilic et al., 2001). Marked depression of respiratory function may be treated by either removal of mucus or blood from the upper airway, extension of the head and neck and/or assisted ventilation by manually squeezing the chest between the finger and thumb at a rate of approximately 90 breaths/min (Flecknell, 1993). In addition, administration of an α_2-adrenoceptor antagonist (atipamezole) or opioid receptor antagonist (naloxone) will help to reverse respiratory depression but will also decrease analgesia. In order to decrease mortality, oxygen supplementation via a face mask has been recommended for injectable and inhalation anesthetic regimen.

Cardiovascular monitoring may include recording of the heart rate (ECG, pulse oximeter), pulse rate (apical pulse through the chest wall), arterial blood pressure and skin color. However, many ECG monitors designed for the use in humans may be unable to detect the low amplitude ECG signals in mice. Additional fluid support is beneficial in terms of raising arterial blood pressure and s.c. or i.p. administration of warmed saline, lactated Ringer's solution or glucose 5% in normal saline solution at 0.5–1 ml per 30-g mouse or at 0.2 ml/h have been recommended (Flecknell, 1993; Wiersema et al., 1997; Wixson and Smiler, 1997). Special care must be taken to avoid hemorrhagic shock which may already occur after a rapid loss of 0.5 ml of blood in mice (Flecknell, 1993).

Assessment of *depth of anesthesia* is based on a number of clinical signs such as loss of the righting reflex, recumbency and loss of purposeful movements, muscle relaxation, respiratory rate and loss of the tail-pinch, pinnae and pedal reflexes (Green, 1979; Gardner et al., 1995; Zeller et al., 1998; Zuurbier et al, 2002). Of these reflexes, loss of the pedal reflex is the most reliable indication of the development of surgical anesthesia in mice (Green, 1979).

Analgesia

The detection of pain perception in individual rodents is based on subjective evaluation of behavioral and attitudinal changes as well as on objective analysis of physiologic parameters (Morton and Griffith, 1985; Wixson and Smiler, 1997). Behavioral signs indicative of pain perception include reluctance to move, abnormal posturing, decreased appetite, vocalization, anxiety, apprehension, hypersensitivity, depression, aggression and polyphagia of bedding. The pain intensity of a given type of procedure and location of the lesion, based on the discomfort level of that procedure in larger, companion animal species or in humans, should be considered for pain assessment in mice. This information may be also very helpful if investigators choose to design an analgesic protocol for a group of mice.

The recognition of attitudinal changes largely depends on preexisting knowledge of the temperament and behavior of each individual animal. Therefore, veterinary and animal husbandry technicians and research staff are crucial members of the veterinary care team.

In addition to behavioral changes physiological signs of pain perception include fluctuations in blood pressure, heart rate, respiratory rate, body temperature and food and water consumption resulting in changes in body weight.

These behavioral, attitudinal and physiological signs of pain perception are currently the best methods available to detect individual rodents that are in some degree of pain as well as to assess the success/failure of ongoing analgesic therapy. However, in mice body weight and food and water consumption may not be greatly affected by invasive surgeries (e.g. thoracotomy) leading to the conclusion that, unlike rats, it is doubtful that these parameters are useful as objective measures of pain (Liles and Flecknell, 1993).

For analgesic treatment in mice nonsteroidal antiinflammatory drugs (NSAIDs) and opioids are most commonly used (Table 34.3). While NSAIDs such as acetaminophen, asprin, carprofen, diclofenac, flunixin, ibuprofen, paracetamol and piroxicam are primarily indicated for the treatment of low to moderate intensity inflammatory pain, opioids are used to combat moderate to severe pain. The class of opioid analgesics used in mice include the pure agonists morphine and meperidine, the partial agonist buprenorphine and the mixed agonist/antagonist pentazocine, nalbuphine and butorphanol. Opioid partial agonist and mixed agonist/antagonists may exhibit a ceiling effect such that further increases in dosage produce no further analgesia but may escalate its adverse effects such as respiratory depression. Administration of an opioid mixed agonist/antagonist to an animal initially receiving a potent opioid agonist may mitigate the remaining analgesic activity of the agonist. In rodents, buprenorphine was found to be 25–40 times more potent than morphine after parenteral (e.g. s.c.) injection (Cowan et al., 1977). Furthermore, buprenorphine may cause tolerance and positive Straub tail effect (tail elevated >45° to the horizontal; ED50: 0.06–0.75 mg/kg s.c.) in mice (Cowan et al., 1977).

Tricyclic antidepressants (e.g. Amitriptyline) are primarily indicated for the treatment of neuropathic pain. They may decrease behavioral signs of pain perception, including digital irritation and autotomy, for up to 2 weeks with no tolerance or overt adverse effects (Abad et al., 1989).

Analgesics may be administered to mice either parenterally or orally. Oral formulations are available for morphine, butorphanol, oxycodone, codeine, meperidine and pentazocine. However, when adding drugs to

TABLE 34.3 : Analgesics used in mice

Drug	Dosage (mg/kg)	Route	Interval	Reference
Acetaminophen	300	i.p.		Jenkins (1987)
Amitrytiline	5–10	i.p.	24 hourly	Abad et al. (1989)
Aspirin	20	s.c.		Flecknell (1984)
	120	p.o.	once	Dobromylskyi et al. (2000)
	120–300	p.o.		Jenkins (1987)
Buprenorphine	0.01	s.c.		Liles and Flecknell (1993)
	0.05	s.c.	8–12 hourly	Flecknell (1993)
	0.05–0.10	s.c.	8–12 hourly	Dobromylskyi et al. (2000)
	2.0	s.c.		Flecknell (1987), Wiersema et al. (1997)
	2.5	i.p.		Harvey and Walberg (1987)
Butorphanol	0.05–5	s.c.		Jenkins (1987)
	1–2	i.m., s.c.	4 hourly	Dobromylskyi et al. (2000)
	1–5	s.c.	4 hourly	Flecknell (1993)
	5.4	s.c.		Harvey and Walberg (1987)
Carprofen	5	s.c., p.o.	daily	Dobromylskyi et al. (2000)
Codeine	20	s.c.	4 hourly	Flecknell (1987), Jenkins (1987)
	60–90	p.o.		Flecknell (1987), Jenkins (1987)
Diclofenac	8	p.o.	daily	Dobromylskyi et al. (2000)
Fentanyl	0.0125–1.0	i.p.		Thurmon et al. (1996)
Flunixin	2.5	s.c.	12–24 hourly	Dobromylskyi et al. (2000)
Ibuprofen	7.5	p.o.		Jenkins (1987)
	30	p.o.	daily	Dobromylskyi et al. (2000)
Meperidine	20	s.c., i.m.	2–3 hourly	Jenkins (1987)
	2–5	i.m., s.c.	2–4 hourly	Flecknell (1993), Dobromylskyi et al. (2000)
	10–20	s.c.	2–4 hourly	Flecknell (1984, 1987)
Nalbuphine	1.0	s.c.		Liles and Flecknell (1993)
	2–4	i.m.	4 hourly	Dobromylskyi et al. (2000)
	4–8	i.p., s.c.		Flecknell (1993)
Paracetamol	300	p.o.	4 hourly	Flecknell (1987)
Pentazocine	5–10	s.c., i.m.	3–4 hourly	Flecknell (1993), Dobromylskyi et al. (2000)
	10	s.c.	3–4 hourly	Flecknell (1987), Jenkins (1987)
Pethidine (meperidine)	10–20	s.c., i.m.	2–3 hourly	Flecknell (1993), Dobromylskyi et al. (2000)
Phenacetin	200	p.o.	4 hourly	Flecknell (1987)
Piroxicam	3	p.o.	daily?	Dobromylskyi et al. (2000)

Intraperitoneally (i.p.), intramuscularly (i.m.), intravenously (i.v.), subcutaneously (s.c.), orally (p.o.).

drinking water the risks are inaccurate dosing, lack of consumption due to palatability and degradation of the agent over time due to hydrolysis (Wixson and Smiler, 1997). Furthermore, a marked first-pass metabolism may rapidly degrade oral opioids, thus making it difficult to achieve efficacious blood and tissue drug levels. Moreover, oral opioids are quite expensive and thus may not be cost effective for 'herd' analgesic therapy.

Euthanasia

When euthanising an animal the method used must be painless, provide rapid unconsciousness and death, require minimum restraint, avoid excitement, should be appropriate for the age, species and health of the animals, must minimize fear and psychological stress in the animal, must be reliable, reproducible, irreversible, simple to administer and safe for the operator (Working Party, 1996).

Recognition and confirmation of death

The cessation of respiration and heart beat, and the absence of reflexes are good indicators of irreversible death in rodents. In addition, death may be confirmed by additional methods such as exsanguination, extraction of the heart or evisceration (Working Party, 1997).

Embryos, neonates

If a fetus is removed from a deeply anaesthetized dam, then it may be killed by decapitation or removal of the heart (Working Party, 1997). Newborn rodents up to 10 days old may be killed by decapitation or concussion while carbon dioxide is *not* recommended (Working Party, 1997).

Adult Rodents

Recommended methods for euthanasia of adult rodents include the use of barbiturates, carbon dioxide, cervical dislocation, decapitation and stunning (National Research Council, 1992).

Barbiturates

Sodium pentobarbital, injected i.v. or i.p., is considered the agent of choice for most euthanasia. Barbiturates are safe and humane. Other barbiturates such as thiopental and thiamylal must be administered i.v. For nervous and intractable animals, sedation with xylazine or preinduction with ketamine plus xylazine might be appropriate (National Research Council, 1992).

Carbon dioxide

Carbon dioxide is a well-accepted, commonly used gas for euthanasia of laboratory animals but *not* in newborns (!). Inhalation of at least 70% CO_2 in oxygen has a rapid anesthetic effect that proceeds to respiratory arrest and death if exposure is prolonged (Working Party, 1997). Animals become unconscious within 45–60 s and should remain in the chamber for at least 5 or 6 min and then examined closely to determine that all vital signs have ceased (National Research Council, 1992).

Cervical Dislocation

This technique consists of a separation of the skull and brain from the spinal cord by anteriorly directed pressure applied to the base of the skull. Cervical dislocation causes almost immediate unconsciousness because of cerebral shock. All voluntary motor and sensory functions cease because of damage to the spinal cord. However, considerable involuntary muscle activity may occur (National Research Council, 1992).

Stunning

A sharp blow delivered to the central skull bones must be of sufficient force to produce massive hemorrhage and thus immediate depression of the central nervous system (CNS). When done properly, unconsciousness is immediate. Stunning should be used only by properly trained persons and when other means are inappropriate or unavailable. After stunning, the animal must be killed immediately by another procedure, such as exsanguination or decapitation (National Research Council, 1992).

Decapitation

Decapitation with a guillotine is used primarily when pharmacological agents and carbon dioxide are contraindicated (e.g. pharmacological and biochemical studies). This method causes rapid death if properly performed. The animal needs to be properly restrained, and its head must be completely severed from its body at the atlanto-occipital joint. The guillotine must be kept in good operating condition, and the blade must be sharp (National Research Council, 1992).

Other acceptable methods for euthanasia in rodents include the use of inhalation anesthetics (halothane, enflurane, isoflurane), T-61 (only to be injected i.v. !) and rapid freezing in liquid nitrogen (only in small neonates; Working Party, 1997).

References

Abad, F., Feria, M. and Boada, J. (1989). *Neurosci. Lett.* **99**, 187–190.

Branson, K.R. and Gross, M.E. (1994). *J. Am. Vet. Med. Assoc.* **204**, 1888–1890.

Calderone, L., Grimes, P. and Shaley, M. (1986). *Exp. Eye Res.* **42**, 331–337.

Chaves, A.A., Weinstein, D.M. and Bauer, J.A. (2001). *Life Sci.* **69**, 213–222.

Cowan, A., Lewis, J.W. and MacFarlane, I.R. (1977). *Br. J. Pharmacol.* **60**, 537–545.

Cruz, J.I., Loste, J.M. and Burzaco, O.H. (1998). *Lab. Anim.* **32**, 18–22.

Dazert, S., Schomig, P., Shehata-Dieler, W.E., Aletsee, C. and Dieler, R. (2000). *Laryngorhinootologie* **79**, 26–29.

Dobromylskyi, P., Flecknell, P.A., Lascelles, B.D., Pascoe, P.J., Taylor, P. and Waterman-Pearson, A. (2000). In *Pain Management in Animals* (eds P.A. Flecknell and A. Waterman-Pearson), pp. 81–145. W.B. Saunders, London.

Dörr, W. and Weber-Frisch, M. (1999). *Lab. Anim.* **33**, 35–40.

Eger, E.I. (1994). *Anesthesiology* **80**, 906–922.

Erhardt, W., Hebestedt, A., Aschenbrenner, G., Pichotka, B. and Blümel, G. (1984). *Res. Exp. Med.* (Berl) **184**, 159–169.

Flecknell, P.A. (1984). *Lab. Anim.* **18**, 147–160.

Flecknell, P.A. (1987). *Laboratory Animal Anaesthesia.* Academic Press, London.

Flecknell, P.A. (1993). In *Guide to Techniques in Mouse Development* (eds P.M. Wassarman and M.L DePamphilis), *Meth. Enzymol.* **225**, 16–33.

Flecknell, P.A. and Mitchell, M. (1984). *Lab. Anim.* **18**, 143–146.

Gardner, D.J., Davis, J.A., Weina, P.J. and Theune, B. (1995). *Lab. Anim. Sci.* **45**, 199–204.

Glen, J.B. (1980). *Br. J. Anaesth.* **52**, 731–742.

Green, C.J. (1975). *Lab. Anim.* **9**, 161–178.

Green, C.J. (1979). *Animal Anaesthesia. Laboratory Animal Handbooks* 8, pp. 147–154. Laboratory Animals Ltd, London.

Green, C.J., Halsey, M.J., Precious, S. and Wardley-Smith, B. (1978). *Lab. Anim.* **12**, 85–89.

Green, C.J., Knight, J., Precious, S. and Simpkin, S. (1981a). *Lab. Anim.* **15**, 163–170.

Green, C.J., Knight, J., Precious, S. and Simpkin, S. (1981b). *Lab. Anim.* **15**, 171–175.

Hart, C.Y., Burnett, J.C. Jr. and Redfield, M.M. (2001). *Am. J. Physiol. Heart. Circ. Physiol.* **281**, H1938–H1945.

Harvey, R.C. and Walberg, J. (1987). In *Principles and Practice of Veterinary Anesthesia* (ed C.E. Short), pp. 380–392. Williams & Wilkins, Baltimore.

Jenkins, W.L. (1987). *J. Am. Vet. Med. Assoc.* **191**, 1231–1240.

Jones, D.M., Arters, J. and Berger-Sweeney, J. (1999). *Lab. Anim. Sci.* **49**, 316–318.

Kilic, N., Henke, J., Pragst, I. and Erhardt, W. (2001). *Vet. Anaesth. Analg.* **28**, 211–212.

Köhler, I., Meier, R., Busato, A., Neiger-Aeschbacher, G. and Schatzmann, U. (1999). *Lab. Anim.* **33**, 155–161.

Koizumi, T., Maeda, H. and Hioki, K. (2002). *Exp. Anim.* **51**, 119–124.

Lewis, G.E. Jr. and Jennings, P.J. Jr. (1972). *Lab. Anim. Sci.* **22**, 430–432.

Liles, J.H. and Flecknell, P.A. (1993). *J. Vet. Anaesth.* **20**, 38.

Lovell, D.P. (1986a). *Lab. Anim.* **20**, 85–90.

Lovell, D.P. (1986b). Variation in pentobarbitone sleeping time in mice. 2. Variables affecting test result. *Lab. Anim.* **20**, 91–96.

Lovell, D.P. (1986c). *Lab. Anim.* **20**, 307–312.

Maekawa, T., Tommasine, C. and Shapiro, H.M. (1986). *Anesthesiology* **65**, 144–151.

Markovic, S.N., Knight, R.P. and Murasko, D.M. (1993). *Anesthesiology* **78**, 700–706.

Mazze, R.I., Wilson, A.I., Rice, S.A. and Baden, J.M. (1985). *Teratology* **32**, 339–345.

Morton, D.B. and Griffith, P.H.M. (1985). *Vet. Rec.* **116**, 431–436.

Murray, W.J. and Fleming, P.J. (1972). *Anesthesiology* **37**, 620–625.

National Research Council (1992). Recognition and Alleviation of Pain and Distress in Laboratory Animals. National Academy Press, Washington.

Papaioannou, V.E. and Fox, J.G. (1993). *Lab. Anim. Sci.* **43**, 189–192.

Park, C.M., Clegg, K.E., Harvey-Clark, C.J. and Hollenberg, M.J. (1992). *Lab. Anim. Sci.* **42**, 508–513.

Peuler, M., Glass, D.D. and Arens, J.F. (1975). *Anesthesiology* **43**, 575–578.

Puig, N.R., Amerio, N., Piaggio, E., Barragan, J., Comba, J.O. and Elena, G.A. (1999). *Reprod. Toxicol.* **13**, 361–367.

Roth, D.M., Swanye, J.S., Dalton, N.D., Gilpin, E.A. and Ross, J. Jr. (2002). *Am. J. Physiol. Herat. Circ. Physiol.* **282**, H2134–2140.

Sakai, T., Ichiyama, T., Whitten, C.W., Gieseke, A.H. and Lipton, J.M. (2000). *Can. J. Anaesth.* **47**, 1019–1024.

Sedgwick, C.J., Erhardt, W., Krobel, R. and Lendl, C. (1992). In *Anästhesie bei Kleintieren* (eds R.P. Paddleford and W. Erhardt), pp. 359–384. Schattauer, Stuttgart.

Short, C.E. (1987). In *Principles and Practice of Veterinary Anesthesia* (ed. C.E. Short), pp. 28–46. Williams & Wilkins, Baltimore.

Silverman, J., Huhndorf, M., Balk, M. and Slater, G. (1983). *Lab. Anim. Sci.* **33**, 457–460.

Spiegel, K., Kalb, R. and Pasternak, G.W. (1983). *Ann. Neurol.* **13**, 462–465.

Tarin, D. and Sturdee, A. (1972). *Lab. Anim.* **6**, 79–84.

Thompson, J.S., Brown, S.A., Khurdayan, V., Zeynalzadedan, A., Sullivan, P.G. and Scheff, S.W. (2002). *Comp. Med.* **52**, 63–67.

Thurmon, J.C., Tranquilli, W.J. and Benson, G.J. (1996). In *Lumb & Jones' Veterinary Anesthesia*, 3rd edn (eds J.C. Thurmon, W.J. Tranquilli and G.J. Benson), pp. 686–735. Williams & Wilkins, Baltimore.

Virtanen, R. (1989). *Acta. Vet. Scand. Suppl.* **85**, 29–37.

Weiss, J. and Zimmermann, F. (1999). Letters to the editor. *Lab. Anim.* **33**, 192–193.

Whelan, G. and Flecknell, P.A. (1994). *Lab. Anim.* **28**, 70–77.

White, W.J. and Field, K.J. (1987). *Vet. Clin. North Am. Small Anim. Pract.* **17**, 989–1017.

Wiersema, A.M., Dirksen, R., Oyen, W.J.G. and van der Vliet, J.A. (1997). *Lab. Anim.* **31**, 151–156.

Wixson, S.K. and Smiler, K.L. (1997). In *Anesthesia and Analgesia in Laboratory Animals* (eds D.F. Kohn, S.K. Wixson, W.J. White and G.J. Benson), pp. 265–203. Academic Press, San Diego.

Working Party (1996). *Lab. Anim.* **30**, 293–316.

Working Party (1997). *Lab. Anim.* **31**, 1–32.

Zeller, W., Meier, G., Bürki, K. and Panoussis, B. (1998). *Lab. Anim.* **32**, 407–413.

Zuurbier, C.J., Emons, V.M. and Ince, C. (2002). *Am. J. Physiol. Heart Circ. Physiol.* **282**, H2099–H2105.

Glossary

acrosome: The caplike, membrane-bound structure derived from Golgi elements found at the anterior portion of the nucleus of a spermatozoon

adnexa: Appendages or adjunct parts

alloantigens: Isoantigens; antigens subject to intraspecies genetic variations

allorecognition: Recognition of alloantigens

amplicons: Small, replicating DNA fragments; PCR products

angiectasis: Gross dilation and often lengthening of a blood vessel

anisocytosis: Presence in the blood of erythrocytes showing excessive variation in size

anosmia: Absence of the sense of smell

apnea: Cessation of breathing

apoE deficient: Apoliproprotein E deficient

ataxia: Failure of muscular co-ordination

audiogenic: Induced by sound e.g.seizures

axenic: Refers to experimental animals raised under sterile conditions; axenic or germfree animals are not contaminated by or completely free of the presence of other organisms

biliary atresia: Congenital absence or closure of the ducts that drain bile from the liver

biotope: The smallest unit of habitat with uniform environmental conditions

cachexia: General ill health and malnutrition

capacitate (sperm): Enabling sperm to become capable of fertilising an ovum after it reaches the ampullary portion of the uterine tube

ceroid: Lipid pigment similar to lipofuscin

chaperones: See molecular chaperones

clonogenic: Giving rise to a clone of cells

congenic: Two inbred stains of animals that are genetically identical except at a single or few specified loci so that their known genetic differences are expressed in the same 'genetic background'

consomic: A strain of animals in which a single, full-length chromosome has been transferred (substituted) from another strain

coprophagy: Ingestion of faeces; normal nutritional behavior for rodents

cosmid libraries: Includes types of vectors used to clone large DNA fragments

decidual: Pertaining to the decidua i.e. the endometrium of the pregnant uterus

dinucleotide repeats: The most common of the microsatellite tandem repeats, consisting of two nucleotides repeating hundreds or thousands of times. (GT)n is the most frequently seen.

doppler: An ultrasound method of examining moving objects (usually red blood cells) in tissues

echocardiography: A method of graphically recording details of the heart – position and motion

ectactic: Morbid condition of dilatation

ectasia: Dilation, expansion or distension

ectromelia: Ectromelia virus is a highly virulent natural pathogen of mice that causes mousepox, a severe disease with high mortality rate. The ectromelia virus is a member of the poxvirus family

EFS: Electrical Field Stimulation

electroporation: The application of an electric field to cause a reversible creation of pore-like openings in the plasma membrane of a cell

enzootic: Present in an animal community at low incidence at all times

epitope: A molecule or portion of a molecule capable of binding to the combining site of an antibody

epizootic: A disease of high morbidity only occasionally present in an animal community; the veterinary equivalent of an epidemic

ethogram: A list and description of an animal's discrete pattern of behaviour

exophthalmus: Abnormal protusion of the eyeball

formites: Object able to harbour pathogenic micro-organisms

fragilised: Made fragile or sensitive

gametogenesis: The development of the male or female sex cells or gametes

gastrulation: The process by which a blastula becomes a gastrula

gene ontologies: Defined, structured classification of genes

gene trap: A type of DNA construct containing a reporter gene sequence downstream of a splice acceptor site that is capable of integrating into random chromosomal locations in the mouse

gene trap mutagenesis: Random mutation of the genome by random insertion of DNA pieces

gnotobiotic – gnotoxenic: Denoting germ-free or formerly germ-free organisms in which the composition of any associated microbial flora, if present, is fully defined.

haemosiderin: An iron-storage protein derived from phagocytosed erythrocytes

HI test: Haemagglutinin inhibition test

haplotype: A set of alleles of a group of closely linked genes usually inherited as a unit

helicase: An enzyme that unwinds the DNA helix at the replication fork to allow the resulting single units to be copied

heteroduplex: A double-stranded DNA molecule formed by renaturation of PCR products from two different alleles

heterozygous: Having two different alleles of a specified gene

hirsute: Shaggy; having abundant or excessive hair

homeostasis: A tendency to stability in the normal body states of the organism

homolog: A gene related to a second gene by descent from a common ancestral DNA sequence.

homozygous: Possessing a pair of identical alleles at a given locus

hyalinosis: Hyaline degeneration presenting as acidophilic intracytoplasmic accumulations

hyperhomocysteinemia: Elevated levels of homocysteine in blood

hyperphagia: Ingestion of a greater than optimal quantity of food

hypertrophic: Relating to a general increase in bulk of a part or organ, due to increase in size, but not in number, of the individual tissue elements

iatrogenic: Disease or symptom produced (inadvertently) as a consequence of medical or surgical treatment

ICU: Intensive care unit

immortalised: Enabled to overcome the normal limitations on growth or life span

immunocompetence: The ability or capacity to develop an immune response following exposure to antigen

infundibulum: A general anatomical term for a funnel-shaped structure e.g. the abdominal extremity of the uterine (fallopian) tube.

isogenic – syngeneic: Denoting individuals or tissues that have identical genotypes and thus could participate in a syngraft

isohistogeneity: Identical tissue components compared with another mouse (inbred)

knock-in mice: Mice in which the wild type copy of a gene is replaced by one with a specific modification (homologous recombination)

knockout mice: Where the function of a gene has been completely eliminated

kobner's reaction: Isomorphic reaction to skin injury at previously uninvolved sites

lacrimal: Pertaining to the tears

lentiviral vectors: HIV-virus-like carriers

LLN test: Local lymph node test (primary response in a draining lymph node)

lordosis: Anterior concavity in the curvature of the lumbar and cervical spine

meningoencephalitis: Inflammation of the brain and meninges

MHC haplotypes: Major histocompatibility complex – set of alleles (see also haplotype)

microsatellites: Simple sequence repeats, involving one, two or more nucleotides, present in noncoding regions of the genome

mitogen: A substance that is able to induce mitosis in certain eukaryotic cells e.g. concanavalin A

molecular chaperones: A family of cellular proteins, e.g. heat shock proteins that assist in the normal assembly or folding process of other proteins but will not be part of the resulting functional complexes

myelosuppression: Suppression of bone marrow activity

myofibres: Muscle fibres

narcosis: A non-specific and reversible depression of function of the central nervous system

nongenotoxic: Does not cause damage to DNA

oncosuppressive: Tumour suppressive

orphan virus: Virus not known to be associated with a disease

orthologs: Genes in two species that have evolved from a common ancestor without duplication

ostium (sing` ostium): Opening of or into a tubular organ or between two distinct cavities within the body

osteopetrosis: Solidified bone; congenital disease that prevents formation of bone marrow and results in abnormal bone development

pantropic infection: Having an affinity for many tissues

paralogs: Paralogs are genes related by duplication within a genome and which may have evolved new functions

pelage: The hairy coat of mammals – truncal hairs

pleiotropic: Multiple, often seemingly unrelated, phenotypic effects caused by a single altered gene or pair of altered genes

plethysmography: A test measuring the systolic blood pressure of a leg compared to the arm. The test is usually performed to rule out blockages in the arms or legs (usually legs)

poikilothermic: Organisms unable to use their metabolism to heat or cool themselves, thus having body temperatures depending on the environment

postgastrulation: See gastrulation

primary tropism: First site of infection

prion: A poorly characterised proteinaceous slow infectious agent lacking detectable nucleic acid and eliciting no immune response, now believed to be responsible for several neurodegenerative diseases

progeria: A syndrome of uncertain pathogenesis, characterised by accelerated aging.

proteome: The full complement of proteins found in a cell or tissue at a particular time and under specific conditions

psammoma bodies: Mineralised bodies occurring in the meninges, choroids plexus and certain meningiomas

pulsatility index (PI): A measure of resistance in carotid artery stenosis

rales: An abnormal discontinuous non-musical sound heard on auscultation over the lungs, primarily during inhalation

rederivation: Elimination of non-acceptable agents from animal colonies via embryo transfer or cesarian section

rete ridges: Downward thickenings of the epidermis between the dermal papillae

retroorbital: At the back of the eye orbit

Robertsonian chromosomes: A fusion of two nonhomologous telocentric chromosomes to form a single metacentric chromosome

rouleaux: A roll of red blood corpuscles like a stack of coins

sebocytes: Sebum secreting cells

seroconversion: Change of a serological test from negative to positive

seropositivity: Showing positive results on serological examination; a high level of antibody

serositis: Inflammation of a serous membrane

spike(s) antigens: A major viral antigen known to neutralise virus infectivity (a glycoprotein)

steatorrhea: Excessive amount of fats in the faeces

sustenticular cells: Supporting cells; tall cells resting on the basement membrane and serving as a support to the shorter specialised cells e.g. in the auditory or olfactory epithelium

synteny: Conservation of gene order, without demonstrable linkage, between divergent lineages

telemetric transmitters: A mobile substation transmitting information

telomere: End region of eukaryotic chromosomes

tetanic: Pertaining to tetanus – a continuously contracted state of muscle resulting from high-frequency stimulation

thermogenesis: Production of heat within the animal's body

tidal volume: Normal respiratory volume of the lung

tinctorial: Pertaining to dyeing or staining

tolerance (immunological): Failure to mount an immune response to an antigen

tonic-clonic spasms: Said of a spasm or seizure consisting of a convulsive twitching of the muscles

transactivator: Diffusible molecule that can induce transcription on both homologous and heterologous molecules of DNA

transcriptome: The full complement of expressed gene transcripts in a particular cell or tissue at a particular time and under specific conditions

transgene: A mutant or other gene experimentally introduced to an animal (usually mouse)

transgenesis: Process of inserting a transgene into an animal

transgenic animals: Animals containing a transgene

transponder: A device that can be inserted under the skin or in the abdomen for field tracking

turgor: The pressure within cells or organs

typhlocolitis: Inflammation of colon and caecum

vascular impedance: Reduction in blood flow in the regional blood vessels

xenoorgans: Organs transplanted from one species into another species

zeitgeber: Time setter

zoonotic: Transmissible from animals to man under natural conditions

Index

Plate 1 Mice of the *Mus spretus* species (left, with an agouti coat color) and C57BL/6 (right, with a non agouti or solid black coat color). In spite of their great similarity in size and body shape these mice are distantly related species but can still produce viable and fertile hybrids (female only). Mice of the *Mus spretus* species have been extensively used for the development of the mouse genetic map.

Plate 2 BALB/c mouse engaging in grooming behaviour and defecation (see Table 4.5, behavioural parameters) close to the partition in a version of the modified hole board test which also measures social behaviour. Courtesy of Dr. Sabine Hölter, German Mouse Clinic, GSF Neuherberg, Germany.

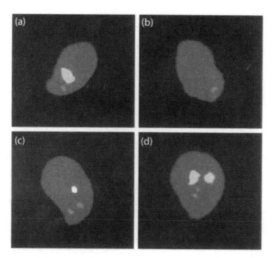

Plate 3 Mouse sperm with fluorescence signals for chromosomes to detect aneuploidy. (a) Normal sperm with chromosomes 8 (labelled with CY3, red) and Y (labelled with FITC, green); (b) hypohaploid sperm with only chromosome 8 (red); (c) hyperhaploid (disomic) sperm with two chromosomes 8 (red) and one X chromosome (labelled with FITC plus CY3, white); (d) diploid mouse sperm with two chromosomes 8 and two Y chromosomes. Images digitized with the software ISIS (MetaSystems, Altlussheim, Germany).

AAD161
40,XY,t(4;10;15)

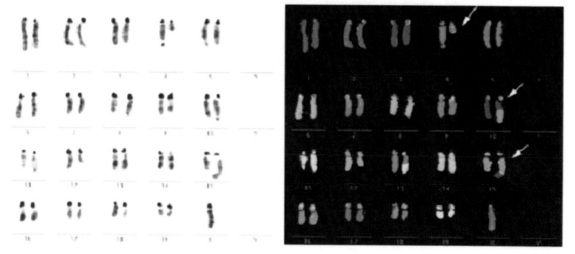

Plate 4 DAPI-banded and multicolour karyotype (M-FISH) of a mouse carrying a complex translocation involving 3 chromosomes T(4;10;15) (obtained from Isabelle Jentsch, GSF-Institute of Human Genetics).

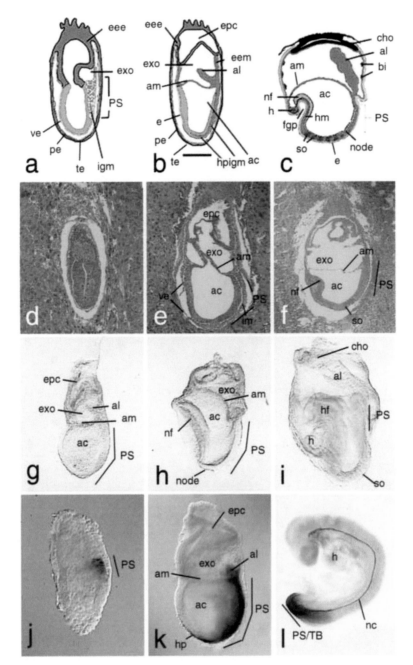

Plate 5 Mouse development from early primitive streak to early organogenesis stages. (a–c) Schematic representation of early postimplantation development. (a) 6.75 d p.c. embryo; (b) 7.5 d p.c. embryo; (c) 8.25 d p.c. embryo. Colour code. Yellow – (visceral, parietal, definitive) endoderm, red – mesoderm, grey – extraembryonic ectoderm, blue – embryonic ectoderm/epiblast, black – chorion. (d–f) Histological sections of early gastrulation stage mouse embryos with surrounding uterine tissue. 10 μm paraffin sections stained with hematoxylin and eosin. (d) 6.25 d p.c. embryo; (e) 7.0 d p.c. embryo; (f) 8.0–8.25 d p.c. embryo; (g–h) whole gastrulation stage embryos; (g) 7.5 d p.c. embryo; (h) 7.75 d p.c. embryo; (i) 8.5 d p.c. embryo. The black bar represents 200 μm. (j–l) Gastrulation stage embryos marked for the primitive streak/tail bud region. Immunohistological detection of Brachyury protein expression in whole embryos (Kispert and Herrmann, 1994). (j) 6.5 d p.c. embryo. Brachyury expression marks the primitive streak which has formed at the future posterior end of the embryo proper. (k) 7.5 d p.c. embryo. The primitive streak has extended to the distalmost tip of the egg cylinder. Brachyury protein expression also indicates the extraembryonic mesoderm (allantois) and the head process. (l) 9.0 d p.c. embryo. The primitive streak or tail bud marks the posterior pole of the embryo. Brachyury protein is detected all along the notochord. Anterior is to the left, posterior to the right, proximal is to the top, distal is to the bottom. ac – amniotic cavity, al – allantois, am – amnion, bi – blood islands, cho – chorion, e – (definitive) endoderm, eee – extraembryonic ectoderm, eem – extraembryonic mesoderm, epc – ectoplacental cavity, exo – exocoelom, fgp – foregut pocket, h – heart, hp – head process, igm – ingressing mesoderm, nc – notochord, nf – neural folfs, pe – parietal endoderm, PS – primitive streak, so – somites, TB – tail bud, te – trophectoderm, ve – visceral endoderm.

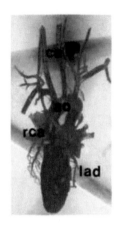

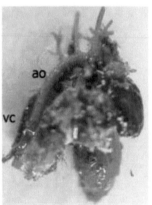

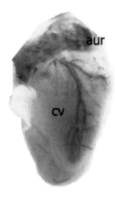

Plate 6 Top Panel: Photograph of latex cast of mouse arterial system including coronary arteries. ca = carotid artery; lad = left anterior descending coronary artery; ao = aortic arch; rca = right coronary artery. (Reprinted with permission from *Am. J. Physiol. Heart Circ. Physiol.* **269**, H2147–H2154.) Middle Panel: Latex cast of major arteries and veins in and around heart and lungs; ao = aortic arch; vc = posterior vena cava. Bottom Panel: Latex cast of major cardiac veins (cv) on the lateral wall of left ventricle.

(a)

(b)

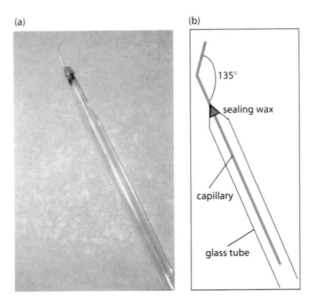

135°

sealing wax

capillary

glass tube

Plate 7 Embryo transfer capillary. (a) Photograph; (b) schematic view.

(a)

(b)

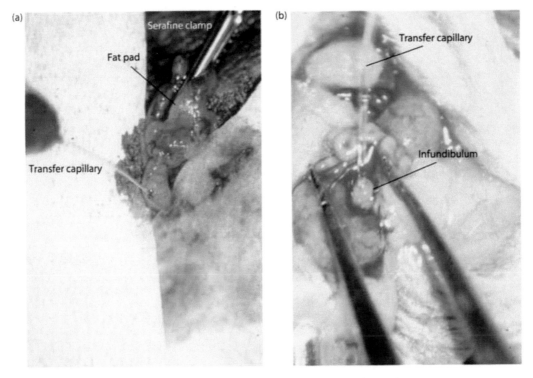

Serafine clamp

Fat pad

Transfer capillary

Transfer capillary

Infundibulum

Plate 8 Embryo transfer (a) to the uterus; (b) to the oviduct.

(a)

(b)

a   b   c   d b e

(c)

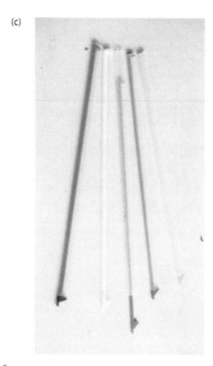

Plate 9 Plastic straws (Minitübs®) used as cryocontainers for embryo freezing. (A) Photograph; (B) schematic view; (a) glass bulb, (b) air space, (c) freezing medium containing the embryos, (d) seeding point, (e) metal bulb; (C) magazines for 10 straws each.

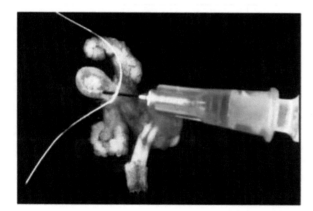

Plate 10 Urinary bladder fixation technique.

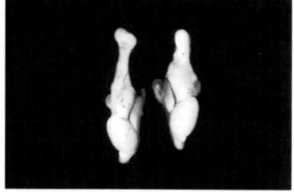

Plate 11 Orientation of the testes and epididymides.

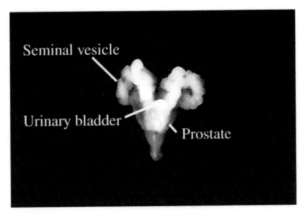

Plate 12 Male reproductive tract with urinary bladder.

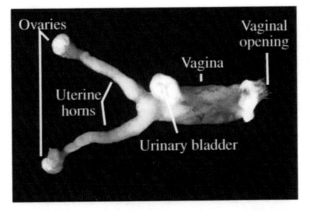

Plate 13 Female reproductive tract with urinary bladder.

Plate 15 'Swiss roll' of intestines. It is important to make the rolls smaller than the cassettes (a), and minimize loss of fixative while rolling the segments (b).

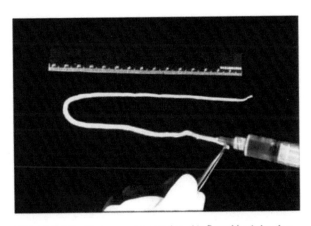

Plate 14 Intestines are removed and inflated by injection with fixative.

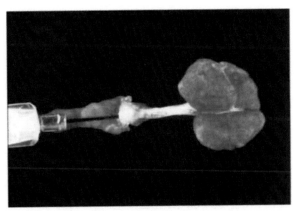

Plate 16 Lung inflation with fixative.

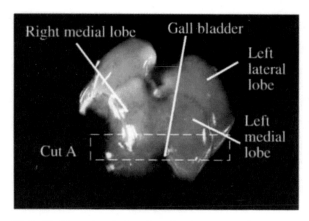

Plate 17 Example of trimming sites for a mouse liver. (a) left lateral lobe, (b) left medial lobe, (c) right medial lobe, and (d) gallbladder.

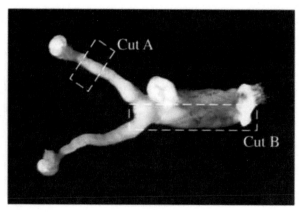

Plate 18 Female reproductive tract. Dotted lines indicate where to cut tissue for histologic processing.

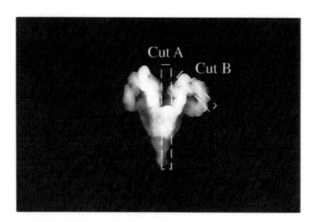

Plate 19 Male reproductive tract. Dotted lines indicate sampling sites for histology.

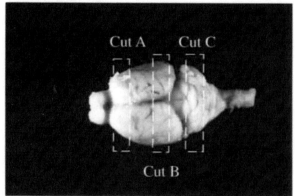

Plate 20 Trimming sites marked for sectioning a mouse brain.

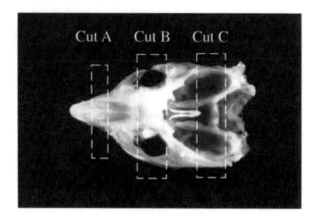

Plate 21 Three sections are cut in the decalcified skull. This exposes: (A) the nasal cavity, (B) eyes and associated glands, and (C) inner, middle, and external ear as well as the pituitary gland.

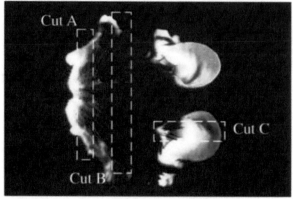

Plate 22 Skin of the head trimmed to study (A) muzzle and vibrissae, (B) eyelids, cilia, Meibomian gland, and conjunctiva, and (C) pinna of the ear.